Disease Problems in Vegetable Production

2nd Revised and Enlarged Edition

S.K. Gupta
Prof. & Head
Department of Plant Pathology
Dr. Y.S. Parmar University of Horticulture and Forestry, Solan

T.S. Thind
Adjunct Professor (Plant Pathology)
Punjab Agricultural University, Ludhiana

Published by
SCIENTIFIC PUBLISHERS (INDIA)

Jodhpur –
5 A, New Pali Road
P.O. Box 91
Jodhpur - 342 001 INDIA

Delhi –
4806/24, Ansari Road
Daryaganj
New Delhi - 110 002 INDIA

ISBN: 978-93-86652-23-2

Visit the Scientific Publishers (India) website at
http://www.scientificpub.com
Printed in India

PREFACE
(2nd Edition)

India has diverse climatic conditions which are suitable for cultivation of various crops like cereals, millets, pulses and horticultural crops. Agriculture still accounts for about 54.6 per cent of total employment and contributed approximately 17 per cent Gross Value Added (GVA) at current prices during 2014-15. During cultivation, these crops are attacked by various diseases of fungal, bacterial and viral nature causing 10 -16 per cent losses of global produce. However, timely and effective management of these diseases has played a key role in making India self sufficient in food production.

For the integrated development of agriculture in the country, diversification of cropping systems is must and cultivation of vegetables plays an important role in this direction. However, vegetables are highly perishable commodity and are ravaged by various diseases that cause substantial yield losses both during cultivation and at post harvest stages. The production of good quality vegetables can only be obtained if the ravages of diseases are kept under check. Moreover, new diseases are emerging and the diseases which were earlier less important are becoming major constraints due to changing climatic conditions. Keeping in view the large scale acceptance of the first edition of the book published in 2006 and suggestions received from the reviewers and also the advances in the subject it was considered imperative to bring out the revised and updated second edition of the book for the benefit of readers, especially post graduate students. In the revised edition, two new chapters on 'Diseases of Ginger' and 'Diseases of Vegetables under Protected Cultivation Conditions' have been incorporated. Some new diseases like bacterial canker of tomato, French bean rust, cauliflower mosaic, Dasheen mosaic of colocasia, blue mold and brown rot of onion etc. have also been added under the respective vegetables. Overall, the book will provide a comprehensive account of all aspects of diseases of commercial vegetables for both under graduate and post graduate students, teachers, extension workers and elite growers.

The help rendered by Dr. (Mrs.) Monika Sharma, Asstt. Professor, College of Horticulture and Forestry, Neri, Hamirpur (HP) and Ms.

Banita and Dr. (Ms.) Adhikshita, Research Fellows in the Department of Plant Pathology in updating the text of this book is gratefully acknowledged. We are thankful to Dr. YSP University of Horticulture & Forestry, Solan and Punjab Agricultural University, Ludhiana for granting permission to revise this book. We are also indebted to our family members for allowing us to take up this time consuming task.

S.K. Gupta
T.S. Thind

PREFACE
(1st Edition)

India is endowed with varied climatic conditions suitable for the production of various vegetables in different areas round the year. Vegetables occupy an important place in Indian economy and are grown on about 30 million ha area. Globally, India is ranked second after China as far as production of vegetables is concerned. In recent years, the trend of commercialization of horticulture sector has been found to be increasing due to opening up of the world economy. Vegetables being a rich source of vitamins, minerals and the much-needed fibre in our daily diet constitute a major component in balanced nutrition. There is considerable increase in consumption of fresh and processed vegetables in India due to improved economy of its people and the estimated demand in future can only be met by increasing productivity through greater technology adoption. Vegetables like other crop plants are affected by various diseases, which result in both qualitative and quantitative losses. The introduction of hybrids and various high yielding varieties of vegetables has resulted in epidemics of many diseases. These have also added to the introduction of several new disease problems and some of the earlier minor diseases have become more important. It is essential that one should have knowledge about the causes of plant diseases, their perpetuation and spread so that timely management practices are implemented. Knowledge of various factors leading to epidemic build-up of diseases is necessary for prediction and timely application of various management methods. Without proper knowledge of diseases, farmers are indiscriminately using agrochemicals to combat the ravages of insect-pests and diseases, which may lead to an increase in the environmental pollution. In the WTO regime, for the export of vegetables and fruits to the other countries, care has to be taken to keep the pesticide residues below the permissible limits. Keeping this in view, it is necessary to use integrated disease management practices involving most effective and economical methods so that quality produce without pesticide residue is obtained.

Over the past decade, a lot of information has been generated by plant pathologists on different aspects of diseases of vegetables and their management options. The present book on *Disease Problems in Vegetable Production* is an effort to compile available information on the latest developments in the symptomatology, etiology, epidemiology and management including host resistance, cultural, biological and chemical methods. The integration of more than one management practice, where ever feasible, has been suggested to make their use more effective, economical and eco-friendly. A total of seventeen chapters dealing with important diseases of vegetables like potato, tomato, crucifers, cucurbits, pea, French bean, chilli and bell pepper, onion and garlic, egg plant, carrot, sugarbeet, colocasia, okra and leafy vegetables have been compiled in this book. The book chapters also include common pathogens of vegetable crops, disease problems in nurseries, post-harvest diseases and nematode disease problems of vegetables. The information has been augmented with coloured photographs of disease symptoms to provide an easy way of identification of different diseases. Disease cycles of important diseases further help in understanding the perpetuation and spread of these pathogens. The book has been written in simple and easily understandable language. It is hoped that the information contained in this book will prove useful to researchers working on vegetable diseases, teachers and students of plant pathology, extension workers and elite growers.

The help rendered by Dr. K.R. Shyam, Professor of Plant Pathology (Retd.) in preparing various disease cycles is gratefully acknowledged. Our sincere thanks are also due to Sh. Mohinder Singh Chauhan, Artist, for drawing diagrams of various disease cycles. We are also thankful to Dr. Y.S. Parmar University of Horticulture and Forestry, Nauni, Solan and Punjab Agricultural University, Ludhiana for the permission to write this book. We are indebted to our family members for allowing us to spare time for this arduous task.

S.K. Gupta
T.S. Thind

CONTENTS

1. Common Pathogens of Vegetable Crops 1
2. Disease Problems in Vegetable Nurseries 11
3. Diseases of Potato 19
4. Diseases of Tomato 62
5. Diseases of Cruciferous Vegetables 123
6. Diseases of Cucurbits 180
7. Diseases of Pea 213
8. Diseases of French Bean 265
9. Diseases of Chillies and Bell Pepper 326
10. Diseases of Onion and Garlic 367
11. Diseases of Egg Plant 389
12. Diseases of Carrot and Sugarbeet 413
13. Diseases of Colocasia 437
14. Diseases of Okra 449
15. Diseases of Leafy Vegetables 462
16. Diseases of Ginger 499
17. Post-Harvest Diseases of Vegetables 517
18. Diseases of Vegetables under Protected Cultivation Conditions 534
19. Nematode Problems in Vegetable Production 570

1

Common Pathogens of Vegetable Crops

India has gained self-sufficiency in the production of cereal foods. Ever increasing human population also warrants sustainable production of other food materials like pulses, oil seeds, vegetables, fruits and spices. Vegetables are short duration crops and are rich source of vitamins, minerals and the much needed fibers in our daily diet.

Plant diseases, caused by fungi, bacteria, viruses, phytoplasma, nematodes etc., have become serious constraints in the production of vegetables. The Irish famine due to potato late blight caused by *Phytophthora infestans* in 1840s created much more awareness about the importance of plant diseases. The introduction of high yielding cultivars/hybrids coupled with higher doses of fertilizers, assured irrigations and intensive agriculture have resulted in the higher susceptibility of plants to diseases causing severe qualitative and quantitative losses, thus making the produce less competitive in the national/ international markets. These problems are further compounded by the changing climatic conditions. In some of these crops, the cumulative losses have been estimated to 26 to 40 per cent and even complete crop failures in certain cases. The ultimate aim of the science of plant pathology is the management of these diseases which can be kept under check by applying various management practices *viz.,* cultural, biological, chemical and through resistance breeding.

For working out management practices for various diseases, it is important to understand the nature, life-cycle (etiology) and disease-cycle (epidemiology) of pathogen(s). These aspects are dealt under various chapters of this book on important diseases of different vegetable crops. Brief description of plant pathogens responsible for causing diseases in vegetable crops is given here under:

Fungal and fungi-like pathogens: Fungi can be defined as nucleated, spore bearing, achlorophyllous organism which generally produce asexually and sexually and whose usually filamentous, branched somatic structures are typically delineated by cell walls containing cellulose or chitin or both. Microscopic examination of fungi shows that all fungal structures are formed of filament like structures called hyphae which when together are called mycelium. The fungi and fungal-like organisms cause diseases on crops. The fungal-like organisms, often referred to as lower fungi, are now considered to belong to the kingdom **Protozoa** or to the kingdom **Chromista** (also known as **Stramenopiles**). True fungi belong to the kingdom **Fungi.** Few typical characteristics of some phyla are given in Table 1.

Table 1. Characteristics of fungal and fungi-like pathogens

Kingdom	Phylum	Vegetative form	Asexual reproduction	Sexual reproduction
Protozoa	Plasmodiophoromycota	Endoparasitic slime molds	Sporangial plasmodium, zoospores	Zoospores
Chromista (Stramenopiles)	Oomycota	Non-septate elongated mycelium	Sporangia produce zoospores or behave like conidia	Oospores
	Chytridiomycota	Non-septateg-lobose or ovoid mycelium	Motile zoospores	Sporangium
	Zygomycota	Non-septate elongated mycelium	Sporangia produce sporangiophores, conidia	Zygospore
Fungi	Ascomycota	Septate mycelium	Conidia, chlamydospores, sclerotia	Ascus containing ascospores (Sexual reproduction and structure are in some fungi are rare, lacking or unknown)
	Basidiomycota	-do-	Aeciospores, Uredinospores, Teliospores	Basidia, Basidiospores

Different types of perfect forms in fungi are shown in Fig. 1.

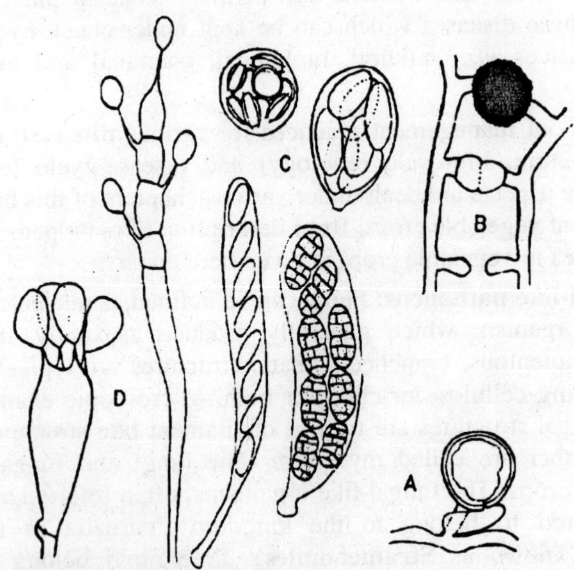

Fig. 1. Perfect forms of fungi (A) Oospore, (B) Zygospore, (C) Asci, (D) Basidium

Typical features of different classes of fungal pathogens are as follows:

1. Oomycota

The phylum Oomycota contains 74 genera and includes important pathogenic genera such as *Pythium, Phytophthora,* downy mildew fungi, *Albugo* etc. These are coenocytic, water loving fungi and are characterized by

- production of biflagellate motile zoospores with two types of flagella, the shorter flagella of the tinsel type directed forward and the whiplash type directed backwards,
- cell walls are composed of glucan-cellulose and lack chitin,
- an advance type of oogamous reproduction takes place by the passage of gamete nuclei,
- meiosis is gametangial rather than zygotic and the vegetative thallus is diploid,
- the life cycle is of haplobiontic type.

Among oomycota, Peronosporales are the most advanced and include biotrophic parasites of higher plants such as the 'downy mildews' and 'white rusts' that complete their entire life cycle on living hosts. In some of these genera, the sporangia germinate directly by germtubes (instead of forming zoospores) when the weather is dry (*Albugo*) while in other (*Peronospora*) the sporangia always germinate by germtubes and are therefore called conidia. This indicates that the conidia might have originated from sporangia. Oomycetes produce sexual stage known as oospores (Gr. oon= egg + spora= seed).

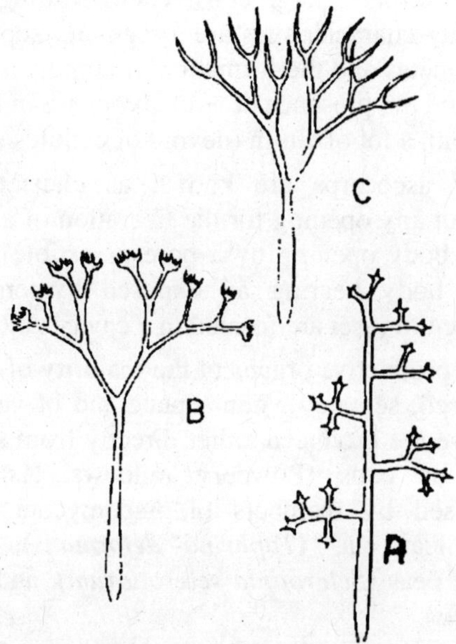

Fig. 2. Types of sporangiophores in Peronosporaceae (A) *Plasmopara,* (B) *Bremia,* (C) *Peronospora*

The Peronosporales are the highly advanced order of the oomycota fungi. They cause diseases of economic importance such as late blight of potato (*Phytophthora infestans*), white rust (*Albugo candida*) of crucifers, damping off of vegetables (*Pythium* spp.). Pythiaceae, Albuginaceae and Peronosporaceae are the three classes in this order.

In Pythiaceae, sporangiophores are usually undifferentiated from the mycelium, branched and indeterminate in growth while in other two classes these are easily differentiated from the mycelium. Sporangia are borne in chains (Basipetal manner) produced by a short unbranched sporangiophore in Albuginaceae while in Peronosporaceae, sporangia are borne singly, or in clusters at the tip of characteristically branched tree like sporangiophores of determinate growth (Fig. 2).

2. Zygomycota

The phylum Zygomycota is characterized by the presence of coenocytic hyphae, absence of motile cells and sexual reproduction that occurs by gasmetangial copulation and results in the formation of Zygospores. The asexual reproduction occurs typically by non-motile sporangiospores. Some of the Zygomycota fungi like *Mucor, Rhizopus*and *Choanephora* are weak parasites of plants including stored vegetables.

3. Ascomycota

Ascomycota includes a wide range of diverse fungi some of which are responsible for causing destructive plant diseases. These are characterized by the production of a distinctive structure, the ascus (Gr. askos = goat skin, sac) within which sexual spores, ascospores are formed by free cell formation after karyogamy and meiosis. Most of the ascomycetes produce large conspicuous fruiting bodies (ascocarp) that enclose the asci. The other characteristics are a regularly septate mycelium, absence of motile zoospore or gametes and the formation of conidia as the typical non-sexual reproductive structures and the presence of a dikaryophase in their life cycle. The cell walls of these fungi contain a lot of chitin (devoid of cellulose).

Different types of ascocarps are known as cleistothecium (globular and completely closed without any opening for the liberation of ascospores), perithecium (a flask shaped fruiting body opening by a pore or ostiole), apothecium (a cup or saucer shaped fruiting body bearing an exposed hymenium also called open ascocarp) and pseudothecium (asci are formed in a cavity within a stroma).

Conidia are chief reproductive organs of the majority of ascomycetes. These are variously shaped and sized, septate or non-septate and of various colours produced singly or in chains. These are produced either directly from somatic hyphae or from specialized conidiogenous cells (Powdery mildews, *Aspergillus, Penicillium*). Important diseases caused by members of Ascomycota are powdery mildews (*Erysiphe* sp.), peach leaf curl (*Taphrina deformans*), apple scab (*Venturia inaequalis*), white rot of peas (*Sclerotinia sclerotiorum*), and stem gall of coriander (*Protomyces macrosporus*).

4. Basidiomycota

Basidiomycota are named after a special structure the Basidium (Gr. basidion = a small base) with in which nuclear fusion and meiosis occur resulting in the production of four basidiospores exogenously on an extension (sterigma) of its wall, in contrast to the endogenous production of ascospores in the ascus of Ascomycota. The basidia bearing bodies of the basidiomycota vary from the minute pustules of the species of *Puccinia* (Uredinales) to the large basidiocarps of the species of *Ganoderma*. Other diagnostic characters are the presence of clamp connections, dolipore septa, and a double layered wall. The most predominant dikaryotic phase is interspersed between plasmogamy and karyogamy and represented by extensive dikaryotic mycelium and producing the basidiocarps. In this phyllum no differentiated sex organs are produced except in the rust fungi and the sexual function (somatogamy) is performed by the less specialized structures usually the somatic hyphae. Rusts and smuts in several crops are caused by members of Uredinales and Ustilaginales, respectively. In Uredinales (eg. *Puccinia*) the teliospore is terminal and germinates to give a promycelium bearing basidiospores on sterigmata from which they are forcibly discharged. In Ustilaginales (*Ustilago* spp.), the teliospores (smut spores) are usually intercalary and the basidiospores are produced either terminally or laterally but are sessile (not borne on sterigmata) and are not forcibly discharged from promycelium.

5. Mitosporic fungi (Anamorphic fungi)

This group comprises of imperfect, asexual or conidial states of Ascomycota and Basidiomycota and those asexual fungi with which no perfect states have been correlated so far. These are also known as 'Imperfect Fungi' because of the lack of the perfect sexual state. It lacks any common phylogenetic origin or relationship. Most mitosporic fungi are conidial states of ascomycetes while a few of basidiomycetes and sometimes, of certain zygomycetes. For some, the perfect state (teliomorph) is known while for some it is still undescribed e.g. several aspergilli, penicilli, fusaria, powdery mildews, rust etc. Several fungi of this group are serious plant pathogens and are responsible for causing serious diseases such as leaf spots, anthraxcnoses, leaf blights, seed rots, damping-off, root rots and wilts of various plants. A large number of them produce a variety of toxins in foods, feeds and grains in storage.

These fungi produce typically well-developed septate, branched hyphae. The cells are usually multinucleate. Reproduction may occur from mycelial fragments, budding or from spores of various types. Spore types produced by the Fungi Imperfecti may be oidia (Gr. oidion = small egg) also called arthrospores (*Geotrichum candidum*), chlamydospores which are thick walled thallic conidia that generally function as resting spores or conidia which are specialized non-motile asexual propagules. The conidia are produced on the simple or branched hyphae called conidiogenous cells or conidiophores. The conidia may be produced either at the tip or side of conidiogenous cells. These may be produced singly or in chains or in a droplet of mucus. The conidiophores may occur singly or may be aggregated to form specialized structures such as sporodochia and synnemata or produced indefinite fructifications known as pycnidia or acervuli. Sections of different fruiting bodies are shown in Fig. 3.

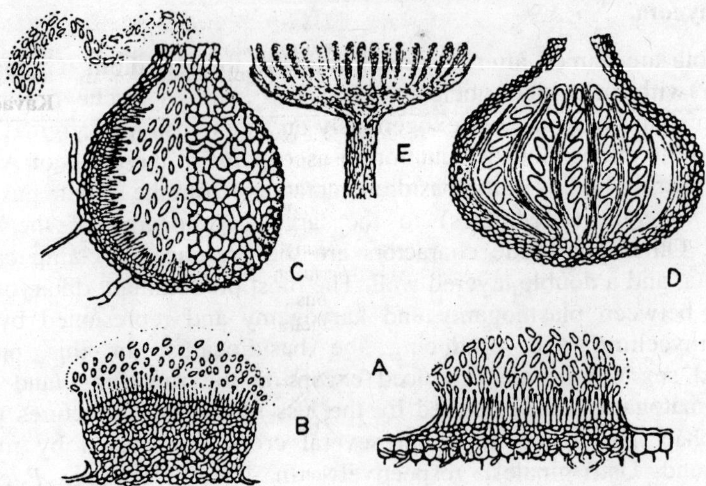

Fig. 3. Vertical sections of complex fructications. (A) Acervulus, (B) Sporodochium, (C) Pycnidium, (D) Perithecium, (E) Apothecium.

Bacteria

Bacteria are simple microorganisms having cell size smaller than the smallest spore of fungi (0.5-1 x 1-3 μm). These are prokaryotic in nature having differentiated nucleus and variable physiological behavior. These are distinguished on the basis of their cell wall characteristics, colony growth, production of diffusible pigments, position of flagella and also certain biochemical characters as given in Table 2.

Table 2. Some fundamental characters of bacterial pathogens

Division	Family	Genus	Colony colour	Flag-ellum	Diffu-sible pigments	Glucose meta-bolism	Oxid-ase test of Kavacs	Gram Reac-tion
Gracillicu tes	Pseudomo-nadaceae	*Pseudo-monas*	Creami sh white	One or many polar flagella	+ (fluo-rescent)	Oxida-tive	+	- tive
		Xantho-monas	Yellow	One polar flage-llum	+	Oxid-ative	-	- tive

Division	Family	Genus	Colony colour	Flag-ellum	Diffu-sible pigments	Glucose meta-bolism	Oxid-ase test of Kavacs	Gram Reac-tion
		Ralstonia	Creamish white	One or many polar flagella	-	Oxidative	+	- tive
	Enterobacteriaceae	*Erwinia*	Creamish whit	Many peritrichous flagella	-	Fermantative	+	- tive
Firmicutes	- (Class: Thallobacteria)	*Clavibacter* (*Cornybacterium*)	Pale orange	Non-motile	-	Weak or nil	-	+tive
Tenericutes (Mollicutes)	Spiroplasmataceae	*Spiroplasma*	Have only cell membrane and lack cell wall					
	-	*Phytoplasma*						

Viruses and Viroids

Viruses and viroids are the smallest entities that have either DNA or RNA as their genomic constituent and a protein coat. These reproduce and multiply in the living cells after infection by using host machinery and are transferred to adjoining cells to manifests different types of symptoms.

Viruses

Viruses represent significant threats to modern agriculture and their biological control remains a challenge in the twenty-first century. Although viruses behave like microorganism in that they cause disease, can function genetically and are able to reproduce, they also behave like a chemical molecule, as they are inactive outside the living system. They consist of nucleic acid and a protein, which forms a protective coat around the nucleic acid. The nucleic material is either RNA or DNA, both can be single stranded or double stranded. Therefore, a virus is a nucleoprotein that has the ability to cause disease. It multiplies in a living host and is very small to be seen under light microscope. The shape of the viruses can vary from rod shape to polyhedral. Viruses cannot reproduce themselves by producing any reproductive structures. Instead, they infect and induce host cell to produce more virus particles. They cause disease utilizing the host machinery during multiplication, make space for itself into the host cell and disrupt all the cellular processes. This upset the host cellular metabolic processes showing various symptoms in the infected host cells. There are hundreds of plant viruses which are classified into 7 orders and 25 families with additional 71 families which are not assigned to any order yet (Table 3) (ICTV webpage, 2013).

Table 3. Classification and nomenclature of viral pathogens

Order	Families
Caudovirales	*Myoviridae, Podoviridae, Siphoviridae*
Herpesvirales	*Alloherpesviridae, Herpesviridae, Malacoherpesviridae*
Ligamenvirales	*Lipothrixviridae, Rudiviridae*
Mononegavirales	*Bornaviridae, Filoviridae, Paramyxoviridae, Rhabdoviridae*
Nidovirales	*Arteriviridae, Coronaviridae, Mesoniviridae, Roniviridae*
Picornavirales	*Dicistroviridae, Iflaviridae, Marnaviridae, Picornaviridae* *Secoviridae* Unassigned
Tymovirales	*Alphaflexiviridae, Betaflexiviridae, Gammaflexiviridae, Tymoviridae*
Virus families not assigned to an order	71 families

However, different genera of virus(s) along with their genome organization and particle shape infecting vegetable crops are described in Table 4.

Table 4. Classification and nomenclature of viral pathogens infecting vegetable crops

Genome organization	Family	Genus	Type species	Particle shape
ssRNA	Potyviridae	*Potyvirus*	*Potato virus Y*	Filamentous
	Closteroviridae	*Crinivirus*	*Lettuce inf. yellows virus*	
	Tombusviridae	*Tombusvirus*	*Tomato bushy stunt virus*	
(+) ssRNA	Luteoviridae	*Polerovirus*	*Potato leafroll virus*	Isometric viruses
		Enamovirus	*Pea enation mosaic virus*	
	Bromoviridae	*Cucumovirus*	*Cucumber mosaic virus*	
(-) ssRNA	*Rhabdoviridae* (Order: *Mononegavirales)*	*Nucleorhabdovirus*	*Potato yellow dwarf virus*	Bacilliform particles
	Bunyaviridae	*Tospovirus*	*Tomato spotted wilt virus*	Membranous circular particles
dsRNA	Partitiviridae	*Varicosavirus*	*Lettuce big-vein virus*	Filamentous
ssDNA	Geminiviridae	*Begomovirus*	*Bean golden mosaic virus*	Isometric
dsDNA	Caulimoviridae	*Caulimovirus*	*Cauliflower mosaic virus*	

Viroids

Viroids are small, circular, low molecular weight ribonucleic acid that can infect plant cells, replicate themselves in plant nucleus and cause disease. Viroids have not yet been detected in any kind of organism besides plants. Because of their small size (250-370 nucleotides), viroids lack sufficient information to code for even one protein, even for a replicase enzyme required to replicate the viroid. Viroids uses the substances and enzymes of plant cells. Viroids have single-stranded RNA molecules with extensive base pairing in parts of RNA strand (Fig. 4). The base pairing results in some sort of hairpin structure with single-stranded and double-stranded regions of the same viroid and contributes to the stability of RNA, given that it lacks a protein coat. Few examples of viroid plant diseases are Potato spindle tuber, Chrysanthemum stunt, Apple scar skin, Citrus exocortis, Cadang-cadang disease of coconut, Tomato apical stunt, Tomato chlorotic dwarf etc.

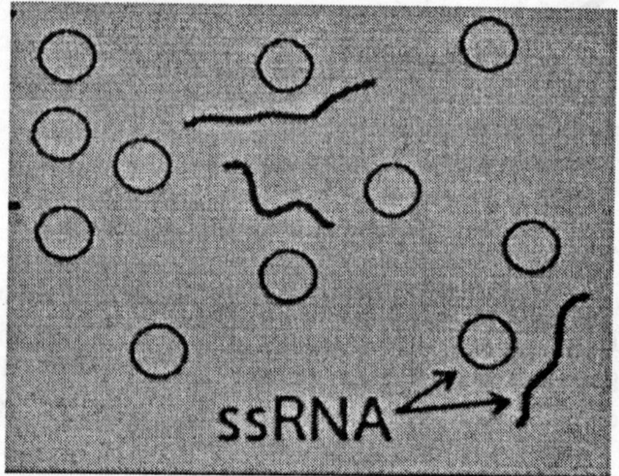

Fig. 4. Circular and linear viroid molecules

Phytoplasma

Phytoplasmas lack cell walls, are bounded by a "unit" membrane, and have cytoplasm, ribosomes, and strands of nuclear material. The size of their chromosomes varies from 530 kilobases of DNA to 1130 kilobases. Their shape is usually spheroidal to ovoid or irregularly tubular to filamentous, and their sizes are comparable to those of the typical mycoplasmas. Figure 5 shows size variations in bacterial cell, phytoplasma and different groups of viruses.

Diverse types of symptoms are produced by different plant pathogens on affected host plants and symptoms may also vary on different parts of the host plants attacked (Agrios, 1997). Proper understanding of the pathogen biology, disease cycle and factors affecting disease development and spread is necessary for devising suitable disease control measures for diverse diseases affecting vegetable crops. Since desirable host resistance is not easily available in vegetable crops, farmers generally rely on the use of fungicides and insecticides for managing fungal diseases and virus vectors. To reduce environmental pollution, integrated disease management involving different management practices provides better options for economical and ecologically safe methods for effective disease management in production of vegetable crops.

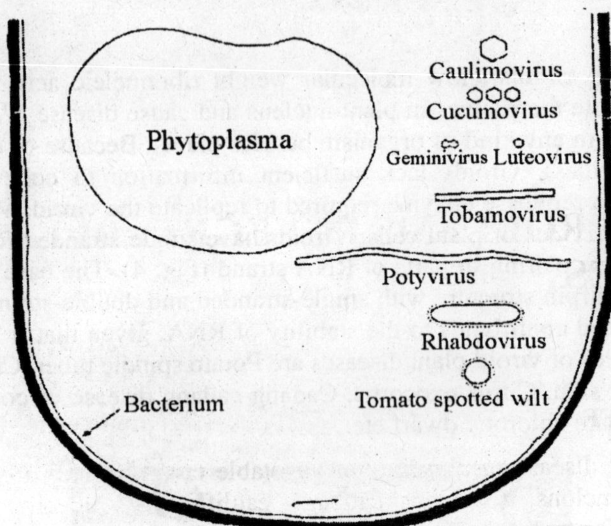

Fig. 5. Showing size variation in bacterial cell, phytoplasma and different groups of viruses.

REFERENCES

Agrios, G.N. 1997. *Plant Pathology.* Academic Press Inc. London, 803 pp.

Kirk, P.M., Cannon, P.F., Minter, D.W. and Stalpers, J.A. 2008. Ainsworth and Bisby's Dictionary of Fungi (10th edition) CAB International, Oxon, UK. 84 pp.

Messiaen, C.M., Blancard, D., Rouxel, F. and Lafon, R. 1991. *Les Maladies Des Plantes Maraicheres.* 3rd Edition. Institut National De Le Recherche Agronomique, Paris, 552 pp.

■■■■

2

DISEASE PROBLEMS IN VEGETABLE NURSERIES

1. DAMPING OFF

It is an important disease in nurseries of vegetable crops like tomato, bell pepper, chillies, brinjal, melons, cucumber, cabbage, cauliflower, broccoli etc. for which nursery is raised for transplanting. The disease is fairly common in poorly managed nursery beds, incited often by the seed and soil borne pathogens. It is responsible not only for the poor seed germination and stand of seedlings but also for carryover of the pathogens to the field where transplanting is done for raising the crop. The infected seedlings provide an ideal avenue for rapid and efficient spread of diseases caused by soil and seed borne fungal and bacterial pathogens (Shyam, 1991).

Symptoms

The disease manifests in two phases (i) pre-emergence damping-off, and (ii) post-emergence damping-off.

(i) Pre-emergence damping-off

In this phase, there is failure of seedling emergence from the soil either due to seed rots or killing of young seedlings before their emergence from the soil, hence resulting in patchy appearance of seedlings in the nursery in early stages of growth (Plate 1a).

(ii) Post emergence damping-off

In case of post emergence damping-off, the disease outbreak is characterized by toppling over of infected seedlings at any time after their emergence from the soil. The infected tissue initially appears to be water-soaked and soft, and subsequently, the stem at the infection point gets constricted resulting in toppling over and mortality of the seedlings (Plate 1). In the nurseries, the disease usually radiates from initial infection points and thus gives the appearance of islands wherein almost all seedlings are killed (Plate 1b, 1c).

The Pathogen(s)

Soil borne pathogens, such as *Pythium aphanidermatum, P. butleri, P. inflatum, Phytophthora parasitica, Fusarium oxysporum, F. solani* f.sp. *minus*, *Phomopsis vexans, Colletotrichum capsici, Rhizoctonia solani* and *Sclerotium rolfsii* are the most commonly associated pathogens with this disease complex. Among these fungi, the species of Pythium and Rhizoctonia solani are more common than others. *P. aphanidermatum* (Eds.) Fi., *P. butleri* Subram. and *P. ultimum* Trow. are common species.

The mycelium of *Pythium* is intracellular in the host. It consists of slender, coenocytic and much branched hyphae. Asexual reproduction is mostly by means of zoospores, which are always differentiated in a vesicle extruded by the zoosporangium. The sporangium forms a beak, the tip of which softens and out of which flows the protoplasmic mass to form a spherical vesicle. In this vesicle, protoplasm cleaves to form zoospores, which are kidney shaped and biflagellate. They rupture the wall of the vesicle and escape in the surrounding film of water. After swimming for a while, the zoospores encyst and then germinate by forming germtube. In sexual reproduction, the oogonium is plurinucleate, arises terminally (less often intercalary) on long or short unmodified hyphae with homogeneous protoplasm and only one nucleus migrates into the periplasm. The antheridium is also plurinucleate at first but only one nucleus functions in the fertilization of egg. The branch bearing the terminal antheridium may arise at a distance from the oogonium on a separate branch or from the same branch. Through a conjugation tube, one male nucleus passes into the egg, which then forms a thick wall and becomes oospore.

Pythium aphanidermatum and *P. butleri* are closely related species. The mycelium of *P. aphanidermatum* is white, well developed and woolly, but not as copious or robust as of *P. butleri*. The hyphae measure 2.8 to 7.5 μm in diameter. The zoosporangia are lobbed, toruloid, branched, but smaller and less profusely branched than those of *P. butleri*. The swellings sometimes only form bud like outgrowths from the main sporangia. Encysted zoospores measure 7.6 to 12.0 μm. Oogonia are spherical, smooth walled, terminal on lateral hyphae and 20 to 29 μm in diameter. Antheridia are monoclinous, intercalary or terminal. One or two antheridia attach to each oogonium. They are usually barrel or dome shaped and measure 9–11 × 10–14 μm in size. Oospores are aplerotic, single, moderately thick walled and 17–19 μm in diameter.

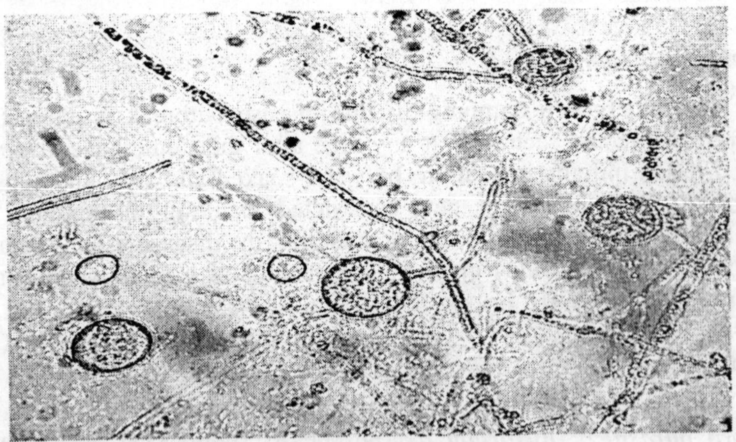

Fig. 1. Sporangia of *Pythium ultimum*

P. butleri differs from *P. aphanidermatum* in terms of more copious robust aerial mycelium, the larger, more swollen and more branched sporangia, larger zoospores, antheridia, oogonia and oospores.

The mycelium of *P. ultimum* is white and cottony. Hyphae are hyaline and coenocytic ranging from 4.65–8.67 μm in size Sporangia are mostly round and somewhat ovoid in shape without any papilla. Oospores are round, smooth, thick walled, plerotic and hyaline and measure 23.09 × 22.91 μm in size (Fig. 1, 2).

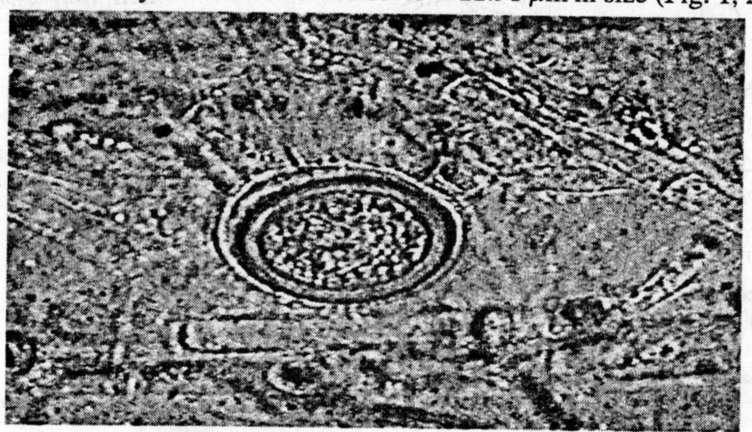

Fig. 2. Oospore of *Pythium ultimum*

Disease cycle

Most of the fungi responsible for this disease are either seed or soil borne. *Pythium* species are more common pathogens associated with this disease. The species of *Pythium* survive from one season to the other as mycelium in plant debris or oospores in the soil (Fig. 3), which germinate in favourable weather and produce zoosporangia. The zoosporangia produce zoospores, which after encystment either cause infection of the germinating seedlings below the soil surface or at collar portion of the seedling near soil line. The mycelium grows inside the diseased tissues and after appearance of symptoms, the infected plants topple down on the soil surface. The sporangia produced on these toppled plants cause secondary infection in the presence of favourable weather. If climate becomes unfavourable, the sexual reproduction sets in and the fungus produces oospores, which can survive under adverse climatic conditions.

Rhizoctonia solani survives in soil in the form of sclerotia. Hyphae grow from these sclerotia and infect seedlings in the nursery beds. The pathogen can survive in soil for several years.

Fusarium sp. are soil borne in nature and survive in the form of chlamydospores. In favourable weather, they germinate and produce sporodochia which in turn produce micro and macro conidia. The conidia germinate by germtube and cause infection. Though *Fusarium* sp. by themselves are capable of causing infection, but it is further enhanced if the nematodes are already present in nursery soil because they provide avenues for entry of this pathogen in the roots.

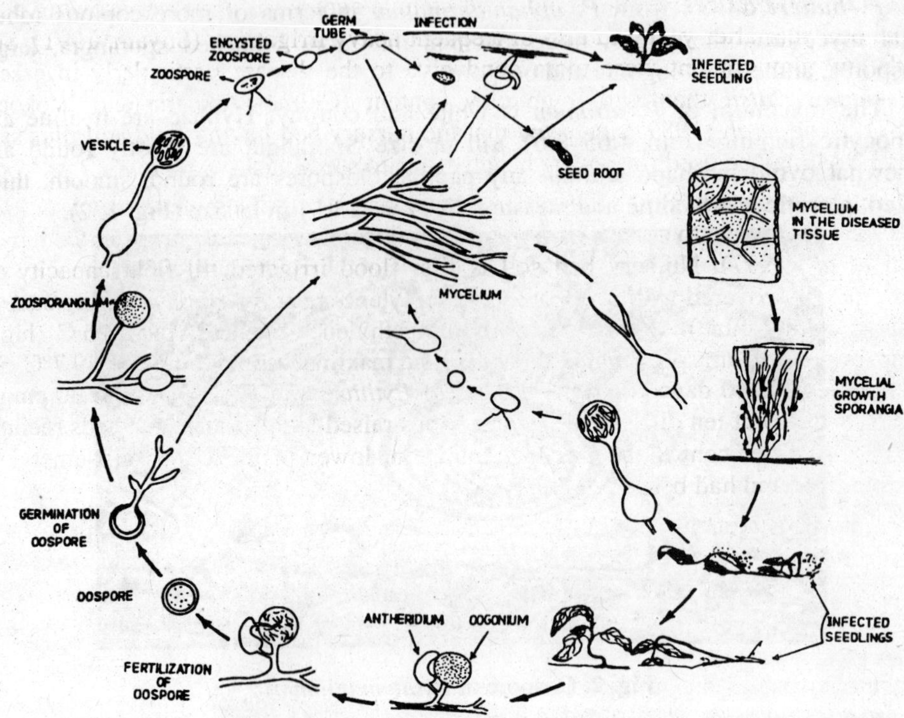

Fig. 3. Disease cycle of damping off of vegetables due to *Pythium* sp.

Epidemiology

The pathogens associated with this disease are either seed or soil borne or both. Species of *Pythium* attack the plants when they are in juvenile stage. Soil moisture near saturation stimulates mycelial growth as well as asexual reproduction. Propagules capable of causing infection of tomato include vegetative mycelium, germinating sporangia, zoospores and oospores. Some propagules, particularly zoospores may come into contact with tomato plants by chemotaxis. Infective propagules are capable of direct penetration, but wounds enhance penetration and infection. High soil moisture, pH 6.0 and heavy soils favour disease development. Pre-emergence damping-off is maximum at 20–25ºC, while post-emergence damping-off is more severe at 30–40ºC (Tripathi and Grover, 1976). Shyam (1991) also emphasized that higher soil moisture and relatively higher soil temperature favour the rapid development of the disease. Kumar (2017) also observed high soil moisture (60–75%) and temperature (25ºC) as optimum for the development of this disease. The disease is further aggravated in ill-aerated soils with poor drainage having thick stand of the seedlings. Seedlings supplied with balanced nutrients are less susceptible to the damping-off (*P. butleri*).

Management

The management of damping-off has been attempted by the adoption of cultural, chemical and biological methods.

(i) Cultural practices: Cultural practices include avoidance of nursery sowing in the same bed year after year and also of frequent heavy irrigations (Shyam, 1991). Soils with low clay content were more conducive to the disease particularly in case of *Sclerotium rolfsii* than with high clay content (Chima-Wokocha and Wokocha, 1987). Rana *et al.* (1989) suggested that the nursery bed having field soil plus sand, plus FYM may be used for raising healthy tomato seedlings.

(ii) Soil solarization: Soil solarization with transparent polyethylene mulch (25 μm) was found effective to control damping-off of different vegetable crops in the nursery (Raj *et al.*, 1997). Nursery bed soil is first flood irrigated till field capacity and subsequently covered with transparent polyethylene sheet for 40 days during summer months (May–June). Mulching with polyethylene resulted in 13.5°C higher temperature at 8 cm soil depth with an average maximum temperature of 49.7°C. Soil solarization for 40 days killed the species of *Pythium* and *Fusarium* upto 30 cm soil depth. Nursery of ten different vegetable crops raised later in solarized beds recorded 18.3 to 42.0 per cent higher seed germination, lower incidence of post-emergence damping-off and had better seedling vigour.

(iii) Chemical control: Soil sterilization and seed treatment with fungicides are the most frequent chemical methods used for the management of this disease. Soil sterilization and soil application of mercuric chloride (1:1500) was recommended for the control of this disease earlier, but due to persistence of mercury in soil for long time, this recommendation stands withdrawn now. Efficacy of soil fumigation with metham, formalin and methyl bromide in reducing soil borne fungi has also been reported (Shukla *et al.*, 1977; McCarter *et al.*, 1978). Pre-sowing application of dazomet (Basamid G @ 30 g/ m^2) was also found effective in the management of damping-off caused by *P. aphanidermatum* (Patel and Patel, 1998). Seed treatment with cuprous oxide, copper oxychloride, carbendazim, captan, thiram, carbendazim + thiram, carboxin, have been recommended for the management of pre-emergence damping-off, in vegetables (Shukla *et al.*, 1977; Vyas *et al.*, 1981; Shyam, 1991; Kaushal and Sugha, 1995) which improved seedling stand in the nurseries. Captan when applied as seed dresser (0.25%) and soil drenching (0.3%) 15 days after planting exhibited highest total healthy plant stand in chilli, cabbage and onion crops (Champawat and Sharma, 2003). Seed treatment with Metalaxyl resulted in minimum pre-emergence damping-off while drenching of Apron XL, Curzate M-8, Metalaxyl, Ridomil Gold and Sectin caused complete reduction of post emergence damping-off under pot conditions (Kumar and Gupta, 2017).

(iv) Biological control: Khara and Hadwan (1990) studied tomato rhizosphere fungi for their antagonism against *R. solani*. *Trichoderma koningii* and *T. pseudokoningii* showed maximum antagonism. Soil inoculation with *T. viride, T. harzianum* and *Laetisaria arvalis* gave good control of *P. indicum*, and there was manifold increase in seed germination in comparison to control (Krishnamoorthy and Bhaskaran, 1990). Seed treatment with these bio-agents was comparable to soil drenching with fenaminosulf or copper oxychloride or seed treatment with captan or thiram. Although soil and seed applications of *Gliocladium virens* and *T. longibrachiatum* against *R. solani* and *P. aphanidermatum* were also efficacious, yet, in general soil application of these bioagents conferred greater protection than the seed dressings (Sreenivasaprasad and Manibhushanrao, 1990), while seed coating with conidial suspension of antagonists like *T. harzianum, T. viride* and *G. virens* was found more effective than soil application in protecting the tomato seedlings from damping-off (Hazarika *et al.*, 2000, Jayraj *et al.*, 2006; Basharat *et al.*, 2011). Seed and soil treatments with *T. viride* and *T. harzianum* alone or in combinations with FYM and *G. virens* reduced the damping-off and other root diseases (Hazarika *et al.*, 2000; Basharat *et al.*, 2011). Bacterial antagonists like *Bacillus subtilis* Bs2 and *Pseudomonas fluorescens* effectively decreased the incidence of damping-off of brinjal and increased root and shoot length (Pandey and Simon, 2010). This may be due to the induction of defence related enzymes involved in the phenyl propanoid pathway, collectively contributed to enhance resistance to the pathogen. Seed treatment with consortium of native biological agents like *Pseudomonas* and *Azotobacter* species exhibited an enhanced level of protection as compared to native individual antagonistic bacteria to control pre- and post-emergence damping-off of tomato and brinjal (Champawat and Sharma, 2003; Dutta and Dutta, 2007; Pandey and Simon, 2010). Inoculation of mycorrhizal fungus *Glomus fasciculatum* either simultaneously or 2 weeks before inoculation with *P. aphanidermatum* reduced damping-off and increased plant height in comparison to control (Hegde and Rai, 1984).

(v) Plant extracts: Kurucheve and Padmavathi (1997) studied the fungitoxicity of five plant products against *P. aphanidermatum*, the causal agent of damping-off of chillies and found garlic (*Allium sativum*) bulb extract as most effective and recorded minimum mycelial growth followed by *Lawsonia inermis* leaf extract. Maximum percentage of seed germination, growth and vigour of chilli seedlings were also observed when seeds were treated with garlic bulb extract. Hooda *et al.* (2011) reported the efficacy of *Lantana camara*, neem cake, cow urine, *Urtica parviflora, Sapium* sp., *Ligustrum nepalensis, Eucalyptus* sp. and Azadirachtin in managing pre-emergence damping-off, while *Curcuma longa* and *Thuja compacta* were effective against post-emergence damping off.

(vi) Integrated disease management: Shyam (1991) recommended the integration of cultural and chemical methods like change of nursery site every year along with pre-sowing soil sterilization with formalin (5%), seed treatment with captan or thiram and drenching of beds after emergence with captan (0.3%) or mixture of mancozeb (0.25%) and carbendazim (0.1%) at 8–10 days interval coupled with avoidance of heavy irrigations for the management of damping off in summer and winter vegetables. Soil solarization with transparent polyethylene mulch during April–May, coupled with seed treatment and soil application of *T. harzianum* were found effective in reducing the incidence of both pre- and post emergence damping-off of tomato under field conditions (Kumar and Gupta, 2017).

2. Foliar pathogens

Several fungal, bacterial and viral diseases infest the foliage of nursery plants after their emergence from soil surface. Sometimes, seed borne fungal and bacterial pathogens like leaf spots (*Alternaria* sp.) and blight in tomato (*Phytophthora infestans*), downy mildew (*Peronospora parasitica*) and black rot (*Xanthomonas campestris* pv. *campestris*) in crucifers, bacterial spot (*X. vesicatoria*), and bacterial canker (*Clavibacter michiganensis* pv. *michiganensis*) in chilli and tomato etc.

appear in the nurseries and multiply rapidly. Keeping this in view, suitable seed treatments and foliar sprays should be applied as suggested in the subsequent chapters.

Viral diseases like leaf curl and mosaics may also appear in the nursery of some vegetables and are carried from the nursery to the fields/protected structures. Vectors like white fly and aphids transmit these viruses. To avoid these diseases in nursery, it is always better to raise the nursery under a net house having insect proof net (40 mesh size) so that the entry of insect vectors is restricted. Insecticides like dimethoate or imidaclorpid should be applied at weekly interval to control these insect vectors. Besides this, some insecticide dust like methyl parathion should be applied around the borders of the nursery bed to avoid the entry of ants and other soil borne insects.

3. Nematode infestation

Sometimes nematodes like *Meloidogyne* sp. can infest seedlings in the nursery beds of vegetable crops, such as tomato, brinjal and cucurbits. This can facilitate to spread the nematodes to the fields where this problem does not exist and can result into severe yield losses at later stages, if remained unchecked. It is advisable to apply carbofuran at the time of nursery bed preparation for management of nematode population below economic threshold.

REFERENCES

Basharat, N., Sobita Simon, Das, S. and Soma, R.V. 2011. Comparative efficacy of *Trichoderma viride* and *T. harzianum* in management of *Pythium aphanidermatum* and *Rhizoctonia solani* causing root rot and damping-off diseases. *J. Plant Dis. Sci.* **6**: 60–62.

Champawat, R.S. and Sharma, R.S. 2003. Integrated management of nursery diseases in brinjal, chilli, cabbage and onion. *J. Mycol. Pl. Pathol.* **33:** 290–291.

Chima-Wokocha, R. and Wokocha, R.C. 1987. Effect of soil type on the damping-off of tomato seedlings caused by *Sclerotium rolfsii* in the Nigerian savanna. *Pl. Soil* **98(3):** 443–444.

Dutta, S. and Dutta, D. 2007. Evaluation of biocontrol potentiality of native plant growth promoting bacteria against *Rhizoctonia solani* mediated damping-off disease of tomato. *J. Mycopathol. Res.* **45:** 201–206.

Hazarika, D.K., Sarmah, R., Paramanick, T. Hazarika, K. and Phookan, A.K. 2000. Biological management of tomato damping-off caused by *Pythium aphanidermatum. Indian J. Plant Pathol.* **18(1&2):** 36–39.

Hegde, S.V. and Rai, P.V. 1984. Influence of *Glomus fasciculatum* on damping-off of tomato. *Curr. Sci.* **53(11):** 588–589.

Hooda, K.S., Joshi, D., Shri Dhar and Bhatt, J.C. 2011. Management of damping-off of tomato with botanicals and bio-products in North Western Himalayas. *Indian J. Horti.* **68:** 19–23.

Jayaraj, J., Radhakrishnan, N.V. and Velazhahan, R. 2006. Development of formulations of *Trichoderma harzianum* strain M1 for control of damping-off of tomato caused by *Pythium aphanidermatum. Arch. Phytopathol. Plant Protec.* **39:** 1–8.

Kaushal, N. and Sugha, S.K. 1995. Role of *Phomopsis vexans* in damping-off of seedlings in eggplant and its control. *Indian J. Mycol. Pl. Pathol.* **25(3):** 189–191.

Khara, H.S. and Hadwan, H.A. 1990. *In vitro* studies on antagonism of *Trichoderma* spp. against *Rhizoctonia solani* the causal agent of damping-off of tomato. *Pl. Dis. Res.* **5:** 144–147.

Krishnamoorthy, A.S. and Bhaskaran, R. 1990. Biological control of damping-off disease of tomato caused by *Pythium indicum* Balakrishnan. *J. Biol. Control* **4(1):** 323–324.

Kumar, Rajesh, 2017. Studies on damping-off of tomato caused by *Pythium* sp. M. Sc. thesis submitted to YSP University of Horticulture and Forestry, Solan, 60p.

Kumar, Rajesh and Gupta, S.K. 2017. Etiology and integrated management of damping-off of tomato in Himachal Pradesh. National Symposium "Biorational Approaches in Plant Disease Management" YSP University of Horticulture and Forestry, Solan, 27–28 Oct., 2017 (Abst.).

Kurucheve, V. and Padmavathi, R. 1997. Fungitoxicity of selected plant products against *Pythium aphanidermatum. Indian Phytopath.* **50(4):** 535–539.

McCarter, S.M., Jaworski, C.A. and Johnson, A.W. 1978. Effect of continuous plant culture and soil fumigation on soil borne plant pathogens and on growth of tomato transplants. *Phytopathology* **68(10):** 1475–1481.

Pandey, A. and Simon, L.S. 2010. Bacterial antagonists for the management of damping-off of brinjal. *Nat. Acad. Sci. Lett.* **33(11/12):** 365–368.

Patel, H.R. and Patel, B.N. 1998. Evaluation of Dazomet (Basamid G) against rove beetles, damping-off, weeds and nematodes in bidi tobacco nursery. *J. Mycol. Pl. Pathol.* **28:** 134–139.

Raj, H., Bhardwaj, M.L. and Sharma, N.K. 1997. Soil solarization for the control of damping-off of different vegetable crops in the nursery. *Indian Phytopath.* **50(4):** 524–528.

Rana, R., Gangopadhyay, S. and Banerjee, M.K. 1989. Incidence of damping-off of tomato in different types of nursery beds. *Pl. Dis. Res.* **4(2):** 179.

Shukla, P., Singh, R.P. and Prasad, M. 1977. Studies on damping-off disease of pea caused by *Rhizoctonia solani* I. Control measures. *Indian J. Mycol. Pl. Pathol.* **7(2):** 157–158.

Shyam, K.R. 1991. Important diseases of summer and winter vegetable crops and their management. In *Horticultural techniques and post harvest management* (eds. Jain *et al.*), Directorate of Extension Education, UHF, Nauni, pp. 62–69.

Sreenivasaprasad, S. and Manibhushanrao, K. 1990. Biocontrol potential of fungal antagonists *Gliocladium virens* and *Trichoderma longibrachiatum. Z. Pflanzenkr. Pflanzenschutz* **97(6):** 570–579.

Tripathi, N.N. and Grover, R.K. 1976. Inoculum potential and soil factors affecting the pathogenesis of *Pythium butleri* in causing damping-off of tomato. *Proc. Indian Natl. Sci. Acad. B.* **41(5):** 466–474.

Vyas, S.C., Andotra, P.S. and Joshi, L.K. 1981. Effect of systemic fungicides on control of root rot on vegetables caused by *Rhizoctonia bataticola* and plant growth. *Pesticides* **15(11):** 22–24.

■■■■

3

DISEASES OF POTATO

Potato (*Solanum tuberosum* L.) is an important vegetable crop and is cultivated in an area of about one million hectares in India, with the total production of more than 1.5 million tons. Important potato growing states are Uttar Pradesh, West Bengal, Bihar, Assam, Himachal Pradesh, Madhya Pradesh, Punjab and Karnataka. One to three crops per year are taken under varied agroclimatic conditions. In India, about 80 per cent of potato crop is cultivated under subtropical and 20 per cent under temperate to sub-temperate climate. The cultivation of this crop is often affected due to attack of various diseases caused by fungi, bacteria, viruses and nematodes etc. (Khurana *et al.*, 1998a). This chapter deals with the important disease problems in potato, their etiology, epidemiology and different strategies for effective management.

FUNGAL DISEASES

1. LATE BLIGHT

Late blight is the most destructive disease of potato throughout the world and was recorded in the beginning of nineteenth century in the Andes Mountains of South America. The disease is believed to have been introduced into Europe from South America in early 1840s from where it spread to other countries and caused a famine in Ireland (known as Irish famine) during 1845-46 resulting into deaths of millions of Irish people and an equal number migrating to other parts of the world, particularly North America. The systematic development of Plant Pathology as a discipline started after this famine. Late blight has the tremendous potential to cause up to 70% reduction in the yield in a susceptible cultivars. The losses ranging from 20-25% in Punjab, 40-50% in Haryana, 15-20% in Uttar Pradesh and 5-10% in Bihar and West Bengal have been reported (CPRI, 1987). In India, annual losses due to late blight have been reported to range between 10-75% (Dutt, 1979). The extent of damage depends largely at which stage the crop is attacked and it is usually heavy when infection occurs on the young crop stage before tuberization that sometimes can result into complete yield losses in epidemic years. The losses vary between 15 to 100 per cent depending on cultivar, crop stage and the fungicides sprayed (Thind and Mohan, 1998).

Geographical distribution

Late blight has been reported from almost all parts of the world, wherever potato is grown. In India, the disease was first reported from Nilgiri hills between 1870 and 1880, from Darjeeling in 1883 and from Shimla hills in 1902. In the plains, the disease was first recorded from Hoogly district of West Bengal between 1898 and 1900 and caused an epidemic in Indo-Gangetic plains in 1913. Later the disease appeared in plains of Uttar Pradesh and other north Indian states including Punjab and Haryana and is now prevalent in all the areas under cool and wet weather conditions.

Symptoms

The symptoms of late blight appear on leaves, stem and tubers. Initially small, water-soaked lesions develop near the tips and margins of the leaves which rapidly grow into large, brown to purplish black, necrotic lesions (Plate 1a) under favourable weather conditions (Fig. 1). During morning hours, whitish downy growth of the pathogen consisting of sprangiophores and sporangia can be seen on the edges of the lesions mostly on the underside of the leaves (Plate 1b). Light brown to dark brown lesions appear on stem and petioles which may elongate later and girdle the affected parts. The affected stems become weak and collapse, causing death of the plant parts above the lesions. Sometime the tender growing stem tips and nearby young foliage are attacked in the young crop and the plants stop their growth and wither soon. Since the disease is polycyclic in nature, the entire crop in a field may be killed in one or two weeks and the field gives blighted appearance (Plate 1d). The tubers may get infected by rain-washed sporangia from the diseased foliage. The infected tubers show irregular, small to large, slightly depressed areas of brown to purplish skin which extend deep into the internal tissue of the tubers (Plate 1c). The infected tuber tissue which is firm and dry in the beginning, is often invaded by secondary pathogens, mainly bacteria, in the field or poorly ventilated storage places resulting into soft rot of tubers.

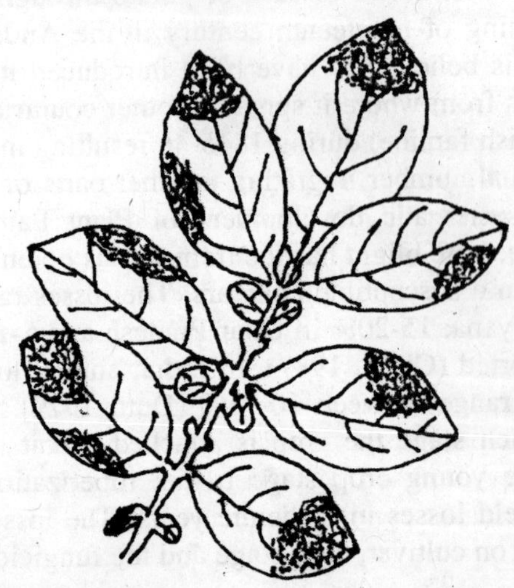

Fig. 1. Symptoms of Late blight on leaves.

Diagnostics

In India, Hussain *et al.* (2013) validated PCR based protocol for detection of latent infection in commercial seed tuber stocks. The pathogen could be detected up to 25 mm away from the diseased lesions. A species specific, SCAR marker (PITH2-F/PITH2-R) has been developed using RAPD markers for detection of *Phytophthora infestans* which showed specificity to *P. infestans* only (IISR, 2013). Besides, duplex PCRs have been developed for simultaneous detection of late blight and early blight, and late blight and dry rot of potato (IISR, 2014). Real time PCR quantification was carried out to determine the *P. infestans* load in the host tissues. Sampling was performed at different distance levels and different time intervals. The studies indicate that the *P.infestans* mycelium can penetrate the host tissue up to 25 mm away from the symptomatic area (IISR, 2013).

The pathogen and variability

Late blight is caused by *Phytophthora infestans* (Mont.) de Bary which belongs to kingdom *Chromista*, phyllum *Oomycota,* class *Oomycetes*, order *Peronosporales* and familiy *Pythiaceae*. The mycelium is coenocytic, hyaline, branched having simple to club shaped haustoria. Sporangiophores originate from internal mycelium and emerge through stomata on the leaves and tubers. These are slender, hyaline, sympodially branched, indeterminate, septate having side branches with swollen bases. Sporangia are formed on the tip of sporangiophores and are hyaline, thin walled, lemon shaped, papillate having up to 30 nuclei, measuring 21-28 × 12-24 μm. The sporangium, turns to the side on maturity and sporangiophore growth continues with characteristic swollen nodes. This fungus is heterothallic and requires two mating types A1 and A2 for sexual reproduction.

Before 1980, the worldwide population of *P. infestans* outside Mexico appearaed to be asexual and to consist of a single clonal lineage of A1 mating type characterized by a single genotype. It is widely believed that new strains migrated into Europe in 1976 and that this led to subsequent population changes including the introduction of the A2 mating type. The population characteristics of the existing population in NW Europe show a diverse population including both mating types, sexual reproduction and oospores and the aggressiveness of *P. infestans* has increased in the past 20 years (Cooke *et al.*, 2011).

In India, although A1 mating type is of common occurrence, prevalence of A2 type has been recorded in Shimla hills after 1984 (Singh *et al.*, 1994). Antheridia and oogonia are produced after A1 and A2 mating types come in close contact. Antheridia are amphygynous while oogonia are spherical. Oospores are 24-46 μm in diameter, thick walled and develop after fertilization. Oospore formation varies depending upon particular A1 and A2 combinations and also on infected potato genotypes (Cohen *et al.*, 1997) but self production of oospores in A1 population is also reported. There has been shift from simple 0 and 4 races to complex 4-8 gene races of *P. infestans* after 1965 in India. Arora (1990) reported that development and fitness of complex races is dependant on cultivars planted and favourable weather conditions.

In India, A2 mating type of *P. infestans* has stabilized in temperate hills while the A1 is dominating in sub-tropical plains (CPRI, 2013, 2014, 2015). Recently, the ploidy status of *P. infestans* isolates collected from different parts of the country was determined and results revealed that population in the country consist of diploids, triploids and tetraploids. The frequency of diploids, triploids and tetraploids was 86,

10 and 4 per cent, respectively (Sharma *et al.*, 2014). Earlier, there were simple single gene races and now there is prevalence of compelx multigene races of the pathogen in most of the potato growing countries. In Kenya, mating type tests with A1 tester isolates coupled with DNA analysis revealed that the fungal isolates from Uganda, Kenya and W. Africa are of A1 mating type (US1 clonal lineage). Variation and lack of consistency in oospore formation have been detected among the isolates from Uganda and Kenya. Isolates from potato were found to be more virulent than isolates from tomato (Olanya *et al.*, 2001).

Sharma *et al.* (2017) studied phenotypic (mating type, metalaxyl sensitivity), pathotypes and gentic (RAPD, AFLP, URP and SSR markers) diversity among 120 isolates of *P. infestans* collected from hills and subtropical plains of India and reported genetic diversity among pathogen population which should be considered while breeding for resistant cultivars or other disease management programme.

Disease cycle

P. infestans over-summers as dormant mycelium in the infected tubers in the cold storage. The extent of tuber infection is higher in the hills (6-39%) than plains (0.1 to 4%) and is more in a susceptible cultivar than in a resistant cultivar. These infected tubers serve as primary source of inoculum for the next crop. The volunteer plants and cull piles containing infected tubers near cold stores may also serve as a source of inoculum. The infected tubers generally give rise to diseased haulms on which the pathogen sporulates during suitable weather conditions, thus providing initial sporangial inoculum for further infection and spread of the disease (Fig. 2).

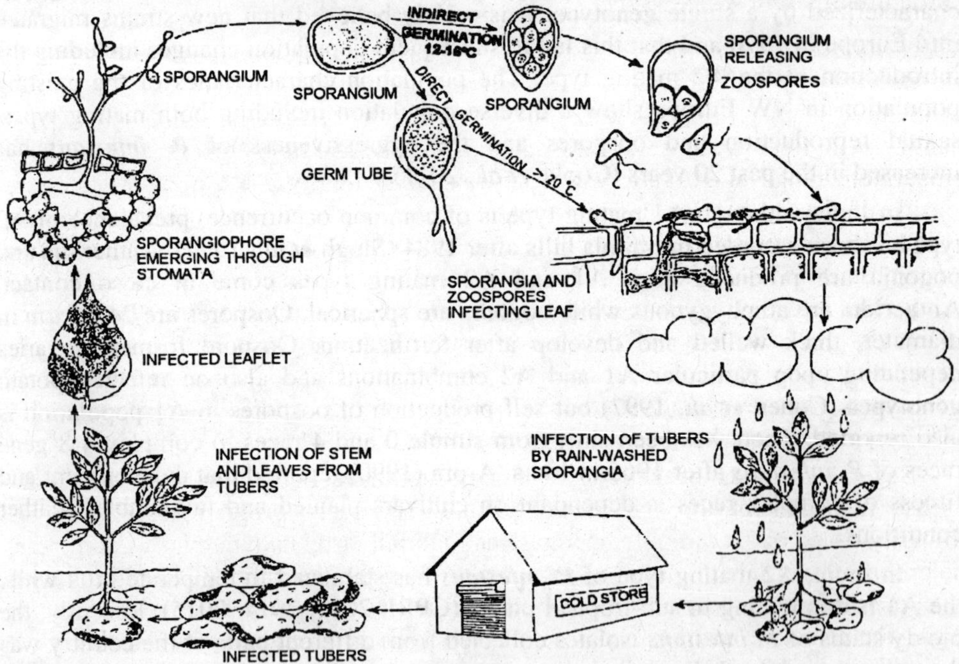

Fig. 2. Disaese cycle of late blight of potato in northern plains of India

The role of oospores in serving as primary source of inoculum is common in temperate countries as both A1 and A2 races of the pathogen are prevalent. In India also formation of oospores has been reported in nature in hilly areas of H.P. (Singh *et al.*, 1994).

The presence of late blight infection in apparently healthy tubers collected from diseased field provides new dimension to the epidemiology of this disease (Anonymous, 1991). Generally, heavily infected tubers rot after planting in the field. The number of infected tubers which emerge above ground are reported to be 0.5% in England and 10% in India. A stock with 5% tuber infection produced 14-15% diseased plants in the field (Anonymous, 1991).

There are several cycles of secondary infection and interval between two cycles depends on weather conditions. Normally a cycle is completed in 5-7 days under favourable weather conditions but it may take 10-12 days if temperature is too low (less than 10ºC) or there is not enough relative humidity. A leaf wetness for minimum period of four hours is required for sporangial germination and infection. Sporangia after landing on the host surface produce zoospores in a free film of water which soon encyst and germinate to cause infection through stomatal openings. Rarely, sporangia may germinate directly by producing a germ tube and cause infection. After proliferating inter-cellularly and getting nourishment through haustoria, the hyphae damage the tissue which form first water-soaked and then necrotic blight lesions. Later, hyphae emerge through stomata as sporangiospores, mainly on lower leaf surface, and produce sporangia which are disseminated by wind and cause subsequent infections in nearby fields.

Epidemiology and disease forecasting

Much work has been done on epidemiology of late blight of potato in other countries and also in India. Development of late blight in the field is generally associated with the prevailing environmental conditions. It appears that the temperature at the time of planting is an important factor in the epidemiology of late blight disease. Generally, a temperature range of 10-20ºC with high relative humidity and cloudy weather are considered favourable for the rapid development of the disease (Singh and Bhat, 2003). The fungus grows best at 20ºC while sporangia formation and germination occur best at 15ºC. Wind and rain splashes help in the dispersal of sporangia and heavy rains and dew wash them down into the soil where tubers may be infected. Sporangia are sensitive to dessication and require a free film of water for germination. Dey *et al.* (2010) reported that sporangial production in *P. infestans* accelerated when temperature gradient between 15-20ºC and low temperature (10-13ºC) supports enormous production of zoospores.

Several methods of late blight forecasting have been developed for better disease management through proper planning of spray programmes. These are primarily based on record of temperature and rainfall or relative humidity and predict the probability of late blight development assuming the presence of inoculum. Based on climatological factors many workers have laid down certain rules to predict late blight for effective management of the disease at minimum cost. 'Blite Cast', a forecasting model, is being used in north-eastern U.S.A. for timing fungicide applications against potato blight (Krause *et al.*, 1975). Such blight prediction rules have also been developed in other countries like Netherlands and England.

In India, Bhattacharyya *et al.* (1983) have achieved good success in predicting outbreaks in the hills. According to them if (i) the 7 day moving rainfall (30 mm for Shimla, 28.9 mm for Ootacamund and 38.5 mm for Shillong) and mean temperature of 23.9ºC or less continue for 7 days, the disease would appear within 3 weeks and if (ii) hourly temperature ranging between 10-20ºC associated with 80% or more relative humidity continues for 18 hours for 2 consecutive days, the blight would appear within a week. In north Indian plains including Punjab state, the disease generally remains at low levels and rarely occurs in epidemic form. Epidemics of late blight have been observed in the plains in 1986-87, 1991-92, 1992-93, 1997-98, 2004-05 crop seasons when frequent winter rains were experienced with foggy weather. Infected tubers stored in the cold stores serve as the primary source of infection. The disease generally starts appearing in the Punjab state from mid-November onwards when rainy spells occur in this period. Rain has been found to be the single most important factor in development and spread of this disease and regression based prediction system has been developed taking into consideration rainfall, temperature, sunshine hours and relative humidity in the plains. Singh *et al.* (2000) have developed a forecasting system 'JHULSACAST' for plains of Western U.P. based on temperature, relative humidity and rainfall.

Recently in India, late blight forecasting systems have been evolved both for temperate highlands and sub-tropical plains (Singh and Sharma, 2013). Predictions for its appearance are done every year using these systems. Using 'JHULSACAST' template, the model has been calibrated for Punjab (Arora *et al.*, 2012), tarai region of Uttarakhand (Pundhir *et al.*, 2014) and plains of West Bengal (Chakraborty *et al.*, 2015). Based on 'JHULSACAST' model, Decision support system (DSS) has also been developed which has three components i.e. (i) prediction for first appearance of disease, (ii) decision rules for need-based fungicide application, and (iii) yield loss assessment model (Arora *et al.*, 2014). Besides, Indo-Blightcast-a pan India model has also been developed for the prediction of late blight across the agro-ecologies (Govindakrishnan *et al.*, 2015). Recently, a wireless sensor network was used for validation of 'JHULSACAST' with other forecasting late blight models in western Uttar Pradesh using human participatory sensing approach and results revealed that 'JHULSACAST' is more accurate than the other tested models (Jagayasi *et al.*, 2015).

For plains of Punjab, Thind *et al.* (2013) developed a dynamic risk assessment model based on duration of relative humidity of more than 90 per cent at a range of temperatures for forecasting late blight of potato. They further developed a web-based decision support system for real time evaluation of weather parameters from automatic weather stations installed in potato growing aeras of the state. The DSS has helped in timing fungicide application against late blight and provided economical disease control.

Studies on potential impact of climate change on potato late blight in Punjab and western Uttar Pradesh indicated delay in late blight appearance by 0 to 6 days in 2020 and 12 to 14 days in 2055 in Punjab and 0 to 8 days in 2020 and 10 to 21 days in 2055 in western Uttar Pradesh. The number of sprays required to control late blight in potato seed crop would be 7.3 and 8 in future scenarios (2020 and 2055) in comparison to 6.5 in baseline scenario in Punjab. In contrast, there would be no change in number of sprays in year 2020 in western Uttar Pradesh, however, due to further rise in temperature in year 2055, it is likely to be reduced by two number of sprays over baseline year 2000 (Dua *et al.*, 2015).

Management

Various measures which aim to reduce the primary inoculum and spread of the disease can help in effective management of potato late blight. These include sanitary measures, use of healthy seed, cultural practices, host resistance and use of fungicides.

(i) Sanitary measures: Infected seed tubers, diseased haulms left in the field and cull piles, which may serve as primary source of inoculum, should be destroyed by burning or properly burying them deep in the soil. Application of certain chemicals such as sodium chlorate or chlorthiamid granules is also useful for this purpose. Volunteer or self sown plants and weeds which may harbour late blight infection can be destroyed by application of some suitable weedicide.

(ii) Use of healthy seed: Since infected seed is the major source of primary inoculum for late blight, apparently infected tubers should be sorted out before and after storage and the cull piles destroyed by burning or burying deep in soil. Potato tubers to be used for seed should be taken from a disease free crop. The use of healthy tubers for seed will ensure healthy and disease free crop.

(iii) Cultural practices: Some alterations in the cultural practices may prove beneficial in reducing late blight incidence in the field. Proper hilling and earthing up of potato tubers should be practiced to minimize chances of tuber infection. Wider row to row and plant to plant spacing ensures better penetration of air and sunshine to lower parts of the plants thus creating unfavourable conditions for development of the disease. Cultivation of the blight resistant and susceptible varieties together reduces the rate of spread of the disease epiphytotic (Bhattacharyya and Singh, 1990). Cultivation of potato with some other crop like wheat also slows down spread of the disease. In the north-western plains of India like Punjab and Haryana, early planted crop (in September) generally harvested pre-maturely after about two months remains free from late blight. Main season autumn crop planted in early October is harvested in January-February and is often attacked by late blight (Sekhon and Sokhi, 1999). In Himalayan hills, the disease generally appears in late June and in plains in mid-November to mid December. So planting of early maturing varieties in early April in hills and in late September or early October in plains followed by dehaulming of plants at 1% disease intensity should prove beneficial. Early identification and quick destruction of infected plants help in the management of late blight effectively and can economize on fungicidal sprays (Wani *et al.*, 2010).

Host resistance

Use of resistant varieties is regarded as the best means of late blight management. Initially the resistance was derived from *Solanum demissum* which soon broke down due to development of specialized races of the pathogen. Early studies had shown that the resistance in potato plants to *P. infestans* was inherited as dominant character and the R genes control hypersensitive response in a race specific manner. Later, it was realized that the pathogen exists as race complexes and has the ability to produce new races rapidly. In India, earlier simple races like race 0 and 4 were prevalent, but now with the introduction of new varieties multigene races are occurring both in hills and plains and eleven multigene races are reported using 0 to 12 R gene differentials. Now the efforts are directed to produce varieties having polygenic field resistance. Such resistance sources are available in species like *S. demissum* and *S. stoloniferum* and such resistance is governed by a number of minor and independent genes. Field resistance is associated with characters like longer incubation period, reduced pathogen growth within the host tissue, reduced sporulation and late plant maturity. Louwes *et al.* (1992) have also reported *S. circaefolium* sub sp. *circaefolium* to be a good resistance source against late blight.

Bjor and Mulelid (1991) reported cultivar Pimpernal without `R' genes to possess resistance to tuber blight. Das *et al.* (1999) have developed blight resistant clones of commercial varieties using gamma radiation. Mahajan *et al.* (2005) reported that some of the leaf biochemical constituents like starch, total phenols and lignin play an important role in conferring pre-infectional resistance against late blight of potato. Total sugars and cellulose was higher in susceptible varieties while starch, total phenols and lignins were higher in resistant varieties.

A large number of varieties possessing resistance to late blight have been developed by Central Potato Research Institute, Shimla. The prominent among them are Kufri Jyoti, Kufri Muthu, Kufri Badshah, Kufri Khasigaro, Kufri Giriraj, Kufri Chipsona 1, Kufri Chipsona 3, Kufri Girdhari, Kufri Himalini, Kufri Shailja etc. (Kumar *et al.*, 2014).

CPRI has characterized its available germplasm for presence of *R1*, *R2* and *R3a* genes and have identified genotypes possessing these genes alone or in combinations (Sharma *et al.*, 2013). Further these genotypes were tested for late blight resistance in the lab by detached leaf method to corroborate the results of molecular marker analysis and found that most of the genotypes having multiple *R*-genes were either resistant or moderately resistant. Attempts have been made to combine *R1*, *R2* and *R3a* genes in single host background and 15 clones have been identified possessing all *R1, R2* and *R3a* genes through MAS and encouraging results with regard to late blight resistance have been obtained under lab screening.

Molecular mapping of the diploid wild species *Solanum chacoense* has been done using AFLP, SSR markers for the identification of major QTLs for late blight resistance (Chakrabarti *et al.*, 2014). Efforts are in progress to isolate novel genes for late blight within these QTLs and develop molecular markers. In India, late blight resistant somatic hybrids between dihaploids of *Solanum tuberosum* L. and wild species like *S. pinnatisectum, S. cardiophyllum* etc. have been developed (Sarkar *et al.*, 2011; Chandel *et al.*, 2015).

Transgenic approach: Transgenics have become alternative and stable approach for potato where the traditional breeding programme is not able to introgress the desired trait. *Avr3a*, the *P. infestans* gene responsible for pathogenicity in potato by suppression of host cell death, has been targeted for development of RNAi gene construct by putting the gene sequence in inverted repeat orientation intervened by an intron sequence. The transgenics have resulted in suppressed disease progression and a reduction in pathogen load (Thakur *et al.*, 2015; Sanju *et al.*, 2015). This indicates that siRNA targeted against *Avr3a* can confer resistance to *P. infestans* in transgenic potato plants.

Biocontrol: Biosurfactants can be used as alternatives to chemical surfactants as their capability of reducing surface and interfacial tension with low toxicity, high specificity and biodegradability make them important for inhibiting pathogens. The metabolite of biosurfactant producing microorganism (*Pseudomonas aeruginosa*) has shown high efficacy against *P. infestans* under *in vitro* conditions (Tomar *et al.*, 2013, 2014).

Chemical control: Fungicides form an important input for managing late blight of potato because host resistance in commercial varieties is often broken down by multigene races of *P. infestans*. Several fungicides have been reported to provide good control of late blight in India and other countries. Bordeaux mixture and Burgundy mixture which were used earlier were later replaced by copper oxychloride, captafol, mancozeb, zineb, propineb and other dithiocarbamates, chlorothalonil etc. (Dutt *et al.*, 1982; Malathi and Jeyarajan, 1995). These have to be used as prophylactic applications before infection is established. Mancozeb is now an accepted fungicide for late blight control in India and is reported effective by several workers. A preventive fungicide application schedule appears more effective (Thind *et al.*, 1989a) for contact fungicides and spraying with these fungicides before usual dates of disease appearance when the crop is about six weeks old in north Indian conditions gives better control.

In the past two decades, several systemic fungicides with significantly better efficacy have been found promising for late blight control. Metalaxyl, an acylalanine fungicide, has proved most effective against this disease (Schwinn and Staub, 1987). Oxadixyl, banalaxyl, ofurace and dimethomorph are other specific fungicides having high efficacy against this disease (Thind and Mohan, 1997; Mohan and Thind, 1999). Dipping of sprouted tubers in 0.1 per cent metalaxyl for 30 minutes suppresses the disease (Singh and Bhattacharyya, 1990). However, excessive and irrational use of these phenylamide fungicides may lead to the development of resistance in the pathogen. Metalaxyl resistant populations of *P. infestans* have been reported in many countries (Dowley and O'Sullivan, 1991; Deahl *et al.*, 1993) including India (Arora, 1991; Singh *et al.*, 1988; Thind *et al.*, 2001). To avoid the development of fungicide resistance, mixture formulations of systemic and contact fungicides are generally recommended. Ridomil MZ (metalaxyl + mancozeb) is one such mixture being commonly used against late blight (Alam *et al.*, 1991; Thind and Mohan, 1997). Other mixtures which are in use for late blight control are ofurace + mancozeb, cymoxanil+ mancozeb, dimethomorph + chlorothalonil, dimethomorph + mancozeb, oxadixyl + mancozeb, oxadixyl + copper oxychloride, banalaxyl + mancozeb and few others. Recently, newly developed fungicides like azoxystrobin, fenamidone, famoxadone, iprovalicarb and mefenoxam (a new enantiomer of metalaxyl) and their combinations with contact fungicides like mancozeb have shown promising efficacy against potato late blight (Thind *et al.*, 2004a,b).

Inducer chemicals like salicylic acid, benzothiadiazol, fosetyl-Al, calcium chloride, ethylene, hydrogen peroxide, potassium hydrogen phosphate were studied to induce the resistance in potato against *P. infestans* and found that salicylic acid provided complete control of the disease when used as tuber dip treatment for ten minutes followed by spray application (Kumar *et al.* 2010). Banyal and Sugha (2010) did the refinement in fungicidal spray schedule for effective management of the disease in north-western hills and reported that the spray schedule consisting of Ridomil MZ followed by Kavach and again Ridomil MZ is highly effective in controlling late blight with maximum disease control of 72.5 per cent and higher yield (36.4%) over the unprotected control.

Kumar *et al.* (2012) studied the field efficacy of seven novel action fungicides against the disease. Fungicides evaluated were Cabrio Top (0.2%), Ergon (0.2%), Moximate 72 WP (0.25%), Sectin 60 WG (0.12%), Valiphenal (IR 5885) (0.25%), Benalaxyl 8% + Mancozeb 64% (0.25%) and Cabrio Team 18.7%WG (0.12%) and the results revealed that the new fungicides proved highly effective in controlling the disease with a corresponding increase in tuber yield.

Prophylactic sprays of chlorothalonil/mancozeb followed by systemic/ translaminar fungicides were found effective than post symptom sprays. This practice will be useful to minimize the yield losses due to late blight and assist in reducing development of resistance against fungicides in pathogen (Lal *et al.*, 2015).

Forecasting based fungicide applications reduce the cost of disease management by reducing number of applications. A preventive spray schedule of contact fungicides like dithiocarbamates is recommended as soon weather conditions like temperature and relative humidity become favourable. Phenylamide fungicide mixtures need to be sprayed only when heavy disease risk situation prevails in the event of intermittent rains and cloudy weather.

2. EARLY BLIGHT

It is a very common disease of potatoes and is called early blight because of its early appearance after sowing. The disease is reported to cause 10-25 per cent annual losses in yields in USA. In Germany and Denmark it has been reported to cause more than 50 per cent losses where as in Israel, the yield losses vary from 20-24 per cent. The pathogen has a wide host range and can infect tomato, chilli, brinjal and few other solanaceous plants.

Geographical distribution

Early blight of potato is prevalent throughout the potato growing regions of the world such as USA, Central America, European and Asian countries including India. In India, it was first reported from Farukhabad in UP in 1903 and later from other parts of the country. The disease is more severe in dry and warm parts particularly in Deccan plateau and Central India.

Symptoms

The disease mainly affects foliage and may uncommonly occur on stem and tubers. The symptoms first occur on lower leaves as dark brown, oval or angular spots scattered on the leaf surface. The spots are usually surrounded by a chlorotic zone which may extend much beyond the lesion due to the presence of toxin 'alternaric acid' produced by the pathogen. Under suitable climatic conditions, they enlarge rapidly to 3-4 mm in diameter, become irregular and may involve entire leaf lamina (Plate 2). Concentric rings may appear on the lesions giving them target board appearance (Fig. 3). Large number of spots on a single leaf results in senescence and dropping. During frequent rains, large dark brown lesions are produced resembling those of late blight but without apparent whitish growth of the fungus. Infection on petioles and stems shows small or elongated dark brown necrotic lesions and the tissue may break at the infection point. Lesions on tubers are dark sunken, round to irregular in shape and surrounded by raised borders with underlying flesh giving dry and leathery appearance. Tuber infection is uncommon in plains.

Fig. 3. Symptoms of early blight on leaves

The Pathogen

Early blight is caused by the demataceous *Alternaria solani* (Ellis & Mart.) Jones & Grout belonging kingdom *Fungi,* phyllum *Ascomycota,* class *Deuteromycetes*, order Moniliales and family Dematiaceae. It has pale to olivaceous brown septate mycelium. Conidiophores may arise singly or in small fascicles through stomata from the mycelium present in the dead centres of spots and are up to 110 x 6-10 μm in size. Conidia are usually solitary, obclavate, oblong to ellipsoid, tapering to a beak almost of the same length or even longer as that of the conidial body. Conidia are muriform, pale to olivaceous brown, smooth, 150-300 x 15-19 μm having 9-11 transverse and 0-few longitudinal septa (Fig. 4). No perfect stage of the fungus is known.

Disease cycle

Although tubers may be infected but they, in general, rarely serve as primary source of infection. Infected plant debris left in the fields after harvest is the main source of primary inoculum. The pathogen can survive as conidia or mycelium in dried infected leaf bits in the soil and is capable of producing fresh crop of conidiophores and conidia even after a year under moist conditions. The pathogen also infects other hosts like tomato and Capsicum as well as weeds like *Solanum nigrum* which may play an important role in its survival (Singh and Nagaich, 1977). Initial infection starts from the lower older leaves early in the season producing a large number of conidia which are disseminated by wind, water and rain splashes and cause secondary spread of the disease. The conidia germinate in the presence of moisture giving rise to 5-10 germ tubes which enter the host tissue directly or through stomata. Dark brown spots appear within 3-4 days after the infection has taken place.

Fig. 4. Conidia of *Alternaria solani*

Epidemiology

Warm and humid climate with frequent rains followed by dry weather is most favourable for disease development. Dew and frequent rains accompanied with high temperature result in profuse sporulation. Continued dry spell and high temperatures, however, may retard disease development and its further spread. The optimum temperature for the germination of conidia is around 25ºC and the fungus can grow well at temperatures between 23-25ºC (Tong *et al.*, 1994). A temperature range of 26-28ºC with RH of 70-80% is favourable for disease development. The disease is often more severe when the plants are predisposed by injury, nematodes (Bombawale *et al.*, 1982), poor nutrition or other types of stress. In hills, the disease becomes severe with the onset of monsoon rains in June-July. In northern plains, the disease remains at low profile in early autumn crop, however, it may appear in high proportions on late as well as spring planted crop owing to frequent winter rains.

Management

Different methods such as sanitary measures, cultural practices, use of fungicides and host resistance can be used for managing early blight.

Field sanitation and cultural practices : Diseased haulms and crop debris should be removed from the infected fields after harvest and destroyed in order to reduce the primary source of inoculum for the next crop. Balanced fertilization of the crop ensures healthy and vigorous plant growth which contracts less disease. Poorly fertilized plants are liable to be attacked more by the disease. Tubers should be allowed to mature in the field before harvest and injury to the tubers should be avoided. Use of disease free tubers for planting, removal of diseased crop debris from the fields and 3-4 years crop rotation help in reducing the primary inoculum.

Host resistance: Resistance to early blight has been identified in *Solanum* species such as *S. phureja* and *S. chacoense* that has been exploited for breeding of resistant cultivars (Stewart *et al*., 1994). Among various cultivars being grown in the plains of India, Kufri Sindhuri shows good resistance against the attack of early blight.

Chemical control : Most of the fungicides recommended against late blight can effectively control early blight of potato as well. Sprays of dithiocarbamate fungicides such as mancozeb, zineb, ziram, propineb and others like chlorothalonil, captan, copper oxychloride, Bordeaux mixture, captan etc. have been reported to be effective against this disease (Thind and Jhooty, 1982; Shteinberg *et al*., 1996). These fungicides give better control when used as preventive applications at the initiation of the first symptoms. Tebuconazole, a triazole fungicide, has also demonstrated good efficacy against early blight. Recently developed strobilurin fungicide azoxystrobin has been reported to provide good efficacy of both late blight and early blight of potato and tomato.

Integrated disease management : More than one management practices can be integrated for achieving better level of disease control. Genotypes having moderate level of resistance can give desired control of early blight with fewer fungicide applications compared to susceptible cultivars and result into economical yield returns (Shteinberg *et al*., 1993). The intensity of fungicide applications can be adjusted in such a way that moderately resistant cultivars can be sprayed at longer intervals while in case of varieties with age related resistance it would be better to increase the frequency of sprays towards the end of the season. Some antagonistic organisms can also be used in the IDM programme. Two strains of *Trichoderma* species have been reported antagonistic to *Alternaria solani* (Martinez and Solano, 1995). A strain 679-2 of *Pseudomonas* species can also inhibit this pathogen through production of a water soluble metabolite and there was no adverse effect of copper fungicides on this antagonist that can be easily integrated with other disease control measures (Casido and Lukezic, 1992).

3. BLACK SCURF

Black scurf is a common disease of potato tubers and is of world wide occurrence. It is prevalent where potatoes are cultivated. Other names of the disease are stem canker, top rosette and web blight. Black scurf causes both qualitative and quantitative damage as it decreases the market value of the tubers both for table as well as for seed purpose. Lakra (1992) reported that black scurf reduced seed germination by 11 per cent and affected yields by 68.2 per cent in the potato cultivar Kufri Chandramukhi.

Geographical distribution

Black scurf is occurring in all the countries where potatoes are grown. Its incidence has been reported from USA, Mexico, Canada, England, France, Germany, Ireland, Asian countries including Japan and India and other potato growing countries. In India, the disease is observed both in hilly areas and plains.

Symptoms

Stem canker and the scurf phase are the two distinct phases of the disease. There may be pre-emergence death of the plants or the germination may be delayed when the pathogen attacks growing tips of the sprouts. If the infection is less, a large number of sprouts may arise and the affected hills show yellowing of plants which generally remain stunted. In stem canker phase, the growing tips of sprouts show browning. Sunken, circular or elongated brown necrotic spots may also be observed on the mature sprouts at the point of infection near mother tuber. Severely affected sprouts are killed. Later, when shoots emerge, similar necrotic lesions are observed on the stem which may extend downwards and may completely girdle the stem. In case of severe cankerous lesions, the tops show stunting, rosetting and purpling of leaves. Green aerial tubers may also appear on the branches and petiole axes and the disease phase is called 'aerial tubers'.

The most prominent symptom of black scurf is the presence of black crust on tubers due to the formation of sclerotia of the fungus (Plate 3). The pathogen produces a large number of sclerotia superficially on the surface of growing tubers. These sclerotia may be hard or spongy forming a black crust of scurf on the tuber surface. These are normally seated on the skin and do not cause any damage to the tuber inside. Black scurf phase is more common than stem canker in India. Dry rot symptoms on tubers are also seen if the pathogen enters through wounds.

The Pathogen

Black scurf is caused by the fungus *Rhizoctonia solani* Kuhn. (teleomorph *Thanatephorus cucumeris* (Frank) Donk). The mycelium is septate, brownish, dolipore type, 5-14 μm wide. The lateral branches of the mycelium are constricted at branching point and possess septum near the junction. The sclerotia are small, brown to black and unlike other fungal sclerotia are undifferentiated into rind and medullae. Outer cells of sclerotia are thick walled and dark in colour. Basidial stage of the fungus has also been reported in some parts of the country on dead potato haulms, which appears as flaky pellicle on the substrate under cool and humid environment. The basidiospores appear on sterigmata arising from barrel shaped to clavate basidia and are usually four in number. These are ellipsoid to oblong and flattened on one side. Different strains of the pathogen have been identified based on cultural and sclerotial characters. Based on anastmosis grouping (AG), *R. solani* infecting potato has been assigned to AG-3 type.

Disease cycle

The fungus persists and perpetuates in the form of sclerotia on the seed tubers, in soil and in plant residues. On germination in the presence of moisture, the sclerotia produce mycelium which may cause infections on the sprouts, stem or young tubers. Black scurf incidence and severity are significantly higher when the inoculum was present in both seed tubers and soil, compared with either of them separately. Thus both major sources of inoculums, seed tubers and soil, are important in disease development. However, when both sources are present, black scurf incidence and severity are increased, leading to economical damage to tuber yield and quality (Tsror and Peretz-Alon, 2005). The fungus can survive as sclerotia in the soil for many years. Basidiospores when formed are capable of causing new infections and help in the spread of the disease. The fungus although has a wide host range but shows some level of specificity depending on host crops. In a cross inoculation experiment, potato isolates of *R. solani* failed to cause any infection on rice plants, while isolates of the fungus from rice appeared to be weekly pathogenic on potato (Thind and Aggarwal, 2008). The strains infecting members of graminaceous crops do not generally cause infections on solanaceous crops including poato and *vice verse*.

Epidemiology

Both soil and infected seed tubers serve as primary sources of inoculum. The pathogen survives as sclerotia in the soil and its viability is greatly reduced by exposure to hot summer temperatures. In northern plains of India, the fungus has restricted survival due to hot summers, however, in hills, it can overwinter in soil. In addition, the pathogen has a wide host range having both solanaceous and non-solanaceous plants. The fungus can grow at temperatures between 8-35ºC with optimim at 25-30ºC. The sclerotia germinate between 8-30ºC with optimum at 23ºC. Maximum infection takes place at around 24ºC.

Soil moisture and temperature are the two important factors affecting scurf development. Low temperature and high soil moisture are conducive for infection on sprouts and stems. This phase is not common on early autumn crop in North Indian plains as the temperature remains high. The more common phase is black scurf where the sclerotia formation is favoured at temperature of 28ºC. The number of sclerotia on tubers may vary with soil temperature and moisture. Black scurf on tubers is more common in crops grown in sandy to sandy loam and moderately wet soils.

Management

The black scurf pathogen survives in soil as well as through seed potato tubers and has a wide host range. Successful management of the disease can be achieved by following proper cultural practices and chemical treatment of the seed. Cultivar Cardinal has been reported as resistant to black scurf by Rauf *et al.* (2007).

(i) Cultural practices: In general, planting of healthy and disease free tubers is helpful in reducing the incidence of the disease. Change in the date of planting is another important criterion. If the time between planting of tubers and sprout emergence is reduced, less infection is observed. In the hills, the crops should be sown in spring to reduce stem canker and necrosis of sprouts. On the other hand, in the plains, early plantation of main crop followed by early harvest is recommended. Two to four years rotation with cereals, brassicas and legumes is helpful for the management of this disease. Green manuring with Dhaincha (*Sesbania cannabina*) reduced the black scurf on tubers by 40 per cent while maize and sunhemp (*Crotolaria juncea* L.) were reported effective by Bhattacharrya *et al.* (1977). The increase in organic matter content of the soil helps in reducing the population of the fungus through enhanced activity of the antagonists such as *Trichoderma* spp. Likewise, soil amendment with neem or margosa cake or with saw dust has been found to reduce black scurf (Sikka *et al.*, 1971; Singh *et al.*, 1972). Other practices like shallow planting and good drainage also help in disease reduction.

(ii) Seed tuber treatment: Treatment of seed tubers with chemicals before planting or before storing results in effective disease control. Dipping of seed tubers in 0.25 – 0.5 per cent solution of organo-mercurials such as Emisan-6, Agallol or Aretan for 5-

10 minutes is widely practiced by the farmers. But these mercurial fungicides are being banned now because of toxicity and pollution problems. Certain oxathiin fungicides are also reported to be effective against stem canker phase of the disease. Tanni *et al.* (1982) reported significant control of the disease by dusting or spraying of tubers with Validamycin. Khanna *et al.* (1991) reported good efficacy of carbendazim to control the disease.

Dipping of seed tubers with 1% acetic acid plus 0.05% zinc sulphate or with 3% boric acid has also been reported effective (Somani, 1986). The treatment with boric acid is more effective, less toxic and safer than organo-mercurials which are not only hazardous but also affect the emergence of sprouts especially when the temperature is above 30°C. Thind *et al.* (2002) have reported almost complete control of black scurf by dipping infected seed tubers in 0.25 per cent solution of Monceren (pencycuron), a phenyl urea based fungicide, for 10 minutes before sowing. Dithane M-45 has been reported to provide good control of black scurf in Pakistan (Rauf *et al.*, 2007). It significantly checked eyes germination inhibition and decreased sprout killing, stem girdling, stem canker, black scurf incidence and severity.

(iii) Biological control: Field experiments showed that *Trichoderma harzianum* and cattle manure compost amendment applied in furrow could reduce black scurf incidence in organically grown potatoes (Tsror *et al.*, 2001). Incorporation of *T. harzianum* to the soil surface had a relatively small effect compared to the in-furrow treatment. Although, the treatments significantly reduced disease incidence and severity on daughter tubers, total yield was unaffected in the experiment using naturally infested soil and diseased seed tubers. Fermentor biomass preparations of *T. viride* (T-1-89) and *Gliocladium virens* (GI-21) applied as dusts to seed potatoes infested with sclerotia of *R. solani* before planting, reduced disease incidence in the field by 50 and 55%, respectively. Validity of sclerotia from seed pieces retrieved from the field 54-89% by specific antagonists (Beagle-Ristaino and Papavizas, 1985). In an integrated management approach evaluated by Arora (2008), the tuber seed treatment with 1.5% boric acid followed by application of *T. viride* formulation containing 1×10^7 cfu/g @ 4.5 g per kg at planting reduced the disease to level achieved with 3% boric acid spray. Soil application with *T. harzianum* at the time of sowing followed by two dosages at 20 days interval have been reported to provide good disease control with better crop stand and yield increase in Pakistan (Rauf *et al.*, 2015).

4. POTATO WART

Potato wart was first reported to occur in Hungary in the year 1895 and since then it has been known to cause serious losses in temperate countries of the world. Besides potato, it has also been reported from tomato and other *Solanum* species.

Geographical distribution

The disease is quite serious in temperate regions of the world such as Western Europe, Russia, Peru, Canada, USA and some other temperate countries. In India, the disease was reported for the first time by Ganguly and Paul (1953) from Darjeeling and is presently restricted to that area only due to strict domestic quarantine on this disease.

Symptoms

All the underground plant parts except roots are attacked. The plant cells multiply rapidly at the infection site and produce hypertrophied tissue masses resulting into wart type symptoms on tubers (Plate 4b) and beads like projections on stems or stolons varying in size from small protuberances to large intricately branched structures. Warts are normally soft, pulpy, spherical and similar in colour to tubers. The colour may change to green on exposure to sunlight. Sometimes whole tuber may be covered with warts (Plate 4a). The colour of warts changes to dark in shade as the tubers mature. Sometimes secondary micro-organisms invade the wart tissues causing their decomposition.

The Pathogen

Potato wart is caused by the fungus *Synchytrium endibiotichum* (Schill.) Perc. which is holocarpic and endobiotic in nature. The fungus produces sporangia which release a large number of uniflagellate planospores. When planospores originate from different sori, these behave as planogametes and copulate to form zygospores which act as resting sporangia of the pathogen. One race and one biotype of the pathogen have been reported from Darjeeling hills in India. Phadtare and Sharma (1971) have reported certain other hosts to be infected by the fungus.

Disease cycle

The resting sporangia are released into the soil on disintegration of warty structures where they can remain dormant for several years. When soil becomes wet after irrigation or rainfall in the next spring, these resting sporangia germinate and cause fresh infections. Numerous planospores which are uniflagellate are released after germination of resting sporangia. These swim in a film of water and penetrate the host cell after dissolving a minute pore in the epidermal wall leaving the flagellum outside the cell. Planospores grow in size in the host cell and the nucleus is also enlarged and forms a thick brown wall around it thus forming prosorus or summer spore. The host cell also increases in size and becomes pear shaped. Host cells around the infected cells are also stimulated and due to hypertrophy and hyperplasia, wart like structures are formed on the host surface. The host cell dies and the prosorus occupies the lower half of it. After maturing the prosorus releases the protoplast in the upper half of the infected cell which is surrounded by the thin membrane. The nucleus divides through mitosis and their number reaches around 32 after which the cytoplasm cleaves to 5-9 segments and again the nucleus starts dividing mitotically till their number in each segment is 200-300. The protoplasm around each nucleus rounds up forming planospores or gametes. When there is sufficient moisture, these segments behave as planospores while under dry conditions, these act as gametes which fuse in pairs to form zygotes. After maturity of planospores on host surface, sporangia are released from the sorus that cause fresh infections in the presence of water film on the host surface. There are repetitive cycles of planospore production in the crop season. Copulation between planogametes from different sori takes place on the host surface or in the soil and after karyogamy biflagellate zygote swims for sometime before resting on the surface of the host. These then penetrate the epidermal cells and the process of hypertrophy and hyperplasia is initiated burying the infected cells deep inside. The zygote increases in size and a thick wall is formed around it. This then acts as resting sporangium and remains dormant in the winter season. The disease is spread to new areas by infected soil or infected host tissue. The disease has been kept restricted in India because of quarantine measures.

Epidemiology

The longevity and viability of the resting spores are affected by soil type and crop rotation. Other factors like soil moisture and pH also affect the process of infection. For germination of resting spores and sporangia a free film of water is necessary. When moisture conditions are available, infection can occur at temperatures ranging from 10-24ºC and at soil pH between 4-8.5. The optimum temperature for infection is 21ºC and slightly acidic or neutral pH is preferred.

Management

The disease mainly being soil borne in nature is difficult to be managed. Various measures including quarantines, cultural practices, host resistance and some chemicals have been found effective in reducing the disease level.

(i) Quarantine measures: Since the disease is soil borne, its eradication is very difficult. Strict domestic quarantine measures enforced in India have helped to restrict the disease in Darjeeling district of West Bengal only where it was first reported. Considering the risk of spread of the disease through contaminated soil particles, the seed as well as table potatoes are not allowed to be taken out from this district and this has helped to contain the disease effectively.

(ii) Cultural : The pathogen can remain viable in the infested soil for several years. This warrants to practice crop rotation with non-host crops for 8-10 years where the disease has become endemic. The pathogen population in the soil can also be reduced through steam sterilization of the soil.

(iii) Host resistance: Some potato lines such as Ronda, Edina, Mira from European countries have shown considerable resistance to potato wart. Varieties developed by CPRI, Shimla like Kufri Jyoti, Kufri Sherpa, Kufri Jeevan, and Kufri Muthun have demonstrated resistance to late blight as well as wart and can play significant role in reducing the disease level in endemic areas.

(iv) Chemical treatments: The use of fungicides like mercuric chloride, copper sulphate, formaldehyde etc. as soil application is, though effective, but is quite expensive and can result into some non-target effects. These can be applied to limited extent to check the establishment of the pathogen into those areas where it has recently been introduced.

(B) BACTERIAL DISEASES

5. COMMON SCAB

Common scab of potato is world wide in distribution. Initially the disease was considered to be of minor importance in India and was confined to hilly tracts but gradually it became wide spread in the plains of Uttar Pradesh, Bihar, Haryana and

Punjab. It is widely prevalent and important in neutral or slightly alkaline and light sandy soils during dry periods. Though the disease does not account for much direct loss in yield yet it drastically reduces the market value of the tubers due to rough and blemished skin. The affected tubers fetch low price in the market and are rejected for seed purposes as well. Under severe root infection, there may be considerable reduction in tuber yield also. Disease incidence as high as 100 per cent has been reported in hilly areas (Shekhawat, 1990). Seed producing states like H.P. and Punjab are badly affected due to this disease as infected tubers become unfit for seed purpose.

Geographical distribution

Common scab of potato was first reported from UK in 1825 and now occurs throughout the world. Its occurrence has been reported from all the major potato growing countries like USA, Mexico, England, Ireland, France, Germany, Japan and other Asian countries including India. In India the disease was first reported from Khasi hills in Assam and since then the disease has become endemic in all the potato growing states of the country (Shekhawat *et al.*, 1993)

Symptoms

Initial symptoms of common scab appear on tubers as small brownish and slightly raised spots which later on enlarge, merge with each other and give corky appearance. Scab symptoms on tubers are categorized as shallow and deep pitted. In shallow scab, the affected tubers show superficial roughened areas (Plate 5b), sometimes raised above but often slightly below the skin of the tubers. The lesions of scab consist of corky tissue which is the result of abnormal proliferation of the cells of tuber epiderm due to invasion by the pathogen. The lesions may vary in shape and size and the colour is generally light to dark brown.

In deep pitted scab, the lesions are dark brown or almost black and measure 1-3 mm or more in depth (Plate 5a). They are also corky in appearance and may join together involving entire surface of the tuber. Other forms of scab are proliferated lenticels with hard corky deposition and concentric series of wrinkled layers of cork around central black core.

The Pathogen

Common scab of potato is generally considered to be caused by *Streptomyces scabies* (Thaxter) Waksman and Henrici. However, many species of *Streptomyces* are now reported to be responsible for causing common scab, though *S. scabies* is generally recognized as its major cause. Other species of *Streptomyces* responsible for causing mild common scab are *S. griseus* (Krainsky) Waksman & Hewici, *S. collinus* Lindenbein, *S. aureofaciens, S. longisporoflavus, S. griseoflavus, S. alborgriseolus, S. violanceorber* and *S. flaveolus* (Dey and Singh, 1983). However, names of some of these species are not included in the Advanced Lists of bacterial names. Glucose asparagine agar, Czapek's agar and Tyrosine-caesin-agar are suitable media for the isolation of *Streptomyces* species. The mycelium of the pathogen is slender, branched with few or no cross walles. The spores are cylindrical or ellipsoid produced on special hyphae that develop cross walls from tip and spores are pinched off. These *Streptomyces* species are gram positive and aerobic.

Among other species in India, Singh and Singh (1991) reported *S. aureofaciens* for inciting various types of shallow or deep scab symptoms on potato tubers.

Disease cycle

Main source of primary infection is either through infected tubers or soil borne inoculum where the pathogen can survive for longer periods. Fallow fields or crop rotation have less influence on the pathogen except in submerged soils. Soil or irrigation water and infected seed tubers are the main means of its spread. The role of diseased plant debris and infested soil in survival of *S. albus* and *S. aureofaciens* and also in initiating disease in succeeding crop of potato has been established in India (Singh and Singh, 1992). Young tubers in early formation stage are most susceptible when infection takes place through unsuberized lenticels, stomata, wounds or even through the cuticle directly. As a result of infection and certain metabolites secreted by the pathogen, the cells around the infected site divide rapidly to produce several layers of corky cells giving scabby appearance. The infected tubers stored for seed purpose serve as inoculum source for the next season.

Epidemiology

S. scabies and other species can survive in soil for several years under favourable conditions. The pathogen can be distributed through movement of soil by wind, water and cultural operations.

Various soil factors influence the severity of the disease. *S. scabies* is active in the pH range of 5.2-8.0 and the severity of disease increases with the rise in pH. A temperature range of 20-30°C is most suitable for the growth of the organism. Optimum temperature for infection is slightly below 20°C and for lesion development slightly about 20°C. The pathogen is active in dry soils and is suppressed by watering.

Management

As the disease is both tuber and soil borne, it is difficult to control. However, some management practices have been adopted to check this disease. Healthy and blemish free seed tubers should be selected for sowing in order to reduce the primary source of inoculum.

(i) Chemical: The tubers can be disinfected by dipping them for 10-30 minutes in 0.25 per cent aqueous solution of mercurial fungicides such as Emisan or 0.5% solution of Agallol or Aretan (Singh and Singh, 1982; Singh and Soni, 1987). The treatment has been commonly accepted by farmers in India. However, due to environmental pollution and health hazards, mercurial fungicides are now being banned. Other tuber treatment chemicals like 3 per cent boric acid and certain antibiotics like Streptocycline and Plantamycin have also given satisfactory control.

(ii) Cultural: Since soil pH and soil moisture are important factors in the development of disease, they have been exploited for minimizing the losses. Successful control of potato scab has been achieved by frequent irrigations of the field at weekly intervals from tuberisation until maturity (Singh and Singh, 1981).

Green manuring and cultivation of certain legumes before planting potato has been emphasized as important cultural practices in controlling common scab. Green manuring probably increases the activity of certain actinomycetes and other bacteria which are antagonistic to *S. scabies*. Delay in sowing also reduces disease incidence. Four year crop rotation with wheat-oat followed by potato-onion-maize also helps in reducing scab in the infested fields (Singh and Jeswani, 1987).

In Russia, *S. scabies* was reduced by application of high rates of potassium in combination with nitrogen and phosphorus fertilizers. In India, Singh and Singh (1993) confirmed the role of nitrogenous fertilizers in the control of common scab (*S. aureofaciens*) of potato. Growing potatoes in acidic soils also reduces scab and addition of sulphur can bring the pH to near 5.5. Application of gypsum is reported effective to reduce common scab (Singh and Soni, 1987).

(iii) Host resistance: Limited information is available on the use of resistant varieties to manage this disease as no commercial variety is known to possess high resistance to scab. Good level of resistance to scab has been reported in *Solanum chacoense* and *S. phureja*. *S. andigena x S. tuberosum* hybrids and red skinned varieties have been found to possess considerable resistance to scab (*S. scarius*) and are being used in the resistance breeding programme (Nagaich, 1983). Singh and Singh (1992) have reported Kufri Sindhuri and some other accessions as resistant to scab caused by *S. aureofaciens*.

6. BLACK LEG AND SOFT ROT

Black leg and soft rot of potato caused by *Erwinia* species are widely prevalent wherever potatoes are grown and cause considerable loss to the standing crop. These bacteria not only cause black leg and wilt of plants in the field but also damage the tubers during transit and storage causing pulpy soft rot. The bacteria are prevalent wherever potatoes are grown.

Symptoms

Symptoms can occur at any stage of plant development. In the field, typical black leg is characterized by appearance of brown black or jet black lesion on the stem above the point of attachment of the seed tuber. The lesion may extend up the stem to a few millimeters or along the entire length. The affected plants are commonly stunted and plants show tendency to grow erect. Leaves may roll and become chlorotic. Sometimes young seedlings arising from diseased seed tubers are destroyed before or soon after emergence. Tubers from field carry the bacterium to storage godowns.

The tubers may develop soft rot in the soil if the moisture and temperature are high. They may also develop sot rot during transit and storage and are transformed partly or totally into a decayed mass of pulp which remains intact by corky epidermis which is not attacked by the pathogen. When attacked by secondary bacteria, a strong repulsive smell is emitted.

The Pathogen(s)

Two bacteria namely *Erwinia carotovora* sub-species *atroseptica* (van Hall) Dye and *Erwinia caroto*vora sub-species *carotovora* (Jones) Bergey *et al.* are responsible for causing black leg and soft rot of potato. *E. chrysanthemi* Bubkholder *et al.* is also sometimes associated with bacterial soft rot of stored tubers.

Disease cycle and epidemiology

The lenticels of seed tubers provide harbouring places for survival of causal bacterial pathogens. They also live saprophytically on decaying plants debris including diseased potato leaves and on tubers left over in the field after harvest. The bacteria survive for longer duration in cool and moist conditions than in warm climate. They may also persist in the rhizosphere of certain weeds. The bacteria from tubers carrying latent infections or from soil infect the haulms at various times during the season. These bacterial pathogens colonise shoots and if vascular tissue is colonized they may infect stem and daughter tubers. These infected tubers serve as source of devastating soft rot in the cold storages. Infection of seed tubers is favoured by moist soil and relatively cool temperature lower than 18-19°C.

Management

(i) Cultural practices: Cultural control measures are the main approach to reduce the black leg and soft rot. These include use of healthy and certified whole tubers as seed, planting at less depth, use of adequate N fertilizers, avoiding injury to tubers at harvest, harvesting tubers during dry weather and sorting out and destroying tubers showing signs of soft rot. These practices will help to minimize the infection. In addition, curing or healing potatoes after harvest at high temperature (16°C) and 80% relative humidity for 10 days will protect potatoes from soft rot infection. Hot water treatment also gives some success in controlling the rot in seed tubers (Wale *et al.*, 1986).

(ii) Chemical treatment: As the bacterium is deep seated, dipping of tubers in fungicidal or bactericidal solution is not of much use. However, there are some reports regarding use of certain organo-mercurials for treatment of cut tubers and treating the seed pieces with antibiotics to check the disease.

7. BACTERIAL WILT

Bacterial wilt of potato is a destructive disease in tropical, sub tropical and warm temperate regions of the world. It was first recorded in India from Pune, Maharashtra. Bacterial wilt causes premature wilting of standing crop and rotting of tubers both in fields and stores. Heavy yield losses may occur due to premature wilting before tuber setting. Up to 75% losses have been reported in some areas of Karnataka (Gadewar *et al.*, 1991). Brown rot bacterium has been listed as a quarantine pathogen for export of potato to EU countries.

Geographical distribution

It has extensive host range and more than 200 plant species including solanaceous crops are affected world wide. In India, it is a serious problem in potato and is endemic in distribution in west coast of Trivandrum, Karnataka, Maharashtra, Madhya Pradesh, eastern plains of Assam, Orissa and West Bengal. The disease is absent in North-western hills and North –western and North-central plains. The bacterium apparently does not occur In plains of North Indian states of Punjab and

Haryana due to high temperature prevalent during summer months which does not allow the bacterium to survive. After extensive surveys of the potato fields in Punjab in the past 10 years and laboratory testing of potato tubers and soil samples using molecular techniques, no incidence of bacterial wilt pathogen has been observed and the state has been declared as a pest free area for brown rot (Hunjan *et al.*, 2016)

Symptoms

Both above and underground plant parts are affected. There is a sudden wilting and death of infected plants or collapse of one or more branches is observed (Plate 6a). Wilting appears during July in the hills about two weeks after the onset of monsoon. Many wilted plants also show stem rot at soil level. Greyish brown discolouration appears through the stem of affected tubers. Cross-section of such tubers reveals a distinct brown colouration in the vascular ring. A slight pressure on the cut tuber causes oozing of typical greyish white bacterial slime out of the vascular ring. Vascular browning is a characteristic symptom of the disease. In North-eastern hills, lenticel infection of potato tubers in the form of water-soaked lesions is quite common (Plate 6b).

The Pathogen

Bacterial wilt is caused by *Pseudomonas solanacearum* Smith (Syn. *Ralstonia solanacearum*) which was first described as *Bacillus solanacearum* by E.F. Smith in 1896 which was later changed to *Pseudomonas solanacearum* (Smith, 1914). The name was later changed to *Burkholderia solanacearum* (Yabuuchi *et al.*, 1992) and is currently known as *Ralstonia solanacearum* (Yabuuchi *et al.*, 1995). It is a rod shaped Gram negative bacterium measuring 0.5-0.6 x 0.8-1.2 μm and having 1-4 polar flagella. Three races of the bacterium have been identified and potato is affected by race 1 and 3. Based on phylogenetic divergence at molecular level using a series of probes based on regions of the genome concerned with virulence, *R. solanacearum* has been split into five biovars comprising two Divisions (Cook *et al.*, 1989). This bifurcation has been confirmed on the basis of sequencing of tRNA genes in the bacterium (Taghavi *et al.*, 1996).

Disease cycle

R. solanacearum is both soil and seed borne and can survive in infected crop debris, potato tubers, wild host plants and weeds. Infested soil and infected seed potatoes serve as source of primary inoculum. Bacterial masses released in the soil by brown rot affected plant parts help in the survival of *R. solanacearum* (Shekhawat *et al.*, 1991). Inoculum threshold of 10^7 cfu/g soil is required for the infection (Devi *et al.*, 1982).

For short and long distance dispersal of the pathogen, infected seed tuber is the most effective inoculum source. Tubers may carry the pathogen in vascular tissue on the tuber surface and in lenticels. In Shillong, 3.2% and 16% tubers of cvs. Kufri Jyoti and Kufri Megha, respectively, carried infection. The pathogen can survive in potato tuber lenticels and in vascular tissues for 8 months at 10-15°C. The pathogen can also survive on tomato, brinjal and chilli seedlings grown in infested area (Keshwal, 1980). After infection, the bacterium reach the xylem vessels and may spread into the plant. These also enter the intercellular spaces of the parenchyma cells in the cortex and pith, dissolve the cell walls and create cavities filled with slimy masses of bacteria. The enzymes secreted by the bacteria break down cell wall substances which when carried in the transpiration stream, collect at vessel ends, form gels and gums which help clog the xylem vessel pores and thus block movement of water. The enzymes also cause softening and weakening of the cell walls which then collapse and the tissues droop and wilt. Secondary spread takes place by root contacts, irrigation water and farm implements. In the hilly areas of Uttarakhand. the pathogen from infected plants is carried through river water used for irrigating potato crops (Shekhawat *et al.*, 1988a).

Epidemiology

A high temperature regime (28-30ºC) and relatively high soil moisture are favourable for wilt development. Temperature and soil moisture have a synergistic effect on disease development. Singh (1980) observed that disease incidence was maximum during July-August when both soil moisture and temperature were comparatively higher. Infection of susceptible host plants generally occurs through the root system. The bacteria multiply in the plant roots and are released into the soil thus increasing the pathogen population. In a study, Sunaina and Gupta (1998) have worked out relationship between seasons and bacterial wilt severity in Kumaon hills of Uttarakhand In summer season (March-July), the bacterial incidence was high, presumably the high temperature and rainfall during these months promote bacterial wilt. In autumn, no wilted plants were observed.

The time taken for symptom development depends on age of plant and on environmental factors. The population build up of *R. solanacaerum* was higher in Kufri Jyoti and Kufri Chandramukhi while it was low in Kufri Lalima (Chakrabarti and Shekhawat, 1991).

R. solanacearum can survive in diverse situations. It has the ability to survive in soil, plant debris and on alternate/collateral hosts belonging to different families like Amaranthaceae, Solanaceae, Compositeae, Leguminoseae etc.

Management

It is a difficult disease to manage as the commercial cultivars are highly susceptible and chemical control is not feasible. The reduction of the initial inoculum is of prime importance to manage bacterial wilt of potato. Different methods are used to manage bacterial wilt of potato among which cultural practices and biological control measures carry much emphasis.

(i) Soil disinfection: Application of stable bleaching power @ 12 kg/ha has been found to reduce bacterial wilt by 80% when applied in furrows at the time of planting (Shekhawat *et al.*, 1988b). Seed bed used for raising seed-lings can be disinfected by maintaining the soil temperature at 85ºC for 15 minutes by steam application.

(ii) Cultural practices: Use of disease free seed is important for avoiding introduction of the wilt pathogen in the new fields. The incidence of bacterial wilt was far less in the whole seed than when cut tubers were planted. In early (February) planted crop, wilt incidence and tuber rot are almost negligible in North-western hills. Crop rotation, the most generally used control measure, is employed world wide to check bacterial wilt. Two years rotation with wheat-maize or wheat-fallow reduced bacterial wilt incidence to 7% from 80% (Shekhawat *et al.*, 1988b). By shifting dates of sowing to avoid periods of high soil moisture due to rainfall and high temperature can also help to decrease the disease incidence.

(iii) Chemical treatment: Plantomycin alone or in combination with Blitox-50 proved effective in controlling bacterial wilt upto 95% (Ojha *et al.*, 1986). Emisan 6, Thiram, Blitox-50 and chlorinated water also checked the disease to the extent of 80, 80, 75 and 60 per cent, respectively. Pretreatment of seed tubers with 500 ppm of an antibiotic C-6 followed by two foliar sprays controlled the disease and increased the yield three times.

(iv) Host resistance: The genetic resistance to *R. solanacearum* is not general to all the strains. The resistance is usually strain specific and is broken down under warm climates. The resistant clones are susceptible to latent infection (Granda, 1988). The resistant cultivars should be grown under cultural practices identified for control of the disease and the seed should be replaced before the latent infection reaches threshold level. *Solanum phureja* has been a valuable source of resistance to bacterial wilt but this is temperature sensitive and best suited for higher elevations or cooler climates. Some of the somatic hybrid lines developed by Laferriere *et al.* (1998) can serve as bridges for the eventual introgression of novel bacterial wilt resistance into potato plants.

(v) Biological control: Microorganisms that can grow in the rhizosphere are ideal for use as biocontrol agents. Certain bacteria like *Pseudomonas fluorescens*, *Bacillus polymyxa*, *Bacillus* spp. and actinomyctes have been found to delay the development of *R. solanacearum* and reduce the incidence of bacterial wilt (Sivamani *et al.*, 1987). Bacteria which are antagonistic to *R. solanacearum* have been isolated from various sources e.g. suppressive soils and rhizosphere of host plants. Studies have shown that isolates of *Bacillus* spp. (BSN-1 and BSN-2) reduced wilt by 87% and 89% and tuber rot by 83% and 84%, respectively and an increase in yield by 129% and 124% was also recorded.

(C) VIRAL DISEASES

Commercial propagation of potato is normally done vegetatively using 'seed' tubers. Therefore, 'Degeneration' of seed stocks due to viruses is common as vegetative propagation results in the continuity of several viral pathogens. There are over 30 viruses infecting potatoes (Table 1), but only five or six are actually important in India/ S. Asia (Khurana, 1992, Singh and Khurana, 1993). The recent advances in healthy seed potato production have overcome this major problem and greatly enhanced both the area and yields of potato. The losses in potato yield due to one or more viruses infecting potatoes vary from low to very high. Generally, severe mosaic caused by potato virus Y (PVY) and potato leaf roll virus (PLRV) alone can reduce the yields up to 70-80% while mild viruses, like PVX, PVS, PVM also depress the yields by 10-30%. If the seed stocks are not maintained well or frequently replaced with fresh ones, the virus infiltration reaches 100% level within 3-4 successive crop seasons/years resulting in almost only half to one third yields which is rather uneconomical (Table 2). The annual losses in potatoes due to viral diseases in India may reach between Rs. 5000-6000 millions (Khurana, 1992).

Table 1. Distribution of viruses by their association on potato

Viruses dependent on potatoes for survival and spread	Viruses not dependent on potatoes for survival and spread	Undetermined
Potato virus X (PVX)	Tobacco rattle virus (TRV)	Potato black ringspot virus (PBRV)
Potato virus Y (PVY)	Tobacco mosaic virus (TMV)	Potato spindle tuber viroid (PSTVd)
Potato virus S (PVS)	Tomato spotted wilt virus (TSWV)	Potato yellow vein virus* (PYVV)
Potato virus M (PVM)	Cucumber mosaic virus (CMV)	Potato deforming mosaic virus (PDMV)*
Potato mop top virus (PMTV)	Tomato black ring virus (TBRV)	
Potato aucuba mosaic virus (PAMV)	Tobacco ring spot virus (TRSV)	
Wild aucuba mosaic virus (WAMV)	Solanum apical leaf curling virus (SALCV)	
Potato virus T (PVT)	Potato yellow dwarf virus (PYDV)	
Potato leafroll virus (PLRV)	Beet curly top virus (BCTV)	
Andean potato latent virus (APLV)	Alfalfa mosaic virus (AMV)	
Andean potato mottle virus (APMV)	Tobacco necrosis virus (TNV)	
Potato virus V (PVV)	Tobacco streak virus (TSV)	
Potato stunt virus (PSV)		

* Their virus origin has not been yet determined.

Planting of infected seed tubers is the main reason for spread of potato viruses from one place to another. Contagious viruses, namely PVX, PVS. etc. rapidly spread upon contract of plant foliage and roots in field as well as upon injury/cutting of the tubers upon harvest, in stores and while planting. Besides, most other potato viruses (viz. PVY, PVA, PLRV, PVM, etc.) spread either through winged aphids from far off fields or within the field by aphid colonies especially that of *Myzus persicae*. Many factors govern aphid transmission of potato viruses such as host preference of aphid species, efficiency of the vector, crop age, climate, vector form and/ or migration behavior etc. (Van Harten, 1983; Khurana and Singh, 1995). Models have been developed for forecasting the likely impact of virus spread and to determine whether or not to spray insecticides for control of aphids. At present, seed quality is determined by the presence and population of aphids in any plot before cutting of haulms.

Characteristics and spread

Generally, potato viruses are pandemic, still mode of their spread and distribution varies from place to place. Out of a large number of viruses infecting potatoes, potato virus Y, A, M, X, S, and leaf roll are the most important ones. Besides, the viroid is also economically important in America and several warm areas. Mosaics and leaf roll are the most common and severe potato viral problems in subtropical climate mainly due to high vector activity and also lack of availability of virus free seed. Table 2 lists the key symptoms for important potato viruses found in India while biophysical characteristics of the major viruses are summarized in Table 3 to further help in identifying them after isolation / culture etc.

Table 2. Symptoms guide for field inspection of important viruses on potatoes in India

Viruses	Main Diagnostic Symptoms
PVX	-*Interveinal or faint mosaic*: Causing mild or barely perceptible mosaic with light and dark green patches, mottling with stunting, crinkling with virulent strains.
PVS*	-*Latent*: It is barely perceptible as mottle or faint vein bending. Some cultures show severe bronzing, necrotic spotting and even leafdrop.
PVM	-*Leafrolling mosaic*: Often causes mild to severe mosaic with slight leaf deformations, necrosis of petioles and veins under certain conditions.
PVA*	-*Supermild Mosaic*: Causes faint mottling with leaf distortion sometimes, top necrosis in some vars. Sometimes severe, invokes rugosity and crinkling of leaves that turn shiny.
PVY	-*Severe/Rugose mosaic*: Symptoms vary with strain and cultivar. Mild-to-severe mosaic and venial necrosis and leaf drop streaking. Plants stunted. Severe mosaic (Plate 7a), rugosity, bunching or twisting of leaves, down-turning of leaflet margins and stunting, necrosis in some cultivars. Only some veins develop necrosis, older leaves collapse and drop, mild mosaic in less sensitive cultivars or some remain even symptomless. Rarely some tubers also show 'ringspots' at harvest or in stores.
PVA+PVX	-(*Crinkle*): Heavy blotching, distorted leaves with wavy margins, severe stunting, mottling, deformity, puckering, necrotic spotting and streaking of leaves.
PVX+PVY	-(*Rugose*): Severe mosaic and rugosity. Plants stunted. Leaves curl downward, necrotic veins in lower leaves and severe mosaic in upper ones. Stunting leaf blotching, leaf banding, necrosis, leaf drop, rugosity and leaf crinkling.
PLRV	-*Leafroll*: Invokes rolling of lower leaves in primary infections and of the upper leaves in the secondary infection with the leaves turning leathery, brittle, pink or brown pigmented (Plate 7b). Infected plants remain stunted, chlorotic and make rattling noise when shaken.
TSWV	-*Stem necrosis*: Characterised by scattered, necrotic lesions or 'ringspots' on shiny leaves followed by necrosis of petioles and stems. The plants have chlorotic and blighted appearance.

*Individually this virus can not be a problem but enhances effect of mild viruses upon combination and is responsible for breaking late blight resistance.

Detection

It is now well established that potato viruses are either contagious or aphid transmitted. Normally, we find combined infections of two or more viruses and such infections can be broadly grouped as either mild/ severe mosaics and/or leafroll. In field, one can easily recognize viral disease symptoms, if perceptible. But neither all viruses cause severe (clear) symptoms nor it is true in varieties for the same virus and at all times. In fact, symptoms depend mainly on the host, weather and virus(es) involved. It is also important to detect viruses in low concentration prior to their symptom expression. This is normally done either by "grow-out" test which was earlier popular for selection of healthy seed stocks in the off-season locations (de Bokx and Want, 1987) or by adopting one or more serological detection tests (Khurana and Garg, 1993).

Inspection of visual field symptoms is the oldest and easiest way to rogue out the diseased plants. The Seed Inspectors still certify the crop health (standard of seed) by counting the number of plant showing symptoms. Apart from field inspection of crop for plants showing symptoms, biological detection of viruses and allied diseases has been in vogue since long and viruses are detected most often by employing one or other serological tests (Khurana and Garg, 1993). Among field tests, chloroplast agglutination test is the simplest and quickest while latex flocculation is sensitive and also economical. Recently, ELISA variants have revolutionized virus detection because ELISA is highly sensitive, specific and extremely successful even for low concentration viruses such as PVS, PVY, PVT etc. or even phloem restricted viruses (PLRV). Immuno specific electron microscopy (ISEM), though expensive and sophisticated, has limited yet definite role in virus diagnosis especially for pre-nucleus mother plants/buffer stocks used for rapid multiplication of seed.

Symptoms

Most common potato viruses (viz. PVX, PVS) are latent and contagious causing faint to mild mosaics while others are aphid transmitted (viz. PVY, PVA, PVM, PLRV) which invoke severe mosaics and leafroll. Tubers from such infected plants produce rather weak plants and form an initial virus source in the fields. Nevertheless, the virus, varieties and plants' age determine the rate of tuber transmission. A combined infection of more than one virus was recorded in many cases in the tubers. Coinfection of two or more viruses, in a cultivar susceptible to a particular virus, augmented the concentration of the other virus with or without aggravation of symptoms but it was not true in the resistant cultivars (Singh and Khurana, 1993). The infected plants do show varied types of symptoms, however, the tubers rarely show symptoms such as necrotic spots/ rings, cracks and/or spindling (Table 3).

Table 3. Biophysical properties of some important viruses affecting potato

Virus	Acronym	Genus	1	3	4	5	6
Cucumber mosaic	CMV	Cucumovirus	Iso	30	N/A	Aphids	Non-persistent
Potato A	PVA	Potyvirus	Fil		730	Aphids	Non-persistent
Potato black ringspot	PBRSV	Nepovirus	Iso	26	N/A	Nematode	-
Potato leafroll	PLRV	Luteovirus	Iso	24	N/A	Aphids	Persistent
Potato mop-top	PVM	Carlavirus	Fil	12	650	Aphids	Non-persistent
Potato S	PVS	Carlavirus	Fil	12	660	Aphids	Non-peristent

Virus	Acronym	Genus	1	3	4	5	6
Potato X	PVX	Potexvirus	Fil	13	515	-	-
Potato Y	PVY	Potyvirus	Fil	11	740	Aphids	Non-persistent
Potato yellow dwarf	PYDV	Rhabdovirus	Bac	75	380	Leaf-hoppers	Persistent
Potato yellow mosaic	PYMV	Geminivirus	Seg.	18-20	?	Whitefly	-
Potato yellowing	PYV	Alfamovirus	Bac	21	60	Aphids	Semi-persistent
Tobacco black ring	TBRV	Nepovirus	Iso	28	N/A	Nematode	Persistent
Tobacco mosaic	TMV	Tobamovirus	Rod	18	300	-	-
Tomato mosaic	ToMV	Tobamovirus	Rod	18	300	-	-

1= Particle morphology 2= No. of distinct particles 3=Particle diameter 4= Particle length 5=Vector type 6=Persistence of virus in vector

Detection of Potato Viruses

Detection by using indicator plants

The viruses usually invoke clear or specific symptoms, such as PVY causes mosaic while PLRV infection results in leaf roll symptoms. However, depending upon the location, virus, host and weather, not all plants express viral disease symptoms, which may be checked biologically, on a set of indicator plants (Table 4) by mechanical, sap, aphid or graft inoculations. There are two types of indicators hosts, viz. systemic and local lesion hosts. The latter are effective being sensitive, quick in response and show clear cut reactions such as chlorotic or necrotic spots. Unfortunately, not many viruses can be routinely detected by using such test plants. There is also a great deal of variation(s) in their reaction to the virus(es), their isolates or strains, viz. *Gomphrena globosa* does not give local lesions for XHB strain of PVX. *Chenopodium quinoa* reacts systemically to PVS-An strain. Providing moderate temperature (18-24°C), high humidity (>80%) and constant light (4000 lux) to test plants increases the host sensitivity to virus(es). However, for the spindle tuber viroid (PSTVd), sap inoculation on *Scopolia sinensis* or tomato variety 'Rutgers' is recommended under low light (1000-2000 lux) and higher temperature (27-31°C). Detached leaf testing for the viruses in laboratory is also useful and avoids cross contaminations and reduces glasshouse space requirement.

Table 4. Plant species reacting to important potato viruses and viroid

Host indicators	APLV	PLRV	PSTV	PVA	PVS	PVT	PVX	PVY	TRV
Chenopodium amaranticolor	L	-	-	-	(L)	S	L	(L)	L
C. quinoa Willd	L	-	-	-	L/S	S	L	(L)	L

Host indicators	APLV	PLRV	PSTV	PVA	PVS	PVT	PVX	PVY	TRV
Datura metel L.	-	-	-	S	-	-	L/S	S	-
D. stramonium L.	-	S	-	-	-	-	S	-	-
Gomphrena globosa L.	-	-	-	-	S	-	L	-	-
Lycopersicon esculentum Mill	-	-	S	-	-	-	S	S	-
Nicandra physaloides Gaert	-	-	(L)	(L)	-	-	-	-	-
Nicotiana clevelondi Gray	S	-	-	S	-	-	S	S	S
N. debneyii Domin	S	-	-	S	S	S	S	S	-
Phaseolus vulgaris L.	-	-	-	-	-	(L)	-	-	-
Physalis floridana Rydb.	-	S	-	L	-	-	S	L/S	-
Scopolia sinensis Hemsl.	-	-	L	-	-	-	-	-	-
S. demissum' AG'	-	-	-	L	-	-	L/S	L	(L)

L-Local symptoms, (L)- Haphazard symptoms, S - Systemic symptoms

Serological detection

Serological virus detection techniques because of their rapidity, cost, reliability, etc. have a large-scale field application. The commonly adopted immunodiagnostic tests are briefly mentioned along with the latest developments and the merits of each test.

Chloroplast agglutination: It is an old yet effective field test for viruses occurring in high concentrations like PVX and PVS.

Microprecipitin test : It is simpler and economical as far requirement of antisera is concerned and has higher sensitivity but the sample (virus) extracts need to be first partially clarified.

Gelimmuno-diffusion: It is also called Ouchterlony double diffusion technique and has been extensively used to determine serological relationship between different strains of one virus or different viruses. Single radial immuno-diffusion test had been effectively employed for some time in detecting potato viruses X, S, M, under lab/field conditions (Shephard, 1972).

Latex agglutination test (LAT): It is based on antibodies adsorbed on to latex polystyrene particles to agglutinate in case of positive reaction. The test can be scored easily through naked eye after 30-60 min. This test has been widely adopted because of high specificity, rapidity and economy of antisera both in the field and laboratory (Fribourg and Nakashima, 1984).

ELISA or enzyme immuno assays (EIA): This helps greatly in testing for viruses, specially for large scale indexing which has changed greatly with the introduction of enzyme immunoassays (EIA) in the mid 1970s. There are many types of EIA. Some are direct and others are indirect and they all incorporate antibodies which have been covalently coupled with one of the enzymes so that the activity of both molecules is retained while the virus detection and quantification is based on the enzymatic (colour) reaction (Khurana and Garg, 1993).

(a) *das-ELISA:* For plant viruses, standard double antibody sandwich-ELISA (das-ELISA) protocol developed by Clark and Adams (1977) has been widely adopted with minor modifications depending upon the virus –host combinations. Recent developments in plant virus serodiagnositcs with ELISA, especially due to their high specificity, sensitivity, automation and convenience in testing, have changed the entire scenario (Khurana, 1990). ELISA is about 4-10 times more sensitive than latex agglutination in detection of potato viruses X and S (Franc and Banttari, 1986).

(b) *Indirect ELISA*: It is an advantageous variant of ELISA employing two species of antibodies, *i.e.* coating antibody is from one animal species (e.g. sheep, goat and chicken) while the second antibody is from another animal species (e.g. rabbit) or vice-versa to reduce the background signals and amplify the positive reactions (Cooper and Edwards, 1986). It also helps in establishing serological relationship among diverse strains of the same virus or among different viruses.

(c) *Dot-ELISA (Dot immunobinding ELISA):* Basically it is similar to standard das-ELISA except that it employs nitro-cellulose membranes and even cellulose filter paper in place of microtitre plate (Smith and Banttari, 1987). It is rather quick and handy if spotted membranes need to be mailed for probing. As low as 1-5 pg of the PVY/PLRV can be detected in very small samples.

Besides, application of ELISA with fluorogenic substrates or enzyme amplify-ication system can greatly enhance tests sensitivity and it is readily possible to detect PLRV in single aphids (Khurana and Garg, 1993; Stanley *et al.*, 1995). Micro-das ELISA (Reichenbacher *et al.*, 1984) and also magnetic microsphere based ELISA have been developed (Banttari *et al.*, 1991) to economize on antibody/conjugate, time and labour yet maintaining high sensitivity of the test for detecting low concentration viruses.

A serosurvey on incidence of three common potato viruses in samples from different locations/ cultivars revealed wide variation, being highest in case of PVS followed by PVX and least for PVY. No direct correlation was, however, observed to exist between the virus titres and their incidence/ susceptibility of a cultivar at one or more locations (Singh and Khurana, 1994).

Immunosorbent electron microscopy (ISEM): It is a highly sensitive method for virus detection and is either at par or better than ELISA. ISEM detected PLRV even in single aphids. However, it is essential to use the ISEM with optimum conditions, viz. PVX/S samples may be extracted in phosphate buffer (O, IM, pH 7) but for PVY it is important to add 0.1M EDTA to prevent loss of virions through aggregation (Garg and Khurana, 1992). Higher titres of the antiserum used for PVY detection in ISEM caused virion degradation and also combination of antisera with that of PVS was responsible for loss of virion through dislodging even after 'trapping' (Garg and Khurana, 1993; 1994).

Molecular detection: Recombinant DNA technology allows reliable detection of nucleic acids of the infectious agents by spotting them on to a nitrocellulose membrane followed by hybridization with labelled probes, 'tailored' specifically against any targeted pathogen such as a virus or viroid.

***Nucleic acid spot hybridization (NASH)*:** Since viroids are without a protein coat, they can not be detected by immunological techniques. NASH is now most widely used in potato pathology for detection of PSTVd. Originally, it was able to detect only as low as 30 ng/g in tuber sprouts. An improvement in NASH has been made by using single stranded DNA probes. A highly specific cDNA probe capable of detecting 10 to 15 pg of PSTVd from as little as 0.75 mg samples is reported (Bernardy *et al*., 1987). Further, they indicated that oligonucleotide primed labelling is a more efficient and reproducible method of incorporation of 32P into the probe than the nick-translation.

The molecular cloning and hybridization procedures are also successfully applied for basic investigation of viral genome structure and resistance mechanisms. A number of PVX strains/serotypes have been found to have differences in specificity of hybridization. These have also helped in understanding the molecular mechanism involved in the breaking of PVX-HB strain.

***Polymerase chain reaction (PCR)*:** PCR has proved useful in the identification and diagnosis of several viroids, viruses, and mycoplasma-like organisms directly from nucleic acid extracts of infected hosts. It has been found effective in detecting PVY in tubers even after storage at 10°C (Barker *et al*., 1993).

Epidemiology

Potato leaf roll virus (PLRV) is transmitted through infected tubers. Aphids like *Myzus persicae* and *Aphis gossypii* transmit it in circulative manner. The vectors require about 2 hr feeding in infected plants to acquire the virus after which they become viruliferous and transmit to the other plants that is favoured by moderate temperature and dry weather. Likewise, PVY that causes streak and severe mosaic spreads mainly through infected tubers and is transmitted further by the above mentioned aphid species. The virus is stylet borne and infected volunteer plants also serve as good source of perpetuation of the virus. PVX causing faint interveinal mosaic is highly contagious and spreads through rubbing of leaves, contact with infected plants and also through tubers, seed cutting knife, farm implements etc. Root clubbing of healthy and diseased plants in the field can also transmit the virus apart from some other host plants. Potato aucuba mosaic virus is sap inoculated and transmitted by *M. persicae.*

Management of viral diseases

Viral diseases of potato apart from reducing crop yields are also the main causes for degeneration of seed stocks because of vegetative propagation. Since different viruses spread and perpetuate in different manners, different control measures have to be adopted for them. However, it is normally recommended to use them as an integrated control package.

(i) Indirect measures: Indirect measures for control of viruses and viroids include selecting vector free period and location, isolation of the crop, sanitation, use of healthy stocks, roguing, use of pesticides for control of aphid vectors, inspection and certification. Dehaulming the seed crop at the right time for multiplication of quality nucleus seed is also recommended (Mandahar *et al*., 1990).

(ii) Direct measures: Direct control measures include cultivation of virus resistant varieties but not many are available in our country especially having combined resistance for two or more important viruses, except for Kufri Jyoti, Kufri Sindhuri, Kufri Ashoka, etc. Breeding for virus resistant variety is the method of choice for developing countries to cut down on the cost of seed production/ certification systems. The preferred form of resistance is either hypersensitivity or extreme resistance/immunity to the viruses and vectors (Banttari *et al.*, 1993). It has been rather easy to breed varieties with combined resistance to PVX and PVY and also PVA but not PVS and leafroll virus.

Thermo-therapy and use of apical meristem culture technique is in vogue to free the important germplasm and hybrids from the viruses. It is easy to eliminate PLRV, PVY, PVX, MLOs, etc. but PVS is probably the most difficult to get rid of. A combination of chemo-and thermo-therapy coupled with meristem culture, at alternating high (36°C/16hr) and low (29°C/8 hr) temperature, has assured greater virus freedom in the mericlones (Conrad, 1991; Slack and Tuford, 1995; Khurana *et al.*, 1996).

(iii) Non-conventional measures: Non-conventional approaches for incorporating virus resistance are now being employed either by using viral-coat protein or replicase or protease genes or even antisense RNA or anti-viral proteins of non-viral origin, for enhancing host resistance.

(iv) Use of true potato seed: Use of true (Botanical) potato seed (TPS) also holds promise for areas where quality seed is not available. Production of potato crop from TPS obtained from homogeneous populations has been tried in many countries including China, India, Mongolia, New Zealand, Vietnam, etc. It is also possible to either use them directly for sowing a potato crop or by transplanting the seedlings or even going in for production of tuberlets in the first stage and then using them for raising a ware crop. TPS offers various advantages, viz., reduced cost of seed material, easy storage and transpor-tation, low incidence of diseases but requires more labour, lacks uniformity, etc.

(v) Preventive measures: Prevention is the best policy for virus management. Consequently, reliable detection methods have a great significance in the production of high quality virus free seed potatoes. To achieve this goal, detailed information on various aspects such as (a) nature of virus; (b) mode of transmission; (c) health standards of the planting materials, which may act as internal sources of the virus spread in the crop, (d) weed hosts which may act as external sources of the viruses, and (e) factors affecting the build up of the vector and virus diseases must be available (Singh and Boiteau, 1988). Based on the information about above mentioned factors, regulating the build up of viruses and allied diseases in the field, following measures may be adopted in a combined way for minimizing their spread.

Quarantine inspection for exclusion of unknown pathogens (especially latent viruses) through plant propagation stocks into any country is an important step in disease management. Post-entry quarantine (PEQ) of tubers, true potato seed (TPS) and even *in vitro* plantlets can help avoid introduction of new virus(es) or allied pathogens which may get into importing countries. The potato viroid (PSTVd) and viruses (PVT and APLV) and also tobacco ringspot virus-calico strain (TRSV-Ca) and Aracacha virus Oca strain (AVB-Oca) from potato in South America get readily transmitted through true seeds (Jones, 1982). PVY, Solanum apical leaf curl virus etc., restricted to Andes, can be easily excluded. They need to be checked and eliminated before allowing any vegetative or true potato seed stocks into any country.

Controlling spread of viruses is possible by avoiding contact and injury of seed/plants. Cutting of seed and mechanical damage to tuber sprouts has a major role in spreading viruses like PVX, PVS and viroid (PSTVd) (Khurana and Garg, 1998). Besides, movement of men, farm machinery and plant protection operations in fields also result in spread of the contagious viruses / viroid.

Avoiding insect vector(s) for seed production

The viruses spread from far off fields by the migrant winged aphids but within the crop by the developing aphid colonies. *Myzus persicae* is the main aphid vector followed by *Aphis gossypii* in readily spreading both PVY and PLRV during the spring season when the aphid population is higher than in autumn/winter seasons. Therefore, seed potatoes are best multiplied as per 'Seed Plot' technique in the autumn or main winter seasons when the vector pressure is very low or negligible in N-W plains of India (Nagaich *et al.*, 1969) following the essentials of the technique as listed below:

1. Ensure planting the crop with reliable disease-free stocks.
2. Presprouting of seed tubers ensures thick and multiple sprouts and avoids 'sickly' late germinating plants.
3. Timely planting and proper combination of seed size and spacing help in achieving maximum number of seed tubers.
4. Use of herbicides to control weeds and to avoid spread of contact transmitted viruses upon manual weeding.
5. Apply recommended plant protection measures.
6. Timely inspection, roguing and killing of haulms.
7. Withhold irrigation 10-15 days before killing the haulms.
8. Harvest the crop 10-15 days after haulms cutting followed by grading, treating the seed tubers and cold store the produce for use in next season.

Seed potato certification (SPC)

Potatoes are normally planted using "Seed" tubers. A "diseased" tuber carries many disease organisms within or over it. Certified seed for planting potatoes is the foundation of maintaining and producing a healthy crop. Seed certification in potatoes is very important in maintaining purity of variety, yields of the crop and health standards over a period of time (Khurana and Singh, 1997). SPC has to be done by state level agency having set state/national standards of SPC by inspecting crop(s) in field/seed lots, checking mainly for cultivar purity and roguing of all diseased and off type plants. Such basic seed may be multiplied for a number of years (usually 5 years) with rigorous inspection and field testing in designated/ suitable primary seed producing areas for preparation of Basic/Breeders seed (normally by govt. agencies/research organizations) until Foundation/ Certified (F I/II) (CI/II) seed stages are reached in sixth or seventh year.

REFERENCES

Alam, S.M.K., Elias, M. and Miah, M.R.V. 1991. Efficacy of foliage fungicides in controlling late blight of potato. *J. Pl. Path.* **7:** 21–22.

Anonymous 1991. Annual Scientific Report, CPRI, Shimla, pp. 77–78.

Arora, R.K. 1990. Fitness variations in isolates of *Phytophthora infestans. Indian Phytopath.* **43:** 248.

Arora, R.K. 1991. Occurrence and frequancy of metaloxyl resistant isolates of *Phytophthora infestans. Indian Phytopath.* **45:** 90.

Arora, R.K. 2008. Management of black scurf of potato with the integrated use of *Trichoderma viride* and boric acid. *Potato J.* **35 :** 130–133.

Arora R K, Islam Ahmed and Singh B P. 2012. Forecasting late blight of potato in Punjab using JUSLACAST model. *Potato J.* **39**: 173–176

Arora R K, Shrama S and Singh B P. 2014. Late blight disease of potato and its management. *Potato J.* **41**:16–40

Banthari, E.E., Ellis, P.J. and Khurana, S.M. Paul, 1993. Management of diseases caused by viruses and virus like pathogens. In: *Potato Health Management* (Ed. R.C. Rowe). APS Press, St. Paul, U.S.A., pp. 127–133.

Banttari, E.E., Clapper, D.L., Hu, S.P., Daws, K.M. and Khurana, S.M. Paul 1991. Rapid magnetic microsphere enzyme immunoassay for potato virus X and potato leafroll virus. *Phytopathology* **81:**1039–1042.

Banyal, D.K. and Sugha, S.K. 2010. Refinement of fungicidal spray schedule for effective management of potato late blight in north-western hills. *Pl. Dis. Res.* **25(1):** 30–34.

Barker, H., Webster, K.D. and Revy, B. 1993. Detection of potato virus Y in potato tubers: A comparison of polymerase chain reaction and enzyme linked immunosorbent assary. *Potato Res.* **36:** 13–20.

Beagle–Ristaino. J.E. and Papavizas, G.C. 1985. Biological control of Rhizoctonia stem canker and black scurf of potato. *Phytopathology* **75 :** 560–564.

Bernardy, M.G., Jacoli, G.G. and Ragetli, H.W.J. 1987. Rapid detection of potato spindle tuber viroid (PSTV) by dot blot hybridization. *J. Phytopath.* **118:** 171–180.

Bhattacharrya, S.K. and Singh, B.P. 1990. Current trends in epidemiology and control of late blight of potato. In : *'Current Facets in Potato Research'* (Eds. J.S. Grewal, G.S. Shekhawat, S.K. Bhattacharrya, R.A. Singh), Central Potato Research Institute, Shimla pp. 121–131.

Bhattacharrya, S.K., Phadtare, S.G. and Sharma, V.C. 1977. Fungal diseases of potato. In : *'Recent Technology in Potato Improvement and Production'*. (Ed. B.B. Nagaich), CPRI, Shimla, pp 414–424.

Bhattacharrya, S.K., Raj, S., Singh, D.S., Khanna, R.N. and Sharma, S.R. 1983. Forecasting late blight of potato in Indian hills. In : *'Potato in Developing Countries'* (Eds. B.B. Nagaich *et al.*) Indian Potato Association, CPRI, Shimla, pp. 20–23.

Bjor T. and Mulelid, K. 1991. Differential resistance to tuber late blight in potato cultivars without R–genes. *Potato Res.* **34 :** 3–8.

Bombawale, O.M., Bedi, P.S., Sharma, N.K. and Sharma, S.K. 1982. Early blight–root knot–nematicide interaction in some potato cultivars. In *"Potato in Developing Countries"* (Eds. B.B. Nagaich *et al.*), CPRI, Shimla, pp 374–378.

Casido, L.E. (Jr.) and Lukezic, F.L. 1992. Control of leaf spot diseases of alfalfa and tomato with application of bacterial predator *Pseudomonas* strain 679–2. *Plant Dis.* **76 :** 1217–1220.

Chakrabarti, S.K. and Shekhawat, G.S. 1991. Population build up *in vivo* of *Pseudomonas solancearum* in three Indian potato cultivars. *Indian Phytopath.* **44(4):** 499–500.

Chakrabarti S K, Singh B P, Thakur G, Tiwari J K, Kaushik S K, Sharma S and Bhardwaj V. 2014. QTL mapping underlying resistance to late blight in a diploid potato family of *Solanum spegazzinii* × *S. chacoense*. *Potato Res.* **57**:1–11.

Chakraborty A, Singh B P, Ahmad I and Sharma S. 2015. Forecasting late blight of potato in the plains of West Bengal using JHULSACAST model. *Potato J.* **42**: 50–57.

Chandel P, Tiwari J K, Ali Nilofer, Devi S, Sharma S, Sharma S, Luthra S K and Singh B P. 2015. Interspecific potato somatic hybrids between *Solanum tuberosum* and *S. cardiophyllum*, potential sources of late blight resistance breeding. *Plant Cell Tissue and Organ Culture* DOI 10.1007/s11240–015–0862–8.

Clark, M.F. and Adams, A.N. 1977. Characteristics of the microplate method of enzyme–linked immunosorbent assay for the detection of plant viruses. *J. Gen. Virol.* **34 :** 475–483.

Cohen, Y., Farakash, S., Reshit, Z. and Baides, A. 1997. Oospore production of *Phytophthora infestans* in potato and tomato leaves. *Phytopathology* **87:** 191–196.

Conrad, P.L. 1991. Potato virus S– free plants obtained by using antiviral compounds and nodal segment culture of potato. *Am. Potato J.* **68:** 507–513.

Cook, D., Barlow, E. and Sequeira, L. 1989. Genetic diversity of *Pseudomonas solanacearum*: detection of restriction fragment length polymorphosins with DNA probes that specify virulence and the hypersensitive response. *Mol. Plant Microbe Interac.* **2:** 113–121.

Cooke, L.R., Schepers, H.T.A.M., Hermansen, A., Bain, R.A., Bradshaw, N.A., Ritchie, F., Shaw, D.S., Evenhuis, A., Kessel, G.J.T., wander, J.G.N., Andersson, B., Hansen, J.G., Hannukkala, A., Naerstad, R. and Nielsen, B.J. 2011. Epidemiology and integrated control of potato late blight in Europe. *Potato Research* **54(2) :** 183–222.

Cooper, J.I. and Edwards, M.L. 1986. Variations and limitations of enzyme amplified immunoassays. In: *Developments and Applications in Virus Testing* (Eds. R.A.C. Jones and L. Torrance). The Lavenham Press, Suffolk, U.K. 300 pp.

CPRI 1987. Annual Scientific Report. Central Potato Research Institute, Shimla, India.

CPRI. 2013. Annual Progress Report. Central Potato Research Institute, Shimla, India.

CPRI. 2014. Annual Progress Report. Central Potato Research Institute, Shimla, India.

CPRI. 2015. Annual Progress Report. Central Potato Research Institute, Shimla, India.

Das, Aparna, Minocha, J.L., Thind, T.S. and Dhaliwal, H.S. 1999. *In vitro* induction and selection for late blight resistance in potato. *Indian Phytopath*. **52(2):** 169–171.

Deahl, K.L., Inglis, D.A. and Demuth, S.P. 1993. Testing for resistance to metalaxyl in *Phytophthora infestans* isolates from North–Western Washington. *Am. Potato J.* **70(11):** 779–795.

DeBokx, J.A. and Van der Want, J.P.H. (Eds.) 1987. Viruses of potatoes and seed potato production, PUDOC, Wageningen, 259 pp.

Devi, L.R., Menon, M.R. and Aiyer, R.S. 1982. Population threshold of *Pseudomonas solanacearum* at the onset of first symptoms of wilt in tomato. *Indian J. Microbiol.* **22:** 41–43.

Dey, S.K. and Singh, N. 1983. Studies on variability of seven isolates of *Streptomyces* species causing common scab of potato in Punjab (India). *Phytopath. Mediterr.* **22 :** 59–66.

Dey, T.K., Hossain, M., Kadian, M.S., Hossain, S., Bonierbale, M. and Mahmud, A.A. 2010. Prevalence, epidemiology and management of potato late blight in Bangla Desh. *Potato J.* **37(3) :** 99–102.

Dowley, L.J. and O'Sullivan, E. 1991. Changes in the distribution of phenylamide resistant strains of *Phytophthora infestans* (Mont.) de Bary in Ireland. *Potato Res.* **34(1):** 67–69.

Dua VK, Singh BP, Ahmad Islam and Sharma Sanjeev. 2015. Potential impact of climate change on late blight outbreak in western Uttar Pradesh and Punjab using Jhulsacast model. *Potato J.* **42(1):** 58–71

Dutt, B.L. 1979. Bacterial and fungal diseases of potato. ICAR, New Delhi, 196 p.

Dutt, B.L., Mathur, P.N. and Shiv Ram 1982. Control of late blight, a serious disease of potatoes in the Shimla hills. In: *'Potato in Developing Countries* (Eds. B.B. Nagaich, G.S. Shekhawat, P.C. Gaur, S.C. Verma), Central Potato Research Institute, Shimla, pp. 319–324.

Franc, G.D. and Banttari, E.E. 1986. Comparison of latex agglutination, enzyme–linked immunosorbent assay and indicator plants for detection of potato viruses S and X in potatoes. *Amer. Potato J.* **63:**357–362.

Fribourg, C.E. and Nakashima, J. 1984. An improved latex agglutination test for routine detection of potato viruses. *Potato Res.* **27:**173–249.

Gadewar, A.V., Trivedi, T.P. and Shekhawat, G.S. 1991. *Potato wilt in Karnataka*. Tech. Bull. No. 17. Central Potato Research Institute, Shimla, 33pp.

Ganguly, A. and Paul, D.K. 1953. Wart disease of potatoes in India. *Sci. & Cult.* **18 :** 605–606.

Garg, I.D. and Khurana, S.M. Paul. 1992. Factors influencing immune electron microscopy of flexuous potato viruses. *Acta Virol.* **36:** 435–442.

Garg, I.D. and Khurana, S.M. Paul. 1993. Morphological changes in the flexuous potato viruses upon decoration in immunosorbent electron microscopy. *Acta Virol.* **37:**407–411.

Garg, I.D. and Khurana, S.M. Paul. 1994. ISEM: Dislodging of PVY/PVA virions upon incubation with antisera to PVS/PVM. In : *'Potato: Present and Future'* (Eds. G.S. Shekhawat *et al.*) Indian Potato Assoc. Shimla, pp. 325–327.

Govindakrishnan P M, Singh B P, Ahmad Islam, Rawat Shashi and Sharma Sanjeev. 2015. Indoblightcast– A simple generic DSS for late blight management. pp 32–33, in *Abstracts: 3rd International Symposium on Phytophthora: Taxonomy, Genomics, Pathogenicity, Resistance and Disease Management* edited by Chowdappa P, Hegde V and Muralikrishna H. ICAR–IIHR, Bengaluru, 9–12 September, 2015

Granada, G.A. 1988. Latent infection by Pseudomonas solanacearum in potato and symptomless plants. In: 'Bacterial Diseases of the Potato – Report of the Planning Conference on Bacterial Diseases of Potato, 1987". International Potato Centre, Lima, Peru, pp. 93–108.

Hunjan, M.S., Singh, P.P., Kaur, S., Kaur, P., Kumar, A. and Sagar, V. 2016. Punjab is a pest free area with respect to brown rot disease of potato. Paper presented in 6th International Conference on *'Plant, Pathogens and People – Challenges in Plant Pathology to Benefit Humankind*', 23–27 February, 2016, New Delhi, Abstract p. 158.

Hussain, T., Sharma, S., Singh, B.P., Jeevalatha, A., Sagar, V., Sharma, N.N., Kaushik, S.K., Chakrabarti, S.K. and Anwar, F. 2013. Detection of latent infection of *Phytophthora infestans* in potato seed tubers. *Potato J.* **40:** 142–148

IISR. 2013. PhytoFuRa Annual Report 2012–13. Indian Institute of Spices Research, Kozhikode

IISR. 2014. PhytoFuRa Annual Report 2013–14. Indian Institute of Spices Research, Kozhikode, pp 20–26

Jagyasi, B., Kumar, V., Pande, A., Singh, B.P., Lal Mehi, Ahmad Islam and Lohia, P. 2015. Validation of Jhulsacast model using human participatory sensing and wireless sensor networks. *Potato J.* **42:**44–49

Jones, R.A.C. 1982. Tests for transmission of four potato viruses through potato true seed. *Ann. Appl. Biol.* **100:** 315–320.

Keshwal, R.L. 1980. Studies on bacterial wilt of solanaceous crops caused by *Pseudomonas solanacearum* E.F. Smith with special reference to its control. Ph.D. Thesis, Univ. of Jabalpur, Jabalpur, M.P., India, 201pp.

Khanna, R., Roy, S., Rai, R.P., Sharma, J., Jeswani, M.D. and Singh, P.H. 1991. Annual Scientific Report, CPRI, Shimla, pp. 74.

Khurana, S.M. Paul 1990. Modern approaches for detection and management of the potato viruses and viroid. In: *'Current Facets in Potato Research'* (Eds. J.S. Grewal *et al.*), Shimla, pp. 98–108.

Khurana, S.M. Paul 1992. *Potato Viruses and Viral Diseases.* Tech. Bull No. 35, CPRI, Shimla, 23pp.

Khurana, S.M. Paul and Garg, I.D. 1993. New techniques for detection of viruses and viroid. In: '*Advances In Horticulture. Vol. 7. Potato* (K.L. Chadha and J.S. Grewal, Eds.), MPH, N.Delhi, pp. 529–566.

Khurana, S.M. Paul and Garg, I.D. 1998. Present status of controlling mechanically and non–persistently aphid transmitted potato viruses. In: *'Plant Virus Disease Control.* (A. Hadidi *et al.*, eds) pp. 593–609.

Khurana, S.M. Paul, Garg, I.D., Singh, B.P. and Gadewar, A.Y. 1998a. Major diseases of potato and their management. In: *Integrated Pest and Diseases Management* (R.K Upadhya *et al.*, Eds.), APH Corpn., New Delhi pp 11-64.

Khurana, S.M. Paul and Singh Sarjeet 1997. Potato seed production and certification. In: '*Plant Pathology: 50 Years of Research in India'* (J.P.Verma and A.Verma, eds.), pp. 109–141, ICAR, New Delhi.

Khurana, S.M. Paul and Singh, M.N. 1995. Factors governing aphid transmission of the potato viruses. In: `*Proc. 2nd Agricultural Science Congress* (M.S. Swaminathan *et al.*, Eds), NAAS, New Delhi, pp. 233–252.

Khurana, S.M. Paul, Chandra, R. and Dhingra, M.K. 1996. Potato viruses and their eradication for production of virus free seed stocks. In: *'Disease Scenario in Crop Plants – I. Fruits and Vegetables'* (V.P. Agnihotri *et al.*, Eds.), IBPSS, Delhi, pp. 195–218.

Krause, R.S., Massie, L.B. and Hyre, R.A. 1975. Blite cast: A computerized forecast of potato late blight. *Pl. Dis. Reptr.* **59:** 95–98.

Kumar, P., Thind, T.S., Bala, A., Amrate, P.K. and Singh, A. 2012. Relative effectiveness and persistence of novel action fungicides against late blight (*Phytophthora infestans*) of potato. *Pl. Dis. Res.* **27(1):** 45–48.

Kumar, S., Thind, T.S., Bala, A. and Gupta, A.K. 2010. Induced resistance in potato against *Phytophthora infestans* using chemical and bio–agents. *Pl. Dis. Res.* **25(1):** 12–18.

Kumar V, Luthra S K, Bhardwaj V and Singh B P. 2014. *Indian potato varieties and their salient features.* Technical Bulletin No. 78 (Revised), ICAR–Central Potato Research Institute, Shimla, HP, India

Laferriere, L.T., Helgeson, J.P. and Allen, C. 1998. *Solanum tuberosum – S. commersonii* somatic hybrids are resistant to brown rot caused by *Ralstonia solanacearum.* Pp. 313–320. In: *Bacterial Wilt Disease : Molecular and Ecological Aspects.* (Prior, P., Allen, C. and Elphinstone, J. (eds.)) Report of the Second International Wilt Symposium held in Gosier, Guadeloupe, France, 22–27 June, 1997. INRA, Springer, 1998, 447 p.

Lakra, B.S. 1992. Correlation of infection intensities of black scurf with yield components of potato. *Indian J. Mycol. Pl. Pathol.* **22 :** 203–204.

Lal Mehi, Yadav Saurabh, Chand Subhash, Kaushik S K, Singh B P and Sharma Sanjeev. 2015. Evaluation of fungicides against late blight (*Phytophthora infestans*) on susceptible and moderately resistant potato cultivars. *Indian Phytopath.* **68**: 345–347

Louwes, K.M, Hoekstra, R. and Mattrheij, W.M. 1992. Interspecific hybridization between the cultivated potato *Solanum tuberosum* sub sp. *tuberosum* L. and the wild species *S. circaefolium* sub sp. *circaefolium* Bitter exhibiting resistance to *Phytophthora infestans* and *Globodera pallida. Theor. Appl. Genet.* **84** : 362–370.

Mahajan, Monica, Thind, T.S. and Dhillon, M. 2005. Pre–infectional biochemical basis of resistance to late blight (*Phytophthora infestans*) in potato (*Solanum tuberosum*) *Indian J. Agric. Sci.* **75(7):** 454–456.

Malathi, P. and Jeyarajan, R. 1995. Evaluation of chlorothalonil for potato late blight in Nilgiris. *Indian J. Mycol. Pl. Pathol.* **25(3):** 211–213.

Mandahar, C.L., Khurana, S.M. Paul and Garg, I.D. 1990. Disease management. In: *'Plant Viruses Vol. II. Pathology'*, (C.L. Mandahar, Ed.), CRC Press, Boca Raton, Fl. (USA), pp. 273–293.

Martinez, B. and Solano, T. 1995. Antagonism of *Trichoderma* species to *Alternaria solani* (Ell. & Mart.) Jones and Grout. *Revista de Protecion Vegetal* **10** : 221–225.

Mohan, C. and Thind, T.S. 1999. Persistence and relative performance of some new fungicides for effective management of potato late blight in Punjab. *Indian J. Mycol. Pl. Pathol.* **29(1):** 32–37.

Nagaich, B.B. 1983. Disease resistance in potato in India, *Indian Phytopath.* **36**:1–10.

Nagaich, B.B., Pushkarnath, Bhardwaj, V.P., Giri, B.K., Anand, S.R. and Upreti, G.C. 1969. Production of disease free seed potatoes in the Indo–Gangetic plains. *Indian J. Agric. Sci.* **39:** 238–2431.

Ojha, K.L., Yadav, B.P. and Bhagat, A.P. 1986. Chemical control of bacterial wilt of ginger. *Indian Phytopath.* **39:** 600–601.

Phadtare, S.G. and Sharma, K.P. 1971. Additional hosts of *Synchytrium endobio–ticum. Indian Phytopath.* **24** : 389–392.

Pundhir V S, Singh B P, Ahmad Islam, Sharma Sanjeev, Kushwaha H S, Singh V K and Joshi Varsha. 2014. Forecasting late blight of potato in Tarai region of Uttarakhand using JHULSACAST model. *Potato J.* **41**: 95–104.

Rauf, C.A., Ashraf, M. and Ahmad, I. 2007. Management of black scurf disease of potato. *Pak. J. Bot.* **39(4)** : 1353–1357.

Rauf, C.A., Naz, F., Ahmad, I., Haque, I.U. and Riaz, A. 2015. Management of black scurf of potato with effective microbes, biological potassium fertilizer and *Trichoderma harzianum. Int. J. Agric. Biol.* **17** : 601–606.

Reichenbacher, D., Kalinnia, I., Schultze, M., Hom, A. and Kleinhemple, H. 1984. Ultramicro–ELISA with a fluorogenic substrate for detection of potato viruses. *Potato Res.* **27:** 353–364.

Sanju, S, Siddappa S, Thakur Aditi, Shukla P K, Srivastava N, Pattanayak D, Sharma S and Singh B P. 2015. Host–mediated gene silencing of a single effector gene from the potato pathogen *Phytophthora infestans* imparts partial resistance to late blight disease. *Functional and Integrated Genomics* **15**:697–706

Sarkar, D, Tiwari J K, Sharma S, Poonam, Sharma S, Gopal J, Singh B P, Luthra S K, Pandey S K and Pattanayak D. 2011. Production and characterization of somatic hybrids between *Solanum tuberosum* L. and *S. pinnatisectum* Dun. *Plant Cell Tissue and Organ Culture* **107**: 427–440

Schwinn, F.J. and Staub, T. 1987. Phenylamides and other fungicides against Oomycetes. In: *"Modern Selective Fungicides: Properties, Applications, Mechanism of Action"* (ed. H. Lyr), Longman Group, UK Ltd., pp. 259–273.

Sekhon, P.S. and Sokhi, S.S. 1999. Effect of date of planting on development of late blight of potato in Punjab. *Indian Phytopath.* **52(4):** 267–269.

Sharma, R, Kaushik S K, Bhardwaj V, Sharma S, Bhatt A K and Singh B P. 2013. Molecular characterization of potato genotypes for late blight resistance. *Potato J.* **40**: 164–172

Sharma, Sanjeev, Guleria Anupama and Singh B P. 2014. Ploidy status and its role in aggressiveness of *Phytophthora infestans.* In National Seminar on *"Emerging Problems of Potato"* held at CPRI, Shimla during 1–2 November 2014. p162.

Sharma, Snajeev, Singh, B.P., Jeevalatha, A. and Chakrabarti, S.K. 2017. Molecular characterization and pathogenicity of Indian *Phytophthora infestans* isolates reveals no correlation between phenotypes and their genetic origin. *J. Mycol. Pl. Pathol.* **47(1):** 1–12.

Shekhawat, G.S. 1990. Emerging trends in epidemiology and management of bacterial diseases of potato. In : *Current Facets in Potato Research.* (J.S. Grewal, G.S. Shekhawat, S.K. Bhattacharyya and R.A. Singh, eds.). Indian Potato Association, CPRI, Shimla, pp. 109–120.

Shekhawat, G.S., Gadewar, A.V., Bahal, V.K. and Verma, R.K. 1988a. Cultural practices for managing bacterial wilt of potato. In *"Bacterial Diseases of Potato: Report of the Planning Conference of Bacterial Diseases of Potato, 1987"*. International Potato Centre, Lima, Peru, pp. 65–84.

Shekhawat, G.S., Kishore, V., Gadewar, A.V., Sunaina, V., Verma, R.K., Bahal, V.K. and Chakrabarti, S.K. 1991. Ecology, detection and management of *Pseudomonas solanacerarum.* Annual Scientific Report, Central Potato Research Institute, Shimla, pp. 76–82.

Shekhawat, G.S., Kishore, V., Sunaina, V., Bahal, V.K., Gadewar, A.V., Verma, R.K. and Chakrabarti, S.K. 1988b. Latency and management of *Pseudomonas solanacerarum.* Annual Scientific Report, Central Potato Research Institute, Shimla, pp. 117–121.

Shekhawat, G.S., Singh, B.P. and Jeswani, M.D. 1993. *Soil and Tuber Borne Diseases of Potato.* Tech. Bull. No. 41. Central Potato Research Institute, Shimla, 47 pp.

Shepard, J.F. 1972. Gel–diffusion methods for the serological detection of potato viruses X, S and M. *Mont. Agric. Exp. Sta. Bull.* 622, 72 pp.

Shtienberg, D., Blachinsky, D., Ben–Hador, G. and Dinoor, A. 1996. Effects of growing reason and fungicide type on the development of *Alternaria solani* and on potato yield. *Plant Dis.* **80:** 994–998.

Shtienberg, D., Blachinsky, D., Kremer, Y., Ben–Hador, G. and Dinoor, A. 1993. Integration of genotype and age related resistance to use fungicide use in management of Alternaria disease of cotton and potato. *Phytopathology* **85 :** 993–1002.

Sikka, L.C., Srivastava, S.N.S., Singh, A.K. and Bhardwaj. V.P. 1971. An integrated approach to the control of *Rhizoctonoa solani* on potato. *Indian Phytopath.* **24 :** 544–547.

Singh A. and Singh, D.V. 1991. Common scab of potato caused by *Streptomyces aureofaciens*: A new record from India. *Indian Phytopath.* **44:**558.

Singh A. and Singh, D.V. 1992. Sources of resistance to common scab of potato. *Indian Phytopath.* **45:** 225–226.

Singh, A. and Singh, D.V. 1993 Influence of different sources of nitrogen on the development of common scab of potato and its yield. *Indian J. Agric. Sci.* **63:**597–598.

Singh, B.P. and Bhat, N. 2003. Emerging trends in the epidemiology of late blight of potato. *Ann. Rev. Plant Pathol.,* **2 :** 43–84.

Singh, B.P. and Bhattacharrya, S.K. 1990. Control of late blight tuber infection by systemic fungicides. *Indian J. Plant Prot.* **18:** 119–123.

Singh, B.P. and Jeswani, M.D. 1987. Managing common scab of potato through chemical control and cultural practices. *J. Indian Potato Assoc.* **14 :** 26–32.

Singh, B.P., Ahmed Islam, Sharma, V.C. and Shekhawat, G.S. 2000. JHULSACAST : A computerized forecast of potato late blight in western Uttar Pradesh. *J. Indian Potato Assoc.* **27 :** 25–34.

Singh, B.P., Chakrabarty, S.K., Roy, S. and Bhattacharrya, S.K. 1988. Adaptability of *Phytophthora infestans* to fungicides. *Indian J. Pl. Prot.* **16:** 277–280.

Singh, B.P., Roy, S. and Bhattacharrya, S.K. 1994. Occurrence of A2 mating type of *Phytophthora infestans* in India. *Potato Res.* **37** :227–231.

Singh B P and Sharma Sanjeev. 2013. Forecasting of potato late blight. *Int. J. Innovative Hortic.* **2(2):** 1–11

Singh, D.S. and Nagaich, B.B. 1977. *Nicandra physaloides* Gaertn, an additional host of *Alternaria solani* (Ellis & Mart.) Jones & Grout causing early blight of potato. *Sci. & Cult.* **43:** 454.

Singh, H. and Singh, N. 1982. Efficacy of different chemicals in controlling common scab of potato. *Indian J. Mycol. Pl. Pathol.* **12:** 337–338.

Singh, H. and Soni, P.S. 1987. Chemical control of common scab of potato. *Pl. Dis. Res.* **2:** 77–79.

Singh, N. and Singh, H. 1981. Irrigation, a cultural practice to control common scab of potato. *J. Indian Potato Assoc.* **8 :** 35–36.

Singh, R.A. and Khurana, S.M. Paul 1993. Viral and allied diseases of poatato. In: *Advances In Horticulture Vol. 7. Potato* (K.L. Chadha and J.S. Grewal, Eds.). MPH, N. Delhi, pp. 491–528.

Singh, R.P. and Boiteau, G. 1988. Control of aphid borne diseases: Nonpersistent viruses. In: *'Potato Pest Management in Canada'* (G. Boiteau *et al.*, Eds.), pp. 30–53, Canada, Agric., NB, 384 pp.

Singh, R.S., Chaube, H.S. and Singh, N. 1972. Studies on the control of black scurf disease of potato. *Indian Phytopath.* **25 :** 343–349.

Singh, Raj Pal 1980. Studies on *Pseudomonas solanacearum* the incitant of bacterial wilt of solanaceous crops. Ph.D. Thesis, Univ. of Gorakhpur, Gorakhpur, India.

Singh, Shatrughna and Khurana, S.M. Paul 1994. Incidence and relative concentration of common potato viruses in five cultivars. *Indian J. Virol.* **10:** 44–50.

Sivamani, E., Anuratha, C.S. and Ganamnickam, S.S. 1987. Toxicity of *Pseudomonas fluorescens* towards bacterial plant pathogens of banana (*Pseudomonas solanacearum*) and rice (*Xanthomonas campestris* pv. *orzae*). *Curr. Sci.* **56:** 547–548.

Slack, S.A. and Tuford, L.A. 1995. Meristem culture for virus elimination. In: *'Plant Cell, Tissue and Organ Culture – Fundamental Methods.* (O.L. Gamborg and G.C. Phillips, Eds.), Springer Verlag, N. York, pp. 117–128.

Smith, F.D. and Banttari, E.E. 1987. Dot–ELISA on nitrocellulose membranes for detection of potato leafroll virus. *Plant Dis.* **71:** 795–799.

Somani, A.K. 1986. Non–hazardous chemical control of black scurf of potato. *Indian J. Agric. Sci.* **56:** 366–369.

Stanley, C.J., Johnson, A. and Self, C.H. 1995. Enzyme amplification can enhance both the speed and sensitivity of immunoassay. *J. Immunol. Methods.* **83:** 89–95.

Stewart, H.E., Bradshaw, J.E., Wastie, R.L., Mackay, G.R., Erlich, O., Livescu, L. and Nachmias, A. 1994. Assessing progenies of potato for resistance to early blight. *Potato Res.* **37:** 237–269.

Sunaina, V. and Gupta, P.K. 1998. Relationship between seasons and severity of bacterial wilt of potato under field conditions. *Indian Phytopath.* **51(1):** 92–94.

Taghavi, M., Hayward, C., Sly, L.I., Fegan, M. 1996. Analysis of the phylogenetic relationships of strains of *Burkholderia solanacearum, Pseudomonas syzygii* and the blood disease bacterium on 165 tRNA gene sequences. *Int. J. Syst. Bacteriol.* **46 :** 10–15.

Tanni, A., Akai, J. and Aota, T. 1982. Chemical control of several potato diseases transmitted through seed tubers. In: *'Potato in Developing Countries* (Eds. B.B. Nagaich *et al.*), CPRI, Shimla, pp. 271–278.

Thakur Aditi, Sanju S, Siddappa S, Shukla P K, Srivastava N, Pattanayak D, Sharma S and Singh B P. 2015. Artifical microRNA mediated gene silencing of *Phytophthora infestans* single effector *Avr3a* gene imparts moderate type of late blight resistance in potato. *Plant Pathol. J.* **14**:1–12.

Thind, T.S. and Aggarwal, R. 2008. Characterization and pathogenic relationship of *Rhizoctonia solani* isolates in potato–rice system and their sensitivity to fungicides. *J. Phytopathol.* **156 (10):** 615–621.

Thind, T.S. and Jhooty, J.S. 1982. Studies on spray schedule of Dithane M–45 and Blitox against early blight of potato. *J. Indian Potato. Assoc.* **9(1):** 28–31.

Thind, T.S. and Mohan, C. 1997. Management of potato late blight by rational use of fungicides. *Proc. Third National Agric. Sci. Cong.* (V. Beri, V.K. Dilawari and M.S. Bajwa, Eds.). pp. 307.

Thind, T.S. and Mohan, C. 1998. Severity of late blight and assessment of yield losses in potato during 1997–98 epiphytotic in Punjab. *Pl. Dis. Res.* **13:** 204–207.

Thind, T.S., Bala, A., Mukerjee, J., Jolly, R.K., Kumar, P., Singh, A. and Sekhon, P.S. 2013. Development of a web–based decision support system for late blight of potato. *Pl. Dis. Res.* **28(1)** : 64–70.

Thind, T.S., Mohan, C. and Kaur, S. 2002. Promising activity of pencycuron, a phenyl urea–based fungicide, for effective management of black scurf of potato. *Indian Phytopath.* **55 (1)**: 39–44.

Thind, T.S., Mohan, C., Arora, J.K. and Raj, Prem 2004b. Potential of some recently developed fungicides and theit combination products for effective manage–ment of late blight of potato. *Indian Phytopath.* **57(1)** : 95–98.

Thind, T.S., Mohan, C., Bedi, J.S., Sokhi, S.S. and Grewal, R.K. 1989a. Role of application time of fungicides for the control of late blight of potato. *Pl. Dis. Res.* **4:** 113–117.

Thind, T.S., Mohan, C., Raj, Prem and Arora, J.K. 2004a. Activity spectrum of storibilurin, a new generation of eco–fungicides, against some fungal pathogens. *Indian Phytopath.* **57(1)** : 104–106.

Thind, T.S., Mohan, C., Sokhi, S.S. and Bedi, J.S. 1989b. A detached leaf technique for maintenance and multiplication of *Phytophthora infestans* and evaluation of fungicides. *Curr. Sci.* **58:** 388–389.

Thind, T.S., Singh, Lakhvir, Mohan, C. and Pal, Joginder 2001. Monitoring for metalaxyl resistance in populations of *Phytophthora infestans* from 1996 to 1999 in Punjab and their characteristics. *Indian Phytopath.* **54 :** 68–74.

Tomar S, Singh B P, Khan M A, Kumar S, Sharma S and Lal M. 2013. Identification of *Pseudomonas aeruginosa* strain producing biosurfactant with antifungal activity against *Phytophthora infestans. Potato J.* **40**:155–163

Tomar Sonica, Singh B P, Lal Mehi, Khan M A, Hussain Tauseef, Sharma Sanjeev, Kaushik S K and Kumar Satish. 2014. Screening of novel microorganisms for biosurfactant and biocontrol activity against *Phytophthora infestans. J. Environ. Biol.* **35**:893–899

Tong, Yen Hui, Liang Ji Nong, Ku–Jug You, Tung, Y.H. and Xu, Jy. 1994. Studies on the biology and pathogenicity of *Alternaria solani* on tomato. *J. Jiangsu–Agric. College* **15:** 29–31.

Tsror, L., Barak, L. and Sneh, B. 2001. Biological control of black scurf on potato under organic management. *Crop Prot.* **20(2)** : 145–150.

Tsror, L. and Peretz–Alon, I. 2005. The influence of the inoculums source of *Rhizoctonia solani* on development of black scurf on potato. *J. Phytopath.* **153 (4)** : 240–244.

Van Harten, A. 1983. The relation between aphid flights and the spread of potato virus Yn(PVYn) in the Netherlands. *Potato Res.* **266:** 1–15.

Wale, S.J., Robertson, K., Robinson, L. and Foster, G. 1986. Studies on the relationship of contamination of seed potato tubers with *Erwinia* sp. to black leg incidence and large scale hot water dipping to reduce contamination. In : *'Crop Protection and Quality of Potatoes'*. Wellsbourne, Warwick, AAB, pp. 285–291.

Wani, A.A., Rewaland, H.S. and Mohan C. 2010. Management of late blight of potato through early detection and destruction of disease foci. *Pl. Dis. Res.* **25(1):** 26–29.

Yabuuchi, E., Kosako, Y., Oyaizu, H., Yano, I., Hotta, H., Hashimoto, Y., Ezaki, T. and Arakawa, M. 1992. Proposal of *Burkholderia* gen. nov. and transfer of seven species of the genus *Pseudomonas* homology group II to the new genus, with the type species *Burkholderia cepacia* (Paaleroni and Holmes, 1981) comb. nov. *Microbiol. Immunol.* **36** : 1251–1275.

Yabuuchi, E., Kosako, Y., Yano, I., Hotta, H. and Nishiuchi, E. 1995. Transfer of two *Burkholderia* and an *Alcoligenes* species to *Ralstonia* gen. nov. Proposal of *Ralstonia pickettii* (Ralston, Palleroni and Doudoroff, 1973) comb. Nov. and *Ralstonia eutrapha* (Davis, 1969) comb. Nov. *Microbiol. Immunol.* **39** : 897–907

■■■■

4

DISEASES OF TOMATO

Tomato (*Solanum lycopersicum* L.) is an important vegetable crop grown globally and consumed as salad, ketchups, sauce, soup, pickles etc. It is also cooked alongwith other vegetables to provide taste and is the basic need of Indian kitchen. Ripened fruits are good source of ascorbic acid and minerals. The crop is generally grown during winter months (Oct.-April) in the plains of India while in the hills of Himachal Pradesh, Uttar Pradesh and Bangalore area of Karnataka it is mostly grown during summer and rainy season, thus fresh fruits are available round the year in the market. Prevalence of high humidity and warm temperature during the growing season of tomato not only favours the luxuriant growth of the crop but also favours the development of various fungal, bacterial and viral diseases. Under favourable environmental conditions, epiphytotics of certain diseases have often reduced the yields considerably in certain years. Some of the important fungal, bacterial and viral disease problems of tomato are described below along with management practices.

(A) FUNGAL DISEASES

1. LATE BLIGHT

Late blight is a serious disease problem affecting cultivation of tomato, the world over where the temperature and moisture permit the pathogen to thrive. Though the fungus has a long history and lot of literature is available on the management of this disease but it still remains a threat to successful cultivation of the crop.

Geographical distribution

Late blight originated from Andes hills of Central and South America, the place origin of potato and then it spread to other parts of the world. The disease on tomato is prevalent in all those areas of the world having cool, moist conditions during growing conditions. On tomato this disease has been reported from all those areas like Europe, north-eastern parts of North America, Mexico, eastern slopes of mountain ranges on the Pacific ocean countries in the West Indies, Central and South America and specific tomato producing areas in Asian, African and Australian countries of the world where late blight on potato occurs (Sokhi *et al.*, 1993). In Australia, it is known as Irish blight on tomato.

In India, late blight was first observed on tomato in 1900 (Butler, 1903). In Himachal Pradesh, this disease on tomato was not observed previously because the then prevailing race 'O' of *Phytophthora infestans*, did not infect tomato but with the appearance of potato races 4 and 1, 3 in 1971, tomato crop was severely attacked at Solan and Dhaulakuan (Bhattacharyya and Singh, 1990). Now the disease is prevalent in all tomato growing tracts of Himachal Pradesh particularly where rainy season crop is grown. In Punjab, the disease appears on tomato when the preceding potato crop is heavily attacked by late blight in January-February and may continue up to March-April if favourable weather conditions persist. In Karnataka, a severe disease outbreak was recorded in 2009 and 2010 crop seasons (Chowdappa *et al.*, 2013).

Losses

The losses from this disease can be more than from any other disease under favourable environmental conditions. Taubenhaus and Ezekiel (1931) observed that in infected fields, the average reduction in fruit yield is atleast 50 per cent while in Argentina, 95 per cent loss in fruit yield has been estimated (Taboada and Montanola, 1959). Truszkowska and Pudelkowa (1966) observed heavy losses from this disease under field conditions than glass house. The disease can also cause serious losses during storage and transit. Chowdappa *et al.* (2013) recorded up to 100 per cent crop losses due to this disease in Karnataka. They also recorded the spread of this disease from Karnataka to neighbouring states like Tamil Nadu and Andhra Pradesh.

Symptoms

Late blight affects leaves, stems and fruits on tomato. Apart it can also infect tomato nursery if it is raised adjacent to infected potato crop. On leaves, it appears as pale, green, irregular spots on the tips and margins, which in moist weather enlarge rapidly with central tissue turning necrotic and dark brown or black. On the lower surface of the leaves, a white downy growth of the fungus appears around the dead areas (Plate 1a), which is more prevalent in morning hours. In dry weather, the growth of the fungus is checked and the water soaked areas dry up and turn brown. Brown streaks also develop along the stems. On tomato fruits, symptoms appear only on green fruits as olivaceous greasy spots, which gradually cover the entire fruit surface (Plate 1b). Initially, the tissue remains firm with varying depths of discolored tissue below the skin but when blight is followed by soft rot, the fruit soon disintegrates.

The Pathogen

The disease is caused by *Phytophthora infestans* (Mont.) de Bary. The fungus belongs to kingdom *Chromista*, phylum *Oomycota*, class *Oomycetes*, order *Peronosporales* and family *Pythiaceae*.It is identified by its characteristic lemon shaped, hyaline sporangia which emerge through leaf stomata. Sporangia (21-38 x 12-23 μm) may germinate directly through germtube or under cool moist conditions through zoospores, which are upto eight in number. Each biflagellate zoospore swims in film of water for sometime, sheds flagella, become encysted and initiates new infections.

Sexual reproduction in *P. infestans* is not very common in nature because *P. infestans* is heterothalic and requires two different mating types, A1 and A2, for sexual reproduction. Uptill 1984, distribution of A2 mating type was restricted to Mexico whereas in rest of the world only A1 mating type was observed. However, of late presence of A2 mating type and oospore formation in nature has been reported from the European countries like Switzerland, England and the Netherlands. In India also, frequent occurrence of A2 mating types of *P. infestans* has been reported from hills in the recent years (Singh *et al.*, 1994). The antheridium is amphigynous and the oogonium is spherical. The germination of oospores has not been seen and is presumed that it germinates by producing germtube.

Vartanian and Endo (1985) identified two races viz., T_0 and T_1 of *P. infestans* by the reaction types on tomato cultivars. In India on tomato, 3 races namely 1.3.4.7.11, 3.11 and 4.7 were identified when tested on R0 to R11 gene differentials of potato (Sokhi *et al.*, 1993). Race 1.3.4.7.11 was occurring both on potato and tomato in Punjab. There is gaining in aggressiveness of potato isolates of *P. infestans* when successively transferred on tomato. A distinct pathotype specific on tomato fruit was also identified in the Punjab State. Chowdappa *et al.* (2013) reported the migration of new population similar to 13_A_2 genotype from Europe which was responsible for severe outbreak of this disease on tomato in Karnataka and neighbouring states. Moreover, all isolates belonging to 13_A_2 clonal lineage were metalaxyl resistant, A2 mating type, mtDNA halotype 1a and had characteristic simple sequence repeats (SSR) and RG57 fingerprints and highly aggressive on tomato.

Host range

Besides potato and tomato, the fungus also attacks *Nicandra physaloides* Gertn., *Petunia hybrida* Vilm., *Datura stramonium* L., *Aster thomsoni* Clarke, *Erigeron multicaulis* D.C., *Galinsoga parviflora* Jav., *Sonchus oleraceus* Linn., *Ipomoea hederacea* Jacq., *I. purpurea* Lam., *Geranium nepalense* Sweet, *Polygonum alatum* Buch. Ham., *Ramix acetosa* Linn., *Bupleurum meddeni* Clark, *Schizanthus grahmi*, *Lycopersicon peruvianum* Mill. and *Solanum melongena* L. (Phadtare, 1976; Raj *et al.*, 1976; Dutt, 1979; Raj *et al.*, 2000).

Disease relationship in potato and tomato

Mills (1940) showed that *P. infestans* pathogenic to potatoes might be weakly pathogenic on tomato and that the continuous culture on that host led to the establishment of a strain markedly pathogenic on tomatoes. Earlier reports suggested that fungus attacking tomato was physiologically different from that on potato. The pathogenicity gradually increased during successive transfers on tomato and after 7th transfer there was no difference between tomato isolates and 'transferred' potato isolates, a phenomenon called as adaptive parasitism. It was further indicated that tomato strains originate in nature from successive passages of the potato strain through tomato foliage. Kishy (1962) grouped the potato isolates into three categories according to their pathogenecity to tomato as potato type, intermediate type and tomato type with respect to their slight, moderate and severe attack on tomato cv. Ponderosa. Gunther *et al.* (1970), on the other hand showed that *P. infestans* was able to pass from potato to tomato without any loss. Sokhi *et al.* (1993) have also shown that late blight fungus can pass from potato to tomato though less virulent in the beginning and becomes equally virulent after successive generations on tomato under field conditions. Sokhi *et al.* (1993) reported that *P. infestans* can move from potato to tomato, though less virulent in the beginning, but gains virulence after successive generations on tomato under field conditions.

In hilly areas both tomato and potato crops are grown simultaneously during April to September-October and the crop can be affected any time after July due to the availability of sufficient moisture and moderate temperatures. All the commercial cvs/hybrids of tomato grown in Himachal Pradesh were found susceptible to this disease (Sharma, 1999).

Disease cycle

The pathogen mainly perpetuates from one season to another through infected seed tubers of potato (Hector, 1926). In the hills the temperatures in the country stores are quite favourable for survival in seed tubers upto next season. In the plains, the pathogen overwinters in seed tubers, which are stored in cold stores (Pushkarnath and Paharia, 1963). It also survives in association with volunteer and abandoned potato and tomato plant materials in fields, cull piles and gardens. Inoculum produced on infected plant material/self grown plants raised from leftover seed tubers from previous crop under moist conditions is carried by wind to nearby plants. Once primary infection has occurred, further spread takes place by air borne or rain splashed sporangia. Successful infection normally proceeds when sporangia land on wet leaves, tender branches or green fruits and the temperature is in between 10 and 25°C. Process of zoospore release from sporangia and penetration of germtube into the host tissue takes 0.5 to 2.5 h depending on the temperature and host plant conditions. The mycelium of the fungus grows between the cells and produces haustoria which feed off the interiors of the cell causing cell death. The mycelial growth into healthy tissue continues, leaving dead discoloured cells. The mycelium emerges through stomata on the underside of the leaf and produces sporangiophores. The sporangia then develop on these sporangiophores. Under favourable environmental conditions, large number of sporangia is produced which infect leaves and stems of the same plant or neighbouring plants. Sporangia and zoospores can also move via the soil water and cause infection of potato tubers which serve as a source of primary inoculum for the disease in ensuing season for both potato and tomato crops.

Epidemiology

The fungus survives in infected potato seed tubers, cull piles, infected plant debris in hills and other host plants that serve as source of infection. Cool and moist conditions favour the dispersal of inoculum and successful field infection. For development of the disease, low temperature and high humidity are of fundamental importance. The most favourable temperature for successful germination of sporangia either via zoospores or directly by germ tubes is 12 and 24°C, respectively (Sokhi *et al*., 1993). Higher temperature and low relative humidity favour direct germination of sporangia while low temperature and high humidity facilitate indirect germination with formation of zoospores. After infection, subsequent development of disease is most rapid at 21°C. Sporangial production is most rapid at 100 per cent relative humidity and 21°C temperature. Sporangia are sensitive to desiccation and after dispersal by wind or splashing rain, these require free water for germination. The fungus remains alive in host tissues at air temperature between 0 and 28°C. Zoosporangia develop at temperature between 9 and 22°C. Development of the fungus within the leaf takes place at 16-18°C. Temperatures above 30°C are considered unfavorable for disease development. The shortest incubation period and latent period were recorded at 22° C for two races of *P. infestans* (Maziero *et al.*, 2009).

Host tissue is little affected by air-humidity, however, zoosporangia are only formed when the relative humidity in the leaf conopy is >95 per cent. When the temperature is optimum, about eight hours of high humidity are required for production of zoosporangia, liberation of zoospores and penetration. Water in the form of dew or rain must be present on the leaf surface for atleast two hours to allow zoospore formation, germination and penetration. Lesion formation on leaves occurs within 3-5 days of infection at 17-25°C. Under optimum conditions, a disease cycle takes only three days. Splashing rain disseminates the sporangia and zoospores. Continued spread of disease is dependent on both host and climatic conditions. Leaf age and position on the plant affects the degree of susceptibility, with older and lower leaves being the most susceptible. The rate of disease development, including sporulation, is affected by degree of general resistance and maturity. Rate of sporulation decreases with the increase in age of plant and leaf. Continuous high temperature has been found to be more damaging than intermittent periods. Light intensity affects development, with high light intensities reducing mycelial invasion and lesion size as compared to low light intensities. Disease spread is more rapid when canopies are dense, which results in altered microclimatic conditions of higher humidities and moderate temperatures. These factors also affect sporulation on infected tissue and thus the continued epidemic development. The fungus survives under normal temperature and moisture conditions required for potato tuber survival in cold stores enabling it to regenerate the disease cycle when the tuber is planted or grows as a volunteer plant in fields and cull piles.

Management

Disease can be managed effectively by using cultural, chemical and host resistance methods.

(i) Cultural methods: Cultural practices like subterranean irrigation, autumn ploughing, rotation, avoiding tomatoes in succession to the same crop, potatoes or other solanaceae plants, wider spacing, avoid direct contact of the fruits and foliage with soil, staking of indeterminate varieties like Solan Gola and hybrids like Naveen and BSS-99 can be staked either with bamboo stick or bushes available in forests to avoid direct contact of fruits with soil and mulching with pine needles and dried grass can be helpful in reducing the disease. Diseased fruits should be collected regularly and dumped in the pits and while spraying the fields with fungicides pits should also be sprayed. Improve drainage of the field and keep field free from weeds. Immediate heating of harvested tomatoes from *P. infestans* infected field for 4 h at 40-45°C and then their transfer to 25°C for quick ripening is also useful. Application of 75 per cent recommended dose of fertilizers along with vermicompost (4 ton/ha) inoculated with biofertilizer recorded significantly lowest per cent disease index (13.84%) and produced highest marketable fruit yield (26.14 ton/ha) as compared to highest PDI of 31.62 per cent and lowest fruit yield (15.42 ton/ha) by the treatment containing 100 per cent sole inorganic fertilizers (Chatterjee and Khalko, 2013).

(ii) Host resistance: Tomato cvs/lines viz., Heinemannas Jubilaum (Czyzewska and Bialkowska, 1973), Rossol VFN, Piacenza 0164 and Roma (Erinle and Quinn, 1980),

GP 12, GGP 13 and GP 18 (Petrovskaya and Stolpwikova, 1989) Ottawa-30, Ottawa-31, PI 270407, W-Va-700, Pierline New Yorker, Nova, FI 814, Hossiline and Hybrid TH 2312 (Sokhi *et al.*, 1993) behaved as resistant. They can be utilized for further breeding programme. Some of varieties like New Hampshire Sure Crop, New Yorker, Nova and Virginia known as resistant elsewhere (Yang, 1979) behaved as susceptible under Punjab conditions (Kaur *et al.*, 1988). Two of the best performing lines with enduring resistance to late blight, LBR19-2 and LBR44-2, were released as new varieties in Tanzania under trade names 'Meru' and 'Kiboko' in 2007 and 2008, respectively (Ojiewo *et al.*, 2010).NC 1 CELBR and NC 2 CELBR, determinate large-fruited, fresh-market tomato breeding lines with a high level of resistance to late blight have been described by Gardner and Panthee (2010).

(iii) Chemical control: In the wake of unstable host resistance various systemic and non-systemic fungicides have been found effective in the management of late blight of tomato and potato. Earlier attempts to control the disease with Bourdeaux mixture were made which was later replaced by copper oxychloride, dithiocarbamates (mancozeb, zineb), captafol and chlorothalonil. Mancozeb is now a widely accepted protectant fungicide for the control of this disease. These preventive fungicides protect the plants principally by inhibiting sporangial germination and preventing penetration. Once infection takes place, these conventional fungicides are rendered ineffective. Timing of protectant fungicide application is very important, as they do not become distributed throughout the foliar tissues to halt previously established infection. New growth after application remains unprotected and these fungicides can be removed by rain and weathering. Protective spray programme before disease appearance in a field proves quite effective.

The introduction of systemic fungicides with specific activity against the members of oomycetes has given a big success in the fight against *P. infestans.* Metalaxyl, one of the acylalanines, has proved very effective against late blight (Schwinn and Staub, 1987) and is translocated within the plant. Oxadixyl, benalaxyl, ofurace, cymoxanil and dimethomorph are other anti-oomycete fungicides having high efficacy against this pathogen and possess systemic activity. These systemic fungicides are considered superior to the traditional protectants since they enter and become distributed within the host tissues and thus less susceptible to weathering, requiring less frequent applications. A spray schedule including alternate applications of protectant and phenylamide fungicides has been developed for effective and economical control of late blight of tomato in Punjab (Mohan and Thind, 2001). Metalaxyl provides disease control by reducing sporulation, germination and intercellular fungal growth within the host tissues and is an effective eradicant stopping previously established infection. These fungicides, however, have to be used with caution and should be applied to the minimum due to risk of development of resistance in the pathogen population. It is always better to use these fungicides in rotation or in combination with protectant fungicides. Pre-prepared mixture products viz., Ridomil MZ (8% metalaxyl + 64% mancozeb), Galben M8-65 (8% benalaxyl + 65% mancozeb), Ridomil-45 (5% metalaxyl + 40% copperoxychloride), Curzate M-8 (8% cymoxanil + 64% mancozeb) etc. are a few formulated products available during these days (Thind and Chahal, 2002). Infinito 68.75 SC @1500 g/ha significantly reduced late blight severity and increased fruit yield (Debnath and Nath (2009).

A mixture composed of 5 parts of plant activator BABA (DL-3-aminobutyric acid) and 1 part of mancozeb was found highly effective in the management of late blight of tomato and potato (Baider and Cohen, 2003). The results showed enhanced effect of mancozeb in BABA-induced plants, suggesting, therefore, that lower dosages of this fungicide may be sufficient to control late blight of tomato and potato under field conditions (Baider and Cohen, 2003). The fungicides dimethomorph-mancozeb, fosetyl-Al, azoxystrobin, and dipotassium phosphonate/phosphate and the plant defense activator acibenzolar-*S*-methyl have been reported to suppres late blight development (Becktell *et al.,* 2005). Peruch *et al.* (2008) reported that Bordeaux mixture @ 0.5% reduced AUDPC by 67, 35 and 49 per cent, as well as disease progress rate were 50, 16 and 30 per cent compared with control in spring 2004, spring 2005 and autumn 2006, respectively, and the yield was increased by 114 percent as compared to other treatments.

Rani *et al.* (2011) studied the persistence of both contact and systemic fungicides against *Phytophthora infestans* in tomato and reported that combination products (systemic and contact fungicides) had better persistence than contact fungicides alone, thereby, giving them an advantage of longer action resulting into better disease control.

(iv) Biological control: Yamaguchi *et al.* (1992) reported tomato seedlings from *Fusarium oxysporum* MT0062 pre treated soil showed reduced development of *P. infestans* on the aerial parts. This may be due to induction of non-specific systemic resistance in tomato seedlings because *F. oxysporum* can sucessfully infect this and other solanaceous crops. *Bacillus* sp. str. C 18 inhibited mycelial growth of seven plant pathogenic fungi including *P. infestans* under *in vitro* conditions and this bacterium also protected tomato plants from *P. infestans* under *in vivo* conditions. Prophylactic sprays of aqueous and ethanolic leaf extracts of *Reynoutria sachalinensis* has also been claimed to provide 40-50 per cent control of this disease (Herger *et al.,* 1988). Pre-infectional treatment (2-14 days before inoculation) with ethanol extracts of *Hedera helix* and *Paeonia suffruticosa* extracts suppressed late blight development drastically (Rohner *et al.*, 2004). Both extracts also exhibited a high degree of activity against late blight on entire tomato plants and inhibited the release of zoospores from sporangia and the germination of zoospores under *in vitro* conditions.

Antifungal activities of essential oils obtained from aerial parts of aromatic plants such as oregano (*Origanum syriacum* var. *bevanii*), thyme (*Thymbra spicata* subsp. *spicata*), lavender (*Lavandula stoechas* subsp. *stoechas*), rosemary (*Rosmarinus officinalis*), fennel (*Foeniculum vulgare*), and laurel (*Laurus nobilis*) were investigated against *P. infestans*. All essential oils were found to inhibit the growth of *P. infestans* in a dose dependent manner. Volatile phase effect of oregano and thyme oils at 0.3 μg/ml air was found to completely inhibit the growth of *P. infestans*. Complete growth inhibition of the pathogen by essential oil of fennel, rosemary, lavender and laurel was, however, observed at 0.4–2.0 μg/ml air concentrations (Soylu *et al.*, 2006).

2. BUCKEYE ROT

This is an important disease in those areas where fruiting time coincides with the rainy season.

Geographical distribution

The disease has been reported from all the tomato growing areas of the world where high relative humidity coupled with warm temperature is prevalent. For the first time, it was recorded in the year 1915 in Florida (Sherbackoff, 1917). Since then it has been reported from U.S.A., Mexico, Canada, England, Cuba, Ireland, Holland, Denmark, Italy, Trinidad, South Africa, Rhodesia, China, Malaysia, Germany and France (Shyam and Gupta, 1998).

In India, the disease was first time reported from Solan area (1200 to 1500 m.a.s.l.) of Himachal Pradesh by Jain *et al.* (1961) and subsequently, it has been reported from Bangalore (Karnataka), Ooty (Tamil Nadu), Haryana and Uttar Pradesh (Kumaon region) by various workers (Sohi, 1978; Srivastava, 1981; Sharma, 1992).

Losses

The losses due to buckeye rot are direct. The infected fruits develop small brown water soaked spots, which increase in size, and as such infected fruits fall down prematurely. Heavy losses to green fruits due to this disease have been reported from Denmark (Gram *et al.*, 1927) and Eastern Trans Vaal (Wager, 1935). In India, losses from this disease have been reported to range from 35-40 per cent (Jain *et al.*, 1961) in Himachal Pradesh conditions. Sokhi and Sohi (1982) reported that the disease was most serious in the months of July - August under Bangalore conditions and caused 34 per cent fruit rot in unsprayed plots in comparison to only 1.17 per cent in sprayed ones. The fruit rot incidence may go upto 90 per cent under high humidity and good rainfall conditions while the annual economic loss to the cultivators has been estimated to be about rupees 3 million (Sharma, 1971).

Symptoms

Immature fruits of green colour irrespective of their development stages are susceptible to the attack of the pathogen and such fruits, which are in contact with soil or near to soil level, are the most vulnerable to the disease and the symptoms appear as water soaked light brown discoloured spots (Plate 2a). These spots increase readily showing concentric dark brown rings slightly resembling the markings on a buckeye nut (horse chestnut). The lesions rapidly enlarge and in 3-4 days whole of the fruit surface turns dark brown and feels soft to touch (Plate 2b). White flocculent superficial growth of the fungus consisting of sporangia and sporangiophores develops profusely on the diseased fruits in warm and humid weather. Later, these fruits may drop down in the field. With the invasion of saprophytic microorganisms, diseased fruits decompose soon. Besides, the zonated fruit spots, the spots without zonation are also observed.

The Pathogen

The fungus responsible for this disease is *Phytophthora nicotianae* B. de Hann. var. *parasitica* (Dastur) Waterhouse.The mycelium of the pathogen is hyaline, coenocytic with finely granular protoplasm, branching typically at right angles with slight constriction at the point of branching. In very old mycelium, septa may form at a very irregular interval. The hyphae are upto 9 μm wide, irregular in width but without marked hyphal swellings. The sporangiophores arise from hyphal threads and produce sporangia generally at their tips in nature, while in culture, the sporangia are terminal as well as intercalary. The sporangiophores vary from 100-230 μm in length. The sporangiophore proliferates and produces new sporangia by pushing aside the old sporangium. The sporangia vary in shape and size under natural epiphytotic conditions. Generally, they are broadly ovoid to globose in shape, but elongated to barrel shaped sporangia are also occasionally found. The sporangia usually have one hemispherical papilla at the tip, though, occasionally sporangia with two papillae are also found. The size of sporangium varies from 25.5 - 58.0 x 20.0 - 52.0 μm on the host while in culture they are 24.0 - 52.5 x 19.5 - 48.5 μm. The sporangia usually germinate by releasing 8-16 zoospores through the papillae opening and occasionally by germ tube (at >27°C). After release, zoospores become kidney shaped with two lateral flagella and move about actively in water for about 30 minutes and then become encysted (6.8 x 10.2 μm in size). Chlamydospores are smooth, globose, slightly yellowish with thick brown walls (20-60 μm in dia.), produced abundantly in culture and germinate by producing zoospores or germtubes.

Antheridia are amphigynous, spherical or oval and 10 x 12-16 μm in size. Oogonia are usually produced in single cultures though often very sparsely or not until after some weeks. They are readily produced when opposite strains are grown together in culture. Oogonia measure less than 28 μm, rarely 31 μm in diameter. On maturity, oogonia become rough, thick walled and yellowish brown. Oospores are aplerotic, 18 to 20 μm in diameter with 2 μm thick wall.

Host range

The pathogen *Phytophthora nicotianae* var. *parasitica* has a very wide host range. Obrero and Aragaki (1965) reported it from pineapple, parsley, hibiscus, carnation, eggplant and papaya besides tomato. Sharma (1971) studied the host range including those plants, which were generally found growing in Solan region either cultivated or wild, and reported eight new hosts viz., *Euphorbia hypericifolia, Physalis minima, Mirabilis jalapa, Commelina obliqua, Tegetes minor, Ipomea muricata* and *Galinsoga parviflora.* It also attacks beans, maize, brinjal, chillies, onion, potato and pumpkin (Singh, 1987).

Variability

Obrero and Aragaki (1965) reported that *P. parasitica* isolates from tomato plants were more pathogenic to tomato fruits than isolates from pineapple, parsley, hibiscus, carnation, eggplant and a soil isolate. An isolate from papaya was non-pathogenic. Cultural and pathogenic variation among isolates of *P. parasitica* from different host plants have been reported (Siradhana *et al.*, 1968; Sharma and Sharma, 1987). Sharma (1971) also studied the cultural and pathogenic variation among 20 isolates collected from buckeye rot affected tomato fruits and reported significant differences among the isolates for cultural and pathogenic characters but no correlation could be established between mycelial growth and virulence. In the past, high yielding variety like Money Maker reported to be resistant to buckeye rot (Sharma *et al.*, 1974) has lost its resistance (Dodan *et al.*, 1995) probably due to appearance of more aggressive strains of the pathogen. Dodan and Shyam (1994) also studied the pathogenic variation in 26 isolates of the pathogen representing different agroclimatic zones of Himachal Pradesh. Isolates showed considerable variation in cultural and pathogenic behaviour but no correlation was found between linear growth, size of sporangia, chlamydo-spores, zoospores and pathogenic behaviour of these isolates. Although the isolates did not exhibit the existence of assemblage(s) of virulence/ avirulences but could be grouped into 18 pathogenic strains on the basis of their aggressiveness pattern and non-differential host-isolate interaction.

Disease cycle

In areas where the crop is grown regularly in the same field, the fungus overwinter in the soil in the form of oospores or chlamydospores and can remain active in soil for atleast one year without the support of a susceptible host. The fungus has also been reported to be internally seed borne and remain viable for more than one and a half year (Sharma and Sohi, 1975). With the onset of monsoon rains in the second fortnight of June, in the presence of high soil moisture and moderate temperatures (20-25°C), the oospores and clamydospores start germinating by producing mycelium and sporangia. The sporangia in turn produce biflagellate zoospores, which after swimming for sometime in film of water loose flagella and become encysted. These encysted zoospores are the source of primary infection, which are splashed by the rain to the immature green fruits. After the establishment of contact between zoospores and fruit surface, the zoospores start germinating within 3 h by germ tube and give rise to appressoria on the fruit surface. Infection hyphae develop from the appressoria, which penetrate the cuticle directly. The fungus invades intracellularly, downward as well as radially. The hyphae invading intracellularly are smooth with granular protoplasm, branching profusely in the walled parenchymatous cells of the host. These hyphae penetrate deep into the fruit tissue and invade seeds. The symptoms develop on fruits after 3-4 day of infection. When most of the area of the fruit is covered with the spot, the infected fruit become mummified and falls down on the ground. The mycelium in the epidermal layer of the affected portion of the fruit agglomerate, pushes through the epidermis and cuticle by rupturing it and grows on the fruit surface under humid conditions, producing a crop of sporangia on simple branched sporangiophores. No sporangia are produced inside the host tissues. These sporangia are again splashed by rain and cause secondary infection. When the conditions become unfavourable, the fungus either produces chlamydospores or goes in sexual reproduction and produces oospores (Fig. 1).

Epidemiology

The pathogen survives in the form of chlamydospores and oospores in soil as well as in infected seed (Sharma *et al.*, 1978). The fungus remains viable for more than two years under ordinary field conditions in the soil and a depth of 10 cm is more favourable for the perennation of the fungus and it can remain viable for a longer period under wet conditions as compared to dry conditions (Sharma, 1971). Fruits which touch or are very near to the ground, are most susceptible to this disease and abundant soil moisture is belived to be an important factor for the initiation of this disease (Kendrick, 1923). Wager (1935) indicated the role of rainy weather in the development of this disease while the disease is totally absent in dry weather (Baker, 1939). Inoculum surviving in soil is splashed from the soil with spattering rains and after infection, sporangia are produced on fruits and are carried by moist wind and water to the adjoining fruits.

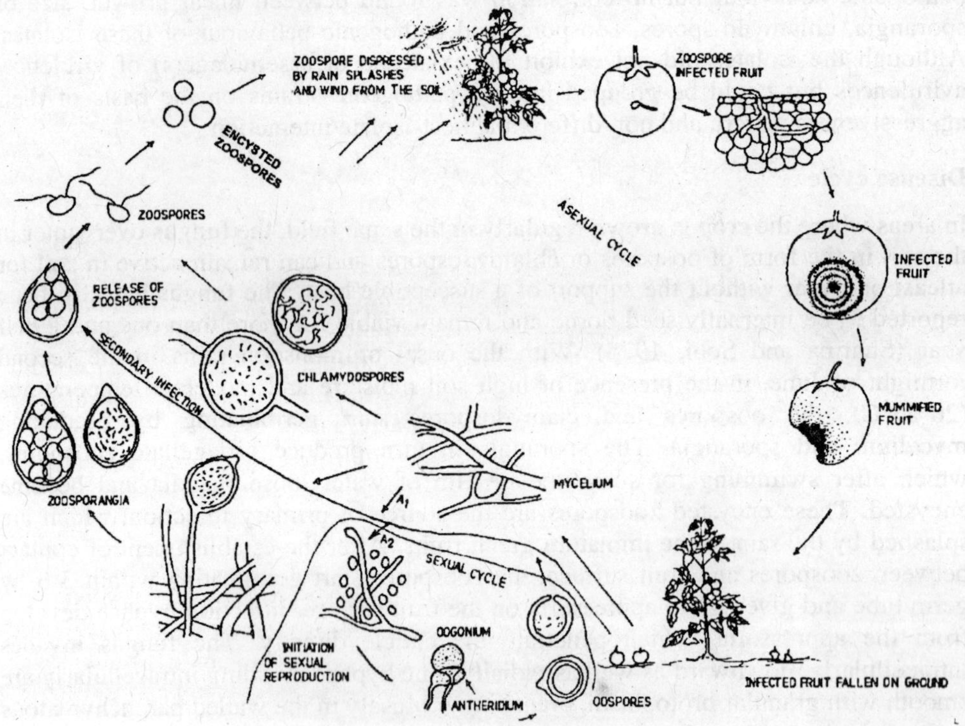

Fig. 1. Disease cycle of Buckeye rot of tomato

The fungus shows optimum growth at 25-30°C temperature. Maximum fruit infection under field conditions occurs at a temperature range of 20-25°C, RH > 60 per cent and high rainfall conditions (Thareja *et al.*, 1989).

The height from the soil at which fruits are attached on the plants appears to play an important role on the severity of the disease. Thareja *et al.* (1989) recorded maximum number of infected fruits when they were in direct contact with soil or near the ground, upto the level of irrigation water (5 cm high), while the infection upto the height of 45 cm has also been reported by various workers (Sharma *et al.*, 1976). Sharma and Sohi (1983) evaluated different levels of N, P and K fertilizers and reported that higher doses of N resulted in higher yield of healthy fruits but fruit rot was also increased significantly while higher levels of P resulted in more yield of healthy fruits and less fruit rot.

Forecasting

Sharma *et al.* (1978) developed a short term forecasting method for this disease based on weekly temperature and cumulative rainfall with June 20 as the zero date. They reported that the disease is not expected to occur at temperatures at or below 20ºC, though at temperatures of 22.5ºC or above even a slight rainfall will result in disease appearance, which is expected to appear after 4 days (48 hours for sporangia production and 48 hours for incubation). The disease is not expected to be serious above 25°C as the sporangia at these temperatures germinate by germtubes and not through zoospores thus reducing the amount of inoculum.

Management

(i) Cultural practices: Use of cultural practices is of immense value during these days due to pollution and health hazards resulting from deposits of chemical sprays. Sherbakoff (1917) advocated staking of plants in the field for the control of the disease and this was also found effective by Wager (1935) but Wilson (1956) found the disease rather frequently on staked plants as was observed by Sharma *et al.* (1976). Wider spacing should be adopted in unstaked commercial plantings to control this disease (Loest, 1948) as the development of disease was slow in widely spaced (90 cm) crop then 45 to 60 cm (Kumar and Sugha, 2000). Foliage and fruit upto a height of 30 cm should be removed to avoid moist and stagnant air conditions (Sharma *et al.*, 1976, Kumar and Sugha, 2000). Mulching of the field was also found effective for the control of this disease. Similarly, intercropping of of tomato with maize in double rows showed less incidence as compared to intercropped with beans or *toria* (Kumar and Sugha, 2000). Application of light irrigation and ensuring optimum crop water use were found important in reducing the incidence of buckeye rot (Bhardwaj and Masand, 1995; Gupta *et al.*, 2005).

(ii) Host resistance: Despite the continuous efforts in evaluation and breeding for disease resistance against buckeye rot of tomato, nothing concrete has been found so far, though a few cvs./lines having tolerance to the disease have been identified. Fruits of eleven tomato varieties proved highly resistant to natural and artificial infections by *P. parasitica* at Tennessee Agricultural Experiment Station while Sherbakoff 57 x Oxhert was most promising with respect to fruit size and resistance (Felix, 1948).

Several varieties/hybrids like hybrid N-64 (STEP 390 x HES 6578), Rossal, EC 129171/PI-2, EC 128-969, EC 129-166, EC 729-1, EC 122-62-1, EC 130035 and EC 129603 are reported to be resistant from different parts of world (Obrero and Aragaki, 1965; Thareja and Srivastava, 1989; Perez *et al.*, 1986; Singh *et al.*, 1997). Dodan *et al.* (1995) screened 110 cvs./lines at Solan against this disease and only one small fruited line viz., *Lycopersicon esculentum* var. *ceresiformae* was found resistant under natural epiphytotic conditions and that also succumbed to inoculations under laboratory conditions. Similar observations regarding this line were recorded in Rajasthan (Fageria *et al.*, 1998). Obrero and Aragaki (1965) who reported that more zoospores and a longer incubation period is required for infecting a resistant variety than susceptible variety by *P. parasitica* var. *nicotianae*. Under field conditions, sufficient number of zoospores might not be reaching to the infection court due to small size of fruits. Of late, line 'EC 129602' was reported as moderately resistant to this disease under Solan conditions (Joshi *et al.*, 2004). Devi and Tiwari (2001) reported that the gene reaction is controlled by polygene and that the resistance was dominant to the susceptible trait and based on analysis of gene action, non-allelic interaction was found in all crosses except Kt-15 x KT-10.

The factors affecting the buckeye rot resistance are plant height, ascorbic acid content, total acidity and total nitrogen content of green fruits (Sharma *et al.*, 1974).

Higher levels of phenolics like total and ortho di-hydroxy phenols and phenol oxidizing enzymes like polyphenol oxidase and peroxidase was observed in resistant varieties than in susceptible ones (Sharma and Bhardwaj, 1977, Singh *et al.*, 1997). Tomato fruits are known to produce antifungal compounds such as lycopene whose quantity vary with cultivars and fruit size. Smaller the fruit size, lower the lycopene contents and stronger the reaction (Trique, 1977).

(iii) Chemical control: In an attempt to manage this disease, various systemic and non-systemic fungicides have been recommended. However, due to continuous rains which are essential both for good growth of tomato plants and rapid development and spread of the disease, routine application of fungicides is sometimes delayed which often results in heavy disease outbreak and poor disease control. Proper timing of initial sprays is of prime importance to achieve desirable level of disease control.

Earlier attempts to control the disease were made by spray application of Bordeaux mixture, tribasic copper sulphate, captafol, copper oxychloride, zineb, mancozeb, chlorothalonil etc. (Sharma *et al.*, 1976). Sequential sprays of captafol, mancozeb and copper oxychloride 40, 55 and 70 days after transplanting increased yield by 50.5 per cent by reducing the incidence of buckeye rot and defoliation diseases (Bhardwaj, 1991). Ridomil MZ and cymoxanil + mancozeb provided good control of the disease when applied before infection (Gupta and Bharat, 2008).

Use of plant extracts

Krishan and Paul (2014) evaluated various botanicals as aqueous extracts and found that marigold leaf extract completely inhibited the growth of *P. nicotianae* var. *parasitica* while crude extracts of *Eupatorium adenophorum* (kali basuti) and *Melia azederach* (Darake) also resulted in 100 per cent inhibition at 1.0 per cent or more concentration.

Integrated disease management

Tomato being a long duration crop is under prolonged exposure to the diseases resulting in greater yield loss. Management of major diseases like buckeye rot is, therefore, an important component in the tomato production technology. This may be achieved through reducing the initial population of the pathogen and then decelerating the rates of increase in these populations. Use of chemicals, resistant varieties and modified cultural practices may help in achieving these objectives. Considering the soil borne nature of the pathogen and the infection caused by the rain splashed inoculum, only single management practice may not achieve the desired goal. During the last two decades, considerable information has been generated on the identification of resistance sources and breeding of improved varieties, yet no variety/hybrid with desirable resistance and desired fruit characters has been developed. However, efforts have been made to integrate the cultural and chemical methods to obtain the desired results.

Spraying of Dithane M-45 or Bordeaux mixture or captafol along with staking and removal of foliage and fruits upto 30 cm from soil level was found to be most effective and economical (Thapa and Sharma, 1978; Sharma, 1992). Integration of fungicide (Indofil M-45, 0.25%) sprays, cultural practices like periodical clipping of lower leaves (15-20 cm), weeding, mulching with white polythene sheet and regular removal of affected fruits not only reduced disease incidence drastically but also increased fruit yield appreciably (Dodan *et al.*, 1994).

Gupta and Shyam (1998) conducted studies to determine the role of pine needle mulch and spray of fungicides, alone and in combination with systemic (Ridomil MZ, 0.25%), and non-systemic (Indofil M-45, Blitox-50 and Captaf- 0.25%) fungicides, on pine needle mulch. They found pine needle mulch and spray applications of Ridomil MZ and Blitox-50 on mulch and foliage as most efficacious in managing this disease but pine needlle mulch impregnated with floor application of Blitox -50 was also quite effective without having the residue deposits on fruits with slightly less benefit : cost ratio (12.9) as compared to its foliage application (16.7).

3. DEFOLIATION DISEASES

(a) ALTERNARIA LEAF SPOTS

It is a commonly occurring disease of tomato throughout the world and may occasionally assume destructive status at any stage of plant growth. Though, *Alternaria solani* has been reported to be the most common pathogen causing leaf spot diseases (early blight). There are reports that other species like, *A. alternata* and *A. alternata* f.sp. *lycopersici* also cause leaf spots and stem canker in tomato (Grogan *et al.* 1975).

Geographical distribution

Early blight caused by *A. solani* occurs wherever tomatoes are grown throughout the world. The occurrence of the disease has been reported from Southern England, Columbia, Bulgaria, Antioqria, North Florida, South Africa and Cameroon (Fontem *et al.*, 1999). It is a destructive disease in England, India, Australia and United States (Jones *et al.*, 1991). In India this disease was first time recorded from Delhi by Chona *et al.* (1958). Since then it has been reported from almost all the states where tomatoes are grown.

Douglas (1922) reported species of *Alternaria* causing spots on fruits and leaves of tomato in U.S.A. The fungus produced spores similar in morphology to that of *A. alternata* causing spots of leaves and green fruits. Subsequently, the occurrence of this pathogen causing seedling diseases and leaf spots of tomato was reported from Egypt (Ashour and El-Kadi, 1956) and India (Tandon and Chaturvedi, 1965). The occurrence of this pathogen causing leaf spots of tomato in Himachal Pradesh has also been reported (Sahi and Shyam, 1993).

Grogan *et al* (1975) were the first to record the occurrence of *A. alternata* f.sp. *lycopersici* on tomato from San Diego County of southern California, U.S.A., causing serious stem canker disease. This pathogen was considered to be a distinct pathotype capable of initiating primary infection of leaves and stem of susceptible tomato cultivars. Subsequently, it was reported from Greece causing stem and leaf lesions of tomato under glasshouse conditions (Malathrakis, 1983). Later, Tagami *et al* (1984) reported the occurrence of stem canker of tomato caused by a distinct pathotype of *A. alternata* from Japan, which was similar to the pathotype earlier reported from U.S.A.

Several workers have reported the occurrence of *A. alternata* on ripe fruits of tomato. However, Grogan *et al.* (1975) maintained that *A. alternata* causing leaf spots of tomato was different from the essentially saprophytic pathotypes of *A. alternata* that were capable of causing rot of ripe fruits of tomato.

Losses

Variable losses have been reported in literature from this disease. The disease generally starts appearing on the lower older leaves mostly 40 days after transplanting and reaches at its maximum in later part of the season, reducing fruit yields considerably. The reduction in fruit yield was 6 per cent when the disease intensity was 25 per cent while there was 40 per cent reduction in yield when disease intensity was 100 per cent. Datar and Mayee (1981) observed up to 78 per cent losses in tomato fruit yield under severe epidemic conditions of early blight (*A. solani*) having coefficient of disease index of 71.6 per cent.

Symptoms

Alternaria solani generally attacks the main season crop grown in plains. The fungus can infect seedlings but generally is a problem of older plants. Lowest leaves are attacked first and then the disease progresses upwards. Dark brown spots with concentric rings develop on the leaves, which give target board effect, the most characteristic symptom of the disease. In humid weather, the affected areas coalesce and form dark brown patches. In severe attacks, affected leaves shrivel and fall down prematurely resulting in early defoliation.

On leaves, the spots appear as small angular scattered, light brown in colour, progressing between veins towards the leaflet margin (Plate 3a). Concentric rings may appear on the lesions giving them target board appearance (Fig. 2). The severely infected leaflets curl and dry out prematurely. Subsequently, the pathogen invades the adjoining healthy leaflets and gradually progresses on the foliage upward from the lower most infected leaves. Elongated dark brown lesions also appear on the stem and branches (Plate 3c). On fruits the infection takes place at the stalk end in the form of dark brown spots near the attachement with the fruit (Pate 3b). The lesion enlarges in size with concentric zonations becoming evident at later stage (Plate 3c). The infected fruits get easily detached from the stalk.

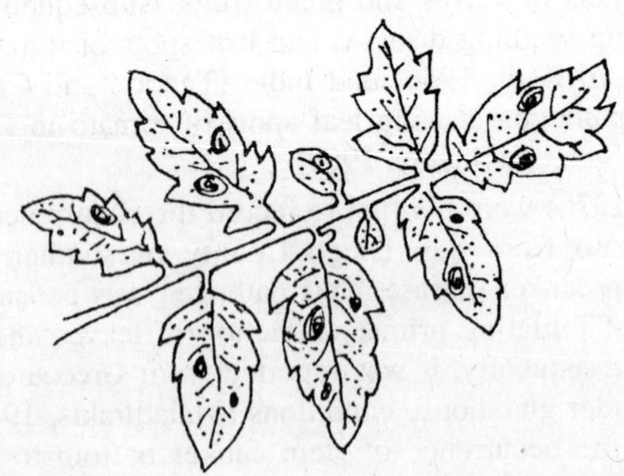

Fig. 2. Symptoms of Alternaria leaf spot of tomato (*A. solani*)

In case of infection caused by ***Alternaria alternata***, the spots on the lower most leaves start appearing about one month after transplanting as minute water soaked specks which soon turn into small, circular scattered, dark brown spots (Plate 4a). Older spots are surrounded by yellow halo and the leaflets become chlorotic. The affected leaves dry prematurely.

In case of ***A. alternata* f.sp. *lycopersici*** the symptoms also appear on stems (Plate 4c) and branches as light to dark brown spots. On leaves, the spots are circular to irregular, often elongated, and enlarge rapidly when the humidity is high due to frequent rains or continuous dew formation (Plate 4b). The older lesions become dark grey in colour often encircling the branches.

The Pathogen(s)

The disease is caused by three pathogens viz. *Alternaria solani* (Ell. and Mart) Jones and Grout, *A. alternata* (Fr.) Keissler and *A. alternata* f.sp. *lycopersici* Grogan Kimble and Misaghi. The morphological characteristics of each pathogen is described below:

The mycelium of *A. solani* consists of septate, branched, light brown hyphae, which become darker with age. The hyphae in the host are at first intercellular, later penetrating into the cells of the invaded tissues. Conidiophores emerge through the stomata from the dead centers of the spot. They are relatively short, 50 to 90 μm and dark coloured. Conidia (120-296 x 12-20 μm) are beaked, muriform, dark and borne singly or in chains of two. The conidia develop from a bud that arises from the terminal cell of the conidiophore. Five to ten transverse septa and a few longitudinal septa are present in each conidium. Germination of conidia occurs within 35 to 45 minutes in water at 28 to 30°C producing 5 to 10 germtubes.

The young colonies of *A. alternata* possessed much fluffy margins with off white centres which turned into dusky neutral grey within 96 h. Later these colonies become appressed and nearly greyish black in colour. The conidia occurring on leaflet lesions under field conditions are 1-6 septate measuring 13.0 - 46.0 x 7.0 - 10.0 μm (av. 29.50 x 7.40μm). The size of beak range from 3.0 - 10.0 μm with an average length of 7.9 μm.

The young colonies of *A. alternata* f.sp. *lycopersici* are fluffy at the margins and off white in the center, subsequently turning dusky and neutral greenish grey within 96 h of growth. Later these become appressed and nearly black in colour. The conidia occurring on leaflet and stem lesions under field conditions are 1-5 septate measuring 17.0 - 50.0 x 10.0 - 13.0 μm (av. 34.40 x 12.30 μm) and beak length ranged from 3.0 - 7.0 μm with an average of 6.13 μm.

Host range

The host range of *A. solani, besides Solanum tuberosum* and *S. lycopersicum* includes *Capsicum annuum, Atropha belladona, Cymphomondra betacea, Hyoscmus albus, H. niger, Nicotiana alata* and *Solanum nigrum* (Singh, 1995). *A. alternata* has also got a wide host range. However, *A. alternata* f. sp. *lycopersici* is limited to tomato cultivars only.

Disease cycle

Alternaria species infecting tomato overwinter in diseased plant debris and can persist for one to two years. Grogan *et al.* (1975) also reported that *A. alternata* can perenate in cankered tomato stems in soil upto 13 months which is sufficient for carry over of the pathogen from one growing season to another. Seed borne nature of *A. solani* has also been established (Khulbe and Sati, 1987 and Shahida and Abdul, 1995) and can be introduced through infected seed. Primary infection of lower leaves first takes place through conidia formed on crop debris in soil. Secondary spread of the disease occurs through conidia developed on primary spots. These conidia are blown by wind, water and insects to the neighbouring leaves\plants. Infection generally occurs through the stomata but the *Alternaria* spp. are also capable of direct penetration. Prevailing environmental conditions such as frequency of rains, amount and duration of dew deposits on leaves, atmospheric humidity and temperature etc further influence disease intensity.

Epidemiology

Datar and Mayee (1981a) reported that the disease severity was maximum in June-July sown crop as compared to September-October and January-April sown crops. This might be due to the prevalence of high humidity during July, August and September months. High soil moisture, humidity, dew and rainfall favour disease development. The optimum temperature for *A. solani* is 28 to 30ºC (Singh, 1995) while for *A. alternata f.* sp. *lycopersici* and *A. alternata* it ranges between 25-30 and 20-25ºC, respectively (Sahi, 1990). Once infection has occurred, conidial dispersal continues throughout the growing season of the crop. Maximum dispersal of conidia occurs in advanced stages of disease development and between 9 a.m and 12 noon (Datar and Mayee, 1982).

Management

The disease can be managed effectively through cultural practices, chemical sprays and host resistance methods

(i) Cultural practices: Since the fungus survives in infected plant material left in the field, cultural practices like destruction of infected plant debris, use of healthy seeds and crop rotation are desirable. Seed borne infections in tomato can be eliminated by exposure of seed to high temperature (dry heat) for 5-7 days, but seed germination was adversely affected (Takano and Takayana, 1985). Substantial reduction in losses of southern grown tomato seedlings was achieved by early planting or by avoiding use of overage plants. Periodical clipping of lower leaves up to 20 cm height from the ground level has been reported to reduce the disease severity and increased the fruit yield (Sahi, 1990). Mohan *et al.* (2009) studied the effect of plant age, leaf wetness and date of transplanting on early blight of tomato and reported that disease appeared at lowest rate with disease index of 19 per cent when tomato seedlings were transplanted after 30 days. Mminimum of 8 hours wetness was required for infection and appearance of symptoms and with the delay in transplanting date disease severity also decreased.

(ii) Host resistance: The information on the availability of the promising cultivars exhibiting resistance to *Alternaria* diseases is very meager being necrotrophic pathogens. Grogan *et al.* (1975) found that out of the several varieties tested, only 6718 VF, Rutgers, Marvel, Marglobe, Pritchard, Ace-55, Improved Garden State and Victor showed varying resistance to *A.alternata* f.sp. *lycopersici*. Genotypes/ cvs. like EC-174069, EC-174076, EC-174025, EC-174057, LA 1783, LA 2992, Marglobe and a mutant SGP_8M_2 were reported resistant to both pathogens like *A. alternata* and *A.alternata* f.sp. *lycopersici* (Sahi, 1990; Fageria *et al.*, 1998). Biezen *et al.* (1995) reported two wild species of tomato viz. *Lycopersicon pennellii* and *L. peruvianum* highly resistant to *A. alternata* f.sp. *lycopersici* and also insensitive to its toxins. Genetic analysis showed that high insensitivity to *A.a.* f.sp. *lycopersici*-toxins from *L. pennellii* is inherited in tomato as a single complete dominant locus. Near complete resistance in some accessions of the wild tomato species *L. hirsutum* (PI 126445 and LA 2099) has also been reported against *A. solani* (Foolad *et al.*, 2000).

Fernandez *et al.* (1998) studied polyphenoloxidase and chitinase isoenzymatic patterns in leaves of two varieties with different degrees of resistance and found chitinase isoenzymes related to defence mechanisms. Resistance or susceptibility of tomato cultivars to *A. solani* is not influenced by level of ascorbic acid present prior to infection but vitamin 'C' content in case of susceptible cultivars was adversely affected after infection as compared to resistant ones (Chhabra *et al.*, 2000). Of late, several genotypes/lines viz., LE-415 Arka Alok, Arka Abha, Arka Meghali, Arka Sourabh, IIHR-305, IIHR-308, IIHR-2266, IIHR-2285 and IIHR-2288 were found resistant (Pandey and Pandey, 2002; Prasad and Naik, 2003) to these spots and can be utilized in breeding programmes.

(iii) Chemical control: Sprays of fungicides like captafol, mancozeb, maneb, zineb, chlorotha-lonil, metiram, copper oxychloride + dichlofluanid, copper oxych-loride + folpet, iprodione and difenconazole have been reported effective for the control of this disease (Alexanderi *et al.,* 1981; Rajgopal and Vidyase-kharan, 1983; Sattar and Kassem, 1991; Follas *et al.*, 1992; Deora *et al.*, 2004) but mancozeb sprays were more cost-effective (Sinha and Prasad, 1991). Soil application of captafol + folpet prevented the stem infection caused by *A. alternata*, f. sp. *lycopersici* for more than 25 days (Malathrakis, 1983). The efficacy of EBI compound like flusilazole against this disease has also been reported (Liu and Wu, 1997). Activity of indole derivatives like 5-fluoroindole against this pathogen has also been suggested (Dogo *et al.,* 1999). Sprays of mancozeb followed by chlorothalonil effectively controlled this disease (Deora *et al.*, 2004). Though iprodione (0.2%) and mancozeb (0.2%) were effective in controlling early blight of tomato under field conditions but best cost: benefit ratio of 1: 11.4 was obtained in mancozeb treatment (Prasad and Naik, 2003, Chand *et al.*, 2014). Gupta and Bharat (2008) observed good control of the disease with combi-fungicides like metalaxyl + mancozeb and cymoxanil + mancozeb. Arunakumara *et al.* (2010) reported good efficacy of propiconazole (0.1%) and pyraclostrobin (0.2% sprays against this disease and increased fruit yields. Bharat and Gupta (2011) used strobilurin fungicides for the management of disease and reported azoxystrobin (18.2%) + difenoconazole (11.4%) (Amistar Top 325 SC) to be highly effective for the management of the disease.

(iv) Biological control: Casida and Lukezic (1992) reported that application of *Pseudomonas* strain 679-2 on tomato leaves reduced severity of *A. solani* leaf spots. Applications of copper fungicide to the leaves did not interfere with the activity of bacterial strain 679-2. Three isolates of *Bacillus* like *B. megaterium* MBS 4, *B. brevis* OBS 11 and *B. subtilis* var. *globigii* CBS 10 were found antagonistic to *A. solani* both in dual and concomitant cultures (Liu and Wu, 1997). The cell-free metabolites produced by these 3 antagonists significantly reduced spore germination.

Use of plant extracts

Extract of *Clerodendrum aculeatum* when applied as foliar spray one day before inoculation of *A. alternata* completely inhibited infection process on tomato (Sharma and Sharma, 1992). Garlic bulb extract and Prosopsis leaf extracts showed considerable amount of inhibition on the growth of *A. solani* under *in vitro* conditions (Prasad and Naik, 2003). Purified antimicrobial metabolite (stigmasterol) isolated from methanol extract of *Muntingia calabura* roots increased the tomato seed germination and seedling vigour and reduced seed infection by *A. solani,* thereby, controlling early blight of tomato under pot culture conditions (Rajesh *et. al*., 2014).

Integrated disease management

Integration of captan sprays with periodical clipping of lower leaves upto 20 cm height plus weeding was found most effective in controlling this disease but the highest yield was obtained in Bordeaux mixture sprayed plots in combination with clipping of lower leaves upto 20 cm height plus weeding (Sahi, 1990).

(b) SEPTORIA LEAF SPOT

Among foliar diseases of tomato, it is one of the most destructive diseases, which is widely distributed throughout the world. It is particularly severe in areas where wet and humid weather conditions persist for long periods. In seasons of moderate temperatures and abundant rainfall, the disease can cause considerable loss by premature defoliation.

Geographical distribution

Leaf spot of tomato caused by *Septoria lycopersici* is of worldwide occurrence and has been reported for the first time from Argentina by Spegazzini (1882). Subsequently, it was reported from Italy, U.S.A., Australia, France, Germany, England, Fiji, Russia, Czechoslovakia, Phillipine, Ceylon and East Africa.

In India, Ramakrishanan and Sundaram (1954) first time reported its occurrence from Coonor (South India). Later, it was reported from Bombay (Thirmulachar *et al.*, 1956), Kumaon Hills of Uttarakhand (Gupta, 1960) and Himachal Pradesh (Sohi and Sokhi, 1969).

Losses

Losses from this disease results because of premature defoliation of the plants, reduction in fruit yield and predisposition of the fruits to sun scald due to less foliage. Premature defoliation is the most damaging phase that makes the plant week due to less photosynthetic activity ultimately leading to drastic reduction in fruit yields.

Sohi and Sokhi (1973) reported that during favourable weather (July to September) under Solan (H.P.) conditions, it causes complete defoliation, thereby, resulting into the failure of the crop.

Symptoms

The plants may be attacked at any stage of their growth. The initial symptoms appear on the under surface of lower leaves as water soaked spots. As the spots grow larger, these become more or less circular in outline (Fig. 3) and show definite brown coloured margins with grey centers (Plate 5a). A few black glistening pinhead sized pycnidia may be seen in the center of spots. The disease on the stem appears as small, slightly elongated, dark spots containing numerous black pycnidia. Small-elongated specks intermediate in size as those found on the leaves and stems appear on calyx. Black circular spots may also be seen on young as well as mature fruits (Plate 5b).

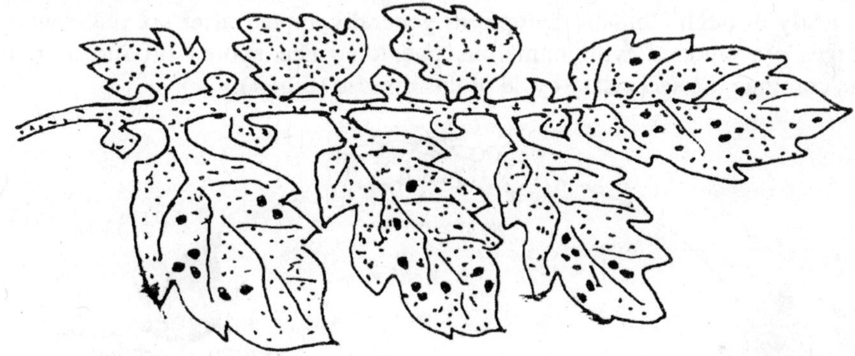

Fig. 3. Septoria leaf spot of tomato (*Septoria lycopersici*)

The Pathogens

The disease is incited by *Septoria lycopersici* Speg. On the host, young mycelium is hyaline, thin walled and sparingly septate. The protoplasmic contents are densely granular, vacuolated, contain numerous small oil globules and the diameter ranges from 1.3 -5.6 μm while the older mycelium is brown, infrequently branched and septate, ranging from 2.5-5.9 μm in diameter. In the initial stage of pycnidial formation, several hyphae aggregate at certain point and after intertwining; form a mass of hyphal tissue that assumes the appearance of pseudoparenchyma. Pycnidia are subglobose, composed of 2-3 layers of brown cells and range between 53-165 μm in width. Pycniospores are filiform, slightly curved, hyaline and septate with pointed or rounded ends and range from 34.5-69.0 × 2.2-4.8 μm in size. No sexual stage has been reported. Two physiologic races I.E. I & II have been identified.

Host range

Host range of this fungus includes potato, eggplant, *Datura tatula, D. stramonium* current tomato, *Solanum carolinense, S. nigrum, Physalis* sp. Sohi and Sokhi (1973) tried six hosts for host range of *S. lycopersici* by artificial inoculation, only *Solanum melongena* got infected but no pycnidia were formed on this host.

Disease cycle

The pathogen overwinters in infected plant debris in the form of mycelium or spores or in debris of solanaceous weed hosts, such as horsenettle (*Solanum carolinense* L.). Progressive increase in the age of pycnidia results in decreased germination of spores. These remained viable up to 6 months after harvest on plant debris stored under laboratory and screenhouse conditions (Kumar and Sugha, 1999). They further observed that plant debris buried near the soil surface helped in overseasoning of the pathogen and successful initiation of the disease. Sohi and Sokhi (1973) and Kumar and Sugha (1999) reported that the pathogen is not carried through the seeds while Walker *et al.* (1937) considered seed and transplant beds as source of infection. The conidia overwintering in soil or plant debris are splashd by rain to the lower leaves of the plants. Germtubes can originate from any cell of the conidia, which penetrate the leaves mostly through stomata. Symptoms generally appear after six days, which in later stages are studded with numerous pycnidia. The spores produced in these pycnidia continue the secondary cycle of this disease (Fig. 4).

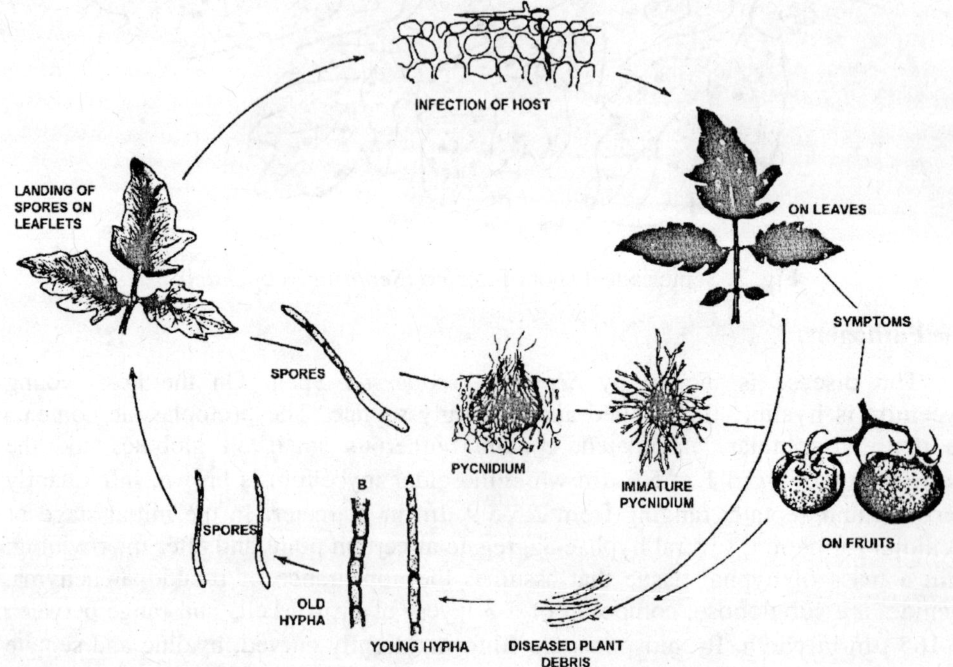

Fig. 4. Disease cycle of Septoria leaf spot of tomato

Epidemiology

The pathogen perennates in diseased plant debris left in the field and also on or in the seed. The fungus has wide host range including solanaceous crops and weeds growing in and around tomato fields. Inoculum load of 525-1050 spores per ml and leaf wetness in the range of 48-60 h were found optimum for normal disease development under artificial inoculation test while inoculum in leaf extract and leachate and dew accelerated development of the disease (Sugha and Kumar, 2000). Temperature of 20-25ºC with 75-92 per cent relative humidity is congenial for disease development. Raina and Razdan (2010) found that relative humidity coupled with rainy days/percentage of total rainfall elicit maximum disease build up.

Management

(i) Cultural practices: The destruction of infected plant material and use of clean seed is useful in checking this disease. Various cultural practices such as early sowing and dry season planting, 1 to 2 years crop rotation between tomato crops, trimming and staking of plants, wider spacing (90 × 50 cm), intercropping with maize, avoiding walking among plants, ploughing under all residual plant material, pruning of lower leaves and staking of plants in case of indeterminate varieties/ hybrids (Kumar and Sugha, 2000), careful attention to the timing of irrigation to reduce the length of time during which the plants remain wet and reduce the contact between foliage and soil and keeping workers and equipment out of field with wet foliage (Jones *et al.,* 1991) have been recommended for its management.

(ii) Host resistance : Several varieties/lines/wild species such as EC2750, EC7785, HP2453, KT-4 and Natural Cross (Sohi and Sokhi, 1969), Sweet 72, Sel. 152, S-12, Marglobe and Gambde (Chauhan, 1987), *L. chilense*, *L. parviflorum*, *L. racimigerum*, *L. peruvianum* vars. *humifusum* and *glandulosum*, *L. pimpnellifolium* forms PI126444 and PI 126936, *L. esculentum* accessions 85602009 and 85602015 and *L. esculentum* var. *ceraciforme* (Sotirova and Rodeva, 1989), EC 262, EC1085, EC 2, EC 13, EC 35337, EC 9, EC 17, EC42555, EC 18, EC 130040P_1-2 (Bedi *et al.*, 1990), *L. esculentum* var. *cerasiforme*, *L. pimpinellifolium* acc. PI 1269205, *L. hirsutum* f. *typicum* acc. PI-127826 and *L. hirsutum* f. *glabratum* acc. PI 19931 (Sotirova *et al.,* 1992), have been reported resistant to this disease. IPA-5 and Rio Fuego were resistant while Caribe and Contessa showed intermediate resistance (Moretto *et al.* 1993) and a higher level of horizontal resistance was shown by IPA-5 than Caribe and Contessa. Pandey and Pandey (2002) reported line LE-415 as resistant genotype having multiple disease resistance including Septoria leaf spot while Lines EC 401927 and CLN 5915-206-D_4 2-2-0 showed moderate degree of resistance to this disease (Joshi *et al*., 2004). Out of 41 cultivars, two varieties (S-12 and SH-12) screened for disease resistance against Septoria leaf spot showed considerable resistance at test locations (Raina and Razdan, 2010).

(iii) Chemical control: Spray applications of various fungicides, such as benomyl, carbendazim, mancozeb, copper oxychloride, folpet, captafol, metiram, copper oxychloride + dichlofluanid, copper oxychloride + folpet, captan, folpet + captafol, cuprous oxide, chlorothalonil, metiram zinc + copper oxychloride have been reported effective against Septoria leaf spot with economical yield gain (Tedla, 1985; Szekely *et al.,* 1990; Bharadwaj, 1991; Raina and Razdan, 2010). Three sprays of Bayleton (0.05%) effectively controlled this disease and provided 99.4% control when applied at 10 days interval (Kumar and Sugha, 1998). Spraying of tomato crop with Dithane M-45 or Dithane Z-78 at 0.2 per cent proved highly effective in restricting disease spread under field conditions thereby stabilizing yield.

4. FUSARIUM WILT

Fusarium wilt is an important disease of this crop particularly in those parts of the world where soil temperature remains high during the period of plant growth in field or under protected cultivation.

Geographical distribution

The disease was reported for the first time from India, Pune (Bihar) by Butler (1918). This disease has been distributed all over the world including British Guiana, U.S.A., Surinam, Chile, Russia, Latvia, Germany, Austria, Italy, Bulgaria, Egypt, Rhodesia, South Africa, Mauritius, China and India.

Losses

The disease inflicts heavy losses, particularly in susceptible cultivars especially when grown under favourable weather conditions. In India, losses ranging from 10 to 80 per cent have been reported. On an average, annual loss due to this disease is 10 to 20 per cent, which may go up to 80 per cent in rare cases. Kapoor (1988) found that the incidence of disease in germplasm/ breeding material ranged from 30 to 100 per cent.

Symptoms

Two of the initial symptoms on young plants are clearing of vein lets and epinasty (drooping of the petioles). The disease first appears in the field as yellowing of the lower leaves (Fig. 5a). The affected leaves die prematurely and the symptoms continue to appear on successively younger leaves. One or more branches may be affected while the others remain symptomless. On any given petiole, leaflets on one side may be affected showing yellowing. Browning of the walls of the vascular bundles (Fig. 5b) follows rather promptly, and is best seen by cutting of stem or petiole and examining a cross section.

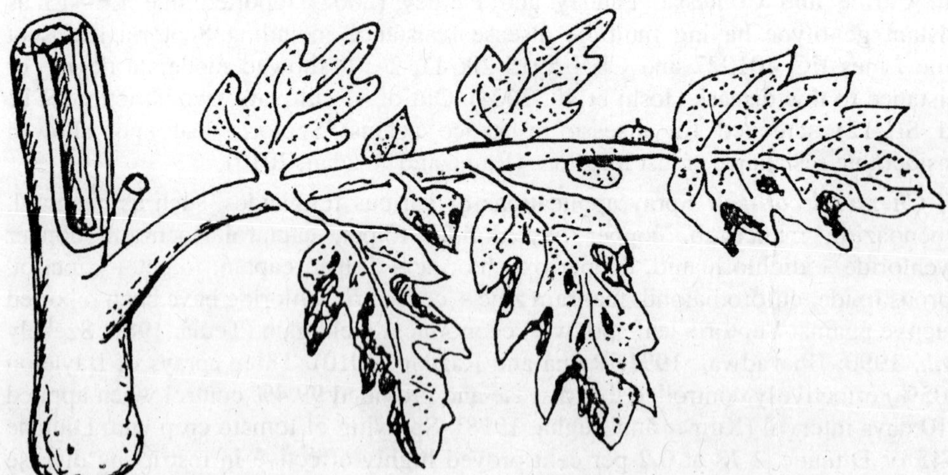

Fig. 5a. Symptoms of Fusarium wilt of tomato on leaves (*F. oxysporum* f.sp. *lycopersica*)

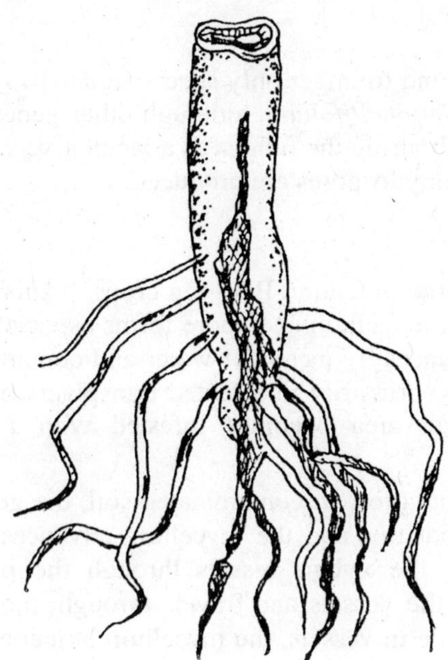

Fig. 5b. Symptoms of fusarium wilt showing browning of vascular bundles

The Pathogen

The disease is caused by the fungus *Fusarium oxysporum* Schlechtend ex Fr. f.sp. *lycopersici* (Sacc.) Snyder and Hans. The old synonyms are *F. lycopersici*, *F. bulbigenum* and *F. bulbigenum* f.sp. *lycopersici*.

The fungus is a facultative saprophyte, which grows readily on various solid or liquid media. The mycelium is hyaline at first becoming cream coloured with age and has branches and septations, and some isolates produce blue or red pigments. The micro conidia are ellipsoidal, straight to curved, 5-12 x 2.2-3.5 μm, one to two celled and hyaline. Macro conidia are the typical 'Fusarium spores', hyaline, 3-5 septate, falcate having gradually pointed and curved ends and appear on sporodochia or without formation of such fruiting bodies, and measure 27-46 x 3-5 μm if three septate or 35-60 x 3-5 μm if five septate. Three septate spores are more common.

Chlamydospores, both rough and smooth walled, characterized by thick walls are terminal or intercalary serving as the chief means of long-term survival in soils in the absence of host (Singh, 1995). They are often solitary but occasionally form in pairs or chains. Three physiological races viz., race 1, 2 and 3 have been reported world over (Jones *et al.*, 1991).Genetic diversity among *F. oxysporum* f.sp. *lycopersici* isolates collected from different tomato growing regions of India was studied by using RAPD primers. RAPD profiles revealed up to 20 per cent genetic diversity among ten isolates of this pathogen (Mishra *et al.*, 2010).

The fungus is regarded as an invader of the epidermis or hair of the fibrous root system of the host, finally reaching the vascular bundles where it is confined largely to xylem vessels. There is enough evidence that invasion is enhanced by wounding of the root system.

Host range

The fungus has been found to infect only species related to *Lycopersicon*, especially *L. esculentum* and *L. pimpinellifolium*, although other genera have also been shown to be latent hosts by harbouring the fungus as a facultative saprophyte in their fibrous root systems where chlamydospores are produced.

Disease cycle

The pathogen is soil borne in nature. Between crops, it survives in the infected plant debris and in the soil as mycelium and spore forms especially as chlamydospores. It spreads over small distances by means of water and contaminated farm equipments, and over long distances, primarily in infected transplants or in the soil carried with them. Usually, once an area becomes infested with *Fusarium*, it remains so indefinitely.

When healthy plants grow in contaminated soil, the germ-tube of spores or the mycelium penetrates root tips and the mycelium advances through the root cortex intercellularly reaching the xylem vessels through the pits. The mycelium then remains exclusively in the vessels and travels through, mostly upward in stem and crown of the plant. While in vessels, the mycelium branches produce micro conidia, which are detached and carried upward in the sap stream. The micro conidia germinate at the point where their upward movement is stopped. The mycelium penetrates the upper wall of the vessel and more micro conidia are produced in the next vessel. The mycelium also advances laterally in to adjacent vessels, penetrating them through pits.

Presumably, a combination of processes, namely, vessel clogging by mycelium, spores, gels and gums is responsible for the break down of water economy of the infected plant. When the leaves transpire more water than the roots and stem can transport to them, the stomata close and the leaves wilt, and finally die followed by death of the plant. The fungus then invades all tissues of the plant extensively reaches the surface of dead plant and then sporulates profusely. The spores may be disseminated to new plants or areas by wind or water.

Sometimes, when the soil moisture is high and the temperature relatively low, infected plants may produce good yields, however, in such cases, the fungus may reach the fruit of the plant and penetrate or contaminate the seed. Usually, the infected fruits decay and drop.

Epidemiology

The fungus is soil borne and can persist in soil for many years. The optimum temperature for growth of the pathogen is 28°C. The symptoms of the disease vary in relation to the soil temperature in which the host plant is grown. The optimum soil temperature for disease development has been found to be 25-31°C, characterized by sudden wilting, usually without yellowing starting from the lower leaves. The rapid development of the disease occurred at 31°C. The temperature conditions of soil and air most favourable to the disease are 27 and 28°C, respectively. Hot dry weather favoured wilt development.

The disease is more severe in acidic soils than in alkaline ones. Hydrogen ion concentration has been found to play a marked role in controlling the growth of causal organism. A pH of 4.0 to 4.5 is best for growth of the pathogen and the disease develop more in sandy soils and least in clay soils. Kapoor (1988) observed that the disease was most serious in warm sandy soils. He also stated that presence of root knot nematodes and monoculture enhanced the disease further. Low nitrogen and high potassium levels in soil predispose the plant to this disease. Application of potassium (100 kg/ ha) or in combination with nitrogen (100 kg N + 100 kg K/ha) decreased the disease incidence to 62 and 67 per cent, respectively while nitrogen (150 kg/ha) alone resulted in 83 per cent disease incidence (Raj and Kapoor, 1995).

Management

(i) Cultural practices: Fusarium wilt can be efficiently controlled by careful crop rotation and the use of healthy seed. Combined application of inorganic fertilizers and organic manures is considered effective in reducing disease incidence. Application of phosphate fertilizers like super phosphate helps to reduce wilt and increase yield. Nitrate fertilizers reduce disease severity and ammonium fertilizers increase disease development. Sarhan and Kiraly (1981) obtained complete control of *Fusarium oxysporum* f.sp. *lycopersici* in cultivar Primset of tomato by application of N as 420 ppm nitrate + raising the soil pH to 7.6 with $Ca(OH)_2$ + drenching the roots with benomyl @ 33 mg/l. Micronutrients like Cu, Zn, Fe, Mg, B and I have given encouraging results when sprayed at transplanting, flowering and fruit formation by increasing resistance in the plants to the disease (Ahmed *et al.*, 1987).

Nirwato *et al.* (1994) reported that application of leguminous foliage such as groundnut and *Sesbania grandiflora* reduced the percentage of infected plants of tomato variety Money Maker to 22.17 and 11.68 per cent, respectively compared to control. Application of organic amendments (crab shell powder and chicken manure) greatly reduced the severity of *F. oxysporum* f.sp. *lycopersici* through reduction in propagules and inhibition of germination.

(ii) Physical methods: Mulching of soil with transparent polyethylene sheet has proved effective against tomato wilt. A 70 per cent reduction in population of the pathogen at a depth of 5 cm, following 30 days of soil solarization in the mulched plot was recorded (Dwiwedi, 1991). The maximum temperature (55°C), recorded in the mulched plot was sufficient to kill the pathogen under Varanasi conditions. El Shami *et al.* (1990) reported that soil solarization by using single layer of transparent polyethylene was much more efficient in reducing the wilt than fumigation with methyl bromide. Soil solarization is thus recommended for management of tomato wilt under Egyptian conditions.

Hot water treatment of roots at 48-49°C for 30 seconds has proved beneficial in delaying the symptoms (Archisi *et al.*, 1985). Steam sterilization of soil is also quite effective under greenhouse conditions for the control of *Fusarium oxysporum* f.sp. *lycopersici* in greenhouse.

(iii) Host resistance: Different sources having varying degrees of resistance have been summarized in the table on next page:

Resistance source (s)	Remarks	Name of worker
EC 26104, Gambde, Globe, HS 101, HS 107, Improved Meerut, Marglobe and Pusa Early Dwarf	Resistant	Chauhan, 1988
Lycopersicon pimpinellifolium: EC 65992, EC 65993, EC 86522 (USA); EC 7353 (Australia) and EC 121453 (Netherlands). *L. hirsutum*: EC 486, EC 65999 and EC 127985 (USA). *L. peruvianum*: EC 2630, EC 65984, EC 86512 and EC 106293 (USA). *L. pimpinellifolium* x *L. hirsutum*: EC 66007, EC 86516, EC 106295, EC 116039 and EC 128013 (USA).	Wild species having resistance	Bisht *et al. 1989*
L. peruvianum: EC 148897 and EC 148808	Immune	Banerjee *et al.* 1990
L. hirsutum var. *glabratum*: EC 144975 and B 6013	Highly resistant	
EC 144981	Moderately resistant	
F_{1s} and backcrosses of Indian varieties (HS 101, HS 102, S 17 and Pusa Ruby) with *L. pimpinellifolium* cv. Pan American	Showed highly resistant reaction	Malhotra and Vashistha, 1993
T 27 and BT 62	Resistant	Mishra and Mishra, 1993

(iii) Chemical control: Soil fumigation with chloropirin can provide an effective control of tomato wilt in highly susceptible varieties. Soil fumigation with methyl bromide (98%) + chloropicrin (2%), applied under polythene covers for 2 days proved very effective against *F. oxysporum* f.sp. *lycopersici* for two seasons.

In greenhouse tomatoes, Atkinson and Adamson (1977) observed consistent reduction in the number of wilted or dead plants by a single pre-plant drench of benomyl or thiophanate-methyl. Etebarian (1992) also reported good efficacy of benomyl, iprodione plus carbendazim and carbendazim alone against tomato wilt. Furadan (carbofuran) has been reported to reduce the incidence of Fusarium wilt when applied to soils at pH 6.5-7.5 @ 10 lb/acre (Overman and Jones, 1975). Since the presence of nematodes in wilt-affected field further aggravates the situation, it is better to apply some nematicide like Furadan in soil before transplanting. Chebil *et al.* (1980) found that soil disinfestation with Basamid (dazomet) ensured better plant growth in soils infested with *Fusarium oxysporum* f.sp. *lycopersici*. Seed treatment with thiram and ceresan @ 4 g/kg seed can also decrease infection by the wilt pathogen. Benlate as seedling dip and soil drench, followed by Bavistin gave highest reduction of the disease incidence (Reddy *et al.*, 2007). Prochloraz and bromuconazole were the most effective fungicides against the pathogen both *in vitro* and *in vivo*, followed by benomyl and carbendazim (Amini and Sidovich, 2010).

(iv) Biological control: Several antagonistic microorganisms have been reported effective in suppressing *Fusarium* wilt of tomato. *Cephalosporium* species may result in a reduction of tomato by reducing the severity of *F. oxysporum* f.sp.

lycopersici. Severity of wilt symptoms was suppressed by non-pathogenic species like *F. oxysporum* f.sp. *melongenae*, *F. oxysporum* f.sp. *cucumerinum* and *F. oxysporum* f.sp. *batatas* as root dip treatment in spore suspension before transplanting (Amemiya *et al.*, 1986). Kroon *et al.* (1991) demonstrated that *F. oxysporum* f.sp. *dianthi* was able to induce resistance against the wilt disease. Additive effects were obtained in most cases when tomato plants were inoculated with non-pathogenic *F. oxysporum* str. F 047 at seeding stage (Fuchs and Defago, 1991).

Pseudomonas isolates from rhizosphere of different plants can reduce infection of *F. oxysporum* f.sp. *lycopersici* in tomato. Mutants from avirulent strain of *Pseudomonas solanacearum* are capable of detoxifying fusaric acid and thus possess the ability to protect tomato plants from wilt.

Coating of tomato seed with *Trichoderma viride*, *Streptomyces griseus* or *Bacillus subtilis* prior to sowing in wilt affected soil significantly reduced damping-off and increased seedling survival (Yehia *et al.*, 1981). Khakimov and Abdullaer (1992) also observed a good control of the pathogen through soil, seed or seedling treatment with trichodermin. Tomato plants infected with *F. oxysporum* f.sp. *lycopersici* wilted less, if they had previously been inoculated with the endotrophic mycorrhizal fungus, *Glomus mosseae*. The fungus enhanced lignin deposition in the cell wall of the epidermis and stele, which is considered to increase resistance to the wilt pathogen. Other organisms like *Streptomyces* spp., *Penicillium* spp. and *Acremonium* spp. have also been tried in this direction.

Sulfhydryl proteinases such as papain, bromelian, ficin and shoot extract of *Inula viscosa* have shown potential as biopesticides against Fusarium wilt of tomato (Qusan *et al.*, 1995). Tomato seeds coated with *T. viridae, T. harzianum and Chaetomium globosum* not only showed greater disease controlling ability but also reported better seed germination in addition to improved growth and vigour of plants (Mehra, 2006). Mehra (2008) tested five different fungal antagonists like *Aspergillus terreus, Penicillium cyclopium, Chaetomium globosum, T. viride* and *T. harzianum* as seedling dip, soil drench and as joint application against Fusarium wilt pathogen. Highest disease control of 42.6 per cent and 44.6 per cent was reported with *T. viride* in seedling dip and soil drench methods. Seed treatment, seedling dip and soil application of *B. subtilis* EPCO 16 in liquid formulation upon challenge inoculation with *F. oxysporum* f.sp, *lycopersici* resulted in lower disease incidence (68.4% reduction) with increased plant growth and fruit yield of tomato relative to control (Ramyabharathi and Raghuchander, 2014).

Integrated disease management

A combination of *T. harzianum* and sheep manure amendment at concentrations (w:w) of 6 and 10 per cent reduced tomato wilt by 21-36 and 29-36 per cent, respectively, after 0-28 months of soil incubation (Barakat and Al-Masri, 2009). The combined use of the biocontrol bacterium, strain LRB3W1, and a fungicide, benomyl, decreased the disease incidence and should be an attractive approach for suppressing tomato wilt (Someya *et al.*, 2006). Among the integrated control treatments, *T. harzianum* + carbendazim showed the best effect (Singh *et al.*, 2007). Carbendazim + *T. viride* and *T. viride* significantly reduced wilt incidence by 73.91, 69.56 and 68.11 per cent, respectively, as against 71.88 per cent wilting in control (under epiphytotic condition i.e. wilt sick soil) (Musmade *et al.*, 2009). Barnwal *et al.* (2011) recorded minimum wilt incidence of 14.9 per cent by seed priming with *T. viride* @4g/ kg seed + soil application of neem oil cake at 250 kg/ha and recorded > 38 per cent increase in fruit yield.

(B) BACTERIAL DISEASES

5. BACTERIAL SPOT

It is a devastating disease of tomato and pepper throughout the world, affecting all the above ground parts. In favourable weather conditions, the yield is significantly reduced.

Geographical distribution

Bacterial spot of tomato was first reported from South Africa by Doidge (1921). Subsequently, it has been reported from many other parts of the world including Indiana, Delaware, Queensland, Israel, Siberia, Bulgaria, Austria, Italy, Chile, Cuba, Southern Georgia, Barbados and Poland.

In India, the disease was recorded on chilli for the first time at Agricultural College Farm, Poona in 1948 (Patel *et al.*, 1950). Since then, it has been reported on tomato from other parts of the country viz. Kerala, Bangalore, Varanasi, Jalgaon and north parts of India (Chand *et al.*, 1994; Borkar, 1995).

Losses

Losses from this disease results from the yield reduction occurring due to premature defoliation causes and fruit spots rendering them unmarketable Fruit yield losses due to this disease may go up to 50 per cent under adverse climatic conditions (Pohronezny *et al.*, 1990). Pernezny *et al.* (1996) conducted several field trials and found that marketable fruit yield was reduced by 30 per cent in control plots as compared to treated plots. Up to 52 per cent losses in marketable fruits have been reported due to this disease under favourable weather conditions (Sharma *et al.*, 2011)

Symptoms

The pathogen affects all above ground plant parts. On leaves and stems, the spots are generally brown circular and up to 3 mm in diameter. During rainy periods or in the presence of dew, the spots are watersoaked, which are darker in colour, less uniformly distributed and not surrounded by halo. In favourable weather conditions, spots coalesce together and a general yellowing occurs on leaflets with many lesions (Plate 6a). Later blightening of foliage takes place that gives a scorched appearance. On young fruits, the lesions appear as small, dark slightly raised dots, sometimes with a narrow watersoaked borders. On mature green fruits, the spots appear as brownish-black elevated scab like areas 2 to 6 mm in diameter with feathered or irregular margins (Plate 6b). For this reason, the disease is also known as scab. At harvest time, they are generally sunken and appear grey or bleached, owing to the dry paper like conditions of the affected skin.

The Pathogen

The disease is caused by *Xanthomonas vesicatoria* (ex. Doidge) Vauterin *et al.* The bacterium is rod shaped, gram negative, strictly aerobic, motile by a single polar flagellum, forming yellow colonies, liquefies gelatine, attacks starch and digests protein and measures 0.7-1.0 × 2.0-2.4 μm. The colony characters on potato dextrose agar medium have been described as smooth, glistening, raised, 1.0-1-2 cm after 4 days and apricot yellow in colour. However, on nutrient agar, it grows relatively slowly and the colonies are smooth, flat glistening, convex, circular with fringed margins, measuring 1 cm after 4 days. Sijam *et al.* (1992) recorded the colony characteristics on CKTM medium as circular, raised, yellow surrounded by a clear ring formed after 1-2 days of spot inoculation. Recently Ramayashreedevi *et al.* (2017) also reported the existence of *X. euvesicatoria* on tomato in five states of India viz. Himachal Pradesh, Karnataka, Uttarakhand, Tamilnadu and Haryana.

Host range

Under field conditions, it is found only on tomatoes, peppers and one of the nightshades, although some weed hosts like *Physalis angulata, P. minima, P. peruviana, Amaranthus dubius, Acalypha virginica, Solanum nigrum, S. americanum* and *Lagascea mollis* have also been reported (Munoz and Granada, 1990).

Disease cycle

The pathogen is mainly seed borne which give rise to infected seedlings and also overwinters in plant debris (Jones *et al.*, 1986). Besides this the pathogen also survives for long periods on dry seeds, inoculated sandy soil, dried leaves and the rhizosphere of pepper and of other non-host plants. In addition, small number of pathogen propagules survived for 18 months in a field previously cropped with pepper affected with bacterial spot (Bashan and Okon, 1986). The bacterium can also overwinter in wheat roots. The disease spreads through wind splashed rains, infected seed and planting material. The bacterium penetrates through stomata and wounds created by wind driven sand, insect punctures or mechanical means.

Epidemiology

The bacterium is disseminated within fields by wind driven rain droplets, the clipping of transplants and aerosols (Jones, 1991). Moderate temperatures and high humidity favour disease development. Longer periods of high humidity (RH 100%) increase susceptibility to bacterial spot. Disease is also affected by day and night temperature fluctuations and at continuous 24°C temperature, it developed most rapidly while a night temperature of 16°C suppressed the disease regardless of day temperature. Nakamura (1990) observed that between the extremes of 19 ± 2°C and 29.5 ± 0.5°C, the disease incidence was proportional to temperature. Later, Jones (1991) reported that the disease is favoured by temperatures of 24-30°C and by high precipitation. However, shortest incubation period of 7 days was recorded at 28°C and 100 per cent relative humidity (Shukla, 2001). Duration of leaf wetness also affect the severity of the disease and with the increase in leaf wetness duration from 6 to 84 h, there was a progressive increase in disease severity.

Shukla and Gupta (2005) also studied the influence of environmental factors on disease development under field conditions and reported that the disease was initiated in the second week of July after the commencement of monsoon rains in Himachal Pradesh. Mean Temperature of 28°C, >90% RH and high soil moisture favoured disease development. Higher rates of nitrogen and potassium application reduce the epiphytic populations of the bacterium (McGuire *et al.*, 1991). The relative density of lesions was influenced by temperature and leaf wetness duration and tomato bacterial leaf spot was more severe at 25° C (Marcuzzo *et al.*, 2009).

Management

Keeping in view the seed borne nature of the disease, attempts have been made to manage the disease through application of various types of seed treatments, seedlings dips and foliar application of chemicals, biocontrol agents, host resistance and cultural practices.

(i) Cultural practices: Rotation of fields to avoid carry over of the bacterium on volunteer plants and crop residue, production of disease free plants, elimination of potential volunteers, avoidance of cull piles near greenhouse or field operations, use of disease free seed have been suggested to reduce the disease (Jones, 1991; Jindal, 1992). Disease to some extent can also be kept under check by application of pine needle mulch (Shukla, 2001).

(ii) Host resistance: Although a large number of tomato cvs./ accessions have been screened against the disease by various workers, only a few have been found resistant such as *Lycopersicon peruvianum* var. *humifuscum*, *L. hirsutum typicum*, *L. hirsutum* var. *glandulosum* and *L. minutum*-913 are reported to be resistant (Beleva and Sotirova, 1977). Roma VF group was less susceptible to this pathogen than cultivars belonging to VC 134 group.

Several cvs./ lines of tomato such as Ohio 4012-3, Ohio 4014-4, Heinz 1568-F_3, L 556, Campbell-28, PI-127813, Heinz 2990, PI-324708, Heinz 603-F_{11}, PI-224573 and Monense have been reported resistant to this disease (Scott and Jones, 1986). Fruits of genotype EC 129171-PI-3 were found to be resistant (Borkar, 1995).

Accessions of subgenus *Eriopersicon* were more resistant than those belonging to subgenus *Eulycopersicon* of wild tomatoes (Sotirova and Bogatsevska, 1998)

The resistance in Hawaii 7998 is reported to be associated with a hypersensitive reaction and controlled by more than 1 gene (Wang *et al.*, 1994). In case of tomato race 3, an avirulence gene Avr-Xv3 was found responsible for elicitation of hypersensitive reaction in leaves of tomato near isogenic line (NIL) 216 (Astua, 2000). Six tomato genotypes and their hybrid diallel combinations, without the reciprocals, were investigated for their reaction to bacterial spot caused by three *Xanthomonas* races. Resistance was rated on a grade scale, and by the area under disease progress curve. The accession UENF 157 was found to be resistant to all pathogen races (Souza *et al.*, 2008). Sharma *et al.* (2011) reported that the resistace in bacterial spot was governed by polygenes. The additive and dominance effect of genes played an important role in the inheritance of this disease.

Chemical control

(i) Seed treatment: Since the pathogen perpetuates through infected and or contaminated seed, attempts have been made by various workers to reduce possible transmission of the bacterium through hot water or chemical seed treatment either alone or in combination (Jones, 1991). Hot water treatment at 50°C for 25 minutes or five minutes dip in mercuric chloride (1: 2000) or treatment with Ceresan, Agrimycin-100 or acetic acid have been found effective against this disease. Soaking of the seeds in sodium hypochlorite either alone or in combination with chlorohydric acid for 20 minutes has also been found effective (Hernandez and Medina, 1991). Kritzman (1993) reported an effective control of bacterial spot when tomato seeds were immersed in a solution containing cupric acetate, acetic acid, 5-ethoxy-3 (trichloromethyl)-1, 2,4-thiadiazole and triton-X 100 in the ratio of 1:4 (w/v) for 30 minutes at 25°C, without any adverse effect on seed germination or seedling vigour.

Bacterized tomato seeds when treated with aqueous extracts of garlic (30 g/ 100 ml of distilled water) provided 35 per cent decrease in number of leaf spots caused by *X. vesicatoria* as compared to tomato plants raised from untreated bacterized seeds (Mangamma and Sreeramulu, 1991)

(ii) Seedling dip: Basymbekov and Firsov (1978) reported that root dip in 0.05 per cent BF-3 (dimethyl-(2 oxy-4-phenyl butone-3yl-2) for 30 minutes or 1 hour before planting reduces the disease. Among the cyclic compounds tested only BF-6 gave best results when roots were soaked for 2 hour in 0.01 per cent solution before planting which resulted in reduced leaf and fruit infections.

(iii) Foliar application: Several copper compounds, antibiotics and various other compounds have been evaluated by different workers all over the world to keep the ravages of this disease under check. Efficacy of various antibiotics/ fungicides viz., streptomycin, streptomycin + terramycin, Agrimycin-100, Agrimycin-100 + tribasic copper sulphate, 0.5 per cent Caffaro powder or copper oxychloride, 0.5 per cent Dithane-cupromix, Kocide 101 and mixture of basic copper sulphate and maneb or mancozeb (Conover and Gerhold, 1981; Jones and Jones, 1985; Maringoni *et al.*, 1986) have been reported against this disease. Jones and Jones (1987) reported that twice a week application of copper-mancozeb combination gave better control than individual fungicide but premixing of copper compounds and mancozeb for upto 4 hours had no effect on disease control. Under field conditions, sprays of streptomycin sulphate (0.01%) + copper oxychloride (0.2%) and copper oxychloride (0.2%) + mancozeb (0.25) were found most effective against this disease both as protective and in eradicative spray applications (Shukla and Gupta, 2004).

Biological control

Ahn and Cho (1996) found glycinecin; a bacteriocin produced by *Xanthomonas campestris* pv. *glycine* to inhibit the growth of *X. vesicatoria*. Campbell *et al.* (1998) tested 5 isolates (*Pseudomonas. fluorescens* A 506, *Erwinia herbicola* B_2, *P. mendocina* B 41, *P. putida* B 56 and *P. syringae* Cit 7) against bacterial spot and found *P. putida* and *P. syringae* effective to reduce the disease incidence under field conditions. PGPR strains 89B-61 and *Bacillus pumilus* SE34 both and their combined application provided significant suppression of bacterial spot in the two field trials conducted in Alabama (Ji *et al.*, 2006). Seed bacterization with *Bacillus subtilis* strain GB03 showed increased plant growth and reduced the bacterial leaf spot in tomato and the rate of reduction in the bacterial spot disease incidence in plants raised from bacterized seeds was directly proportional to the increased levels of PAL, PPO and phenol contents (Umesha and Hariprasad, 2010).

6. BACTERIAL WILT

Bacterial wilt is one of the most devastating bacterial diseases of solanaceous plants particularly in many warm temperate, subtropical and tropical regions of the world. The disease is also called southern bacterial wilt, solanaceous wilt, and southern bacterial blight and by many other common names in countries wherever it occurs. Losses from this disease are tremendous particularly during warm humid weather. In Phillippines, average losses of 15 per cent from this disease in tomato crop have been reported while in improved hybrids in Taiwan it caused upto 55 per cent losses (Hartman *et al.*, 1991). In India, Devi and Menon (1980) observed maximum incidence in October – November and minimum in February under Kerala conditions. Upto 90.6 per cent loss in tomato yield due to this disease has been reported in some badly affected fields in India (Kishun, 1987).

Geographical distribution

Kelman (1953) was the first to review the literature and indicate the worldwide economic importance of bacterial wilt on its many hosts. Since then the occurrence of this disease has been reported from Atkinson fields of British Guiana, Israel, Canada, Bulgaria, Philippines and Taiwan.

In India, the disease is known to occur in Punjab, Himachal Pradesh, Madhya Pradesh, Bihar, West Bengal, Assam, Orissa, Kerala, Karnataka and Tamil Nadu on tomato and other crops like potato, eggplant, chilli etc. (Khan *et al.*, 1980). In Himachal Pradesh, the occurrence of the disease was first time recorded from Kangra Valley in mid 1981 and later the existence of this disease was also reported from those areas of Solan district that are about 700 a.m.s.l. (Gupta *et al.*, 1998). Upto 4-60 per cent incidence of this disease was recorded in different north and eastern states of India (Singh *et al.*, 2010).

Symptoms

Characteristic symptoms of bacterial wilt on most of the hosts are rapid and complete wilting of normal grown up plants (Fig. 6). During wet and warm climatic conditions, the most conspicuous symptom seen on tomato plants is sudden drooping of leaves, without yellowing often accompanied by rotting of the stem (Plate 7). The roots appear healthy and often are well developed. Pathogen is mostly confined to vascular region. In advance cases, it may invade the cortex and pith and cause yellow brown discolouration of tissues. Bacterial polysaccharides mechanically block the vascular system, which checks the trans-location of water and minerals resulting in wilting of plants.

The Pathogen

The bacterium responsible for wilt was first named as *Bacillus solanacearum* and then changed to *Pseudomonas solanacearum*. The nomenclature of the bacterium has again been changed to *Ralstonia solanacearum* (Yabuuchi *et al.*, 1995). The bacterium is gram negative, rod shaped, frequently occurs in pairs, motile with 1 to 4 polar flagella and measures 0.5-0.7 x 1.5–2.5 μm. It is aerobic and catalase and oxidase-positive and forms nitrites from nitrates. It, like some other non-fluorescent pseudomonads, produces intracellular, refractile, sudanophilic inclusions consisting of polyhydroxy-butyric acid. Some strains produce a brown diffusible pigment on complex media and a black pigment on autoclaved potato slants. The pathogen is negative for levan production and starch hydrolysis. It causes weak hydrolysis of gelatin. Five races (race 1,2,3,4,5) and four biovars (biovars I, II, III and IV) are known to occur in different parts of the world. In India, race 1 and biovar III and IV are prevalent (Kalha and Sood, 199;, Singh *et al.*, 2010). PCR with *R. solanacearum* specific primers Y2 and OLII were used for molecular detection and all isolates were amplified at 288 bp.The primer was also able to distinguish the *R. solanacearum* from other saprophytic bacteria through colony PCR within 2 days (Singh *et al.*, 2010).

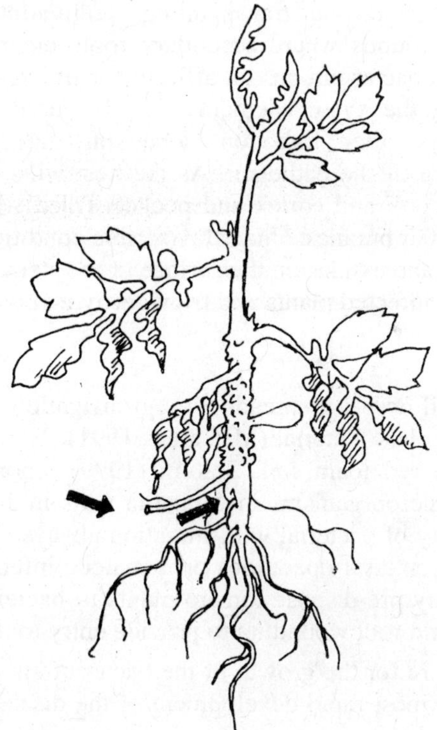

Fig. 6. Symptoms of Bacterial wilt of tomato (*Ralstonia solanacearum*).

Host range

The pathogen infects over 200 plant species including grasses, herbs, shrubs and trees (Gadewar and Shekhawat, 2000). Apart from tomato several other plants such as tobacco, *Nicotiana rustica*, egg plant, *Solanum nigrum*, *Datura stramonium, Phaseolus vulgaris, P. lunatus, Croton glandulosus* and *Crotolaria striata* are the host plants for bacterial wilt while race 3 and biovar II attacks potato, tomato,

Capsicum annuum var. *gronum*, aubergine, *Nicotiana glutinosa, Solanum capsicaotrum, Datura stramonium* (Fernandez, 1986). Muhammad (1987) also observed symptoms of bacterial wilt on *Crassocephalum crepidioides, Cosmos* sp., *Solanum khasianum* and *Syzigium aromaticum*. The important crops affected in India are potato, tomato, brinjal, chillies, ginger, jute, cotton, cumin, fennel, sesame etc. (Gadewar and Shekhawat, 2000).

Disease cycle

The pathogen is soil borne and can survive atleast for 2 years in the absence of any host in the diverse agro-climates of India (Shekhawat *et al*., 1979). The bacteria from tomato and brinjal survive up to 3 years in fallow and upto unlimited period in cultivated land. It is found upto 45 cm depth in cultivated and fallow lands but its population decreases with increase in depth. Although the pathogen is not considered as seed borne, but Devi and Menon (1980) were able to produce bacterial wilt in tomato plants grown from seeds collected from partially wilted plants. The bacterium enters through wounds made by transplanting, cultivation, insects or certain nematodes and natural wounds where secondary roots emerge (McCarter, 1991). Once inside the host, the bacterium has an affinity for the vascular system where it multiplies rapidly filling the xylem elements with bacterial cells and slime. Wilt occurs within 2-5 days after infection depending upon host susceptibility, temperature and virulence of the pathogen. As the host wilts, the pathogen invades parenchyma cells in the pith and cortex and pockets filled with masses of bacteria develop around the vascular bundles. Under favourable conditions the bacterium may move through the cortex and exudes on the surface of the stem. It is released in to the soil from the roots of the infected plants and from decaying host material.

Epidemiology

Since the pathogen is soil borne, it spreads through irrigation water, soil movement, implements and infected plant material (McCarter, 1991). The disease is more severe in sandy loam soil than red loam soil. Shiomi (1999) reported that the naturally existing population of microorganisms in Mutsumi soils in Japan was significantly able to reduce the severity of bacterial wilt. Infection always occurs through wounds caused by transplanting, cultural operations or nematode infection (Keshwal, 2000). Root knot nematodes may pre-dispose tomato plants to bacterial wilt by interfering with nutrition of plants and root wounding to provide entry to the pathogen.

Optimum temperature for the growth of the bacterium is 35-37°C. Disease does not develop at 15°C and most rapid development of the disease takes place at 37°C. No infection occurs at or below 50 per cent soil moisture. The temperature should be accompanied with the plentiful supply of moisture. Bacterial wilt is significantly correlated with temperature, relative humidity and mainly rainfall (Bora *et al*., 1996). Small variation in an environmental parameter caused variation in wilt incidence.

The pathogen can grow over a wide range of pH of the soil and optimum pH is 6.2-6.6. The disease is prevalent in soils near neutrality but absent from those that were strongly acidic or alkaline. Organic manures promote the activity of bacterium while inorganic fertilizers decrease its activity. There is decrease in infection with the increase in the age of the plants (Singh, 1995).

Management

Control of bacterial wilt in infested soil is very difficult. Different workers throughout the world have made efforts to combat this disease through cultural, chemical, biological and host resistance either alone or in combination.

(i) Cultural practices: Attempts to minimise losses due to bacterial wilt by shifting dates of transplanting and harvesting had limited success in certain parts of the world. These changes in planting dates are generally designed to avoid either periods of high temperature or heavy rainfall or both. The disease can be kept under check if the crop rotations are followed and in rotation crops like maize or ragi is included (Sharma and Kumar, 2004) because these crops reduce bacterial population in soil significantly.

Although soil solarization has been exploited for the control of numerous soil borne pathogens, its role for the control of soil borne bacterial pathogens has not been much studied. Flourescent pseudomonads and other microorganisms recolonized in solarized soil are also involved in increased growth response of tomato plants in solarized soils (Gamliel and Katan, 1991). The fluorescent pseudomonads like *P. putida, P. fluorescens* and *P. alcaligenes* were recovered from solarized soil, which stimulated growth of tomato plants. Additional reduction in the density of *P. solanacearum* was achieved when solarization was combined with fumigation (Chellemi *et al.*, 1994).

P. solanacearum, is polyphagous in nature and as such attacks not only plants of economic importance but number of other ornamentals and diverse group of weeds are also known to harbour this pathogen. Moffett and Haywards (1980) studied the role of weed hosts in the survival of *P. solanacearum*. Devi and Menon (1980) listed nine hosts including two crop plants, three ornamentals and four weeds. Destruction of weeds, collateral and other off-season hosts should help in minimising the inoculum potential of the pathogen.

Application of various soil amendments such as saw dust, peat moss also decreased wilt incidence. Suppression of pathogen in amended soil was observed by stimulating the growth of saprophytic micro-organisms especially actinomycetes and *Bacillus* species or by the release of toxins during decomposition by soil micro-organisms (Yao *et al.*, 1994). When different nitrogen forms (NH_4Cl, $NaNO_3$ and $NaNO_2$) substituted urea in the components of S-H mixture, only the soil amended with $NaNO_2$ was found effective in suppressing the wilt pathogen (Chou *et al.*, 1995). Rapid bacterial population decline was mainly caused by the rapid increase in nitrite.

(ii) Host resistance: At least two types of resistance seem to exist in tomato, one is polygenic, originating from *Lycopersicon esculentum* var. *ceresiformae* and needs to be sustained by additional crop management measures. The second type is a dominant resistance present in cv. Hawaii, which is of simpler heredity, like the monogenic resistance described in Hawaii 7998 (Scott *et al.*, 1988). Most of the resistant sources against *P. solanacearum* were found in *L. pimpinelli-folium* or ecotypes of *L. esculentum* var. *ceresiformae*, which are wild tomato germplasm with small fruits and resistant to elevated temperatures. The Asian Vegetable Research and Development Centre (AVRDC), Taiwan together with institutions in many countries including Australia, India, South Africa and U.S.A. have played a key role in the development of cvs. with stable resistance (Wang *et al*., 1996).

Tomato lines/cultivars *viz.,* VC 9-1UG and VC 11-1UG (Bedekar, 1977), Scorpio (Peterson *et al*., 1983), Musashikyoryoku 560, Matsurao, eootstocks LD 89 and BFM-R (Yoneyama, 1985), Redlands Summertaste and Redlander (Mofett, 1986), Arka Abha (BWR-1), Arka Alok (BWR-5) (Sinha *et al*., 1988), Sonali (Sel-2) (Patil *et al*., 1990), DPT 38 (Kumar *et al*., 1997) are resistant.

Graft of susceptible cultivars of tomato on resistant rootstocks belonging to *Solanum* spp. Has been reported to be effective in management of this disease. Rajput *et al.* (1982) controlled this disease by grafting susceptible tomato cultivars on to resistant eggplant (pusa Purple Cluster) rootstocks.

Variations in pathogen strains and environment are the main reasons, commonly given to explain the breakdown of resistance in tomatoes. High soil temperature and humidity were reported to markedly decrease the resistance of tomato (Mew and Ho, 1977). Resistance also depends on nematode population in soil (Jatala *et al.,* 1975). At present Hawaii 7996, Hawaii 7998 and CRA 66 seem to have acceptable stability of resistance with variation in strains and temperature. Various aspects of resistance to bacterial wilt have been reviewed (Kelman, 1953; Hayward, 1991; Prior *et al*., 1994). The combi-nation of host resistance with location specific additional measures for control is probably one of the cheapest strategies available and also one of the most satisfactory approaches from an ecological point of view (Prior *et al*., 1994). F_1 cross viz., EC-339074 x EC-386021 (Swarna Sampada) was found superior than others with respect to resistance and yield in bacterial sick plot (Sharma *et al.,* 2006). Tomato lines/ cultivars viz., EC 369060A (Swarna Naveen), BT-10, CKVT-17, Sikkam Local, BB-6, JC-8, Arka Keshav, Arka Nidhi, Singnath, BB-40, Green Round have been reported resistant to this disease (Singh *et al*., 2014).

(iii) Chemical control: Soil treatment with various chemicals like chloropicrin, methyl bromide and methyl bromide-chloropicrin mixture (67-33%) are effective in retarding the wilt development. Application of bleaching powder (15 kg/ ha) has also been found effective against this disease (Ho, 1988). The disease can also be controlled effectively if dazomet application is combined with soil solarization (Yamada *et al*., 1997).

Bacterinol – 100, when applied as dry seed dresser, a nursery spray and field application resulted in 11.5 per cent yield increase over untreated control (Rao, 1990).

Streptomycin sulphate/ oxytetracycline when sprayed @ 200 ppm gave good control of tomato wilt (Mazumdar, 1998) and Agrimycin (streptomycin) @ 200 ppm application suppressed the wilt upto 30 per cent (Gunawan, 1989). Terlai, a bactericide also gave good control of this disease but caused chlorosis of tomato if used at higher concentration (Hanudin and Machmud, 1994). Applications of bleaching powder (10g and 5g/l) also effectively reduced the disease incidence accompanied with increase in yield (Mazumdar, 1998).

Incorporation of salts like KNO_3, $NaNO_3$, KCl and NaCl to soil (1: 99) reduced the soil population of *P. solanacearum* after four days of application but $NaNO_3$ was the most effective salt and reduced the bacterial population upto 65 per cent

(Hanudin and Machmud, 1994). In nutrient solution the population of wilt bacterium decreased with the increased calcium concentration in the nutrient solution (Yamazaki *et al.*, 2000). They further suggested that the calcium concentration in soil should be increased to have better control of the wilt. Root dipping of tomato seedlings for 6 h in a solution of 200 ppm streptomycin, chloramphenicol, oxytetracycline hydrochloride, norfloxacin, amoxicillin + cloxacillin, chloroquine phosphate and 2000 ppm dichlorophen before planting, reduced the incidence of bacterial wilt to a great extent (Mondal *et al.*, 2005).

(iv) Biological control: Soil microflora that can grow in the rhizosphere are ideal for use as biocontrol agents, since the rhizosphere provides the frontline defence for roots against attack by pathogens. Several biological control agents like *Pseudomonas fluorescens* (Hartman *et al.*, 1993), and *Bacillus* spp. (Anuratha and Gnanamanickam, 1990), *B. licheniformis, B. cereus, B. subtilis* (Phae *et al.,* 1992), vesicular arbuscular mycorrhiza (Suresh and Rai, 1991) have been found to delay wilt development and reduce disease incidence in tomato and potato.

Tomato seedlings root dipped in *Pseudomonas glumae* selected strains provided good control of wilt (Wakimoto, 1987). In addition to production of antibiotics, other factors like occupation of attachment sites and induction of resistance in host tissues might also play role in biocontrol process. Antibiosis of fluorescent Pseudomonas against wilt bacterium was mainly due to their production of fluorescent siderophores.

El-shanshoury *et al.* (1996) has reported about the possibility of using biocontrol agent along with herbicides like pendimethalin and metribuzin for increased control of bacterial and Fusarium wilt in tomato. Minimum disease incidence (9.27%) observed in *Bacillus subtilis* isolate BS-5 treated soil, which has performed significantly better than untreated control after 35 days of inoculationwith *R. solanacearum* and *B. subtilis* inoculation. Whereas BS-9 isolate exhibited maximum phosphorus solublization, siderophore production, indole acetic acid production and ammonia production in vitro while under glasshouse conditions it increased height of tomato plants and biomass production which was better than untreated control (Singh *et al.,* 2012). Ashlesha and Paul (2012) evaluated seven organic inputs viz., biosol, matka khad, fermented butter milk, panchgavya, biodynamic compost tea, cow urine and vermiwash against *R. solanacearum* both *in vitro* and in polyhouse conditions. Combination of biosol, cow urine, fermented butter milk, biodynamic compost tea, vermiwash and homa ash enhanced the survivability of seedlings up to 13 days when tested at 1: 10 ratio after 1 h duration dip duration and seedlings kept in Hoagland's solution. Drenching of mixture of biosol, fermented butter milk, cow urine, vermiwash and homa ash before and after transplanting along with capsicum seedling deep in biosol for 10 min before transplanting in polyhouse provided 86.64, 85.22 and 84.64 per cent disease control at Rajol, Palampur and Sungal areas, respectively.

Use of plant extracts

Garlic cloves suspension (10ml) or ground garlic bulb (6g) when added to rhizosphere not only reduced the disease but also resulted in normal plant growth and increased fruit weight (Hutagalung, 1988). Two organic formulations comprising the mixture of asafoetida (*Ferula foetida* Rege), turmeric powder (*Curcuma longa* Linn.) and water (ATW) in the ratio of 1.5g: 5g: 10 l and 1g: 5g: 10 l also recorded maximum disease control of 70.3 per cent and 69 per cent respectively with increased yields (Mazumdar, 1998). Garlic exhibited the strongest antibacterial activity against bacterial wilt *in vitro* and *in vivo*, followed by Datura and Nerium. In greenhouse experiments, the application of plant extracts at the time of inoculation resulted in the highest reduction of disease index.

Integrated disease management

Integration of soil solarization for 8 and 10 weeks during March to May alongwith incorporation of antagonistic rhizobacteria i.e. *Pseudomonas fluorescence* and *Bacillus cereus* in soil prior to solarization resulted in increased soil temperature at 5 cm depth in wilt sick plots by 5.2 to 6.2°C under Palampur (Himachal Pradesh) conditions and caused up to 73.4 per cent suppression of wilt incidence which was attributed to enhanced population of the antagonists and reduced *Ralstonia solanacearum* population (Kumar and Sood, 2001). No wilt incidence was observed when bacterized (*Pseudomonas fluorescens* and *Bacillus cereus*) tomato seedlings of moderately resistant line EC 191536 were planted in 10 weeks solarized soil (Kumar and Sood, 2005). Sharma and Kumar (2009) reported that karanj cake alongwith bleaching powder application was more effective in reducing the bacterial growth at flowering stage thereby increasing the yield of tomato.

7. BACTERIAL CANKER

It is a serious disease of tomato both under open and protected conditions and cause substantial yield losses in favourable climatic conditions. Losses are associated with stunting, wilting, cankering, and sometimes death of systemically infected plants along with fruit spotting resulting from secondary infections.

The disease is described in detail in the chapter Diseases of vegetables under protected cultivation.

Symptoms

The disease is observed in midseason, especially when plants are in flowering and at the mature and ripe stage. The disease symptoms appear on almost all the above ground plant parts *viz.*, leaves, petioles, stem, branches and fruits. Young plants show poor growth and wilting of branches. The lower leaves of affected plants become yellow, shrivelled with dark brown spots usually at the edges and surrounded by yellow halo. In some cases one half of the affected plant shows wilt symptoms while the other half appears healthy. The stem of the affected plant also develop scars or cavities at some locations. Sometimes splitting of the affected stem was also noticed with longitudinal cankers (Plate 8a). The pith of affected stem appears reddish brown, especially at the nodes. The pith later turned somewhat mealy in appearance and hollow. Fruit symptoms appear as bird's eye lesions on severely affected plants (Plate 8b). The lesions on fruits were yellow to brown, slightly raised, about 3 mm in diameter and surrounded by a persistent white halo.

The Pathogen

The disease is caused by bacterium *Clavibacter michiganensis* subsp. *michiganensis* (Cmm). The bacterium is aerobic non-sporulating, gram positive, rod shaped. The colonies of bacterium are mostly creamish to yellow in colour, roundish and semi fluidal. The colonies became deeper yellow and glistering with the period of incubation. All the bacterial isolates were found positive for Gram stain, esculin hydrolysis, gelatin liquefaction, catalase and 2,3,5- triphenyl tetrazolium chloride (TTC), and were negative for starch hydrolysis, oxidase, levan production and casein hydrolysis.

Disease cycle and epidemiology

The bacterium is seed borne and known to spread through contaminated seed to new areas. Due to seed borne nature of the pathogen, disease spread to the areas where it was not reported earlier. The pathogen spread within plantings by splashing rain and human activities. Once the bacterium enters through leaf stomata and/ or small wounds, systemic infection occurs. Moisture condition of the plant and temperature are found to be important factors in the development of disease. During summers particularly in hilly areas, the congenial conditions prevail for the development of bacterial canker in the month of June when the temperature remains high and pre-monsoon showers provide sufficient humidity. Warm temperatures (23°C to 28°C) along with high moisture or relative humidity (>80%) favours disease development.

Management

Management of this disease is very difficult after the appearance of symptoms. Some practices are suggested to keep the ravages of this disease in check which are described in chapter diseases of vegetables under protected cultivation.

(C) VIRAL DISEASES

8. LEAF CURL

Leaf curl is one of the most important and devastating diseases of tomato causing heavy losses in yield and quality of fruits. Leaf curl caused by Tomato Yellow Leaf Curl Virus (TYLCV) is one of the devastating diseases of this crop and inflicts heavy losses in yield depending on the severity and stage of the infection (Kalloo, 1988).

Geographical distribution

Tomato leaf curl also refereed to as tomato yellow leaf curl disease was first reported to be major constraint to tomato production in the Near East in the mid-1960s (Cohen and Nitzany, 1966). Since then the virus was found to affect tomato plants in three large geographical regions: the Mediterranean Basin (Cyprus, Egypt, Jordan, Israel, Lebanon, Turkey), Western Africa (Cape Verde, Senegal), Eastern Africa (Sudan, Tanzania) and Indian sub-continent (Czosnek *et al*., 1991). In India, the occurrence of leaf curl disease on tomato was first reported by Vasudeva and Samraj (1948).

Losses

The disease caused 50 to 80 per cent losses in Cyprus (Ioannou, 1985) and led to complete crop failures in Mexico and Florida during 1989-90 winter growing season

(Watterson, 1993). Up to 92.3 per cent losses have been estimated if the plants are infected within 20 days of transplanting and up to 74.1 and 28.9 per cent if the plants are affected 35 and 50 days after transplanting, respectively (Sastry and Singh, 1973). Of late a severe epidemic of this disease has been observed in Jammu area and the incidence ranged in between 80.00 to 98.00 per cent (Kumar *et al.*, 2004).

Symptoms

The virus causes mosaic, interveinal yellowing, vein clearing and crinkling and puckering of the leaves accompanied by rolling of the leaf margins (Fig. 7). Initially young leaves show slight yellow discolouration which later become leathery and brittle. Significant reduction of nodes and internodal lengths occurs giving the plant a bushy appearance. In advanced stages of infection severe stunting and partial to complete sterility occurs. Infected plants bear few or no fruits.

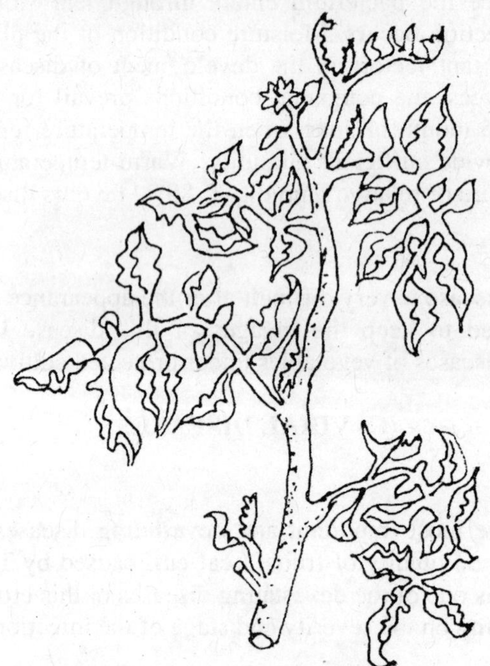

Fig. 7. Symptoms of leaf curl of potato.

The Virus

The disease is caused by *Tomato Yellow Leaf Curl Virus* (TYLCV). The virus TYLCV belongs to the Gemini Virus group (Chatchawan-kanphanich *et al.*, 1993). The particle size of this virus is varying from 15-20 x 25-30 nm (Chenulu and Giri, 1985). The genome of TYLCV cloned and sequenced from Israel has been found to be monopartite and contain only one single-stranded DNA of approximately 0.63 x 10^6 daltons corresponding to 2,787 nucleotides (Czosnek *et al.*, 1991). Indian tomato leaf curl virus (ITmLCV) has been reported to react serologically with whitefly-transmitted gemini viruses namely, African cassava mosaic, squash leaf curl, tomato golden mosaic and bean golden mosaic but not Indian cassava mosaic virus and the tobacco leaf curl virus from Japan (Muniyappa *et al.*, 1991). The ITmLCV is similar in size to the old world geminiviruses, but is slightly larger than the DNA of the geminiviruses from Americas. The common region of the ITmLCV is more similar to that of the tomato leaf curl viruses from Taiwan and Australia and less similar to TYLCV from Israel, the Thailand tomato yellow leaf curl virus, tomato golden mosaic from Brazil and tomato mottle virus from Florida. ITmLCV is a distinct geminivirus because its AL 1 and AR 1 open reading frames showed less than 80 per cent nucleotide sequence identity with seven other tomato infecting geminiviruses (Chatchawankanphanich *et al.*, 1993).

Host range

The virus has a wide host range including cultivated species, weed hosts and ornamentals. Under natural conditions tomato leaf curl can be transmitted to *Acanthospermum hispidum, Ageratum conyzoides, Abelmoschus esculentus, Abutilon variegata, Althea rosea, Bidens biternata, Blansvella rhombosida, Carica papaya, Crotolaria juncea, C. mucronata, Capsicum annum, C. frutescens, Conyza stricta, Datura stramonium, Euphorbia geniculata, Flavena australensis, Gossypium orboreum, Galinsoga parviflora, Hibiscus sabdariffa, Lycopersicon pimpinellifolium, L. peruvianum, L. hirsutum, L. glandulosun, Nicandra physaloides, Nicotiana tabacum, N. benthamiana, N. glutinosa, Oxalis acetosella, Physalis minima, P. peruviana, P. ungulata, Petunia hybrida, Phlox drumondii, Sida mysoriensis, Solanum nigrum, S. melongena, Syndrella nodiflora, Withiana* spp., *Xanthium strumarium* and *Zinia elegans* (Sastry *et al.*, 1978; Saikia and Muniyappa, 1989).

Transmission

It is neither seed nor sap transmissible. The sole agency of its transmission is the white fly (*Bemisia tabaci*) that transmits the virus in a persistant manner. For successful transmission of the virus by vector, minimum acquisition and inoculation feeding periods are 15 to 30 minutes and the latent period of atleast 21 h. Transmission appears to be in a circulative, non-propagative manner. B biotype whitefly was an efficient vector for ToLCTWV. Transmission was achieved to 50-60 per cent of the plants by one adult per plant, and the frequency of transmission approached or reached 100 perc ent when the vector density increased to five per plant (Zhang *et al.*, 2010).

Epidemiology

Presence of weed hosts, perennials etc. serve as the main source of primary inoculum for the virus. Moderately high temperature (33-39°C) favours transmission efficiency and acquisition threshold of the virus. The latent period of the virus in tmato plants varies from 8 days in summer to 90 days in winter that accounts for its higher incidence in the summer months (Mayee *et al.*, 1974).

Management

Although no highly effective methods have been developed for its management, however, adoption of an integrated management strategy taking into account crop sanitation, cultural practices, host resistance and timely use of insecticides have been found effective to reduce tomato leaf curl virus.

(i) Cultural practices: Careful selection of sowing dates helps to check the incidence of the disease. Since the activity of whilte flies is mostly affected by temperature, rainfall and humidity, the worst effects of the disease could be avoided by following July to December plantings. Care should be taken to destroy all weeds, which may serve as alternate or collateral hosts for the virus. Soil mulching with straw has been found effective to repell whiteflies (Nitzany *et al.*, 1964). Planting of barrier crops like maize or jowar around the main tomato plot, atleast 30 days prior to tomato transplanting helps to check the entry of viruliferous whiteflies. The crops grown under mulching *viz.,* rice straw mulch and black plastic mulch had lesser disease severity compared with those without mulch (Saha *et al.*, 2010).

(ii) Host resistance: Resistance to TYLCV has been observed in wild relatives of cultivated tomatoes like *Lycopersicon cheesmanii, L. chilense, L. hirsutum* and *L. peruvianum* (Kasrawi *et al.*, 1988) and its successful transfer to high yielding tomato varieties would contribute much in disease control. The resistant varieties 'Hisar Anmol' and 'Hisar Gaurav' are presently undergoing multi-locational trials to test the stability of resistance (Kalloo and Banerjee, 1993). Among various cvs. /lines EC-65968, EC-65980, EC-148897 and EC-148898 from *L. peruvianum* (Varma *et al.*, 1980) 49-B-295 from *L. esculentum* (Joshi and Choudhary, 1981) and LA 386, LA 177 from *L. hirsutum* (Muniyappa *et al.*, 1991) have been identified with resistance/tolerance to tomato leaf curl virus in India. Genotypes viz., H 88-78-1, H 88-78-2 and H 88-87 were found highly resistant under field conditions, however, under artificial inoculation conditions only H 88-78-1 showed highly resistant while H 88-78-2 and H 88-87 were moderately resistant (Singh *et al.*, 2010).

Twelve tomato genotypes, two resistant checks (56-14-7 and 58-11-1-1), two susceptible checks (Punjab Upma and Punjab Chhuhara) and eight advanced lines (11-2-2, 2-20-2, 6-2-3, 4-4-1-1, 7-4-5, 8-3-2, 8-5-2-1 and 1-2-3) were used for identification of resistance against tomato leaf curl virus (ToLCV) based on disease incidence, scanning electron microscopic and molecular studies. Results revealed that advanced lines which showed zero percentage of disease incidence were having smaller and lesser stomata, larger and higher number of trichomes. Further, the absence of virus in advanced lines through molecular studies proved that the advanced lines were resistant to ToLCV (Arora *et al.*, 2011).

(iii) Chemical control: Pre-sowing treatment of tomato nursery beds with Furadan or Disystox @ 1 kg a.i./ ha helps to check whitefly population (Singh and Sastry, 1976). Application of Cycocel (200-500 ppm) at seedling stage significantly reduced the incidence of disease (Banerjee *et al.*, 1991). Growing of tomato nurseries under protected conditions and spraying with Nuvacron after transplanting substantially increased the yields (Muniyappa and Saikia, 1983). Sprays of Krishi oil (2%) after fruit formation were effective in reducing the spread of the virus by preventing acquisition and inoculation of the virus by whiteflies (Singh *et al.*, 1973).

9. TOMATO MOSAIC

Tomato mosaic is one of the ubiquitous and dangerous diseases of tomato especially under glasshouse/polyhouse conditions. This is a common disease of this crop throughout the world and cause considerable yield losses. Although not usually a severe problem in greenhouse tomatoes because of the wide use of resistant cultivars, the disease is still a serious threat worldwide where resistant cultivars are not available and the cultivation of plants includes frequent handling (Jones *et al.*, 1991).

Geographical distribution

The association of this virus with the disease was first reported from Connecticut (Clinton, 1909). Since then it has been reported from different parts of the world. The virus is also reported to be prevalent in India (Kulkarni, 1924).

Symptoms

The characteristic symptoms vary from conspicuous yellowing, necrotic streaks to light and dark green mottling often accompanied by wilting depending upon the strains of the virus causing infection (Plate 9). At higher temperatures, severe mottling accompanied by slight stunting of plants occur whereas at lower temperatures plants are severely stunted accompanied by slight mottling and distorted leaves showing fern leaf (Fig. 8b) symptoms. Depending on the strain shoe string symptoms also appear (Fig. 8a). Necrotic sunken lesions usually occur on fruits. Internal browning of tomato fruits on infected plants is sometimes observed (Boyle and Wharton, 1957). This type of symptoms is also evident in mature but unripe fruits and usually involves the fleshy parenchyma whose cells often collapse (Boyle, 1994).

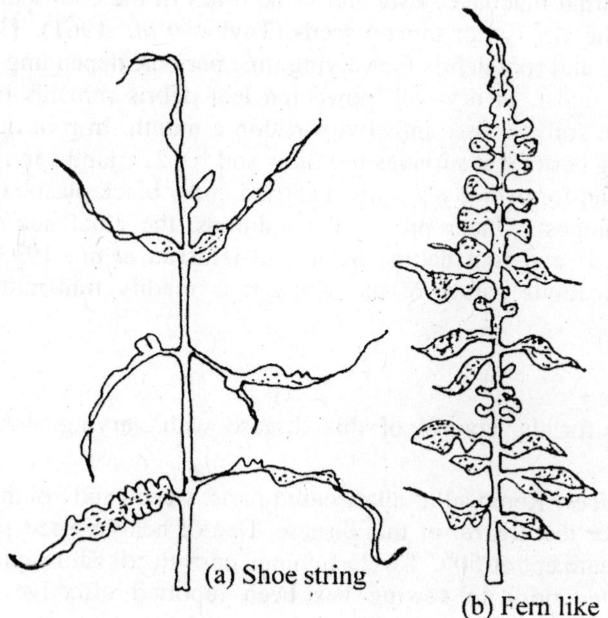

Fig. 8. Symptoms of Tomato mosiac

The Virus

Tomato Mosaic Virus (ToMV) is a member of *Tobamovirus* group. ToMV is an RNA-containing virus with straight rod particles about 300 x 18 nm. The amount of nucleic acid is 5 per cent and its molecular weight is 2 million Daltons. ToMV is distinctive in that it remains active in extracted host plant juice up to 50 years, longer than any other known virus. It is infectious in dilutions up to 1-10,00,000 and withstands heating up to 93°C for 10 minutes. It can be readily distinguished from tobacco mosaic virus (TMV) by differences in their host range, serological affinities and protein composition and therefore, both are now considered as distinct viruses (Gibbs, 1977).

Host range

The virus has a fairly wide host range, including many solanaceous species and most species tested in the families Aizoaceae, Amaranthaceae, Chenopodiaceae and Serophulariaceae.

Transmission

The virus is readily sap transmissible and is one of the most infectious and persistent viral agents known. ToMV is readily transferred from plant to plant by workers hands, tools and clothing and in rare cases by insects (Jones *et al.*, 1991). The virus is seed borne and upto 94 per cent of the seed contains virus.

Epidemiology

Infected seed and leaf and root debris are common sources of inoculum. The virus is present in the external mucilage, testa and some times in the endosperms of tomato seeds but not in the embryo of tomato seeds (Taylor *et al.*, 1961). The ToMV can also survive in leaf and root debris for varying time periods, depending upon whether the soil is dry or moist. In dry soil, powdered leaf debris remains infective for 2 years, but in moist soil it looses infectivity within a month. In root debris the virus can persist for long periods, it survives in fallow soil for 22 months in root debris at a depth of 120 cm and for over 2 years when carried under black plastic covered with a layer of clean compost. Under protected conditions, the chief source of primary inoculum was seeds and root debris in the soil (Hassan *et al.*, 1993). Due to the persistent and infectious nature of the virus, it is readily transmitted by human activities.

Management

Various methods for the control of this disease with varying success has been reported.

(i) Cultural practices: Removal of all diseased parts, weeds and volunteer plants has been advocated for the control of this disease. Use of healthy seed for sowing and their hot water treatment at 50°C for 25 minutes or with trisodium phosphate @ 90 g/l for 20 minutes prior to sowing has been reported effective (Marathe and Summanwar, 1995). Since the virus is easily carried from one crop to another on clothing, workers hands/ tools and chewing/smoking of tobacco. Care must be taken to ensure that workers wash their hands in 3 per cent solution of trisodium orthophosphate and then scrub well with soap and water. Steam sterilization of solids helps to inactivate ToMV especially in case of root infections.

(ii) Host resistance: Resistant cultivars having Tm-1, Tm-2 and Tm-22 genes conditioning resistance to ToMV should be used since they retain their resistance over long periods of time. Cherian and Muniyapa (1998) have reported resistance to tomato mosaic tobamovirus in *Lycopersicon peruvianum* accessions. The lines showing resistance under natural conditions to this disease have been found to be 12-80, 13-80, Kristi and Revermun Uluchshenny.

(iii) Cross protection: Inoculation of tomato seedlings with attenuated strains of the virus (Broadbent, 1976) by mixing 100 ml of diluted virus suspension with 1 g carborundum powder helps to protect against severe strains. This method has been successfully used on a large scale and has been referred to as vaccination (Jeyarajan, 1996).

REFERENCES

Ahmed, B.Y., Sharif, F.M. amd Sarhan, A.R.T. 1987. Effect of certain micronutrientson Fusarium wilt of tomato. *I. J. Agric. Water Resour. Res. Pl. Food* **6:** 13–28.

Ahn, E.J. and Cho, Y.S. 1996. Cloning of the bacteriocin gene from *Xanthomonas campestris* pv. *glycines. Korean J. Pl. Pathol.* **12:**169–175.

Alexandri, A.A., Eliade, E. and Nagle, R.M. 1981. Some aspects of the effect of some fungicides on tomato plants. *Analele Institut. de Cercetari pentru–Protectia–Planletor* **16** : 331–350.

Amemiya, Y., Yamaguchi, K., Hirano, K. and Iida, W. 1986. Suppression of Fusarium wilt of tomato by use of crop protection. *Tech. Bull. Fac. Hortic., Chiba Univ.* **37:** 79–83.

Amini, J. and Sidovich, D.F. 2010. The effects of fungicides on *Fusarium oxysporum* f.sp. *lycopersici* associated with Fusarium wilt of tomato, *J. Plant Prot. Res.* **50(2):** 172–178.

Anuratha, C.S. and Gnanamanickam, S.S. 1990. Biological control of bacterial wilt caused by *Pseudomonas solanacearum* in India with antagonistic bacteria. *Pl. Soil* **124:** 109–116.

Araki, K., Hidefumi, I. Iwasaki, S., Kanamori, T., Yasada, T. and Nonoyama 1985. *Bull. Veg. Ornam. Crops Res. Stn., Ano, Mie.* **13:** 93–108.

Archisi, M., Gennori, M., Matta, A. 1985. Retardations of *Fusarium* wilt symptoms in tomato by pre and post inoculum treatments of the roots and aerial part of host in hot water. *Physiol. Pl. Pathol.* **26(2):** 175–183.

Arora, N., Kaur, S. and Sharma A. 2011. Identification of tomato leaf curl virus (ToLCV) resistant genotypes based on disease incidence, scanning electron microscopic and molecular studies. *Pl. Dis. Res.* **26(1):** 76–81.

Arunakumara, K.T. and Kulkarni, M.S., Thammaiah, N. and Hegde, Yashoda 2010. Fungicidal management of early blight of tomato. *Indian Phytopath.* **63(1):** 96–97.

Ashlesha and Paul, Y.S. 2012. Evaluation of organic inputs against *Ralstonia solanacearum*–a bacterial wilt pathogen of Capsicum under protected cultivation. *Pl. Dis. Res.* **27(1):** 66–70.

Ashour, W.A. and El–Kadi, M.M. 1956. Damping off disease of tomato seeds and its control. *Ann. Agric. Sci. Cairo* **1:** 111–126.

Astua, M.G. 2000. Xv4–Vr XV4: a new gene for interaction identified between *Xanthomonas campestris* pv. *vesicatoria* race T 3 and the wild tomato relative *Lycopersicon pennellii. Mol. Pl. Microbe Interac.* **13(12):** 1346–1355.

Atkinson, R.G. and Adamson, R.M. 1977. Effects of fungicides on wilt and yield of green house tomatoes grown in *Fusarium* infected saw dust. *Can. J. Pl. Sci.* **57(3):** 675–680.

Baider, A. and Cohen, Y. 2003. Synergistic interaction between BABA and mancozeb in controlling *Phytophthora infestans* in potato and tomato and *Pseuudoperonospora cubensis* in cucumber. *Phytoparasitica* **31:** 399–409.

Baker, R.E.D. 1939. Notes on the diseases and fruit rots of tomatoes in the British West Indies. *Trop. Agric. Trindad* **16(1):** 252–257.

Banerjee, M.K., Arora, S.K. and Kalloo, G. 1991. Effect of Cycocel on the development of leaf curl disease in tomato. *Crop Res.* **4:** 218–221.

Banerjee, M.K., Kalloo, G. and Jalali, I. 1990. Studies on Fusarium wilt resistance in *Lycopersicon* screening and method of assessment. *Veg. Sci.* **17(2):** 167–174.

Barnwal, M.K., Maiti, D. and Pandey, A.C. 2011. Management of tomato wilt caused by *Fusarium oxysporum* f.sp. *lycopersici* by using bioagents and organic amendments. *Indian Phytopath.* **64(2):** 194–196.

Bashan, Y. and Okon, Y. 1986. Internal and external infection of fruits and seeds of peppers by *Xanthomonas campestris* pv. *vesicatoria. Can. J. Bot.* **64:** 2865–2871.

Basymbekov, M.B. and Firsov, V.F. 1978. Trials with some new niprations against bacterial spot of tomato. *Vestn. S. Kh. Nauk. Kazakhstana.* **21:** 38–41.

Becktell, M.C., Daughtrey, M.L. and Fry, W.E. 2005. Epidemiology and management of petunia and tomato late blight in the greenhouse, *Plant Dis.* **89:** 1000–1008.

Bedekar, H.S. 1977. Reaction of four tomato cultivars of different mixtures of isolates of *Pseudomonas solanacearum. Veg. Sci.* **4:** 1–5.

Bedi, J.S., Sokhi, S.S., Munshi, G.D. and Mohan, C. 1990. Identification of resistance to some diseases in tomato lines. *Pl. Dis. Res.* **5(2) :** 188–190.

Beleva, L. and Sotirova, V. 1977. Reaction of wild species, subspecies and cultivars of tomato to strains of *Xanthomonas vesicatoria* (Doidge) Dowson. *Gradinarska i Lozarska Nauka.* **14:** 84–93.

Bharadwaj, C.L. 1991. Efficacy and economics of fungicides spray schedules for the management of foliar and fruit diseases in tomato. *Indian J. Mycol. Pl. Pathol.* **21(1) :** 56–59.

Bharat, N.K. and Gupta, S.K. 2011. Management of tomato buckeye rot and Alternaria leaf spot through strobilurin fungicide. *Pl. Dis. Res.* **26 (1):** 97–98.

Bhardwaj, C.L. and Masand, S.S. 1995. Effect of mulching and irrigation schedules on buckeye rot (*Phytophthora nicotianae* var. *parasitica*) of tomato (*Lycopersicon esculentum*). *Indian J. Agric. Sci.* **65(3):** 223–225.

Bhattacharyya , S.K. and Singh, B.P. 1990. Symptomatology and host range. In: *Late Blight of Potato* (Eds. Bhattacharyya, S.K., Shekhawat, G.S. and Singh, B.P.). *Tech. Bull. No.* 27, CPRI, Shimla, pp. 4–5.

Biezen, E.A., Van der Glagostskaya, T., Overduin, B., Nijkamp, H.J.J., Hille, J. and Vander, B.E.A. 1995. Inheritance and genetic mapping of resistance to *Alternaria alternata* f. sp. *lycopersici* in *Lycopersicon pennellii. Mol. Gen. Genet.* **247(4):** 453–461.

Bisht, I.S., Umesh Chandra, Thomas, T.A. and Chandra, I. 1989. Field reactions of wild tomato species to various diseases. *Veg. Sci.* **16(1):** 104–106.

Bora, L.C., Gogoi, P.K. and Das, B.G. 1996. Bacterial wilt of tomato in relation to few environmental parameters in Assam. *J. Agric. Sci. Soc. Northeast India* **9(2):** 185–186.

Borkar, S.G. 1995. Differential sensitivity of tomato tissues to infection by *Xanthomonas campestris* pv. *vesicatoria. Indian J. Mycol. Pl. Pathol.* **25:** 208–210.

Boyle J.S. and Wharton, D.C. 1957. The experimental reproduction of tomato internal browning by inoculation with strains of tobacco mosaic virus. *Phytopathology* **47:** 199–207.

Boyle, J.S. 1994. Abnormal ripening of tomato fruit. *Plant Dis.* **78:** 936–944.

Broadbent, L. 1976. Epidemiology and control of tomato mosaic virus. *Annu. Rev. Phytopathol.* **14:** 75–96.

Butler, E.J. 1903. Potato diseases of India. *Agril. Ledg. Crop Dis. Pest. Ser.* **7** : 87–124.

Butler, E.J. 1918. *Fungi and Diseases in Plants.* Thacker Spink & Co. Calcutta, India. 300 pp.

Campbell, L., Moss, B., Byrne, J. Dianese, A. and Wilson, M. 1998. The good, the bad and the ugly: biocontrol of bacterial spot of tomato. *Highlights Agric. Res. Albama Agric. Exp. Sta.* **45:** 8–10.

Casida, L.E. Jr. and Lukezic, F.L. 1992. Control of leaf spot diseases of alfalfa and tomato with applications of the bacterial predator *Pseudomonas* strain 679–2. *Plant Dis.* **76(12):** 1217–1220.

Chand, G. Kumar, R., Kumar, A., Kumar, S. and Srivastava, J.N. 2014. Field evaluation of different fungicides against Alternaria leaf blight of tomato in eastern Bihar. *Pl. Dis. Res.* **29(2):** 213–214.

Chand, R., Singh, R. anu Singh, P.K. 1994. Distribution of pathogenic groups and races in *Xanthomonas campestris* pv. *vesicatoria* in the peninsular India. *Indian Phytopath.* **47:** 251–252.

Chatchawankanphanich, O., Chiang, B.T., Green, S.K., Singh, S.J. and Maxwell, D.P. 1993. Nucleotide sequence of a geminivirus associated with tomato leaf curl from India. *Plant Dis.* **77:** 1168.

Chatterjee, R. and Khalko, S. 2013. Influence of organic amendments and inorganic fertilizers on late blight incidence and yield of tomato (*Lycopersicon esculentum* Mill.), *Int. J. Geol. Agric. Environ. Sci.* **1**:36–38.

Chauhan, M.S. 1987. Reaction of tomato varieties to leaf spots and fruit rots in Haryana. *Indian J. Mycol. Pl. Pathol.* **17(3):** 355.

Chauhan, M.S. 1988. Reaction of varieties/lines of tomato to wilt. *Indian J. Mycol Pl. Pathol.* **18(1):** 72–73.

Chebil, A., Margari, A. and Verlodt, H. 1980. Preliminary results with early tomato cultivation in plastic bags with a substrate of sea weed *Posscidonia oceanica. 32nd Int. Symp. Phytopharm. Phytiatry Part II.* **45:** 775–781.

Chellemi, D.O., Olson, S.M. and Michell, D.J. 1994. Effects of soil solarization and fumigation on survival of soil borne pathogens of tomato in Northern Florida. *Plant Dis.* **78:** 1167–1172.

Chenulu, V.V. and Giri, B.K. 1985. Viral and mycoplasmal diseases of tomato and their management. In: *Perspectives in Plant Virology* (Eds. Gupta, B.M., Singh, B.P., Verma, H.N. and Srivastava, K.M.), Lucknow, India. 322pp.

Cherian, S. and Muniyapa, V. 1998. Screening of tomato hybrids and *Lycopersicon* species to tomato mosaic tobamovirus. *Pl. Dis. Res.* **13(1):** 8–13.

Chhabra, M.L., Garg, A.P., Banerjee, M.K. and Gandhi, S.K. 2000. Influence of Alternaria blight on vitamin C content of tomato plants. *Pl. Dis. Res.* **15(2):** 223–224.

Chona, B.L., Lall, G. and Kakaria, N.C. 1958. *The fungi of Delhi.* ICAR Publ., New Delhi, pp.42.

Chou, C.C., Hou, S.T. and Tzeng, K.C. 1995. Suppression of the bacterial wilt pathogen, *Pseudomonas solanacearum* in soil by S–H amendment and relation of its inorganic nitrogen products to the suppressive effects. *Pl. Pathol. Bull.* **4:** 163–172.

Chowdappa, P., Nirmal Kumar, B.J, Madhura, S., Mohan Kumar, S.P., Kevin Meyers, Fry, W.E. and Cooke, D.E.L. 2013. Emergence of 13_A2 blue lineage of *Phytophthora infestans* was responsible for severe outbreaks of late blight on tomato in India. *J. Phytopathol.* **168:** 48–59.

Clinton, G.P. 1909. Notes on fungal diseases etc. for 1908. *Conn. State Exp. Stn. Rep.* 1907–1908: pp. 857–858.

Cohen, S. and Nitzany, F.E. 1966. Transmission and host range of the tomato yellow leaf curl virus. *Phytopathology* **56:** 1127–1131.

Conover, R.A. and Gerhold, N.R. 1981. Mixtures of copper and maneb or mancozeb for control of bacterial spot of tomato and their capability for control of fungus diseases. *Proc. Fl. State Hortic. Soc.* **94:** 154–156.

Czosnek, H., Navot, N., Zamir, D. and Laterrot, H. 1991. Diagnosis of tomato yellow leaf curl virus with cloned DNA probes: A geographical assessmentof the disease. P. 40–43 In: *Resistance of the tomato to TYLCV. Proc. Sem. EEC contract DGXII–TS2–A–055 F(CD) partners.* (Eds. Laterrot, H. and Trousse, C.) INRA–Station d' Amelioration des Plantes Maraicheres, Montfavet–Avignon, France.

Czyzewska, S. and Bialkowska, H. 1973. Observations on the soundness of fruits in a collection of field tomatoes. *Biul. Warzywniczy* **15 :** 223–248.

Datar, V.V. and Mayee, C.D. 1981. Assessment of losses in tomato yield due to early blight. *Indian Phytopath* **34(2):** 191–195.

Datar, V.V. and Mayee, C.D. 1981a. Epidemiology of early blight of tomato foliage to early blight caused by *Alternaria solani. Indian Phytopath.* **33:** 83–86.

Datar, V.V. and Mayee, C.D. 1982. Conidial dispersal of *Alternaria solani* in tomato. *Indian Phytopath* **35:** 68–70.

Debnath, S. and Nath, P.S. 2009. Control of late blight (*Phytophthora infestans*) disease of tomato in plains of West Bengal. *J. Crop and Weed* **5(2):** 122–123.

Deora, A., Randhawa, H.S. and Sharma, R.C. 2004. Incidence of Alternaria leaf blight in tomato and efficacy of commercial fungicides for its control. *Ann. Biol.* **20:** 211–218.

Devi, A.K.B. and Tiwari, R.N. 2001. Inheritance of resistance to fruit rot, *Phytophthora nicotianae* var. *parasitica* Dast. in tomato *Lycopersicon esculentum* Mill. *Orissa J. Hort.* **29(1)**: 91-94.

Devi, L.R. and Menon, M.R. 1980. Seasonal incidence of bacterial wilt of tomato. *Indian J. Microbiol.* **20(1):** 13–15.

Dodan, D.S. and Shyam, K.R. 1994. Pathogenic variation within *Phytophthora nicotianae* var. *parasitica* isolates from tomato. *Indian J. Mycol. Pl. Pathol.* **24(3):** 196–201.

Dodan, D.S., Shyam, K.R. and Bhardwaj, S.S. 1995. Relative response of tomato cultivars/ lines against buckeye rot under field conditions. *Pl. Dis. Res.* **10(2):** 135–136.

Dodan, D.S., Shyam, K.R. and Gupta, S.K. 1994. Integrated management of buckeye rot of tomato. *Pl. Dis. Res.* **9(2):** 141–144.

Dogo, M., Toyoda, H., Tampo, H., Matsuda, Y., Nonomura, T. and Ouchi, S. 1999. Antimicrobial activities of indole derivatives against soil–borne pathogens of tomato. *Bull. Inst. Comp. Agril. Sci., Kinki Univ.* **7:** 91–95.

Doidge, E.M. 1921. A tomato canker. *Ann. App. Biol.* **7:** 407–430.

Douglas, B. 1922. A new *Alternaria* spot of tomatoes in California. *Phytopathology* **12:** 146–148.

Dutt, B.L. 1979. *Bacterial and fungal diseases of potato*, ICAR, New Delhi, 196pp.

Dwivedi, S.K. 1991. Studies on population dynamics of *Fusarium oxysporum* f.sp. *lycopersici* in solar heated soil. *Natl. Acad. Sci. Newsl.* **14(6):** 235–237.

El Shami, L., Salem, D.E., Fadl, F.A., Ashour, W.E. and El Zayat, M.M. 1990. Soil solarization and plant disease management. II. Effect of soil solarization in comparison with soil fumigation on the management of Fusarium wilt of tomato. *Agric. Res. Rev.* **68(3):** 601–611.

El–Shanshoury, A.R., Abu–El–Sououd, S.M., Awadal, A. and El–Bandy, N.B. 1996. Effects of *Streptomyces corchorusie, Streptomyces* mutabilis, pendimethalin and metribuzine on the control of bacterial and *Fusarium* wilt of tomato. *Can. J. Bot.* **74(7):** 1016–1022.

Erinle, I.D. and Quinn, J.G. 1980. An epiphytotic of late blight of tomatoes in Nigeria. *Plant Dis.* **64 (7):** 701–702.

Etebarian, H.R. 1992. Studies on Fusarium wilt of tomato and its chemical control in Varamin area. *Indian J. Agric. Sci.* **23:** 1–14.

Fageria, M.S., Dhaka, R.S. and Jat, R.G. 1998. Field evaluation of tomato genotypes for resistance to buckeye rot and *Alternaria* blight. *J. Mycol. Pl. Pathol.* **28(1):** 52–53.

Felix, E.L. 1948. Buckeye rot resistant tomatoes. *Phytopathology* **38(7)**: 569.

Fernandez, A., Solorzano, E., Peteira, B. and Leon, O. 1998. Pathogenesis related proteins in tomato plants infected with *Alternaria solani*. *Rev.de Prot. Veg.* **13(3):** 157–163.

Fernandez, M.C. 1986. Some hosts of *Pseudomonas solanacearum* in Chile. *Agric. Tecnica* **46(1):** 101–105.

Follas, G.B., Beetz, H.M. and Popay, A.J. 1992. Control of leaf spots in carrots and early blight in potatoes and tomatoes with difenconazole. *Proc. 45th N Z Pl. Prot. Conf., Wellington, New Zealand,* Aug. 11–13, 1992, pp. 46–49.

Fontem, D.A., Gumedzoe, M.Y. and Nono–Wondim, R. 1999. Biological constraints in tomato production in the Western highlands of Cameroon. *Tropicultura* **16–17**: 89–92.

Foolad, M.R., Ntahimpera, N., Christ, B.J. and Lin, G.Y. 2000. Comparison of field, greenhouse and detached leaflet evaluations of tomato germplasm for early blight resistance. *Plant Dis*. **84(9):** 967–972.

Fuchs, J. and Defago, G. 1991. Protection of tomatoes against *Fusarium oxysporum* f. sp. *lycopersici* by combining non–pathogenic Fusarium with different bacteria on untreated soil. *Bull. SROP* **14(8):** 51–56.

Gadewar, A.V. and Shekhawat, G.S. 2000. Bacterial wilt and brown rot. In: *Diseases and Pests of Potato–a Manual* (Ed. S.M. Paul Khurana), CPRI, Shimla pp. 22–26.

Gamliel, A. and Katan, J. 1991. Involvement of fluorescent pseudomonads and other microorganisms in increased growth response of plants in solarized soils. *Phytopathology* **81:** 494–502.

Gardner, R.G. and Panthee, D.R. 2010. NC 1 CELBR and NC 2 CELBR: early blight and late blight–resistant fresh market tomato breeding lines, *HortSci*. **45(6):** 975–976.

Gibbs, A.J. 1977. Tobamovirus group. In: *Description of Plant Viruses No*. 184. CMI/ AAB, Kew, Surrey, England.

Gram, E., Jogerson, C.A. and Rostrup, S. 1927. Oversight over Sydomme hos. Landbrugets og havebrugets Kultur planter i. 1927.. *Tidsskr. Planteaue* **35:** 778–836.

Grogan, R.G., Kimble, K.A. and Misaghi, I.J. 1975. A stem canker disease of tomato caused by *Alternaria alternata* f. sp. *lycopersici*. *Phytopathology* **65 :** 880–886.

Gunawan, O.S. 1989. Control of bacterial wilt (*Pseudomonas solanacearum* E.F. Smith) with Agrimycin 15/1.5 WP on tomato in Dago, Bandung. *Bull. Penelitian Hortic*. **17(3):** 41–44.

Gunther, E., Eggert, D. and Grummer, G. 1970. Resistance of tomatoes to late blight (*Phytophthora infestans*). *Ann. App. Biol.* **65 (2):** 255–262.

Gupta, S.C. 1960. *Septoria Lycopersici* Speg.– aserious leaf spot disease of tomato in Nainital area. *Proc. Natl. Acad. Sci. India* B, **30(1):** 98–100.

Gupta, S.K. and Bharat, N.K. 2008 . Management of buckeye rot and late blight of tomato through combi fungicides. *Pestology* **32:** 17–19.

Gupta, S.K. and Shyam, K.R. 1998. Role of non–foliage application of fungicides in the management of buckeye rot of tomato. *J. Mycol. Pl. Pathol* **28(1):** 67 (Abstr.).

Gupta, S.K., Dohroo, N.P. and Shyam, K.R. 1998. Occurrence of bacterial wilt of tomato in Himachal Pradesh. *Pl. Dis. Res*. **13(2):** 174.

Gupta, S.K., Upmanyu, Sachin and Sharma, R.C. 2005. Biology, epidemiology and management of buckeyr rot of tomato. In *Challenging Problems in Horticultural and Forest Pathology*. (Eds. R.C. Sharma and J.N. Sharma). Indus Publishing Co., New Delhi, pp. 183–199.

Hanudin and Machmud, M. 1994. Prospects for the use of soil amendments for bacterial wilt control on tomato. *ACIAR Bacterial Wilt Newsl.* **10:** 12–13.

Hartman, G.L., Hong, W.F. and Wang, T.C. 1991. Survey of bacterial wilt on fresh market hybrid tomatoes in Taiwan. *Pl. Prot. Bull., Taiwan* **33(2):** 197–203.

Hartman, G.L., Hong, W.F., Hanudin and Hayward, A.C. 1993. Potential of biological and chemical control of bacterial wilt. In: *Bacterial Wilt* (Eds. Hartman, G.L. and Hayward, A.C.). *Proc. Int. Conf. held at Kaohsiung Taiwan, 28–31 Oct. 1992. ACIAR Proc.* **45:** 322–326.

Hassan, S., Arif, M. and Defoer, T. 1993. Epidemiological studies of tomato viruses in Malakand Agency of North West Frontier province of Pakistan. *Sarhad J. Agric.* **9:** 37–43.

Hayward, A.C. 1991. Biology and epidemiology of bacterial wilt caused by *Pseudomonas solanacearum. Annu. Rev. Phytopathol.* **29:** 65–87.

Hector, G.P. 1926. Appendix II. Annual Report of the economic Botanist to the Govt. of Bengal for the year 1924–25. *Annu. Rept. Dept. Agric. Bengal, 1924–25*, pp. 5–9.

Herger, G., Klingauf, F., Mangold, D., Pommer, E.H. and Scherer, M. 1988. Efficacy of extracts of *Raynoutria sachalinensis* (F. Schmidt) Nakai against fungal diseases, especially powdery mildews. *Nachrichtenbl. Disch Pflanzenschutz–dienst.* **40:** 56–60.

Hernandez, A.M. and Medina, E. 1991. Chemical control of *Xanthomonas campestris* pv. *vesicatoria* in tomato seeds. *Prot. de Plantas.* **1:** 63–72.

Ho, B.L. 1988. Bacterial wilt control in tomato using household disinfectants. *MARDI Res. J.* **16:** 73–76.

Hutagalung, L. 1988. Garlic bulb as a material for suppressing the incidence of bacterial wilt (*Pseudomonas solanacearum* E.F. Smith) on tomatoes. *Bull. Penelitian Hortic.* **16:** 84–93.

Ioannou, N. 1985. Yield losses and resistance to tomato strains of tomato yellow leaf curl and tobacco mosaic viruses. pp.11. *Agric. Res. Inst. (Cyprus) Tech. Bull.* 66, Nicosia, Cyprus.

Jain, S.S., Sharma, S.L. and Juneja, S.L. 1961. Studies on buckeye rot of tomato – new to Himachal Pradesh. *Proc. Indian Sci. Cong. Assoc.* **50:** 351–352.

Jatala, P., French, E.R. and Gutassa, L. 1975. Inter–relationships of *Meliodogyne incognita acrita* and *Pseudomonas solanacearum* on potatoes. *J. Nematol.* **7:** 325.

Jeyarajan, R. 1996. Virus disease management by cross protection. *Indian J. Mycol Pl. Pathol.* **26(2):** 147–153.

Ji, P., Campbell, H.L., Kloepper, J.W., Jones, J.B., Suslow, T.V. and Wilson, M. 2006. Integrated biological control of bacterial speck and spot of tomato under field conditions using foliar biological control agents and plant growth–promoting rhizobacteria, *Biol. Contr.* **36(3):** 358–367.

Jindal, K.K. 1992. The versatile bacteria: how they show up on vegetable crops, the remedies thereof. *Indian Hortic.* **36:** 18–21.

Jones, J.B. 1991. Bacterial spot. In: *Compendium of tomato diseases.* (Eds. Jones, J.B.; Stall, R.E.; Zitter, T.A.) APS Press, St. Paul, Minnesota pp. 27.

Jones, J.B. and Jones, J.P. 1985. The effect of bactericides, tank mixing time and spray schedule on bacterial leaf spot of tomato. *Proc. Fl. State Hortic. Soc.* **98:** 244–247.

Jones, J.B. and Jones, J.P. 1987. The effect of bactericides, tank mixing time and spray schedule on bacterial leaf spot of tomato. *Citrus Veg. Mag.* **50:** 34–35.

Jones, J.B., Jones J.P., Stall, R.E. and Zitter, T.A. 1991. *Compedium of tomato diseases*. APS Press, St. Paul, Minnesota, U.S.A. 73 p.

Jones, J.B., Pohronezny, K.L., Stall, R.E. and Jones, J.P. 1986. Survival of *Xanthomonas campestris* pv. *vesicatoria* in Florida on tomato crop residue, weeds, seeds and volunteer tomato plants. *Phytopathology*. **76:** 430–434.

Joshi, A., Gupta, S.K. and Kohli, U.K. 2004. Resistance to Septoria leaf–spot (*Septoria lycopersici*) and buckeye fruit–rot (*Phytophthora nicotianae* var *nicotianae*) of tomato (*Lycopersicon esculentum*). *Indian J. Agric. Sci.* **74:** 194–198.

Joshi, G.C. and Choudhary, B. 1981. Screening of *Lycopersicon* and *Solanum* species for resistance to leaf curl virus. *Veg. Sci.* **8:** 45–50.

Kalha, C.S. and Sood, A.K. 1994. Occurrence of biovar III of *Pseudomonas solanacearum* on tomato in Himachal Pradesh. *ACIAR Bacterial Wilt Newsl.* **10:** 10.

Kalloo, G. 1988. Disease resistance in vegetable crops. *Vegetable Breeding*. Vol. 2, CRC Press Inc. Florida, USA. Pp. 1–94.

Kalloo, G. and Banerjee, M.K. 1993. Tomato leaf curl virus resistant varieties: Hisar Anmol (H–24) and Hisar Gaurav (H–36). *Hortic. Sci.* (in press).

Kapoor, I.J. 1988. Fungi involved in tomato wilt syndrome in Delhi, Maharashtra and TamilNadu. *Indian Phytopath.* **41:** 208–213.

Kasrawi, M.A.M., Sarwan, A. and Mansour, A. 1988. Sources of resistance to tomato yellow leaf curl virus (TYLCV) in *Lycopersicon* species. *Euphytica* **37:** 61–64.

Kaur, S., Singh, S., Kanwar, J.S. and Cheema, D.S. 1988. Variability in tomato for resistance to late blight under field conditions. *Indian Phytopath.* **41:** 486–487.

Kelman, A. 1953. The bacterial wilt caused by *Pseudomonas solanacearum*. *North Carolina Agric. Exp. Stn. Tech. Bull.* **99:** 194pp.

Kendrick, J.B. 1923. Phytophthora rot of tomato, eggplant and pepper. *Proc. Indian Acad. Sci.* **38:** 299–306.

Keshwal, R.L., Khare, U.K. and Singh, R.P. 2000. Effect of physical properties of soil on wilt incidence and population of *Ralstonia solanacearum*. *Ann. Pl. Prot. Sci.* **8(1):** 40–43.

Khakimov, A.Kh. and Abdullaev, B.Ya. 1992. *Trichoderma* against fusariose of tomato. *Zaschita Rastenii Moskva* **8:** 25.

Khan, A.N.A., Shetty, K.S. and Patil, R.B. 1980. Occurrence of bacterial wilt of chilli in Karnataka and its relationship to the wilts of other solanaceous vegetables. *Indian Phytopath.* **32:** 507–512.

Khulbe, R.D. and Sati, M.C. 1987. Seed transmission of *Alternaria solani* in tomato. *Indian Phytopath.* **40:** 106–107.

Kishun, R. 1987. Loss of yield of tomato due to bacterial wilt caused by *Pseudomonas solanacearum. Indian Phytopath.* **40:** 152–155.

Kishy, K. 1962. Studies on physiological specialization of *Phytophthora infestans* on tomatoes. I. On the tomato–intermediate and potato type of *Phytophthora infestans* and on the tomato races of *Phytophthora infestans*. *Ann. Phytopathol. Soc. Jpn.* **27 (4):** 172–179, 180–188.

Kritzman, G. 1993. A Chemi-thermal treatment for control of seed-borne bacterial pathogens of tomato. *Phytoparasitica*. **21:** 101–109.

Kronn, B.A.M., Scheffer, R.J. and Elgersma, D.M. 1991. Induced resistance in tomato plants against Fusarium wilt induced by *Fusarium oxysporum* f. sp. *dianthi*. *Neth. J. Plant Path.* **97(6):** 401–408.

Kulkarni, G.S. 1924. Mosaic and other related diseases of crops in the Bombay Presidency. *Poona Agric. College Mag.* **16:** 6–12.

Kumar, K., Mahajan, R., Chib, H.S. and Kalha, C.S. 2004. An epidemic of tomato leaf curl disease in Jammu. *Indian Phytopath.* **57(2):** 181–184.

Kumar, P. and Sood, A.K. 2001. Integration of antagonistic rhizobacteria and soil solarization for the management of bacterial wilt of tomato caused by *Ralstonia solanacearum. Indian Phytopath.* **54(1):** 12–15.

Kumar, P. and Sood, A.K. 2005. An ecofriendly approach for the management of bacterial wilt of tomato. *Pl. Dis. Res.* **20:** 55–57.

Kumar, S. and Sugha, S.K. 1998. Fungitoxic management of Septoria leaf spot of tomato. *Pl. Dis. Res.* **13(1):** 87–88.

Kumar, S. and Sugha, S.K. 1999. Overseasoning of *Septoria lycopersici* an incitant of leaf spot disease in tomato. *J. Mycol. Pl. Pathol.* **29(1):** 83–86.

Kumar, S. and Sugha, S.K. 2000. Role of cultural practices in the management of Septoria leaf spot of tomato. *Indian Phytopath.* **53:** 105–106.

Kumar, V., Singh, B.M. and Sugha, S.K. 1997. Reaction of tomato genotypes to bacterial wilt and Phytophthora fruit rot. *Pl. Dis. Res.* **12(1):** 90–94.

Liu, C.H. and Wu, W.S. 1997. Chemical and biological control of tomato early blight. *Pl. Pathol. Bull.* **6(3):** 132–140.

Loest, 1948.

Malathrakis, N.E. 1983. Alternaria stem canker of tomato in Greece. *Phytopathol. Mediterr.* **22 :** 33–38.

Malhotra, S.K. and Vashistha, R.N. 1993. Genetics of resistance to Fusarium wilt race 1 in currant tomato (*Lycopersicon pimpinellifolium*). *Indian J. Agric. Sci.* **63(4):** 246–247.

Mangamma, P. and Sreeramulu, A. 1991. Garlic extract inhibitory to growth of *Xanthomonas campestris* pv. *vesicatoria*. *Indian Phytopath.* **44:** 372–374.

Marathe, T.S. and Arvind Summanwar, S. 1995. Elimination of seed borne tobacco mosaic virus from hybrid tomato seeds by trisodium phosphate seed treatment. *Indian J. Mycol. Pl. Pathol.* **25:** 99.

Maringoni, A.C., Kurozawa, C., Barbosa, V. and da Silvaneto, J.M. 1986. Chemical control of bacterial spot (*Xanthomonas campestris* pv. *vesicatoria* (Doidge) Dye) of tomato (*Lycopersicon esculentum* Mill.) *Summa Phytopathol.* **12:** 92–101.

Mayee, C.D., Kanwar, J.S. and Nandpuri, K.S. 1974. The comparative performance of different genotypes of tomato vis–à–vis leaf curl and mosaic. *Punjab Agric. Univ. J. Res.* **11:** 362–364.

Maziero, J.M.N., Maffia, L.A. and Mizubuti, E.S.G. 2009. Effects of temperature on events in the infection cycle of two clonal lineages of *Phytophthora infestans* causing late blight on tomato and potato in Brazil. *Plant Dis.* **93(5):** 459–466.

Mazumdar, N. 1998. Managing *Ralstonia solanacearum* wilt of tomato. *J. Mycol. Pl. Pathol.* **28(2):** 189–192.

McCarter, S.M. 1991. Bacterial wilt. In: *Compendium of tomato diseases*. (Eds. Jones, J.B., Jones, J.P., Stall, R.E. and Zitter, T.A.). APS Press, St. Paul, Minnesota, pp. 28–29.

McGuire, R.G., Jones, J.B., Stanley, C.D. and Csizinszky, A.A. 1991. Epiphytic populations of *Xanthomonas campestris* pv. *vesicatoria* and bacterial spot of tomato as influenced by nitrogen and potassium fertilization. *Phytopathology* **81:** 656–660.

Mehra, R. 2006. Biocontrol of Fusarium wilt of tomato using antagonistic microorganisms. *Pl. Dis. Res.* **21 (2):** 146–147.

Mehra, R. 2008. Biocontrol of Fusarium wilt of tomato caused by *Fusarium oxysporum* f.sp. *lycopersici. Pl. Dis. Res.* **23 (2):** 51–54.

Mew, T.W. and Ho, W.C. 1977. Effect of soil temperature on resistance of tomato cultivars to bacterial wilt. *Phytopathology* **67:** 900–911.

Mills, W.R. 1940. *Phytophthora infestans* on tomato. *Phytopathology* **30 :** 830–839.

Mishra, N.C. and Mishra, S.N. 1993. Perfomance of tomato varieties against wilt (*Fusarium oxysporum* f.sp. *lycopersici*) and fruit borer (*Helicoverpa armigera*) in the North Eastern Ghat Zone of Orissa. *Indian J. Pl. Prot.* **21(2):** 183–186.

Mishra K.K., Kumar, A. and Pandey, K.K. 2010. RAPD based genetic diversity among different isolates Fusarium oxysporum f.sp. lycopersici and their comparative biocontrol. *World J. Microbiol. Biotechnol.* **26(6):** 1079–1085.

Mofett, M.L. 1986. A note on the bacterial wilt resistance breeding program in Queensland. *ACIAR Bacterial Wilt Newsl.* **1:** 1–2.

Moffett, M.L. and Hayward, A.C. 1980. The role of weed species in survival of *Pseudomonas solanacearum* in tomato cropping land. *Aust. Pl. Pathol.* **9(2):** 6–8.

Mohan, C. and Thind, T.S. 2001. Relative performance of spray schedules of systemic and non–systemic fungicides against late blight of tomato. *Pl. Dis. Res.* **16(2):** 247–250.

Mohan, C., Singh, N. and Sharma, V.K. 2009. Effect of plant age, leaf wetness and date of transplanting on early blight of tomato. *Pl. Dis. Res.* **24(1):** 65–68.

Mondal, B., Chattopadhyay, S.B., Bhattacharya, I. and Khatua, D.C. 2005. Efficacy of different antibiotics and anti–bacterial compounds against bacterial wilt disease of tomato in West Bengal. *Environ. Ecol.* **23(Spl–1):** 73–77.

Moretto, K.C.K., Nakamura, K. and Barreto, M. 1993. Resistance of some tomato cultivar to Septoria leaf spot (*Septoria lycopersici*) as evaluated by some monocyclic parameters. *Summa Phytopathol.* **19:** 161–164.

Muhammad, M. 1987. New hosts for *Pseudomonas solanacearum. ACIAR Bacterial Wilt Newsl.* **2:** 8.

Muniyappa, V. and Saikia, A.K. 1983. Prevention of the spread of tomato leaf curl disease. *Indian Phytopath.* **36:** 183 (Abstr.).

Muniyappa, V. Swanson, M.M., Duncan G.H. and Harrison, B.D. 1991. Partial purification, properties and epitope variability of Indian tomato leaf curl geminivirus. *Ann. App. Biol.* **118:** 595–604.

Muniyappa, V., Jalikop, S.H., Saikia, A.K., Chennarayappa, S.G., Ishvara, B.A. and Ramappa, H.K. 1991. Reaction of *Lycopersicon* cultivars and wild accessions to tomato leaf curl virus. *Euphytica* **56:** 37–41.

Munoz, C.C.A. and Granada, G.A. 1990. The study of hosts of the causal agent of tomato bacterial spot, *Xanthomonas campestris* pv. *vesicatoria. Fitopatol. Colombia* **14:** 52–55.

Musmade, N., Tini, P. and Thakur, K.S. 2009.Biological and chemical management of tomato wilt caused by *Fusarium oxysporum* f.sp. *lycopersici. J. Soils Crops* **19(1):** 118–121.

Nakamura, K. 1990. Influence of temperature on the infection of tomato plants by *Xanthomonas campestris* pv. *vesicatoria. Summa Phytopathol.* **16:** 106–110.

Nirwato, H., Djajati Horjodi, S.S., Tjtrosomo, S., Horjadiw Widodo, W.D., Sudorsono. 1994. Study on the use of soil amendment with leguminous foliage and liming in effort to control *Fusarium* wilt disease of tomato. *Acta Hortic.* **369:** 155–162.

Nitzany, P.E., Geisciberg, H. and Koch, B. 1964. Tests for the protection of cucumber from a whitefly–borne virus. *Phytopathology* **54:** 1059–1061.

Obrero, F.C. and Aragaki, M. 1965. Some factors influencing infecton and disease development by *Phytophthora parasitica* Dast. on tomato. *Pl. Dis. Reptr.* **49(4):** 327–331.

Ojiewo, C.O., Swai, I.S., Oluoch, M.O., Silue, D., NonoWomdim, R., Hanson, P., Black, L. and Wang, T.C. 2010. Development and release of late blight–resistant tomato varieties 'Meru' and 'Kiboko'. *Int. J. Veg. Sci.* **16(2):** 134–147.

Overman, A.J. and Jones, J.P. 1975. Effect of Furadan on yield and expression of Fusarium wilt in tomato planted as bare root and containerised transplants. *Proc. Fl. State Hortic. Soc. 1975.* **88:** 214–217.

Pandey, P.K. and Pandey, K.K. 2002. Field screening of different tomato germplasm lines against *Septoria, Alternaria* and bacterial disease complex at seedling stage. *J. Mycol. Pl. Pathol.* **32:** 234–235.

Patel, M.K., Kulkarni, Y.S. and Dhande, G.W. 1950. Bacterial leaf spot of chillies. *Indian Phytopath.* **3:** 95–97.

Patil, B.P., Rajput, J.C., Palve, S.B., Jambhale, A.S. and Diwaker, M.P. 1990. New tomato 'Sonali' high yieldingand wilt resistant. *Indian Hortic*. **35(3):** 6–9.

Perez, M.C., Torres, F., Molina, F. and Verona, G. 1986. Resisistence to buckeye rot (*Phytophthora parasitica* Dastur) in varieties of tomato (*Lycopersicon exculentum* Mill.). *Centro Agricola* **13(3):** 3–16.

Pernezny, K., Datnoff, L.E., Muller, T. and Collins, J. 1996. Losses in fresh market tomato production in Florida due to target spot and bacterial spot and the benefits of protectant fungicides. *Plant Dis.* **80:** 559–563.

Peruch, L.A.M., Silva, A.C.F.da and Rebelo, A.M. 2008. Effect of Bordeaux mixture and alternative products in the management of late blight of tomato under organic system at the Southern Coast of Santa Catarina. *Agropecuaria Catarinense* **21(2):** 62–67.

Peterson, R.A., Inch, A.J., Herrington, M.E. and Saranah, J. 1983. Scorpio–a tomato resistant to bacterial wilt biovar III. *Aust. Pl. Pathol*. **12(1):** 8–10.

Petrovskaya, N.N. and Stolpnikova, T.V. 1989. Resistance of mutant lines of tomato to diseases. *Nauchno–Tekhricheskii Byulleten Vsesoyuznogo ordena Leninai ordema Druzhby Narodov Nauchno Issledovatelskogo Instituta Rasteni–evodstva Ineni N.I. Vavilova* **187:** 32–34.

Phadtare, S.G. 1976. *Datura stramonium* as collateral host of *Phytophthora infestans* in India. *J. Indian Potato Assoc.* **3** : 74.

Phae, C.G., Shoda, M., Kita, N., Nakano, M. and Ushiyama, K. 1992. Biological control of crown and root rot and bacterial wilt of tomato by *Bacillus subtilis* NB22. *Ann. Phytopathol. Soc. Jpn* **58:** 329–339.

Pohronezny, K., Moss, M.A., Dankers, W. and Schenk, J. 1990. Dispersal and management of *Xanthomonas campestris* pv. *vesicatoria* during thinning of direct–seeded tomato. *Plant Dis*. **74:**800–805.

Prasad, Y., Naik, M.K. 2003. Evaluation of genotypes, fungicides and plant extracts against early blight of tomato caused by *Alternaria solani. Indian J. Pl. Prot.* **31(2):** 49–53.

Prior, P., Grimault, V. and Schmit, J. 1994. Resistance to bacterial wilt (*Pseudomonas solanacearum*) in tomato: Present status and prospects. In: *Bacterial wilt: The disease and its causative agent, Pseudomonas solanacearum* (Eds: Hayward, A.C. and Hartman, G.L.). pp. 209–223.

Pushkarnath and Paharia, K.D. 1963. Survival of *Phytophthora infestans* on infected tubers in cold storage in the plains of India. *Indian Potato J.* **5** : 48–51.

Qusan, J.R., Al Abed, A.S. and Abu Blan, H.A. 1995. Antifungal activity of clanny mula (*Inula viscose*) on *Helminthosporium sativum* and *Fusarium oxysporum* f. sp. *lycopersici. Phytopathol. Meditr.* **34(1):** 7–14.

Raina, P.K. and Razdan, V.K. 2010. Influence of weather factors and management of Septoria leaf spot of tomato in sub–tropics. *Indian Phytopath.* **63(1):** 26–29.

Raj, H. and Kapoor, I.J. 1995. Influence of different plant nutrients on severity of tomato wilt (*Fusarium oxysporum*). *Indian J. Mycol Pl. Pathol.* **25:** 290.

Raj, P., Thind, T.S. and Mohan, C. 2000. Addition to host range of *Phytophthora infestans* from Punjab. *Pl. Dis. Res.* **15(1):** 119–120.

Raj, S., Bhattacharyya, S.K. and Sharma, S.R. 1976. New hosts of *Phytophthora infestans* in nature. *Indian Phytopath.* **29 :** 342–43.

Rajagopal, R. and Vidhyasekaran, P. 1983. Effect of fungicidal control of leaf spot diseases of tomato on the quality of fruits. *Indian Phytopath.* **36 :** 352–354.

Rajesh, R., Jaivel, N. And Marimuthu, P. 2014. Bioefficacy of *Muntigia calabura* botanical formulation against *Alternaria solani* early blight of tomato. *J. Mycol. Pl. Pathol.* **44(2):** 166–171.

Rajput, J.C., Palve, S.B. and Limaye, U.P. 1982. Control of bacterial wilt of tomato. *J. Maharastra Agric. Univ.* **7:** 191.

Ramakrishnan, T.S. and Sundaram, N.V. 1954. Notes on some fungi from South Indian–IV. *Indian Phytopath* **7**: 140–151.

Ramyabharathi, S.A. and Raghuchander, T. 2014. Efficacy of secondary metabolites produced by *Bacillus subtilis* EPCO 16 against tomato wilt pathogen *Fusarium oxysporum* f.sp. *lycopersici. J. Mycol. Plant Pathol.* **44(2):** 148–153.

Ramayashreedevi, G.S., Singh, D., Srivastava, Arpita, Biswas, K.S. and Gupta, A.K. 2017. Characterization of *Xanthomonas* species causing bacterial leaf spot disease of pepper (*Capsicum annuum*) in India. *Indian J. Agric. Sci.* **87(12):** 1679–1686.

Rani, R., Sharma, V.K. and Rattan G.S. 2011. Persistence of fungicides on tomato foliage following application of Indofil M–45 and Ridomil MZ. *Pl. Dis. Res.* **26 (2):** 152–155.

Rao, A.N.S. 1990. Protection of tomato from bacterial wilt (*Pseudomonas solanacearum* Smith.) Bactrinol–100. *Curr. Res. Univ. Agric. Sci. Bangalore* **19:** 100–101.

Reddy, T.B.M., Govindappa, M.R., Padmaja, A.S., Shankarappa, K.S. and Siddaramaiah, A.L. 2007. Chemical management of Fusarium wilt in tomato. *Environ. Ecol.* **25(4):** 929–930.

Rohner, E., Carabet, A. and Buchenauer, H. 2004. Effectiveness of plant extracts of *Paeonia suffruticosa* and *Hedera helix* against diseases caused by *Phytophthora infestans* in tomato and *Pseudoperonospora cubensis* in cucumber. *Z. Pflanzenkr. Pflanzenschutz* **111:** 83–95.

Saha, G., Das, D., Lepcha, L., Deka,N., Nanda, M.K., Chowdhury, S. and Khan, S.A. 2010. Studies on leaf curl disease infection pressure and fruit yield of tomato as influenced by mulching and different meteorological variables, *J. Agrometeorol.* **12(1):** 138–140.

Sahi, H.P.S. 1990. Epidemiology and management of *Alternaria* leaf spots of tomato in Himachal Pradesh. Ph.D. Thesis submitted to Dr YS Parmar Univ. of Hort. and Forestry, Solan, 97pp.

Sahi, H.P.S. and Shyam, K.R. 1993. Occurrence of *Alternaria* leaf spots of tomato in Himachal Pradesh–a new record *Pl. Dis. Res.* **8(2):** 140–141.

Saikia, A.K. and Muniyappa, V. 1989. Epidemiology and control of tomato leaf curl virus in Southern India. *Trop. Agric.* **66(4):** 350–354.

Sarhan, A.R.T. and Kiraly, Z. 1981. Control of Fusarium wilt of tomato with an integrated nitrate lime fungicide regime. *Acta Phytopathol. Acad. Sci. Hung.* **16(1–2):** 9–14.

Sastry, K.S.M. and Singh, S.J. 1973. Assessment of losses in tomato leaf curl virus. *Indian J. Mycol. Pl. Pathol.* **3:** 50–54.

Sastry, K.S.M., Singh, S.J. and Sastry, K.S. 1978. Studies on epidemiology of tomato leaf curl virus. *Indian J. Hortic.* **35:** 269–277.

Sattar, M.H.A. and Kassem, N.A. 1991. Screening of fungicides for control of early blight of tomato. *Arab J. Pl. Prot.* **9(1):** 9–13.

Schwinn, F.J. and Staub, T. 1987. Phenylamides and other fungicides against Oomycetes. 1987. In: *Modern Selective Fungicides – Properties, Applications, Mechanism of action* (Ed. H. Lyr.) Longman Group UK Ltd. London pp. 259–273.

Scott, J.W. and Jones, J.B. 1986. Sources of resistance to bacterial spot in tomato. *Hortic. Sci.* **21**: 304–306.

Scott, J.W., Somodi, G.C. and Jones, J.B. 1988. Bacterial spot resistance is not associated to bacterial wilt resistance in tomato. *Proc. Fl. State Hortic. Soc.* **101:** 390–392.

Shahida, P. and Abdul, G. 1995. Seed borne mycoflora of tomato. *Pakistan J. Bot.* **27(1):** 201–208.

Sharma, A.K. 1992. Integrated management of buckeye rot of summer tomatoes in the hills. *Indian J. Mycol. Pl. Pathol.* **22(3):** 287–288.

Sharma, Deepa, Gupta, S.K. and Jarial, Kumud 2011. Studies on the inheritance pattern of bacterial spot (*Xanthomonas vesicatoria*) in tomato. *Indian Phytopath.* **64(2):** 128–130.

Sharma, J.P. and Kumar, S. 2004. Effect of crop rotation on population dyanamics of *Ralstonia solanacearum* in tomato wilt sick soil. *Indian Phytopath.* **57(1):** 80–81.

Sharma, J.P. and Kumar,S. 2009. Linear reduction of propagules of *Ralstonia solanacearum* in soil by cake and chemicals. *Indian Phytopath.* **62(1):** 49–53.

Sharma, J.P., Jha, A.K., Singh, A.K., Pan, R.S., Rai, M. and Kumar, S. 2006. Evaluation of tomato against bacterial wilt (*Ralstonia solanacearum*) in Jharkhand.*Indian Phytopath.* **59(4):** 405–409.

Sharma, M. 1999. Studies on late blight of tomato (*Lycopersicon esculentum* Mill.) caused by *Phytophthora infestans* (Mont.) de Bary. M.Sc. Thesis, Dr. Y.S. Parmar Univ. of Hort. & Forestry, Nauni, 52pp.

Sharma, N. and Sharma, N. 1992. Control of *Alternaria alternata* by *Clerodendrum aculeatum. Natl. Acad. Sci. Lett.* **15(4):** 107–109.

Sharma, R.C. and Sharma, S.L. 1987. Variability in *Phytophthora nicotianae* var. *parasitica. J. Res. Punjab Agric. Univ.* **24:** 449–452.

Sharma, S.L. 1971. Buckeye rot of tomato caused by *Phytophthora parasitica* Dast. Ph.D. Thesis, Panjab University, Chandigarh, 168pp.

Sharma, S.L. and Bhardwaj, S.S. 1977. Relation of polyphenol oxidase activity in tomatoes to buckeys rot (*Phytophthora nicotianae* var. *parasitica* (Dast.) Waterh.) *Indian J. Exp. Biol.* **15(8):** 688.

Sharma, S.L. and Sohi, H.S. 1975. Seed borne nature of *Phytophthora parasitica* causing buckeye rot of tomato. *Indian Phytopath.* **38:** 130.

Sharma, S.L., Chowfla, S.C. and Sohi, H.S. 1976. Control of buckeye rot of tomato by cultural Practices. *Indian J. Mycol. Pl. Pathol.* **6:** 51–54.

Sharma, S.L., Chowfla, S.C. and Sohi, H.S. 1978. Epidemiology and forecasting of tomato buckeye rot. *Indian Phytopath.* **31:** 424–429.

Sharma, S.L., Sohi, H.S., Chowfla, S.C. and Peshin, S.N. 1974. Notes on the reaction of tomato varieties in field to the causal organisms of buckeye fruit rot. *Indian J. Agric. Sci.* **44:** 323–324.

Sharma, S.R. and Sohi, H.S. 1983. Buckeye rot of tomato as influenced by different levels of N, P and K fertilizers. *J. Turk. Phytopathol.* **12(1):** 19–26.

Shekhawat, G.S., Kishore, V., Singh, D.S., Khanna, R.N., Singh, R. and Bahal, V.K. 1979. Survival of *Pseudomonas solanacearum* under diverse agroclimates in India. *Indian J. Agric. Sci.* **49:** 736–738.

Sherbakoff, C.D. 1917. Buckeye rot of tomato fruit. *Phytopathology* **7:** 119–129.

Shiomi, Y., Nishiyama, M., Onizoka, T. Maromoto, T. 1999. Comparison of bacterial community structure in the rhizoplane of tomato plants grown in soils suppressive and conducive towards bacterial wilt. *App. Environ. Microbiol.* **65(9):** 3996–4001.

Shukla, Arti 2001. Studies on bacterial spot of tomato caused by *Xanthomonas vesicatoria* (Doidge) Dye. Ph.D. thesis submitted to Dr. Y.S. Parmar University of Horticulture & Forestry, Nauni, Solan, 87pp.

Shukla, Arti and Gupta, S.K. 2004. Management of bacterial spot (*Xanthomonas campestris* pv. *vesicatoria*) through chemicals. *J. Mycol Pl. Pathol.* **34:** 88–90.

Shukla, Arti and Gupta, S.K. 2005. Role of epidemiological factors on the development of bacterial spot (*Xanthomonas vesicatoria*) of tomato. *Indian Phytopath.* **58:** 319–322.

Shyam, K.R. and Gupta, S.K. 1996. Fungicidal control of buckeye fruit rot (*Phytophthora nicotianae* var. *parasitica*) and Alternaria leaf spot (*Alternaria alternata, A. alternata* f.sp. *lycopersici*) of tomato (*Lycopersicon esculentum*). *Indian J. Agric. Sci.* **66(5):** 309–312.

Shyam, K.R. and Gupta, S.K. 1998. Buckeye rot of tomato. Research Information Bulletin No.1. Dr. Y.S. Parmar University of Hort. & Forestry, Nauni, 20pp.

Sijam, K., Chang, C.J. and Gitaitis, R.D. 1992. A medium for differentiating tomato and pepper strains of *Xanthomonas campestris* pv. *vesicatoria*. *Can. J. Bot.* **14:** 182–184.

Singh, B.P., Roy, S. and Bhattacharyya, S.K. 1994. Occurrence of A2 mating type of *Phytophthora infestans* in India. *Potato Res.* **37:** 227–231.

Singh, C.B., Srivastava, S., Jain, P. and Yadava, F.D. 2007. Integrated approach for control of tomato wilt caused by *Fusarium oxysporum* f. sp. *lycopersici*. *Flora and Fauna Jhansi* **13(2):** 340–342.

Singh, D, Sinha, S., Yadav, D.K., Sharma, J.P., Srivastava, D.K., Lal, H.C. and Mondal, K.K. 2010. Characterization of biovar/ races of *Ralstonia solanacearum*, the incitant of bacterial wilt in solanaceous crops. *Indian Phytopath.* **63(3):** 261–265.

Singh, D., Chowdappa, P. and Sharma, Pratibha 2014. *Diseases of Vegetable Crops: Diagnosis and Management*. Indian Phytopathological Society, IARI, New Delhi and Today and Tomorrow's Printers and Publishers, New Delhi, 734pp.

Singh, D., Yadav, D.K., Sinha, Shweta and Upadhyay, B.K. 2012. Utilization of plant growth promoting Bacillus subtilis isolates for the management of bacterial wilt incidence in tomato caused by *Ralstonia solanacearum* race 1biovar 3.*Indian Phytopath.* **65(1):** 18–24.

Singh, R.K., Rai, N. & Singh, S.N. 2010. Response of tomato genotypes to tomato leaf curl virus, *Indian J. Agric. Sci.* **80(8):** 755–758.

Singh, R.S. 1987. *Diseases of Vegetable Crops.* 2nd ed. New Delhi: Oxford and IBH Publ. Pvt. Ltd. pp. 86–88.

Singh, R.S. 1995. *Diseases of Vegetable Crops.* 3rd ed. New Delhi: Oxford and IBH Publ. Pvt. Ltd. pp. 107–109.

Singh, S.J. and Sastry, K.S. 1976. Tomato leaf curl virus and its control. *Indian Hortic.* **21:** 9–10.

Singh, S.J., Sastry, K.S.M. and Sastry, K.S. 1973. Effect of oil spray on the control of tomato leaf curl virus in field. *Indian J. Agric. Sci.* **43:** 669–672.

Singh, Y., Jamwal, R.S., Pathania, N.K. and Kalia, P. 1997. Inheritance of biochemical traits in relation to buckeye fruit rot (*Phytophthora nicotianae* var. *parasitica*) in tomato. *Indian J. Hill Fmg.* **10(1/2):** 10–15.

Sinha, P.P. and Prasad, R.K. 1991. Evaluation on fungicides for control of early blight of tomato. *Madras Agric. J.* **78(1–4):** 141–143.

Sinha, S.K., Sinha, A.N. and Jain, B.P. 1988. Reaction of wilt–resistant tomato varieties and lines to Pseudomonas solanacearum. *ACIAR Bacterial Wilt Newsl.* **4:** 3.

Siradhana, B.S., Ellett, C.W. and Schmitthenner, A.F. 1968. Pathogenic and cultural variation in *Phytophthora nicotianae* var. *parasitica* from green house plants. *Phytopathology* **58:** 718–719.

Sohi, H.S. 1978. Some important diseaes of vegetables in India and losses caused by them. *The Lal Baugh J.* **23:** 1–11.

Sohi, H.S. and Sokhi, S.S. 1969. Relative resistance and susceptibility of different varieties of tomato to *Septoria lycopersici* Speg. *Indian J. Hortic.* **26 :** 83–86.

Sohi, H.S. and Sokhi, S.S. 1973. Morphological, physiological and pathological studies in *Septoria lycopersici. Indian Phytopath.* **25:** 666–673.

Sokhi, S.S. and Sohi, H.S. 1982. Assessment of losses in tomato caused by buckeye rot *Phytophthora nicotianae* var. *parasitica. Indian Phytopath.* **35:** 675–676.

Sokhi, S.S., Thind, T.S. and Dhillon, H.S. 1993. Late blight of potato and tomato. Punjab Agricultural University, Ludhiana. DR/Publ. –93/26 p.19.

Someya, N., Tsuchiya, K., Yoshida, T., Noguchi, M.T. and Sawada, H. 2006. Combined use of the biocontrol bacterium *Pseudomonas fluorescens* strain LRB3W1 with reduced fungicide application for the control of tomato Fusarium wilt. *Biocont. Sci.* **11(2):** 75–80.

Sotirova, V. and Rodeva, R. 1989. Genetic sources of resistance to *Septoria lycopersici* (Speg.) in tomato. *Genet. Sel.* **22:** 480–484.

Sotirova, V., Stoimenova, E. and Rodeva, R. 1992. Source of combined resistance to diseases in wild species of *Lycopersicon* Mill. *Genet. Sel.* **25:** 38–44.

Sotirova, V.G. and Bogatsevska, N.S. 1998. Sources of resistance to bacterial spot fruit infection in wild tomato species. *Proc. Int. Symp. 18–19 Nov. 1998. Aschersliben, Germany* **5:** 29–31.

Souza, De. M.F.M., Rodrigues, R., Amaral, Do.Jùnior. A.T. and Pombo, S.C. 2008. Resistance to *Xanthomonas* spp. in tomato: Diallel analysis and gene effects estimative in a breeding programme carried out in Brazil. *J. Phytopathol.* **156:** 660–667.

Soylu, E.M., Soylu, S. and Kurt, S. 2006. Antimicrobial activities of the essential oils of various plants against tomato late blight disease agent *Phytophthora infestans, Mycopathologia* **161 (2):** 119–128.

Spegazzini, C. 1982. Fungi Argentini. Pugillus IV. Buenos Aires.

Srivastava, M.P. 1981. Buckeye rot of tomato in Haryana. *Haryana Agric. Univ. J. Res.* **11:** 563–564.

Sugha, S.K. and Kumar, S. 2000. Factors affecting the development of Septoria leaf spot of tomato. *Indian Phytopath.* **63(2):** 178–180.

Suresh, C.K. and Rai, P.V. 1991. Interaction of *Pseudomonas solanacearum* with antagonistic bacteria and VA–mycorrhiza. *Curr. Res.* **20:** 26–37.

Szekely, I., Baicu, T. and Alexandri, A.A. 1990. Investigations of the efficacy of new fungicides for control of late blight (*Phytophthora infestans* (Mont.) de By.) and other leaf diseases of tomato. *Acad. de Stinte Agricola si silvice* **23:** 245–254.

Taboda, V. and Montanola, M. 1959. Outbreaks and new records. *F.A.O. Pl. Prot. Bull.* **7(6):** 80.

Tagami, Y., Tomikava, A., Yamamoto, T., Katagiri, N., Kohmoto, K. and Nishimura, S. 1984. Occurrence of stem canker of tomato caused by a distinct pathotype of *Alternaria alternata* in Japan. *Proc. Kanshi Pl. Prot. Soc.* **26:** 29–32.

Takano, T. and Takayana, M. 1985. The disinfecting effectiveness of dry heat on two species of seed borne pathogens of quarantine significance. *Res. Bull.Pl. Prot. Sci., Jpn* **21:** 1–9.

Tandon, R.N. and Chaturvedi, C. 1965. *Alternaria* leaf spot of tomato. *Proc. Natl. Acad. Sci., India Sec.B* **35:** 286–290.

Taubenhaus, J.J. and Ezekiel, W.N. 1931. Late blight of tomatoes and potatoes. *Texas Agric. Exp. Stn. Cir.* 60.

Taylor, R.H., Grogan, R.G. and Kimble, K.A. 1961. Transmission of tobacco mosaic virus in tomato seed. *Phytopathology* **51:** 837–842.

Tedla, T. 1985. Effect of captafol and Ridomil (R) MZ in the control of late blight (*Phytophthora infestans*) and Septoria leaf spot (*Septoria lycopersici*) on tomato. *Acta Hortic.* **158:** 389–399.

Thapa, C.D. and Sharma, S.L. 1978. Chemical control of fruit rot diseases of tomato. *Indian J. Pl. Prot.* **6(2):** 76–78.

Thareja, R.K. and Srivastava, M.P. 1989. Varietal resistance of tomato against buckeye rot. *Indian Phytopath.* **42:** 297 (Abstr.).

Thareja, R.K., Srivastava, M.P. and Singh, S.1989. Epidemiology of buckeye rot of tomato caused by *Phytophthora nicotianae* var. *parasitica. Indian Phytopath.* **42:** 582–584.

Thind, T.S. and Chahal, S.S. 2002. Perspectives in use of fungicides in plant disease control. *Annu. Rev. Pl. Pathol.* **1:** 177–212.

Thirumalachar, M.J., Bhatt, V.V., Dhande, G.W. and Patel, M.K. 1956. Additions to the fungi of Bombay. *Indian Phytopath.* **9:** 9–14.

Trique, B. 1977. Biochemical reactions of isolated fruits of Lycopersicon esculentum Mill. experimentally infected by *Phytophthora nicotianae* var. *parasitca. Comptes Rendus Hebdomadaires des Scances del Acad. Des. Sci.* **284(16):** 1533–1536.

Truszkowska, W. and Pudelkowa, Z. 1966. Observations on some tomato diseases in the glasshouse and in the field. *Acta Mycol. Warszawa* **2:** 183–201.

Umesha, S. and Hariprasad, P. 2010. Plant health improvement by *Bacillus subtilis* strain GB03 in tomato against bacterial spot disease, *Indian Phytopath.* **63(2):** 127–130.

Varma, J.P., Hayati, J. and Poonam. 1980. Resistance in *Lycopersicon* species to tomato leaf curl disease in India. *J. Pl. Dis. Prot.* **73:** 137–144.

Vartanian, V.G. and Endo, R. M. 1985. Overwintering hosts, compatibility types, and races of *Phytophthora infestans* on tomato in southern California. *Plant Dis.* **69:** 516–519.

Vasudeva, R.S. and Samraj, J. 1948. A leaf curl disease of tomato. *Phytopathology* **38:** 364–369.

Wager, V.A. 1935. Brown rot of tomato fruits due to *Phytophthora parasitica* Dast. *S. Afr. J. Sci.* **32:** 235–237.

Wakimoto, S. 1987. Biological control of bacterial wilt of tomato by non–pathogenic strains of *Pseudomonas glumae. Korean J. Pl. Pathol.* **3:** 300–303.

Walker, J.C. 1937. Plant diseases. *Rep. Wis. Agric. Expt. Stn. 1935–36.* (Bull. 438), pp. 106–120.

Wang, J.F. *et al.* 1994. *Phytopathology* **84:** 702–706.

Wang, J.F., Hanson, M.P. and Barnes, J.A. 1996. Preliminary results of worldwide evaluation of international set of resistance sources to bacterial wilt in tomatoes. *ACIAR Bacterial Wilt Newsl.* **13:** 8–10.

Watterson, J.C. 1993. Development and breeding of resistance to pepper and tomato viruses. pp 80–101. In: *Resistance to Viral Diseases of Vegetables: Genetics and Breeding*. (Ed. Koyte, M.M.). Timber Press, Portland, Oregon.

Wilson, J.D. 1956. Comparative control of buckeye rot of tomato by various fungicides. *Phytopathology* **46:** 511–512.

Yabuuchi, E., Kosako, Y., Yano, I., Hotta, H. and Nishiuchi, Y. 1995. Transfer of two *Burkholdria* and an *Alcaligenes* species to *Ralstonia* gen. nov.: Proposal of *Ralstonia picketti* (Ralston, Palleroni and Doudoroff, 1973) Comb. nov. and *Ralstonia eutropha* (Davis 1969) Comb. nov. *Microbiol Immunol.* **39:** 897–904.

Yamada, M., Nakazawa, Y. and Kitamura, T. 1997. Control of tomato bacterial wilt by Dazomet combined with soil solarization. *Proc. Kanto–Tosan Pl. Prot. Soc.* **44:** 75–78.

Yamaguchi, K., Kida, M., Arita, M. and Takashashi, M. 1992. Introduction of systemic resistance by *Fusarium oxysporum* MT 0062 in solanaceous crops. *Ann. Phytopathol. Soc. Jpn.* **58(1):** 16–22.

Yamazaki, H. and Hushina, T. 1995. Calcium nutrition affects resistance of tomato seedlings to bacterial wilt. *Hortic. Sci.* **30:** 91–93.

Yamazaki, H., Kikuchi, S., Hoshina, T. and Kimuka, T. 2000. Effect of calcium concentration in nutrient solution on development of bacterial wilt and population of its pathogen *Ralstonia solanacearum* in grafted tomato seedlings. *Soil Sci. Pl. Nut.* **46(2):** 535–539.

Yang, C.Y. 1979. Bacterial and fungal diseases of tomato. *Proc. First Int. Symp. Trop. Tomato.* 23–37 Oct., held at Shanhua Taiwan, Republic of China. 111–123pp.

Yao, G., Zhang, F. and Li, Z. 1994. Control of bacterial wilt with soil amendment. *ACIAR Bacterial Wilt Newsl.* **11:** 8.

Yehia, A.H., Fl Hassan, S.A., Ismail, F.K. 1981. Studies on damping–off disease of tomato seedlings and its biological control. *Meso. J. Agric.* **16(2):** 115–124.

Yoneyama, S. 1985. Resistance of some cultivars and rootstocks to bacterial wilt of tomato. *Proc. Kanto Tosan Pl. Prot. Soc.* **32:** 85.

Zhang, R.R., Li, M., Xu, F.C., Li, F.F., Zhou, X.P. and Liu, S.S. 2010. Effects of vector density, plant age and temperature on the transmission of Tomato leaf curl taiwan virus by the B biotype *Bemisia tabaci, Acta Phytophylacica Sinica* **37(1):** 1–6.

■■■■

5

DISEASES OF CRUCIFEROUS VEGETABLES

Cruciferous vegetables are important Kharif vegetable crops, which are grown both for table and seed purpose. This vegetable group constitutes crops like cauliflower, cabbage, radish, turnip, broccoli, brussel's sprouts, knol-khol and rutabaga. These crops are grown throughout the country and are attacked by number of diseases, which not only reduce the quantity but also quality of the produce. The diseases, which are of common occurrence, are described in this chapter along with their management.

(A) FUNGAL DISEASES

1. DOWNY MILDEW

Amongst the downy mildews affecting vegetable crops, the one caused by *Peronospora parasitica* on members of the genus *Brassica* is the most commonly occurring disease worldwide. It is widespread in those regions of the world, which have cool and wet climate.

Geographical distribution

The disease has worldwide distribution. The disease was first recorded in 1883 in USA on *Brassica* sp. (Farlow, 1883). Subsequently, it was reported from many countries of the world (Channon, 1981). In India, this disease was first reported by Butler (1918). Now it is prevalent in almost all states of the country wherever crucifers are grown.

Losses

The losses from this disease can be enormous because it causes infection in nursery beds as well as in the field after transplanting. Systemic infection may occur in the curds/ heads and flourish in storage. Under favourable weather conditions, *P. parasitica* may infect up to 50–60 per cent of cabbage seed crop and reduce yields by 16–20 per cent (Vasileva, 1976). Downy mildew can significantly affect the yield and developmental characters like size and weight of silique, number of silique/plant, number of seeds/silique and weight of seeds of radish (Achar, 1992). Seed yields loss can be as high as 58 per cent while infection also affects the size and weight of roots. Gaikwad *et al* (2004) reported 75–90 per cent seedling mortality due to this disease in Pune, Maharahtra. In addition to the damage caused by the downy mildew itself, the affected tissues become susceptible to secondary bacterial (*Xanthomonas* sp.) and fungal (*Alternaria* and *Rhizopus* sp.) invasions that induce more rapid decay of the heads.

Symptoms

(i) Cauliflower : This disease more seriously attacks young plants than the adult plants. The characteristic symptoms appear as angular translucid spots in the inter-veinal spaces of the leaves which are purplish brown on the underside, while corresponding upper surface is tan to yellow in colour (Plate 1a). In wet weather, the downy growth of the fungus bearing conidia and conidiophores also appears on the under surface of the leaves. In many situations, partial or complete infection of the leaves may be the total expression of the disease in the field. Infection of cauliflower curds has also been observed in fields (Bains *et al.*, 1981). The affected curds look brownish at the top, which later turn dark brown to black (Plate 1b).

(ii) Cabbage: When the fungus enters the stalk at the leaf base of an old head of cabbage, a greyish-black discolouration occurs. In some storage lots of cabbage this discolouration has been found extending up through the stalk to the innermost bud leaves. On cabbage heads, the pathogen may cause numerous sunken black spots, varying in size from minute spots to an inch or more in diameter (Sherf and Mcnab, 1986).

(iii) Broccoli: Downy mildew appears first on the lower leaves of the broccoli plants. The first symptoms of leaf infection are small water-soaked spots surrounded by a halo of light green tissue on the under surface of the leaf. Under favourable conditions for development of infection, the spots enlarge to form indefinite yellow areas. Later, the tissues within these infected areas collapse and become light brown and parchment-like. The mildew lesions vary in shape and size. The largest lesions are usually bounded by leaf veins. The initial spots of infection may also be localized. The tissues of the spot collapse to form a small brown lesion. Systemic infections are usually confined to the upper portion of the main stalk and to the branches leading to the florets of the head. Infected lesion develops brown to black netted lesions, and in others as long strands of discoloured tissues. In some plants systemic invasion of the head can be detected by diffuse blue to purple areas on the stalk and branches of the head.

The Pathogen

The fungus responsible for this disease is *Peronospora parasitica* (Pers.) Fr. which is an obligate parasite. The mycelium of the fungus is strictly intercellular with large, finger-shaped intracellular haustoria, which become clavate and branched, and nearly fill the cell cavity. Bains *et al.* (1981) reported the size of haustoria from cauliflower as 19.5–51.7 × 9.9–27.0 μm and are mostly bulbolus rarely lobbed. The conidiophores are produced during darkness, initially as unbranched hyphae from aggregations of intercellular mycelium beneath the host epidermis. They are divided into primary and secondary branches (dichotomously branched), which ultimately bifurcate to form sterigmata which bear the single conidium at the tip (Fig. 1). The conidia are broadly oval, ellipsoidal and hyaline, and they fall off mainly by hygroscopic twisting of the conidiophores in response to change in atmospheric humidity. In cauliflower leaves, conidia measure 25.6–33.8 × 22.5–29.5 μm.

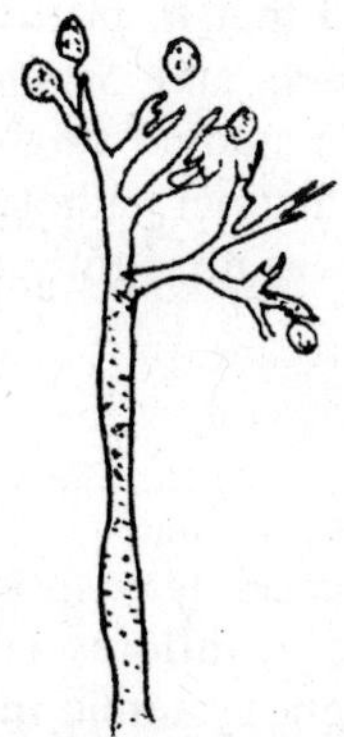

Fig. 1. Sporangiophore and sporangia of *Peronospora parasitica*

Oospore formation is favoured by conditions which induce senescence of the host tissues. There is evidence of both homothalism and heterothalism in *P. parasitica.* Sheriff and Lucas (1989) studied the cytogenetics of heterothalic and homothalic isolates of this fungus and reported that self fertility of the isolates may be due to the presence of a third mating type allele on the 5th chromosome, a condition known as secondary homothalism. The oospores are globose measuring 26 to 43 µm in diameter. They are enclosed in crest like folds and appear pale yellow in colour and germinate by germtube.

Host range

The pathogen attacks several species of the genus Brassica, Other alternate hosts are *Arabidopsis* spp., *Chenopodium album, Camelina sativa, Cheiranthus allionii, C. cherri, Capsella bursa-pasteris, Coronopus didymus, Eruca sativa, Iberis amara, Lepidium sativum, Malcomia africana, Matthiola incana, Nasturtium officinale, Raphanus sativus, Radicula nasturtium-aquaticum, Rorippa* spp., *Sinapsis alba, Sisymbrium officinale* and *S. irio* (Channon, 1981; Bains and Jhooty, 1983; Verma and Thakur., 1989).

Variability

Specificity in the downy mildew fungus on crucifers is very complex, since it ocurs on a wide range of wild hosts as well as cultivated species. *P. parasitica* is highly specialized and seldom occurs in the same biological form on more than one crucifer. A turnip isolate is able to infect seedlings of turnip but not rutabaga or radish. Earlier *P. parasitica* was divided into three biological strains, namely f.sp. *brassicae*, the chief hosts of which are *B. oleracea, B. napus, B. rapa, B. nigra, B. juncea, B. tournefortii* and *B. fructiculosa;* f.sp. *sinapidis*, the principle hosts of which are *Sinapsis arvensis* and *S. alba;* f.sp. *raphani*, the chief hosts of which are *Raphanus raphanistrum* and *R. sativus*. Felton and Walker (1946) differentiated races of *P. parasitica* found on *R. sativus* and *B. oleracea* on the basis of their host specificity. Natti *et al.* (1967) reported that the predominant physiologic races of *P. parasitica* pathogenic to broccoli and other types of *B. oleracea* grown commercially in New York were race 1 and race 2. The latter race was pathogenic to plants resistant to race

1. Bains and Jhooty (1983) reported that *P. parasitica* isolates from different hosts i.e. from *Brassica, Raphanus, Eruca* and *Sisymbrium* vary and were not cross infective. Mehta and Saharan (1994) studied stagheads of 6 host species on 17 host differentials and classified into two distinct pathotypes, one from cauliflower and the other from oilseed *Brassica*. There was no significant difference between the isolates in percentage of spore germination.

Disease cycle

The downy mildew pathogen perennates as oospores in senescenced host tissues and in *Albugo candida* induced malformed inflorescence. It also survives as conidia on leaves and inflorescence, and as latent systemic mycelium in seeds or infected plant debris (Sahran *et al.*, 1997). Infections are favoured by low temperatures and high atmospheric humidity following rain or dew. The penetration is usually direct, but occasionally also occurs through stomata. Primary infection occurs due to soil borne oospores, while conidia released from conidiophores found on the cotyledons or hypocotyls favour secondary spread. Water droplets also help pathogen dispersal over short distances (Fig. 2).

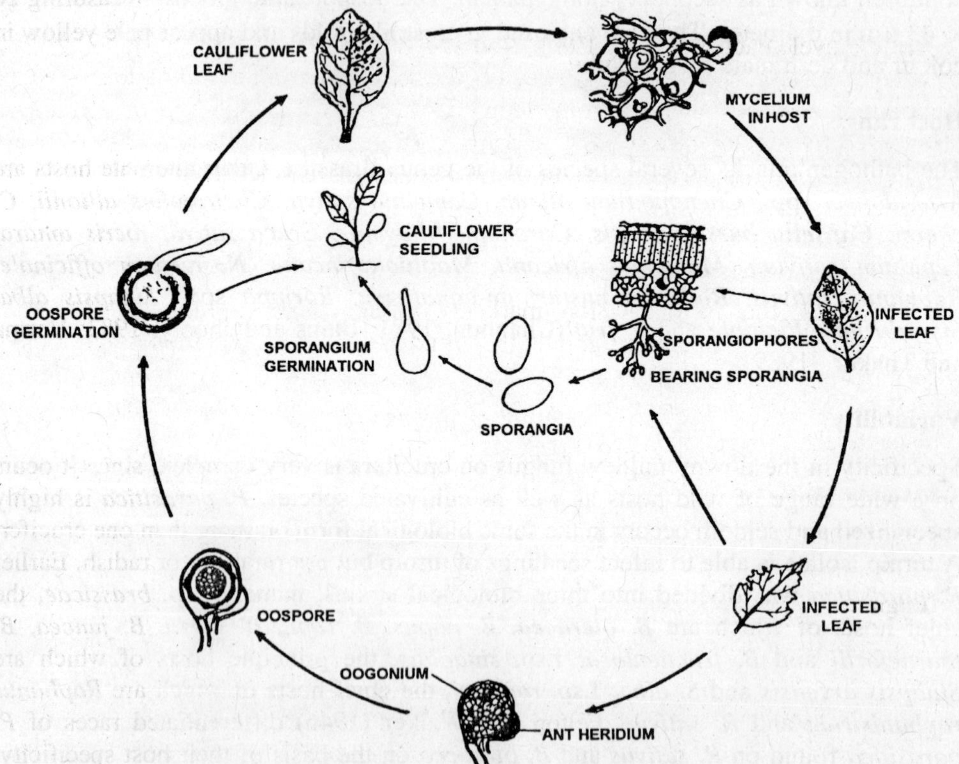

Fig. 2. Disease cycle of downy mildew of crucifers

Epidemiology

This is a polycyclic disease, which initiates from the low level of initial inoculum and then increases exponentially through successive cycles on the host during the growing season. Epidemics of this disease are governed by the prevalent environmental factors like air temperature and relative humudity. At 15ºC, conidia germination, appressoria formation and penetration of the host take place in 4, 6 and 12 h, respectively. Felton and Walker (1946) found that in pre-penetration stage, conidial germination was most rapid at 8–12ºC while penetration of the host by infection hyphae and formation of the haustoria occurred most rapidly at 16 and 20–24ºC, respectively. For disease development a similar temperature optima of 8–14ºC has also been reported by d'Ercole (1975). Saharan *et al.* (1997) observed that lower temperature of 15–16ºC results in slower growth of both the host and the pathogen, less damage to the host, more prolific sporulation, more reinfection and consequently more profuse disease development.

Cauliflower plants, which are deficient in potash, suffer most from this disease. Cabbage plants grown in soil fertilized with less potash and more phosphorus are more prone to the attack of downy mildew than in unfertilized soil. However, no consistent effect of fertilizers on development of downy mildew on crucifers has also been observed (Felton and Walker, 1946).

The discharge of sporangia from diseased leaves of Chinese cabbage showed a periodic cycle each day (Lin, 1981). Sporangial release increases steadily after 2 a.m. each day and reaches at peak between 6–8 a.m., decreasing rapidly after 8 a.m. Few sporangia can be detected from noon to 10 p.m. The discharge of sporangia is favoured by temperature of 18ºC and RH < 75 per cent. Shao *et al.* (1990) reported that in favourable weather conditions (temperature and RH), sporangia release is three times higher in the morning than afternoon. Dew periods and microclimates within cauliflower seedling canopy play an important role in downy mildew epiphytotics compared to environmental conditions outside (Pandey *et al.*, 2001)

Among the factors which markedly influence the development of downy mildew are air temperature and high humidity. In pre-penetration stage, conidial germination was most rapid at 8–12°C while penetration of the host by the infection hyphae and formation of the haustoria occurred most rapidly at 16°C and 20–24°C, respectively.

Management

Since the fungus survives in the form of oospores or on perennial hosts, the management can be attempted through integration of cultural practices with fungicidal cover and host resistance.

(i) Cultural practices: Destruction of crop debris and perennial weed hosts, crop rotation with non-cruciferous crops, avoidance of dense planting to reduce relative humidity around plants, use of clean seed (without fragments of pods which may contain oospores) are some of the cultural practices recommended for the control of this disease (Sherf and Macnab, 1986; Singh, 1987). Thakur and Mathur (2002) suggested other cultural practices like manipulation of planting dates, host nutrition, crop rotation as well as inter and mixed cropping to reduce the primary inoculum and secondary spread of the disease.

(ii) Host resistance: Clear evidence has been obtained on the physiologic specialization (pathogenic variability) of *P. parasitica* at the generic and specific levels of the host. A list of cultivars/lines of different crucifer vegetables reported resistant are given below:

(a) Cauliflower: Cultivars/lines like Igloo, Snowball Y, Dok Eglon, RS-355, PI 181860, PI 188562, PI 204765, PI 204768, PI 181860, PI 188562, PI 204779, PI 241612, PI 264656, 291567, PI 373906, KPS-1. PI 231210, PI 189028 PI 208474, CC 3-5-1-1, PI 246077, Early Winters White Head (Kontaxis *et al.,* 1979; Thomas and Jourdam, 1990; Sharma *et al.* 1991; Hoser *et al*, 1995; Mahajan *et al*., 1995; Sharma *et al*., 1995) have been reported as resistant to this disease. However, none of the variety/lines was found resistant to this disease under Varanasi conditions, while lines Kuwari-17, Kuwari-8, Kuwari-4 and First Early Laxmi exhibited moderately resistant reaction (Pandey *et al*., 2001). Cvs./lines like KJ-31, KT-25, KPS-1, PSBK-1, IIHR-260-1, 265, 302 and 305-1 were found resistant, while KT-9, Pusa Snowball-1 and Punjab Katki were found less susceptible at different locations (Mehra and Batra, 2005, Thakur *et al.*, 2009, Varalakshmi *et al.,* 2009). Resistance and susceptibility to downy mildew was highly correlated with an increase/decrease in number of stomata, particularly on the dorsal side of the leaf surface (Mahajan and Gill, 1998). Phenols and sugar level in different varieties also play an important role in the initiation of the disease (Mahajan, 1999).

(b) Cabbage: Cultivars/lines like January King, Balkhan, Spitz Kool, PI 246063, Copenhagen Market Express, PI 263054, PI 263057, Pi 357374, PI 418784, PI 418985, PI 418986, PI 418987, PI 418988, PI 246077, PI 245013, Tromchuda cabbage, Algarvia (ISA 207), PI 245015, Geneva 145-1 (Elenkov, 1979; Sherf and Macnab, 1986; Hoser *et al.,* 1991; Carvalho and Monteiro, 1996) have been reported resistant to this disease.

(c) Broccoli: Cultivars/lines *viz.,* Calabrese, Grand Central, PI 231210, Italian Green sprouting, Hybrid 1230, Green surf, 2804, Hyb. 2805, Hyb. 2803, GSV 82-4310, XPH 1117, Hyb 288, AVX 7361, PI 263056, PI 263057, PI 3573, PI 3574, PI 418984, PI 418985, PI 418986, PI 418987, PI 418988, OSU CR*, Citation, Excalibur, Nancy (Laemmlen and Mayberry, 1984; Sherf and Macnab, 1986; Hoser *et al*., 1991), have been reported resistant to this olisease.

(d) Radish: Cultivars like Okura, Tokinishi (All season), Baniba, NoirLon, d'Horloge, Rave a Forcer (Bonnet and Blancard, 1987; Silvae *et al*., 1996) have been reported resistant to this disease.

(iii) Chemical control

(a) Seed treatment : Fungicidal seed treatment followed by foliar sprays is a common practice to control this disease. Metalaxyl seed treatment @ 0.3–0.6 g a.i./kg reduced downy mildew infection on broccolli (Paulus and Nelson, 1977) and a significant yield increase was observed when plants raised from such treated seed were sprayed once or twice with the same compound. Seed treatment with Apron SD (35% metalaxyl) controlled downy mildew of cauliflower for more than 2 weeks after sowing (Crute, 1984). White *et al.* (1984) observed that seed treatment with

Apron SD (1g metalaxyl/kg seed) gave complete control of this disease on cauliflower when inoculated 10 days after sowing. Following seed treatment, metalaxyl was detectable in the cotyledons, true leaves and roots of cabbage seedlings upto 4 weeks after sowing.

In summer cauliflowers, good control was achieved by drenching the compost with propamocarb, fosetyl-aluminium foliar sprays and by applying a dichlofluanid foliar spray programme (Davies and Wafford, 1987)

(b) Foliar sprays: During early period (1920 to 1960's) reliance for control of this disease rested on frequent application of spray/dust of chemicals, such as chloranil, copper based material (Bordeaux mixture), zineb, spergon, dichlofluanid and propineb (Whitwell and Griffin, 1967). Later other non-systemic fungicides like captafol, mancozeb, copper oxychloride, have been found to be superior to other fungicides on a large number of crucifers at several locations.

The introduction of systemic fungicides like metalaxyl during 1980's, provided effective control of this disease either alone or in combination with non-systemics. Ryan *et al.* (1977) reported the efficacy of fosetyl-aluminium, metalaxyl + mancozeb, cyprofuram and propamocarb in the management of this disease on cauliflowers. Later oxadixyl plus mancozeb, cupric hydroxide and chlorothalonil gave significantly better protection from this disease than mancozeb alone (Brophy and Liang, 1992). Neutralized phosphoric acid (2 applications of 2.4 kg a.i./ha, 21 and 7 days before harvest) sprays applied on to the cauliflower in the field within 3 weeks of harvest reduce curd infection of downy mildew under storage conditions. There was no effect of phosphoric acid on crop appearance and maturity. The maximum phosphonate residue in curds at harvest was 12 μg/g which was considered a safe limit (Mckay *et al.*, 1992).

Three sprays of Ridomil 25 WP @ 2kg/ha at weekly intervals begining from 28 days after transplanting controlled this disease effectively on chinese cabbage and provided 65 per cent more marketable yields (Yang *et al.*, 1983). Due to risk of resistance development, metalaxyl in prepacked mixture with mancozeb (Ridomil MZ) is preferred for the control of downy mildew on various crops. Four sprays with Daconil (0.1%), Dithane M-45 (0.2%), Ridomil MZ (0.2%) or Alliette (0.3%) at 8–10 day intervals were highly effective in controlling this disease on radish and root yield was significantly higher in sprayed plots besides significant reduction in the apparent infection rate (r) and the basic infection rate (R) of downy mildew in treated plots (Sharma and Sohi, 1982; Sharma, 1983). Gupta and Shyam (1994) studied the erradicant activity of eight fungicides and found metalaxyl + mancozeb (0.25%) and cymoxanil (0.03%) + mancozeb (0.2%) more effective over a 14 days period on cabbage seedlings.

(iv) Induced Resistance

Various defence inducing elicitors of either biotic or abiotic origin has been used to check plant diseases. These compounds have been used at a very low concentration and might have no direct action on the pathogens and were capable of inducing resistance in plants.CGA 245704, also known as benzothiadiazole (BTH) has shown potent systemic acquired resistance (Godard *et al.*, 1999). Preinoculation application of BTH induced resistance in cauliflower against *P. parasitica* (Sharma, 2002).

Induction of resistance was maximum after three days of treatment and it persisted upto 28 days.

2. ALTERNARIA LEAF SPOT

Alternaria diseases are the most common and serious menace to the cultivation of cruciferous vegetables world over. Infection on young leaves, stems and siliquae generally results in heavy yield losses as well as quality of seed.

This disease is known by many names like black spot of cabbage, brown spot of cauliflower, Alternaria storage rot, Alternaria blight, Alternaria sooty spot, black pod blotch, brown rot, curd blight (Verma and Saharan, 1994). These diseases are generally caused by three species of *Alternaria* namely *A. brassicae* and *A. brassicicola*, which are common on most of the cruciferous vegetables and *A. raphani*, which mainly attacks radish.

Geographical distribution

Both *Alternaria brassicae* and *A. brassicicola* are common on vegetable crucifers and have been reported from almost every continent on Brassicaceae hosts (Anonymous, 1969). *A. brassicae* was first reported in 1909 from U.S.A. on cabbage (Fawcett, 1909), while *A. brassicicola* was first recorded in 1924 from Holland on cabbage (Bolle, 1924). Since then both species have been reported from almost all the continents of the world (Verma and Saharan, 1994).

A. raphani was first reported in 1945 from Denmark on *Brassica* sp. (Neergaard, 1948). Since then it has been reported from Canada, Denmark, Egypt, Greece, India, Japan, Netherland and U.S.A. (Ellis, 1971). *A. brassicicola* causes pod blight of cabbage and due to summer rains result in heavy losses in arid dry temperate zone (Kinnaur district) of Himachal Pradesh (Bharat and Gupta, 2011)

Losses

Alternaria infection in cruciferous vegetables reduces both quality and quantity of the produce. In nursery it causes damping-off, while in older plants it causes spotting of lower leaves and heads of cabbage, cauliflower and broccoli. A 50 per cent reduction in the yield of cabbage, cauliflower and broccoli has been reported from this pathogen in wet season in U.S.A. Market losses are due to decay, which develops in transit and storage (Ramsey and Smith, 1961). Smith *et al* (1988) reported that the reduction in seed yield due to *A. brassicicola* infection may be as high as 80 per cent and the fungus may severely depress germination to the extent that infected seed may be un-saleable. Losses upto 30 per cent have also been reported from Columbia (Tamayo *et al.*, 2001). Upto 55 per cent yield losses in mid maturing varieties of cauliflower have been recorded in cv. Pant Shubhra under Uttrakhand conditions and there was a direct relationship between curd area infection and reduction in seed yield (Prasad and Vishunavat, 2006). Similarly, Arora *et al.* (2011) reported 80 per cent reduction in seed yield and rotting of curds of cauliflower rendering them unmarketable due to *Alternaria* infection.

Symptoms

The symptoms produced by two species (*A. brassicae* and *A. brassisicola*) are often similar and indistinguishable. In nursery beds, minute dark spots appear on the cotyledons and seedling stem, which lead to a severe damping off or stunting of the young plant. On older plants, all above ground parts are attacked including leaves, stems/heads, inflorescence, siliquae and seed. On leaves, symptoms appear as minute dark brown to black spots, which may enlarge by forming concentric rings (Plate 2). Each spot is surrounded by a yellow halo of chlorotic tissues. Older lesions are circular, often zonzte with a papery thin center which in humid weather may be covered with a mat of spores which are light brownish in case of *A. brassicae* and dark olive brown in case of *A. brassicicola.* In later stages, the center of the lesions may drop down to give a shot hole effect. Severe defoliation by *A. brassicae* has been reported in some cultivars of stubble turnips. The lesions are linear on stem, petiole and siliquae (Verma and Saharan, 1994).

Browning of cauliflower/broccoli heads occurs which generally begin at the margin of the individual flower or flower cluster. Plants grown for seed show, dark necrotic lesions on the main axis, inflorescence, branches and on the siliquae. In favourable weather conditions, the lesions coalesce and result in pre-mature ripening and splitting of the siliquae, which may result in high level of seed infection. Infected seeds remain small in size, shrunken, discoloured or covered with fungal growth and have low viability.

Turnip and rutabaga foliage and roots may also become infected in field but root symptoms develop after being placed in storage. The leaf spots are nearly circular, often zonate and are of various shades of brown to black. If storage temperature is high, spores of the fungus also appear on the spots (Sherf and Macnab, 1986).

Radish plants kept for seed purpose are severely affected by *A. raphani.* On leaf, the spots are raised, spherical to elliptical and upto 1 cm in diameter. Black sporulation may also appear on the lesions. The center soon dries and may drop down (Singh, 1987). The natural infection of radish seed with *A. raphani* may result in a lack of germination, pre or post emergence damping off, appearance of distinctive lesions on table radish and in the spotting and blighting of leaves, stalks and siliquae.

The Pathogen(s)

Three species of *Alternaria* are responsible for this disease. *A. brassicicola* (Schw.) Wiltshire and *A. brassicae* (Berk.) Sacc. attack most of the crucifer vegetables, while *A. raphani* Groves & Skolko usually attacks radish. The morphological characters of these three species are described in detail below:

A. brassicae

The morphological characters as described by Ellis (1971) are colonies amphigenous, effused, rather pale olive, hairy. Mycelium immersed, hyphae branched, septate, hyaline, 4–8 µm thick. Conidiophores arising in groups of 2–10 from the hyphae, emerging through stomata, usually simple, erect, frequently geniculate, more or less cylindrical, septate, up to 170 µm long, 6–11 µm thick, bearing one to several small, but distinct conidial scars. Conidia solitary or occasionally in chains of up to 4, acropleurogenous, arising through small pores in the conidiophore wall, straight or singly curved, obclavate, rostrate with 16–19 (usually 11–15) transverse septa and 0–8 (usually 0-3) longitudinal or oblique septa, pale or greyish olive, 75–350 µm long and usually 20–30 µm (sometimes up to 40 um) thick in the broadest part, the beak about 1/3 to 1/2 the length of the conidium and 5–9 µm thick (Fig. 3). This specie produces chlamydospores. It is mostly confined to the cruciferae family.

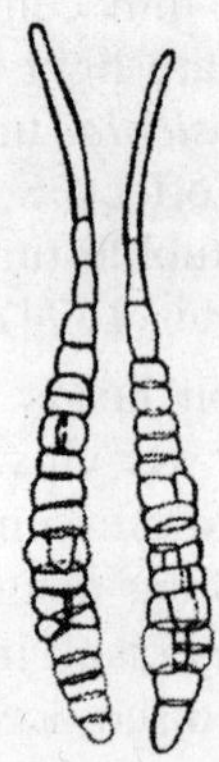

Fig. 3. Conidia of *A. brassicae*

A. brassicicola

The morphological characters as described by Ellis (1971) are colonies amphigenous, effused, dark olivaceous brown to dark blackish brown, velvety. Mycelium immersed, hyphae branched septate, hyaline at first, later brown or olivaceous brown, inter- and intra-cellular, smooth, 1.5–7.5 µm thick. Conidiophores arising singly or in groups of 2–12 or more, emerging through stomata, usually simple, erect occasionally geniculate, but often slightly swollen at the base, septate, pale to mild olivaceous brown, smooth upto 70 µm long, 5–8 µm thick. Conidia mostly in chains of up to 20 or more, sometimes branched, arising through small pores in the conidiophore wall, straight, nearly cylindrical, usually tapering slightly towards the apex or obclavate, the basal cell rounded, the beak usually almost non-existent, the apical cell being more or less rectangular or resembling a truncated cone, occasionally better developed, but then always short and thick, with 1–11, mostly less than 6, transverse septa, often constricted at the septa, pale to dark olivaceous brown, smooth or becoming slightly warted with age, 18–30 µm long, 8–30 µm thick in the broadest part with the beak 1/6 the length of the conidium and 6–8 µm thick.

A. raphani

According to Ellis (1971) conidiophores simple or occasionally branched, septate, olivaceous brown, upto 150 µm long, 3–7 µm thick, sometimes swollen slightly at the tip and usually with a single conidial scar, conidia commonly in chains of 2–3, straight or slightly curved, obclavate or ellipsoidal, generally with a short beak, mid to dark golden brown or olivaceous brown, smooth or sometimes minutely verruculose, with 3–7 transverse and often a number of longitudinal or oblique septa; constricted at the septa, 50–130 (70) µm long, 14–30 (22) µm thick in the broadest part. Chlamydospores formed abundantly in culture, sometimes in chains at first 1 celled, round, finally many celled and irregular, brown. Conidiophores often develop from them. Unlike *A. alternata* and *A. brassicicola*, this specie forms chlamydospores and the beak of the conidium is smaller than that of *A. brassicae* but longer than the almost nonexistent beak of *A. brassicicola*.

Host range

Alternaria species attack a large number of brassicaceous crops. Specices wise host range according to Verma and Saharan (1994) is as follows:

(i) A. brassicae

Among the common hosts of economic importance are horse radish (*Armoracia rusticana*), kale (*Brassica oleracea* var. *viridis*), Kohlrabi (*B. oleracea* var. *gongylodes*), rutabaga (*B. napa brassica*), brussels sprouts (*B. oleracea* var. *gemmifera*), cauliflower (*B. oleracea* var. *botrytis*), cabbage (*B. oleracea* var. *capitata*), rapeseed (*B. napus* ssp. *oleifera*), wild mustard (*B. tournifortii*), black mustard (*B. nigra*), chinese cabbage (*B. chinensis*), white mustard (*B. hirta*), white mustard (*B. kabar*), canola (*B. rapa* ssp. *oliefera*), turnip (*B. rapa* ssp. *rapifera*), abyssinian mustard (*B. carinata*), mustard (*B. juncea*), guar (*Cyamopsis psoraloides*), crambe (*Crambe abyssinica* and *C. maritima*), wallflower (*Cheiranthus cheirii*), taramira (*Eruca sativa*), wormseed mustard (*Erysimum cheiranthoides*), candytuft (*Iberis amara*), garden cress (*Lepidium sativum* and *L. latifolium*), lettuce (*Lactuca sativa*), honesty (*Lunaria annua*), stock (*Matthiola incana*), radish (*Raphanus sativus*), wild mustard (*Sinapis alba* and *S. arvensis*), hedge mustard (*Sisymbrium officinale*), tumbling weed (*Sisymbrium altissimum* and *S. orientale*), and stinkweed (*Thlaspi arvense*).

(ii) A. brassicicola

The host range includes cabbage, cauliflower, mustard, chinese cabbage, turnip, abyssinian mustard, crambe, taramira and radish.

(iii) A. raphani

Among the economic hosts reported susceptible to this pathogen are Indian mustard, cauliflower, cabbage, swedes, wallflower, taramira, candytuft, lettuce, stock and stinkweed.

Variability

Cultural and pathogenic variations in different isolates of *Alternaria* species infecting Cruciferae have been reported. Saharan and Kadian (1983) distinguished three isolates of *A. brassicae* infecting *Brassica* sp. in India and designated them as RM_1, RM $_2$ and V_3. All three isolates were virulent on rapeseed and mustard group of crops. RM_1 was avirulent on *B. oleracea* var. *capitata,* while RM_2 was avirulent on both *B. oleracea* var. *capitata* and *B. oleracea* var. *botrytis*. Isolate V_3 was virulent on eight *Brassica* sp. tested. Mridha (1983) tested thirteen isolates of *A. brassicae* on selected cultivars of winter oilseed rape and reported difference in their virulence. Cultural and morphological variation in the isolates of *A. brassicae* has also been indicated. Spontaneous occurrence of albino mutants of this fungus has also been reported (Campbell *et al.*, 1968). Atkinson (1953) obtained 312 isolates of *A. raphani* from different geographical areas and classified them as wild type and variant type. He considered the former as being less virulent than the latter.

Genetic variability among different isolates of *Alternaria* infecting cole crops has also been studied. RAPD and ISR primers showed mean similarity coefficient of 0.73 and 0.84, respectively among 32 isolates of *A. brassicae* from cauliflower (Sharma *et al.*, 2013). Further internal transcribed spacer analysis showed all isolates were 90–100 per cent similar to each other.

Disease Cycle

The three *Alternaria* sp. perennate on infected plant debris, in or on infected seed and cruciferous weeds. Infection of host plants begins directly from infested seed, from spores produced on crop debris, from infected cruciferous weeds or possibly from microsclerotia and chlamydospores if produced on infected debris. Spores are produced abundantly in wet weather and dispersed locally by rain splashes and wind currents. Spore release which is passive, is stimulated by falling humidity and restricted by high humidity exhibiting a distinct diurnal periodicity, with maximum spore concentration occurring in the early afternoon and minimum in the early morning. Although all the three species of *Alternaria* occurring on cruciferous vegetables are reported to be seed borne (Maude and Humpherson-Jones, 1980), yet it hardly plays any role in survival and dissemination of this pathogen in tropical countries like India where summer temperatures often reach >35ºC (Chahal, 1981). However, infected seed is the most important source of primary inoculum in cooler areas of the world. *Alternaria* conidia from their courts of survival germinate in free water, by giving rise to germtube, which emerges, from any cell of the conidia. Germtubes generally penetrate the undamaged host tissues directly, although indirect penetration through stomata has also been reported in *A. brassicae* (Tsunedo and Skoropad, 1978). Penetration is preceded by formation of appresoria from the tip of germtubes. Once within the host, the epidermal cells are fully invaded and mycelia ramify through and between the mesophyll and palisade cells. Early in the post-infection phase, the invaded epidermal cells become necrotic, and the parenchyma tissues ahead of the advancing hyphae often collapse (Dixon, 1981). Under favourable environmental conditions, lesions develop and produce wind borne conidia, which may cause secondary infections on the same or neighbouring plants. The cycle continues under favourable weather conditions.

Epidemiology

Moderate temperatures, high humidity and splashing rains favoured Alternaria diseases. Relative humidity between 95–100 per cent at least for 18 h and temperature of 21–27ºC for three successive days are essential for epidemic development. *A. brassicae* forms abundant spores after a succession of wet and dry periods. Susceptibility of the host increases with the age of the plant. Reduced light intensity is more favourable for lesion development.

The optimum temperature for sporulation is 18–24 and 20–30ºC for *A. brassicae* and *A. brassicicola*, respectively, at which both fungi produce spores in 12–14 h. At temperature >24ºC, sporulation of *A. brassicae* is inhibited. *A. raphani* infection progresses rapidly at 22–26ºC.

Management

Since the pathogen(s) overwinters either in infected plant debris or through seed, it is essential to integrate all the available control methods like cultural, biological, chemical and host resistance.

(i) Cultural Practices: Closer spacing, high doses of nitrogen and frequent irrigation rapidly increase severity of Alternaria blight. The pod blight of radish seed crop was significantly less in the normal season transplanting of 15th December in the Punjab conditions, where twig blight is significantly less in the early transplanting of 15th November (Sandhu *et al.,* 1984, 1985). Cultural practices like use of clean, large, healthy seed, long crop rotation, sanitation, weed control, planting at recommended times, use of balanced nutrients, proper plant density and proper drainage of the fields have been suggested by various workers from time to time (Dixon, 1981; Singh, 1987). Application of fertilizers containing humus and organic manure reduces the disease better. To reduce storage and transit losses in cabbage and cauliflower, heads should be handled carefully to avoid bruising and surface moisture allowed to evaporate before storage. The temperature of the storage house should be kept at 33–34ºF alongwith ample ventilation to reduce humidity.

(ii) Seed treatment

(a) Hot water seed treatment : Heating of white cabbage and cauliflower seed by 30 minutes immersion in hot water maintained at 45ºC or for 20 minutes at 50ºC provided excellent control of *A. brassicae* and improved seed germination by 13 per cent, besides elimination of other fungi. Ellis (1968 a,b) reported complete elimination of *Alternaria* infection from cruciferous seeds by hot water treatment for 25 minutes at 50ºC.

(b) Chemical seed treatment: Various fungicides like mancozeb, zineb, captafol, dichlofluanid, captan, fenpropimorph, quinolate, iprodione, thiram, and carboxin have been found effective as seed dresser for the control of seed borne *Alternaria* spp. (Chahal, *et al.*, 1977, Ellis, 1968a,b, Maude and Suett, 1986). Soaking *Brassica* seed in an aqueous suspension of thiram (0.2%) for 24 h at 30ºC eradicates *Alternaria* infection. Seed treatment with iprodione (1.25 g a.i/kg seed) drastically reduced seed borne infection of *A. brassicae* and *A. brassisicola* (Maude and Suett, 1986). *A. brassisicola* was effectively controlled by dressing of chinese cabbage seed with thiram or powdered *Streptomyces griseoviridis* (Valkonen and Koponen, 1990).

(c) Seed treatment with bioagents: Cruciferous seeds treated with either *Gliocladium virens*-19, *Trichoderma harzianum*-22, *T. harzianum*-50, *Penicillium corylophilum*-36 and *P. oxalicum*-76 provided significantly improved emergence of healthy seedlings (Wu and Lu, 1984). The hyphae of antagonistic fungi are able to adhere to conidia or coil around or penetrate germ tubes or hyphae of *A. brassicicola*. The conidia of *A. brassicicola* in the presence of antagonistic fungi get shriveled and plasmolyzed. *Chaetomium globosum, T. harzianum, T. koningii* and *Fusarium* sp.

treated radish seed infected with *A. raphani* and *A. brassicicola* showed consistently more number of healthy seedling as compared to untreated ones (Vannacci and Harman, 1987). White *et al.* (1990) obtained good control of damping off of cauliflower and cabbage (*A. brassicicola*) by seed treatment with Mycostop, a powdery formulation of spores and mycelium of *Streptomyces griseoviridis*.

iii. Host Resistance

Very limited sources of resistance have been reported in cruciferous vegetables against *Alternaria* sp. Cvs./lines like 1-6-1-2, 1-6-1-4, Pusa Shubhra, PI's 231208-209, 217934, 231209, 267725, KT-9 of cauliflower and Cambridge No. 5 of brussels sprouts have been reported resistant to *A. brassicae* and *A. brassicicola* (Braverman, 1971; Williams and Pink, 1987; Sharma *et al*., 1991). Sharma *et al* (1991) reported that KT-9 had more deposition of cuticular wax on its foliage than other cultivars, which is inhibitory to the deposition of water borne inoculum on leaf surface. Wax is also known to reduce the rate of conidial germination and the number of germ tubes formed by each conidium (Verma and Saharan, 1994). Dhingra *et al.* (2012) studied the role of phenols and oxidative enzymes in imparting resistance against Alternaria blight and reported that presence of higher levels of phenols and oxidative enzymes in moderately resistant genotypes provide efficient antioxidant system to fight against this pathogen and may be responsible for resistance against Alternaria blight in cauliflower genotypes.

iv. Chemical control

Various systemic and non-systemic fungicides like Baycor, Benlate, Blitane, Blitox-50, Bordeaux mixture, Brestan, Captaf, Difolatan, Cuman L, Dithane M-45, Dithane Z-78, Daconil, Duter, Folicur, Rovral, Procymidone and Prochloraz have been reported effective against *Alternaria* sp. both under *in vitro* and *in vivo* conditions (Chand and Jatian, 1969; Maude and Dudlay, 1972; Lapis and Ricaforte, 1974; Chahal *et al*., 1977; Singh, 1986; Singh, 1987; Verma and Saharan, 1994).

A. brassicae and *A. brassicicola* on cauliflower were successfully controlled and yield increased by 2–22 times after spraying with a mixture of thiram + benomyl alternated with mancozeb + insecticide (Kudela *et al*., 1978). Dithane M-45 (0.25%) proved most effective in controlling Alternaria blight of radish followed by Bordeaux mixture (0.4%) in increasing seed weight, improving seed germination and in reducing seed infection (Hussaini and Singh, 1989). Efficacy of mancozeb was also reported superior to other fungicides in controlling Alternaria blight of five different cruciferous vegetables (Thind and Jhooty, 1988). However, Mondal *et al.* (1989) found Rovral (0.2%) best followed by Dithane M-45, Dithane Z-78 and captan for the control of this disease in radish.

Three sprays of iprodione @ 0.5-1 kg a.i./ha, applied at 3 week intervals from the young green pod stage until cutting, controlled pod infection of *A. brassicicola*, and increased yields (Humpherson-Jones and Maude, 1982). Singh *et al* (1989) found three sprays of mancozeb, ziram and zineb effective against *A. brassicicola* leaf spot of cabbage. Sprays of chlorothalonil were adjudged as best against pod blight of cabbage followed by mancozeb (Bharat and Gupta, 2011). Foliar application of potassium and phosphoric salts at low dose has also been found to induce resistance to varying levels in different cole crops which offers an ecofriendly and cheap approach to combat the ravages of these leaf spots.

Storage rot of white cabbage due to *A. brassicicola* can be effectively controlled with iprodione mixed in talc powder (Kear *et al.*, 1977). Similarly, a post-harvest iprodione drench on stored winter cabbage controls *Alternaria* sp. (Geeson and Browne, 1979). Pre-harvest sprays of mancozeb (0.2%) or folpet (0.2%) protected cabbage heads from *A. brassicae* storage rots (Tasca and Trandaf, 1984). Ronilan-smoke, a vinclozoline formulation used as smoke pellets, in Brassica storage rooms also controls *A. brassicae* apart from *Botrytis cinerea* infections (Jennrich, 1985).

3. STALK ROT

This is an important disease of cauliflower seed crop (late varieties). Once this disease gets introduced in a field, it is very difficult to raise a healthy crop in the same field without taking appropriate control measures.

Geographical Distribution

Sclerotinia sclerotiorum was first recorded in England in the year 1837 on vegetables (de Bary, 1886), and thereafter it was reported on lettuce from Mediteranean sea board of France (Roselle, 1937). Later it was found on cauliflower in California, on cabbage seed plants in Washington, on peas from Kincardineshire in Britain, on bean, cauliflower and lettuce in W. Australia and also reported to be widespread on cabbage and cauliflower in Denmark.

In India, on cauliflowers this disease was recorded for the first time in 1973 in few field of Saproon valley in Solan district of Himachal Pradesh (Anonymous, 1974).

Losses

The disease has the potential to cause 80–90 per cent reduction in potential seed yields of untreated cauliflower fields (Sharma, 1979). Plant mortality due to this disease has been reported upto 70 per cent in various fields in Himachal Pradesh and corresponding losses in seed yield were 280 kg/ ha (Shyam *et al.*, 1994).

Symptoms

The symptoms of the disease appear on all above ground parts of the plants and start appearing in the months of December and January with the earthing up of plants. The leaves of the affected plants lose their turgor during daytime and droop down to the ground, but regain turgidity during night or early morning. The bright green sheen of the affected plants is replaced by the dull whitish green colour, ultimately becoming pale yellow. The yellowing starts from tips of the older leaves downwards which shed, prematurely. In most of the cases, mid-rib and petioles at a point touching the soil, show small, discrete to large irregular dark brown to black necrotic lesions (Plate 3a). The lesions are covered with fluffy growth of extrametrical fungus under cool and humid weather. Rotting from petioles advances to the stalk where dark brown to black spots is produced girdling whole of the stem at ground level. Thereafter, the stalk rot progresses towards the curd and occasionally whole of the pith portion up to the forks of curd branching get completely rotten (Plate 3b). The pith and the curd also rot, giving way to large cavities lined inside with fluffy mycelium and numerous sclerotia of the causal fungus. Under cool humid conditions, the mycelium emerges out and can be seen sticking to affected portion of the plant.

With the progress of the disease, curds are also affected and show brown to dark brown rotting which may start from any portion of the curd, but generally from the center. The affected tissue becomes soft and mushy bearing numerous sclerotia. Inflorescence is also affected during the months of April-May. If there are plenty of rains during bolting and seed setting, the fungus progresses fast and engulfs whole of the branches and inflorescence where mycelium can be seen hanging out with sclerotia sticking to it. However, if the weather is dry, the mycelial growth is restricted only up to branches. Affected branches become dry and bear shriveled seeds.

The Pathogen

The disease is caused by *Sclerotinia sclerotiorum* (Lib.) de Bary. Mycelium in culture as well as on host surface is white, cottony, branched, consisting of closely septate hyphae, which are 6.0 to 16.0 μm broad and filled with dense granular protoplasm. Micro-conidia are produced on short lateral branches of the vegetative mycelium in chains and measure 1.5 to 3.5 μm. When the food supply is exhausted and the vegetative growth ceases, the hyphae with granular protoplasm collect in small dense masses and form sclerotia. These sclerotia are first white in colour, then turn black and are round to irregular in shape measuring 2.5 to 15 mm. The sclerotia in culture are morphologically similar to those produced on the host.

The sclerotia also germinate by producing stalked apothecia, which range from one to five per sclerotium when proper humidity and light conditions are provided to overwintered sclerotia. The best appothecial development takes place at 7–11°C (Kruger, 1976) and their production ceases at 16–28°C with low soil moisture. Length of the stipe of the apothecia ranges from 8–20 mm, whereas diameter of the apothecial discs ranges from 2–9 mm. Apothecia are brown in colour and are round or lobate type. Asci are cylindrical, 100–153 x 10–15 µm in size. Each ascus contains eight ascospores, which are released in clouds. Ascospores are elliptical and ranged from 11–13 x 6–6.5 µm in size.

Host range

The pathogen is polyphagous in nature and has a wide host range of over 350 species in 60 families including plants of uneconomical importance (Gupta and Dohroo, 1993). Roy (1973) proved its pathogenecity on 27 hosts including that on cauliflower (*B.c.* var. *botrytis*). Khan (1976) reported that all the vegetable hosts *viz.*, knol khol (*Brassica oleracea* var. *caulorape)*, lettuce (*Lactuca sativa*), cabbage (*Brassica oleracea* var *capitata*), brussel's sprout (*B. oleracea* var. *gemmifer*), spinach (*Spinacia oleracea*), fennel (*Foeniculum vulgare*), tomato (*Lycopersicon esculentum*), pea (*Pisum sativum*), French bean (*Phaseolus vulgaris*), capsicum (*Capsicum annuum*), mustard (*Brassica oleracea* var. *sarson*), brinjal (*Solanum melongena*), turnip (*Brassica rapa*), beans (*Phaseolus coccineus*), carrot (*Daucus carota*), celery (*Apium graveolens* var. *dulca*) were found susceptible to the pathogen. Corato and Baviello (2000) reported the occurrence of this pathogen on *Brassica carinata* and *Hibiscus cannabinus*.

Disease cycle

In the areas where crop is grown regularly year after year in the same field, the fungus overwinters in soil in the form of sclerotia in the plant debris. They possess a considerable degree of resistance to drought, cold and other destructive agencies. In the presence of proper humidity and light conditions, sclerotia germinate by forming apothecia, which in turn form asci and ascospores. If the spores, upon escaping from the ascus, lodge on a susceptible host, a new infection may originate. Mycelium from sclerotia is also capable of infecting cauliflower. After infection, the fungus produces oxalic acid, which kills the cells in advance. During cool and moist weather fluffy mycelium appears on the host surface and when food is exhausted, the mycelium coagulates and starts forming sclerotia, which are initially white in colour, but later turn black (Fig. 4).

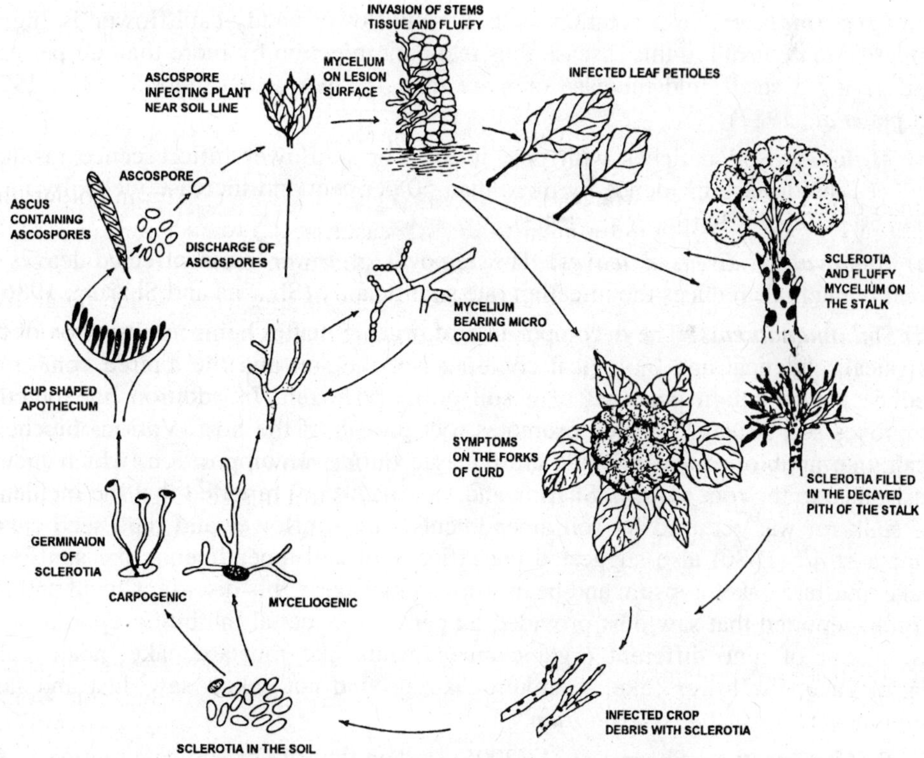

Fig. 4. Disease cycle of stalk rot of cauliflower

Epidemiology

The fungus responsible for this disease is soil borne in the form of sclerotia, which can infect the susceptible host over a wide range of temperatures *i.e.,* from 0 to 25ºC with an optimum at 15 to 20ºC. The fungus can tolerate wide pH range, but is best adapted to an acidic substrate. Sharma and Sharma (1985) also studied the effect of weather parameters on the development of this disease in cauliflower and reported that a temperature regime of 20–25ºC and 90–95 per cent relative humidity are congenial for disease development. A multiple correlation between disease incidence and three independent variables *i.e.,* temperature, rainfall and humidity was found significant for the disease development and 73 per cent of the disease has been found to be affected by these factors. Application of nitrogenous fertilizers significantly enhanced this disease (Sandhu, 1986). Doses of CAN (calcium ammonium nitrate) gave 66.32 per cent disease incidence compared with urea (47.8%) and zero nitrogen (33.4%).

Management

Since, the fungus responsible for this disease is a persistent soil inhabitant, many cultural, biological and fungicidal control methods besides host resistance have been tried for the control of this disease.

i. Cultural methods

***(a) Crop rotation*:** Crop rotation such as cauliflower-paddy-cauliflower is highly effective in controlling this disease thus reducing infection by more than 60 per cent and giving a seed yield increase of more than 161 per cent (Sharma *et al*., 1983; Gupta *et al*., 1987)

***(b) Mulches*:** Soil mulches with pine needles or sunflower inflorescence residues reduces the disease incidence by more than 50 per cent and increase yields by more than 50 per cent (Sharma *et al*., 1983).

***(c) Removal of diseased leaves*:** The removal of lower most affected leaves at weekly intervals reduces the infection rate significantly (Sharma and Sharma, 1986).

(d) Soil amendments: The decomposition of organic matter helps in alteration of the physical, chemical and biological conditions of the soil and the altered conditions reduce the inoculum potential of a soil borne pathogen. In addition, the practice improves soil structure, which promotes root growth of the host. Various biochemicals like antibiotics and phenols are released during decomposition, which induce resistance to the root system. Sharma and Sharma (1986) reported that the incidence of stalk rot was reduced by soil amendments with sunflower and rape seed cakes. Gupta *et al.* (1986) also suggested the efficacy of soil amendments like sunflower cake, mustard cake, gypsum and bean straw in managing this disease. Chand and Rai (2008) reported that saw dust provided 52 per cent mycelial inhibition upon *in vitro* evaluation of nine different organic amendments like mustard cake, neem cake, castor cake, sunflower cake, sesamum cake, ground nut cake, saw dust and farm yard manure against *S. sclerotiorum*.

(e) Soil solarisation: Sharma *et al.* (2005) studied the role of soil solarization in the management of this disease and reported solarization of soil during May and June reduced the populations of *S. sclerotiorum* in soil (89.2–98.3%) and ability of surviving sclerotia to form apothecia (90.0–98.6%). The maximum reduction (100%) occurred in the top 7.5 cm layer of soil after 8 weeks, but significant effects of solarization were seen at 15 and 30 cm soil depths after 9 weeks (100.0 and 93.3%). They further reported that these reductions were due to microbial colonization by

Trichoderma harzianum and *T. viride* and degradation of sclerotia weakened by sublethal temperature produced by solarization.

(ii) Host resistance

Cultivars/lines *viz.*, Master Osena, Avans, Janavon, Early Winters Adam's Head, EC 162587, EC 103576, KN- 81, EC 131592, SN-445, RSK, RSK-1301, BR, KN-81, KM-1, KM-1 x KJ-38, RSK-130 x KJ-38 and Olimpus have been reported resistant (Sharma *et al.*, 1982; Dohroo, 1988; Sharma *et al.*, 1995; Kapoor, 1996; Sharma *et al.*, 2000; Sharma *et al.*, 2003). These cvs./lines can be utilized in breeding programmme against stalk rot of cauliflower. However, Janavon reported resistant against this disease (Dohroo, 1988; Kapoor, 1996) was found moderately resistant later (Sharma *et al.*, 2000). The resistance to the disease is inherited polygenically and under the control of recessive genes, and primary to additive gene action (Baswana *et al.*, 1991), while the best cross of non-heading cabbages PI 206942 x cabbage and cauliflower revealed major recessive gene for resistance plus modifiers (Dickson and Petzoldt, 1996).

(iii) Chemical methods

Different fungicides show differential activity on various stages in life cycle of *Sclerotinia* sp., but benzimidazoles have significant impact on the reduction of this disease. Letham *et al.* (1976) reported that foliar sprays of benomyl gave excellent control of both aerial and basal infection of cauliflower. Sharma and Sharma (1984) evaluated fourteen fungicides against this disease and reported that all the fungicides reduced the incidence of disease and improved seed yields, but Benlate (benomyl) and MBC (carbendazim) performed best. Benomyl has been reported to be the most cost effective chemical. Pre-sowing soil application of carbendazim (5%) granules effectively controlled *S. sclerotiorum* (Sugha, 1999). Maxiumum reduction in disease incidence was caused by carbendazim followed by metalaxyl-MZ, mancozeb and thiophanate methyl (Zewain *et al.*, 2005).

(iv) Biological control

The biological control aims at eradication and control of pathogens through the activity of other microorganisms. In this method, such procedures are included which enhance the activity of microorganisms in soil or at the infection site. *Trichoderma viride* was reported to have maximum antagonistic activity against the pathogen (Gupta and Agarwala, 1988). Suppression of *S. sclerotiorum* by culture filterates of rhizosphere fungi like *Fusarium solani, Aspergillus terreus, T. viride* and *Rhizopus arhizus,* has also been reported (Gupta and Dohroo, 1989). *Gliocladium virens* and isolates of *T. harzianum* and *T. viride* were found highly effective in causing considerable growth inhibition of *S. sclerotiorum* at three incubation temperatures (15, 20 and 25°C) tried by agar plate method, however, the lowest incubation temperature of 15°C gave enhanced growth inhibition of the test pathogen (Kansal *et al.*, 1990). Sclerotial germination was reduced when they were dipped for 1 h in spore suspension of *T. harzianum* spore suspension or in *Azadirachta indica* leaf extract (Sharma and Basandhari, 1997). Application of culture filterate of *T. harzianum* ware also found effective in reducing the incidence of this disease and increased seed yield (Zewain *et al.,* 2005). Under field conditions, application of Nemazal and *T. harzianum* reduced the disease incidence to 40 per cent (Cuong and Dohroo, 2007).

4. WHITE RUST

It is one of the most common diseases of radish and other crucifers throughout the world and can cause considerable loss when it occurs alongwith downy mildew. In India, among vegetables it is more serious on radish than other vegetables. The disease is more serious in seed crops of *Brassica* sp. and once this disease is introduced in any area, its incidence increases continuously.

Geographical distribution

The disease is widely prevalent on cultivated *Brassica* and other cruciferous species through out the world. The occurrence of the disease has been reported from U.K., USA, Brazil, Canada, Germany, Japan, Pakistan, Palestine, Romania, Turkey, Fiji, New Zealand and China.

In India, the disease was first time reported from Assam (Chowdhary, 1944). Since then the disease is prevalent in almost all the states of the country wherever cruciferous vegetables are grown.

Symptoms

Local infection results in to isolated pustules on stem and leaves which show characteristic raised glossy white blisters 1–2 mm in diameter (Fig. 5). Systemic infection stimulates hypertrophy and hyperplasia resulting in enlarged and distorted organs. The inflorescence may become enormously thickened (12–15 times the normal thickness), fleshy and greenish leading to sterility (Plate 4). Blisters after rupturing the host epidermis release white powdery mass comprising of sporangia.

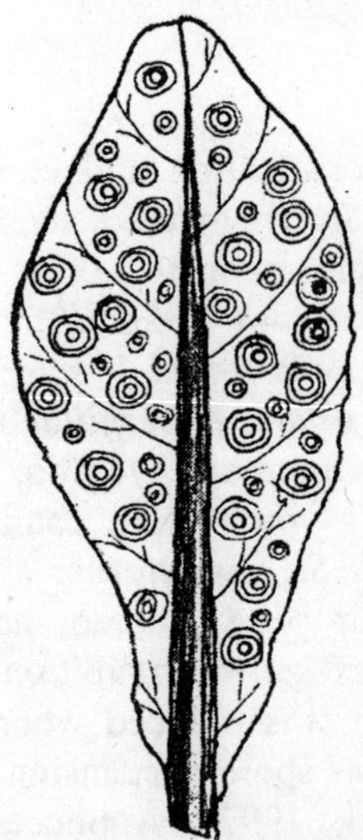

Fig. 5. Symptoms of white rust

The Pathogen

The fungus responsible for this disease is *Albugo candida* (Pers. Ex Chev.) Kuntze. It is an obligate parasite reproducing both sexually and asexuallay. Mycelium is intercellular and non-septate which produces knob shaped haustoria inside the host cells to draw nourishment. Numerous short conidiophores arise from the mycelium, which are arranged in rows beneath the host epidermal layer press on the epidermis to cause the pustule/blister. Inside the pustules, sporangia are produced in chains arising from the sporangiophores at the base of the cavity. The sporangia are spherical, thick walled hyaline and interconnected by a pad of gelatinous disclike tissue. In the presence of moisture these pads swell and disintegrate freeing the sporangia from the chains. The sporangia are hyaline, nearly spherical and measure 14–16 x 16–20 μm in size. They germinate either by germtube or by zoospores.

The oogonia and anthredia are formed in the intercellular spaces and they join by means of fertilization tube, resulting in oospore formation. The oospore is spherical with a thick irregular wall, deep yellow to dark brown in colour. The oospore germinates by formation of a vesicle in which zoospores are formed. The zoospores are elliptical to kidney shaped and are biflagellate. They soon loose the flagella become encysted and germinate by germtube.

Host range

A. candida infects a large number of host plants belonging to Aizoaceae, Cruciferae, Capparidaceae and Cleomaceae families, including leaves of *Reseda alba* (Mukerji, 1975; Waterhouse, 1975). It infects plants of 241 species in 63 genera of Cruciferae family alone (Biga, 1955).

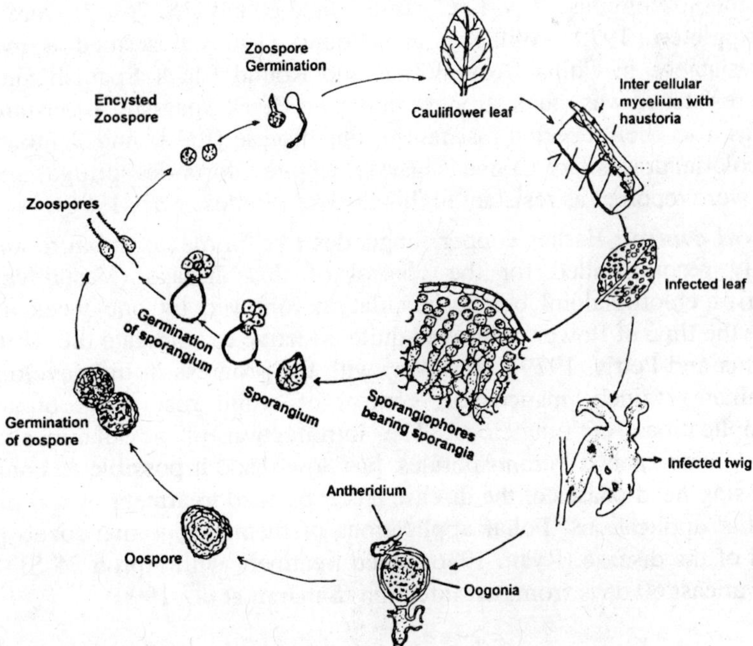

Fig. 6. Disease cycle of white rust of crucifers

Disease cycle

The fungus survives either as oospores in the infected plant debris or mixed with seed or as perennial mycelium in some weed hosts. These oospores may germinate in the ensuing crop season to produce zoospores, which serve as primary inoculum. Secondary spread is through sporangia produced on the spots caused by primary inoculum (Fig. 6).

Epidemiology

The fungus subsists from one season to another by means of oospores and perennial mycelim in cruciferous weeds. Cool and moist weather favours the disease development and prevented by low relative humidity. Moisture is essential for germination of sporangia through zoospores. The optimum temperature for germination of sporangia by zoospores is 20°C. White rust progressed at a faster rate with dewfall, mean relative humidity >65 per cent, mean temperature 10 to 18°C and potential evapo-transpiration < 60 (Lakra and Saharan, 1989).

Management

(i) Cultural practices: Practices like collection and burning of diseased plant material in order to prevent oospore formation, isolation of cauliflower from other crucifers, reduction of humid conditions around the plants by maintaining proper spacing, weed free cultivation, application of fertilizers like phosphorus and potassium and avoidance of excess application of nitrogen help in reduction in the disease severity.

(ii) Host resistance: Brussel's sprout cultivars like Topscore, Ulysses, Tetrarch, Wanhoe, Troika, Romulus, AW-1, Corvette, Jade-6 and RS 76055 were rated as resistant (Appleton, 1979). Williams and Pound (1963) described a monogenic dominant resistance in China Rose Winter and Round Black Spanish varieties of radish. The resistance was much stronger in Round Black Spanish. A mustard variety RH 8541 has also been reported resistant to this disease (Lakra and Saharan, 1988). Portugese cole landraces like Couve Algarvia, Couve Gloria de Portugal and Couve Portuguess were reported as resistant to this disease (Santos *et al.*, 1996).

(iii) Chemical control: Earlier copper fungicides like Bordeaux mixture and Blitox were mainly recommended for the control of this disease (Vasudeva, 1958). Application of chlorothalonil before inoculation followed by one week after first spray and at the time of flowering proved quite effective to eradicate the white rust of turnip (Verma and Petrie, 1979). However, with the progress in the development of dithiacar-bamates (zineb, mancozeb), control of white rust was obtained with multiple applications of fungicides. The introduction of acylalanines, specific systemic fungicide against Peronosporales, has now made it possible to control both the leaf and stag-head phases of the disease either by seed treatment or soil drenching and few foliar applications. Foliar applications of metalaxyl + mancozeb provided best control of the disease (Ryan, 1986). Seed treatment with Apron 35 SD protects the crop for atleast 60 days from rust infection (Saharan *et al.*, 1990).

5. CLUB ROOT

This is one of the important diseases of cruciferous vegetables particularly in temperate climates. Once the disease appears in a field it is very difficult to grow a profitable crop in that field and the affected plants are complete loss because of premature wilting.

Geographical distribution

This is one of the earliest known diseases of crop plants in Europe. The occurrence of the disease has been reported from Russia, England, USA, Korea, Turkey, Sri Lanka, India and Austria.

Although the disease is not common in India but it has been reported from Tamil Nadu (Loganathan *et al.*, 2001). The disease has also infested other temperate seed belts of other states.

Losses

The disease incidence in Turkey has reported as high as 91.4 per cent (Gulsoy, 1978). In Tamil Nadu, the incidence of the disease ranged from 13–51 and 19–35 per cent in cabbage and cauliflower, respectively (Loganathan *et al.*, 2001).

Fig. 7. Symptoms of club root of crucifers

Symptoms

The symptoms of the disease are recorded only after the pathogen has progressed considerably in root system. The earliest symptoms include the drooping of the leaves during hot of the day. The other above ground parts of the affected plant show stunting, chlorosis and reduction in the size of head. When such plants are pulled out and the roots are examined, hypertrophy of root system can be seen (Fig. 7). The roots are malformed into club-like structures due to thick, fleshy growth of roots. In general the root tips are malformed, leaving the basal portions of the root mostly normal. The hypertrophy causes malfunctioning of the xylem, which results in drooping of the leaves. In advanced stages, the affected tissues are invaded by secondary saprophytic organisms mostly soft rot bacteria, which rot such tissues and finally turn black in colour.

The Pathogen

The fungus responsible for this disease is *Plasmodiophora brassicae* Woronin. The fungus produces several resting spores inside the host cells, which are lying in soil or diseased plant debris. The spores are hyaline, spherical, contain chitin in walls and measure about 4 μm in diameter. In favourable conditions, they germinate by swelling to several times their original size, giving rise to a single biflagellate zoospore from each resting spore which emerges through a rift in the cell wall. The zoospores are naked, devoid of cell wall, with uninucleate protoplasm, irregular or amoeboid in shape. However, the shape changes frequently due to the movement of the protoplasm. In contact with root hair or epidermal cells of the host roots, the zoospore shed flagella and bodily enters the root through a hole dissolved in the cell wall. The hole is then closed by host action. Once inside the host, the pathogen can be seen in a plasmodial stage multiplying and spreading within host cells. With the progress of the disease, the plasmodium divides in to unicellular units and around each such unit a wall is formed, resulting in a resting spore. The resting spores are capable of germinating immediately after formation.

Host range

The fungus can attack several cruciferous plants and weeds, which are commonly found in and around the field.

Disease cycle

The pathogen perennates in soil as resting spores in the infected plant debris. The pathogen once introduced in a field, may survive in soil for more than 10 years. When the susceptible crop plants are planted in such fields, resting spores infect the roots. The spores multiply and spread in the plasmodial stage, causing hyperplasia of the host cells and later forms the resting spores (Fig. 8).

Epidemiology

The club root organism is soil borne. The disease spread from one field to another either through seedlings raised on infested soil or wild species of mustard which act as reservoir for the inoculum. The disease can develop at a wide range of temperature and the disease is not restricted to any region on account of temperature. The organism is sensitive to alkaline conditions and the spores germinate poorly under these conditions. Soil moisture also play important role in spore germination. If the soil is dry, the spores of the fungus fail to germinate and the disease development is checked. Moist soil permits spore germination though it may be for short periods only. Movement of infested soil from one field to other can also take place on farm implements and through surface irrigation water.

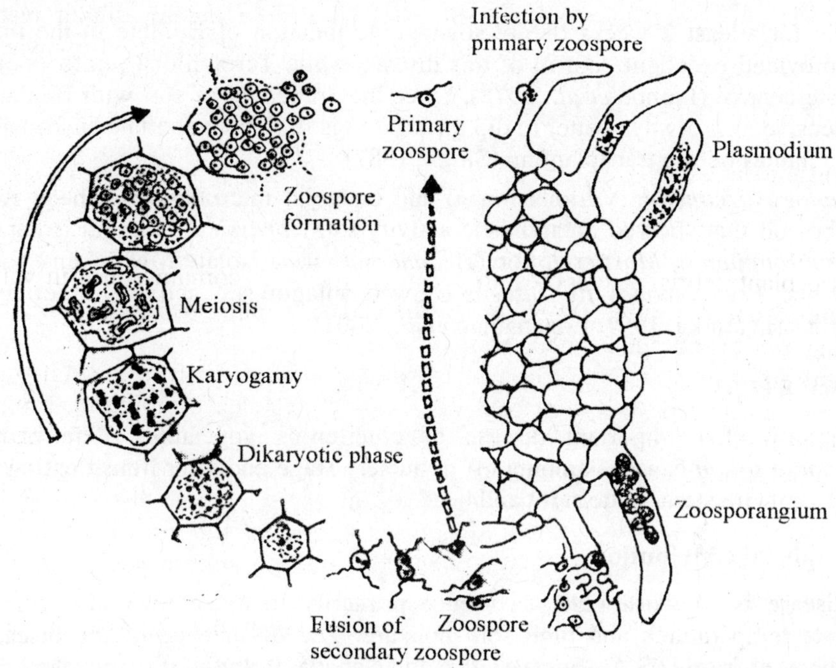

Fig. 8. Disease cycle of club root of crucifers

Management

(i) Cultural practices: The pathogen is soil borne and once introduced in soil it is very difficult to eradicate. Some preventive measures like destruction of infected plant debris, removal of cruciferous weeds from in and around the field, and raising of nursery in disease free soil should be followed. Since the pathogen is sensitive to alkaline conditions, the pH of the soil should be raised to 7.2 by application of lime. After liming, the field should be irrigated immediately so that the movement of alkaline particles to neutralize the acidity can take place. Manipulation of supplies of calcium, boron, nitrogen and soil pH values is associated with adantagenous changes to soil microflora and diminution of the efficiency of stages in the life cycle of pathogen (Dixon, 2008) cultivation of short duration resistant *Brassica* hosts such as radish and toria sarson reduced clubroot disease intensity and enhanced the yield of cauliflower, cabbage and leafy vegetable rayosak in Darjelling area of West Bengal (Bhattacharya and Dixon, 2010).

(ii) Host resistance: Many workers screened various cvs./lines/accessions of cabbage, kale, cauliflower, brussel's sprouts and broccoli to find out club root resistant sources (Crete and Chang, 1980; Monteiro and Williams, 1989; Dias *et al.*, 1993). Monteiro and Williams (1989) recorded that kale type was more resistant than cabbage. No resistant cauliflower or broccoli has been detected but several land races of cabbage and kale have been identified. Broccoli and radish lines OR-1 and Sexafire, respectively were found completely resistant.

(iii) Chemical control: Club root can virtually be eliminated by soil fumigation with methyl bromide (Lambe *et al.*, 1978) or by treating the soil of the nursery bed with formalin for atleast 2 weeks before sowing. Application of Benlate in the infected fields provided excellent control of this disease while Terrachlor (quintozene) gave moderate control (Lambe *et al.*, 1978). Since the drenching of soil with Benlate will be expensive, it is always better to dip the seedlings in (0.5%) Benlate suspension for 15–20 minutes before transplanting (Singh, 1987).

(iv) Biological control: Various fungal and bacterial microbes have been isolated from the soil that showed antagonistic activity to *P. brassicae*. Fungal isolates like *Haplosporangium* or *Mortierella* or *Trichoderma viride* isolate MNT 7 and bacterial isolate like *Pseudomonas fluorescens* showed antagonistic activity under *in vitro* conditions (Djatnika, 1991; Loganathan *et al.*, 2001).

6. WIRE STEM

This is an other important disease of cruciferous vegetables. The pathogen, *Rhizoctonia solani* causes damping off in nursery stage and after transplanting plants suffer from wire stem, bottom rot and head rot.

Geographical distribution

The disease is of worldwide occurrence primarily in those regions, which have moderate temperatures and high soil moisture. The occurrence of the disease has been reported from U.S.A., Europe, Asia and Canada. In India, it is prevalent in hilly areas of Himachal Pradesh, Jammu and Kashmir, Uttranchal and North Eastern States. Gupta and Shyam (1995) reported the occurrence of this disease from various locations of Himachal Pradesh.

Losses

In nursery bed the pathogen causes considerable losses and often more than 50 per cent plants are killed in seedling stage. Yield losses due to wire stem are estimated as 12 per cent loss of heads when the plants are infected slightly, whereas under severe infection losses increase to 36 per cent (Dixon, 1981). In certain fields under Himachal Pradesh conditions, the incidence of the disease was more than 50 per cent while in general it ranged from 15–20 per cent (Gupta and Shyam, 1995). In Kullu valley, the estimated loss has been reported upto 60 per cent in transplanted cauliflower crop due to this disease (Kapoor, 1999).

Symptoms

In nursery, the disease kills the young seedlings, sometimes even before their emergence. The pathogen can invade the seedlings beyond juvenile stage, may be up to 15–20 cm high. Symptoms on such seedlings appear as light brown to dark brown areas on the stem of the infected plants extending 1–2 cm below the soil line. The lesions girdle the stem, which become sunken and at later stage turned in to tough and woody structure. Such affected areas are relatively thinner than the stems of the healthy plants and this phase is called as wire stem (Plate 5). In older plants, it causes bottom rot in which lower leaves droop and dry out and turn dark. Sometimes the infection may progress to upper leaves and eventually reaches the head and causes head rot.

The Pathogen

The fungus responsible for this disease is *Rhizoctonia solani* Kuhn (Tel: *Thanatephorus cucumeris* (Frank) Donk). Mycelium is septate. Hyphae are up to 12 μm wide and 250 μm long, branches develop from near to the distal end of the cell, with a constriction at the point of origin and septa forming above it. Conspicuously, the septa are of the dolipore type. In culture, sclerotia develop within 5–6 days. The size of sclerotia ranges from 1 to 3 mm in diameter. The sclerotia consist of brown barrel shaped monilioid cells measuring up to 30 μm in width.

The basidial stage of the fungus (*Thanatephorus cucumeris*) also occurs but very rarely. The sexual hymenium consists of a thin sheet composed of barrel-shaped or subcylindrical basidia 10–25 x 6–19 μm, from which between four and seven sterigmata develop which are 5.5–36.5μm long with occasional septa. Borne on the sterigmata are basidiospores, which are hyaline, oblong to broad ellipsoid and unilaterally flattened. They are smooth and thin walled, 6–14 x 4–8 μm in size.

Host range

The fungus is polyphagous in nature and has been recorded on a number of hosts from various parts of the world. It is most widely distributed and attacks 287 genera belonging to dicotyledonous and monocotyledonous families of Angiosperms and also three each of Gymnosperms and Pteridophytes (Ahuja, 1976).

Disease cycle

The pathogen is soil borne. It survives from one season to another usually as mycelium or through sclerotia in the soil and in or on infected perenial plants or propagative material such as potato tubers. The sclerotia are resistant to drought and to large fluctuations in temperature. It is present in most soils and once established, remains their indefinately. It spreads through rain, irrigation water, tools etc.

Epidemiology

The pathogen is soil borne in nature. The optimum temperature for disease development is 22–24ºC, while higher soil moisture enhanced disease development (Iqbal *et al.*, 1988). Cauliflower plants up to the age of 20 days are more prone to attack than 35 days old and sandy loam soil favoured maximum incidence while it was minimum in clay soils (Chauhan *et al.,* 2000). The incidence, of the disease was lower in those fields, which have neutral pH, clay loam soil and received balanced doses of NPK fertilizers (Sharma and Chaudhary, 1984). Application of NH_4 form of nitrogen decreases the disease incidence as compared to NO_3 form (Chernyaeva and Bisuas, 1980).

Management

(i) Cultural practices: Practices like use of healthy seed, change of nursery site every year, disinfection of the plant bed either with soil solarization (Pullman *et al.*, 1981) or by fumigants like formaldehyde, deep ploughing during summer months, shallow seeding to encourage rapid emergence and selection of the healthy seedlings at the time of transplanting should be followed to keep the disease under check. Soil amendment with cellulose powder was also found effective in reducing the disease incidence, which may be due to increase in C : N ratio in soil. This results in the decrease of fungal population, but actinomycetes and bacterial populations increased (Kundu and Nandi, 1985).

(ii) Host resistance: Cultivars like South Pacific, Chou-fleur d' orgevale and Autumn Giant (D'- Ercole, 1974; D'-Ercole and Nipoti, 1978) were reported resistant to this disease.

(iii) Chemical control: The seed should be treated either with captan, carbendazim, iprodione or thiophanate methyl (Yada and Imoto, 1983) before sowing. Foliar application of carbendazim or mancozeb alone or in combination has also been recommended (Jana *et al.*, 1989).

(iv) Biological control: Applications of bioagents either as seed treatment or soil application were found effective in reducing the incidence of the disease. *Trichoderma harzianum* and *Arachniotus* sp. applied as seed coating in the form of paste and with and without certain chemical amendments in the pot soil gave significant control of the disease (Iqbal and Khalid, 2002). The combination of seed treatment and soil application with *Trichoderma* sp. reduced pre- and post-emergence damping off of cabbage in nursery and enhanced seedling vigour (Roy *et al.*, 1998).

A spore mixture of *Streptomyces arenae* and *S. chibaensis* at 1.5 x 10^7 spores/ml proved to be most effective in controlling the disease besides improving shoot length and dry mass of seedlings (Kundu and Nandi, 1984). An antibiotic and siderophore producing *Pseudomonas* strain (RRLJ 008) displayed antibiosis against this pathogen and seed bacterization improved the germination, shoot length, root length, fresh and dry mass and enhanced the yield and chlorophyll content of leaves (Boruah and Kumar, 2002). Soil treatment with water dispersible granules or spore suspension and seed treatment with *Bacillus subtilis* strain B2g resulted in increased emergence rates in cabbage plants inoculated with *R. solani* (Marten *et al.*, 1999). Cauliflower seedling dip and soil application of *T. harzianum* significantly reduced the disease incidence (Bhagat and Pan, 2008)

Cauliflower plants preinoculated with VA mycorrhizae developed less infection by *R. solani* as compared to uninoculated plants (Iqbal *et al.*, 1988). VAM infection in preinoculated plants was higher at 18-20°C temperature and 15-20 per cent soil moisture.

(v) Integrated disease management: Soil amended with neem cake (1 kg/m^2) and solarized by covering with white polyethylene sheet for two weeks and treatment of seeds with *T. harzianum* proved effective in the management of this disease (Singh *et al.*, 2002).

7. YELLOWS

Erwin F. Smith first called attention to this disease on cabbage in the Hudson valley of USA in 1899 (Dixon, 1981). Since then the disease has been reported from warmer areas of the world. This disease is not significant in winter crops.

Geographical distribution

Besides U.S.A, the occurrence of the disease has also been reported from Iowa, New Mexico, Japan, Ukrain, Greece and India. In India, The occurrence of the disease on cabbage, turnip and radish has been reported from West Bengal, Himachal Pradesh and Kashmir Valley (Chattopadhyay and Mustafee, 1967; Gangopadhyay and Kapoor, 1975).

Losses

The yellows disease can cause heavy losses to the cabbage crop and 50 to 100 per cent yield reduction is common in contaminated fields. The disease caused about 90 per cent losses in Iowa and Kansas, USA (Harter and Jones, 1923). About 50 per cent cabbage yield losses were recorded at Sonu Poulo State (Galli and Tokeshi, 1962).

Symptoms

The disease affects the plants at any age. The first sign is the lifeless yellow green colour of the foliage of the affected plants. Sometimes the yellowing is uniform, but more often it is more intense on one side of leaf and plant causing curling of the leaves and stem. The lower leaves become yellow first and then the disease progresses upward. Affected leaves droop prematurely and growth of plant is retarded. The vascular system becomes yellow to dark brown. The yellowing and drying of the plants are chiefly due to the toxic effects of the byproduct of the fungus in the vessels. Severely infected plants die prematurely.

The Pathogen

The disease is caused by *Fusarium oxysporum* Schlechtend ex Fr. f.sp. *conglutinans* (Wr.) Snyder and Hansen. The mycelium is septate, initially hyaline, becoming cream coloured with age and finally showing ocherous strands throughout the colony. Microconidia are single celled, hyaline, ovoid to ellipsoid, measuring 6–15 x 2.5–4 μm in size. Macroconidia are usually relatively rare. They are fusiform, hyaline and mostly 2–3 septate and measure 25–33 x 3.5–5.5 μm in size. In older mycelium, chlamydospores are formed. In all, five physiological races like race 1, 2, 3, 4 and 5 of the pathogen have been distinguished (Armstrong and Armstrong, 1952; Ramirez-Villapudda *et al.*, 1985).

Disease cycle

The pathogen is soil borne first as saprophyte on diseae plant debris and then as chlamydospores in the absence of the host. The organism thrives in variety of soil types and once it is established, it remains there indefinitely. The disease is disseminated by means of soil on implements and through irrigation water. The chlamydospores penetrate the host through root tips and invade the root cortex.

Epidemiology

The pathogen is soil borne in nature and survives in the form of chlamydospores. It is not seed borne. The disease is favoured by high soil temperature and below 17°C very little infection can take place. Excessive soil moisture and higher doses of nitrogen favour the disease. The pathogen can be disseminated through seedlings raised on contaminated soil, implements and surface irrigation water.

Management

(i) Cultural practices: The disease can be kept under check by following crop rotation, deep ploughing during summer months, application of well rotten manure combined with potassium fertilizers, use of disease free soil for raising the nursery, addition of vermicompost in the bed soil (Szczech and Brzeski, 1994) before sowing. Disease can be greatly reduced by soil solarization and dried cabbage residues amendment (1% w/w). The solarization will not only increase the temperature of the soil to a critical level but also trap the fungitoxic gases emanating from the cabbage amendments (Ramirez- Villapudda and Munnecke, 1987).

(ii) Host resistance: Wilt and yellows diseases are best controlled through resistant varieties. Yellows resistant varieties like Wisconsin All Seasons, Wisconsin Ballhead and Wisconsin Hollander, all of which contain single gene for resistance to *F.o.* f.sp. *conglutinanas* (Pound and Walker, 1951). Two varieties Hybelle and Sanibel were developed as multiple disease resistant F_1 cabbage hybrids with major genes resistant to this disease (Williams *et al.*, 1968) while cabbage Harmat and F_1 hybrid Drumhead were partially resistant (Pesti *et al.*, 1987).

(iii) Chemical control: The soil of nursery bed can be sterilized by application of either formaldehyde or dazomet (Takeuchi *et al.*, 1995). Dip the seedlings before transplanting in the solution of carbendazim or roots in zinc dimethyl dithiocarbamate (1%) for 30 minutes before transplanting (Gangopadhyay and Kapoor, 1975).

(iv) Biological control: Several biological control agents like *Penicillium* sp., *Bacillus subtilis* and Actinomycetes (Reyes, 1984; Kobayashi, 1991) have been reported effective in reducing the incidence of the disease in heavily infested fields.

8. BLACK LEG

It is an important disease of cruciferous vegetables, oilseed and fodder brassica crops which causes considerable damage to these crops particularly in favourable environmental conditions.

Geographical distribution

The occurrence of this disease has been reported in several continents of the world wherever these crops are grown (Anonymous, 1978). The disease was first recorded in France in 1849 and in USA during 1910. In 1973, there was a severe outbreak of the disease in Midwestern and eastern cabbage growing areas of the United States (Gabrielson and Maguire, 1977). In India, the disease has not been a serious problem, though its occurrence on cabbage has been reported from Kullu Valley in 1961 (Munjal and Gill, 1961).

Losses

The disease causes serious losses in cruciferous vegetables and oil seed crops, the world over. In Canada, the disease is responsible for significant yield losses in oil seed and turnip rape (Chen and Seguin Swartz, 1996). Yield losses up to 70 per cent due to this disease have been reported from Mexico in cauliflower crop (Moreno-Rico *et al.*, 2001).

Symptoms

The plants are susceptible to the disease from nursery stage to maturity. All plant parts like stems, leaves, pods and seeds are infected. The symptoms first appear as oval, depressed, light brown to grey cankers near the base of the stem, which gradually enlarge to girdle the stem and are often surrounded with a purplish border (Fig. 9). Lesions on leaves are round to irregular shaped, white to buff in colour that become papery with age and are covered with numerous black dots on them, representing the fruiting bodies on them. Severely girdled cabbage plants often survive until the weight of the mature heads cause the plant to fall over due to the weakened conditions of the stem. Cankers on the stem restrict the uptake of water and nutrients, which cause premature ripening and shrivelled seeds and pods.

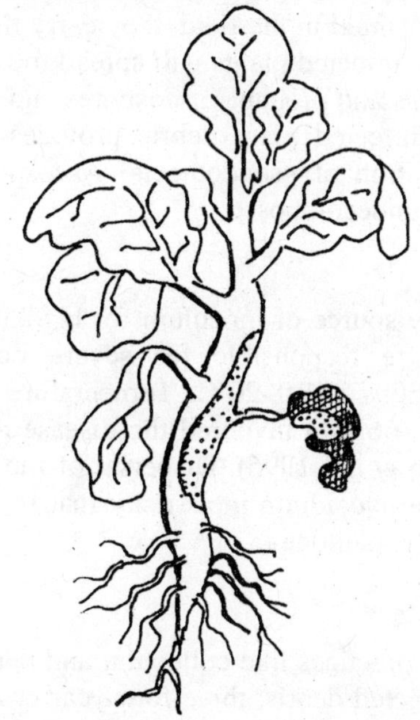

Fig. 9. Symptoms of black leg of crucifers

The Pathogen

The disease is caused by *Leptosphaeria maculans* (Desm.) Ces & de Not. while asexual stage is *Phoma lingam* (Tode ex Fr.) Desm. Asexual stage is common in nature. The hyphae are septate, branched, hyaline when young and become dark walled with age. Two types of pycnidia *i.e.,* sclerotioid and globular type exists. Both have black walls, which are several cell layers thick. Sclerotioid type has walls of several layers of thick walled pseudosclerenchymatous cells and is 200 to 500 μm in diameter and is found on dead, woody tissues and may serve as overwintering structure. A globular type has cell walls that are thickened in the outer cell layer only and is found on living tissue and in cultures only. Pycniospores are hyaline, cylindrical, mostly straight, one celled and 3–5 x 1.5–2 μm in size. When they are in abundance, they appear pink in colour. The sexual stage consists of pseudothecia which are immersed, becoming erumpent, globose, black with protruding ostioles and are 300 to 500 μm in diameter. The pseudoparaphyses are hyaline, segmented and filiform. Asci are bitunicate, cylindrical to clavate and are 80–125 x 15–22 in size. They contain eight biseriate, multiseptate, and cylindrical to ellipsoidal ascospores with rounded ends that are yellow brown in colour and measure 35–70 x 5–8 μm in size.

Disease cycle

The pathogen perennates in the infected seed as dormant mycelium (Bedlan, 1985) or in diseased plant debris. If the infected seed germinates, the fungus also becomes activated and cause infection of the hypocotyls. The infected seedlings are killed and the fungus on such dead seedlings continued to produce pycniospores, which cause secondary infections. Raindrops and irrigation water disperse the pycniospores. If the infected seedlings are transplanted in the field, they carry the disease to the field and the spores produced on such infected plants will spread the disease in field. Pods can be infected with pycniospores and probably ascospores and can lead to seed infection (Gabrielson, 1983). Dead infected plant debris provides the substrate for sexual reproduction and the formation of pseudothecia. Ascospores are produced in the debris as long as it remains undecomposed.

Epidemiology

Infected seed is the primary source of inoculum for blackleg epidemics in cabbage. Only the severe strains are responsible for severe epidemic. Ascospores are discharged in moist conditions at 10–20°C. Temperature (18.3°C), rainfall (122.3 mm) and relative humidity (66.2%) favoured the disease development, its dispersal and infection (Moreno-Rico *et al.*, 1996). Dispersal of the pathogen is by splashing of spores that ooze out of a pycnidium in a sticky matrix. Insects feeding on stems can create entry points for the pathogen.

Management

(i) Cultural practices: The practices like collection and destruction of infected plant debris, plough down of infected debris, three-four year crop rotation, use of healthy seed produced in dry climate, seed bed hygiene, delay in sowing to avoid the period of maximum ascospore discharge and destruction of crucifer weed hosts are generally recommended to delay the onset of the disease early.

(ii) Seed treatment: Since the pathogen is seed borne and express itself in the nursery bed. It is always better to use the treated seed. The seed can be treated either with hot water at 50°C for 30 minutes without significant losses of germination (Nega *et al.*, 2003) or thiabendazole applied as slurry at 2.5 g a.i/kg seed (Maude *et al.*, 1984) or seed soaking in benomyl (Harman and Nash, 1978).

(iii) Chemical control: Soil applied zineb @ 5g/m^2, 3 days before sowing reduced disease infection (Klinkovskaya, 1976). Applications of various systemic fungicides like benomyl (Lambe *et al.*, 1978), fluazinam (Mitani *et al.*, 1998), combination of difenoconazole and iprodione (Bedlan, 1998) have been reported effective in the management of this disease.

9. RING SPOT

This disease is important in those areas where cool and moist climatic conditions persist during crop growth period.

Geographical distribution

In 19th century, the occurrence of this disease was reported from France. Since then this disease has been reported from northwestern Europe, southern England, parts of the USSR, the western coastal valleys of California, USA, Peru, Equador, New Zealand (Dixon, 1981), Venezuela (Tortolero and Carrasco, 1982), Germany (Rudnick, 1986) and Netherlands (Ende and Van-den- Ende, 1993).

In India, the disease has been reported from Andeman Islands and Himachal Pradesh (Ramesh and Rao, 1989; Kapoor, 1999).

Losses

The disease causes considerable yield losses if the environmental conditions are favourable for the development of the disease and result in premature defoliation if the mature foliage is heavily affected and reduces seed yields. Early planting (April) of cabbage in Netherlands was severely damaged by this pathogen (Ende *et al.*, 1999). In India up to 40 per cent incidence of this disease was recorded in Andaman Islands (Ramesh and Rao, 1989).

Symptoms

The symptoms of the disease appear on all aerial parts of the plants from the cotyledonary leaves to the seed pods, but usually mature foliage is most frequently attacked. On leaves, brownish grey, small, circular dark spots, visible on both leaf surfaces, which expand up to 2–3 cm in diameter appear. Often a number of clearly delimited, black-specked concentric zones (ring spots) are conspicuous on which fructifications appear. Each lesion usually has a definite edge bounded by a narrow water-soaked area and chlorotic zone where lesions are numerous and the entire leaf becomes yellowish, showing curled, cracked and ragged edges. On leaf midribs, stems and siliquas lesions may be rectangular or oval-shaped. The pathogen affects cabbage in storage also and symptoms appear as dark brown to black lesions that eventually become shrivelled and leathery.

The Pathogen

The disease is incited by *Mycosphaerella brassicicola* (Fr. Ex Duby) Lindau. Two types of spores are produced on the host. In early stages pycnidia like spermogonia appear as black dots. They are flask shaped and contain several one celled, hyaline spermatia measuring 2.5–4.5 x 1.5–2.5 μm in size. These spores do not germinate, but help in spermatisation and development of the sexual fruiting bodies (pseudothecia), which develop later. The pseudothecia are mostly found on leaves, but sometimes on stems and seed pods, as globular dark brown structures with apical papillate ostioles measuring 100–130 μm in diameter. Asci are bitunicate eight spored, 30–45 x 12–18 μm in size. Ascospores are hyaline, cylindrical, bicellualar with rounded ends and no constriction at the septum and 18–23 x 3–5 μm in size.

Host range

The pathogen is wide spread on *Brassica* crops. Besides cauliflower and cabbage, it also affects broccoli, brussel's sprouts, rape, rutabaga and kale.

Disease cycle

The fungus mainly overwinters from one season to another in diseased crop debris and on continuously growing Brassica crops. The pathogen can also survive in infected seed. The ascospores are released from the overwintering inoculum and cause infection. They enter through the stomata and the hyphae established between and within host cells throughout the leaf giving rise to the appearance of lesions on both surfaces of the leaf.

Epidemiology

The primary inoculum mainly consisting of ascospores emanated from the plant debris or from the *Brassica* crop growing nearby. Cool and humid weather favours disease development. Ascospore discharge was found to be stimulated by light and to occur within a broad temerature range *i.e.,* 5–20°C under humid conditions (Ende and Van-den-Ende, 1993). Inoculum containing mycelial fragments required a long duration of leaf wetness (96–120 h) for infection to occur which is uncommon in nature whereas ascospores caused infection after a much shorter duration (<48 h).

Management

(i) Cultural practices: Cultural practices like destruction of infected plant debris, crop rotation, deep ploughing of the soil during summer months, selection of nursery bed site well away from existing crops and free from infected plant debris, wider plant spacing, use of healthy seed, treating the seed with hot water (45°C for 20 minutes) and addition of potassium fertilizers are considered as means of avoiding infection.

(ii) Host resistance: White cabbage cultivars *viz.,* Bentley, Bingo and Glaxy were reported as tolerant to *Mycosphaerella brassicicola* (Biesheuvel, 1991) the cabbage moth.

(iii) Chemical control: The disease can be kept under check by spraying fungicides like benomyl, chlorothalonil, carbendazim, thiophanate methyl, mancozeb and cyperoconazole (Broadhurst and Popay, 1994). However, devlopment of tolerant strains of *M. brassicicola* from Pukekohe region of New Zealand has been reported (Broadhurst and Popay, 1994). This problem of resistance can be overcome by rotation of the fungicides and by mixing of fungicides with different modes of action.

(B) BACTERIAL DISEASES

1. BLACK ROT

It is one of the most destructive vascular diseases of crucifers and occurs in all parts of the temperate and subtropical zones of the world where rainfall or heavy dew are plentiful and average temperatures are in between 25 and 30°C.

Geographical distribution

The disease was first observed at Lexington, Kentucky (German, 1891). Since then it has been reported on various cruciferous hosts from different parts of the world including Philippines, Norway, Mississippi, Delaware, Australia, USA, Melnik, New York, Mauritius, Bulgaria, Holland, South Africa, Rhodesia, Queensland, Mozambique, California, Greece, Israel, Florida, Yugoslavia, Hawaii, Thailand, and Basilicata.

In India, the disease was first reported on cabbage from Bombay (Patwardhan, 1928). On cauliflower, it was first reported by Patel *et al* (1949) and subsequently it was reported from Katrain (Kullu district) area of Himachal Pradesh (Rao and Srivastava, 1964). Since then the occurrence of this disease has also been reported from other parts of the country including Manipur and Kashmir (Gupta, 1991; Gupta and Chaudhary, 1995; Bhat and Masoodi, 1999).

Losses

The disease is very destructive and can cause considerable losses by premature defoliation. It affects the quality of curd and heads in cauliflower and cabbage, respectively, during transit and renders them unmarketable. A crop loss upto 50 per cent in cabbage and 50 to 70 per cent in cauliflower has been recorded (Jorgenson and Walter, 1955). About 50 per cent severity of this disease was recorded in cabbage susceptible cultivars like Golden Acre and Pride of India (Gupta, 1991).

Symptoms

The plants may be affected at any time during their growth period *i.e.,* from the stage of young seedling till maturity. In cauliflower the earliest symptoms on young seedling occur as necrotic lesions on cotyledons or the lower leaves of the seedlings which show blackening and later die and drop off, leaving a symptomless plant for several weeks while the pathogen apparently develops systemically. Later, infection of leaves takes place through water pores at the margins or occasionally through wounds caused by chewing insects like diamond back moth. The causal bacterium colonizes the vascular system and when temperature exceeds 25°C, the bacterial cells move rapidly through the vascular system and produce an extracellular polysaccharide, which plugs the xylem vessels. With the plugging of xylem vessels, water flow is restricted resulting in the formation of characteristic 'V' shaped chlorotic lesions usually originating from the margins of leaves and progressing towards midrib (Plate 6). The veins show a brown to black discolouration. In turnip and rutabaga, the fleshy roots are invaded from the leaves, which show black discolouration and internal breakdown. The vascular discolouration extends to the main stem and proceeds upward and downward. Affected leaves turn yellow and fall down prematurely due to formation of abscission layer. Curds are also invaded and discoloured. Invasion of the flowering stalks may lead to infection of the seed (Cook *et al.,* 1952). Systemic infection during storage often renders the product unmarketable. The infection of this pathogen may be followed by attacks of soft rot bacteria or *Sclerotinia sclerotiorum* (Agrios, 1997).

The Pathogen

The bacterium responsible for this disease is *Xanthomonas campestris* pv. *campestris* (Pammel) Dowson. It is a small, rod shaped, aerobic, Gram negative, non-spore forming bacterium. The bacterium has a single polar flagellum and it is catalase positive, hydrogen sulphide positive, oxidase negative and does not produce nitrate or indole. It produces a yellowish extracellular polysaccharide (EPS) called *Xantham* on media containing glucose. *X. c.* pv. *campestris* hydrolyses starch, a characteristic that is used for easy recognition of the bacterium on Schad's selective medium. Its growth is inhibited or retarded in acidic range of pH. The *in vitro* survival temperature range is very wide (5 to 38°C), with the optimum temperature being 30°C and lethal at about 50°C. Six races of the pathogen are present worldover (Taylor *et al.*, 2002) and out of these, race 1 and 4 are most predominant (Vicente *et al.*, 2001).

Genetic and pathogenic variability among Indian isolates of *X. c.* pv. *campestris* and other species of *Xanthomonas* was studied by Singh *et al.* (2011) and reported that three distinct groups of pathogenic variability *i.e.,* Group 1, 2 and 3 were formed with *X. c.* pv. *campestris* clustering under Group 1 except one strain. The strains of *X. c.* pv. *campestris* further clustered in to 6 subgroups like A, B, C, D, E and F based on disease reaction on host. The hrpf gene sequence analysis also showed distinctness of *X. c.* pv. *campestris* from other *Xanthomonas*.

Host Range

The bacterium is found to attack all members of Cruciferae family (Mahiar and Khalif, 1999) including rutabaga, cabbage, savoy cabbage, broccoli, turnip, radish, kohlrabi, *Lapidium sativum, Raphanus sativus* var. *caudatus, Brassica juncea*, fodder cabbage, *Capsella bursa-pastoris* and *Cantella asiatica,* but failed to infect any of the non-cruciferous plant species tested (Bhat *et al.*, 2010).

Disease cycle

The pathogen survives in infected seed, diseased plant debris and on cruciferous weeds (Kuan *et al.*, 1986). When the infected seed germinates, the seed coat that comes out alongwith the cotyledons serves as the source of primary inoculum. The bacterium enters cotyledons through the marginal stomata and passes into true leaves. In the leaves, the entry is usually through hydathodes. On entry, the bacteria move to the xylem vessels where they multiply rapidly and move up and down the plant. Another important source of inoculum is infected plant debris in soil (Williams, 1980). The pathogen can better survive from season to season when protected by host. The bacterium can also perennate in stems of cabbage and brussel sprouts. Dispersal of *X. c.* pv. *campestris* from cruciferous weeds through air has also been reported (Kuan *et al.*, 1986) which may also serve as important source of primary inoculum. When the disease inoculum is present in the field, secondary spread is facilitated by splashing of contaminated soil by irrigation water and/or rains, blowing of fallen leaves and occasionally by insects (Fig. 10). Potential of flea beetles *Phyllotreta cruciferae* to transmit this bacterium is reported, but to a limited extent (Shelton and Hunter, 1985).

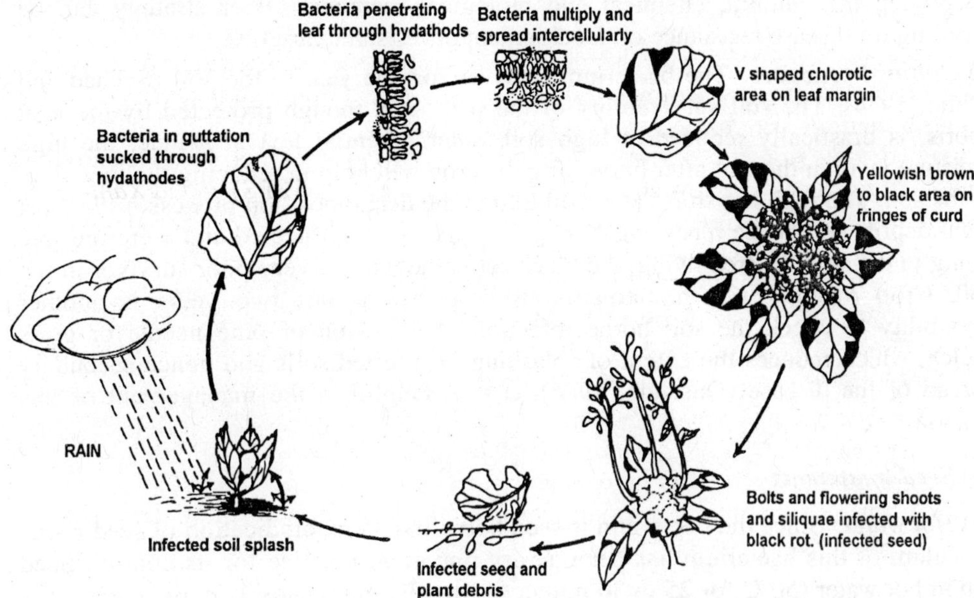

Fig. 10. Disease cycle of black rot of cauliflower

Epidemiology

The infection which may occur at any stage of plant growth generally initiated either by the germination of infected seed or by the entry of the pathogen in the leaves through hydathodes or wounds with rain splashes coming from the soil containing infected plant debris. The bacterium under *in vitro* conditions can survive at a very wide range (5 to 38°C) of temperature with optimum being about 30°C. In regions where rainfall is very low during raising of seedlings, the disease is not common due to restricted dispersal. Disease development is influenced by temperature, relative humidity and rainfall (Jarial and Shyam, 2004). Under field conditions, the optimum temperature for the disease development is about 26.5 to 30°C, minimum being 5°C and maximum 36°C. Definite role of boron in the susceptibility of cauliflower to this disease has been reported (Kotur and Kumar, 1989). A strong correlation between boron uptake in leaf tissues, residual boron in soil and calcium/boron ratio in leaf tissues and disease incidence showed that boron imbalance enhances the host susceptibility. Heavy rains have been reported to be responsible for the fast spread of the pathogen through splashes. Although a relative humidity above 85 per cent has been reported to be conducive for disease development (Lin, 1981), but it does not seem to be a very important factor for disease development as severe outbreaks of the disease have also been reported under hot and dry weather conditions.

Management

Management of black rot can be achieved through eradication of the pathogen in the seed and reduction or eradication of the inoculum from soil. Attaining one of the two will not be sufficient. The ravages of the disease can be kept under check by integrating the cultural, chemical and biological methods. Other strategy can be breeding for disease resistance or development of resistant varieties.

(i) Cultural practices: The bacterium survives for one year in the soil (Schaad and White, 1974). The soil survivability of the pathogen, though protected by the host debris, is drastically reduced at high soil water potential levels. Hence, the high moisture level in the soil after harvesting the crop will help in reducing the inoculum in the soil. The high moisture level will hasten the decomposition process of the host plant debris, and hence rendering *X. c.* pv. *campestris* unprotected and therefore free living in the soil. As free living, the black rot pathogen is a very poor survivor in the soil. Crop rotation with non-cruciferous crops for atleast two years is another possibility to check the soil borne infection. Application of pine needle or grass mulch, which reduces the extent of splashing of infested soils and hence secondary spread of the disease (Onsando, 1988), can be helpful in the management of this disease.

(ii) Seed treatments

(a) Hot water: Since the bacterium is seed borne in nature, eradication of seed borne inoculum of this bacterium has been a very important practice for its control. Seed dip in hot water (50°C for 25 to 30 minutes) virtually eliminates *X. c.* pv. *campestris* without affecting seed vigour. However, it may also cause a reduction in seed germination and seed vigour. Sharma (1981) recommended hot water treatment at 52°C for 30 minutes followed by 30-minute dip in Streptocycline (100 ppm) solution. This treatment causes very limited reduction in seed germination, if the seed is well developed, and provides good control of seed borne inoculum.

(b) Chemical seed treatment: Several antibiotics like streptomycin, aureomycin and terramycin have been tried to eradicate or greatly reduce *X. c.* pv. *campestris* in infected brassica seeds. The use of antibiotics, however, had phytotoxic effects on the seeds. Use of antibiotics followed by sodium hypochlorite did not cause phytotoxicity, reduction of seed vigour or viability of the sort induced by hot water treatment (Humaydan *et al.*, 1980). Other chemical seed treatments recommended are soaking of seed in cupric acetate at 40°C and slurry treatments of calcium hypochlorite at 10 to 20 g a.i./kg seed (Schultz *et al.*, 1986). Seed treatment with the mixture of streptocycline (100 ppm) + captan (2000 ppm) has also been reported effective in controlling the disease and restoring the seed viability (Shekhawat *et al.*, 1982).

iii. Host resistance

(a) Cabbage: Reports from India, Australia, Japan, South America, Portugal and various regions of U.S.A. indicate that the Japanese cabbage cultivar Early Fuji can resist a variety of indigenous strains of *X. c.* pv. *campestris* (Williams, 1980). Screening of introductions and local germplasm of cabbage revealed resistance in some cultivars like 294 BC, Puakea, MGS-2-3, Sn 445, Brunswick Pie Crto, P.I. 436606, Matsunami, YR Isobue, Sachikaze 2 from Japan, Pusa Mukta, Shotoku Uniao, Master AG 325, Globe Master and Washe Shivssei, Penca da Merandela, Pencade Chaves, Gloria de Portugal, Silveirnha and Couve da crrocha cultivars from Portugal, Early Flat Dutch, Master AG 235, Empax F_1, Norma Red, AC 16525, IIHR

380, IIHR 386, IIHR 416, IIHR 431, IIHR 466, IIHR 472, B 295, Delicious, Irodori, YR Kinshu, YR Aoba and Ayahikari have been reported to be good source of resistance to this disease (Fukaya *et al.*, 1988; Kalda *et al.*, 1991; Castellane and Braz, 1991; Gupta, 1991; Henz and Melo, 1994; Pandey *et al.*, 1995; Abiko *et al.*, 1998; Yang *et al.*, 1998). The F_1 hybrid between the cabbage AC 16525 and Broccoli IIHR 104 showed resistance to *X. c.* pv. *campestris* in cabbage be governed by a dominant gene (Pandey *et al.*, 1995). Of late, Cv. Symphony has also been reported as resistant (Sharma *et al.*, 2003).

(b) Cauliflower: Several cvs./lines *viz.,* Ivans and Glory, Master 325 and Empax F1 (Henz and Melo, 1994), RBSI, EC 162587, Sn 445 and Lawyana (Sharma *et al.*, 1995), IIHR, 103, Broccoli IIHR 104, IIHR 54, IIHR 55, IIHR 82, IIHR 106 (Pandey *et al.*, 1995), KN-81 and BR-2 (Sharma *et al.*, 2000), Leamington and Late enterprisel (Sharma *et al.*, 2003), Kt-2 (Thakur *et al.*, 2004) have been reported to be resistant to black rot while Aritos F_1, Maxim F_1, YR Rampo, K.K. Cross, Kenzan, Y.R. 265, Y.R. Ranshu Gold, Y.R. Ranzan, Y.R. Ranju and Punjab Katki have been reported to be moderately resistant to black rot (Henz and Melo, 1994, Thakur *et al.,* 2009). Resistance to black rot is inherited polygenically and governed by dominant genes (Sharma *et al.*, 1992; Sharma *et al.*, 1995).

(iii) Chemical control: Field management of black rot through chemicals has not been attempted much, mainly because of restrictions on the indiscriminate use of antibiotics on the crops. Three sprays of streptocycline (100ppm) at transplanting, curd formation and at pod formation alongwith hot water (52°C) seed treatment followed by dip in streptocycline (100 ppm) for 30 minutes each has been found effective in the management of this disease (Sharma, 1981). Chemical seed dip with Streptocycline (100 ppm) + mancozeb (0.25%) gave almost complete control, while three sprays with same chemicals resulted in 56 per cent disease control when tested under glasshouse conditions (Jarial and Shyam, 2003). Applications of 10–12.5 kg/ha stable bleaching powder, as soil drench is also found effective (Saini and Parashar, 1981).

(iv) Biological control: The biological control of this disease is still in its infancy. When protozoa are associated with *X. c.* pv. *campestris* in soil, the population of the bacterium declines considerably. Endophytic bacterial strains isolated from cabbage leaves showed antagonistic activity both under green house and field conditions (Assis *et al.*, 1998). Strains EN_4 (*Kluyvera ascorbata*) and EN_5 (*Alcabegenes piechaudii*) showed 70.8 and 41.7 per cent reduction in disease incidence, respectively, while highest disease severity reduction was shown by EN_4 (77%) in application of antagonist 4 day before and simultaneously and antibiosis was considered as the mode of action of the antagonists. Seed treatment, soil application and slurry culture drenching with *Azotobacter chroococcum* provided moderate effect on this disease, whereas seed treatment, seedling dip and foliar spray with *Pseudomonas fluorescens* provided encouraging control of this disease (Beura *et al.,* 2006, Singh *et al.,* 2010).

Plant origin pesticides: As many as 70 plant species possessing strong antibacterial properties against *X. c,* pv. *campestris* have been reported (Tiwari *et al.*, 2004). Important among them is the seed extracts of Aparajta, Chitrak, Dhatura and Jangli sem. The bacterium was most sensitive to the leaf extracts of *Camellia sinensis* (Bhardwaj and Laura, 2009).

(v) Integrated disease management: Since the pathogen is both seed and soil borne, a combination of seed treatment (streptocycline, 0.01% + mancozeb, 0.25%, for 3 h) and foliar sprays of the same chemicals coupled with *Eucalyptus hybrida* leaves mulch were found effective against this disease and resulted in 59 per cent disease control under field conditions (Jarial and Shyam, 2003).

2. SOFT ROT/ CURD ROTS

This is a common disease of almost all vegetables in field, transit and storage. The bacterium responsible for this disease affects succulent plant parts especially the heads, curds and roots.

Geographical distribution

The disease occurs on a wide range of vegetables all over the world including India. In northern India, where cabbage and cauliflower are grown for seed production, the disease causes rotting of curds and heads. Occurrence of this disease on maturing cabbage heads from Udaipur (Rajasthan) and Kashmir Valley has also been reported (Chakravarti and Hegde, 1972; Bhat *et al.,* 2010).

Losses

The losses due to infection are quite serious. The affected curds and heads of cauliflower and cabbage fail to throw flowering shoots followed by stump rot. Mortality due to curd rot ranged from 1.5 to 27 per cent and corresponding losses in seed yield were 6 to 108 kg/ha (Shyam *et al.*, 1994). In Norway, this disease caused serious problems in Chinese cabbage cultivation (Baivoll, 1995).

Symptoms

Based on the characteristics of curd rots and pathogen associated with these rots, the curd rot syndrome is classified into four broad groups (Shyam ***et al.***, 1994).

(a) Erwinia rot: The rot initially appears as small water soaked areas often in the centre of the curd which soon assumes light brown to tan colour (Plate 7). During cool and wet weather, the affected area increases rapidly. The centre of the lesions turn dark brown to black assuming wet rot symptoms emitting foul smell which often attract flies. Rot exceeding 50 per cent of the curd surface usually leads to mortality of the plants.

(b) Sclerotinia curd rot: This rot is characterized by appearance of dark brown areas on the forks of the branches of the base of the curd. This is followed by appearance of dark brown to black lesions on the forks engirdling completely the branches and the rotting extends down to core. At this stage, the symptoms are quite conspicuous and curd turns fade white, lumps droop down exposing the central necrotic rotten surface. Subsequently, the rot proceeds downward to the core of a stem leading to premature mortality of the plants.

***(c) Xanthomonas* rot:** This rot is characterized by the appearance of yellowish brown to black areas on the fringes of the bolt surface often coupled with bacterial ooze in wet weather. The rot spreads down towards the base of the branches forming dark brown to black streaks. The affected branches may lodge down near the forks. Though, this rot does not cause premature death of the plants, it reduces the seed yield substantially.

***(d) Alternaria* rot:** This is characterized by the necrosis of the curds forming a brownish black discolouration. The affected curds do not die prematurely. Under high humidity conditions, these curds may however, be invaded by saprophytic microbes.

Curd rot due to *Pythium aphanidermatum* is characterized by a brownish discolouration of stem and branches of curd mostly starting from the base of the peduncle. The severely infected curds developed dark brown soft rot and failed to bolt.

The Pathogen

The bacterium responsible for this disease is *Erwinia carotovora* (Jones) Bergey *et al.* It is a weak parasite and facultative anaerobe. The bacterium is Gram-negative rod, with peritrichous flagella. It produces grayish white colonies on nutrient agar medium and cloudy growth in nutrient broth. It produces gas in the nutrient medium utilizing several carbohydrates as carbon source. Besides *Erwinia caratovora,* association of 13 fungal pathogens like *Alternaria alternate, A. brassicae, A. brassicicola, Botrytis cinerea, Chatomium globossum, Cladosporium cladosporioides, Fusarium equiseti, F. lateritium, Peronospora parasitica, Phialophora* sp., *Phoma tropica, Pythium aphanidermatum, Rhizoctonia solani* and *Sclerotinia sclerotiorum* and one more bacteria *i.e., Bacillus polymyxa* have been found associated with curd rot in different areas of Himachal Pradesh (Chakrabarty *et al.,* 1989). Later, Chakrabarty (1993) reported that the association of fungi causing leaf blight (*Alternaria brassicae*), grey mold (*Botrytis cinerea*), seedling rot (*Fusarium equiseti*), leaf spot (*Phoma tropica*), damping-off (*Pythium aphanidermatum*), root rot (*Rhizoctonia solani*) and stalk rot (*Sclerotinia sclerotiorum*) and soft rot bacterium (*Erwinia caratovora*) with curd rot complex.

Host range

The host range of this bacterium is very wide and includes cabbage, cauliflower, swiss chard, common mallow, gladiolus, lettuce, onion, pepper, potato, spinach, squash, sweet melon, tomato etc. (Rajeh and Hamed, 2000). Bhat *et al.* (2010) also studied the host range and reported that this bacterium could cause rotting of *Dacus carota, Cucumis sativus, Allium cepa, Solanum tuberosum, B. oleracea* var. *botrytis* and *Lycopersicon esculentum.*

Disease cycle

The bacterium is omnipresent and survives saprophytically on dead and succulent plant tissues in soil and stores. It enters the host through wounds to reach the intercellular spaces. After multiplying slowly in the intercellular spaces, it builds up and spreads rapidly by producing pectolytic enzymes that act on the middle lamellae making them collapse. The loosened cells become easily susceptible to the spread of the bacteria, which rapidly cover the entire tissue.

Epidemiology

The pathogen survives in infected plant debris in soil and in storage. The bacterium overwinters in soil at depth of 10 to 50 cm but not at the soil surface (Kikumoto, 1980). In valley areas of Himachal Pradesh and Jammu and Kashmir where cabbage and cauliflowers (Snowball group) are grown for seed production, the rotting of heads and curds start just after frosting, chilling and melting of snow which are responsible for cold injury and provides avenues for the entry of the bacterium (Kapoor, 1987). The disease is favoured by high humidity and optimum temperature for disease development lies in between 25 to 30°C (Kapoor, 1999). Infection mainly spreads through the contact with adjacent leaves (Kikumoto, 1987).

Different abiotic stresses also play an important role for the initiation of this disease under field conditions. Low temperature due to heavy rains during February–March predisposed the curds to freezing injury and subsequently to disease development in March. Mean air temperature of 12–16ºC favours this disease in Himachal Pradesh (Shyam *et al.*, 2001). Frost occurrence and boron deficiency also predispose the cauliflower curds to rot. Higher doses of nitrogen increase the succulence of curd tissue making them more prone to frost damage, thereby enhance vulnerability to pathogens, while increase in Mg and Boron increased curd weight and reduced disease incidence (Chakrabarty and Taneja, 1990). Application of boron imparted considerable resistance to infection by curd rot pathogen *F. equiseti* which was attributed to its affinity in catalyzing the uptake of calcium, an important constituient of cell wall and middle lamella, thereby, conferring resistance to infection (Chakrabarty and Shyam, 1990). They also reported that the protective effect of boron against infection was decreased with the increase in the level of nitrogen irrespective of whether the plants were subjected to or protected from frost. This may be possibly due to interference of higher levels of nitrogen in the uptake of boron or immobilization of boron by calcium.

The incidence and severity of of cabbage soft rot due to *E. carotovora* sub sp. *carotovora* was highest during summer months i.e. June to August in Kashmir (Bhat *et al.*, 2010).

Management

(i) Cultural practices: Since the pathogen is wound parasite, care must be taken to avoid causing wounds and bruises while conducting cultural operations and at the time of harvest. Wider row spacing significantly reduces disease incidence (Fritz and Honma, 1987).

(ii) Host resistance: Several cultivars of Chinese cabbage like Beijing 75, Lu Chun Bai 1, Baiyang 60, Japanese Xiayangbai 60 days, Mingyue (Xu *et al.*, 1993; Li *et al.*, 1993; Wang and Wang, 2001) and hybrid Weibai 6 (Han *et al.*, 2003) have been reported resistant to this disease. *Brassica rapa* accessions like CC-23, CC-1, Majica and Arcadia were having partial tolerance to the pathogen (Ren *et al.*, 1995). Cabbage cultivars *viz.*, Azerbaidzhan, Globe Yellow Resistant, Round up and Horano S E (Mamataliev, 1991) were reported resistant, while cauliflower hybrids like Carillon x Sel.-12, Pyramis x K-1 and Pyramis x Carrilon showed field resistance to the disease (Sharma *et al.*, 1988).

(iii) Chemical control: Applications of streptocycline in combination with copper oxychloride have been recommended for the control of this disease (Kapoor, 1999). Spraying should be initiated immediately with the initiation of the symptoms. In cauliflower seed crop, the curd rot can be kept under check when the curds are pasted with a slurry of chloramphenicol + captafol (1 : 25) and subsequently sprayed with an aqueous suspension of this mixture at 0.01 + 0.25 per cent during the season (Chakrabarty, 1993). Application of boron and lime have also been found effective in reducing the incidence of Erwinia rot in cauliflower (Kotur, 1992)

(C) VIRAL DISEASE

1. CAULIFLOWER MOSAIC

This is a wide spread disease occurring on cauliflower and on related plants.

Geographical distribution

Occurrence of this disease on *Brassica campestris* and *B. oleracea* was first time reported from U.S.A. in 1937 and since then it is distributed worldwide.

Symptoms

Many strains of the virus induce conspicuous mosaic and mottling of leaves, the severity of which is dependent on the varietal tolerance of susceptible species and environmental conditions (Plate 8). Symptoms are usually much more severe in plants simultaneously infected with turnip mosaic potyvirus or broccoli necrotic yellows rhabdovirus.

The Virus

The disease caused by *Cauliflower Mosaic Virus*. Synonyms of the virus are *Brassica Virus 3, Broccoli Mosaic Virus, Cabbage Mosaic Virus* and *Cabbage Virus B*. The virus is a plant pararetrovirus belonging to the genus *Caulimovirus* and family *caulimoviridae*. The virions are isometric, not enveloped, 50 nm in diameter, rounded in profile without a conspicuous capsomere arrangement. This is a DNA containing virus which is double stranded, circular (Yasaka *et al*., 2014). The capsid shells of virions are composed of multilayers. Capsid appears round.

Host Range

Host range of the virus spreads in three host families like Cruciferae, Rosaceae and Solanaceae. Susceptible hosts are *Arabidiopsis thaliana, Brassica capestris, B. c.* ssp. *chinensis, B. c. napus, B. c.* ssp. *pekinensis, B. c.* ssp. *rapa, B. oleracea, B. o.* var. *botrytis, Datura stramonium, Nicotiana clevelandii, Raphanus* etc.

Transmission

Most isolates of the virus are transmitted in a semi-persistent manner by the aphid vectors. *Brevicoryne brassicae* and *Myzus persicae* and atleast 25 other species of Aphididae are the vectors of this virus. The virus is lost by the vector when it moults;

does not multiply in the vector; not transmitted congenitally to the progeny of the vector and also transmitted by the mechanical inoculation.

Management

Adequate weed control and sanitation should be practiced. Plough-down of previous crop residues should be done immediately. Transplant beds should be isolated from commercial crop fields and overwintering cruciferous weeds. To keep the vector population in check, insecticidal sprays can be also done.

REFERENCES

Abiko, K., Kubota, m., Yamada, K. and Hida, K. 1998. Varietal resistance to black rot caused by *Xanthomonas campestris* pv. *campestris* (Pammel 1895) Dowson 1939. *Proc. Kansai Pl. Prot. Soc.* **40:** 47–49.

Achar, P.N. 1992. Yield loss in *Raphanus sativus* from systemic infection by *Peronospora parasitica. Phyton (Buenos Aires).* **53:** 89–94.

Agrios, G.N. 1997. *Plant Pathology.* Academic Press INC. London, 803 pp.

Ahuja, S.C. 1976. Banded leaf and sheath blight of maize. Ph.D. Thesis, Post-graduate School, IARI, New Delhi. 153p.

Anonymous, 1969. CMI distribution map of *Alternaria brassicicola*, Kew, Commonwealth Mycological Institute.

Anonymous, 1974. Annual report of the Department of Botany and Plant Pathology, HPU, Solan, 1973.

Anonymous, 1978. Commonwealth Mycological Institute, Distribution maps of plant diseases, *Leptosphaeria maculans* (Desm.) Ces & de Not. CMI Map No. 73.

Appleton, J. 1979. Lancs sprouts trial rates 55 varieties for white blister. *Grower* **91:** 52–58.

Armstrong, G.M. and Armstrong, J.K. 1952. Physiologic races of the *Fusarium* causing wilts of the crucifers. *Phytopathology* **42:** 255–257.

Arora, N., Madan, Jasleen, Dhingra, Madhu and Aujla, I.S. 2011. Identification of morphological and biochemical markers for screening of cauliflower (*Brassica oleracea* L.) genotypes against Alternaria blight. *Pl. Dis. Res.* **26:** 26–34.

Assis, S.M.P., Silveira, E.B. da, Mariano, R. de L.R. and Menezes, D. 1998. Endophytic bacteria – method for isolation and antagonistic potential against cabbage black rot. *Summa Phytopathol.* **24(3/4):** 216–220.

Atkinson, R.G. 1953. Survival and pathogenicity of *Alternaria raphani* after five years in dried cultures. *Can. J. Bot.* **31(5):** 542–547.

Bains, S.S. and Jhooty, J.S. 1983. Host range and morphology of *Peronospora parasitica* from different sources. *Indian J. Mycol. Pl. Pathol.* **13 :** 372–375.

Bains, S.S., Sokhi, S.S. and Jhooty, J.S. 1981. Out–break of *Peronospora parasitica* on cauliflower curd in Punjab. *Indian Phytopath.* **34 :** 389–390.

Baivoll, G. 1995. Introduction of Chinese cabbage in Norway: problems and possibilities. *J. Veg. Crop Prod.* **1:** 3–18.

Baswana, K.S., Rastogi, K.B. and Sharma, P.P. 1991. Inheritance of stalk rot resista– nce in cauliflower (*Brassica oleracea* var. *botryrtis* L.). *Euphytica* **57(2):** 93–96.

Bedlan, G. 1985. Phoma disease of cabbage plants. *Pflanzenschutz.* **12:** 6–7.

Bedlan, G. 1998. Application–management to control post–harvest *Botrytis* infections of chinese cabbage. *Gesunde Pflanzen* **50(6):** 162–164.

Beura, S.K., Mohapatra, K.B., Paul, P.K. and Nandi, A. 2006. Integrated management of black rot disease of cauliflower in Orissa. *J. Mycopathol. Res.* **44:** 293–296.

Bhagat, S. And Pan, S. 2008. Biological control of root and collar rot of cauliflower (*Rhizoctonia solani*) by talc based formulation of *Trichoderma harzianum* Rifai. *J. Biol. Cont.* **22:** 483–486.

Bhardwaj, S.K. and Laura, J.S. 2009. Antibacterial activity of some plant extracts against plant pathogenic bacteria, *Xanthomonas campestris* pv. *campestris. Indian J. Agric. Res.* **43:** 26–31.

Bharat, N.K. and Gupta, B.K. 2011. Occurrence and management of pod blight of cabbage seed crop under arid temperate conditions. *Pl. Dis. Res.* **26:** 92–93.

Bhat, K.A., Bhat, N.A., Masoodi, S.D., Mir, S.A., Zargar, M.Y. and Sheikh, P.A. 2010. Studies on status and and host range of soft rot disease of cabbage (*Brassica oleracea* var. *capitata*) in Kashmir Valley. *J. Phytol.* **10:** 55–59.

Bhat, N.A. and Masoodi, S.D. 1999. Occurrence of *Xanthomonas campestris* pv.*campestris* (Pammel) Dowson incitant of black rot of cabbage in Jammu and Kashmir. *SKUAST J. Res.***1:** 195–197.

Bhat, N.A, Nusrat Sayeed, Bhat, K.A and Mir, S.A. 2010. Pathogeneicity and host range of *Xanthomonas campestris* pv.*campestris*– incitant of black rot of crucifer. *J. Phytol.* **2:** 1–5.

Bhattacharya, I. and Dixon, G.R. 2010. Management of clubroot disease (*Plasmodiophora brassicae*) of brassicas using trap cropping techniques. *Acta Hortic.* **867:** 157–164.

Biesheuvel, A.R. 1991. White cabbage. New cultivars are more tolerant of diseases. *Groenten en Fruit, Vollegrondsgroenten* **1:** 26–27.

Biga, M.L.B. 1955. Review of the species of the genus *Albugo* based on the morphology of the conidia. *Sydowia* **9:** 339–358.

Bolle, P.C. 1924. Plant diseases caused by blackening fungi (Phaeodictyae). *Meded. Phytopath. Lab. Willie Commelin Scholten", Baarn (Holland)* **7:** 77.

Boning, K. 1936. Investigations on horse-radish diseases and their control. *Angew. Bot.* **18(6):** 482–495.

Bonnet, A. and Blancard, D. 1987. Resistance of radish (*Raphanus sativa* L.) to downy mildew, *Peronospora parasitica. Cruciferae Newsl.* **12 :** 98–99.

Boruah, H.P.D. and Kumar, B.S.D. 2002. Plant disease suppression and growth promotion by fluorescent *Pseudomonas* strain. *Folia Microbiol.* **47:** 137–143.

Braverman, S.W. 1971. Reaction of broccoli and cauliflower introductions to *Alternaria brassicicola. Pl. Dis. Reptr.* **55 :** 454–457.

Broadhurst and Popay, 1994. Evaluation of fungicides for control of ringspot on cabbage. *Proc.47 th NZ Pl. Prot. Conf., Waitangi Hotel, New Zealand, 9–11 Aug., 1994*, pp. 178–182.

Broplhy, T.F. and Liang, M.D. 1992. Screening of fungicides for the control of downy mildew on container–grown cabbage seedlings. *Crop Prot.* **11 :** 160–164.

Butler, E.J. 1918. *Fungi and Diseases in Plants*. Thaker Spink & Co., Calcutta, India. pp 297–300.

Campbell, R., Larner, R.W. and Madelin, M.F. 1968. Notes on an albino mutant of *Alternaria brassisicola. Mycologia* **60:** 1122–1125.

Carvalho, T. and Monteiro, A. 1996. Preliminary study on the inheritance of resistance to downy mildew (*Peronospora parasitica* (Pers. Ex. Fr.) at cotyledon stage in Tranchuda cabbage "Algarvia". *Cruciferae Newsl.* **18 :** 104.

Castellane, P.D. and Braz, L.T. 1991. Evaluation of cabbage cultivars during the spring in Jaboticabal, Sao Paulo. *Hortic. Brasileira* **9(1):** 13–14.

Chahal, A.S. 1981. Seed borne infection of *Alternaria brassicae* in Indian mustard and its elimination during storage. *Curr. Sci.* **50:** 621–623.

Chahal, A.S., Kang, M.S. and Chohan, J.S. 1977. *Alternaria* blight –A serious disease of rapeseed and mustard. *Prog. Fmg.* **12** : 21–22.

Chakrabarty, P.K. 1993. Chemical management of curd–rot complex of cauliflower (*Brassica oleracea* convar *botrytis* var. *botrytis*). *Indian J. Agric. Sci.* **63:** 50–55.

Chakrabarty, P.K. and Taneja, N.K. 1990. Effect of nutrients on curd rot of cauliflower (*Brassica oleracea* convar *botrytis* var. *botrytis*). *Indian J. Agric. Sci.* **60:** 74–76.

Chakrabarty, P.K. and Shyam, K.R. 1990. Effect of frost, boron, and different levels of nitrogen on curd rot of cauliflower. *Indian Phytopath*. **43:** 237–239.

Chakrabarty, P.K., Shyam, K.R. and Bhardwaj, S.S. 1989. Some new records of micro-organisms associated with curd rot of cauliflower. *Indian J. Mycol. Pl. Pathol.* **19:** 118–120.

Chakravarti, B.P. and Hegde, S.V. 1972. Occurrence of bacterial soft rot in black rot affected cabbage plants in Udaipur, Rajasthan. *Indian Phytopath*. **25:** 118–122.

Chand, J.N. and Jatian, S.S. 1969. Efficacy of different fungicides against *Alternaria* blight of radish (*Raphanus sativus* L.). *JNKVV Res. J.* **3:** 132–133.

Chand, P. and Rai, D. 2008. *In vitro* evaluation of different organic amendments against *Sclerotinia sclerotiorum* (Lib) de Bary. *Pl. Dis. Res.* **22(2):** 84–85.

Channon, A.G. 1981. Downy mildew of *Brassicas.* In : *The downy mildews*, Ed. D.M. Spencer. Academic Press, London. pp. 321–339.

Chattopadhyay, S.D. and Mustafee, T.P. 1967. Outbreaks and new records, India: occurrence of two new diseases on vegetables. *Pl. Prot. Bull. F.A.O.* **15:** 40–41.

Chauhan, R.S., Maheshwari, S.K. and Gandhi, S.K. 2000. Effect of soil type and plant age on stem rot disease (*Rhizoctonia solani*) of cauliflower (*Brassica oleracea* var. *botrytis*). *Agri. Sci. Dig*. **20:** 58–59.

Chen, C.Y. and Seguin Swartz, G. 1996. Hypersensitive response of *Arabidiopsis thaliana* to an isolate of *Phoma lingam* virulent to oil seed and turnip rape. *Eucarpia Cruciferae Newsl.* **18:** 106–107.

Chernyaeva, I.L. and Bisuas, M. Ch. 1980. Effect of different forms of nitrogen fertilizers and organic additives on *Rhizoctonia solani* Kuhn, the pathogen of black leg of cabbage seedlings. *Mikologiya Fitopatol.* **14:** 453.

Chowdhary, S. 1944. Some fungi from Assam, I. *Indian J. Agric. Sci.* **14:** 230–233.

Chupp, C. and Sherf. A.F. 1960. Crucifer diseases. In: *Vegetable diseases and their control.* Ronald Press Co. New York, Chapter **8:** 237–288.

Cook, A.A., Larson, R.H. and Walker, J.C. 1952. Relation of the black rot pathogen to cabbage seed. *Phytopathology* **42:** 316–320.

Corato, U.de and Baviello, G. 2000. Occurrence of *Sclerotinia sclerotiorum* on *Brassica carinata* in Southern Italy (Brasilicata). *Informatore Fitopatol.* **50(5):** 61–63.

Crete, R. and Chang, M.S. 1980. Screening brassicas for resistance to club root (*Plasmodiophora brassicae* Wor.). *Can. Pl. Dis. Surv*. **60:** 17–19.

Crute, I.R. 1984. Downy mildew of Brassicas. 35th *Annu. Rep. 1983. Natl. Veg. Res. Sta., Wellesbourne, Warwick, UK*, 76 pp.

Cuong, N.D. and Dohroo, N.P. 2007. Study on biologically based management practices of Sclerotinia rot of cauliflower. *Pl. Dis. Res*. **22(2):** 124–128.

d' Ercole, N. 1975. Downy mildew of cauliflower in North–Central Italy. *Informatore Fitopatol.* **25:** 21–23.

d'–Ercole, N. 1974.Trials on the varietal susceptibility of cauliflower to infection by *Rhizoctonia solani* in the greenhouse. *Informatore Fitopatol.* **24:** 11–12.

d'–Ercole, N. and Nipoti, P. 1978. *Rhizoctonia* disease of cauliflower: varietal susceptibility in the greenhouse. *Italia Agricola* **115:** 161–164.

Davies, J.M.L. and Wafford. 1987. Control of downy mildew in module–raised cauliflowers. *Mono. – Brit. Crop Prot. Counc.* **39 :** 379–386.

de Bary, A. 1886. Ubereinige Sclerotinien und Sclerotienkrankheiten. *Bot. Ztg.* **44:** 377–474.

Dhingra, M., Arora, N. and Aujla, I.S. 2012. Field evaluation of cauliflower genotypes against alternaria blight and role of phenols and oxidative enzymes in fungal resistance. *Pl. Dis. Res.* **27(1):** 54–59.

Dias, J.S., Ferreira, M.E. and Williams, P.H. 1993. Screening of Portuguese Cole land races (*Brassica oleracea* L.) with *Peronospora parasitica* and *Plasmodiophora brassicae. Euphytica* **67:** 135–141.

Dickson, M.H. and Petzoldt, R. 1996. Breeding for resistance to *Sclerotinia sclerotiorum* in *Brassica oleracea. Acta Hortic.* **407:** 103–108.

Dixon, G.R. 1981. *Vegetable crop diseases.* AVI Publishing Company, INC. Westport, Connecticut. 404pp.

Dixon, G.R. 2008. Calcium and pH as parts of coherent control strategy for clubroot disease (*Plasmodiophora brassicae*). *Acta Hortic.* 867.

Djatnika, I. 1991. Prospects of biological control of club root. *Eucaripia Cruciferae Newsl.* **14/15:** 142.

Dohroo, N.P. 1988. Germplasm reaction of cauliflower to stalk rot (*Sclerotinia sclerotiorum*). *Indian J. Pl. Pathol.* **6(2):** 144.

Eastburn, D. 1989. Disease management of cabbage and broccoli–an IPM approach. *Trans. Illinois State Hortic. Soc.* **123:** 32–35.

Elenkov, E. 1979. Breeding disease resistant varieties of vegetable crops. *Rastitelna Zashchita* **27 :** 13–16.

Ellis, M.B. 1968a. *Alternaria brassicae.* Commonwealth Mycological Institute. Descri–ptions of Pathogenic fungi and Bacteria. No. 162. CMI, Kew, Surrey, England.

Ellis, M.B. 1968b. *Alternaria brassisicola.* Commonwealth Mycological Institute. Descriptions of Pathogenic fungi and Bacteria. No. 163. CMI, Kew, Surrey, England.

Ellis, M.B. 1971. *Dematiaceous Hyphomycetes.* CMI, Kew Surrey, England **608:** 464–497.

Ende, J.E. van–den and Van–den Ende, J.E. 1993. Seedbed infection of white cabbage by *Mycosphaerella brassicicola. Neth. J. Pl. Pathol.* **99:** 139–148.

Ende, J.E. van–den, Everaarts, A.P. and Van–den Ende, J.E. 1999. More knowledge of the *Mycosphaerella* infection in cabbages. *PAV Bull. Vollegrondsgroenteteelt.* **3:** 35–37.

Farlow, W.G. 1883. Enumeration of the Peronosporeae of the United States. *Bot. Gaz.* **8 :** 305–315, 327–337.

Fawcett, H.S. 1909. Cabbage diseases. *Florida Agric. Exp. Stn. Ann. Rep.*, 59–60.

Felton, M.W. and Walker, J.C. 1946. Environmental factors affecting mildew of cabbage. *J. Agric. Res.* **72:** 69– 81.

Fritz, V.A. and Honma, S. 1987. The effect of raised beds, population densities and planting date on the incidence of bacterial soft rot in Chinese cabbage. *J. Am. Soc. Hortic. Sci.* **112:** 41–44.

Fukaya, M., Miyagawa, T., Koide, H., Ohno, T. and Ichikawa, K. 1988. Study on control of black rot of cabbage. *Res. Bull. Aichi–ken Agric. Res. Stn., Jpn.* **20:** 245–251.

Gabrielson, R.L. 1983. Black leg disease of crucifers caused by *Leptosphaeria maculans* (*Phoma lingam*) and its control. *Seed Sci Tech.* **11:** 749–780.

Gabrielson, R.L. and Maguire, J.D. 1977. The biology and control of *Phoma lingum* in crucifer seed crops. *Search* **14:** 2–8.

Gaikwad, A.P., Kokade, D.S., Nimbalkar, C.A. and Desai, U.T. 2004.Control of downy mildew (*Peronospora parasitica*) of cauliflower (*Brassica oleracea* var. *botryis* subvar. *cauliflora*). *Indian J. Agric. Sci.* **74:** 230–232.

Galli, F. and Tokeshi, H. 1962. Cabbage yellows caused by *Fusarium oxysporum* f.sp. *conglutinans*. *Rev. Agric. Piracicaba* **37:** 24–27.

Gangopadhyay, S. and Kapoor, K.S. 1975. Note on the control of turnip wilt for seed production in Kullu Valley. *Indian J.Agric. Sci.* **45:** 172–174.

Geeson, J.D. and Browne, K.M. 1979. Effect of post harvest fungicide drenches on stored winter white cabbage. *Plant Pathol.* **28 :** 161–168.

German, H. 1891. A bacterial disease of cabbages. *Bot. Gazette* **16(9):** 18–91.

Godard, Jean, Francois, S. Zaidi, Claudie Monto, Daniel le Cora and Drissa Sileu. 1999. Benzothiadiazole (BTH) induces resistance in cauliflower (*Brassica oleracea* var *botryis*) to downy mildew of crucifers caused by *Peronospora parasitica. Crop Protec.* **18:** 397–405.

Gulsoy, H.E. 1978. Research on the distribution of club root (*Plasmodiophora brassicae* Wor.) of cabbage, its relation with soil humidity and pH, chemical control methods and determination of resistant cultivars. *Bitki Koruna Bulteni* **18:** 28–57.

Gupta, D.K. 1991. Studies on black rot of cabbage in Manipur. *Indian J. Mycol. Pl. Pathol.* **21(2):** 203–204.

Gupta, D.K. and Chaudhary, K.C.B. 1995. Infection of radish and rayosag seeds by *Xanthomonas campestris pv. campestris. Indian J. Mycol. Pl. Pathol.* **25:** 232.

Gupta, S.K. and Agarwala, R.K. 1988. Biological control of *Sclerotinia* stalk rot of cauliflower. *Indian J. Plant Pathol.* **6(1):** 71–74.

Gupta, S.K. and Dohroo, N.P. 1989. Suppression of *Sclerotinia sclerotiorum* (Lib.) de Bary causing stalk rot by rhizosphere fungi of cauliflower. *J. Biol. Contl.* **3(2):** 136.

Gupta, S.K. and Dohroo, N.P. 1993. *Sclerotinia* rot of cauliflower. *Int. J. Trop. Pl. Dis.* **11:** 131–138.

Gupta, S.K. and Shyam, K.R. 1994. Antisporulant activity of fungicides on downy mildew (*Peronospora parasitica*) on cabbage. *Indian J. Agric. Sci.* **64:** 891–893.

Gupta, S.K. and Shyam, K.R. 1995. Occurrence of wire stem of cauliflower–a new record. *Pl. Dis. Res.* **10:** 152–153.

Gupta, S.K., Agarwala, R.K. and Sharma, S.L. 1986. Efficacy of soil amendments in managing slalk rot (*Sclerotinia sclerotiorum*) of cauliflower. *Indian J. Plant Pathol.* **4(2):** 129–131.

Gupta, S.K., Agarwala, R.K. and Sharma, S.L. 1987. Effect of crop rotation on the management of stalk rot of cauliflower. *Indian J. Plant Pathol.* **5(1):** 79–80.

Han, T.L., Feng, L.R., Guo, X.M., Cao, Q.C., Tan, J.X., Han, T.L., Feng, L.R., Guo, X.M., Cao, Q.C. and Tan J.X. 2003. Selection of a new Chinese cabbage (*Brassica pekinensis*) F 1 hybrid Weibai 6. *Chinese Veg.* **1:** 28–29.

Harman, G.E. and Nash, G. 1978. Soaking *Brassica* seeds in fungicide solutions to eradicate seed borne fungi: a comparison of aqueous and organic solvent infusion techniques. *Pl. Dis. Reptr.* **62:** 408–412.

Harter, L.L. and Jones, L.R. 1923. Cabbage diseases. *USDA Farmer's Bull.* **1351:** 28pp.

Henz, G.P. and Melo, P.E.de. 1994. Resistance of cabbage culrtivars to *Xanthomonas campestris pv. campestris*. *Pesquisa Agrop. Brasileira* **29(9):** 1411–1415.

Hoser, K.J., Lakowska, R.E. and Antosik, J. 1991. Resistance to some *Brassica oleracea* L. plant introductions to downy mildew (*Peronospora parasitica*). *Cruciferae Newsl.* **14/15:** 144–145.

Hoser, K.J., Lakowska, R.E. and Antosik, J. 1995. The inheritance of resistance to some *Brassica oleracea* cultivars and lines to downy mildew. *J. App. Genet.* **36:** 27–33.

Humaydan, H.S., Harman, G.E., Nedro, B.L. and Dinitto, L.V. 1980. Eradication of *Xanthomonas campestris* the causal agent of black rot from brassica seeds with antibiotics and sodium hypochlorite. *Phytopathology* **70:** 127

Humpherson–Jones, F.M. and Maude, R.B. 1982b. Control of dark leaf spot (*Alternaria brassicicola*) of *Brassica oleracea* seed production crops with foliar sprays of iprodione. *Ann. App. Biol.* **100 :** 99–104.

Hussaini, S.H. and Singh, H. 1989. Efficacy of different fungicides against Alternaria blight of radish seed crop. *Pl. Dis. Res.* **4:** 105–107.

Iqbal, M.A. and Khalid, I. 2002. *In vitro* evluation of various antagonists against mycelial growth of *Rhizoctonia solani* and effect of seed treatment with antagonists on the control of root rot of cauliflower (*Brassica oleracea* L.). *Pakistan J. Phytopathol.* **4:** 11–14.

Iqbal, S.H., Ghazala, N, Mubashir, N., Nasim, G. and Niaz, M. 1988. Role of vesicular – arbuscular mycorrhiza as a deterrent to damping off caused by *Rhizoctonia solani* in *Brassica oleracea. Biologia-Pakistan* **34:** 79–84.

Jana, B.K., Laha, S.K. and Khatua, D.C. 1989. Leaf blight caused by *Rhizoctonia solani* and its control. *Pesticides* **23:** 33–34.

Jang, P. and Safeeula, K.M. 1990. Factors affecting germination of *Peronospora parasitica* in radish. *Proc. Indian Acad. Sci.* **100 :** 113–116.

Jarial, Kumud and Shyam, K.R. 2003. Integrated management of black rot of cauliflower caused by *Xanthomonas campestris* pv. *campestris. Pl. Dis. Res.* **18(2):** 152–158.

Jarial, Kumud and Shyam, K.R. 2004. Black rot of cauliflower in relation to meteorological factors. *Asian J. Microbiol. Biotech. Env. Sci.* **6:** 113–117.

Jennrich, H. 1985. Ronilan–smoke, a new vinclozolin formulation for control of *Botrytis cineria* and other diseases in greenhouses and storage rooms. *Med. fac. Landbouww. Rijksuniv. Gent* **50 :** 1227–1233.

Jorgenson, M.C. and Walter, J.M. 1955. The 1953 outbreak of black rot at Ruskin. *Proc. Fla. Hortic. Soc.* **67:** 109–111.

Kalda, T.S., Tewari, R.N., Ram Singh, Narendra Singh and Verma, T.S. 1991. Evaluation of Pusa Mukta, a resistant variety to black rot disease in cabbage. *Indian J. Hortic.* **(3)**: 247–250.

Kansal, S., Bhardwaj, S.S. and Shyam, K.R. 1990. Antagonistic effect of *Trichoderma* and *Gliocladium* species against *Sclerotinia sclerotiorum* causing stalk rot of cauliflower III–Effect of nonvolatile subsatnces produced by antagonists *Pl. Dis. Rep.*(Supl): 110–114.

Kapoor, K.S. 1987. Bacterial soft rot (*Erwinia carotovora*) of cabbage. *Plant Dis*. **71:** 861.

Kapoor, K.S. 1996. Disease resistance in cauliflower (*Brassica oleracea* var. *botrytis* L.) against *Sclerotinia sclerotiorum* (Lib.) de Bary. *Veg. Sci.* **113(2):** 285–289.

Kapoor, K.S. 1999. Fungal and bacterial diseases of crucifers. In: *Diseases of Horticultural Crops: Vegetables, Ornamentals and Mushrooms* (Eds. Verma, L.R. and Sharma, R.C.). Indus Publishing Co., New Delhi, pp 184–215.

Kear, R.W., Williams, D.J. and Stevens, C.C. 1977. The effect of iprodione on the fungal deterioration of stored white cabbage. *Proc. 1977 British Crop Prot. Conf., Pests Dis., Brighton* : pp. 189–195.

Khan, M.A. 1976. Studies on stalk rot of cauliflower caused by *Sclerotinia sclerotiorum.* M Sc. Thesis, HPU, College of Agriculture, Solan, 80 pp.

Kikumoto, T. 1980. Ecological aspects of the soft rot bacteria. *Bull. Inst. Agric. Rés., Tohoku Univ.* **31:** 19–41.

Kikumoto, T. 1987. Ecological studies on the soft rot bacteria of vegetables. Areas of disease development in Chinese cabbage. *Bull. Inst. Agric. Res., Tohoku Univ.* **38:** 83–90.

Klinkovskaya, I.K. 1976. Effects and after - effects of zineb and heat treatment of seed on the incidence of diseases and yield of cabbage. *Byulleten Usesoyuznogo Nauchno Issledovatel' skogo Instituta Zashchity Rastenii* **38:** 31–35.

Kobayashi, N. 1991. Biological control and integrated control of cabbage yellows. *Proc. Assoc. Pl. Prot. Kyushu.* **37:** 9–14.

Kontaxis, D.G., Mayberry, K.S. and Rubatzky, V.E. 1979. Reaction of cauliflower cultivars to downy mildew in Imperial Valley. *California Agric.* **33:** 19.

Kotur, S.C. 1992. Effect of season and of boron and lime application on curd yield, curd rot and utilization of boron by cauliflower (*Brassica oleracea* convar. *botrytis* var. *botrytis*) in an Alfisol. *Indian J. Agric. Sci.* **62:** 286–289.

Kotur, S.C. and Kumar, S. 1989. Response of cauliflower (*Brassica oleracea* var. *botrytis* var. *botrytis*) to boron in Chotanagpur region. *Indian J. Agric. Sci.* **59:** 640–644.

Kruger, W. 1976. Investigations of effect of apothecial development of *Sclerotinia sclerotiorum* (Lib.) de Bary. *Nacherichtenbl. Pflanzenschutz.* **28(9):** 129–135.

Kuan, T.L., Minsavage, G.V. and Schaad, N.W. 1986. Aerial dispersal of *Xanthomonas campestris* pv. *campestris* from naturally infected *Brassica campestris. Plant Dis.* **70:** 409

Kudela, V., Novak, O. and Skorpik, M. 1978. Effect of chemical control on seed yield of cauliflower. *Ochr. Rostl.* **14 :** 275–284.

Kundu, P.K. and Nandi, B. 1984. Control of cauliflower damping off by using antagonist coated seeds. *Pedobiologia* **27:** 43–48.

Kundu, P.K. and Nandi, B. 1985. Control of *Rhizoctonia* disease of cauliflower by competitive inhibition of the pathogen using organic amendments in soil. *Plant and Soil* **83:** 357–362.

Laemmlen, F.F. and Mayberry, K.S. 1984. Broccoli resistance to downy mildew. *California Agric.* **38:** 17.

Lakra, B.S. and Saharan, G.S. 1988. Influence of host resistance on colonization and incubation period of *Albugo candida* in mustard. *Eucarpia Cruciferae Newsl.* 13: 108–109.

Lakra, B.S. and Saharan, G.S. 1989. Effect of temperature, relative humidity and light on germination *of Albugo candida* sporangia from mustard. *Indan J. Mycol. Pl. Pathol.* **19:** 264–267.

Lambe, R.C., McCart, G., O'–Dell, C., Tabor, T. and Widener, B. 1978. A team approach to cabbage disease control. *Am. Veg. Grow.* **26:** 42–44.

Lapis, D.B. and Ricaforte, E.T. 1974. *Alternaria* leaf spot of crucifers in the Philippines. *Philipp. Agric.* **58 :** 121–137.

Letham, D.B., Huett, D.O. and Trimboli, D.S. 1976. Biology and control of *Sclerotinia sclerotiorum* in cauliflower crops in costal New South Wales. *Pl. Dis. Reptr.* **60(4):** 286–289.

Li, J.K., Zhuang, C.L., Wang, D.S., Sun, D.H. and Zhang, D.Q. 1993. Characteristics and cultural requirements of the spring Chinese cabbage variety LU Chun Bai 1. *Chinese Veg.* **4:** 50–51.

Lin, C.Y. 1981. Studies on black rot of cruciferous crops caused by *Xanthomonas campestris* pv. *campestris* in Taiwan. *Plant Prot. Bull., Taiwan* **23:** 157–167.

Lin, C.Y. 1981. Studies on downy mildew of Chinese cabbage caused by *Peronospora parasitica.* In: *Chinese cabbage Proc. Ist Int. Symp. AVRDC, Shanhua, Taiwan.* Pp. 105–112.

Loganathan, M., Swarnakumari, N. Sivakumar, M. Prakasam, V., Mohan, I., Ramaraj, B. and Samiyappan, R. 2001. Biological suppression of fungal nematode complex diseases of major cruciferous vegetables. *South Indian Hortic.* **49(specl.):** 296–300.

Mahajan, V. 1999. Relationship of phenols and sugars in downy mildew resistance of cauliflower. *J. Maharashtra Agric. Univ.* **24:** 97–99.

Mahajan, V. and Gill, H.S. 1998. Stomatal density in relation to downy mildew disease intensity in cauliflower. *J. Maharashtra Agric. Univ.* **23:** 31–33.

Mahajan, Vijay, Gill, H.S., Moore, T.A. and Mahajan, V. 1995. Inheritance of downy mildew resistance in Indian cauliflowers (group–III). *Euphytica* **86 :** 1–3.

Mahiar, M. and Khalaif, H. 1999. Black rot of crucifers in Jordan: sources of inoculum. *Dirasat. Agric. Sci.* **26:** 329–337.

Mamataliev, M.M. 1991. Resistance of cabbage to *Erwinia carotovora* and *Fusarium* wilt. *Tezisy dokladov*. 124.

Marten, P., Bruckner, S. and Luth, P. 1999. Plant growth promotion of different cultivated plants and biological control of soil borne phytopathogenic fungi by *Bacillus subtilis* strain B2g. *Z. Pflanzenkr. Pflanzenschutz* **106:** 74–81.

Maude, R.B. and Dudley, C.L. 1972. Dark leaf spot (*Alternaria brassicicola*) of cabbage. *22nd Annu. Rep. 1971, Natl. Veg. Res. Stn., Wellesbourne, Warwick, U.K.* p. 76.

Maude, R.B. and Humpherson–Jones, F.M. 1980. Studies on the seed borne phase of dark leaf spot (*Alternaria brassisicola*) and grey leaf spot (*Alternaria brassicae*) of *Brassicas. Ann. App. Biol.* **95:** 311–319.

Maude, R.B. and Suett, D.L. 1986. Application of fungicide to *Brassica* seeds using a film–coating technique. *Proc. 1986 British Crop Prot. Conf., Pests Dis.* **1 :** 237–242.

Maude, R.B., Humpherson–Jones, F.M. and Shuring, C.G. 1984. Treatments to control *Phoma* and *Alternaria* infections of *Brassica* seeds. *Plant Pathol.* **33:** 525–535.

McKay, A.G., Floyd, R.M. and Boyd, C.J. 1992. Phosphonic acid controls downy mildew (*Peronospora parasitica*) in cauliflower curds. *Aust. J. Exp. Agric.* **32:** 127–129.

Mehra, R. and Batra, V. 2005. Evaluation of cauliflower vvarieties for resistance against downy mildew disease and yield. *Res. Crops.* **6:** 154–155.

Mehta, N. and Saharan, G.S. 1994. Morphological and pathological variations in *Peronospora parasitica* infecting *Brassica* species. *Indian Phytopath.* **47:** 153–158.

Mitani, S., Ohhashi, K., Hayashi, H. and Matsuo, N. 1998. Fungicidal activity of fluazinam on *Phoma lingam* (*Leptosphaeria maculans*), causal agent of blackleg on cabbage, *Proc. Kanto Tosan Pl. Prot. Soc.* **45:** 43–46.

Mondal, S.N., Rashid, M.A., Monowar, S.M. and Abdullah, A.M. 1989. Efficacy of fungicides in controlling *Alternaria* blight of radish seed crop. *Thai. J. Agric. Sci.* **22:** 191–196.

Monteiro, A.A. and Williams, P.H. 1989. The exploration of resources of Portuguese cabbage and kale for resistance to several *Brassica* diseases. *Euphytica* **41:** 215–225.

Moreno– Rico, O., Frias–Trevino, A.G., Luna–Ruiz, J.J., Manjano–Flores, D.E., Romero–Cova, S. and Seguin–Swartz, G. 2001. Characterization and pathogenicity of isolates of *Leptosphaeria maculans* from Aguascalientes and Zacatecas, Mexico. *Can. J. Pl. Path.* **23:** 270–278.

Moreno–Rico, O., Moreno–Martinez, E. and Romero–Cova, S. 1996. Etiology, incidence, severity and epidemiology of the canker and/ or foot rot disease of crucifers cultivated in Aguascalientes and Zacatecas States of Mexico. *Fitopatologia* **31:** 184–190.

Mridha, M.A.U. 1983. Virulence of different isolates of *Alternaria brassicae* on winter oilseed rape cultivars. *Sixth Int. Rapeseed Conf., Paris, France* pp. 1025–1029.

Mukerji, K.G. 1975. *Albugo candida*. Descriptions of plant Pathogenic Fungi and Bacteria. No. 460. CMI, Kew, Surrey, England.

Munjal, R.L. and Gill, H.S. 1961. Occurrence of black leg disease of cabbage in India. *Indian Phytopath.* **14:** 16–19.

Natti, J.J., Dickson, M.H. and Atkin, J.D. 1967. Resistance of *Brassica oleracea* varieties to downy mildew. *Phytopathology* **57:** 144–147.

Neergaard, P. 1948. *Eleventh Annu. Rep. J.E. Ohlsen Phytopathological Laboratory,* August 1945 to July 1947. 19 pp.

Nega, E., Ulrich, R., Werner, S. and Jahn, M. 2003. Hot water treatment of vegetable seed– an alternative seed treatment method to control seed borne pathogens in organic farming. *Z. Pflanzenkr.Pflanzenschutz* **110:** 220–234.

Onsando, J.M. 1988. Management of black rot of cabbage caused by *Xanthomonas campestris* pv. *campestris* in Kenya. *Acta Hortic.* **218:** 311–314.

Pandey, K.K., Pandey, P.K., Singh, B., Kalloo, G. and Kapoor, K.S. 2001. Sources of resistance to downy mildew (*Peronospora parasitica*) disease in Asiatic group of cauliflower. *Veg. Sci.* **28:** 55–57.

Pandey, S.C., Naik, G., Ram Kishun and Sridhar, T.S. 1995. Breeding resistant varieties in cauliflower and cabbage. In: *Agroecosystems' management* (eds. Mandal, N.C., Dasgupta, M.K., Ghosh, D.C., Mukhopadhyay, S.K., Gupta, D.D. and Majumdar, D.K.). Sriniketan, India, Palli–Siksha Bhawan: 144–149.

Patel, M.K., Abhyankar, S.G. and Kulkarni, Y.S. 1949. Black rot of cabbage. *Indian Phytopath.* **2:** 58–61.

Patwardhan, G.B. 1928. Field, garden and orchard crops of Bombay Presidency. *Bull. Dept. Agric. Bombay*, 1928. 30pp.

Paulus, A.O. and Nelson, J. 1977. Systemic fungicides for the control of Phycomycetes on vegetable crops applied as seed treatments granular or foliar spray. *Proc. 9th Brit. Insecticide Fungicide Conf., Brighton, England, 21–24 Nov., Vol. 3, London, UK.*

Pesti, M., Viola, M. and Meszaros, A. 1987. *In vitro* selection of *Fusarium* wilt resistance cabbage lines. *Eucarpia Cruciferae Newsl.* **12:** 100–101.

Pound, G.S. and Walker, J.C. 1951. Mosaic resistance in cabbage. *Phytopathology* **40:** 1083–1090.

Prasad, L. and Vishunavat, K. 2006. Assessment of yield losses in cauliflower seed crop due to Alternaria blight. *Indian Phytopath.* **59:** 185–189.

Pullman, G.S., Devay, J.E., Garber, R. and Weinhold, A.R. 1981. Soil solarization: Effect of Verticillium wilt of cotton and soil borne population of *Verticillium dahliae*, *Pythium* spp., *Rhizoctonia solani* and *Thielaviopsis brassicola. Phytopathology* **71:** 954–959.

Rajeh, O.N. and Hamed, K. 2000. Soft rot disease of vegetables in Jordon: host range, reaction of some potato cultivars to the infection and effect of planting date. *Dirasat Agril Sci.* **27:** 149–157.

Ramesh, C.R. and Rao, S.C. 1989. Outbreaks and new records. India. Ring spot (*Mycosphaerella brassicicola*) of cabbage and cauliflower observed in the Andaman Islands. *FAO Pl. Prot. Bull.* **37:** 95.

Ramirez– Villapudda, J., Endo, R.M., Bosland, P. and Williams, P.H. 1985. A new race of *Fusarium oxysporum* f.sp. *conglutinans* that attacks cabbage with type A resistance. *Plant Dis.* **69:** 612–617.

Ramirez–Villapudda, J. and Munnecke, D.E. 1987. Control of cabbage yellow (*Fusarium oxysporum* f.sp. *conglutinans*) by solar heating of field soils amended with dry cabbage residues. *Plant Dis.* **71:** 217–222.

Ramsey, G.B. and Smith, M.A. 1961. Market diseases of cabbage cauliflower, turnips, cucumber, melons and related crops. *U.S.D.A. Handbook* **184:** 49 pp.

Rao, Y.P. and Srivastava, D.N. 1964. Epidemic of bacterial black rot disease of cauliflower at Katrain and measures for its prevention. *Indian J. Hortic.* **21:** 244–246.

Ren, J.P., Dickson, M.H. and Petzoldt, R. 1995. Improving the level of resistance to soft rot (*Erwinia carotovora*) by recurrent selection in *Brassica rapa. Eucarpia Cruciferae Newsl.* **17:** 86–87.

Reyes, A.A. 1984. Suppressive activity of different fungi against cabbage yellows in the soil and *Fusarium oxysporum* f.sp. *conglutinans in vitro. Phytoprotection* **65:** 27–33.

Roselle, E. 1937. The collar disease of lettuce, its causes and its control. *Prog. Agric. Vitic.* **107(2):** 496–697.

Roy, A.K. 1973. Host range of *Sclerotinia sclerotiorum* and *Sclerotium rolfsii* in Jorhat, Assam, *Sci. & Cult.* **39(7):** 319–320.

Roy, S.K., Das, B.C. and Bora, L.C. 1998. Non pesticidal management of damping off of cabbage caused by *Rhizoctonia solani* Kuhn. *J. Agric. Sci. Soc. North East India* **11:** 127–130.

Rudnick, M. 1986. Epidemic occurrence of ring spot disease on cabbage in Schleswig–Holstein, caused by *Mycosphaerella brassicicola* (Duby) Lindau. *Nachrichtenbl. Pflenzenschutz.* **38:** 83–85.

Ryan, E.W. 1977. Control of cauliflower downy mildew (*Peronospora parasitica*) with systemic fungicides. *Proc. Ninth British Insecticide Fungicide Conf., Brighton, England, Vol 1 & 2, Res, Rep., London, U.K. Session 6B. Pests and Diseases of vegetables* 297–300p.

Ryan, E.W. 1986. Research Report–1985. *Horticulture* 35–36.

Saharan, G.S. and Kadian, A.K. 1983. Physiologic specialization in *Alternaria brassicae. Cruciferae Newsl.* **8:** 32–33.

Saharan, G.S., Kaushik, C.D. and Gupta, P.P. 1990. Optimum fungicidal spray schedule for the control of white rust of mustard. In: *Plant Pathological Research Problems and Progress* (Ed. Srivastava, M.P. and Saharan, G.S.). Haryana Agric. Univ., Hissar, India, pp. 25–29.

Saharan, G.S., Verma, P.R. and Nashaat, N.I. 1997. Monograph of downy mildew of crucifers. *Tech. Bull. Agric. – Agri. Food Pl. Prot. Soc.* 1997–01, 197 pp.

Saini, L.C. and Prashar, R.D. 1981. Comparative efficacy of stable bleaching powder with other antibacterial formulations in controlling black and soft rot of cauliflower. *Indian Phytopath*. **34:** 465–469.

Sandhu, K.S. 1986. *Sclerotinia sclerotiorum* a destructive pathogen of cauliflower seed crop in Punjab. *Punjab J. Res.***23(1):** 78–81.

Sandhu, K.S., Singh, H. and Dhillon, T.S. 1984. Effect of *Alternaria* blight on radish varieties sown on different dates. *J. Res. Punjab Agric. Univ*. **21:** 543–546.

Sandhu, K.S., Singh, H. and Kumar, R. 1985. Effect of different nitrogen levels and dates of planting on *Alternaria* blight and downy mildew diseases of radish seed crop. *J. Res. Punjab Agric. Univ.,* **22:** 285–290.

Santos, M.P., Dias, J.S., Monteiro, A.A., Dias, J.S., Crute, I. and Monteiro, A.A. 1996. Screening Portugese cole landraces for resistance to white rust (*Albugo candida* (Pers.) Kuntze). *Acta Hortic.* **407:** 453–459.

Schaad, N.W. and White, W.C. 1974. Survival of *Xanthomonas campestris* in soil. *Phytopathology* **64:** 1518–1520.

Schultz, T., Gabrielson, R.L. and Olson, S. 1986. Control of *Xanthomonas campestris* pv. *campestris* in crucifer seed with slurry treatments of calcium hypochlorite. *Plant Dis.* **70:** 1027.

Shao, X.H., Chen, Q.Q. and Zhang, F.Q. 1990. Research on relation between some ecological factors and *Peronospora parasitica* (Pers.) Fr. on chinese cabbage. *Acta Agric., Shanghai* **6:** 78–81.

Sharma, B.K. and Basandhari, A.K. 1997. Effect of biocontrol agents, fungicides and plant extracts on sclerotial viability of *Sclerotinia sclerotiorum. Indian J. Agric. Sci.* **67(3):** 132–133.

Sharma, B.R., Dhiman, J.S., Thakur, J.C., Singh, A. and Bajaj, K.L. 1991. Multiple disease resistance in cauliflower. *Adv. Hortic. Sci.* **5:** 30–34.

Sharma, B.R., Swarup, V. and Chatterjee, S.S. 1992. Inheritance of resistance to black rot in cauliflower. *Can. J. Genet. Cytol.* **14:** 363–370.

Sharma, P. 2002. Induction of systemic resistance in cauliflower to downy mildew by exogenous application of plant activator. *Ann. Plant Protec. Sci.* **10:** 99–102.

Sharma, P., Swati, D., Sharma, M. and Bhati, D.S. 2013. Genetic variation of *Alternaria brassicae* (Berk.) Sacc., causal agent of dark leaf spot of cauliflower and mustard in India. *J. Gen. Plant Pathol.* **79:** 41–45.

Sharma, Pratibha, Zewain, Q.K., Bahadur, P. and Sain, S.K. 2005. Effect of soil solarization on sclerotial viability of *Sclerotinia sclerotiorum* infecting cauliflower (*Brassica oleracea* var. *botrytis* subvar. *cauliflora*). *Indian J. Agric. Sci.* **75(2):** 90–94.

Sharma, R.C. 1979. Stalk rot of cauliflower caused by *Sclerotinia sclerotiorum* (Lib.) de Bary. PhD Thesis, HPKVV, College of Agriculture, Solan, 146 pp.

Sharma, R.C. and Sharma, S.L. 1984. Evaluation and economics of fungicidal spray against *Scleritinia* rot of cauliflower seed crop. *Seed Res.* **12(2):** 95–97.

Sharma, R.C. and Sharma, S.L. 1985. Correlation of meteorological factors with stalk rot of cauliflower caused by *Sclerotinia sclerotiorum. Indian J. Ecol.* **12(2):** 327–330.

Sharma, R.C. and Sharma, S.L. 1986. Effect of soil amendments and artificial defoliation of infected leaves on *Sclerotinia* stalk rot of cauliflower. *Indian J. Plant Pathol.* **4(2):** 126–128.

Sharma, R.C., Sharma, S.L. and Sharma, I. 1982. Varietal reaction of cauliflower against stalk rot caused by *Sclerotinia sclerotiorum* (Lib.) de Bary. *Crop. Improv.* **9(2):** 167–168.

Sharma, S.K., Kohli, U.K., Kanwar, H.S. and Kumar, S. 2003. Performance of newly introduced temperate cabbage and cauliflower germplasm against black rot and stalk rot diseases. *Pl. Dis. Res.* **18(1):** 90–91.

Sharma, S.K., Korla, B.N. and Thakur, B. 2000. Evaluation of cauliflower genotypes for multiple disease resistance in mid hills of Himachal Pradesh. *Indian J. Plant Pathol.* **18(1&2):** 75–78.

Sharma, S.L. 1981. Control of black rot of cauliflower by hot water seed treatment and field sprays with Streptocycline. *Indian J. Mycol. Pl. Pathol.* **11:** 17–20.

Sharma, S.L., Kaith, D.S. and Khan, M.A. 1983. Cultural control of *Sclerotinia* stalk rot of cauliflower. *Indian Phytopath.* **36(4):** 601–603.

Sharma, S.R. 1983. Effect of fungicidal spray on white rust and downy mildew diseases and seed yield in radish. *Gartenbauwassenschaft* **48 :** 108–112.

Sharma, S.R. and Sohi, H.S. 1982. Effect of fungicides on the development of downy mildew and white rust of radish. *Indian J. Agric. Sci.* **58 :** 521–524.

Sharma, S.R., Gill, H.S. and Kapoor, K.S. 1988. Heterosis and combining ability studies in late cauliflower (*Brassica oleracea* var. *botrytis* L.). *Veg. Sci.* **15:** 55–63.

Sharma, S.R., Kapoor, K.S. and Gill, H.S. 1995. Screening against Sclerotinia rot (*Sclerotinia sclerotiorum*), downy mildew in cauliflower (*Peronpspora parasitica*) and black rot (*Xanthomonas campestris*) in cauliflower (*Brassica oleracea* var. *botrytis* sub var. *cauliflora*) *Indian J. Agric. Sci.* **65(12):** 916–918.

Sharma, Y. and Chaudhary, K.C.B. 1984. Epidemiology of cauliflower wilt in different soils in Jammu. *Indian J. Plant Pathol.* **2:** 78–79.

Shekhawat, P.S., Jain, M.L. and Chakravarti, B.P. 1982. Detection and seed transmission of *Xanthomonas campestris* pv. *campestris* causing black rot of abbage and cauliflower and its control by seed treatment. *Indian Phytopath.* **35:** 442–447.

Shelton, A.M. and Hunter, J.E. 1985. Evaluation of the potential of the flea beetle *Phyllotreta cruciferae* to transmit *Xanthomonas campestris* causal agent of black rot of crucifers. *Can. J. Pl. Pathol.* **7:** 308.

Sherf, A.F. and Macnab, A.A. 1986. Diseases of crucifers. In. *Vegetable diseases and their control* (2 Edn.). The Ronald Press. New York, pp. 251–306.

Sheriff, C. and Lucas, J.A. 1989. Cytogenetical study of hetrothalic and homothalic isolates of *Peronospora parasitica. Mycol. Res.* **92 :** 302–310.

Shyam, K.R., Gupta, S.K. and Mandradia, R.K. 1994. Prevalence of different types of curd rots and extent of yield loss due to plant mortality in cauliflower seed crop. *Indian J. Mycol. Pl. Pathol.* **24:** 172–175.

Shyam, K.R., Gupta, S.K. and Mandradia, R.K. 2001. Role of abiotic stress on the development of bacterial curd rot of cauliflower seed crop in Himachal Pradesh. *Indian J. Mycol. Pl. Pathol.* **31:** 199–201.

Silvae, D., Nashaat, N.I. and Tirilly, Y. 1996. Differential responses of *Brassica oleracea* and *B. rapa* accessions to seven isolates of *Peronospora parasitica* at the cotyledon stage. *Plant Dis.* **80:** 142–144.

Singh, B.P., Singh, S.P., Abu, M. and Mishra, G.M. 1989. Relative efficacy of different fungicides for the control of leaf spot of cabbage. *Veg. Sci.* **16 :** 62–65.

Singh, D., Dhar, S. and Yadav, D.K. 2010. Effect of endophytic bacterial antagonists against black rot disease of cauliflower caused by *Xanthomonas campestris* pv. *campestris. Indian Phytopath.* **63:** 122–126.

Singh, D., Dhar, S. and Yadav, D.K. 2011.Genetic and pathogenic variability of Indian strains of *Xanthomonas campestris* pv. *campestris* causing black rot disease in crucifers. *Curr. Microbiol.* **63:** 551–560.

Singh, D.B. 1986. Evaluation of fungicides for the control of *Alternaria brassicae* and *Drechslera graminea. Iranian J. Plant Pathol.* **22 :** 39–43.

Singh, H.M., Ali, S., Chakraborti, D.K., Singh, V.K., Rajput, S.K.S. and Srivastava, D.K. 2002. Integrated pest management in early cauliflower. *Ann. Pl. Prot. Sci.* **10:** 192–193.

Singh, R.S. 1987. *Diseases of Vegetable crops.* Oxford and IBH Publishing Co. Pvt. Ltd. New Delhi, 362pp.

Smith, I.M., Dunez, J., Philips, D.H., Lelliott, R.A. and Archer, S.A. 1988. *European Handbook of Plant Diseases.* Blackwell Sci. Publ., Oxford. London.

Sugha, S.K. 1999. Efficacy of carbendazim granules against *Sclerotinia sclerotiorum* of pea (*Pisum sativum*), cabbage (*Brassica oleracea* var. *capitata*) and cauli– flower (*Brassica oleracea* var. *botrytis*). *Indian J. Agric. Sci.* **69(4):** 302–303.

Szczech, M. and Brzeski, M.W. 1994. Vermicompost fertilizers or biopesticide? *Sesja Noukowa* **41:** 77–83.

Takeuchi, J., Watanabe, K., Horie, H., Takeda, N., Morozumi, M. and Kowamura, S. 1995. Control of yellows (*Fusarium oxysporum* f.sp. *conglutinans*) of Komatsuma (*Brassica campestris*) by dazomet. *Proc. Kanto–Tosan Pl. Prot. Soc.* **42:** 73–74.

Tamayo, M.P.J., Becerra, V.D.C. and Jaramillo, N.J.E. 2001. *Alternaria brassicae*, causal agent of head rot of cauliflower (*Brassica oleracea* L. var. *botrytis* L.). *ASCOLFI–Informa.* **27(2):** 10–11.

Tasca, G. and Trandaf, F. 1984. Efficacy of pre–harvest treatment with fungicides in reducing cabbage decay during storage. *Lucrari stiintifice Institutul de cercetari si Proiectari Pentru Valorificarea si Industrializarea Legumelor si Fructelor* **15:** 89–93.

Taylor, J.D., Conway, J., Roberts, S.J., Astley, D. and Vicente, J.G. 2002. Sources and origin of resistance to *Xanthomonas campestris* pv. *campestris* in *Brassica* genome. *Phytopathology* **92 (1):** 105–111.

Thakur, J.C, Dhatt, A.S., Lal, Tarsem, Bal, S.S. and Pathak, Mamta 2009. Punjab Katki: a new variety of cauliflower. *J. Res. PAU* **46:** 106–107.

Thakur, J.C., Singh, V.P. and Chander Parkash 2004. Performance of cauliflower (*Brassica oleracea* var. *botrytis*) selections in the temperate climate of India. *Cruciferae Newsl.* **25:** 103–104.

Thakur, R.P. and Mathur, K. 2002. Downy mildews of India. *Crop Protec.* **21:** 333–345.

Thind, T.S. and Jhooty, J.S. 1988. Differential behaviour of fungicides against Alternaria blight of some cruciferous hosts. *Indian J. Mycol. Pl. Pathol.* **18:** 122–127.

Tiwary, R.K.S., Chandravanshi, S.S. and Ojha, B.M. 2004. Evaluation of some medicinal plant species for their antibacterial activity against *Xanthomonas campestris* pv. *campestris,* the black rot pathogen of cabbage. *Indian Phytopath.* **57:** 308–311.

Thomas, C.E. and Jourdam, E.L. 1990. Evaluation of broccoli and cauliflower germplasm for resistance to Race 2 of *Peronospora parasitica. Hortic. Sci.* **25:** 1429–1431.

Tortolero, O. and Carrasco, A. 1982. Cabbage ring spot (*Mycosphaerella brassicicola*) in Venezuela. *Fitopatologia* **17:** 21–24.

Tsuneda, A., Skoropad, W.P. 1978. Phylloplane fungal flora of rapeseed. *Trans. Brit. Mycol. Soc.* **70:** 329–334.

Valkonen, J.P.T. and Koponen, H. 1990. The seed borne fungi of chinese cabbage (*Brassica pekinensis*), their pathogenicity and control. *Plant Pathol.* **39:** 510–516.

Vannacci, G. and Harman, G.E. 1987. Biocontrol of seed borne *Alternaria raphani* and *A. brassicicola. Can. J. Microbiol.* **33:** 850–856.

Varalakshmi, B. Ganeshan, Girija, Gopalakrishnan, C., Pushpalatha, A. And Chethana, B.S. 2009. Identification of sources of resistance to Alternaria leaf spot (*Alternaria brassicicola*), black rot (*Xanthomonas campestris*) and downy mildew (*Peronospora parasitica*) in cauliflower (*Brassica oleracea*). *Indian J. Agric. Sci.* **79:** 482–483.

Vasileva, E.D. 1976. Damage by *Peronospora* disease to cabbage and the effectiveness of polycarbacin in its control. *Referativnyi Zhurnal* (1977), **155:** 1299.

Vasudeva, R.S. 1958. Diseases of rape and mustard. In: *Rapeseed and Mustard.* 16: 77–86. Indian Central Oilseed Comm. Hydrabad, India.

Verma, P.R. and Petrie, G.A. 1979. Effect of fungicides on germination of *Albugo candida* oospores *in vitro* and foliar phase of white rust disease. *Can. Pl. Dis. Surv.* **59:** 53–59.

Verma, P.R. and Saharan, G.S. 1994. *Alternaria brassicae* (Berk.) Sacc., *A. brassicicola* (Schwein.) Wiltsh. and *A. raphani* Groves and Skolko Introduction, bibliography and subject index. *Agric. Can. Res. Stn. Saskatoon Tech. Bull. No.* **93–002:** 81p.

Verma, P.R. and Thakur, P.C. 1989. Comparative field resistance of cabbage collection to downy mildew at seedling stage. *Indian J. Pl. Prot.* **17:** 79–80.

Vicente, J.G., Conway, J., Roberts, S.J. and Taylor, J.D. 2001. Identification and origin of *Xanthomonas campestris* pv. *campestris* races and related pathovars. *Phytopathology* **91(5):** 492–499.

Wang, Z.S. and Wang Z.S. 2001. Occurrence and control of soft rot in summer and autumn Chinese cabbage (*Brassica pekinensis*). *Fujian Agric Sci. Technol.* **3:** 9–11.

Waterhouse, G.M. 1975. Species of *Albugo* in Aizoaceae. *Trans. Brit. Mycol. Soc.* **65:** 504–507.

White, J.G, Crute, I.R. and Wynn, E.C. 1984. A seed treatment for the control of *Pythium* damping off and *Peronospora parasitica* on brassicas. *Ann App. Biol.* **104:** 241–247.

White, J.G., Linfield, C.A., Lahdenpera, M.L. and Uoti, J. 1990. Mycostop a novel biofungicide based on *Streptomyces griseoviridis. Proc. 1990 Brit. Crop Prot. Conf., Pests Dis., Brighton* **1:** 221–226.

Whitwell, J.D. and Griffin, G.W. 1967. Chemical control of downy mildew of seedling cauliflower. *Proc. 4th Brit. Insecticide Fungicide Conf. 1967.* **I:** 1–460.

Williams, J. and Pink, D.A.C. 1987. Development of an *in vitro* screen in brussels sprout for resistance to *Alternaria brassicae. Cruciferae Newsl.* **12:** 91.

Williams, P.H. 1980. Black rot: a continuing threat to world crucifers. *Plant Dis.* **64:** 736–742.

Williams, P.H. and Pound, G.S. 1963. Nature and inheritance of resistance to *Albugo candida* in radish. *Phytopathology* **53:** 1150–1154.

Williams, P.H., Walker, J.C. and Pound, G.S. 1968. Hybelle and Sanibel, multiple disease resistant F_1 hybrid cabbage. *Phytopathology* **58:** 791–792.

Wu, W.S. and Lu, J.H. 1984. Seed treatment with antagonists and chemicals to control *Alternaria brassicicola. Seed Sci. Tech.* **12:** 851–862.

Xu, J.B., Tao, G.H., Chen, G., Yang, R., Zhang, F.L. and Sun, J.Z. 1993. Newly released medseason–maturing Chinese cabbage cultivar Beijing 75 and its extension work. *Chinese Veg.* **4:** 3–5.

Yada, S. and Imoto, M. 1983. Effects of mixing fungicides in seed dressing on the control of damping off of (*Rhizoctonia solani* Kuhn.) in Chinese cabbage. *J. Jpn Soc. Hortic. Sci.* **52:** 27–34.

Yang YauWen, Tsai ChinhChang and Wang TsuTsuen 1998. A heat tolerant broccoli F_1 hybrid, 'Ching-Long 45'. *Hortic Sci.* **33(6):** 1090–1091.

Yang, C.Y., R.L.F. Hu and U. Pupipat 1983. Control of Chinese cabbage downy mildew (*Peronospora parasitica*) by metalaxyl (Ridomil) and other fungicides. *Int. Symp. on Pesticides in Developing countries– Present and Future, Ibaraki, Japan. Trop. Agric. Res. Centre*, 125–130.

Yasaka, R., Nguyen, H.D, Ho, S.Y.W., Duchene, S., Korkmaz, S., Katis, N., Takahashi, H., Gibbs, A.J., Ohshima, K. 2014. The temporal evolution and global spread of *Cauliflower Mosaic Virus*, a plant pararetrovirus. PLoS ONE **9(1):** e85641

Zewain, Q.K., Bahadur, P. and Sharma, P. 2005. Integrated disease management of stalk rot of cauliflower. *Indian Phytopath.* **58(2):** 167–173.

■■■■

6

DISEASES OF CUCURBITS

Cucurbits, member of family cucurbitaceae, form an important group of vegetable crops in different parts of the world. The family cucurbitaceae contains at least 9 genera and of them 16 species are cultivated as vegetables. In India, cucurbits like ash gourd, bitter gourd, bottle gourd, cucumber, sponge gourd, long melon, or snake cucumber, kundru, muskmelon, pointed gourd, pumpkin, ridge gourd, snake gourd, squash, round gourd and watermelon are cultivated round the year in one or the other region of the country. During cultivation, these crops are attacked by number of fungal, bacterial and viral diseases and disease losses are sometimes heavy under ideal disease development conditions. In this chapter some of the important diseases are described along with their management measures.

(A) FUNGAL DISEASES

1. DOWNY MILDEW

It is an important disease of cucurbitaceous crops particularly in those areas of the world having adequate moisture, due to intermittent rains and moderate temperatures during the crop period. The pathogen is confined to cucurbitaceous crops only and during favourable weather conditions, this disease can cause considerable yield losses (Plate 14c). It occurs practically on all the cultivated cucurbits in open field and protected crops under glasshouse/plastic houses (Palti and Cohen, 1980). In North Indian plains also, the disease has assumed serious proportions on cucumber grown under plastic low tunnels (Bhat *et al.*, 2013).

Geographical distribution: The disease was first recorded in Cuba during 1868 (Chupp and Sherf, 1960). Subsequently it was reported from Japan, U.S.A., Germany, Massachusetts, Chechoslovakia, Ceylon, South Africa, Kenya, Porto Rico, Queensland, Palestine, Sierra Leonne, Mozambique, Israel, Serbia and Yugoslavia, Nicaragua, Rhodesia, Turkey, South Australia, Alta Canada, Bulgaria and Poland (Robak, 1995). The distribution of downy mildew pathogen is widest on cucumber (70 countries) in all continents followed by muskmelon (50 countries); on *Cucurbita* and watermelon it extends to about 40 and 25 countries, respectively (Palti and Cohen, 1980).

Losses: Losses are directly related to the length of the lag period between planting and onset of the disease. If the outbreak is earlier, the losses may be enormous. Prevalence of unfavourable weather conditions for plant growth may also increase losses due to this disease. Overhead sprinkling irrigation is usually associated with higher losses than those encountered where trickle irrigation is used (Cohen, 1980). In North Carolina, cucumber production was reduced by 30 per cent due to the infection of downy mildew (Ellis and Cox, 1948), while Mahrishi and Siradhana (1988) reported 79.06 per cent reduction in muskmelon fruit yield in Rajasthan when no sprays of Dithane M-45 were applied. They further reported that the losses are directly proportionate to disease intensity.

Symptoms: The symptoms appear as pale green areas on the upper surface of the leaves, which are separated by islands of darker green colour (Plate 1a). Soon the colour of infected areas changes from pale green to yellow angular spots bounded by leaf veins. During moist weather, the corresponding lower surface is covered with a faint purplish to greyish black fruiting layer of the fungus containing sporangiophores and sporangia (Plate 1 b). Lesions on the upper leaf surface turn necrotic brown from the centre outwards. Mostly the leaves, which are located in the centre of the vine, are attacked first and then disease progresses upward and downward until the entire vine is killed (Plate 1c). Primary lesions on cucumber are angular due to their limitation by leaf veins, but on melon (*Cucumis melo*) and other cucurbits no such limitation of lesion growth is seen. The fruits are rarely affected but heavy disease pressure prevents fruit maturation and reduces flavour and quality.

The Pathogen

The fungus responsible for causing cucurbit downy mildew is *Pseudoperonospora cubensis* (Berk. and Curt.) Rostow. The mycelium is hyaline, coenocytic and intercellular, developing abundantly within the leaf mesophyll layer, but also penetrating to the palisade layer. Individual hyphae have an oval shape, with intercellular haustoria formed as finger-like branches; hyphal size is 5.4–7.2 μm in diameter. Sporangiophores develop in groups of between one and five, produced through individual stomata. These are 180–400 μm long and 5–7 μm wide with a bulbous base and are dichotomously branched in the upper third of their growth (Fig. 1). Sporangia form singly on subacutely shaped sporuliferous tips, are pale gray to purple in colour, ovoid to elliptical in shape, possess a thin wall with a papilla at the distal end and measure 20–40 x 14–25 μm. They germinate to produce biflagellate zoospores 10-13 μm in diameter. Rarely the sporangia of *P. cubensis* germinate directly by forming a germ tube. Oospores are produced rarely and under some specific environmental conditions. The oospores are spherical (19-22 μm diameter), rarely obovoid to ellipsoid, light yellow to pale yellow with a smooth wall of 1.5–3.5 μm thickness (Mahrishi and Siradhana, 1984).

Host range and host specificity

Various workers from different parts of the world reported that *P. cubensis* infecting cucurbits differed in its host range (Ellis, 1951; Hughes and Van Haltern, 1952). The host range includes *Bennincasa hispida* (Thumb.) Cogn., *Bryonopsis laciniosa* Naud. var. *erythrocarpa* Naud., *Citrullus vulgaris* Schrad, *C. vulgaris* var. *fistulosus* Stewart, *Coccinea indica* Wright and Arnold, *Cucumis angulatus* Forsk., *C. anguria* L., *C. callosus* (Rotl.) Cogn., *C. dipsaceus* Ehrenb., *C. melo* L., *C. melo* L. var. *dudaim* Naud., *C. melo* var. *utilissimus* L., *C. sativus* L., *C. sativus* L. var. *angulicus* Bailey, *Cucurbita maxima* Duch., *C. moschata* Duch., *C. pepo* L., *C. pepo* L. var.

ovifera (L.) Bailey, *Echinocystis lobata* (Michx.) T. and G., *Lagenaria siceraria*, *L. vulgaris* Ser., *Luffa aegyptiaca* Mill., *L. acutangula* Roxb., *L. cylindrica* Roem., *Melothria maderaspatana* Cogn., *Melothria scabra* Naud., *Momordica balsamina* L., *M. charantia* L., *Sicyos angulatus* L. *Trichosanthes anguina* L. and *T. cucumerina* L. (Hoerner, 1940; Bains and Jhooty, 1976b; Mahrishi and Siradhana, 1988a).

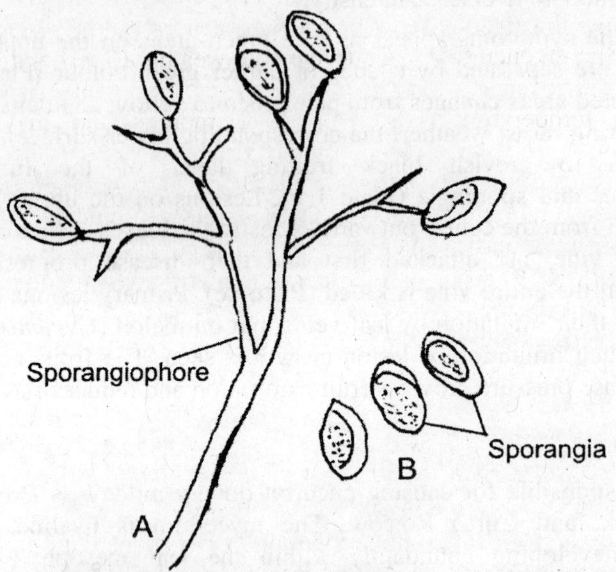

Fig. 1. Sporangiophore and sporangia of *P. cubensis*

Bains and Jhooty (1976b) from Punjab reported pathological specialization in the downy mildew pathogen, as the muskmelon (*Cucumis melo* L.) isolate of *P. cubensis* did not infect ashgourd (*Benincasa hispida* (Thumb.) Cogn.) and pumpkin (*Cucurbita moschata* Dush.) while the reverse was possible. Round gourd (*Citrullus vulgaris* var. *fistulosus* Stewart) was not observed to be infected under natural conditions of Rajasthan, but infection occurred on artificial inoculation with sporangial suspension taken from muskmelon (*C. melo*) and ridgegourd (*Luffa acutangula* Roxb.) (Mahrishi and Siradhana, 1988a). Watermelon and ashgourd were not infected by the muskmelon isolate of *P. cubensis* under Punjab conditions (Bains and Jhooty, 1976b), but they were potential hosts in Rajasthan (Mahrishi and Siradhana, 1988a) while watermelon is a host in Mysore.

Differences in the severity of infection on different crops indicate the presence of physiological races in the *P. cubensis*. Bittergourd, a non-susceptible crop in other regions is heavily damaged in Assam. Five pathotypes infecting watermelon, muskmelon, bottlegourd, *Cucurbita pepo* and *Luffa aegyptica* were distinguished based on high compatibility with specific hosts (Bains and Parkash, 1985). Two or more races exist in Japan and the United States, but not in Europe or the Middle East (Palti and Cohen, 1980). Inoculation of 26 cultivars representing 13 species and subspecies within 7 genera of Cucurbitaceae in Israel, USA and Japan revealed distinct physiological specialization among pathogen isolates (Thomas *et al.*, 1987a). Five isolates of *P. cubensis* could be distinguished based on high compatibility with specific hosts and all the five pathotypes were highly compatible with *Cucumis sativus* and *C. melo* var. *reticulus*.

Disease cycle

The pathogen mainly survives as mycelium and sporangiophores in the off-season on living hosts, as the oospores are not common. The pathogen can survive on cultivated and wild cucurbits throughout the year and between cropping periods in vegetative state and sporangia on cucurbits that survive cold weather (Sherf and Macnab, 1986). In Denmark, in the absence of oospores, zoosporangia have been reported to survive below freezing temperature (-18°C) for 3–4 months and may serve as the overwintering structure in northern latitudes (Lange *et al.*, 1989). Wherever the oospores are produced, they may explain the mode of perenation, but their role in initiating the disease on cultivated cucurbits has not been elucidated (Cohen, 1980). Oospores have been found in cucumber in Japan (Hiura and Kawada, 1933) and China (Chen *et al.*, 1959) and in pumpkin, bottle gourd and sponge gourd from India (Maharishi and Siradhana, 1984; Singh and Sokhi, 1989). They have also been recorded in wild cucurbits from India (Khosla *et al.*, 1973; Bains *et al.*, 1977) and in older infected leaves of greenhouse cucumbers at the end of the season in Austria (Bedlan, 1989). However, their role in acting as primary source of infection for the next season has not been elucidated. The incidence of downy mildew increased five folds over the last three years when the crop was continuously grown in the same area year after year (St. Anand and Weliner, 1991).

In India, Butler (1918) reported the possibility of perpetuation of *P. cubensis* in wild cucurbits in regions with suitable climatic conditions. Similarly, Bains and Jhooty (1976a) also proposed that in Punjab and possibly in North India this pathogen survives as active mycelium on plants of *Luffa aegyptiaca* Mill. since during their earlier studies they could not observe oospores on any of cultivated or wild cucurbit hosts of the fungus. However, Khosla *et al.* (1973) and later Bains *et al.* (1977) were able to record oospores on parwal (*T. diocia*) from Madhya Pradesh and on *Melothria maderaspatana* (a weed) from Punjab, respectively. Later Mahrishi and Siradhana (1984) recorded oospores on pumpkin (*Cucurbita moschata*) and bottlegourd (*Lagenaria siceraria*) from Rajasthan. The oospores may serve as a mode of overseasoning of the fungus in those regions where they are produced, but yet to be proved and confirmed.

In the presence of free moisture on leaf surface, sporangia germinate by producing zoospores. Zoospores become encysted after swimming for some time in film of water. After germination, the germtube penetrates the stomatal openings and this is succeeded by the production of intercellular hyphae and haustoria. Symptoms appear 3–7 days after infection on which secondary inoculum is produced which is blown by wind to the neighbouring plants/fields.

Epidemiology

The pathogen generally overwinters in mycelial form in wild and self sown cultivated hosts under protected conditions. Under favourable weather conditions, sporangia are produced which are blown by wind to cause infection in spring and early summer. Sporangium concentration in the air seems to be the most important biotic factor determining disease onset. Disease appeared at cotyledonary stage, i.e., 8 days after sowing in plants growing nearby infected fields (Cohen and Rotem, 1971). When no such inoculum source was available, disease occurrence was noticed 7 weeks after sowing. Dew is another important factor, which provide free moisture on the leaf, thereby providing ideal conditions for sporangial germination and infection. Apart from leaf wetness essential for infection, the factors determining disease progress are rate of foliage growth and physiological age of the host, amount of the primary inoculums available, light, and the rate at which lesions necrotize (Palti and Cohen, 1980).

Bains and Jhooty (1978a) studied behaviour of *P. cubensis on* muskmelon in Punjab. They reported that the pathogen started producing sporangia before midnight and the sporangia mature by 3 A.M. Maximum dispersal of sporangia occured between 6 and 10 A.M. Maximum sporulation occurred at 20 ± 2°C in the moist saturated atmosphere, and optimum temperature for germination of sporangia and infection was 20°C. Temperature above 35°C arrested infection as well as sporulation and lesions become brown, necrotic. Zoospores are motile for longer periods at 10 to 15°C than at 30 to 35°C (Bains and Jhooty, 1976a, 1978a). Relative humidity above 75 per cent was conducive for disease development. Maximum distribution of sporangia was recorded at a height of 25 cm above the ground level and decreased with the increase in the height from the ground (Sharma *et al.*, 2003). Apparent infection rate (r) showed positive correlation with temperature and moisture conditions (Maharishi and Siradhana, 1988c). Ullasa and Amin (1988) observed that day temperature of 25–30°C, night temperture of 15–21°C and RH >75 per cent favoured the infection of *Luffa acutangula* by *P. cubensis*. Singh *et al.* (1990) observed the influence of post– infection high temperature on disease development in muskmelon cv. Punjab Hybrid and found that continuous exposure of plants at 35 ± 1°C for more than 48 h or even a short exposure of 12 h at 40 ± 1°C was lethal as it eliminated already established infection. Alternation of high temperature with low temperature had less effect on survival of *P. cubensis*. Maximum sporulation was recorded in cucumber leaves when exposed to 20:15°C day: night thermo periods. Exposure to higher temperatures decreased sporulation because of rapid necrotization of the infected tissues (Cohen and Rotem, 1971; Singh and Singh, 1994). Singh *et al.* (1996) reported maximum temperature and wetness duration during SMW 21 to 24 as key limiting factors in disease development. A function of these two parameters, HTR (humid-thermal relation) has been used in developing prediction system for muskmelon downy mildew in Punjab. Moderate temperature (23°C) coupled with high rainfall and relative humidity (>80%) helped significantly in disease development on cucumber (Sharma *et al.*, 2003).

Low nutritional status of the muskmelon plants pre-disposes them to infection by *P. cubensis,* while disease development was less on plants grown in high P, low K and high N nutrition solution (Bains and Jhooty, 1978b). Maharishi and Siradhana (1988b) also observed reduced incidence of disease in plants supplied with higher doses of P, but lower concentration along with N and K increased disease intensity. Trace elements like Zn and Cu also reduced the disease significantly.

On cucumber grown under low plastic tunnel, the disease appeared early in mid-February (compared to first week of April in the crop sown in the open field) was maximum (53.2%) on the crop sown on 30 January. Maximum disease severity (54.4%) was observed when no side ventilation was given and least incubation period (2 days) was recorded at 30ºC and 122 h leaf wetness duration on cv. Punjab Naveen (Bhat *et al.*, 2013).

Long distance dispersal is a fundamental aspect of epidemic development that influences the timing and extent of outbreaks of cucurbit downy mildew. The dispersal potential of *P. cubensis* appears to be limited primarily by sporangia production in source fields and availability of susceptible hosts and less by sporangia survival during transport. Understanding pathogen diversity and population differentiation is a critical aspect of disease management (Ojiambo *et al.*, 2015).

Management

The disease can be kept under check by integration of various management methods like cultural, chemical and host resistance.

i. ***Cultural practices:*** Cultural methods include destruction of plant debris of previous crop and cucurbitaceous weeds, separating mature and newly sown crops, avoiding high crop population densities for preventing the buildup of moist microclimates within leaf canopy and use of irrigation from overhead, which leads to moisture being retained on the leaves.

ii. ***Host resistance:*** Cvs./lines/strains/hybrids like Shipper, Poinsette, Calypso, Addis, Sampson, Israeli lines 17 and 22, Sadao, Russiu, Heiva-K-2841, Schimo-schirazu S 2, Chimochirazu S 2, Belair, County Fair, Nanet, Ninghuang 2, Konkurent, Pasad, Salyut $F_1$249, Progress X169-3, 176 x 182, lines 1144-74, Gibrid 2, Line 22, 371, 196, Levina F_1, Kamon F_1, 6501, 6502, Belcanto F_1, 88-2, 88-7 and 88-8, Ningyang Spiny, Rodnichok F_1, Bngadnyi F_1, Mover F_1, Edinstvo 31, Feniks, Wonye 501 and Wonye 502, Jinchun 4, 86604065-90, S440S, Slicer Astrea F, Nastasja F, Regal F, Xia Qing 4, Zao Qing 2 were reported to be resistant to downy mildew from different parts of the world (Cohen, 1976; Pivovarov *et al.*, 1977; Jiang, 1981; Meyer and Ziesenis, 1982; Yakimenko, 1983; Pivovarov, 1984; Lisitisin and Pluzhnikova, 1990; Wu, 1991; Lai and Zhang, 1992; Stryapkova *et al.*, 1992; Om *et al.*, 1992; Lu *et al.*, 1994; Angelov and Georgiev, 1995; Komnenic *et al.*, 1995 and Zuo *et al.*, 1995). Singh *et al.* (1996) identified KP2, KP7, KP 9, SP 2, SP 3 and EC 163888 as resistant sources of snapmelon against *P. cubensis*. Cv. Chitradurga of cucumber was reported as immune to this pathogen under Solan conditions (Raj *et al.*, 2003). Goswami *et al.* (2011) reported three genotypes viz. IC 274014, IC 267397, MCPS-2 and a wild type Mehna Chibber as highly resistant to downy mildew with the disease severity of 2.0, 2.0, 1.6 and 2.8 per cent, respectively, while four lines viz., MCPS-7, MCPS-17, MCPS-28 and Pusa Sharbati, with disease severity in the range of 3.8–5.0 per cent were resistant.

Thomas (1978) found that plants which produced brown lesions had fewer lesions and ones with yellow lesions had more lesions. Thomas *et al.* (1987b) introduced the use of reaction types 2,3 and 4 representing levels of resistance compared to the susceptible reaction type fixed by RTI. A double digit classification system for the RT of leaves 1 and 2, respectively, in green howal inoculations of two leaf stage plants is an excellent predictor of the disease reaction of older plants under field conditions and provided plant breeders with a technique that will expedite selection for downy mildew resistance in muskmelon. Progeny from test crosses and new accession can be evaluated reliably in the green house-growth chamber for downy mildew resistance as soon as the seed is available. This classification technique for resistance evaluates the specific type of lesion produced rather than estimating foliage damage as per cent chlorotic or necrotic tissue, per cent lesion area, per cent foliage loss etc. The extent of sporulation by *P. cubensis* on muskmelon leaves exhibiting RT 1,2,3 and 4 gives an additional measure of the relative resistance level indicated by each RT. Thus spore production decreases progressively as the level of resistance represented by RT 2, 3 and 4 increased as shown in Table 1.

Table 1. Sporulation of *P. cubensis* on leaves of some muskmelon genotypes

Entry	Reaction type	Sporangia/ cm^2 of leaf area x 10^3	
		Glasshouse	Field
Perlita	1	52.4	41.6
Cinco	2	18.3	12.4
B 8-1	3	6.0	1.1
B 4-1	4	0.3	Traces

Source: Thomas *et al*. (1987b)

Cohen *et al.* (1989) showed that *P. cubensis* produced 10-15 mm lesions in the susceptible cv. Anaras-Yoknean (A-1), but 1–3 mm water-soaked lesions in the resistant cv. PI 124111F (PI). Sporulation on cv. PI was extremely limited compared to the conspicuous sporulation on cv. A-1. Reuveni *et al.* (1992) and Chauhan *et al.* (1994) used peroxidase activity in uninfected musmelon plants to determine resistance and susceptibility to *P. cubensis*. After infection, peroxidase activity increased with time in both susceptible and resistant plants. The ratio of activity in infected to uninfected leaves increased overtime in the susceptible plant.

iii. Chemical control: The disease can be kept under check by spraying or dusting of chemicals. Copper in the form of Bordeaux mixture or fixed copper and mancozeb are the principal fungicides for the control of this disease but care must be taken with copper fungicides since some cultivars exhibit copper phytotoxicity (Dixon, 1981). Jhooty and Munshi (1975) screened nine fungicides for the control of this disease on muskmelon and found Dithane M-45 as most effective. Conventional fungicides like Dithane M-45, Dithane Z-78, Difolatan and Tricop-50 could not eradicate the established infection of *P. cubensis,* but Dithane M-45 significantly reduced sporulation and sprayed leaves were protected for 9 days in case of Dithane M-45, Dithane Z-78 and Difolatan and 5 days in case of Tricop-50 (Bains and Jhooty, 1978c). Mahrishi and Siradhana (1988a) also reported the efficacy of Dithane M-45 in the management of this disease on muskmelon.

These protectant fungicides have to be applied in anticipation of the infection, but had little or no therapeutic effect on disease development after infection had established. During the past two decades highly active systemic fungicides have brought in a new era in practical disease control with considerable benefit to the farmers. Phenylamides is one such group of fungicides which is specifically developed for the control of one of the most threatening types of diseases like downy mildews and late blights caused by Oomycetes on various annual and perennial crop plants. Downy mildew of cucurbits is polycyclic in nature causing serious epidemics within short periods of favourable weather conditions. It was difficult to control these diseases earlier with conventional fungicides, which required repeated applications at high rates. Phenylamides have specific activity against Oomycetes and have significantly improved control of these diseases by post-infection curative effect, low application rates, and high activity under high infection pressure. However, development of resistant strains to metalaxyl application has been reported in several countries (Reuveni *et al.*, 1980). To avoid this problem combination products are now available. Thind *et al* (1991) reported that mancozeb (0.3%) provided good control of downy mildew of muskmelon when used as protective treatment, but failed to check the established infection even when applied 24 h of inoculation whereas Ridomil MZ (0.25%), Galben M8-65 (0.25%) and Fosetyl-Al (0.3%) were remarkable in their protective as well as eradicant activity both under artificial and natural disease conditions. Ridomil MZ (0.25%), Indofil M-45 (0.25%) and Curzate (0.03%) plus Indofil M-45 (0.25%) also provided complete control of this disease on cucumber in prophylactic spray programme (Gupta *et al.*, 1993; Gupta and Shyam, 1998). Gupta and Shyam (1996) reported the antisporulant activity of Ridomil MZ for longer periods that can be utilized in planning various spray schedules for effective management of this disease. Efficacy of cyazofamid against this disease has also been reported (Mitani *et al.*, 2003). Thind and Mohan (2001) and Sharma *et al.* (2003) studied disease-weather relationship in downy mildew of muskmelon and cucumber, respectively, and reported effective control of the disease by fungicide mixtures, such as Patafol (ofurace + mancozeb), Acrobat MZ (dimethomorph + mancozeb), Ridomil MZ (metalaxyl + mancozeb) and Curzate M-8 (cymoxanil + mancozeb) than individual components under high disease risk period. However, Khetmalas and Memane (2003) recommended the sprays of Bordeaux mixture or Alliete followed by mancozeb for the control of this disease and economic returns. Strobilurin fungicides, such as azoxystrobin and kresoxim methyl have also been reported to provide good control of cucumber downy mildew (Ishi *et al.,* 2001). Ridomil Gold (0.25%) and Amistar (0.1%) were found to provide highly effective control of downy mildew of cucumber grown under plastic low tunnel with significant disease increase folllwoed by Curzate M8 and Acrobat (0.25% each) (Bhat *et al.*, 2013).

Thind *et al.* (2010) studied the sensitivity changes to metalaxyl in *Pseudoperonospora cubensis* populations collected from cucumber and muskmelon fields and reported that maximum number of sporangial populations (12 out of 25) exhibited resistant response were collected from district Amritsar with ED_{90} values of metalaxyl ranging in between 30–150 μg/ml and resistance factor in between 6–30. They further reported that resistant populations possessed normal pathogenic potential and exhibited strong competitive fitness when inoculated with sensitive populations. Resistant populations did not show cross resistance to fungicides with different modes of action, such as azoxystrobin, cymoxanil, dimethomorph, fluopicolide, propamocarb, chlorothalonil and mancozeb and these can be utilized for the management of the disease in those areas where metalxyl resistant strains has developed. Ishii *et al.* (2001) observed resistance development in *P. cubensis* to strobilurin fungicides azoxystrobin and kresoxim methyl in cucumber growing areas in Japan. A single point mutation (GGT to GCT) in the cytochrome *b* gene, resulting in substitution of glycine by alanine at position 143, was found in resistant isolates of this pathogen.

Cohen (2015) reported remarkable activity of oxathiapiprolin, a new oomycete fungicide (piperidinyl thiazole isoxazoline class) against *P. cubensis* in cucurbits. The compound inhibited zoospore release, cystospore germination, lesion formation, lesion expansion, sporangiophore development and sporangial production. When applied to the foliage as a preventive spray no lesions developed due to inhibition of zoospore release and cystospore germination, and when applied curatively, at one or two days after inoculation, small restricted lesions developed, but no sporulation occurred. When applied later to mature lesions, sporulation was strongly inhibited. Seed coating was highly effective in protecting the developed cucumber plants against downy mildew

Synergistic effect of plant activator and mancozeb in reducing this disease has been reported (Baider and Cohen, 2003). In green house studies, a mixture composed of 5 parts of plant activator BABA (DL-3-aminobutyric acid) and 1 part of mancozeb was found highly effective in the management of downy mildew of cucumber. The results also showed enhanced effect of mancozeb in BABA-induced plants, suggesting, therefore, that lower dosages of this fungicide may be sufficient to control downy mildew under field conditions.

iv. Plant extracts: Ethanol extracts of *Hedera helix* and *Paeonia suffruticosa* extracts were found highly effective in the inhibition of downy mildew of cucumber when applied as pre-infectional treatment i.e., 2 days prior to inoculation (Rohner *et al.*, 2004). The disease reducing activity was confined to the site of applications within the leaf area and the extracts induced no systemic resistance in higher inserted leaves. Both extracts also exhibited curative effects when applied at 1–4 days after inoculation, sporangial production in *P. cubensis* is drastically reduced.

2. POWDERY MILDEW

Powdery mildew of cucurbits is a wide spread disease which occur both under field as well as indoors i.e., in glasshouses, plastic tunnels or polythene huts, the world over. The disease is reported to be caused by two pathogens viz., *Sphaerotheca fuliginea* and *Erysiphe cichoracearum,* however, the conidial stage of *S. fuliginea* is common. *Leviellula taurica* has also been reported on few cucurbits in some countries. The disease is more severe in dry weather conditions and can cause substantial fruit yield reductions. It has been reported on cucurbitaceous hosts including muskmelon, watermelon, bottle gourd, pumpkin, cucumber, squash, bittergourd, spongegourd, roundmelon, parwal and kundu (Sohi, 1984).

Geographical distribution

The disease is found all over the world occurring in all the continents wherever cucurbits are grown. Hirata (1968) listed 20 countries in Europe, 13 in Pacific, 19 in Asia, 11 in North America and Central America, 6 in South America and 18 in Africa where from occurrence of powdery mildew disease of cucurbits has been recorded. The occurrence of powdery mildew caused by *S. fuliginea* has been reported from Crimea, Northern Iran, Japan, Israel, Moscow, Bulgaria, South Africa, West

Australia, Hungary, Germany, Leningrad, Northern Italy and Sudan while the occurrence of *E. cichoracearum* has been reported from U.S.A., Azerbaijan, Bulgaria, Hungary, USSR, Northern Italy, Switzerland and China. *L. taurica* is known to occur on cucurbits in USSR, Rumania and Bulgaria in Europe, Sudan, Kenya and Libya in Africa and India in Asia (El-Ammari and Khan, 1987). In USSR, Bulgaria, India, Kenya and Libya, *L. taurica* exists on cucurbits alongwith *E. cichoracearum* and in Rumania and Sudan alongwith *S. fuliginea* (Khan, 1989). He further reported that in India, Sudan, Libya and Rumania, *S. fuliginea* predominate over the other species, while in Bulgaria predominanace of *E. cichoracearum* has been claimed.

In India, Butler (1918) reported the occurrence of *S. fuliginea* and *E. cichoracearum* on cucurbits of which *S. fuliginea* is well spread throughout the country. In northern India, *E. cichoracearum* is the major pathogen involved in Jammu and Kashmir (Mir *et al.*, 1988), while *S. fuliginea* is the predominant pathogen on cucurbit crops in Punjab (Jhooty, 1967). In Himachal Pradesh, association of *S. fuliginea* with the disease has been reported (Gupta *et al.*, 2001).

Losses

The disease has a great damaging potential and regularly appears on most of the cucurbits every year and results in substantial fruit yield loss to the growers in tropics, sub-tropics and Mediterranean regions of the world, but exact estimates of losses are not well documented (Khan, 1989). Economic losses are invariable due to this disease and sometimes may lead to yield losses ranging between 12 and 75 per cent to total crop failure (Tafradzhiiski, 1959).

Symptoms

Symptoms of this disease appear on all above ground parts of the plants, such as leaves, petioles, stems and other succulent parts. Symptoms usually develop first on older shades and abaxial leaf surfaces (Zitter *et al.*, 1996). Most common symptoms of the disease caused by *S. fuliginea* and *E. cichoracearum* are the appearance of small white, circular floury spots on the leaf surface, which gradually enlarge to coalesce with each other covering the entire leaf surface with whitish powdery fungal growth (Plate 2a). When the spots are in the advance stages of growth, similar spots start appearing on the stem, petioles etc. The powdery spots generally appear first on the lower surface of the leaves, which gradually spread to the upper surface as well. Fruit infection is very rare. Dixon (1981) described the symptoms of powdery mildew (*S. fuliginea*) as discrete mealy white colonies developing on the upper leaf surface, which gradually coalesce and expand on the petioles and stems. However, Ristic (1985) reported that in case of *E. cichoracearum* symptoms appear on both upper and lower surface of the leaves.

Symptoms caused by *L. taurica* are quite distinguishable from those produced by *S. fuliginea* and *E. cichoracearum*. Leaves attacked by *L. taurica* show drooping and drying. The most conspicuous and common symptoms caused by *L. taurica* are yellowish spots on the upper leaf surface and white powdery covering on the corresponding areas on the lower surface. White powdery spots on the lower surface are mostly angular and margined by leaf veins. The spots are continuous, but separated by normal green leaf areas (El-Ammari and Khan, 1987).

The Pathogen(s)

The disease is caused by three pathogens world over. Earlier, it was thought that *Erysiphe cichoracearum* DC. was the only cause of this disease but later on *Sphaerotheca fuliginea* (Schlecht. ex. Fr.) Poll. was identified and found responsible for causing this disease on many cucurbits except watermelon. Libya is one of the few countries where a third pathogen, *Leveillula taurica* (Lev.) Arnaud is also present along with *E. cichoracearum* and *S. fuliginea*, the first two on indoor cucumbers while *S. fuliginea* is predominant infecting all the crops (Khan and El-Ammari, 1987). Some important differences between *S. fuliginea* and *E. cichoracearum* are mentioned in Table 2. The most important among these is the presence of fibrosin bodies in conidia of *S. fuliginea*.

The conidia of *E. cichoracearum* are single celled, hyaline, barrel- shaped and in long chains (Fig. 2). The conidial dimensions vary with the physiologic race and the host. They measure 30–64 x 13–32 μm. Cleistothecia are globose, dark, with hyaline to dark brown mycelioid appendages (Fig. 3), 80 to 180 μm in diameter depending upon the race (average about 131 μm). They contain 8 to 18, sometimes 25 asci. The wall cells of the cleistothecium are indistinct and 10 to 20 μm wide. The asci are pedicellate, ovate to broadly ovate or ellipsoid, and measure 36–60 or 90 x 22.5–50 (mostly 25 μm). The number of ascospores per ascus is usually two, rarely three. The ascospores are one celled, hyaline oval to sub-cylindrical and measure 18–30 x 12–18μm. The species is hetrothallic.

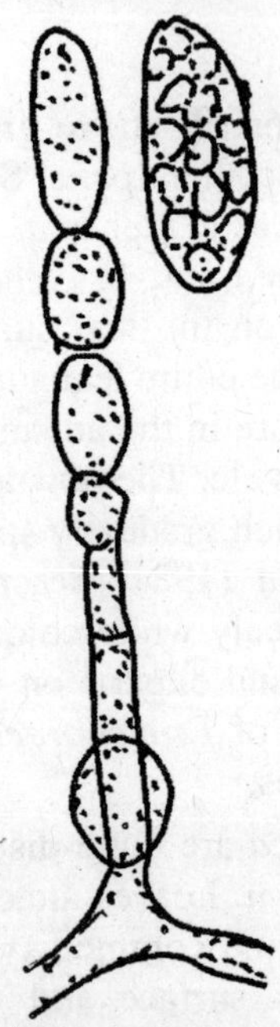

Fig. 2. Conidiophore and conidium of *Erysiphe cichoracearum*

Table 2. Differentiation of cucurbit powdery mildew pathogens

Character	*E. cichoracearum*	*S. fuliginea*
Anamorphic Characters		
Mycelium	Ectophytic	Endophytic
Colour of Mycelium	Floury white	Brownish white
Germ tube	Simple	Forked
Appressoria formation	Present	Absent
Fibrosin bodies	Absent	Present
Shape of conidia	Barrel to cylindrical	Ellipsoidal
Length/Breadth	>2 index	<2 index
Length:Breadth	2:1 ratio	3:2 ratio
Teleomorphic Characters		
Number of asci/cleistothecium	10-25	1
Number of ascospores/cleistothecium	2-3	8

(Source: Dixon, 1981; Khan and Sharma, 1995)

The mycelium of *S. fuliginea* is hyaline, occasionally brown when old, usually evanescent, but sometimes persistent. Conidia are formed in long chains. They often show fibrosin bodies and are ellipsoidal to barrel-shaped, 25–37 x 14–25 μm in size. Cleistothecia are scattered to densely gregarious (Plate 2b), 66 to 98 μm in diameter, usually 85 μm (Fig. 4). The wall cells are usually 25 μm wide. Appendages are variable in number, myceloid, brown, tortuous and as long as the ascocarp. Asci are broadly elleptic to subglobose and measure 50–80 x 37–60 μm. Each ascocarp contains a single ascus. There are eight ascospores in the ascus. They are ellipsoid to nearly spherical and measure 17–22 x 12–20 μm. It is also a hetrothallic fungus (Kabitarani, 1993). The sexual stage (cleistothecia) are not very common in India and hence identity of powdery mildew causing pathogens of cucurbits has been based on size and shape of conidia, presence and absence of fibrosin bodies in the conidia, germination of conidia in relation to moisture stress and shape of germ tube and appresorium of germinating conidia. Lebeda (1983) reported that under Czechoslovakia conditions, *S. fuliginea* was predominant in glasshouse and *E. cichoracearum* under field conditions. El-Kazzar (1983) conducted a survey of cucurbitaceous crops in various parts of Egypt and reported that powdery mildew is caused by *S. fuliginea* and not *E. cichoracearum* as previously believed, mainly on the basis of presence of fibrosin bodies in conidia of *S. fuliginea*.

Physiological specialization

The existence of distinct pathological forms has been reported in both the pathogens (Khan *et al.*, 1976). On the basis of differential reaction on different cucurbit hosts, Kaur and Jhooty (1985) identified 3 distinct form of *S. fuliginea* in Punjab (India). Form 1 infected *Luffa cylindrica*, form 2 infected *Citrullus lunatus* and form 3 infected *Cucumis melo, C. pepo* and *Lagenaria siceraria*. Occurrence of two races in *E. cichoracearum* and 4 (0,1,2,3) in *S. fuliginea* has been reported by Khan and Sharma (1993). Race 1, 2 and 3 occur in most of the cucurbit growing areas of the country. Race 3 is wide spread and infects most of the cultivated cucurbits. Race 2 is restricted to greenmelon (*Citrulus vulgaris* var. *fistulosus*) and race 1 to sponge gourd (*Luffa cylindrica* or *L. aegyptica*) and other cucurbits. *L. aegyptica*, *C. vulgaris* var. *fistulosus* and *Cucumis sativus* can be used as differential hosts for identification of 3 races of *S. fuliginea* isolated from cucurbits. Cohen *et al.* (1996) reported occurrence of race 3 of *S. fuliginea* on cucurbits in Israel. Bardin *et al.* (1997) reported virulence variation and DNA polymorphism in *S. fuliginea*. They chose 28 strains, which were diverse according to host and geographic origin. Cluster analysis on RAPD data did not separate groups within the species of *S. fuliginea*.

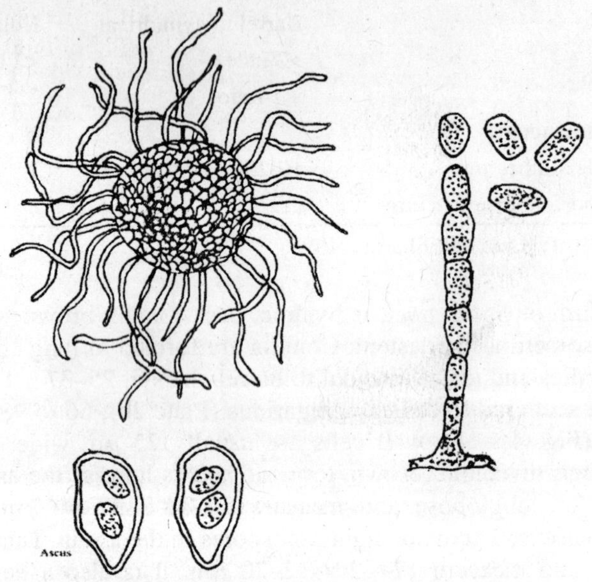

Fig. 3. Cleistothecium, ascus, conidiophore and conidia of *Erysipha cichoracearum*

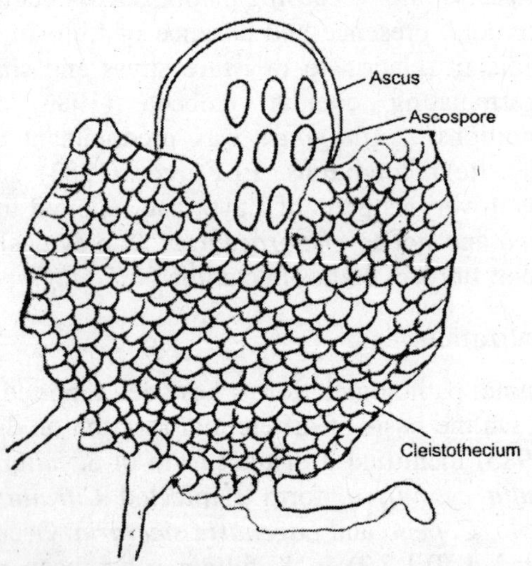

Fig. 4. Cleistothecium of *Sphaerotheca* sp.

Disease cycle

The mode of perpetuation and survival of powdery mildew of cucurbits is perplexing. Whether they over-winter or over-summer as cleistothecia or vegetative mycelium and whether ascospores or conidia initiate primary disease cycle remain to be satisfactorily answered. Some efforts made in this direction have achieved conflicting results and the questions still persist for which answers are speculative or evasive. Cleistothecia in cucurbit podery mildew are formed rarely and from several countries no records of cleistothecia are available. This rarity is more observed for *S. fuliginea* than in *E. cichoracearum*. This excludes the possibility of regular functional role of cleistothecia in survival of the pathogens and there are no experimental evidences to show their role in initiation of this disease. Ballantyne (1975) ruled out the possibility of overwintering or oversummering on non-cultivable cucurbits because most of these cucurbits do not survive in winter conditions. In U.S.A., transfer of inoculum from cold north to warm south in winter and to north from south when season is warm has been suggested (Yarwood, 1957).

In India, Jhooty (1967) indicated that cucurbit powdery mildew can overwinter as active mycelium and conidia in sheltered situations on a number of volunteer or self-grown cucurbits. The powdery mildew fungus (*S. fuliginea*) can thrive well on congenial cucurbits climbing on the tops of houses and other dwelling units especially in rural areas of Indian plains (Khan, 1983). During summer, intense heat creates problem for over-summering of the fungi. But as the cucurbits are grown abundantly during summers on hills, it is quite likely that these pathogens survive on the summer crops on hills and blown to the plains when the intense heat and rains are over and cucurbit hosts are available. The occurrence of cleistothecia of *S. fuliginea* on cucumber leaves in Himachal Pradesh has been reported (Gupta *et al*., 2001).

Powdery mildew can also overwinter on indoor cultivation units and may serve as inoculum source for infection of outdoor cucurbits grown in summer and *vice versa*. Difference in growing time of cucurbits in indoor and outdoor units ensures the availability of the hosts for survival throughout the year.

The primary infection of the leaves is caused by wind blown ascospores (wherever cleistothecia are present) or by conidia formed on the earlier sown crop/indoor cultivation in the neighbouring areas. In the process of infection conidia germinate and after penetration colonization of the entire leaf takes place. The conidia formed abundantly on primary infections are blown by wind and air currents, and cause secondary infections and the cycle is repeated.

Epidemiology

Cucurbit powdery mildews are generally favoured by moderate temperatures, reduced light, fertile dry soil and succulent plant growth (Yarwood, 1957), but two environmental factors viz., temperature and humidity are closely inter-related. The disease is favoured by moderate to warm temperatures (Branzanti and Brunelli, 1992). Munger (1979) observed the development of *S. fuliginea* to be favoured by a night/day temperature regime of 15–20°C whereas about 20–25°C is optimum for *E. cichoracearum* development. Free water film on the host surface has been reported to be detrimental for mycelial growth and conidial germination of powdery mildew pathogens (Yarwood, 1963) but conidial germination of *S. fuliginea* in a saturated atmosphere irrespective of infection has also been suggested (Jhooty, 1967). However, Khan (1989) established an inverse correlation between disease severity and rainfall, which presumably suggests a positive role of high humidity only for initiating conidial germination of *S. fuliginea* (Gupta *et al.*, 2001). Tursumbaev (1976) reported that *E. cichoracearum* requires 60–80 percent relative humidity for the development of the disease and dry conditions were favourable for the growth and sporulation of the fungus (Molot and Lecoq, 1986). The disease develops better in shade than in full light. Thus it is more severe in fields with close plantings. High carbohydrates level in luxuriantly growing plants favours the disease.

Management

i. Host resistance: A perusal of the documented literature revealed that cucumber cultivars/lines viz., Palmetto, Ashley, Poinsett, Cherokee, Polaris, Stone, Natsufushiari, Dal'nevostochnaya, Prescott, Grazhdanka 417, American Poinsett, Japanese Yomaki, Kobus mix, Nimbustol, Hybrid 2343, Hybrid 3219, Nii33, Mp 1 and American PMR 475 have been reported to possess resistance against *S. fuliginea* (Gordeeva, 1961; Yuko, 1974; Imam *et al.*, 1977; Khan, 1989); while Sprint 4405, Castle 2506, C 7-63, BSME 150-30-75, BSME 145-51-150-2 and Bright were resistant to *E. cichoracearum* (Sumner and Pathak, 1987; Rublevskii, 1983 and Leber and Heck, 1991). Some other muskmelon genotypes such as PMR-5, PMR-6, PMR-7, Cantaloupe No. 45, Arka Rajhans, Jacunba, Edisto, Gulf Stream have been reported to be resistant to powdery mildew in North India (Waraitch *et al.*, 1977; Sokhi *et al.*, 1990). McGrath and Staniszewaska (1996) reported that yellow summer squash (*Cucurbita pepo*) hybrids PSX 2287 and HMX 1707 had significantly lower disease severities than the susceptible hybrids Goldbar and Supersett, when evaluated under field conditions in Long Island.

However, inconsistencies in disease reactions to powdery mildew have often been reported. Angelov (1979) failed to elicit a resistant reaction under Bulgarian conditions among 43 cultivars reported to be resistant to *S. fuliginea*. Though in some cases, fungal development and spore production were observed to be restricted among the cultivars evaluated of which line 229-1, Aura, PI 227208 A and *Cucumis anguria* showed high degree of resistance, while Poinsette, Summer Prolific, PI 197088 and PI 197676 were resistant which suggested the role of different strains of the fungus for differences in annual incidences of the disease. Corbaz and Taillens (1994) observed that 3 cultivars resistant to the *S. fuliginea* in USA were found to be highly susceptible elsewhere in the world, which further confirmed the assumption that varietal reactions to powdery mildew are race dependent.

ii. Chemical control: The pathogen(s) responsible for this disease are sensitive to the action of fungicides at all the stages of their lifecycle except cleistothecial stages. Initially sulphur fungicides were generally recommended for the control of this disease but some of the cucurbits like cantaloupe and to certain extent cucumbers are sensitive to sulphur fungicdes. But gourds, pumpkins, squashes and watermelon are sulphur tolerant. Karathane (dinocap), which was initially developed as acaricide, also controls powdery mildews on cucurbits effectively with no phytotoxicity. Among systemic fungicides, high efficacy of carbendazim, benomyl and thiophanate-methyl has been reported against this disease, but the pathogens generally acquire resistance (Khan, 1989), if these fungicides are continuously used. Several EBI fungicides like bitertanol, triadimefon, tridemorph, triforine, penconazole and fenarimol have been successfully used for the management of this disease at a much low dose of 0.05 to 0.1 per cent with significant yield increases. Three sprays of hexaconazole (0.05%) were found effective in the management of this disease in protective spray programme (Gupta and Gupta, 2001).

EBI fungicides, which have longer curative and eradicant action, can be utilized in need based disease management programme depending on condition for disease development in these can be very effective under very high disease pressure situation. Eradicant action of triadimefon, hexaconazole and flusilazole against *S. fuliginea* has been reported (Gupta *et al.*, 1998 and Gupta and Gupta, 2001), which can be utilized in planning spray schedules for the effective management of this disease. However, the use of some EBI fungicide should be avoided as pathogens can develop resistance with their prolonged use over some years. It is advisable to limit their application to two or three sprays per season to avoid build up of resistant strains of the pathogens. Triadimefon and benomyl provided effective control of powdery mildew in summer squash in Long Island (McGrath and Staniszewaka (1996), but isolates of *S. fuliginea* showing resistance to both these fungicides were observed after few years of their use. Goswami *et al.* (2012) reported that triazole fungicides viz., propiconazole and triadimefon, and strobilurin fungicide azoxystrobin (0.5% each) were most effective in controlling powdery mildew on summer squash where disease severity ranged in between 1.5–3.0 per ecnt. Further, plant product mentha oil (5.0%) and cow milk (10.0%) were also found good in disease management.

iii. Biological Control: Application of conidial suspensions of hyperparasite *Ampelomyces quisqualis* to greenhouse cucumbers inoculated with *S. fuliginea* increased cucumber yield at par with the triforine and also when a reduced rate of triforine and the hyper parasite were combined (Sundheim, 1982). He further reported that *A. quisqualis* extensively parasitized *S. fuliginea*. Electron micrographs of leaf sections of diseased plants indicated marked deterioration in the morphological organization of chloroplast membranes, while in *A. quisqualis* treated plants, chloroplast seemed unchanged and like those uninfected plants (Abo-Foul *et al.*, 1996). Ability of two mycoparasite-based products AQ10 (*Ampelomyces quisqualis*) and Mycotal (*Lecanicillium lecanii* as well as three strains of *Bacillus subtilis*) were evaluated to manage powdery mildew on greenhouse-grown melon in Spain (Romero *et al.*, 2007). The strains of *B. subtilis* provided disease control similar to that achieved with the mycoparasites or the fungicide azoxystrobin.

3. ANTHRACNOSE

This is a common disease of the foliage and fruits of various cucurbits like watermelon, muskmelon, bottlegourd, cucumber etc. under humid weather conditions (Hasan *et al.*, 1975).

Geographical distribution

The disease was first recorded in 1867 on gourds in Italy. Since then the disease is appearing in different countries of the world where humid conditions are prevalent during cultivation of various cucurbits. In India, the occurrence of the disease has been reported from Punjab, Haryana, Assam, Karnataka and other states wherever cucurbits are grown in rainy weather (Parkash *et al.*, 1974).

Losses

Early infections may lead to premature fruit dropping and malformed fruits. Cankered fruits are also vulnerable to the attack of many other soft rotting fungi. Such diseased fruits loose market value. Yield losses of marketable fruits upto 99.5 per cent have been reported (Ullasa and Amin, 1986).

Symptoms

Lesions can form on seedlings, leaves, petioles, stems and fruits of susceptible cucurbits. The disease symptoms vary on various hosts and also on different parts of the same host. On leaves, lesions first appear near the veins and are rough, circular, light brown to reddish and can reach more than 1 cm in diameter (Plate 3a). Centre of the lesion may crack or drop out giving shot hole appearance. The lesions are dark coloured in watermelon and brownish to tan in cucumber. On fruits, lesions are circular, sunken, water-soaked, which when coalesced cover larger areas (Plate 3b). Old lesions turn black and are covered with pink spore masses especially in moist weather (Parkash *et al.*, 1974; Zitter *et al.*, 1996). On petioles and stems, lesions are shallow, elongated tan areas. On watermelon leaves, the lesions are brown to black. The margins of lesions on watermelon leaves are typically irregular and associated with leaf veins. Young fruits may have black spots, which result in abortion or malformation.

The Pathogen

The disease is caused by *Colletotrichum orbiculare* (Berk. and Mont.) Arx (Syn. *C. lagenarium* (Pass.) Ellis and Halst.). The conidia are one celled, hyaline, oblong to ovate 4–6 x 13–19 μm in size and are disseminated by water. The teleomorph *Glomerella lagenarium* F. Stevens is rarely found in nature. The pathogen produces black stromata (Acervulus) bearing black setae and hyaline conidiophores on the host surface (Plate 4). Conidia are produced successively by budding at the tips of conidiophores accumulating in pinkish mass (Plate 5). Pathogen specialization has been observed and seven (1 to 7) physiological races have been differentiated in the fungus. The race 1 and 2 are common and virulent (Zitter *et al.*, 1996).

Host range

Besides watermelon, cucumber, muskmelon, it is also reported to attack other cucurbitaceous hosts like snake gourd, ash gourd, round melon, bottle gourd, gourd, long melon etc. from different parts of the country (Prakash *et al.*, 1974).

Disease cycle

The pathogen is seed borne, can also survive in infected plant debris and in volunteer plants. In favourable weather conditions, conidia are disseminated to the healthy plants and cause infection.

Epidemiology

The pathogen overwinters in infected plant debris. The spore masses formed on the lesions are disseminated by rain splashes, irrigation water, beetles and other insects leading to secondary infection. Humid or rainy weather is essential for infection. Optimum temperature for spore germination is 22–27°C alongwith 100 per cent RH for at least 24 h is essential. Invasion takes place up to 72 h after conidial deposition and visible symptoms appear within 72 h after infection.

Management

i. Cultural practices: Deep ploughing of crop residues immediately after harvest reduces the inoculum level effectively. This should be combined with a crop rotation schedule in which no cucurbits are grown for at least one-year period. Remove all the cucurbitaceous weeds from in and around the field.

ii. Host resistance: Due to pathogenic specialization, varieties with resistance to specific races have been developed. Many cucumber cultivars have resistance to races 1, 2 and 3. Several watermelon cultivars are resistant to races 1 and 3. Resistance to anthracnose has been difficult to achieve in melon cultivars. Of late, the work on systemically acquired/-induced resistance in this disease has been initiated which can be used for cross protection. Systemic increase in soluble peroxidase activity has been noted in watermelon and muskmelon induced by *C. orbiculare*. Evidence that salicylic acid is accumulated to high levels prior the onset of the induced resistance of cucumber plants to this pathogen has indicated that this compound as a candidate for the SAR signal molecule (Ryals *et al.*, 1993). Resistance to races 1 and 3 can be found in Africa 8, Charleston Gray and Congo, while African Citron W-695 (Sitterly, 1972), Charleston Gray, Fairfax, Candy Red, Verona, Gray Ball and Super Sweet (Fursa, 1988) were resistant to race 2. In cucumber both monogenic and polygenic types of resistance occur (Sitterly, 1972).

iii. Chemical control: Sprays of benomyl followed by captafol control the disease very effectively. Benomyl or carbendazim applications significantly decreased the disease and increased yields. Peregrine *et al.* (1984) recommended alternate sprays of benomyl and mancozeb to check this disease, whereas the use of disease free seed is essential to reduce the initial inoculum of the pathogen. Treatment of seed with fungicides like thiram and destruction of infected plant debris has been advocated.

4. FUSARIUM WILT

This is an important disease of cucurbitaceous crops and the affected plants wilts when they are in fruiting stage.

Geographical distribution

The disease is prevalent in all the cucurbit growing regions of the world. In India also this disease has been reported from all the states wherever these crops are grown.

Losses

The disease can cause complete crop loss under favourable weather conditions and yield loss up to 80 per cent in the worst affected areas has been reported. Crops like bottle gourd, cucumber, muskmelon and watermelon are worst affected.

Symptoms

The symptoms of the disease mostly appear at flowering and fruiting stage as flaccidity of leaves may be sometimes accompanied by yellowing and/or marginal necrosis. The basal leaves are the first to loose the turgor and the symptoms proceed upwards (Plate 6). Wilting may be sudden or gradual, taking 4–5 days for complete loss of turgor in the latter case. A very striking underground symptom is replacement of tap root by number of lateral roots which try to make up for the clogged xylem vessels of the infected tap root. Vascular discolouration may or may not be present. Sometimes the plants may not show wilting, but remain severely stunted and do not bear fruits.

The Pathogen

The disease is caused by *Fusarium oxysporum* Schlechtend ex Fr. f.sp. *cucumerinum* Owen., *F.o.* f.sp. *melonis* Snyder & Hans., *F.o.* f.sp. *niveum* (E.F. Smith) Snyder & Hans., *F.o.* f.sp. *lagenariae* Matuo & Yamamoto and *F. o.* f.sp. *momordicae* are found on cucumber, muskmelon, watermelon, bottlegourd and bitter gourd, respectively. The fungus produces microconidia, macroconidia and chlamydospores.

Disease cycle

The wilt pathogen is both seed and soil borne. The pathogen survives as chlamydospore in the soil. The pathogen gets introduced in new areas through infected seed. Seed borne nature of the pathogen has been recorded in bottle gourd, cucumber, muskmelon, watermelon and pumpkin (Palodhi and Sen, 1983). The fungus enters the roots either through germinating seeds or through wounds caused by insects and nematodes especially *Meloidogyne incognita.*

Epidemiology

The pathogen survives in soil or in infected plant debris as chalamydospore. The temperature for disease development ranges in between 18 to 22°C, while 21°C is optimum. The disease is promoted by higher light intensity, relative humidity and high evaporation rate. The fungus produces abundant macroconidia in the vessels and in all the infected tissues of the host. The macroconidia and the hyphae transform into thick-walled chlamydospores, which are the surviving structures in the soil. High nitrogen and low potassium and calcium levels favour disease. Nitrogen increases the susceptibility of the host and also virulence of the pathogen (increased production of fusaric acid), while potassium increases the host resistance by extra lignifications of the xylem vessels. Potassium also has an inhibitory effect on the synthesis and activity of enzymes of *Fusarium.*

Management

i. Cultural practices: The disease can be kept under check by following cultural practices like crop rotation with crops like garlic, radish, onion, beetroot and lucerne (Sen, 1986), soil solarization and amendment of soil with cakes like margosa and mustard cakes. The S-H mixture containing organic wastes and inorganic fertilizers has been successful in reducing watermelon wilt in Taiwan (Sun and Huang, 1985).

ii. Host resistance: The most feasible method for the control of wilt disease is through breeding resistant varieties. Several cvs./hybrids of different cucurbits resistant to this disease like Durgapura Madhu, Punjab Sunehri of muskmelon have been bred (Radhakrishnan and Sen, 1985). Melon breeding line MR-1 is resistant to races 0, 1, and 2 of *Fusarium oxysporum* f.sp. *melonis*. Resistance to race 1 and 2 is conferred by dominant genes Fom 2 and Fom 1, respectively, and a single dominant gene also confers resistance to race 0. Genes for resistance to race 1 and race 2 assort independently (Zinc and Thomas, 1990).

iii. Biological control: Non-pathogenic isolates of *Fusarium* from suppressive soils can be used for the control of this disease and the mode of action is induced systemic resistance or a prior colonization and protection of the root from outside. Growth promoting rhizobacteria like *Pseudomonas putida* and *Serratia marcescens* could also incorporate systemic resistance in cucumber against this disease.

5. SEED ROTS AND DAMPING OFF

These diseases have been reported in production areas on crops like melon, cucumber, watermelon, pumpkin and summer squash etc. As these crops are sown directly, if the pathogen(s) is present in the soil may cause these diseases.

Symptoms

Diseased seeds are discoloured, lighter in weight and fail to germinate. Such seeds later rot by the secondary pathogens. Seedlings may die in the soil itself before emergence or may be attacked when the cotyledonary leaves are opening. After emergence of the seedlings from the soil surface, water-soaked lesions appear at the collar region and toppling down of seedlings after 2 to 3 days, are the characteristic symptoms of post-emergence damping-off.

The Pathogens

Cucurbit damping-off and seedling root rots have been ascribed to various species of *Pythium* like *P. aphanidermatum* (Edson) Fitzp., *P. irregulare* Buisman, *P. ultimum* Trow. and *P. spinosum* Swada. *Rhizoctonia solani* Kuhn, *Phytophthora capsici* Leon. and *Phytopahthora drechsleri* Tucker occur independently and also in combination with the species of the *Pythium*. *Fusarium* sp. including *F. oxysporum* f.sp. *cucumerinum* Owen frequently cause rotting of seeds and also damping-off.

Disease cycle and Epidemiology

All the pathogens associated with this disease are either seed borne or soil borne. These pathogens are affected by soil temperature and moisture as well as rainfall during the germination and growth of the seedlings.

Management

Planting on raised beds to allow for maximum water drainage after each irrigation is required to avoid the disease problem. Treat the seed either with captan or thiram @ 2.5 to 3 g/kg seed. After seedling emergence, drench with the mixture of mancozeb (0.25%) and carbendazim (0.1%)

(B) BACTERIAL DISEASES

1. BACTERIAL WILT

This is an important bacterial disease problem and often poses threat to melon and cucumber production in several countries (Latin, 1996). It is not a serious problem in squash, pumpkin and watermelon.

Symptoms

Severe wilt symptoms develop on cucurbit hosts such as cucumber and melon. Only a branch or whole plant may die due to wilt. Plants first appear green and then become necrotic as the wilt becomes irreversible. Wilt may appear at any stage of plant development, but it is more severe early in the season at rapid growth stage. In case of pumpkin and summer squash, plants may show dramatic wilt symptoms during the day when heat is more, but partially recover during next morning. Leaves of affected plants become chlorotic and the margins become necrotic before the plant collapses and ultimately dies. Young pumpkin plants normally lose vigour and may be killed within two weeks of initial wilt symptoms. However, summer squash appears more tolerant and the plants remain vigourous and continue to produce even weeks after the appearance of wilt symptoms.

The Pathogen

The disease is caused by bacterium *Erwinia tracheiphila* (Smith) Bergey *et al.* The bacterium is gram negative, rod-shaped, motile bacterium with peritrichous flagella. Striped cucumber beetle and spotted cucumber beetle transmit it. The pathogen may also be transmitted by western spotted cucumber beetle, banded cucumber beetle and grasshoppers (Leach, 1964).

Disease cycle

Not much is clear about the overwintering of the bacterium causing wilt. Earlier, it was thought to overwinter in the intestinal tract of adult cucumber beetles on fence bushes or nearby trees. However, histological studies of pathogen-vector association at later stage did not agree to this viewpoint. Recent investigations have suggested that asymptomatic weed hosts may play an important role in the survival of this bacterium. It has been observed that the bacterium remain viable in dried plant debris for very short time. Seed or soil does not transmit it. In the stems of host plant, the bacterium dies as the stem dries up and is disintegrated.

Dispersal of bacteria within and across the field is through cucumber beetle vectors. These are mechanically transmitted by contact with contaminated beetle mouthparts when the vector feeds on stems and leaves. These bacteria generally reproduce and multiply at the wound sites and then enter the xylem vessels after which they move down the petioles and stem. Wilting primarily results due to vascular plugging by bacterial masses and formation of gums and resins subsequently. Spread of bacteria to healthy shoots takes place through the dissolved walls of adjacent xylem vessels. Subsequent infections result from infected plants serving as inoculum sources.

Epidemiology

The incubation period of this disease is very long ranging from several days to weeks and is influenced by the plant age and the point where infection has occurred. Young and succulent plants are most vulnerable to wilt (Bassi, 1981). As the plants proceed towards maturity, many factors operate to decrease the spread of the wilt. It has been observed that beetle populations decline in mid summer, as they do not prefer older, brittle leaves and stem as their food. Moreover, bacteria introduced into the new growth at plant tips have to travel greater distances in the plant to induce wilting. Although, there is not much effect of weather on the incidence of wilt, but it may affect the expression of symptoms. Wilt is favoured by warm weather, adequate soil moisture, enough sunlight and balanced nutrient concentration, which are ideal, conditions for growing various cucurbits.

Management

Being a bacterial disease and transmitted by vector, integrated approach including cultural practices, vector control and host resistance is suggested.

*i. **Cultural practices:*** Wilted plants should be rogued out at early stage. Some trap crops have been suggested to slow the progress of wilt. However, these measures are not of much help unless they are integrated with a suitable insecticide spray programme. Some bait formulation containing volatile attractants, a feeding stimulant and a small amount of insecticide have been developed which show good promise for effective control of bacterial wilt in cucurbits.

*ii. **Host resistance:*** Development of bacterial wilt resistant cultivars is an important approach for effective management of this wilt and holds good promise for the time to come. The resistant cultivars should be able to limit the multiplication and movement of the bacteria within the host. However, it is not easy task to get such resistance in cucurbits like melon. There is a need to strengthen research efforts in this direction.

*iii. **Vector control:*** Bacterial wilt can be managed to a large extent by controlling cucumber beetles serving as vector. In disease prone areas, the use of a systemic insecticide applied to soil and timely foliar sprays of contact insecticides are recommended for economical production of cucurbits like cucumber and melon.

2. ANGULAR LEAF SPOT

This is an important disease of cucurbitaceous vegetables and is most serious in cooler areas.

Geographical distribution

The disease is prevalent in all areas wherever cucurbits and related species are grown. In India, Jindal (1994) reported that it is more serious in cooler areas of the country.

Symptoms

The symptoms of the disease appear as small water-soaked spots on leaves, stem, blossoms and fruits (Plate 7). In favourable weather conditions, the spots progress on the leaf lamina, which are delimited by veins of the leaves and become angular (Fig. 5). Later they become tan coloured and may enlarge to irregular shapes giving the leaves a blighted or scorched appearance. On watermelon and squash these spots are black and brown with a yellow halo. A crust of bacterial ooze is formed on them and the dead tissue may drop off leaving shot holes. The spots on fruits are small, circular, brown to black and superficial.

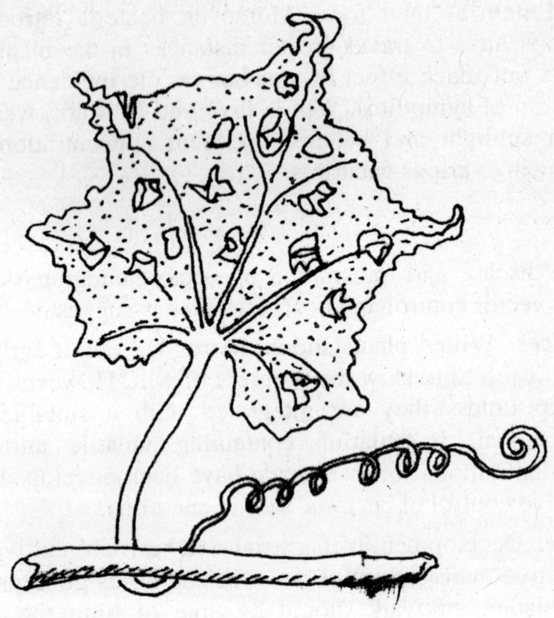

Fig. 5. Symptoms of angular leaf spot

The Pathogen

The bacterium associated with this disease is *Pseudomonas syringae* pv. *lachrymans* (Smith and Bryan) Young *et al.* It is gram negative, aerobic, non-spore forming, straight rod shaped bacterium, motile with 1 to 5 polar flagella (Kagiwata, 1990). In iron deficient medium, it produces slime in the sucrose medium. It is non-nitrate reducing and secrete protease. Optimum temperature for growth is 25 to 30ºC and growth ceases at 41ºC.

Disease cycle

The pathogen is soil borne in infected plant debris and also survives in the infested seed. Bacteria invade cotyledons after the germination of the seed. In leaves, the bacteria enter the leaves either through wounds or stomata in the presence of dew or rain water.

Epidemiology

The bacterium survives in infected plant debris and seed. The bacteria can spread through rain splashes and irrigation water. Moisture is the most important factor governing disease development and its severity. Optimum temperature for the development of the disease is 20 to 24°C.

Management

i. Cultural practices: Cultural practices like collection and destruction of infected plant debris, crop rotations avoiding cucurbitaceous crops in rotation and use of disease free seed produced in the dry areas help in reducing the disease. Hot water treatment of the seed is effective in reducing the seed borne infections, but affects seed germination.

ii. Chemical control: Seed borne infection can be partially reduced by soaking in antibiotics and inorganic salts. In crop fields, spray either streptomycin sulphate (100 ppm) or copper fungicides. Don't spray copper fungicides during hot of the day or to young plants.

(C) VIRAL DISEASES

1. CUCUMBER MOSAIC

This disease is prevalent in cucurbit growing areas of the world and cause severe yield losses.

Geographical distribution

The virus has wide occurrence throughout the world and is known to infect about 800 plant species. Doolittle (1916) and Jagger (1916) described this disease for the first time. Since then it has been reported from various countries of the world, but it is more prevalent in temperate regions of the world. In India, the occurrence of this disease has also been reported.

Losses

The disease can cause widespread epidemics, alone or in combination with other viruses and is responsible for serious losses in yields (Provvidenti, 1996). The infection of cucumber plants at cotyledonary stage reduced the yield by 89 per cent in summer and 96 per cent in winter (Kazada *et al.*, 1975). During 1999 crop season, an epidemic of this disease was prevalent on muskmelon, cucumber and squash in Punjab and in adjoining states causing huge losses (100% in some fields) to the crops. The problem is still very important on early sown cucurbits (Anonymous, 1999).

Symptoms

The disease is more severe and damaging in summer squash, different melons and some pumpkins. The virus causes severe stunting of plants, foliar yellow mosaic, malformation and reduction of leaf size and stem internodes. Systemic symptoms consist of downward leaf curling, mosaic and reduced leaf size in stages of growth (Plate 8). The intensity of symptoms depends on the species and cultivars infected, age of the hosts and environmental conditions. Infected fruits are distorted, become discoloured and gently remain small in size. Seed production is greatly reduced in infected plants.

The Virus

The disease is caused by *Cucumber mosaic virus* (CMV) is a member of cucumovirus group and a number of strains, pathotypes and serotypes of this virus are known, most of which are able to induce severe symptoms in cucurbits (Francki *et al.*, 1979). Based on serological studies, the virus is related to peanut stunt virus.

Host range

CMV has a wide host range including several vegetables like carrot, celery, legumes, lettuce, onion, pepper, tomato, spinach etc., ornamentals like anemone, aster, delphinum, geranium, lilies, petunia, primula, viola, zinnia etc., and woody and semi-woody plants like banana, passion fruit and tree tomato.

Transmission

Since CMV is known to infect diverse plants, several weeds and cultivated crops which may serve as reservoir of the virus for transmission, and spreads into other crops. The virus is transmitted in a non-persistent way by several aphid species including *Myzus persicae* (Sulzer), *Macrosiphum euphorbiae* (Thomas), and *Aulacorthum solani* (Kaltenbach). Acquisition of the virus by aphid instars takes place within one minute. The vectors lose their ability to transmit the virus within a period of two hours. Several factors, such as vector type, virus strain, environmental conditions etc. affect the transmission efficiency. The virus is easily transmitted mechanically. There are divergent views on the seed-borne nature of CMV.

Management

Since the virus has a very wide host range, an integrated approach is required to control cucumber mosaic virus more effectively. Eradication of weed hosts is a difficult task because of extensive host range.

i. Host resistance: The best method to keep the viral diseases under check is the use of resistant varieties. Resistance in cucumber to this virus has been derived from a Chinese cultivar. This resistance has been widely exploited and most American cucumber cultivars and also a large number available in other countries are now resistant to CMV (Provvidentii *et al.*, 1978). In melons, resistance has been derived from Japanese pickling melons and pome Korean and Chinese Melons possess high level of resistance. A number of wild squash species have been found to possess high degree of resistance to to CMV and these have been used in interspecific crosses with *Cucurbita pepo*. Resistance to CMV has also been found in some *C. maxima* lines from South America and in a line of *C. moschata* from Nigeria. Although watermelon is known to be resistant to CMV but is occasionally attacked quite severely by specific pathotype of the virus. In another approach, the coat protein of CMV has been introduced in to cucumber, melon and squash, offering a good level of resistance to several strains of the virus (Gonsalves *et al.*, 1992).

Chemical control

Since aphid vectors transmit the disease, application of insecticides like dimethoate and malathion are widely recommended for the control of vectors to minimize transmission to healthy crop. Mineral oils are also used for the control of aphids.

REFERENCES

Abo-Foul, S., Raskin, V. I., Sztejnberg, A. and Marder, J.B. 1996. Disruption of chlorophyll organization and function in powdery mildew–diseased cucumber leaves and its control by the hyperparasite *Ampelomyces quisqualis*. *Phytopathology* **86:** 195–199.

Angelov, D. 1979. Study of the susceptibility of cucumber cultivars to the pathogen of powdery mildew (*Sphaerotheca fuligenea* Poll.). *Gradinarska I Lozarska Nauka* **16(2):** 98–103.

Angelov, D. and Georgiev, P. 1995. Study of the susceptibility of cucumber accessions to downy mildew, *Pseudoperonospora cubensis* (Berk. et Curt.) Rostov. *Rasteniev" dni Nauki*. **32 (5):** 245–246.

Anonymous, 1999. Annu. Rep. Dept. Plant Pathology, PAU, Ludhiana.

Baider, A. and Cohen, Y. 2003. Synergistic interaction between BABA and mancozeb in controlling *Phytophthora infestans* in potato and tomato, and *Pseuudoperonospora cubensis* in cucumber. *Phytoparasitica* **31:** 399–409.

Bains, S.S. and Jhooty, J.S. 1976a. Over wintering of *Pseudoperonospora cubensis* causing downy mildew of muskmelon. *Indian Phytopath*. **29 :** 213–214.

Bains, S.S. and Jhooty, J.S. 1976b. Host range and possibility of pathological races of *Pseudoperonospora cubensis* – cause of downy mildew of muskmelon. *Indian Phytopath*. **29 :** 214–216.

Bains, S.S. and Jhooty, J.S. 1978a. Epidemiological studies on downy mildew of muskmelon caused by *Pseudoperonospora cubensis*. *Indian Phytopath.* **31:** 42–46.

Bains, S.S. and Jhooty, J.S. 1978b. Relationship between mineral nutrition of muskmelon and development of downy mildew caused by *Pseudoperonospora cubensis. Plant and Soil* **49:** 85–90.

Bains, S.S. and Jhooty, J.S. 1978c. Mode of efficacy of four fungitoxicants against *Pseudoperonospora cubensis* on muskmelon. *Indian Phytopath.***31 :** 339–342.

Bains, S.S. and Prakash, V. 1985. Susceptibility of different cucurbits to *Pseudoperonospora cubensis* under natural and artificial epiphytotic conditions. *Indian Phytopath*. **38:** 138–139.

Bains, S.S., Sokhi, S.S. and Jhooty, J.S. 1977. *Melothria maderaspatana* — A new host of *Pseudoperonospora cubensis*. *Indian J. Mycol. Pl. Pathol.* **7:** 86.

Ballantyne, B. 1975. Powdery mildew on cucurbitaceae: identity, distribution, host range and sources of resistance. *Proc. Linnean Soc. New South Wales* **99(2):** 100–120.

Bardin, M., Nicot, P.C., Normand, P. and Lemair, J.M. 1997. Virulence variation and DNA polymorphism in *Sphaerotheca fuliginea,* causal agent of powdery mildew of cucrubits. *European J. Pl. Pathol.* **103:** 545–554.

Bassi, A. Jr. 1981. Epidemiological study of *Erwinia tracheiphila* with immunofluorescence technique. *Phytopathology* **71:** 859 (Abstr.).

Bedlan, G. 1989 First detection of oospore of *Pseudoperonospora cubensis* (Berk et Curt.) Rost. on glasshouse cucumbers in Austria. *Pflanzenschutzberichte* **50:** 119–120.

Bhat, J.A., Thind, T.S., Bala, A. and Kumar, P. 2013. Factors affecting development of downy mildew of cucumber grown under plastic low tunnel and its management with fungicides. *Pl. Dis. Res.* **28(1) :** 58–63.

Branzanti, B. and Brunelli, A. 1992. Study of aetiology of powdery mildew of cucurbits in Emilia Romagna. *Informatore Fitopatol.* **42(11):** 37–44.

Butler, E.J. 1918. *Fungi and diseases in plants*. Thacker, Spink and Co. Calcutta, pp. 314–315.

Chauhan, A., Singh, P.P. and Dhillon, M. 1994. Studies on peroxidase activity in muskmelon genotypes in relation to downy mildew infection. *J. Pl. Sci. Res.* **10:** 10–11.

Chen, C.P., Sung, C.C. and Ho, C.C. 1959. A brief report of the discovery of oospores of downy mildew of cucumber (*Pseudomonas cubensis* (Berk. & Curt.). *Zhibing Zhishi* **3(6):** 144–145.

Chupp, C. and Sherf, A.F. 1960. *Vegetable diseases and their control*. Ronald Press Co., New York, 693pp.

Cohen, R., Katzir, N., Schneiber, S., Greenberg, R. and Yarden, O. 1996. Occurrence of *Sphaerotheca fuliginea* race 3 on cucurbits in Israel. *Plant Dis.* **80:** 344.

Cohen, Y. 1976. Quantification of resistance of cucumbers and Cantaloups to *Pseudoperonospora cubensis*. *Phytoparasitica* **4(1):** 25–31.

Cohen, Y. 1980. Downy mildew of cucurbits. In: *The downy mildews* (Ed. Spencer, D.M.). Academic Press, San Francisco, pp 341–353.

Cohen, Y. 2015. The novel oomycide oxathiapiprolin inhibits all stages in the asexual life cycle of *Pseudoperonospora cubensis* – causal agent of cucurbit downy mildew. *PLoS One.* 10(10) : e0140015. doi: 10.1371.

Cohen, Y., Eyal., H., Hanamia, J. and Malik, Z. 1989. Ultrastructure of *Pseudoperonospora cubensiis* in muskmelon genotypes susceptible and resistant to downy mildew. *Physiol. Mol. Pl. Pathol.* **34:** 27–40.

Cohen, Y. and Rotem, J. 1971. Field and growth chamber approach to epidemiology of *Pseudoperonospora cubensis* in cucumbers. *Phytopathology* **61:** 736–737.

Corbaz, R. and Taillens, J. 1994. Tolerance of cucumbers to powdery mildew: a confusing situation. *Revue Suisse de Vittculture, d' Arboriculture et d' Horti.* **26(6):** 397–398.

Dixon, G.R. 1981. *Vegetable Crop Diseases*. AVI Publishing Co., Inc., Connecticut, 404 pp.

Doolittle, S.P. 1916. A new infectious mosaic disease of cucumber. *Phytopathology* **6:** 145–147.

El-Ammari, S.S. and Khan, M.W. 1987. Identity of powdery mildew of cucurbits in Libya. *Arab J. Pl. Prot.* **5(2):** 92–98.

El-Kazzar, M.K. 1981. *Sphaerotheca fuliginea* (Schlect. Ex. Fr.) Poll the causal agent of powdery mildew on many cucurbits in Egypt. *Egypt. J. Phytopathol.* **13(1/2):** 65–66.

Ellis, D.E. 1951. Noteworthy diseases of cucurbits in North Carolina in 1949 and 1950. *Pl. Dis. Reptr.* **35(2):** 91–93.

Ellis, D.E. and Cox, R.S. 1948. Dusting cucumbers to control downy mildew. *Bull. N.C. Agric. Exp. Stn.* **362:** 16.

Francki, R.I.B., Mossop, D.W. and Hatta, T. 1979. Cucumber mosaic virus. In: "*Descriptions of Plant viruses*", No. 213. CMI and Association of Applied Botanists, Kew, England.

Fursa, T.B. 1988. Importance of the world gene pool of watermelon for breeding Soviet varieties. *Sbornik Nauchnykh Trudov po Priklandnoi Botanike, Genet. Sel.* **118:** 9–14.

Gonsalves, D., Chee, P., Provvidenti, R., Seem, R. and Slightom, J.M. 1992. Comparison of coatprotein–mediated and genetically–derived resistance in cucumbers to infection by cucumber mosaic virus under field conditions with natural challenge inoculations by vectors. *Biotechnol.* **10:** 1562–1570.

Gordeeva, N.G. 1961. The relationship of ecological groups of melon to fungal diseases. *Sborn Trud. Aspir. Molod Nauch. Sotrud. Vses Inst. Rastenievod* **2(6):** 288–293.

Goswami, S. and Thind, T.S. 2012. Management of powdery mildew (*Sphaerotheca fuliginea*) of summer squash with new fungicides and natural products. *Indian Phytopath.* **65 (2) :** 198–199.

Goswami, S., Thind, T.S., Raheja, S. and Vashisht, V.K. 2011. Screening of muskmelon lines against downy mildew caused by *Pseudoperonospora cubensis* under field conditions. *Indian Phytopath.* **64 (4) :** 374–375.

Gupta, S.K. and Amita, Gupta 2001. Management of cucumber powdery mildew through fungicides. *Pl. Dis. Res.* **16(2):** 186–192.

Gupta, S.K. and Shyam, K.R. 1996. Antisporulant activity of some fungicides against *Pseudoperonospora cubensis* on cucumber. *Indian J. Mycol. Pl. Pathol.* **26(3) :** 293–295.

Gupta, S.K., Amita Gupta, Shyam, K.R. and Bhardwaj, R. 2001. Morphological characterization and effect of meteorological factors on the development of cucumber powdery mildew. *Indian Phytopath.* **54(3):** 311–315.

Gupta, S.K., S.K. Sharma and K.R. Shyam. 1998. Antisporulant activity of ergosterol biosynthesis inhibitor fungicides against powdery mildew (*Sphaerotheca fuliginea*) of cucumber (*Cucumis sativus*). *Indian J. Agric. Sci.* **68(7):** 382–383.

Gupta, S.K., Shyam, K.R. and Dohroo, N.P. 1993. Effect of fungicides on the severity of Downy mildew and yield of cucumber (*Cucumis sativus* L.) in Himachal Pradesh. *Pestology* **17(3) :** 37–39.

Gupta, S.K. and Shyam, K.R. 1998. Protective activity of fungicides against downy mildew of cucumber. *Pl. Dis. Res.* **13(1):** 60–61.

Hasan, A., Khan, W. M. and Khan, A.M. 1975. *Indian J. Mycol Pl. Pathol.* **5:** 214.

Hirata, K. 1968. Notes on the host range and geographic distribution of the powdery mildew fungi. *Trans. Mycol. Soc. Jpn.* **9 :** 73–88.

Hiura, M. and Kawada, S. 1933. On the overwintering of *Peronoplasmopara cubensis* ((Berk. and Curt.) Rostow. *Japanese J. Bot.* **6(4):** 507–513.

Hoerner, G.R. 1940. The species of the genus *Pseudoperonospora* and their recorded hosts. *Pl. Dis. Reptr.* **24(9):** 170–173.

Hughes, M.B. and Van Haltrern, F. 1952. Two biological forms of *Pseudoperonospora cubensis. Pl. Dis. Reptr.* **36:** 365–367.

Imam, M.K., Abobakr, M.A. and Yocoub, H.M. 1977. Inheritance of some characters in cucumber, *Cucumis sativus* L. I. Powdery mildew resistance and mottling of immature fruit. *Libyan J. Agri.* **6(1):** 105–114.

Ishii, H., Fraaije, B.A., Suglyama, T., Noguchi, K., Nishimura, K., Takeda, T., Amano, T. and Hollomon, D.W. 2001. Occurrence and molecular characterization of strobilurin resistance in cucumber powdery mildew and downy mildew. *Phytopathology* **91(12) :** 1166–1171.

Jagger, I.C. 1916. Experiment with the cucumber mosaic disease. *Phytopathology* **6:** 148–151.

Jhooty, J.S. 1967. Identity of powdery mildew of cucurbits in India. *Pl. Dis. Reptr.* **51:** 1079–1080.

Jhooty, J.S. and Munshi, G.D. 1975. Control of downy mildew of muskmelon with fungicides. *Indian J. Mycol. Pl. Pathol.* **5:** 105–106.

Jiang, M.X. 1981. Selection and evaluation of the cucumber F_1 hybrids Ninghuang 1 and Ninghuang 2. *Ningxia Agril. Sci. Techn. Ningxia Nongye Keji.* **5:** 10–11.

Jindal, K.K. 1994. Occurrence of angular leaf spot bacterium *Pseudomonas syringae* pv. *lachrymans* on cucumber plants. *Pl. Dis. Res.* **9:** 66–67.

Kabitarani, A. 1993. Sexuality and pathological specialization in *Sphaerotheca fuliginea. Indian Phytopath.* **46:** 50–53.

Kagiwata, T. 1990. Bacteriological characteristics of cucumber angular leaf spot pathogen *Pseudomonas syringae* pv. *lachrymans. J. Agric. Sci.* (Tokyo) **35:** 116–128.

Kaur, J. and Jhooty, J.S. 1985. Pathological specialization in *Sphaerotheca fuliginea* causing powdery mildew of cucurbits. *Indian Phytopath.* **38:** 302–305.

Kazada, V., Hervert, V. and Polak, Z. 1975. Decrease in yield of greenhouse cucumber infected with cucumber mosaic virus. *Ochr. Rost.* **11:** 1–6.

Khan, A., Khan, M., Khan, W., Akram, M. and Khan, A. 1976. Studies on the cucurbit powdery mildew – II. Varietal response of some cucurbits to *Sphaerotheca fuliginea. Indian Phytopath.* **29:** 306–308.

Khan, M.W. 1983. The identity of powdery mildew of cucurbits–A critical appraisal. *Acta Bot. Indica* **11:** 97–126.

Khan, M.W. 1989. Powdery mildew of cucurbits–A three pathogen disease. *Int. J. Trop. Pl. Dis.* **7:** 107–123.

Khan, M.W. and El–Ammari, S.S. 1987. Outbreaks and new records, Libya. Occurrence of three species of powdery mildew on cucrbits. *FAO Plant Prot. Bull.* **35:** 66–67.

Khan, M.W. and Sharma, G.K. 1993. Races of *Sphaerotheca fuliginea* infecting cucurbits in India and host differentials for their identification. *Indian Phytopath.* **46:** 236–242.

Khan, M.W. and Sharma, G.K. 1995. Taxonomic evaluation of anamorph characters in identification of powdery mildew fungi on cucurbits. *Indian Phytopath.* **48(3):** 314–324.

Khetmalas, M.B. and Memane, S.A. 2003. Management of downy mildew disease of cucumber during rainy season. *J. Maharashtra Agric. Univ.* **28:** 281–282.

Khosla, H.K., Dave, G.S. and Nema, K.G. 1973. Occurrence of downy mildew of parwal (*Trichosanthes dioica* Roxb.) in Madhya Pradesh, India. *JNKVV Res. J.* **7:** 175–177.

Komnenic, M. Obserdovic, A. and Stankovic, L. 1995. Resistance of different cucumber genotypes to *Pseudoperonospora cubensis*, causal agent of downy mildew. *Zastita Bilja.* **46(1):** 69–74.

Lai, L.Y. and Zhang, S.D. 1992. Disease resistance identification of cucumber varietal resources in Shandong. *J. Shundong Agril Sci.* **1:** 20–25.

Lange, L., Eden, U. and Olson, L.W. 1989. Zoosporogenesis in *Pseudoperonospora cubensis*, the causal agent of cucurbit downy mildew. *Nordic J. Bot.* **8(5):** 497–504.

Latin, R.X. 1996. Bacterial wilt. In "*Compendium of cucrbit diseases*" (Eds. Zitter, T.A. Hopkins D.L. and Thomas, C.E.). APS Press, Minnesota, p 36.

Leach, J.G. 1964. Observations on cucumber beetles as vectors of cucurbit wilt. *Phytopathology* **54:** 606–607.

Lebeda, A. 1983. The genera and species spectrum of cucumber powdery mildew in Czechoslovakia. *Phytopathol. Zeitschri.* **108:** 71–79.

Leber, B. and Heck, M. 1991. Cultivars of greenhouse cucumbers for integrated cultivation. *Gemuse–Munchen* **27(3):** 144–197.

Lisitisin, V.N. and Pluzhnikova, L.E. 1990. Results of evaluating cucumber varieties for the purpose of breeding for resistance to bacterial disease and downy mildew. *Ovoshchevodstvo I Bakhchevodstvo* **35:** 82–84.

Lu, S.Z., Ma, D.H., Huo, Z.R., Shen, W.Y., Li, S.J. and Chen, Z.W. 1994. New released cucumber cultivar Jinchun 4–high–quality, disease resistant and high yielding. *China Veg.* **2:** 1–3.

Mahrishi, R.P. and Siradhana, B.S. 1984. On the occurrence of oospores of *Pseudoperonospora cubensis* in Rajasthan. *Indian Phytopath.* **37(2):** 323–325.

Mahrishi, R.P. and Siradhana, B.S. 1988b. Effect of nutrition on downy mildew disease caused by *Pseudoperonospora cubensis* (Berk. & Curt.) Rostow. on muskmelon. *Ann. Arid. Zone Res.* **29(2):** 153–155.

Mahrishi, R.P. and Siradhana, B.S. 1988c. Epidemiology of downy mildew of muskmelom (*Cucumis melo* L.) caused by *Pseudoperonospora cubensis* (Berk. and Curt.) Rostow. *J. Turk. Phytopath.* **17(2):** 67–73.

Mahrishi, R.P. and Siradhana, B.S. 1988a. Studies on downy mildew of cucurbits in Rajasthan: incidence, distributiion, host range and yield in muskmelon. *Ann. Arid. Zone Res.* **27(1):** 67–70.

McGrath, M.T. and Staniszewaka, H. 1996. Management of powdery mildew in summer squash with host resistance, disease threshold–based fungicide programs, or an integrated program. *Plant Dis.* **80 :** 1044–1052.

Meyer, E. and Ziesenis, R. 1982. Downy mildew on outdoor cucumbers. *Gemuse*. **18:**3.

Mir, N.M., Dhar, A.K., Dar, G.H., Zargar, M.Y. and Khan, M.A. 1988. Presence of cleistothecial stage of *Erysiphe cichoracearum* DC. *in vivo* on cucurbits and a new host record from India. *Indian J. Pl. Pathol.* **6(1):** 79–80.

Mitani, S., Kamachi, K., Sugimoto, K., Araki, S. and Yamaguchi, T. 2003. Control of cucumber downy mildew by cyazofamid. *J. Pesticides Sci.* **28:** 64–68.

Molot, P.M. and Lecoq, H. 1986. Powdery mildew of cucurbits. I. Bibliographic data. Preliminary studies. *Agronomie* **6(4):** 335–362.

Munger, H.M. 1979. The influence of temperature on powdery mildew resistance in cucumber. *Rep. Cucurbit Genet. Coop.* **2:** 9–10.

Ojiambo, P.S., Gent, D.H., Quesada–Ocampo, L.M., Hausbeck, M.K. and Holmes, G.J. 2015. Epidemiology and population biology of *Pseudoperonospora cubensis* : A model system for management of downy mildews. *Annu. Rev. Phytopathol.* **53 :** 223–246.

Om, Y.H., Hong, K.H., Oh, D.G. and Kwack, S.N. 1992. Breeding of gynoecious cucumber inbreds 'Wonye 502'. *Res. Rep. Rural Dev. Admn. Hortic.* **34(1):** 1–4.

Palodhi, P.R. and Sen, B. 1983. Perpetuation of cucurbit wilt pathogen in riverbed cultivation. *Indian J. Mycol Pl. Pathol.* **13:** 164–168.

Palti, J. and Cohen, Y. 1980. Downy mildew of cucurbits (*Pseudoperonospora cubensis*)–The fungus and its hosts, distribution, epidemiology and control. *Phytoparasitica* **8:** 109–147.

Peregrine, W.T.H., Ahmed, K. and Momin, M. 1984. *World Crops* **36:** 184–185.

Pivovarov, V.F. 1984. Breeding cucumber for disease resistance using different ecological conditions. *Selektsiya I Semenovodstvo, USSR.* **10:** 20–22.

Pivovarov, V.F., Ushakov, P.P. and Simanca, A.O. 1977. Result of evolutions and selections of cucumber cultivars for improvement of their resistance to downy mildew (*Pseudoperonospora cubensis* R.). *Informe Cientifico Tecnico, Instituto de Investigaciones Fundamentales en Agricultura Triopical* **39:** 20–25.

Prakash, O., Sohi, H.S. and Sokhi, S.S. 1974. *Indian J. Hortic.* **31:** 278–282.

Provvidenti, R. 1996. Cucumber mosaic. In: *Compendium of cucurbit diseases* (Eds. Zitter, T.A.; Hopkins, D.L., Thomas, C.E.), APS Press, Minnesota, p.38–39.

Provvidenti, R., Robinson, R.W. and Munger, H.M. 1978. Resistance in feral species to six viruses infecting cucurbits. *Pl. Dis. Reptr.* **62:** 326–329.

Radhakrishnan, P. and Sen, B. 1985. Efficacy of different methods of inoculation of *Fusarium oxysporum* and *Fusarium solani* for inducing wilt in muskmelon. *Indian Phytopath.* **38:** 70–73.

Raj, D., Gupta, S.K., Shyam, K.R., Sharma, S.K. and Sharma, H.R. 2003. Evaluation of germplasm against downy mildew (*Pseudoperonospora cubensis*). *Pl. Dis. Res.* **18(1):** 67.

Reuveni, M., Eyal, H. and Cohen, Y. 1980. Development of resistance to metalaxyl in *Pseudoperonospora cubensis. Plant Dis.* **64:** 1108–1109.

Reuveni, R., Shimoni, M., Karchi, Z. and Kuc, J. 1992. Peroxidase activity as a biochemical marker for resistance of muskmelon (*Cucumis melo*) to *Pseudoperonospora cubensis. Phytopathology* **82:** 749–753.

Ristic, S. 1985. Contribution to the study of cucumber powdery mildew, *Erysiphe cichoracearum* DC in the glasshouse with special reference to biological and morphological characters. *Zastita Bilja* **36(3):** 303–316.

Robak, J. 1995. Epidemiology and control of cucumber downy mildew *Pseudoperonospora cubensis. Biuletyn Warzywniczy* **2:** 145–149.

Rohner, E., Carabet, A. and Buchenauer, H. 2004. Effectiveness of plant extracts of *Paeonia suffruticosa* and *Hedera helix* against diseases caused by *Phytophthora infestans* in tomato and *Pseudoperonospora cubensis* in cucumber. *Z.r Pflanzenkr. Pflanzenschutz* **111:** 83–95.

Romero, D., De Vicente, A., Zenouh, H., Cazorla, F.M., Femandez–Ortuno, D., Tores, J.A. and Perez–Garcia, A. 2007. Evaluation of biological control agents for managing cucurbit powdery mildew on greenhouse-grown melon. *Plant Pathol.* **56(6)** : 976–986.

Rublevskii, V.V. 1983. Promising cucumber hybrids. *Instituta Rastenievodstva Ilmeni N.I. Vavilova* **128:** 61–62.

Ryals, J., Ward, E. Ukenes, S., Gaffnay, T., Friedrich, L., Alexander, D., Rella, M.G., Gut–Rella, M., Bruening, G., Garcia–Olmedo, F. and Ponz, F. 1993. Toward the genetic engineering of fungal resistance in crops. Workshop on Engineering Plants Against Pests and Pathogens, 11–13 January, 1993, Madrid, Spain, 22–26; Centro de Reuniones Internacionales Sobre Biologia.

Sen, B. 1986. Cultural management of soil–borne diseases. pp. 367–381. In: *Vistas in Plant Pathology*. (Eds. Varma, A. and Varma, J.P.). Malhotra Publishing House, New Delhi, India.

Sharma, D.R., S.K. Gupta and K.R. Shyam. 2003. Studies on downy mildew of cucumber caused by *Pseudoperonospora cubensis* and its management. *J. Mycol. Pl. Pathol.* **33 (2):** 246–251.

Sherf, A.F. and Macnab, A.A. 1986. *Vegetable Diseases and Their Control*, John Wiley and Sons, New York, USA, 728pp.

Singh, P.P., Jhorar, O.P., Singh, R. and Sokhi, S.S. 1996. Predictive model for downy mildew of muskmelon. *Veg. Sci.* **23:** 186–194.

Singh, P.P. and Sokhi, S.S. 1989. First report of occurrence of oospores of *Pseudoperonospora cubensis* on two cucurbitaceous hosts. *Curr. Sci.* **58:** 1330–1331.

Singh, P.P. and Singh, R. 1994. Epidemiological investigations of downy mildew of muskmelon (*Cucumis melo* L.). *Pl. Dis. Res.* **9:** 226.

Singh, P.P., Thind, T.S. and Lal, Tarsem. 1996. Reaction of some muskmelon genotypes against *Pseudoperonospora cubensis* under field and artificial inoculation conditions. *Indian Phytopath.* **49:** 188–190.

Singh, P.P., Thind, T.S., Sokhi, S.S. and Grewal, R.K. 1990. Studies on development of downy mildew of muskmelon under hot weather conditions. *Pl. Dis. Res.* **5(Special):** 104–107.

Sitterly, W.R. 1972. Breeding for resistance in cucurbits. *Annu. Rev. Phytopathol.* **10:** 146–149.

Sohi, H.S. 1984. Present status of our knowledge of important fungal diseases of selected vegetables in India and future needs. *Indian J. Mycol. Pl. Pathol.* **14:** 1–34.

Sokhi, S.S., Munshi, G.D., Grewal, R.K., Thind, T.S. and Sandhu, K.S. 1990. Current fungal disease problems of important vegetables. In: *Basic Research for Crop Disease Management* (Ed. Vidhyasekaran, P.). Daya Publishing House, pp. 382–404.

St. Amand, P.C. and Wehner, T.C. 1991. Crop loss to 14 diseases in cucumber in North Carolina for 1983–1988. *Rep. Cucurbit Genet. Coop. No. 14,* pp. 15–17.

Stryapkova, L.V., Korganova, N.N. and Treeva, M.M. 1992. Losses caused by Peronosporosis can be avoided. *Zashchita Rastenii (Moskova)* **12:** 10–11.

Sumner, D.R. and Pathak, S.C. 1987. Control of foliar disease of cucumber with resistant cultivars and fungicides. *App. Agril. Res.* **2(50):** 324–329.

Sun, S.K. and Huang, J.W. 1985. Mechanisms of control of Fusarium wilt diseases by amendment of soil by S–H mixture. *Pl. Prot. Bull.* **27:** 159–169.

Sundheim, L. 1982. Control of cucumber powdery mildew by the hyperparasite *Ampelomyces quisqualis* and fungicides. *Plant Pathol.* **31:** 209–214.

Tafradzhiiski, I. 1959. Powdery mildew on cucumbers in the greenhouse and pothouse industry and its control. *Gradinarstvo* **1(1):** 22–25.

Thind, T.S. and Mohan, C. 2001. Disease–weather relationship and relative activity of some new fungicides in different application scedules against muskmelon downy mildew. *J. Mycol. Pl. Pathol.* **31:** 174–179.

Thind, T.S, Goswami, S., Kaur, R., Raheja, S. and Mohan, C. 2010. Development of metalaxyl resistance in *Pseudoperonospora cubensis* and its management with novel action fungicides. *Indian Phytopath.* **63(4):** 387–391.

Thind, T.S., Singh, P.P., Sokhi, S.S. and Grewal, R.K. 1991. Application timing and choice of fungicides for the control of downy mildew of muskmelon. *Pl. Dis. Res.* **6(1):** 49–53.

Thomas, C.E. 1978. Reaction of susceptible and resistant contaloupes to *Pseudope–ronospora cubensis. Pl. Dis. Reptr.* **62:** 221–222.

Thomas, C.E., Cohen, Y. Jourdain, E.L. and Eyal, H. 1987b. Use of reaction types to identify downy mildew resistance in muskmelon. *Hortic. Sci.* **22:** 638–640.

Thomas, C.E., Inaba, T. and Cohen, Y. 1987a. Physiological specialization in *Pseudoperonospora cubensis. Phytopathology* **77:** 1621–1624.

Tursumbaev, A. 1976. Effect of sowing dates and weather conditions on the susceptibility of cucumber to powdery mildew. Vestn. S. Skokh. Nauki Kazakistan **19(11):** 49–51.

Ullasa, B.A. and Amin, K.S. 1986. *Trop. Pest Mgt.* **32:** 277–282.

Ullasa, B.A. and Amin, K.S. 1988. Ridgegourd downy mildew epidemics in relation to fungicidal sprays and yield loss. *Mysore J. Agril. Sci.* **22(1):** 62–67.

Waraitch, K.S., Munshi, G.D., Nandpuri, K.S. and Lal, T. 1977. Screening of muskmelon, wildmelon and snapmelon for resistance to powdery mildew (*Sphaerotheca fuliginea*). *Phytopathol. Mediterr.* **16:** 37–39.

Wu, Q.X. 1991. New cucumber varieties. *Bull. Agril. Sci. Techn.* **3:** 34–35.

Yakimenko, L.N. 1983. Promosing cucumber varieties for summer sowing in the Krasnodar area. *Trudy po Prilladnoi Botanike Genetike I Selektsii.* **77:** 22–25.

Yarwood, C.E. 1957. Powdery mildews. *Bot. Rev.* **23(4):** 235–307.

Yarwood, C.E. 1963. Predisposition to powdery mildews. *Phytopathology* **53:** 1144.

Yuko, S. 1974. Melon breeding I. Use of the root dipping technique in screening for *Fusarium* wilt resistance and studies on sources of resistance in melons and cucumbers. *Bull. Veg. Ornmental Crops. Res. Sta. C. (Kurume)* **1:** 15–27.

Zinc, F.W. and Thomas, C.E. 1990. Genetics of resistance to *Fusarium oxysporum* f.sp. *melonis* races 0, 1 and 2 in muskmelon line MR–1. *Phytopathology* **80:** 1230–1232.

Zitter, T.A., Hopkins, D.L. and Thomas, C.E. 1996. *Compendium of Cucurbit diseases*. APS Press, Minnesota, 87 pp.

Zuo, Q.Y., Huang, T., Lin, Y., Chen, Q.H., Xie, S.D.A and Zhu, T.S. 1995. Creation and utilization of breeding parents of cucumber. *Acta Hort.* **402:** 334–339.

■■■■

7

DISEASES OF PEA

Pea (*Pisum sativum* L.) belonging to family Leguminosae includes both ornamental and edible forms – the garden pea and field pea. It is believed to be native of Europe and West Asia, while its wild prototype probably came from Ethiopia. It is one of the most important legume crops and is grown over a greater part of the world. During cultivation, the crop is severely attacked by various fungal, bacterial and viral diseases, which not only reduce the yields but also affect the quality of the produce. A comprehensive account of various diseases affecting pea crop is given in this chapter including management practices.

(A) FUNGAL DISEASES

1. POWDERY MILDEW

Powdery mildew is one of the important fungal diseases and occurs almost every year in severe epidemic form wherever this crop is grown throughout the world. Under favourable weather conditions, it causes considerable yield reductions.

Geographical distribution

Powdery mildew is the most widely distributed disease of pea. Its occurrence has been reported from Afganistan, Alaska, Agrentia, Australia, Canada, China, France, Germany, Hungry, India, Iraq, Italy, Japan, Libya, Malaysia, Mauritius, Pakistan, Philippines, Portugal, Russia, Spain, Siberia, South Africa, Sudan, Switzerland, Tanzania, Tucuman, Turkey, U.S.A., Zambia and many other countries (Singh and Singh, 1988).

In India, the disease was first reported by Butler (1918) from Dehradun (U.P.). But reports of its common occurrence all over the pea growing areas of the country have been received only after 1935 (Uppal *et al.*, 1935).

Losses

The damage due to powdery mildew varies depending on many factors including the varietal susceptibility, sowing time, stage of the crop at the time of infection and prevailing temperature and humidity from middle of February onwards. However, the reduction in yield is mainly due to reduced photosynthetic activity in infected plants. Munjal *et al.* (1963) reported that in a completely infected crop the losses were 21–31 per cent in terms of pod number and 26–47 per cent in terms of pod weight. However, up to 50 per cent reduction in yield due to attack of this disease have been reported (Gritton and Ebert, 1975, Rathi and Tripathi, 1994, Katcch *et al.*, 2010). Epidemic development of powdery mildew is very fast and it progresses as compound interest. Since the disease attains severity at the end of season, seed crops and dry edible types, that mature later than the vegetable types, are more likely to become infected and sustain a heavy loss in productivity.

Symptoms

Symptoms first appear as minute discoloured patches with thin fine meshwork of white mycelium arising at many places on the upper surface of lower leaves. These white patches soon join together to form larger whitish floury areas (Plate 1). As the disease progresses, upper leaves of the plants also express similar symptoms. In severe cases of disease development both upper and lower surfaces of the leaf are infected. The conidia and condidiophores are produced in such profusion that the foliage looks as if dusted with flour and the whole crop in the field appears white from a distance. When the conidial production is nearly over, the leaf surface becomes patchy yellow to brownish and the foliage is then killed.

With progress of the disease, the symptoms are also noticed on tendrils, petioles and stems. In severe cases whole of tendrils and petioles are covered with white floury growth. The basal portion of the stem near the ground level is first infected by the pathogen and shows white floury patches. These patches on petioles and stem later turn light brown to dark brown and necrotic.

Pods are infected in all the developmental stages, but immature pods are more prone to rapid infection. The affected pods show white floury patches consisting of white powdery mass. After infection, the pods remain small in size, shrivelled and pod number and pod weight are also reduced. These floury patches consist of mycelium and spores of the fungus. These patches then turn light brown and finally dark brown in colour and become necrotic. Such infected pods are either unmarketable or fetch minimum price in the open market.

The Pathogen

The causal agent of this disease is *Erysiphe pisi* DC. ex Saint-Amans (sometimes also referred as *Erysiphe polygoni* DC.). *E. pisi* and *E. trifolii* infect pea in the US Pacific Northwest (Attanayake *et al.*, 2010). The fungus is ectoparasitic, spreading on the surface of the host and sending haustoria into the epidermal cells to draw out nourishments. The haustoria lie outside the cytoplasm of the host and special structures are formed between the haustoria and the cytoplasm. These interfacial structures distinguish haustoria from hyphae and therefore, their structures and physiology are important factors in parasitism.

The conidiophores arise vertically from the superficial hyphae on the host surface. Each conidiophore bears conidia either singly or in chains (Plate 2). Of these, only the last one in the chain matures as the spores are formed in regular order from the tip downwards. The mature conidia fall-off quickly and are disseminated by wind. The conidiophores are septate and their cells are not much different from conidia in shape. The conidia are ovate, barrel-shaped, hyaline, and unicellular and measure 25–35 × 13–16 μm (Fig. 1).

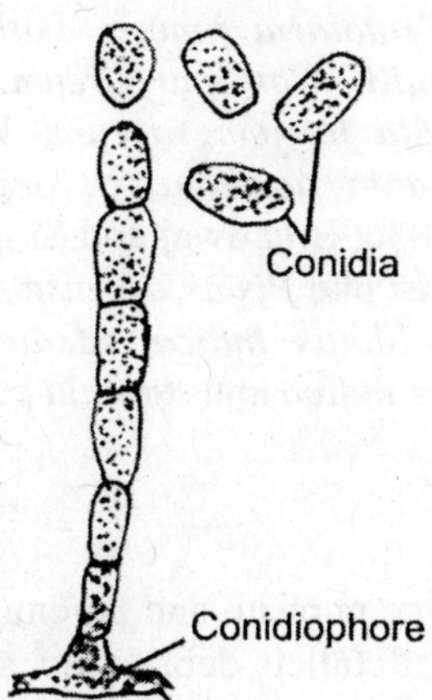

Fig. 1. Conidiophores and conidia of *Erysiphe pisi*

Later in the season, cleistothecia (sexual stage) may appear as dark coloured, round, minute bodies scattered in the mycelial web and measure 85–180 μm in diameter. The peridium is composed of distinct polygonal cells and is provided with a number of myceloid appendages. These outgrowths may be few or many, free or interwoven and resemble the ordinary hyphae. Each cleistothecium contains usually 2–8 asci. These asci originate from a single point in the fruit, appearing in a fanlike manner. They are ovate, nearly sessile and 46–72 x 30–45 μm in size. The asci contain 3–8 ascospores, which are elliptical, hyaline, and unicellular, and measure 10–25 x 9–14 μm. Cleistothecia production appears to be related with host nutrition and environmental conditions.

Although specificity for host species has been reported in *E. pisi*, there is no published information on the occurence of races. Studies have failed to establish existence of physiological races in pea growing areas of Uttar Pradesh (Singh, 1985). However, some variability in 10 isolates from Haryana has been indicated (Rathi and Tripathi, 1992). Banyal *et al.* (2014) studied the virulence pattern of 36 isolates representing diverse pea growing areas of Himachal Pradesh determined on a differential set of pea genotypes viz., JI-2302 (er-1), EC3341160, NIC 11181, EC292164, EC 329561, JI 1559(er-1), JI 2480 (er-2), PB29B and Lincoln (check) and the virulence analysis showed the presence of 19 pathotypes of *E. pisi*. Three pathotypes, namely PMP-12, 14 and 18 were found highly virulent infecting five differentials, whereas PMP 7 was least virulent showing resistant reaction on all the differentials.

Host range

Powdery mildews exhibit some degree of host specialization. There are conflicting reports regarding the range of parasitism of *E. pisi* from peas. According to some workers, it has a narrow host range and confined to *Pisum sativum, P. arvens, Lathyrus* sp. and *Lens esculenta* only (Yarwood, 1957; Singh and Singh, 1983a, Jhooty and Munshi, 1990). On the other hand, different workers have listed a large number of leguminous species as host of *E. pisi*. These include *Cajanus cajan, Cicer arietinum*, *Clitoria biflora*, *Crotolaria juncea*, *Lathyrus aphaca*, *L. odoratus*, *L. sativus*, *L. esculentus*, *Macroptilium atropurpureum*, *Medicago polymorpha*, *Pisum arvense*, *P. sativum*, *Trigonella foenumgraecum*, *Vicia sativa*, *Vigna mungo*, *V. umbellata*, *V. sinensis*, *Sesbania aculeata*, *S. aegyptica*, *Phaseolus mungo*, *P. vulgaris* (Singh and Singh, 1983a; Bhardwaj and Singh, 1984). The pea isolate of *E. pisi* could infect 11 plant species like *Pisum arvense* , *Lathyrus aphaca*, , *L. sativa*, *L. odoratus*, *Lens culinaris*, *Melilotus indica*, *Medicago hispida*, *Vicia sativa*, *V. hirsute*, *V. faba*, *Aeschynomene indica* and *Rubinia pseudoacacia* out of 25 evaluated (Singh *et al*., 2012).

Disease cycle

The causal fungus is an obligate parasite and perennates either through cleistothecia formed mostly on the infected fallen debris and sometimes on standing crop in Bangalore area of Karnataka (Venkatakrishna, 1951) and Lahaul, Spiti and Kinnaur districts of Himachal Pradesh (Kapoor and Chaudhary, 1995) or in seed as dormant mycelium (Uppal *et al*., 1935) or in the form of mycelium and conidia on the host (Jhooty and Munshi, 1990) and on annual and perennial hosts. In Palampur, Himachal Pradesh conditions, during late March or April self grown plants of *Aeschynomene indica* and *Rubinia pseudoacacia* as well as new foliage of *R. pseudocacia* plants get infected from inoculum from long season pea cultivars already in the field (Singh *et al*., 2012). Powdery mildew is present in abundance on *A. indica* and *R. pseudocacia* during September and this forms the source of inoculum for early planted pea crop in the area.

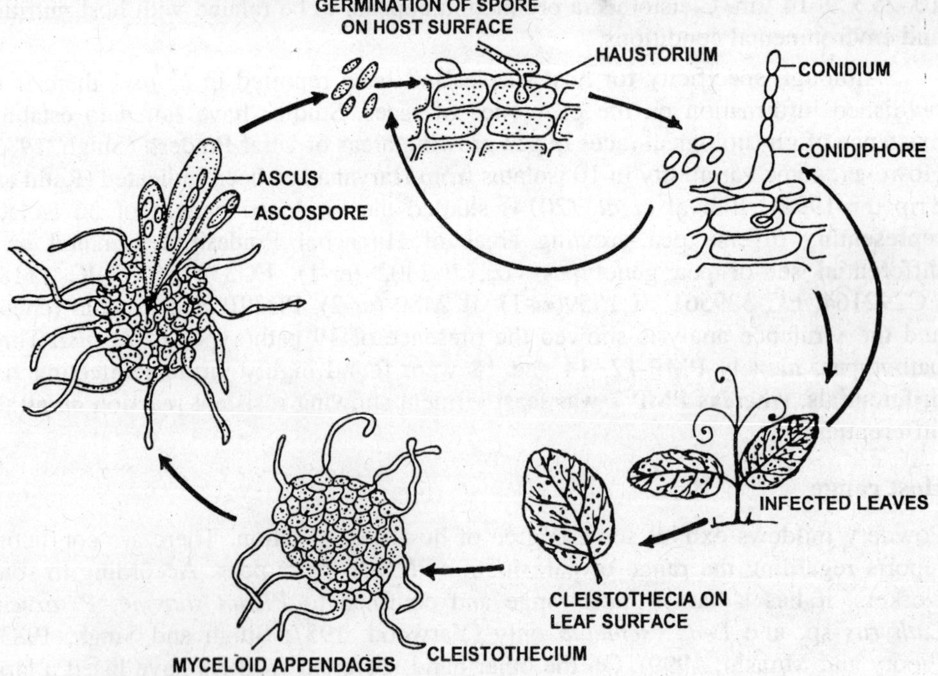

Fig. 2. Disease cycle of pea powdery mildew (courtesy Dr. H.B. Singh)

However, the concept of seed borne nature of the disease could never be confirmed (Yarwood, 1978). The primary infection of the lower leaves is caused by wind blown ascospores (where ever cleistothecia are present) (Griffee, 1969) or by conidia formed on the crop sown earlier in the neighbouring areas (Jhooty and Munshi, 1990). In the process of infection, the germination of conidia occurred after 2.5 h of incubation, while appressoria were formed after 3h on susceptible pea varieties and the germination was evident from both the poles of a conidium by maximum of six germ tubes (Singh *et al.*, 2000). Colonization of the entire leaf may occur in 150 hours under conducive weather conditions. Conidia of *E. pisi* not only germinate without liquid water but can do so even in completely dry environment. The conidia formed abundantly on primary infections are blown by wind and air currents and cause secondary infections. In areas where favourable conditions for cleistothecia formation are prevalent, the fungus reproduces sexually and form ascospores (Fig. 2).

Epidemiology

The powdery mildew of pea becomes a serious problem under warm and dry conditions during the day and cool nights. Irrigation at flowering to pod formation stages can prolong the vegetative phase, thereby increasing period for susceptibility. Crop raised in low and wet areas with high soil moisture is more vulnerable to the disease as these conditions tend to maintain the plants in vegetative phase for relatively longer periods. The disease is more prevalent in dry weather and moderate temperatures. A temperature range of 10–30ºC is favourable for conidia germination with optimum being at 20ºC. Highly significant and positive correlations were observed between temperature and disease severity while high relative humidity was not a pre-requisite for development of pea powdery mildew (Banyal and Tyagi, 1997). A fairly dry soil and heavy application of nitrogenous fertilizers increase the disease incidence (Singh, 1987). Maximum disease occurred in plants supplied with high N and Fe combined with low doses of P, K, Zn and Cu while disease intensity was significantly reduced by double doses of P, K and Zn combined with low N and Fe (Sindhan and Parashar, 1990).

Management

The attack of pea powdery mildew can be managed by using cultural practices, fungicides, plant extracts and oils and host resistance.

(i) Cultural practices: Collection and destruction of plant debris and avoidance of close planting can be helpful in reducing disease incidence (Singh, 1987). Change in date of sowing has also been recommended. Early sown crop (late October and early November) escapes the onslaught of the disease, whereas in the late planting (late November and December) the disease intensity was very high resulting in heavy reduction in yield (Schroeder and Providenti, 1965; Shukla and Amin, 1991). Mid October sown crop despite of having maximum disease severity resulted in minimum apparent infection rate (r) and highest green pod yield, however, late November sown crop escaped the discasc to a considerable extent but it gave minimum green pod yield (Gupta et al., 2014). Plots with raised bed planting and without irrigation had more disease severity as compared to flat bed and with irrigation but the higher yield was recorded in raised bed conditions (Kumari et al., 2013).

(ii) Host resistance: Several workers have investigated varietal resistance of pea to powdery mildew, but very few varieties have been reported to be resistant (Singh *et al*., 1988). Pal *et al.* (1980) screened 292 pea accessions, but found only five entries such as P431, P438, Glorire de Quimper, Sel 18, and Sel 30 resistant against powdery mildew. Cultivars P-185, P-388, P-6583, P 6587, P-6588 and T-10 (Singh, 1985), Pant P-5 (Singh *et al.*, 1988), P-185, P6583, NDPI, JP6, JP886 and HUP-9 (Gupta, 1990) P6588, Rachna, P6587, P383, P388, DPS-3, DPS-6, HUP-2, DMR9, P-185 and P6583 (Mathur *et al.*, 1991), Groach Laser (Mahmood *et al.*, 1991), UVS-1, DPH-13, DPH-19, HFP4, JP4, JP71, Pant P-5, PLP555, PLP582, PLP590, PM5, PMR3, PMR4, PMR334, PLP555, PLP582, PM5, PMR3, PMR4, Sel8-1, Sel8-12-1, Sel35, Sugar Giant Tall, VG1, VG2, (Jindal *et al.* 1995), Pant P-9, Rachana, DMR 11 (Khare and Lakpale, 1997) have been reported resistant to this disease.

Some of the cultivars are reported to have a longer latent period and reduced number and area of mildew specks compared with other cultivars. These parameters are responsible for the lower rate of disease spread in these varieties. Such slow mildewing varieties reported from different parts of the country are IPS-207, IPS-77, IPS-293, IPS-198 and IPS-214 (Krishna and Mishra, 1988) and P185, JP4, JP753, NDP1 and HUP2 (Gupta, 1990a). Krishna *et al.* (1989) also reported less losses in pod length, number of seeds/ pod, 100 seed weight and total seed yield in slow mildewing varieties than the fast mildewing.

Hooda *et al.* (2010) evaluated 117 and 26genotypes of garden and field pea and found high resistance against powdery mildew in 2 genotypes (VP 233 and JP 15) of garden peas. Germplasm comprising of 701 accessions of garden and field pea originating from 60 countries were screened for powdery mildew resistance under natural epiphytotic conditions and 64 accessions found resistant in field screening for 2 years at one location were further screened both in field at two locations and artificially in laboratory to four isolates and 57 accessions showed resistant reaction for 3 consecutive years in field screening, but only 14 accessions originating from 10 countries showed resistant reaction in laboratory screening (Rana *et al.*, 2013). There was sufficient genetic diversity and agronomic superiority in the resistant accessions e.g., EC 598655, EC 598878, EC 598704, IC278261 and IC 218988, which may serve as useful genetic material for breeding resistant varieties for powdery mildew resistance and high yield.

Resistance in powdery mildew is controlled by a single gene pair 'er' (Singh *et al.*, 1983; Azmat *et al.*, 2010). One recessive gene controls resistance in T_{10}, but three pairs of recessive genes have been involved in var. PRS15 and P388 (Ram, 1992). Fondevilla *et al.* (2006) reported that in pea, two single recessive genes, er1 and er2, have been identified for resistance to powdery mildew caused by *Erysiphe pisi.* Combining the resistance provided by gene er1 and by line JI2480 into new cultivars is likely to increase their level of resistance and enhance durability of the protection. Gupta and Mate (2009) reported that conidia of *E. polygoni* varied in size on different cultivars of the same host. Susceptible cultivars showed an increase in size of conidia than resistant and moderately resistant. Similarly on resistant cultivars, the size of conidia reduced when inoculated from susceptible or highly susceptible cultivars. Munshi *et al.* (1987) showed that resistant cultivars contained greater quantities of orthodihydroxy phenols and polyphenols while cuticle thickness, germination of conidia and formation of appresoria were the same in resistant and susceptible ones.

(iii) Chemical control: Although several fungicides have been tested to control this disease, but wettable sulphur in general is still being used as the most effective fungicide for the control of this disease for the last one-century (Spencer, 1978, Yarwood, 1978). Among the large number of fungicides tested by various workers, high level of control of the disease was achieved with the use of Calixin (tridemorph), Karathane (dinocap) and Sulfex (wettable sulphur) as foliar spray (Bhardwaj and Sharma, 1984). However, Sulfex is the most economical fungicide till date.

Foliar sprays of EBI fungicides like Vigil (diclobutrazol), Baytan (triadimenol), Bayleton (triadimefon), Tilt (propiconazole), Rubigan (fenarimol), Punch (flusilazole), Contaf (hexaconazole), Topas (penconazole), Score (difenconazole), Saprol (triforine) and Folicur (tebuconazole) have also shown very good control of this disease at much lower doses than sulphur besides having good antisporulant activity (Rana *et al.*, 1991; Maheshwari *et al.*, 1992; Upadhyay and Gupta, 1994; Gupta and Shyam, 1998; Kapoor and Thakur, 1997; Gupta and Sharma, 2004).

Tilt (propiconazole) was the most effective treatment, which not only increased seed yield but also reduced disease severity significantly as compared to unsprayed check followed by Karathane (0.1%) and Bavistin (0.1%) (Prasad and Dwivedi, 2007). The fungicides like hexaconzole (0.05%) and penconazole (0.05%) were found highly effective and economical for controlling the disease and giving better seed yield (Surwase *et al.*, 2009). Gupta and Kumar (2008) evaluated different modes of fungicides in the management of this disease. In protective spray programme, three sprays of kresoxim methy (0.1%) at fortnightly intervals were found highly effective in checking the disease severity followed by difenoconazole, propiconazole, diniconazole, hexaconazole while wettable sulphur was least effective. Maximum post-symptom antisporulant activity of 10 days was also provided by kresoxim methyl (0.1%) and hexaconazole (0.05%).

Since these are systemic and site-specific fungicides, continuous use of these may result in the development of resistant strains of the pathogen. Rotation of fungicides and their use in mixture with non-systemics will be a good strategy to prolong the life of these fungicides.

(iv) Biological control: Srivastava and Bisht (1983) sprayed the infected plants with eight phyllosphere fungi and maximum suppression of the disease was obtained with *Cladosporium cladosporiodes* followed by *Curvularia lunata* (*Cochliobolus lunatus*). The Coccinellid, *Psyllobora nana* were observed feeding on the mildew *Erysiphe polygoni* in Cuba and can be used as biocontrol agent (Cruz *et al.*, 1990). Shailbala and Khulbe (2009) evaluated different organic preparations of Pyrite, cow dung, cow urine, soapnut, mustard oil + soapnut, cow dung + pyrite + soapnut, mustard oil + cow urine + soapnut and reported that soapnut performed significantly superior in controlling pea powdery mildew disease.

(v) Plant extracts and oils: Control of this disease by plant extracts and oils is superior to chemicals because they are environment firendly and safe from residue point of view. Singh and Singh (1983b) and Singh (1985) screened various plant extracts and oils against conidia germination of *E. pisi* and reported that the conidial germination was significantly reduced in all the extracts and oils at 100 ppm or more whereas, neem and garlic oils proved to be highly effective. Volatile oils and extracts of plant parts also have strong effect on germination of conidia or conidial density (Singh *et al*, 1984). Out of eleven ethanol extracts of botanicals evaluated against *E. pisi* both under laboratory and field conditions, *Acorus calamus* (1%) caused maximum inhibition of conidial germination (98.13%) and germ tube length (82.99%) while, under field conditions though no botanical was as effective as the standard check i.e. hexaconazole but among botanicals three sprays of *Urtica dioica* (2%) showed least disease severity and increased green pod yield (Gupta and Kumar, 2014).

2. DOWNY MILDEW

It is an important disease of pea crop particularly in those areas where cool and wet climate is prevalent during its cultivation. The disease during favourable weather conditions causes considerable yield losses

Geographical distribution

Berkeley in England first discovered the disease in 1846 (Chupp and Sherf, 1960) and since then it has been found in all parts of the world where the crop is grown intensively. The occurrence of the disease has been reported from USA, Ukrain, Italy, Poland and Australia. In India it is prevalent in Indo-gangetic plains including Uttar Pradesh and Punjab (Pavgi and Singh, 1965; Bains and Dhaliwal, 1986).

Losses

In Wisconsin, downy mildew epidemic occurred in 1973 and some fields had 62-85 per cent diseased pods (Hagedorn, 1974). It was also identified as one of the major contributor to the pea yield decline syndrome in south Australia (Davidson and Ramsey, 2000).

Symptoms

The disease generally starts from the lower leaves and then progresses upward. Two types of infection *i.e.* local and systemic occur on pea plants. In local infection, round or less frequently elongated, yellow to brown spots appear on the upper surface of leaves, stipules, and petioles and sometimes on stems. The patches may be of small size or may occupy a larger area of the leaf lamina. During cool and moist weather, the corresponding under surface is covered with a white to grayish violet downy growth of the fungus, which is composed of sporangiophores and sporangia. Systemic infection, results in stunting and distortion of the plants, usually killing them before flowering. On pods, the patches are pale green, more or less elliptical which later turns brown. Seeds developing under these patches remain undersized and aborted. Later in the season, oospore develops in the senescent tissues and can survive in soil for up to 15 years (Van der Gaag and Frinking, 1997).

The Pathogen

Downy mildew is caused by *Peronospora pisi* Syd. The mycelium is aseptate, hyaline, intercellular and branched with finger shaped haustoria. Sporangiophores first appear as simple elongating hyphae from stomata on abaxial leaflet surface, then branching from a single axis (monopodially) to produce multiple, terminal sporangia, which developed synchromously and polyblastically on each sporangiophore (Falloon and Sutherland, 1996). Sporangia had smooth surface during development but finely they were echinulated when mature and are delimited by septa at the ends of terminal sporangiophore branches. Gametangia (oogonia and anthredia) developed extensively on inner surface of field grown pea pods from smooth, bulbous hyphae, adhering to the host epidermis. Each oogonium is surrounded by several anthredia. Oospores with in pod tissue of field grown plants were enclosed by oogonial membranes. Each oospore had a heavily reticulated outer wall enclosing cytoplasm and liquid possibly lipid. The pathogen also shows some variation, and atleast six races have been reported (Hubbeling, 1975).

Disease cycle

The pathogen perenates through oospores lying in soil. Sometimes, it is seed borne also both under the seed coat or as a contaminant with seed lots. These oospores cause primary infection in the ensuing season. Secondary spread of the disease occurs through sporangia disseminated by moist wind.

Epidemiology

Disease occurrence is usually associated with unsusally wet weather and long periods of cool overcast weather (Hagedorn, 1974). Sporangial production requires high humidity for atleast 12 h. The sporangia germinate best at low temperatures between 4 and 8°C and require rain for dispersal. Sporangia germinate readily in free moisture by germtube. The longevity of sporangia decreases significantly as temperature increases. At 5°C no oospores are produced while 20°C is optimum (Van der Gaag *et al.*, 1996). Proximity of pea crops to each other also results in severe outbreak of the disease. Manganese concentration in pea plants is positively correlated with the severity of downy mildew (Davidson and Ramsey, 2000).

Management

(i) Cultural practices: Cultural practices like destruction of infected plant debris, follow atleast 3 year or longer crop rotation, wider spacing, use of healthy seed (may be produced in dry areas), deep ploughing and well-drained fields are useful in reducing the disease.

(ii) Host resistance: Complete immunity among pea cultivars is lacking against downy mildew although some degree of resistance/tolerance have been reported in some cultivars/lines like Gastro, Starcovert, Starnain, Puget, Cobri, 119, 21 and Mukta (Hubbeling, 1975; Stegmark, 1988; 1990 and Annonymous, 1999). Sugar pea was the most susceptible cultivar to *P. pisi* followed by Mange tout and Little marvel cultivars. On the other hand, the lowest infection was found on Master B cultivar followed by Lincoln and Relavil cultivars (El-Marzoky, 2008).

(iii) Chemical control: The pathogen is known to overwinter in the form of oospores in the infected plant tissues and in the seed. Therefore, seed treatment with various systemic fungicides like metalaxyl and A9873 (mixture of metalaxyl, cymoxanil and fludioxanil) has been reported effective against this disease (Forster *et al.*, 2001). Foliar application of metalaxyl (Singh and Dickinson, 1980) or cymoxanil + copper (Bugiani and Govoni, 1996) after the appearance of initial symptoms have been advocated against this disease. However, ineffectiveness of seed treatments with metalaxyl and oxadixyl in New Zealand has been reported (Falloon *et al.*, 2000) while cymoxanil or fosetyl-Al protected seedlings from infection by metalaxyl resistant strains. Recently available strobilurin fungicides alone or in combination with protectants can also be used to keep the ravages of this disease under check. To avoid buildup of fungicide resistant strains, it is always better to rotate the fungicides and apply them in mixture with non-systemics.

3. RUST

Pea rust is an important disease, which occurs quite frequently in pea growing areas in North and North West parts of India. Two rusts have been reported on peas caused by *Uromyces pisi* (Pers.) de Bary and *Uromyces viciae-fabae* (Schroet), of which the latter is of worldwide occurrence.

Geographical distribution

The disease is serious menace to the cultivation of this crop in wet areas, the world over. The identity of the fungus was established long back by Jordi in 1904 (Buchheim, 1922). Since then it has been reported from Morocco, U.S.S.R, Burma, France, Germany, Japan, U.S.A., Afghanistan, Palestine, Switzerland, Norway, Portugal, Pakistan, Lithuania and Israel.

In India, pea rust was first reported on *Vicia faba* by Sydow and Butler from Pusa (Bihar) in 1906. The first report of its occurrence on pea in India was by Butler (1918). Subsequently the disease was reported from Poona where it caused severe damage to sweet pea crop (Patel, 1933) and now is prevalent in severe form in the North and North Western part of India.

Losses

The disease can assume epidemic proportions under favourable weather conditions resulting into considerable yield losses. Some times it becomes very serious, often killing the entire plant during cool and wet weather.

Symptoms

The rust pustules appear on all the green parts of the plant above ground, being most numerous on the underside of the leaves (Fig. 3), less abundant on the pods, and occurring sparingly on the stems. Infection is evident first as minute, raised pustules, which later become distinct, yellowish, circular sori (Plate 3a). The yellow spots have aecia and persist for longer time. The uredo pustules develop on both surfaces of the leaves as well as on other parts and give light brown powdery appearance to the plants in case of severe infections. Late in the season, teleuto pustules also develop on the leaves, but most commonly on stems and petioles (Plate 3b). They are dark brown or almost black in colour.

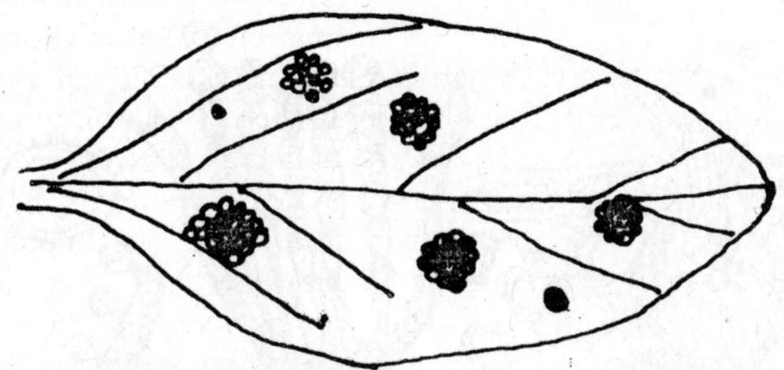

Fig. 3. Symptoms of Pea rust on leaves

The Pathogen

The disease is incited by two species of *Uromyces* viz., *U. pisi* (Pers.) de Bary and *U. viciae-fabae* (Schroet), but the latter one is of worldwide occurrence. *U. viciae-fabae* is autoecious rust with all its spore stages on the same host and no alternate host is required in the life history. The fungus is heterothallic. On peas the pycnia occur in small groups associated with the aecia. The aecia are cupulate and 0.3 to 0.4 mm in diameter. The peridium is short and whitish. The aeciospores are round to angular or elliptical with hyaline wall. The wall of these spores is verrucose. The uredia measure 14 to 26 (more commonly 18 to 22) μm in diameter. The uredia present a powdery appearence. The urediniospoes are round to ovate, light brown, echinulate, with 3–4 equatorial germ pores and measure 20–30 (22–28) x 18–26 (19–22) μm (Fig. 4). The telia occur in the same sorus as the uredia. They are dark brown to black. The teliospores are subglobose, ovate or elliptic with rounded or flattened appex, which is considerably thickened and appears papillate (Plate 4). These spores are smooth, brown and measure 25–38 x 18–27 μm. The wall of the teliospores is 1–2 μm thick at the sides and 5–12 μm at the apex. The pedicel is persistent, yellowish-brown, thick and up to 100 μm long.

Host range

Besides pea, it also infects many species of *Vicia* (*V. amoena* var. *sachalinensis*, *V. cracca* var. *japonica*, *V. faba*, *V. japonica*), *Lathyrus maritimus*, *L. palustris* var. *lineaefolius*, *L. odoratus* and Lentils (Chupp and Sherf, 1960; Mohamed *et al.*, 1983).

Variability

Aeciospores and urediniospores of isolates collected from pea, lentil, faba bean, *Vicia hirsuta* and *V. tetrasperma* showed little variation in morphology, colour or size, whereas teliospores markedly differed in size and shape (Chauhan and Singh, 1995) but isolates of *U. vicia-fabae* from cultivated and wild *Vicia* spp. were cross infective and showed no variability with regard to virulence. However, Seven isolates of *U. viciae-fabae* causing rust of pea were differentiated into six pathotypes, based on differential reaction on 15 pea and lentil hosts (Singh and Sokhi, 1980).

Two biological forms of *U. pisi* were distinguished viz., *U. pisi* f.sp. *pisi* on pea and *U. pisi* f.sp, *lathyri* described as new on *Lathyrus* spp. (Sidenko, 1966).

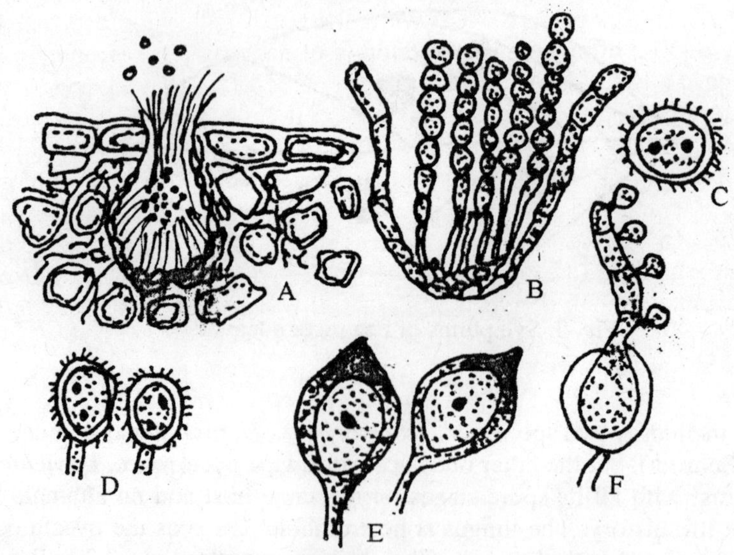

Fig. 4. Different spore stages of *Uromyces viciae fabae* (A. Pycnium, B. Aecium, C. Aeciospore, D. Uradiniospore, E. Telentospores, F. Germination of Teleutospore)

Disease cycle

The rust is seed borne in broad beans, but in peas and lentil it is mainly soil borne as teliospores in crop debris. In India the rust appears to survive on weed hosts belonging to *Lathyrus*, *Vicia* etc. and the spores are wind blown to the main crop (Singh, 1987). Wild hosts may serve as primary or secondary source of infection. After infection the osmotic pressure of germtubes of the urediniospores and haustoria was greater than that of its host thereby increasing the permeability of the plasma membrane of the host cells and resulting in absorbtion of nutrients by the haustoria.

Epidemiology

Very little information is available on the effect of environmental factors on development of pea rust. Prasada and Verma (1948) studied the effect of temperature on the formation and development of aecial, uredinial and telial stages of *U. viciae-fabae* on lentil. They found that temperature range of 17–22ºC is optimum for infection and for development of uredinia, a slightly higher temperature range (20–26ºC) was required. The formation of telia was observed when the crop reached maturity. Teliospores can withstand summer heat in the plains of India and thus play an important role in its perpetuation. However, aeciospores and urediniospores cannot tolerate the summer temperatures in the plains of India. On pea, severity of infection by *U. viciae-fabae* and pustules/plant increased progressively with an increase in the duration of leaf wetness upto 24 h, but did not increase further significantly at optimum tempeature 20ºC (Chauhan and Singh, 1994). The observed relationship between severity of pea rust and duration of leaf wetness at 20ºC may be useful in predicting disease outbreaks if initial inoculum is present. Benda (1967) investigated the role of pH gradient in pea stipules and reported that lower stipules with telia had higher pH than upper ones with uredia and the pH boundary for development for two stages being 6.3–6.5 while the upper most stipules which were most acidic were devoid of symptoms. Production of aeciospores was observed at a temperature range of 10–25°C, with a maximum at 25+2°C. Among the different growth stages of pea, the pod formation stage was highly susceptible and produced the maximum number (744) of aecidia/leaf at 20–25°C. Significant effects of growth stages and temperature were also noticed for pustule number. Urediniospore production mainly coincided with the senescence of the pea plants. Maximum germination (2%) of aeciospores was observed at 25°C, whereas maximum urediniospore germination (3.5%) was at 15°C. Temperatures >15°C decreased urediniospore germination. A relative humidity (RH) of 100 per cent was favourable for aeciospore germination, while 98% RH favoured urediniospore germination. Typical histopathological behaviour of the aeciospores was observed (Kushwaha *et al.*, 2006). Bal and Kumar (2012) studied the influence of weather parameters like relative humidity, temperature and rainfall on the development of pea rust and further its influence on loss in yield and reported that temperatures ranging between 10°C–24°C and relative humidity of more than 70 per cent favoured the development of disease with an average loss of 52.6 per cent in seed yield.

Management

The management of this disease can be attempted by cultural, chemical and host resistance.

(i) Cultural practices: The disease generally appears late in the season and hence the losses can be reduced by planting early. Sangar and Singh (1994) studied the effect of date of sowing on the severity of rust and found that delayed sowing after 5th October increased the disease incidence. Field sanitation to destroy diseased crop debris and long crop rotations avoiding broad beans, *Vicia, Lathyrus* should be followed for minimising losses from this disease (Singh, 1987). Flat bed cultivation alongwith irrigation was found to be more suitable for rust infection and upto 28 per cent severity was recorded (Kumari *et al.*, 2013).

(ii) Chemical control: Hiremath and Pavgi (1971) showed that spore germination in *U. fabae* was completely inhibited by Aureofungin (20 μg/ml). Sugha *et al.* (1994) studied the sensitivity of aeciospores and urediniospores to systemic fungicides and found that aeciospores were most sensitive to thiabendazole followed by benomyl, carbendazim and thiophanate-methyl and least sensitive to bitertanol, while urediniospores were more sensitive to triadimefon, dinobuton and myclobutanil and least sensitive to bitertanol. Differences in sensitivity of the spore types to fungicides were attributed to differences in their morphology and longevity.

Fungicides such as mancozeb, triadimefon, tridemorph, maneb + tridemorph, hexaconazole, penconazole, flusilazole and difenconazole were found effective in reducing the rust severity under field conditions when used as foliar sprays (Singh *et al.*, 1981; Upadhyay and Gupta, 1994; Gupta and Shyam, 1998). Post-infection activity of various EBI fungicides against this disease has also been reported (Gupta and Shyam, 2000). Amin and Ramachander (1989) studied the effect of triforine on epidemic development of rust and found that disease incidence was minimum when applied before 52 days of sowing and was effective only for 7–10 days. Correlations of disease intensities, area under disease progress curve (ADPC) and average apparent infection rate with pod yield were negative and significant. During fruit formation disease intensities were more highly correlated with yields. The lowest disease attack was recorded in plants treated with mancozeb followed by those with Carbendazin (Khan *et al.,* 2009). Singh *et al.* (2008) reported individual fortnightly foliar spray, commenced with the initiation of the disease, of carbendazm difenconazole, tebuconazole and hexaconazole away systemic and at 10 dyas interval of Antracol, Microsoil and share among non-systemic fungicides proved effective in reducing the diseazse severity and increasing pod yields.

(iii) Host resistance: Pea cvs./lines viz., PJ 207508, PJ 222117, EC 109183, Batri Brown 3 and Batri Brown 5, Wisconsin, DPMR 68, DP2 and NDP1 are reported completely resistant (Pal *et al.*, 1980; Narsinghani *et al.*, 1980; Gupta, 1990) while cvs./lines Nosovski 24, Kellervo, Kombainovskii 58, K-4449, Anger yellow fodde, EC 42959, Perf. 326, IC 4604, C12, P 185 and JP 885 were moderately resistant (Golubev, 1965, Sohi *et al.*, 1974; Gupta, 1990). Among the tested pea cultivars, Climax gave the highest yield due to lowest disease severity and Meteor had the lowest yield due to maximum disease attack (Khan *et al.*, 2009).

Studies on inheritance of resistance to rust in peas revealed that resistance was controlled by single dominant genes (Pal *et al.* 1979; Abusaleha *et al.*, 1989). Abusaleha and Pal (1990) studied linkage relationship between rust resistance and seed character in peas and found that gene for resistance to rust is not linked to the gene for green seed colour (on chromosome 4) or the wrinkled seed trait (chromosome 7). Resistant genotypes contained higher levels of phenolics and susceptible ones had more sugar. Phenylalanine ammonia lyase (PAL) activity in rust infected leaves of pea lines resistant to *Erysiphe polygoni* was higher than that in healthy leaves possibly accounting for the higher phenolic content of infected leaves.

4. ASCOCHYTA BLIGHT

Ascochyta blight is a serious disease of this crop particularly under favourable weather conditions and cause significant losses.

Geographical distribution

Jones (1927) and Lawyer (1984) clarified and described the importance of these pathogens involved in various types of symptoms at different pea crop stages and environment. All the three diseases are world wide in occurrence. *Ascochyta* disease complex has been reported from Poland, Germany, Chile, India, Austria, East Africa, Bulgaria, Scandanavia and Netherlands (Guerrero, 1987; Price and Cishahayo, 1986; Oyarzun *et al.,* 1993 and Kaiser *et al.*, 1998).

Mycosphaerella blight and *Ascochyta* foot rot are the most important diseases in the United States. *Mycosphaerella pinodes,* the perfect stage of *Ascochyta pinodes* is an important foliar pathogen than *Ascochyta pisi* (Lawyer, 1984). Two species *i.e., Phoma medicaginis* var. *pinodella* and *M. pinodes* have been reported to be important foot rot pathogens (Wallen, 1965; 1974 and Sarkar *et al.*, 1982).

Srivastava and Gupta (1990) reported pea blight (*A. pisi*) in Sikkim, which caused serious losses from December to March. Singh (1994) isolated *Phoma medicaginis* var. *pinodella* (Jones) Boerema from crops showing foot rot symptoms in 1991 and 1992 in the sub-tropical zone of Jammu. Bhardwaj and Sagar (1996) collected diseased pea samples infected by *M. pinodes*. Initial symptoms were small black flecks and pinhead spots raised on pods, which enlarged into broad, concentric spots leading to foliage blight and yellowing. *Phoma medicaginis* var. *pinodella* associated with *Ascochyta* disease complex of peas was isolated from pea cultivar Lincoln from Bajaura, in Himachal Pradesh by Sagar and Bhardwaj (1997).

Losses

The disease may cause yield losses up to 40 per cent, but sometimes in blight phase alone the losses may go up to 60 per cent (Bretag *et al.*, 1995; Tivoli *et al.*, 1996).

Symptoms

It is difficult to distinguish the three species of the pathogen on the basis of symptoms on the host, but microscopic examination of their spores makes the distinction easy. All the three species attack the shoots, leaves and seeds however *M. pinodes* and *A. pinodella* also attack roots (Bajan, 1968).

(a) Ascochyta or Mycosphaerella blight: This disease is characterized by brown to purplish, irregular areas on the foliage. Sometimes, in the presence of moisture for long periods, the lesions become larger, circular and somewhat zonated. The small, brown to purplish irregular spots, which appear on the pods, enlarge to irregular purplish area or large area could become blotched with the coalescing of lesions (Plate 5a). Developing seeds may be deformed, remain small in size and are sometimes stained. The earliest symptoms on stem appear as black to purplish streaks, more pronounced at the nodes, which could enlarge into brown to purplish irregular areas on the entire stem of 6 to 10 inches above the roots. Pycnidia are usually darker in shade and produced on lesions especially those on leaves and pods in a characteristic ring pattern.

(b) Ascochyta foot rot: The symptoms, manifested by *Phoma medicaginis* var. *pinodella,* are strikingly similar to those described for blight but in this case foot rot lesions predominate (Plate 5b). The dark brown lesions on the stem start soon after germination below the soil line and gradually extend down to roots and up the stem.

(c) Ascochyta leaf and pod spot: Symptoms of this disease appear as definite, circular, sunken, tan to brown spots on leaves and pods while elongated on stems and petioles (Plate 5c). A dark brown margin usually surrounds the tan spot. Pycnidia may form on these spots also. Underlying seeds in the pods may also be affected and remain shriveled.

The Pathogens

Three morphological distinct species of *Ascochyta* cause closely related diseases of pea. *Ascochyta pinodes* (Berk & Blox) Vestgrn. (perfect stage *Mycosphaerella pinodes*), is responsible for causing blight of pea. Another species *Ascochyta pisi* Lib. causes *Ascochyta* blight and pod spot and the third *Phoma medicaginis* var. *pinodella* (Jones) Boerema (syn. *Ascochyta pinodella* Jones.) causes foot rot (Wallen

1965, 1974; Sheridan, 1973 and Sarkar *et al.*, 1982). Of late, the occurrence of a new blight caused by *A. pisarum* has been reported from Himachal Pradesh (Paul and Rathour, 1998).

Historically, Stone (1912) and Vaughan (1913) considered *Mycosphaerella pinodes* (Berk. & Blox) Stone to be the ascigerous stage of *A. pisi* however, Jones (1927) has pointed out that the disease caused by *M. pinodes* differs markedly from that caused by *A. pisi* and that two organisms have been confused in literature. He substantiated the technical descriptions, which were given to separate *M. pinodes* and *Phoma medicaginis* var *pinodella* (microform) from *A. pisi.* Later on, the descriptions given for conidial (*A. pinodes*) and ascosporic stages of *Mycosphaerella pinodes* were given by Punithalingas and Holliday (1972) in CMI publication and further reproduced by Lawyer (1984).

The three species of *Ascochyta* pathogenic on pea can be identified by the microscopic examination of their spores. *A. pisi* has gregarious pycnidia at the centre or in circular zones, somewhat prominent, depressed globose, light to dark brown ostiolate and 75–225 μm in size. Spores are hyaline oblong with rounded ends, straight or somewhat curved 1 (1–3) septate, somewhat constricted 4–10 × 3–5 μm in size with a mean length less than 13.5 μm. *M. pinodes* has the scattered pycnidia, rarely at centre, somewhat prominent, irregularly depressed, globose, light to dark brown, indistinct in spot, ostiolated and 65–175 μm in diameter. Spores are hyaline oblong with rounded ends straight or somewhat curved, 1(1–3) septate, sharply constricted and measure 12–17 × 3.5–5.5 μm in size with a mean length of more than 13 μm. However, Singh (1987) reported that the spores sizes of *A. pisi*, *A. pinodes* and *A. pinodella* are 13.5 × 4.2 μm, 5.5 × 3.7 μm and 12.3 × 4.5 μm, respectively.

Differences in symptoms and morphology were recorded between 2 isolates of *Phoma medicaginis* var. *pinodella* causing characteristic narrow concentric spots on peas from Bajaura and Kukumseri in Himachal Pradesh, India (Bhardwaj and Sagar, 1998). The former caused dark brown spots and the latter caused brown-purple spots. There were also differences in pycnidial and pycnidiospores dimensions, and 2 biotypes were identified. A new micro form of *P. medicaginis* var. *pinodella* causing small raised and dark brown irregular spots on pea leaves at Kukumseri was identified which possessed the smallest pycnidia, pycnidiospores and chlamydospores. The mean sizes of pycnidiospores of 2 macro form biotypes were 11.56 × 3.68 and 11.67 × 4.37 μm, respectively. For micro form biotype this value was 4.71 × 2.40 μm.

The fungus *Mycosphaerella pinodes* has been reported to be homothalic. Rudakov and Lepikhova (1974) stated that the natural populations of *A. pisi* and *M. pinodes* have seven and three biotypes, respectively which differ in morphology, physiology and virulence. There are also reports on the prevalence of five distinct races of *A. pisi* from the United States, USSR and Canada. In Canada the four races identified were generally localized in separate pea growing areas (Lawyer, 1984). Furgal (1986) identified six races of *A.* [*Mycosphaerella*] *pinodes* and race 5 and 6 were most virulent and affected almost all the cultivars of pea. Furgal (1984) identified 6 races of *A.* [*Mycosphaerella*] *pinodes*, 5 of *A. pisi* and one race of *A.* [*Phoma medicaginis* var.] *pinodella*. Ovchinnikova and Andryukhina (1984) described reaction of as many as 125 races of *A.* [*Mycosphaerella*] *pinodes*. Clulow *et al.* (1991) for the first time described genetic variation in *M. pinodes* in the pathogenesis of pea plants. Based on reactions of 9 differential host lines, 45 isolates from a wide range of geographical locations were classified by stem symptoms into 9 groups (Stem pathotypes 1–9) and by leaf symptoms into 16 groups (Foliar pathotypes A-P). The teleomorph of *P. medicaginis* var. *pinodella* was also reported (Bowen *et al.*, 1997). The anamorphic pycnidial fungus *Ascochyta pisi* is one member of a species complex that causes Ascochyta blight of pea, a potentially devastating disease. The teleomorphic state of this fungus was induced under laboratory conditions. Using morphological and molecular characters, the teleomorph was placed within the genus *Didymella* as *D. pisi* and describe a heterothallic mating system using a PCR-based mating type assay and *in vitro* crosses (Chilvers *et al.*, 2009).

Epidemiology

Perpetuation of *Ascochyta pisi* is carried in infected seed and it also overwinters in infected plant debris where pea is grown during summer months (Bajan, 1968, Bedlan, 1985). The rate of seed transmission in case of *Mycosphaerella pinodes* was found to be higher at 15ºC (100%) than at 18° (61.5%) or 25°C (70.0%) and the pathogen was able to survive in seeds for 4 years (Corbiere *et al.*, 1994). Musaev (1967) reported that pycnidia overwinter on plant debris and primary infection occurs by overwintering pycnidiospores. Maude (1966) reported that *A. pisi* infected seed gave 40 per cent infected seedlings showing lesions within 4 weeks of sowing.

Maximum temperatures ranging between 17.7 and 18.9ºC at pre-flowering and a temperature range of 9.8ºC to 26.3ºC at post-flowering coupled with frequent high rainfall and a relative humidity of 85-88% are favourable conditions for *Mycosphaerella* blight of pea (Bhardwaj and Thakur, 2000). Pycnidiospores of *Mycosphaerella pinodes* survived dry periods of up to 21 days after inoculation. Rewetting restored the infection capacity of the pycnidiospore, resulting in high levels of disease. The effects of wet-dry-wet cycles depended on when the dry period occurred during the infection process. No disease symptoms appeared when dry periods occurred during germination. Pycnidiospores formed appressoria but hyphae did not penetrate if a 6-12 h wet period preceded the dry period, and only a few flecks appeared during the dry period. Coalescent necrosis occurred when the dry period followed penetration. The disease was severe in each case when plants were returned to wet conditions after a period of dryness. Lesion development depend on the duration of the initial wet period, and the characteristics (temperature and duration) of both the dry period and the final wet period (Roger *et al.*, 1999).

Development of pycnidia and perithecia of *M. pinodes* was studied in pea cv. Solara under glasshouse and field conditions (Roger and Trivoli, 1996). Development and quantity of pycnidia were related to inoculum concentration and physiological state of the plants. Pycnidia were produced on both green and senescent organs while perithecia on senescent organs only. Spore trapping showed that both pycniospore dispersal and ascospore discharge were initiated by rainfall or dew. Pycniospores and ascospores were dispersed through the growing season, suggesting that ascospores also play an important role in secondary infections

Water-logging of peas already infected with *M. pinodes* may result in more severe infection and greater reductions in plant growth, and cultivars more sensitive to water logging may suffer greater losses from disease (McDonald and Dean, 1996). Susceptibility of the host to the three pathogens *i.e., A. pisi, Mycosphaerella pinodes* and *A. pinodella* was augmented by the application of N, P, Ca and reduced by the application of Potassium.

Spore germination of *A. pisi* was 85–87 per cent in a drop of water or at 100 per cent relative humidity when temperature was 20–21ºC (Susuri, 1976) while 15–25°C temperature was reported optimum for spore germination (Dorozhikin and Kremneva, 1976). Incubation period for *A. pisi* at 25ºC was four days and pycnidia were formed after seven days, whereas at 3ºC, the incubation period was 8 days and the pynidia were formed after 21 days (Malikova, 1973). Singh (1987) reported that the infection did not occur below 80 per cent RH, but it rapidly increased above 90 per cent, Singh *et al.* (2005) also reported significant effect of temperature, moisture duration and their interaction on Ascochyta blight development. Temperature ranging between 20–30ºC with moisture duration of 12–24 h resulted in severe disease development as compared to lower temperature of 10–15ºC. The maximum disease was observed at 30°C, when moisture duration was either 12 or 24 h.

Management

(i) Cultural practices: Use of disease free seed is important for the control of all the three pathogens. Disease free seed can only be produced if the crop is grown in low rainfall areas. The soil borne inoculum can be reduced by following long crop rotation and by destruction of the crop refuse by burning either in the field or after threshing (Bedlan, 1985).

(ii) Host resistance: Kvasnikov and Krotova (1977) found that out of 260 cultivars of pea 13 were relatively resistant to *A. pisi*. Ovchinnikova and Andryukhina (1977) reported that out of 800 cultivars of pea, the garden pea Victoria and Laga and fodder peas Angerskii, Zheltyl, Mestnyi, Melkai and Ranni Zelenyl-33 were found resistant. Out of 266 cultivars tested by Amelina and Chaplygina (1979), Norlii (K-6607 VIR) was the most resistant. When the resistance of 320 cultivars and hybrids of pea was tested by Bakalova (1979) in the field by innoculating with *A. pisi,* no immunity was found however, incidence was low (2%) on K-5778, K-5790, K-5795, K-5810, K-6129, K-5928, K-5574, K-6489, Prekulskii and Yarkandbelye. Furgal-Wegrzycka (1984) reported that mycelial growth and sporulation of *A. pisi* were inhibited by water extracts of resistant but not of susceptible lines. Sandhu and Dhillon (1984) reported that in varietal trials at Ludhiana and Gurdaspur the pea cvs. Pleiofila and ML-21 were moderately resistant to *A. pinodella.* Iqbal *et al.* (2001) found three lines 89P117-5, 88P022-6-28 and 88P0-6-29 of pea as highly resistant to blight (*Ascochyta pisi*).

Pea cvs./lines *viz.,* Bartel, Birte, Bodil, Borek, Jubilat, Karo, Meteor, Rondo, Zolty Pomorski (Furgal-Wegrzycka, 1990); KM 01, KM 02, KM 03, Solara (leaf less), Bohatyr (Ondrej, 1992) and Lu 15/92 (Ondrej, 1994) were reported to be resistant against *A. (Mycosphaerella) pinodes*. Cvs./lines K 1632, K 3055, K 5072, K 5117, K 5513, K 6391, K 7354 and K 8195 (Vladimirtseva *et al.*, 1990), Bartel, Birte, Borek, Cobeco, Finale and Paloma (Furgal-Wegrzycka, 1992) Oscar, Pony

Express and Ru/ 53 (Obradovic *et al.*, 1994) were resistant to *A. pisi*. The determination of varietal resistance against the disease was performed with pathogenicity tests, which clearly showed that the cultivar Bolero expressed a tolerant reaction and the rest of the cultivars tested were susceptible or highly susceptible.

(iii) Chemical control: Seed treatment with fungicides viz., thiram, thiabendazole (Bretag, 1985), Pulsan + oxadixyl and cymoxanil + mancozeb (Rossignol, 1988) were found effective in reducing the seed borne infection of *Ascochyta* sp. The seeds treated with cymoxanil + oxadixyl + carbendazim + thiram showed less rotting than the cymoxanil + oxadixyl treated seeds (Sanssene *et al.*, 1998). Maude (1966) developed a thiram soak technique whereby pea seed is treated with thiram (0.2%) suspension for 24 h at 30°C.

Foliar application of fungicides, such as thiabendazole (Bretag, 1985), copper oxychloride, chlorothalonil and benomyl (Warkentin *et al.*, 1996) have been found effective in reducing the severity of Ascochyta blight and increasing the yield and seed weight of pea. Single application of mancozeb at the early flowering stage was effective in reducing *Mycosphaerella* blight severity and in increasing yield (Warkentin *et al.*, 2000). EBI fungicides like Prochloraz were also found effective in reducing the disease and increasing yield (Nasir and Hoppe, 1997). Combined seed treatment with carbendazim and thiram, and two foliar sprays each with mancozeb and dinocap reduced the severity of the disease and increased yield (Singh *et al.*, 1992).

(iv) Biological control: Bradshaw-Smith *et al.* (1991) studied the effect of seed coating with antagonist *Pythium oligandrum* or fungicides (thiabendazole + fosetyl-aluminium + captan or thiabendazole + metalaxyl + thiram) both under *in vitro* and *in vivo* conditions and found that *P. oligandrum* was an aggressive parasite of pathogens under *in vitro* conditions but failed to control foot rot under field conditions while fungicidal seed treatments significantly reduced foot rot under field conditions. Seed dressing with *Trichoderma koningii* and *Gliocladium roseum* was found effective in protecting the seed from *Ascochyta pisi* (Lacicowa and Pieta, 1996). Four isolates of *Pseudomonas* were tested alone and in combinations for suppression of the disease and promotion of plant growth under field conditions. The individual isolates significantly reduced the Ascochyta blight incidence when applied as either bacterial suspension through foliar and plant debris treatments compared to the respective mixture isolates. The average mean of disease reduction with the most promising isolate 51 was 60 per cent in the foliar and 55 per cent in the plant debris treatment. In addition to disease suppression, pseudomonads promoted plant growth in terms of increased plant height and grain yield. Moreover, under *in* vitro conditions, pseudomonads were compatible with some fungicides at concentrations as high as 100 ppm.

(v) Plant extracts: Crude *Eucomis autumnalis* bulb extract prevented *M. pinodes* spore infection of the leaves under *in vivo* conditions by inhibiting the spore germination and showed no phytotoxic reaction on the leaves (Pretorius *et al.*, 2002)

5. FUSARIUM WILT

This disease is also known as wilt and root rot complex. It is an important disease prevalent in those areas where pea is cultivated in autumn. In northern plains of India, wilt is a serious problem in the early sown (October planted) pea crop and is also associated with stem fly. The crop is susceptible to two similar wilt diseases *i.e.,* true and near wilt caused by *Fusarium oxysporum* f. sp. *pisi*. Wilt affected plants are more difficult to pull from the soil than those damaged or killed by root rots. If wilt affected plants are not examined until wilt symptoms have been apparent for several days, root – rotting fungi in the soil may have so damaged the roots that it is almost impossible to identify the organism responsible.

Geographical distribution

Both wilt diseases occur worldwide wherever peas are grown more or less continuously. Near wilt is more common than true wilt because many pea varieties are available which are resistant to true wilt. Linford first described the disease in 1928 from Wisconsin (Linford, 1928). Subsequently, its occurrence was reported from Rumania, Australia, Holland, NewZealand, England, Polland, British Coloumbia, Canada and Taiwan.

In India, the disease has been reported from Poona and Bombay, Maharashtra (Bhide *et al*., 1955), Punjab (Maheshwari *et al*., 1983), Delhi (Phal and Choudhary, 1983), Madhya Pradesh (Sharma *et al*., 1989), West Bengal (Chakraborty and Chakraborty, 1989) and Himachal Pradesh (Kumar, 1993).

Losses

Losses from this disease are more than any other disease because if the plants are infected in the initial stage of growth, not even a single pod is produced. Losses due to this disease to the tune of 30–40 per cent are common (Vigier *et al*., 1983). Incidence of this disease in six districts of Madhya Pradesh ranged from 12.5 to 30.2 and 19.5 to 37.3 per cent in 1981–82 and 1982–83 crop seasons, respectively (Sharma *et al*., 1989). In Northern India, 13.9 to 95 per cent yield losses have been recorded (Maheshwari *et al*., 1983). Losses up to 60 per cent were observed in the crop during the years of epiphytotics in Himachal Pradesh (Kumar, 1993).

Symptoms

(a) True wilt : Plants are susceptible to the disease at any stage of plant growth. The first symptoms are usually a yellowing of the lower leaves and a stunting or dwarfing of plant growth. The stipules and margins of the leaflets curl downward and inward. The stems at or near the soil line may be slightly thickened and brittle. A cross section of the lower stem often reveals a lemon to orange-brown and finally black discolouration of water-conducting tissue within the stem. Affected plants may wilt and die either slowly or rapidly soon after the first symptoms are noted due to destruction of feeder roots (Plate 6). Affected plants die more rapidly in dry than in moist soils. If infection occurs when the plants are fairly young, they often wither and die without producing any pods. If infections occur late in the season, a few flattened and poorly filled pods may form. The disease generally occurs in circular to oval pattern in the field.

(b) Near wilt: Like true wilt, plants may become infected at any stage of growth. Seedlings may emerge, collapse and die. When slightly older plants are affected, a black canker develops on the underground stem. The canker enlarges both upward and downward. Finally, the entire cortex of the taproot may be affected. Above ground symptoms are quite similar to those of true wilt. However, plants affected with near wilt die more slowly, and the water conducting tissues in some plants are orange to bright brick red. The discolouration may run up the entire plant. Near wilt affects pea varieties, which are resistant to true wilt and mostly kill the plants later in the season, sometimes when the crop is nearly mature. Due to this, the loss from near wilt is less than that from true wilt. Symptoms of this phase are not usually found in the circular spots in a field as in true wilt and only scattered plants are usually affected.

The Pathogen

The fungus responsible for this disease is *Fusarium oxysporum* Schlechtend ex Fr. f. sp. *pisi* (J.C. Hall) Snyder and Hansen. In culture, four different states like septate mycelium (Plate 7a), microconidia (2.5 – 4 μm x 6 – 15 μm), macroconidia (Fig. 5A) that are typically fusiform, hyaline and multiseptate (3.5 – 5.5 μm x 25 – 22 μm) and chlamydospores (Fig. 5B) of this fungus can occur (Plate 7b). Singh *et al.* (2011) studied the effect of different media on growth and enzyme production of *F. oxysporm* f.sp. *pisi* and reported that the best medium for growth and enzyme activity production like pectinolytic (PGTE, PTE, PG, PME and macerating) and cellulolytic (Cx) is Richardís medium.

In 1928, Linford described wilt of pea in Wisconsin and gave *Fusarium orthoceras* App. & Wr. var. *pisi* as the causal agent (Linford, 1928). Under a revised classification, it was described as *F. oxysporum* Schlecht. f. sp. *pisi* (Linford) race 1 Snyd. & Hans. (Snyder and Hansen, 1940). Another race which causes near wilt was also described (Snyder, 1933 and Snyder and Walker, 1935) and was designated as race 2 in the revised classification. This race caused wilt of numerous cultivars that were resistant to race 1. Later on, race 1, 2, 3, 4, 5, 6, 7, 8, 9, 10 and 11 were described (Armstrong and Armstrong, 1974). However, 11-race classification is based more on virulence differences than on true genetic differences in the host and it was advocated that these 11 races should be grouped into either race-1 or race-2 types (Kraft and Haglund, 1978).

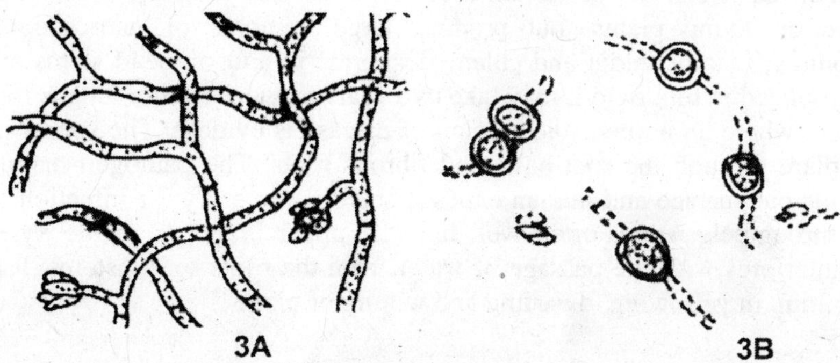

Fig. 5A. Sporodochia with micro- and macro-conidia; **Fig. 5B.** Chlamydospores

Cultural and pathogenic variability is common in the *F. oxysporum* f. sp. *pisi* races. Four isolates of *F.o.* f.sp. *pisi* were divided into two pathotyes on the basis of their growth, pigmentation, morphology of hyphae, microconidia, macroconidia, chlamydospores, sclerotia production and aggressiveness (Verma and Dohroo, 2003a). Among these, pathotype I (*FOP* I and II) resulted in rapid wilt, while pathotype II (*FOP* III and IV) caused slow wilt.

All the American and Dutch wilt resistant pea varieties to race –1 of the pathogen proved susceptible to the isolate of *F.o.* f.sp. *pisi* from Bombay and the authors proposed it as a biotype of American pathogen (Sukapure *et al.*, 1957). Existence of three physiological races of the pathogen from Maharashtra has also been reported (Sulaiman and Utikar, 1967). New Delhi strain of the pathogen is reported different from race-1 (Phal and Chaudhary, 1983).

In Himachal Pradesh, variation in morphological, cultural, pathological and molecular features of *F. oxysporum* f. sp. *pisi* isolates was studied (Gupta *et al.*, 2011) and found that mycelial colour varied from white to dull white with pinkish tinge. Size of microconidia and macroconidia varied from 3.16 × 3.16 to 9.13 × 5.44 and 11.77 × 3.16 × 24.60 × 5.91 μm, respectively. Except one isolate all isolates formed chlamydospores on PDA medium. Isolates also varied in response to carbendazim amended media at 20 mg/l. Studies on pathogenic variability revealed that the disease incidence varied from 0 to 87 per cent after 7 days of inoculation. Genetic analysis of the fingerprints revealed upto 70 per cent variation among isolates DNA banding pattern analysis of gel image. All isolates were resolved into three groups with 2, 7 and 11 isolates, respectively.

Host range

The pathogen has a wide host range, which includes *Vicia gigantia, V. faba*, lucern, sweet clover, vetch, Soyabean, garden bean, red clover, *Cyamopsis tetragonoloba*, lentil, methi and *Clianthus dampieri* (Majumdar and Raychaudhari, 1986; Dutta *et al.*, 2002).

Disease cycle

The pathogen is soil borne in nature and can survive in soil in the absence of a pea crop for 10 years or longer, as thick walled chlamydospores. The pathogen is not seed borne, but it can be carried on the seed coat. The fungus *Fusarium* does not reproduce on living plants, but produce large numbers of microscopic spores (microconidia, macroconidia and chlamydospores) in and on dead stems and roots. Once introduced in to a field it may take two years or more for the fungus to increase in numbers where an appreciable amount of disease is evident. The spores penetrate the pea plant through the root hairs and fibrous roots. The pathogen mainly enters through the cut surface and that an exposed stele is necessary for infection. It grows upward through the stem, often well into the upper branches in the xylem. This process interferes with the passage of water from the roots to the stems, leaves and pods resulting in yellowing, dwarfing and wilting of plants.

Epidemiology

The pathogen is soil borne and also survives on the seed. The spores can be carried from one field to another on farm equipments, on pea refuse and in wind or waterborne soil. They may be introduced in to new areas on the seed. The disease is more common in acidic than in alkaline soils. A soil temperature of 23 to 27°C is most favourable for true wilt and a slightly higher temperature *i.e.,* 27°C or more for near wilt. Plant mortality was maximum in November at soil temperature of 23.8°C and minimum in January at a maximum air temperature of 18.6°C (Maheshwari *et al*., 1982) whereas in Himachal Pradesh, maximum disease development occurred during the second fortnight of October when the soil temperature ranging between 21.8 to 22.5°C and soil moisture between 13.0 to 14.4 per cent (Sharma and Sharma, 2003).

Management

There is no completely satisfactory control of wilt diseases once the causal fungus is introduced and becomes prevalent in the soil. Single management method cannot reduce the magnitude of this disease. Integration of different methods can keep the disease under check.

(i) Cultural practices: Minimum incidence of the disease with maximum grain yield was obtained in the late sown crop *i.e.,* on 1st and 15th November when the soil temperatures are low for the development of wilts (Maheshwari *et al*., 1981a) while it was very severe when the plantings were done in first week of September (Sharma and Sharma, 2003). Higher yields and decreased incidence of wilt was also obtained when November (1st fortnight) sown crop was given three irrigations at presowing, flowering and pod formation stages (Sharma *et al*., 1994). Soils flooded with water develop less severe wilt and root rot than non-flooded soils (Tu, 1991). A mixture (4.4% bagasse, 8.4% rich husk, 4.25% Oyster shell powder, 8.25% urea, 1.04% potassium nitrate, 13.16% calcium superphosphate and 60.5% mineral ash) was also developed for the control of wilt diseases of pea, watermelon and radish (Huang, 1991). Use of balanced fertilizers (nitrogen, phosphorus and potash) also reduces the wilt complex in peas (Sagar and Sugha, 1998). Minimum disease incidence and apparent infection rate were observed when seeds were sown at 8 cm distance compared to other spacings (Verma and Dohroo, 2005a). Late sowing resulted in corresponding increase in Fusarium wilt, as the mid-September sown crop had the highest disease incidence (Verma and Dohroo, 2005b). The incidence of root rot-wilt disease complex and white rot were least where pea-Kuth or pea-onion rotation was followed in Zone IV (dry temperate zone), whereas in Zone II (subtemperate midhills) of Himachal Pradesh, the rice/maize-pea rotation was found effective in comparison to French bean-pea rotation (Kapoor *et al*., 2006). Among different mulches, duck weed (*Lemna minor*) mulching was found to be most effective in reducing wilt incidence and increasing the green pod yield by >150 per cent over control (Devi *et al.,* 2010).

(ii) Host resistance : The only cconomic control in wilt-infected soil is to grow such varieties of pea, which have either resistance or tolerance to wilt. Different cultivars such as Green Admiral, Horal, Prince of Wales, Senator, Bruce, First of All, Dwarf Telephone, Giant Butter, Delwiche Commando, Alderman, Bliss, Morsess Market,

Multiplied, Progress, Queen and Triumph Gruno, Italo, Green Era (Carter, 1965), Kala Nagni (Phal and Choudhary, 1983), inbred lines WSU-28, WSU-31 (Haglund and Anderson, 1987), Malaga, 496 Jarigrasom, L'govskii Zelenozernyl Paloma, Spiket (Vladimirtseva, 1990), DPR-3, DRP-4 (Kumar, 1993 and Kumar *et al.*, 2001) and Franklin (McPhee and Muehlbouer, 2002a) carrying various levels of resistance to Fusarium wilt have been reported.

In China, cv. New Season has been reported to be resistant to race 1 and 2 while it was susceptible to race 6 (Lin *et al.,* 1984). Cultivar Sure Crop derived from a cross of New Era with Laxton's Progress out yielded New Era by 20 per cent and only a trace of disease was found on it (Rich, 1987). Tiwari and Narsinghani (1986) reported that none of the cultivars was found highly resistant to disease but entries JP-501. A/2, JP-1130, JP-1140, JP Batri Brown 6, JP-179 were tolerant and Asauji and Kala Nagini were moderately resistant. Pea cvs./lines viz., DPP 54, KS 268, LP 3 and UN 53-6 as resistant to root rot and wilt complex (Kapoor *et al.*, 2006). Sanjunichi' and 'Tsurunashiakahana', among the main field pea kinds, were resistant cultivars (Mashita and Fukaya, 2006). Inheritance of wilt resistance revealed that single dominant gene is governing the resistance.

(iii) Biological control: The severity of wilt was reduced by simultaneous inoculation with *F. solani* f. sp. *pisi.* The reduction was more when the proportion of *F. solani* f.sp. *pisi* in the inoculum was more and the effect was also more when it was applied 2 or 4 days before application of *F.o.* f. sp. *pisi.* Soil infestation with *Pythium ultimum* reduced the severity of wilt caused by *F.o.* f.sp. *pisi* race 1 and 5 but soil infestation with *F. solani* f. sp. *pisi* had no effect (Kraft, 1978). Pre-inoculation application of *Gliocladium roseum* provided good control of wilt of pea (Saksirirat *et al.*, 1994). Oyarzun *et al.* (1994) observed the efficacy of non-pathogenic isolates of *F. oxysporum* in reducing disease severity and preventing the plant weight loss. Pre-colonization of sterilized soil with non-pathogenic isolate increased the antagonistic effect. Seed treatment with antagonists like *Trichoderma viride* and *T. harzianum* was found effective in the management of this disease both under glasshouse and field conditions (Verma and Dohroo, 2005).

Addition of chitin at 2.4 or 4.8 g/ 500g of inoculated soil on the day of planting pea seedlings in pots reduced the wilt index by 12 per cent and by 25 and 45 per cent, respectively, if added 3 or 8 weeks before planting (Khalifa, 1965). He further observed that when chitin was added to naturally infested soils in a small field experiment at sowing or 4 weeks earlier, the reduction was 82 and 75 per cent, respectively which he attributed to stimulation of microorganisms in the presence of chitin that antagonized or lysed the pathogen. The germination rate of chalamydospores of *Fusarium* can be reduced by raising the soil pH to 7.4 by addition of calcium cynamide, S-H mixture and lime, but not in the range of 4.0–7.0 (Lin and Lo, 1987). Seed treatment with wheat bran based *Trichoderma harzianum* formulation resulted in 57 per cent reduction of the disease under pot conditions and 44 per cent disease control under field conditions.

(iv) Chemical control: Seed treatment with Phytobacteriomycin and dusting with Trichothecin increased seed germination by 5-10 per cent, improved yield and reduced incidence of wilt (Kuzmina, 1966). Seed soaking in 1 : 1 combined suspension of captafol (0.1%) and captan (0.2%) for 30 minutes at 30ºC completely controlled the disease (Gangopadhyay and Kapoor, 1976). Seed dressing with Ceresan, Dexon, captan, carbendazim and benomyl has been found effective for the control of wilt (Lorenzini *et al.*, 1975, Utikar *et al.*, 1978 and Gupta *et al.*, 1989). Seed dressing with Benlate and Dithane M-45 and soil treatment with phorate+ captan reduced plant mortality and increased yield (Maheshwari *et al.*, 1981b). Phospho-organic insecticides are powerful cutinase inhibitors and inhibited cutinase released by *F.o.* f.sp. *pisi* structures for infection (Koller *et al.*, 1982). Bavistin, Topsin-M and Tecto applied as seed treatment were highly effective in increasing fresh and dry weight, root and shoot length, nitrogen content in pea plants and also improved seed germination and plant survival (Neweigy *et al.*, 1985). Maheshwari and Jhooty (1987) evaluated various seed dressing, seed soaking and soil drenching fungicides against mixed inoculum of *F. solani* f.sp. *pisi* and *F. o.* f.sp. *pisi* in pot experiments and found that benomyl was most effective. Pre-sowing treatment of seeds with Aldrin (0.5%) and Bavistin (0.05%) gave 95.2 per cent reduction in plant mortality due to stemfly-wilt complex in Punjab and increased yield by 10 q/ha (Singh *et al.*, 1989). Since Aldrin is banned, now chlorpyriphos (0.4%) alongwith Bavistin is recommended for the control of this complex in Punjab (Anonymous, 2003). Seed treatment with Bavistin (2.5g/kg of seed) followed by one foliar spray each of Bavistin (0.1 %) and Contaf (0.05%) have also been found effective against this disease and increased green pod yield.

Herbicides like trifluralin, ethafluralin, dinitramine and dinoseb have been reported to reduce root rot (complex of *F. oxysporum* f.sp. *pisi* and *F. solani* f.sp. *pisi*) both in glasshouse and field conditions but the yield was not increased (Johnston *et al.*, 1980).

(v) Plant extracts: Garlic (*Allium sativum*) clove extract completely inhibited the growth of *F. o.* f.sp. *pisi* under *in vitro* conditions (Verma and Dohroo, 2003).

(vi) Integrated disease management: Among the different integration levels, amendment with *Lantana camera* (10 t/ha) + bioagent Tricoguard (2.5 kg/65 kg FYM/ha) + spray with carbendazim (0.1%) at pre-flowering stage was found most effective in the management of root rot-wilt complex (Kapoor *et al.*, 2006).

6. WHITE ROT

White rot or mould is the common name of *Sclerotinia* rot, a potentially destructive disease of peas under conditions of heavy vine growth and wet weather. The disease has been reported to be sporadically severe in many areas worldwide.

Geographical distribution

Gray and Findlater (1960) first recorded this disease on pea in Great Britain and since then it has been reported from different parts of the world. Subsequently it was reported from United States, India, New Zealand, Morocco, Netherlands, Bermuda, Brazil, Argentina and Scotland. In India, the disease was first recorded in 1968 from Katrain area of Kullu valley of Himachal Pradesh (Gangopadhyay and Kapoor, 1975). Besides Himachal Pradesh, the disease has also been reported from Uttarakhand, Punjab and Kashmir (Sandhu and Dhillon, 1984; Sharma, 1985).

Losses

Gray and Findlater (1960) reported that the attack of this disease was so severe in Great Britain that one third of the plants were killed and the yield was reduced from 40 to 13 cwt./acre. In India, 20.8 to 70.3 per cent loss in yield has been reported due to this disease (Gangopadhyay and Kapoor, 1975). Due to extensive cultivation of peas during rabi season in northern plains and during summer in dry and wet temperate zones of Himachal Pradesh and Uttranchal, white rot has emerged as a serious disease of this crop, causing 5–40 per cent losses in different areas. In the terai region of north India, where winters are very cold and soil moisture is high, white rot inflicts heavy loss and in Punjab and Himachal Pradesh also, the production is adversely affected by this disease (Singh *et al.*, 2002).

Symptoms

Symptoms are seldom seen at the seedling stage. Usually, symptoms are seen at later stages of growth as water-soaked lesions, which may develop on any part of the plant. The infection may occur at any part of the foliage, but occurs mainly on the stems or branches. The maximum infection develops at the flowering stage of the crop, when petals fall on the ground. These catch infection immediately from ascospores and white fluffy mycelial growth of the fungus invades the stem and branches. At the point of infection, a dry, discoloured spot develops. Later on, fluffy, white mycelium develops and forms dense mycelial mats (Plate 8). Affected areas become slimy. Prominent dark sclerotia, irregular in size and shape are found on the surface and embedded throughout the infected tissues. On pods also fluffy white mycelial growth appears which later develops into dense mycelial mats. Rotting of the flesh takes place and in the rotting tissues, a large number of sclerotia of the fungus can be seen. The diseased tissues become whitish and may be shredded. The seeds also become shriveled and discoloured. Sclerotia appear on affected seeds also.

The Pathogen

The fungus responsible for this disease is *Sclerotinia sclerotiorum* (Lib.) de Bary. The mycelium of the fungus is hyaline, septate and branched. Hyphae are inter and intracellular and invade all the tissues of the affected host portion. The mycelium forms sclerotia, which are generally irregular and elongate and vary in size with age and environmental conditions. Apothecia emerge through pores in rind of the sclerotia when moisture is abundant. Apothecia are tan, fleshy, funnel-shaped cups on thin stalks. Ascospores (5.9–7.3 x 11.7–15.1 μm) are discharged from the apothecia in great abundance to a height of 1-15 cm.

Host range

The pathogen is polyphagus in nature and has a very wide host range of over 350 species in 60 families including plants of uneconomical importance (Gupta and Dohroo, 1993). Its host range includes *Brassica campestris* var. *sarson* and *B.c.* var. *glauca,* knol khol, lettuce, cabbage, brussel's sprout, spinach, fennel, tomato, pea, French bean, capsicum, mustard, brinjal, turnip, beans, carrot, celery. Corato and Baviello (2000) reported the occurrence of this pathogen on *Brassica carinata* and *Hibiscus cannabinus*.

Disease cycle

The pathogen perpetuates in the form of sclerotia left in soil or plant debris. The sclerotia germinate either carpogenically or myceliogenically. Sclerotia on or just below the soil surface can germinate myceliogenically and form mycelium, which invade plant issues directly at the contact points. During the extended periods of cloud and high moisture, ascospores discharged from apothecia are the more common source of infection. Ascospores either directly infect the host or they fall on fallen flowers (petals) and grow in them to produce mycelium. This mycelim can also cause infection and help in secondary spread of the disease. After infection, the fungus produces oxalic acid, which kills the cells in advance. During cool and moist weather fluffy mycelium appears on the host surface and when food exhaust, the mycelium coagulate and start forming sclerotia, which are initially white in colour but later turns black.

Epidemiology

The fungus responsible for this disease is soil borne in the form of sclerotia, which can infect the susceptible host over a wide range of temperatures *i.e.*, from 0 to 28ºC with an optimum at 20 to 25ºC. The fungus can tolerate wide pH range, but is best adapted to an acidic substrate. Ascospores start germinating within 4 h at a temperature range of 5–25ºC (optimum 20ºC). At 15–20ºC, number and dry weight of sclerotia is maximum. Prolonged spells of rain during Feburuary–March play an important role in the primary infection and spread of the disease and the incidence was positively correlated with the percentage of rainy days and total rainfall (Singh and Singh, 1991; Sharma and Munshi, 1992). Maximum disease and lesion size were recorded in March at 14ºC. Low temperature and high moisture on the plant surface favour the initiation of disease in the post flowering stages of the crop (Sharma, 1985). As the period of leaf wetness increases, infection efficiency of the pathogen also increases (Sharma and Kapoor, 1997). The sclerotia can be distributed with the seed, by farm implements or by animals and spread by irrigation water. Ascospores can be blown by the wind from plant to plant or from field to field.

Management

(i) Cultural control: In disease prone areas, cultural practices like destruction of plant debris, deep ploughing, sowing of seed free from sclerotia, crop rotation with paddy (Sharma and Gill, 1996), wider spacing and sowing of the crop late i.e. after November (Singh and Singh, 1995) should be practiced.

(ii) Host resistance: Management of this disease through resistant varieties is limited, as most of the commercially grown varieties were found susceptible (Kapoor and Singh, 1992). Various cultivars/lines *viz.,* Kelvedon Wonder, Mexique-4, EC 8633, EC 3833 (Gangopadhyay and Kapoor, 1975), Dark Skin Perfection, Perfection-132, Wisconsin Perfection, ID-89-1, ID-2 (Blanchette and Auld, 1978), HPPC-77, HPPC-95, HPPC-94, HPPC-22, DMR-3, NDP-8, Dark Skin Perfection (Kapoor and Singh, 1992), DPP-19, DPP-54, DPP-58, DPP-71, DPP-9, DPP-86, HPCC-95, P-3471, P-3496, P-3549, P-3611, P-3641 and P-3673 (Sharma and

Kapoor, 2001) were reported as resistant, while Gwaldam was moderately resistant (Sharma, 1992).

(iii) Chemical control: Since sclerotia are soil borne and their germination continues over a period of months producing ascospore showers, the fungicides may or may not come in proper contact with sclerotia, the primary unit of survival of infection. Hence their effectiveness is reduced (Sharma *et al.*, 1992). Moreover, only a small proportion of fungicides that too diluted concentration may reach the sclerotia buried in the soil, which may not be effective in rendering the sclerotia non-viable. So it is difficult to depend solely on chemical control. However, presowing application of carbendazim (5%) granules @ 10 kg/ha alone and in combination with seed treatment at 2.5 g/kg seed have been reported in controlling this disease effectively (Sugha, 1999). The disease can also be kept under check by the application of Bavistin or Benlate or captan (Sharma, 1987; Sharma and Munshi, 1992). Drenching with Bavistin was found highly effective in the management of this disease as compared to foliar sprays (Sharma *et al.*, 1999). Three sprays of triadimefon, carbendazim, thiophanate methyl given at the time of flowering, 10 and 50 per cent fruiting and furrow application of carbendazim granules (10 kg/ha) at the time of sowing were highly effective in reducing the disease severity and increasing the pod yield (Sharma *et al.*, 1999).

Farmers are using herbicides for the control of weeds in the standing crops. Sometimes the application of herbicides also plays important role in the decomposition of overwintering structures of soil borne pathogens. Herbicides like Alachlor and Basalin caused hundred per cent germination inhibition of sclerotia of *S. sclerotiorum* myceliogenically (Sharma and Kapoor, 1999).

(iv) Biological control: Since chemical control of the disease is inadequate, biological control can be an alternative. Parasitization of mycelium and sclerotia of *S. sclerotiorum* by *Trichoderma harzianum* and the destruction of latter within 15 days have been reported (Sharma and Singh, 1990). Application of wheat bran culture of *T. harzianum* under field conditions gave significant control of the disease plus a yield increase. Mycelial preparations were more effective than spore preparations (Sharma and Singh, 1990; Singh, 1991). Effectiveness of Funginil (*T. viride* formulation) @ 4 g/kg seed was also found effective in reducing the disease severity to 25.83 per cent (Sharma *et al.*, 1999). Reduction in the carpogenic germination of sclerotia by soil treatment with mycoparasites, such as *Coniothyrium minitans* and *Talaromyces flavus* under the canopy of peas has been reported (Huang *et al.*, 2000). Application of *T. harzianum* as sodium alginate pellets resulted in 58 per cent reduction in the disease under pothouse conditions (Kapoor and Kumar, 2004). *Trichoderma atroviridae* when applied as seed treatment was reported equally effective as Bavistin spray (0.1%) at flowering stage in management of the disease (Kaur *et al.*, 2006). Out of 17 isolates, only three isolates (S-1, S-6 and S-12) of *T. harzianum*, exhibited significantly higher sclerotial parasitization and hyphal inhibition activity. These isolates reduced the viability of *S. sclerotiorum* sclerotia to <25 after two weeks of incubation and also exhibited >58 inhibition in the hyphal growth of the pathogen. Under glasshouse conditions also the isolates exhibited significant reduction in post emergence seedling rot and white rot of garden pea as compared to control (Deeksha *et al.*, 2008).

Bacillus cereus strain alf-87A inhibited the vegetative growth and ascospore germination of the pathogen by diffusible metabolites under *in vitro* conditions, while *in vivo* reduced the basal pod rot infection by 39–55 per cent from air borne ascospores when sprayed on pea plants at pod development stage (Huang *et al.*, 1993). Out of 8 organic substrates evaluated for mass multiplication of bioagents, maximum multiplication of bioagents was found in FYM, followed by *Lantana camara* and wheat bran. Population dynamics studies of *T. viride* revealed significant increase in population even after 60 days of application of bioagent (Kapil and Kapoor, 2005).

Kaur *et al.* (2009) tested nine isolates of *Trichoderma* sp., four of *Coniothyrium minitans*, thirty of fluorescent pseudomonads and ten of *Bacillus subtilis* for their bio-efficiency against *Sclerotinia sclerotiorum* following dual culture and poisoned food techniques and reported that three isolates of *T. atroviride* (T1, T2 and T38) inhibited the colony growth and sclerotia formation of *S. sclerotiorum,* two isolates of *C. minitans* (CmCo and Cm20) inhibited pathogen colony growth up to 79.22 and 77.0 per cent, respectively at 50 per cent culture filtrate concentration and Fp6 isolate of *Pseudomonas putida* biotype B and Bs1 isolate of *B. subtilis* inhibited colony growth up to 50.67 and 67.71 per cent, respectively at 50 per cent culture filtrate concentration. All these antagonists reduced the disease incidence under field conditions, but fungal antagonists showed better biocontrol potentiality as compared to bacterial antagonists.

(v) Plant extracts: Application of plant extracts like *Lantana camara*, *Gaussurea lappa*, garlic and neem extract has also been found effective in inhibiting the myceliogenic germination of sclerotia (Sharma and Kapoor, 1999 and Sharma *et al.*, 1999). Six neem-based biopesticides were evaluated at two concentrations (0.5 and 0.1%) against *S. sclerotiorum*. Results revealed that Wanis was the most effective at both concentrations, followed by Neemgold (Kapil and Kapoor, 2005).

(vi) Integrated disease management: Sclerotial viability was reduced when they were dipped for I h in *T. harzianum,* triadimefon and *Azadirachta indica* before incubation at 22°C (Sharma and Basandrai, 1997) which indicated the enhanced efficacy when fungicide, biocontrol agents or plant extracts were used in combinations. Soil application of wheat bran culture of *T. harzianum* @ 120 kg/ha at the time of sowing plus wide row spacing (50–60 cm) plus foliar sprays of Bavistin at flowering stage has been found effective in controlling the disease (Singh *et al.*, 2002). Singh and Banyal (2007) suggested the integrated management practices for best control of white rot by following seed treatment with Bavistin (2.5g/kg seed), soil application of *Trichoderma viride* (2.5kg/62kg FYM/ha) and one foliar spray each of Bavistin (0.1%) and Contaf (0.05%), respectively.

(B) BACTERIAL DISEASES

7. BACTERIAL BLIGHT

It is an important disease of pea in areas where cool and moist climate is prevalent during the growing season.

Geographical distribution

Bacterial blight was first reported from Colorado, United States in 1915 (Sackett, 1916) and subsequently from other parts of USA viz., Monatana, South Carolina and Arizona. Later on the disease was also reported from many other countries *viz.*, Bulgaria, Hungry, Japan, Australia, Greece, Italy, New Zealand, Isreal and Pakistan.

In India, the first report of the disease was made in 1972 at Udaipur, Rajasthan (Chakravarti and Hegde, 1972). Severe outbreak of this disease was observed during 1986–1987 crop season in Kullu valley of Himachal Pradesh (Sood, 1988). Subsequently, Jindal and Bhardwaj (1989) also noticed the disease in Solan, Kullu, Mandi and Sirmaur districts of H.P.

Losses

Under favourable environmental conditions, the disease can cause significant losses. If the crop is attacked early *i.e.,* at seedling stage, the whole crop may be lost. At later stage, defoliation, blasting of blossoms and pods can lower the yield, quality and value of the crop (Kapoor and Singh, 1999). It is also known to cause as much as 70 per cent yield loss in pea (Forbes and Bretag, 1991).

Symptoms

The bacterium produces disease symptoms on all the above ground plant parts. The initial symptoms on leaflets and stipules are manifested as small lesions having water-soaked appearance, usually developing into large, irregularly shaped areas measuring 1–1.5 cm in diameter. Lesions have an angular shape but may have a distinct fan like appearance if infection occurs where a stipule joins the stem or a leaflet joins the midrib (Plate 9). On stem, lesions develop as dark brown, elliptical areas that may encircle the stem and extend upto several centimeters invading the stipules and leaflets to give the distinctive fan like symptoms. Under favourable environmental conditions, the lesions may exude a cream coloured ooze of the bacteria. Infection of stem may kill the plant completely or it may result in development of lateral stems, which continue growing apparently unaffected. On pods, lesions develop as dark green, circular or irregularly shaped areas, 1–1.5 cm in diameter, having water soaked appearance which later on become dark brown and shiny. In some pods, infected areas may show olive coloured water-soaked spots. The flower buds are often killed before they can open. Infection from sepals spreads to the peduncle and then taken to the pods, killing the flowers and causing shrivelling of pods.

The Pathogen

The disease is caused by *Pseudomonas syringae* pv. *pisi* (Sackett) Young, Dye & Wilkie (Sackett, 1916; Young and Dye, 1970; Boelema, 1967; Hunter and Cigna, 1981). However, it has also been shown that *P.s.* pv. *syringae* Van Hall can cause a disease of pea similar to *P.s. pisi* (Hoitink and Hagedorn, 1966; Jindal and Bhardwaj, 1989). Both are gram negative, non-spore forming, and motile rods with one or more polar flagella. However, the two can be distinguished from each other by biochemical, phage, serological and pathogenicity tests.

Different isolates of *P.s. pisi* are known to consist of distinct races, which attack certain cultivars of their host, but not others. Races of the bacterium on pea were first distinguished by the differential reactions of the pea cultivars, Early Onward and Patridge (Taylor, 1972). One hundred and forty six isolates of *P.s. pisi* from UK and overseas were categorized into six races using a set of differentials (Taylor *et al.*, 1989). *P.s.* pv. *pisi* consists of seven identified races. One of these (Race 6) lacks all avirulence genes and is common around the world and in Australia. Globally, Race 2 and Race 6 predominate; however, in Australia, Race 3 predominates due to the widespread cultivation of cultivars susceptible to Race 3, but resistant to Race 2. (Hollaway *et al.*, 2007).

Host range

Alternate hosts of the pathogen include *Lathyrus odoratus, L. latifolius, Vigna* sp., *Dolichos lablab, Vicia atropurpurea, V. bengalensis, V. oillosa, Trifolium prantense, Glycine max, Phaseolus vulgaris, Pisum sativum* var. *arvense* and *Lablab niger* (Tsiolosani and Barbakadze, 1976; Gnanamanickan and Kandaswamy, 1978 and Hagedorn, 1984).

Disease cycle

Neither soil nor infected plant parts are primary sources of infection in any of the pea growing areas of the USA (Hagedorn, 1984). The pathogen overwinters in the infected seed. The seed carries the bacterium both externally and internally. The bacterium can persist in the infected seed for three years (Lawyer, 1984). It colonizes the intercellular and intracellular spaces of the seed coat, but does not penetrate the embryo or cotyledons. Even a very low level of seed infection can cause economic loss, since the disease can spread fast from primary infection foci. The pathogen is carried in irrigation water, splashed by rain or blown in wet winds to other plants and field infection usually occurs through stomata and wounds.

Epidemiology

The bacterium is seed borne in nature. The disease can be transmitted from one lot to another and from one seed to another by threshing equipment or through treatment of seed by flowable fungicides (Kapoor and Singh, 1999). If soil is very wet at the time of emergence, it also favours blight transmission from the infected seed to resulting plant. Even a very low level of seed infection can cause economic loss since the disease can spread swiftly over fields from a single centre of infection. Secondary infection occurs readily because hard rains, hail, wind and sand damage and recurrent frost favour attack and spread of blight. People, insects and birds contribute to dissemination of secondary inoculum. The optimum temperature for this disease is 27.7ºC while minimum is 7.2ºC and maximum 37.7ºC. Frost is an important factor affecting disease development. In Otara (New Zealand), the bacterial blight was affected by alternating periods of frost and warm rain (Watson and Dye, 1971).

Soils with 80 per cent of water holding capacity favoured disease than those with lower percentage *i.e.*, 40 or 60 per cent. If moisture at the time of emergence is limited, blight may not occur and is transmitted to the resulting plants in smaller proportions (Hagedorn, 1984). Transmission of *P.s.* pv. *pisi* was reduced at low soil water extent (Hollaway *et al.*, 1996).

Detection

Often the percentage of seeds infected with bacterium in a sample is so low that detection becomes extremely difficult. At the same time seed borne phytopathogenic bacteria cannot be detected through the standard methods used for the detection and identification of seed borne fungi. One of the simplest techniques for detecting bacterial pathogens in seed is the visual observation. Funiculus of pea seeds harvested from pods inoculated with *P.s.pisi* was deep green in colour and water soaked in appearance. Such seeds were enveloped in bacterial slime and in many instances the seeds showed green water soaked spots only near one or the other end of the hilum.

Serological (Lyon *et al.*, 1995; Blanka *et al.*, 1999 and Mollenbruck and Sander, 1991) and molecular (Bavage *et al.*, 1991; Cournoyer *et al.*, 1993 and Fraaije *et al.*, 1993) detection and characterization of the bacterial blight pathogen have also been reported.

Management

(i) Cultural practices: General cultural practices like proper drainage, wider spacing, weed eradication and restricted irrigation, all reduce humidity and therefore, help in reducing the disease intensity. The use of clean, certified, disease free seed produced in arid areas (Smith and Close, 1966; Young and Dye, 1970; Boelema, 1967 and Hagedorn, 1984) is the most effective method for the control of bacterial blight of pea. Timely sowing and avoidance of frost-pockets (Boelama, 1967), seed production in arid areas, avoiding sprinkler irrigation and using clean equipments during harvesting and post-harvest operations could minimize seed infection (Hagedorn, 1984). The incidence of the disease was affected by sowing date, but not by frost protection (Bhardwaj *et al.*, 1992). They recorded highest incidence on the earliest sown crop and lowest incidence on the late sown crop.

(ii) Host resistance: Very limited success has been achieved on this aspect. Few pea cultivars like Pitica Galbens, Minunea Americii, Victoria Baragan, ICAR 53/54 and Victoria Strube Partridge-73 (Maple) have been reported resistant to the disease (Docea *et al.*, 1954, Crampton and Goulden, 1974). Taylor *et al.* (1982) reported 6 wide groups of pea that had resistance factors in them, while eleven such resistant genotypes were also identified (Taylor *et al.*, 1985). Gerlagh and Vos (1987) reported that dark seeded varieties did not exhibit a hypersensitive reaction. Kokoekova *et al.* (1998) and Blanka *et al.* (1999) reported that SG-L-7 and SG-L-38 showed high susceptibility to some races of the bacterium. Resistance to Race 6 within *P. sativum* has not been found, but attempts are being made to incorporate a race non-specific resistance identified from *P. abyssinicum* into field pea (Hollaway *et al.*, 2007).

(iii) Seed treatment: Heat treatment is generally used to eliminate pathogens from seed particularly when the pathogen is out of reach of non-systemic chemicals. Hot water seed treatment has also been recommended for the control of this disease (Boethinger and Bowers, 1975). Grondeau *et al.* (1992) studied the effect of dry heat in oven and soaking in hot water on peas contaminated with *P. syringae* pv. *pisi,* and reported that germination of seed was more affected by soaking than by dry heat. Seed germination was 5 to 20 per cent less than in control when seed was treated with dry heat for 72 h at 65ºC and was 20 to 45 per cent less than control when seed was soaked for 15 minutes at 60ºC.

Use of surface sterilizing agents, such as one per cent of sodium hypochlorite has also been found to reduce the primary seed borne infection of *P. syringae* pv. *pisi* by 85–90 per cent without adversely affecting the seed germination (Lawyer, 1984). Streptomycin slurry treatment (2.5g a.i./kg of seed) has also been found equally effective for the control of bacterium. However, soaking seeds in streptomycin solution (0.25%) for 2 h gave a higher degree of control than the slurry treatment (Taylor and Dye, 1976). Similar control of seed borne infection was obtained by soaking seed in streptocycline solution (250mg/ml) for 2h. It reduced the disease incidence by 95 per cent and did not reduce the seed germination (Jindal, 1990). None of these treatments mentioned above, however, could eradicate the bacterium completely from the seeds.

(iv) Chemical control: Spray application of streptocycline (100 μg/ml) alone or in combination with copper oxychloride (0.2%) at 7-day interval was significantly superior in reducing the disease severity and also increased the yields (Jindal, 1990). Reduction in disease upto 62 per cent by weekly sprays of streptomycin has also been reported (Forbes and Bretag, 1991).

(v) Integrated disease management: Bacterial blight can be successfully controlled by using an integrated disease management strategy incorporating crop rotation, pathogen-free seed, avoidance of planting in areas prone to frequent frosts or extreme wet weather, crop hygiene and avoiding early sowing. Seed treatment and application of foliar bactericides have limited use in control of this disease (Hollaway *et al.,* 2007).

(C) VIRAL DISEASES

1. PEA ENATION MOSAIC

It's an important disease of this crop and in favourable weather conditions can cause severe yield losses.

Geographical distribution

Pea enation mosaic virus was first described on faba bean in USA in 1935 (Osborne, 1935). Subsequently, it was reported from Germany, Holland, UK, Iran, Sicely, South Africa and India.

Losses

Losses from this disease can be considerable in epidemic years. Kumari *et al.* (2001) reported that this disease was wide spread in Syria and caused 16 to 50 per cent yield losses in lentil and peas.

Symptoms

Initial symptoms on the leaves and stipules of susceptible varieties include typical malformation and plants are usually severely stunted and distorted. Later, infection cause chlorotic, translucent flecks, small brown necrotic lesions and vein clearing on the foliage. Small outgrowths or enations are often formed on the underside of leaves, but this depends on the susceptibility of the variety. Young susceptible plants may be killed if the infection occurs before flowering. Leaves gradually curl and in some cases develop enations near the veins on the abaxial surface. Plants become "bushy" and have shortened internodes. Infection prior to pod formation results in pods that are distorted and stunted (Tornos *et al.*, 2008).

The Virus

The virus responsible for this disease is *Pea enation mosaic virus* (PEMV). It is a unique virus because it is associated with two genera *Enamovirus* and *Umbravirus* based on the two distinct RNAs in its genome, and is referred to as *Pea enation mosaic virus-1* and *Pea enation mosaic virus-2*, respectively (Demler *et al.*, 1996). Virus consists of two types of 28 nm isometric particles and is distinguished by particle density and resistance to degradation by phosphotungstate stain, while biologically and epidemiologically, PEMV mostly resemble with *luteovirus*.

Host range

The virus *Pea enation mosaic virus* has a wide host range but restricted to leguminosae family. The species of *Cicer, Lathyrus, Lens, Lupinus, Medicago, Melilotus, Phaseolus, Pisum, Trifolium* and *Vicia* are infected by this virus in nature (Peters, 1982). *Glycyrrhiza glabra* and *Trifolium alexandrinum* as the new hosts of this virus in Iran (Farzadfar and Izadpanah, 2001).

Disease cycle

PMEV has a wide host range, but overwintering perennial hosts are not well documented, although clover and vetch are suspected. Earlier alfaalfa was considered as its perennial host and inoculum source for this virus (McWhorter and Cook, 1958), but it could not be detected by ELISA or dot blot hybridization in alfalfa samples (Larsen *et al.*, 1996). PMEV is commonly and efficiently transmitted by the pea aphid (*Acyrthosiphon pisum*) in a persistent manner, although several other aphid species have been reported to transmit the virus (Demler and de Zoeten, 1994).

Epidemiology

The PMEV overwinter in leguminous crops or weeds especially vetch and clover species. The pea aphid commonly over winter as eggs or adults on winter annual or perennial legumes and the flights of aphids from PMEV infected plants spread the virus to surrounding plants. Toros *et al.* (1978) reported that when pea aphid nymphs were given acquisition access period of 3 h, the virus had a latent period of 10 h. Following the latent period, the aphid was able to transmit the virus during brief probes into the cytoplasmic tissues.

Management

The use of resistant or tolerant pea cultivars is the most effective method in the management of losses caused by this virus. Pea cultivars like Franklin and Lifter (McPhee and Muehlbauer, 2002 a,b) have been reported to be resistant to this disease. Only fodder cultivars 'Wiato' and 'Zagoba' were significantly less infected by viruses (Marcinkowska, 2007). Srbija a late afila type was found tolerant to pea enation mosaic virus (Orevic *et al.*, 2004). Resistance is governed by dominanat gene *En* and a recessive gene *en/en*. Control of pea aphid by application of insecticides also reduces the incidence of the disease.

2. PEA SEED BORNE MOSAIC

Among pea seed borne viruses, the pea seed borne mosaic virus (PSbMV) is known to be transmitted through seed significantly. This disease assumes special importance because it is mostly in the form of seeds and the germplasm collections are conserved and exchanged internationally. PSbMV is economically important pathogen with worldwide distribution that causes significant losses in pea yield and reduces seed and produce quality (Frew *et al.*, 2002).

Geographical distribution

The virus was first discovered in Czechoslovakia in 1966 (Musil, 1966). Subsequently, it was reported from Japan, USA, Canada, Yugoslavia, Germany, Poland, Switzerland, New Zealand, United Kingdom, India, Morocco and France (Maury and Khetarpal, 1992).

Losses

Losses due to this disease are enormous because the pathogen is seed borne and the seeds are exchanged oftenly. The virus was demonstrated to be seed transmissible (Stevenson and Hagedorn, 1969). Hampton and Braverman (1979) reported that 23 per cent of the 1835 PI lines, originating primarily from India, were infected with PSbMV.

Symptoms

The initial symptoms of the disease in susceptible varieties include leaf roll, mosaic, leaf and stem malformation and stunting that can be severe depending on the cultivar/line and prevailing environmental conditions. Leaflets appear narrow and shortened, and start rolling downward. At later stages of infection plant shows apical malformation in the form of resetting and tendril curling. Symptomless infections of peas are common. Pods on infected plants are frequently distorted or misshapen. Symptoms may also develop on seedcoats in the form of cracks.

The Virus

Pea seed borne mosaic virus (PSbMV) is responsible for this disease and it belongs to the *Potyvirus* in the family *Potyviridae*. The virus consists of 780 nm flexuous particles with a ss-RNA genome. PSbMV also exists as several strains or pathovars. Strains P-1 and P-4 infecting pea, and the lentil strain L, which produces more severe symptoms in pea than do P-1 and P-4 have been reported (Alconero *et al.*, 1986; Safarova *et al.*, 2008).

Host range

Pea is considered to be the natural host of this virus. PSbMV has wide host range and infects 47 plants in 12 families (Aapola *et al.*, 1974). The virus is seed borne in pea, lentil and faba bean.

Disease cycle

The virus is seed borne in pea and lentil. PSbMV is transmitted in a non-persistent manner primarily by aphid vectors like *Acyrthosiphon pisum*, *Myzus persicae* and *Aphis craccivora*. However, 18 other aphid species have been reported to transmit the virus (Khetarpal and Maury, 1987). The aphid can acquire the virus with in five minutes. The aphid acquires the virus from nearby infected field and transmits in healthy fields. Seed transmission results from direct invasion of immature embryos where the virus multiplies in the embryonic tissues and persists during seed maturation (Wang and Maule, 1992).

Epidemiology

PSbMV overwinters in the infected seeds of pea and lentil. In cultivars Assam local, Pant P-5 and KPMR-400, the seed transmission percentage was nil, but the presence of virus was recorded by performing ELISA of same seedlings, which could be explained by means of latent infection virus (Anand *et al.*, 2006). Under field conditions, the disease spread by aphids from the infected seed grown plants to the neighbouring fields. Optimum temperature for the disease is 28–32ºC.

Management

Seed certification programmes can be helpful in reducing the incidence of this disease. Growing resistant varieties will be helpful in the effective management of this disease. Out of eleven cultivars, none of the cultivars was found to be immune to the virus. However, NDVP 250 and VRPMR 9 were tolerant to the virus (Sontakke *et al.*, 2008). In field experiments, of 39 genotypes of field pea (*Pisum sativum*) evaluated, 15 were ranked as highly susceptible, 10 susceptible, 9 moderately resistant, and 5 resistant (Coutts *et al.*, 2008). Application of insecticides for the control of vectors can be another approach.

REFERENCES

Aapola, A.A., Kenesek, J.E. and Mink, G.I. 1974. The influence of inoculation procedure on the host range of pea seed-borne mosaic virus. *Phytopathology* **64:** 1003–1006.

Abusaleha, Pal, A.B. and Vani, A. 1989. Inheritance of rust resistance in peas. *Indian J. Hortic.* **46:** 84–86.

Abusleha and Pal, A.B. 1990. Linkage relationship between rust resistance and seed character in peas. *Indian J. Hortic.* **47(2):** 220–223.

Alconero, R., Provvidenti, R. and Gonsalves, D. 1986. Three pea seed borne mosaic virus pathotypes from pea and lentil germplasm. *Plant Dis.* 70: 783–786.

Amelina, K.V. and Chaplygina, I.N. 1979. The resisitance of pea to Ascochytosis. *Referativnyl Zhurnal* **7:** 256

Amin, K.S. and Ramachander, P.R. 1989. Effect of Saprol on epidemic development of rust and powdery mildew on pea and their influence on yield. *Indian J. Pulse Res.* **2(2):** 140–146.

Anand D., Deepika Rohatgi and Pankaj 2006. Seed–transmission of seed–borne mosaic virus in *Pisum sativum. Ann. Pl. Protec. Sci.* **14(1):** 252–253.

Anonymous, 1999. Variety 'Mukta'. *Pl. Var. J.* **12:** 37–38.

Anonymous, 2003. *Package of Practices for Vegetable Crops.* Punjab Agriculture University, Ludhiana.

Armstrong, G.M. and Armstrong, J.K. 1974. Races of *Fusarium oxysporum* f. sp. *pisi,* causal agents of wilt of pea. *Phytopathology* **64:** 849–857.

Attanayake, R.N., Glawe, D.A., McPhee, K.E., Dugan, F.M. and Chen, W. 2010. *Erysiphe trifolii* – a newly recognized powdery mildew pathogen of pea. *Plant Pathol.* **59(4):** 712–720.

Azmat, M.A., Nawab, N.N., Shahid Niaz., Abdul, R., Khalid M., Khan, A.A. and Khan, S.H. 2010. Single recessive gene controls powdery mildew resistance in pea. *Int. J. Veg. Sci.* **16(3):** 278–286.

Bains, S.S. and Dhaliwal, H.S. 1986. Downy mildew of peas in Punjab. *Indian Phytopath.* **39(3):** 493–494.

Bajan, C. 1968. Investigation on pathogenecity and biology of three species of *Ascochyta* found on pea. *Acta Agrobot.* **21:**17–24

Bakalova, V.V. 1979. Resistance of pea species to *Ascochyta* diseases. *Sel. Semenorodstvo* **2:**29–30

Bal, R.S. and Kumar, A. 2012. Influence of weather parameters on pea rust development and its effect on yield. *Pl. Dis. Res.* **27 (2):** 162–164.

Banyal, D.K. and Tyagi, P.D. 1997. Role of climatic factors in the development pf powdery mildew of pea. *J. Mycol. Pl. Pathol.* **27(1):** 64–66.

Banyal, D.K., Singh, A., Upmanyu, S., Chaudhary, Jaya and Sharma, P.N. 2014. Diversity analysis of *Erysiphe pisi* populations causing pea powdery mildew in Himachal Pradesh. *Indian Phytopath.* **67(3):** 263–267.

Bavage, A.D., Vivian, A., Artherton, G.T., Taylor, J.D. and Malik, A.W. 1991. Molecular genetics of *Pseudomonas syringae* pv. *pisi* plasmid involvement in cultivar specific incompatibility. *J. Gen. Microbiol.* **137:** 2231–2239.

Bedlan, G. 1985. Pea leaf and pod spot. *Pflanzenschutz.* **147:** 14–15.

Benda, J. 1967. The dependence on the pH of the host tissue for the production of uredia and telia in *Uromyces pisi* (Pers.) de Bary. *Ceska Mycol.* **21(2)** : 90–91.

Bharadwaj, S.S. and Sharma, S.L. 1984. Field evaluation of mildewcides against powdery mildew (*Erysiphe polygoni*) of peas (*Pisum sativum*) in Himachal Pradesh. *Indian J. Mycol. Pl. Pathol.* **14:** 103–104.

Bhardwaj, C.L., Sagar, V. 1998. A comparison of macro and micro forms of *Phoma medicaginis* var. *pinodella* infecting peas. *J. Mycol. Pl. Pathol.* **28:** 161–163.

Bhardwaj, C.L. and Sagar, V. 1996. Symptomatological, morphological, cultural and pathogenicity characters of *Mycosphaerella pinodes* from Himachal Pradesh. *Indian Phytopath.* **49:** 176–180.

Bhardwaj, C.L. and Singh, B.M. 1984. Host range of *Oidium* state of *Erysiphe pisi* on some leguminous hosts of Kangra valley of Himachal Pradesh. *Indian Phytopath.* **37:** 732–733.

Bhardwaj, C.L. and Thakur D.R. 2000. Dry grain yield in rainfed garden pea (*Pisum sativum*) in relation to environment and management of *Mycosphaerella* blight. *Indian J. Agric. Sci.* **70:** 231–233.

Bhardwaj, C.L., Sharma, B.K. and Jamwal, R.S. 1992. Effect of date of sowing and frost protection on the incidence of bacterial blight of pea. *Indian Phytopath.* **43:** 221–222.

Bhide, V.P., Hegde, R.K. and Sukapure, R.S. 1955. *Fusarium* wilts of watermelon and peas in Bombay State. *Curr. Sci.* **24(9):** 311.

Blanchette, B.L. and Auld, D.L. 1978. Screening field peas for resistance to white mold. *Crop Sci.* **18(6):** 28–30.

Blanka, K., Iveta, P. and Vaclav, K. 1999. Bacterial blight of pea–detection of pathogen and resistance of varieties. In: *New aspects of resistance research on cultivated plants, bacterial diseases. Proc. Int. Symp. Nov. 18–19, 1998.*

Boelema, B.H. 1967. Bacterial blight (*Pseudomonas pisi*) of peas, 1953–1966. *Agric. Res.* **131:** 3.

Boettinger, C.F. and Bowers, J.L. 1975. Hot water treatment a possible control for seed–borne bacterial blight in southern peas. *Hortic. Sci.* **10:** 48.

Bowen, J.K., Lewis, B.G. and Matthews, P. 1997. Discovery of the teleomorph of *Phoma medicaginis* var. *pinodella* in culture. *Mycolo. Res.* **101:** 80–84.

Bradshaw–Smith, R.P., Craig, G.D. and Biddle, A.J. 1991. Glasshouse and field studies using *Pythium oligandrum* to control fungal foot rot pathogens of peas. *Aspects App. Biol.* **27:** 347–350.

Bretag, T.W. 1985. Chemical control of Ascochyta blight of field peas. *Aust. Pl. Pathol.* **14(2):** 42–43.

Bretag, T.W., Kearne, P.J. and Price, T.V. 1995. Effect of Ascochyta blight on the grain yield of field peas (*Pisum sativum* L.) grown in Southern Australia. *Aust. J. Expt. Agric.* **35:** 531–536.

Buchheim, A. 1922. On the biology of *Uromyces pisi* (Pers.) Winter. preliminary note. *Centralbl. fur Bakt. Ab.* **2(21–24)** : 507–508.

Bugiani, R. and Govoni, P. 1996. Incidence of attacks of downy mildew on pea and control strategies. *Informatore Fitopatol.* **46:** 46–49.

Butler, E.J. 1918. *Fungi and Disease in Plants.* Thacker Spink & Co., Calcutta, 547pp.

Carter, M.V. 1965. Plant Pathology. *Rep. Waiteagric. Res. Inst. 1964 –65.* pp. 57–68.

Chakraborty, U. and Chakraborty, B.N. 1989. Interaction of *Rhizobium leguminosarum* and *Fusarium solani* f. sp. *pisi* on pea affecting disease development and phytoalexin production. *Can. J. Bot.* **67(6):** 698–701.

Chakravarti, B.P. and Hegde, S.V. 1972. Outbreaks and new records. India. Bacterial blight of pea caused by *Pseudomonas pisi* Sackett. *FAO Pl. Prot. Bull.* **20:** 21–22.

Chauhan, R.S. and Singh, B.M. 1994. Effect of different durations of leaf wetness on pea rust development. *Pl. Dis. Res.* **9(2):** 200–201.

Chauhan, R.S. and Singh, B.M. 1995. Morphological variation in teliospores of *uromyces vicia fabae* Schroet collected from different hosts. *Pl. Dis. Res.* **10(1):** 55–58.

Chilvers, M.I., Rogers, J.D., Dugan, F.M., Stewart, J.E., Chen, W.D. and Peever, T.L. 2009. *Didymella pisi* sp. nov., the teleomorph of *Ascochyta pisi. Fungal Biol.* **113(3):** 391–400.

Chupp, C. and Sherf, A.F. 1960. *Vegetable diseases and their control.* The Ronald Press Co., New York, 693pp.

Chupp, C. and Sherf, A. F. 1960. *Vegetable diseases and their control.* John Wiley and Sons, Inc., New York, 693pp.

Clulow, S.A., Lewis, B.G. and Matthews, P.A. 1991. Pathotype classification for *Mycosphaerella pinodes*. *J. Phytopathol.* **131:** 322–332.

Corato, U. de and Baviello, G. 2000. Occurrence of *Sclerotinia sclerotiorum* on *Brassica carinata* in Southern Italy (Brasilicata). *Informatore Fitopatol.* **50(5):** 61–63.

Corbiere, R., Gelie, B., Molinero,V., Spire, D. and Agarwal, V.K. 1994. Investigations on seedborne nature of *Mycosphaerella pinodes* in pea seeds. *Seed Res.* **22:** 26–30.

Cournoyer, B., Sharp, J.D., Astuto, A., Gibbon, M.J., Taylor, J.D. and Vivian, A. 1993. Molecular characterization of the *Pseudomonas syringae* pv. *pisi* plasmid-borne avirulence gene avr Pp1B which matches the P3 resistance locus on pea. *Mol. Pl. Microbe Interac.* **8:** 700–708.

Coutts, B.A., Prince, R.T. and Jones, R.A.C. 2008. Further studies on Pea seed–borne mosaic virus in cool–season crop legumes: responses to infection and seed quality defects. *Aust. J. Agric. Res.* **59(12):** 1130–1145.

Crampton, M.J. and Goulden, D.S. 1974. New peas resistant to wilt. *N. Z. J. Agric. Res.* **128:** 12–13.

Cruz, B., Chianglok, M.L. and Castanedaruiz, R.F. 1990. *Psyllobora nana* (Coleoptera, Coccinellidae), biological control agent. *Ciencias de la Agric.* **40:** 168.

Davidson, J.A. and Ramsey, M.D. 2000. Pea yield decline syndrome in south Australia: the role of diseases and the impact of agronomic practices. *Aust. J. Agric. Res.* **51:** 347–354.

Deeksha Joshi, Hooda, K.S. and Bhatt, J.C. 2008. Biocontrol potential of fungal antagonists from western Himalayan region against sclerotial plant pathogens. *Pestic. Res. J.* **20(1):** 1–5.

Demler, S.A. and de Zoeten, G.A. 1994. Pea enation mosaic virus. In: "*Encyclopedia of Virology*" (Eds. Webster, R.G. and Granoff, A.). Academic Press, San Diego, USA. Pp. 1083–1089.

Demler, S.A., de Zoeten, G.A., Adam, G. and Harris, K.F. 1996. In: "*The Plant Viruses, Vol. 5, Polyhedral Virions and Bipartite RNA Genomes*", (Eds. Harrison, B. D. and Murant, A. F.) Plenum Press, New York, USA, pp. 303–304.

Devi, P.S., Singh, I.M., Elizabeth, M.K. and Gopimohan, N. 2010. Effect of mulching on wilt incidence and yield of pea (*Pisum sativum* L.). *Indian Phytopath.* **63(3):** 348–349.

Docea, E., Radulescu, E., Vulpe, O., Gheorgies, C. and Fratila, E. 1954. Observations of the behaviour of some varieties of pea with respect to attack of principle diseases. *Lucr. Ser. Stunt. Inst. Agron. Nicolae Balcescu Ser. A.* **7:** 275–286.

Dorozhkin, N.A. and Kremneva, A.M. 1976. Effect of temperature and air humidity on the growth and the development of the pathogen of Ascochytosis of pea. *Botanika. Issled., Minsk* **18:**181–184.

Dutta, S., Chakraborty, M. R. and Chatterjee, N.C. 2002. New report of Fusarial wilt of glory pea. *Indian Phytopath.* **53(1):** 117.

El Marzoky, H.A. 2008. Pathological studies on downy mildew of pea caused by *Peronospora viciae* (Berk.) Casp. f. sp. *pisi* (Sydow) in Egypt. *Ann. Agric. Sci. Cairo* **53(2):** 385–394.

Falloon, R.E. and Sutherland, P.W. 1996. *Peronospora viciae* on *Pisum sativum*: morpho-logy of asexual and sexual reproductive structures. *Mycologia* **88(3):** 473–483.

Falloon, R.E., Follas, G.B., Butler, R.C. and Goulden, D.S. 2000. Resistance in *Peronospora viciae* to phenylamide fungicides: reduced efficacy of seed treatments of pea (*Pisum sativum*) and assessment of alternatives. *Crop Prot.* **19:** 313–325.

Farzadfar, S. and Izadpanah, K. 2001. Sources and properties of Iranian isolate of Pea enation mosaic virus. *Iranian J. Pl. Pathol.* **37:** 3–4.

Fondevilla, S., Carver, T.L.W., Moreno, M.T. and Rubiales, D. 2006. Macroscopic and histological characterisation of genes er1 and er2 for powdery mildew resistance in pea. *Euro. J. Pl. Pathol.* **115(3):** 309–321.

Forbes, C.J. and Bretag, T.W. 1991. Efficacy of foliar applied streptomycin for the control of bacterial blight of peas. *Aust. J. Pl. Pathol.* **20(3):** 115–118.

Forster, B., Sztor, E., Burke, R., Follas, G., Falloon, R. and Biddle, A.J. 2001. A9873C, a broad spectrum fungicide seed treatment for peas. *Seed treatment: challenges and opportunities. Proc. Int. Symp. Wishaw, North Warwickshire, U.K.* Feb. 26–27, 2001, pp. 105–110.

Fraaije, B.A., Franken, AAJM, Zouwen, PS-Vander, Bino, R.J., Langerak, C.J., Vander Zouwn, R.S. 1993. Serological and conductimetric assays for the detection of *Pseudomonas syringae* pv. *pisi* in pea seeds. *J. App. Bacteriol.* **73:** 409–413.

Frew, T.J., Russell, A.C. and Timmerman-Vaughan, G.M. 2002. Sequence tagged site markers linked to the sbm1 gene for resistance to pea seedborne mosaic virus in pea. *Pl. Breed.* **121:** 512–516.

Furgal, Wegrzyck, H. 1986. Physiological races of the fungus *Ascochyta pinodes* (Jones) infecting field pea (*Pisum arvense* L.). *Hodowla-Roslin, -Aklimatyzacja-i-Nasiennictwo.* **28:** 83–89.

Furgal, Wegrzycka, H. 1984. Breeding of pea and field pea lines resistant to fungus diseases, with special reference to *Fusarium* and *Ascochyta* species. *Zeszyty–Naukowe–Akademii–Rolniczo-Technicznej-w-Olsztynie,-Rolnictwo.* **39:** 73–125

Furgal-Wegrzycka, H. 1990. Studies of the susceptibility of peas and field pea cultivars to *Ascochyta pinodes* (Jones). *Acta–Agron.* **43:** 131–140.

Furgal-Wegrzycka, H. 1992. Susceptibility of pea and field cultivars to *Ascochyta pisi* (Lib.). *Acta Agron.* **45(1–2):** 39–49.

Gangopadhyay, S. and Kapoor, K.S. 1975. *Sclerotinia* rot of peas in Kullu valley. *Veg. Sci.* **2:** 28–30

Gangopadhyay, S. and Kapoor, K.S. 1976. Wet seed treatment to control *Fusarium* wilt of pea. *Veg. Sci.* **3(1):** 74–78.

Gerlagh, M and Vos, I. 1987. Testing pea cultivars for resistance to *Pseudomonas syringae* pv. *pisi. Prophyta* **41 (5):** 108–111.

Gnanamanickam, S.S. and Kandaswamy, T.K. 1978. Incidence of bacterial blight of *Lablab niger* caused by *Pseudomonas pisi. Madras Agric. J.* **65:** 446–447.

Golubev, A.A. 1965. Pea specimens relatively resistant to diseases. *Sb. Trud. Aspir. molod. nauch. Sotrud. vses. Inst. Rast.* **6(10) :** 121–127.

Gray, E.G. and Findlater, W.T. 1960. *Sclerotinia sclerotiorum* on peas in Kincardineshire. *Plant Pathol.* **9:** 130–132.

Griffee, P.J. 1969. Cleistocarps of *Erysiphe polygoni* on *Pisum sativum* in Zambia. *Trans. Brit. Mycol. Soc.* **53:**491–493.

Gritton, E.T. and Ebert, R.D. 1975. Interaction of planting date and powdery mildew on pea plant performance. *J. Am. Soc. Hortic. Sci.* **100:** 137–142.

Grondeau, C., Ladonne, F., Fourmond, A., Poutier, F. and Samson, R. 1992. Attempts to eradicate *Pseudomonas syringae* pv. *pisi* from pea seeds with heat treatments. *Seed Sci Tech.* **20:** 313–323.

Guerrero, C.J. 1987. Identification of *Phoma medicaginis* (Malbr & Roum) var. *pinodella* (Jones) Boerema, in plants and seeds of pea (*Pisum sativum* L.) in Region IX, Chile. *Agric. Tecnica.* **47:** 80–83.

Gupta, G.P. 1990. Evaluation of pea germplasms for their reaction to powdery mildew and rust. *Indian J. Pulse Res.* **3:** 186–188.

Gupta, P.C., Maheshwari, S.K. and Suhag, L.S. 1989. Efficacy of some seed dressing fungicides in controlling the wilts and root rot complex of pea. *Indian J. Mycol. Pl. Pathol.* **18(2):** 176.

Gupta, R.P. 1990a. Evaluation of slow mildewing resistant varieties of peas in the field. *Indian J. Pulses Res.* **3(1):** 52–55.

Gupta, S.K. and Dohroo, N.P. 1993. *Sclerotinia* rot of cauliflower. *Int. J. Trop. Pl. Dis.* **11:** 131–138.

Gupta, S.K. and Kumar, A. 2008. Management of *Erysiphe pisi* through strobilurin and EBI fungicides. *Indian Phytopath.* **61(2):** 184–191.

Gupta, S.K.a nd Kumar, A. 2014. Management of powdery mildew of pea with botanicals. *J. Res. Punjab agric. Univ.* **51(1):** 45–48.

Gupta, S.K. and Sharma, H. R. 2004. Efficacy of some EBI fungicides against pea powdery mildew. *Pl. Dis. Res.* **19(2):** 190–191.

Gupta, S.K. and Shyam, K.R. 1998. Control of powdery mildew and rust of pea by fungicide. *Indian Phytopath.* **51(2):** 184–186.

Gupta, S.K. and Shyam, K.R. 2000. Post–infection activity of ergosterol biosynthesis inhibiting fungicides against pea rust. *J. Mycol. Pl. Pathol.* **30:** 414–415.

Gupta, S.K., Kumar, Amit and Gupta, Meenu. 2014. Effect of date of sowing and plant densities on the progress of powdery mildew of pea. *Pl. Dis. Res.* **29(2):** 143–147.

Gupta, S.K., Rana, Shakun and Jarial, Kumud 2011. Variation in morphological, cultural, pathological and molecular features of *Fusarium oxysporum* f. sp. *pisi* isolates causing wilt of pea (*Pisum sativum*). *J. Mycol. Plant Pathol.* **41(2):** 275–278.

Gupta, V.R. and Mate, G.D. 2009. Conidial size of *Erysiphe polygoni* influenced by the host reaction. *J. Pl. Dis. Sci.* **4(2):** 215–217.

Hagedorn, D.J. 1974. Recent pea anthracnose and downy mildew epiphytotics in Wisconsin. *Pl. Dis. Reptr.* **58(3):** 226–229.

Hagedorn, D.J. 1984. Bacterial diseases of peas. In: *Compendium of pea diseases.* American Phytopathological Society, Minnesota pp. 8–11.

Haglund, W.A. and Anderson, W.C. 1987. WSU–28 and WSU–31 pea inbred lines with resistance to specific races of *Fusarium* wilt. *Hortic. Sci.* **22(3):** 513–514.

Hampton, R.O. and Braverman, S.W. 1979. Occurrence of pea seed borne mosaic virus and new virus immune germplasm in the plant introduction collection of *Pisum sativum. Pl. Dis. Reptr.* **63:** 95–99.

Hiremath, R.V. and Pavgi, M.S. 1971. *In vitro* assay of Aureofungin against some rust fungi. *Hind. Antibiot. Bull.* **13:** 83–86.

Hoitink, H.A.J. and Hegedorn, D.J. 1966. Bacterial brown spot (*Pseudomonas syringae*) of bean and pea. *Phytopathology* **56:** 881.

Hollaway, G.J., Bretag, T.W. and Price, T.V. 2007. The epidemiology and management of bacterial blight (*Pseudomonas syringae* pv. *pisi*) of field pea (*Pisum sativum*) in Australia: a review. *Aust. J. Agric. Res.* **58(11):** 1086–1099.

Hollaway, G.J., Bretag, T.W., Gooden, J.M. and Hannah, H.C. 1996. Effect of soil water content and temperature on the transmission of *Pseudomonas syringae* pv. *pisi* from pea seed (*Pisum sativum*) to seedlings. *Aust. J. Pl. Pathol.* **25:** 26–30.

Hooda, K.S., Bhatt, J.C., Joshi, D., Shridhar and Mahajan, V. 2010. Resistance in pea genotypes against powdery mildew in north west Himalayas. *Pl. Dis. Res.* **25 (1):** 69–70.

Huang, H.C., Erickson, R.S. and Huang, H.C. 2000. Soil treatment with fungal agents for control of apothecia of *Sclerotinia sclerotiorum* in bean and Pea crops. *Pl. Pathol. Bull.* **9(2):** 53–58.

Huang, H.C., Kokko, E.G., Yanke, L.J. and Phillippe, R.C. 1993. Bacterial suppression of basal pod rot and end rot of dry peas caused by *Sclerotinia sclerotiorum. Can. J. Microbiol.* **39(2):** 227–233.

Huang, J.W. 1991. Control of soil borne crop diseases by soil amendments. *Pl. Prot. Bull. Taipei* **33(1):** 113–123.

Hubbeling, N. 1975. Resistance of peas to downy mildew and distinction of races of *Peronospora pisi* Syd. *Mededelingen–van–de–Faculteit Landbouwwetens–chappen RijRijksuniversiteit Gent.* **40:** 539–543.

Hunter, J.E. and Cigna, H.A. 1981. Bacterial blight of peas in New York State. *Plant Dis.* **65:** 612–613.

Iqbal, S.M., Jamali, A.R., Rauf, C.A. and Akram, A. 2001. Screening of pea (*Pisum sativum* L.) germplasm against blight disease caused by *Ascochyta pisi. Pakistan J. Phytopathol.* **13:** 64–66.

Jhooty, J.S. and Munshi, G.D. 1990. Mode of perpetuation of powdery mildew of pea in North–Western India. *Indian Phytopath.* **43(4):** 455–456.

Jindal, K.K. 1990. Chemical control of bacterial blight of pea. *Pl. Dis. Res.* **5 (Spl.):** 91–93.

Jindal, K.K. and Bhardwaj, S.S. 1989. Etiology of bacterial blight of pea caused by *Pseudomonas syringae* pv. *syringae. Pl. Dis. Res.* **4:** 165–166.

Jindal, K.K., Gupta, S.K., Shyam, K.R. and Rastogi, K.B. 1995. Evaluation of pea germplasm against bacterial blight and powdery mildew. *Pl. Dis. Res.* **10(1):** 35–38.

Johnston, H.W., Ivany, J.A. and Cutcliffe, J.A. 1980. Effect of herbicides applied to soil on Fusarium root rot of processing peas. *Plant Dis.* **64(10):** 942–993.

Jones, L.K. 1927. Studies on the nature and control of blight, leaf and pod spot and foot rot of peas caused by the species of *Ascochyta. NY State Agric. Res. Bull.* 547, 45 pp.

Kaiser, W.J., Muehlbauer, F.J., Hannan, R.M. and Mihov, M. 1998. First report of natural infection of *Pisum sativum* subsp. *elatius* by *Mycosphaerella pinodes* in Bulgaria. *Plant Dis.* **82:** 830.

Kapil, R and Kapoor, A.S. 2005. Management of white rot of pea incited by (*Sclerotinia sclerotiorum*) using *Trichoderma* spp. and biopesticides. *Indian Phytopath.* **58(1):** 10–16.

Kapoor, A.S, Paul, Y.S. and Singh, A. 2006. Integrated management of white root rot and root rot–wilt disease complex of pea. *Indian Phytopath.* **59(4):** 467–474.

Kapoor, A.S. and Chaudhary, H.K. 1995. Mode of perpetuation of *Erysiphe pisi* in dry temperate zone of Himachal Pradesh. *Indian Phytopath.* **48:** 77–78.

Kapoor, A.S. and Kumar, P. 2004. Evaluation of different organic substrates, viability of *Trichoderma harzianum* and management of white rot of pea. *Pl. Dis. Res.* **19(1):** 55–56.

Kapoor, A.S. and Singh, A. 1999. Pea diseases and their management. In: *Diseases of Horticultural Crops – Vegetables, Ornamentals and Mushrooms* (Eds. Verma, L R. and Sharma, R.C.), Indus Publishing Co., New Delhi, pp 278–300.

Kapoor, A.S. and Singh, D. 1992. Varietal resistance of pea to white rot. *Pl. Dis. Res.* **7(1):** 889–890.

Kapoor, A.S. and Thakur, B.R. 1997. Fungicidal spray schedule for the economical management of pea powdery mildew. *Indian Phytopath.* **50(1):** 115–118.

Katoch, V., Sharma, S.D., Pathania, S., Benyal, D.K., Sharma, S.K. and Rathore, R. 2010. Molecular mapping of pea powdery mildew resistance gene *er* 2 to pea linkage group III. *Mol. Breed.* **25(2):** 229–237.

Kaur, J., Munshi, G.D. and Singh, R.S. 2006. Comparative efficacy of biocontrol agents against *Sclerotinia sclerotiorum* (Lib.) de Bary causing white rot of pea. *Pl. Dis. Res.* **21 (2):** 121–124.

Kaur, J., Munshi, G. D., Singh, R. S. and Koch, E. 2009. Selection of biocontrol agents for the management of white rot of peas caused by *Sclerotinia sclerotiorum. Pl. Dis. Res.* **24 (2):** 148–155.

Kavasnikov, B. and Krotova, T. 1977. Resistance of varietiies of garden pea to *Ascochyta* diseases. *Referativnyl Zhurnal* **1:**14–24.

Khalifa, O. 1965. Biological control of *Fusarium* wilt of peas by organic soil amendments. *Ann. App. Biol.* **86(1):** 129–137.

Khan, I. A., Khan H., Ali A., Raziq F., Hussain S., Ahmad M. and Attauddin 2009. Evaluation of various fungicides and cultivars for the control of pea rust under natural conditions. *Sarhad J. Agric.* **25(2):** 261–268.

Khare, N. and Lakpale, N. 1997. Sources of resistance to powdery mildew in field pea. *J. Mycol Pl. Pathol.* **27(2):** 219–220.

Khetarpal, R. K. and Maury, Y. 1987. Pea seed borne mosaic virus: a review. *Agronomie* **7:** 215–224.

Kokoekova, B., Pankova, I. and Kudela, V. 1998. Reaction of Czech pea varieties to races of *Pseudomonas syringae* pv. *pisi. Pl. Prot. Sci.* **34:** 126–130.

Koller, W., Allan, C. R. and Kolattukudy, P. E. 1982. Protection of *Pisum sativum* from *Fusarium solani* f. sp. *pisi* – Inhibition of cutinase with organophosphorus pesticides. *Phytopathology* **72(11):** 1425–1430.

Kraft, J. M. 1978. Effect of root rot pathogen on Fusarium wilt of pea. *Pl. Dis. Reptr.* **62(3):** 216–221.

Kraft, J. M. and Haglund, W. A. 1978. A reappraisal of the race classification of *Fusarium oxysporum* f. sp. *pisi. Phytopathology* **68:** 273–275.

Krishna, A. and Mishra, S. P. 1988. Slow mildewing in pea. *Indian Phytopath.* **41(2):** 204–207.

Krishna, A., Mishra, S.P., Mandloi, K.C. and Khosla, H.K. 1989. Assessment of yield losses in slow and fast mildewing pea cultivars. *Indian J. Mycol. Pl. Pathol.* **18(2):** 177–179.

Kumar, A. 1993. Studies on resistance to *Fusarium* wilt in some cultivars of peas (*Pisum sativum* L.). M.Sc. Thesis, Dr. Y.S. Parmar Univ. Horti. Forestry, Solan, H. P., 71pp.

Kumar, A., Kohli, U.K. and Kumar, A. 2001. Evaluation of garden pea genotypes for horticultural traits and resistance to *Fusarium* wilt. *Haryana J. Hort. Sci.* **30(3–4):** 217–219.

Kumari, S., Kumar, S., Tripathi, H.S. and Singh, V.K. 2013. Effect of planting techniques, irrigation and weed management on major diseases and yield of field pea. *Pl. Dis. Res.* **28 (2):** 132–134.

Kumari, S.G., Makkouk, K.M. and Bayaa, B. 2001. Pea enation mosaic virus–1 infecting lentil in Syria, and further information on its host range, purification, serology and transmission characteristics. *Arab J. Pl. Prot.* **19:** 65–72.

Kumari, Snehlata, Kumar, S., Tripathi, H.S. and Singh, V.K. 2013. Effect of planting techniques, irrigation and weed management on major diseases and yield of field pea. *Pl. Dis. Res.* **28(2):** 132–134.

Kushwaha C., Chand R. and Srivastava, C. P. 2006. Role of aeciospores in outbreaks of pea (*Pisum sativum*) rust (*Uromyces fabae*). *European J. Pl. Path.* **115(3):** 323–330.

Kuzmina, G. 1966. Antibiotics against wilt. *Zashch. Rast. Mosk.* **11(4):** 31.

Lacicowa, B. and Pieta, D. 1996. The efficiency of microbiological dressing of pea seeds (*Pisum sativum* L.) against pathogenic soil–borne fungi. *Roczniki–Nauk–Rolniczych. – Seria–E,–Ochrona–Roslin.* **25:** 15–21.

Larsen, R.C., Kaiser, W.J. and Klein, R.E. 1996. Alfalfa, a non–host of pea enation mosaic virus in Washington State. *Can. J. Pl. Sci.* **76**: 521–524.

Lawyer, A.S. 1984. Diseases caused by *Ascochyta* spp. In: *Compendium of pea diseases* (Ed. D.J. Hegdorn), pp. 11–15. The American Phytopathological Society, Minnosota, U. S. A.

Lin, Y.S. and Lo, C.T. 1987. Control of *Fusarium* wilt and root rot of garden peas by soil amendments. *J. Agric. Res. China* **36(4)**: 435–444.

Lin, Y.S., Sun, W and Wong, P. H. 1984. Fusarium root rot and wilt of garden peas in Taiwan. *J. Agric. Res. China*, **33(4)**: 395–405.

Linford, M.B. 1928. A Fusarium wilt of peas in Wisconsin. *Wis. Agric. Exp. Res. Bull.* **85:** 44pp.

Lorenzini, G, Mannerucci, G. F., Tricolo, E. and Gambogi, P. 1975. Some pathogenic aspects of *Fusarium oxysporum* f. sp. *pisi* isolated from pea seed and efficacy of treatment with systemic fungicides. *Rivista di Patol. Veg.* **11(1/2)**: 25–50.

Lyon, N.F., Taylor, J.D., Roberts, S.J., Maury, Y., Duby, C., Masmoudi, K., Faris Mokaiesh, S., Corblere, E., Mendes Pereira, E., Spire, D., Samson, R., Malandrin, L., Groundeau, C, Franken, AAJM and Fraaiji, B. A. 1995. International programme on the serological detection of bacterial fungal and viral pathogens of protein pea seeds. *Bull. OEPP* **23:** 393–401.

Maheshwari, S.K. and Jhooty, J.S. 1987. Laboratory evaluation of fungicides against the pathogen associated with wilt and root rot complex of pea. *Agric. Sci Dig.* **7(2)**: 89–90.

Maheshwari, S.K., Gupta, J. S. and Jhooty, J.S. 1981a. Effect of various cultural practices on the incidence of the wilt and root rot of pea. *Indian J. Agric. Res.* **51(3)**: 149–151.

Maheshwari, S.K., Gupta, J.S. and Sharma, P. D. 1982. Studies on temperature relation and chemical control of wilt and root rot of pea. *Acta Botanica Indica* **10(1)**: 89–91.

Maheshwari, S.K., Jhooty, J.S. and Gupta, J.S. 1981b. Studies on wilt and root rot of pea I. Efficiency of various chemicals in controlling the wilt and root rot complex of pea. *Agric. Sci Dig.* **1(1)**: 37–38.

Maheshwari, S.K., Jhooty, J.S. and Gupta, J.S. 1983. Survey of wilt and root rot complex of pea in northern India and the assessment of losses. *Agric. Sci. Dig.* **3(3/4)**: 139–141.

Maheshwari, S.K., Gupta, P. C. and Suhag, L. S. 1992. Management of powdery mildew of pea through fungicides. *Legume Res.* **15(1)**: 51–52.

Mahmood, T., Ahmad, I. Haq, I.U. and Aslam, M. 1991. Varietal resistance of pea to powdery mildew. *Indian Phytopath.* **44(1)**: 52–54.

Majumdar, M. and Raychaudhury, S.P. 1986. Screening of pea germplasm against Fusarium root rot and evaluation of host range of the pathogen. *Indian Phytopath.* **36(2)**: 328–330.

Malikova, G.B. 1973. The resistance in pea to Ascochytosis and some questions of the biology of the pathogens. *Genet. Sel.* **51**:103–110

Marcinkowska, J. 2007. Diseases of pea on newly registered cultivars. *Phytopathologia Polonica* **45**: 29–42.

Mashita, N. and Fukaya, M. 2006. Race of Fusarium wilt in the *Pisum* occurred in Aichi and resistant variety to Fusarium Wilt race 2. *Res. Bull. Aichiken Agric. Res. Center* **38:** 65–71.

Mathur, Kamlesh, Shekhawat, K.S. and Bhata, L.K. 1991. Performance of pea varieties against powdery mildew. *Indian Phytopath.* **44(1)**: 137.

Maude, R.B. 1966. Pea seed infection by *Mycosphaerella pinnodes* and *Ascochyta pisi* and its control by seed soaked in thiram and captan suspension. *Ann. App. Biol.* **57**: 193–200.

Maury, Y. and Khetarpal, R.K. 1992. Pea seed borne mosaic virus. In: *Plant Diseases of International Importance.* Vol. II (Eds. Chaube, H. S. and Singh, U. S.), New Jersey, Printice Hall, 376pp.

McDonald, G. K. and Dean, G. 1996. Effect of water logging on the severity of disease caused by *Mycosphaerella pinodes* in peas (*Pisum sativum* L.). *Aust. J. Exp. Agric.* 36: 219–222.

McPhee, K.E. and Muehlbauer, F.J. 2002a. Registration of 'Franklin' green dry pea. *Crop Sci.* **42:** 1378.

McPhee, K.E. and Muehlbauer, F.J. 2002b. Registration of 'Lifter' green dry pea. *Crop Sci.* **42:** 1377–1378.

McWhorte, F.P. and Cook, W.C. 1958. The hosts and strains of pea enation mosaic virus. *Pl. Dis. Reptr.* **42:** 51–60.

Mohamed, H.A., Shatta, H.M., Abdelal, H.R., El–Fahl, A.M. and Ismail, I.A. 1983. Host range and viability of urediniospores of *Uromyces fabae* (Pers.) deBary. *Agric. Res. Rev.* **6(2):** 73–82.

Mollenbruck, G. and Sander, E. 1991. Optimisation of serological detection (ELISA) of the quarantine bacterium and selection of a pea cultivar for biotest. *Z. Pflanzenkr. Pflanzenschutz* **98:** 630–639.

Munjal, R.L., Chenulu, V.V. and Hora, T.S. 1963. Assessment of losses due to powdery mildew (*Erysiphe polygoni* DC.) on pea. *Indian Phytopath.* **16:** 268–270.

Munshi, G.D., Jhooty, J.S. and Bajaj, K.L. 1987. Basis of resistance in peas to powdery mildew caused by *Erysiphe polygoni. Indian J. Plant Pathol.* **17(3):** 280–283.

Musaev, T. 1967. Biology of *Ascochyta* spp. on pea *Uzbek. Biol Zh.* **11:** 14–16

Musil, M. 1966. Uber das Vorkommen des Viruses des Blattrollens der Erbse in der Slowakei (Vorlaufige Mitteilung). *Biologia (Bratislavia)* **21:** 133–138.

Narsinghani, V.G., Singh, S.P. and Pall, B.S. 1980. Note on rust resistant pea varieties. *Indian J. Agric. Sci.* **50(5):** 453.

Nasir, M. and Hoppe, H.H. 1997. Evaluation of foliar fungicides for control of *Mycosphaerella pinodes* on peas. Tests Agrochemicals and Cultivars No. 18, 12–13; *Ann. App. Biol.* 130.

Neweigy, N.A., El–Din, I.F.G., Ziedan, M.I., Hanazy, E.A. and Meame, M.A.A. 1985. Effect of seed dressing of peas and soybean seeds on their growth in soil infested with *Sclerotium rolfsii* or mixture of three pathogenic fungi (*Rhizoctonia solani, Fusarium solani, S. rolfsii). Ann. Agric. Sci. Moshtohor* **23(3):** 1109–1114.

Obradovic, A., Marinkovic, N., Mijatovic, M. and Dordevic, R. 1994. Resistance of pea genotypes to *Ascochyta pisi. Zastita–Bilja.* **45:** 151–154.

Ondrej, M. 1992. Differences in varietal resistance of pea (*Pisum sativum* L.) to root diseases and *Mycosphaerella pinodes. Rostilinna Vyroba* **38(2):** 157–164.

Ondrej, M. 1994. New information in resistance breeding of pea (*Pisum sativum* L.) against Ascochyta leaf and pod spot of pea (*Mycospharella pinodes*). *Rocenka Genet. Zdrojerastlin*, 19–23 pp.

Orevic, R., Zecevic, B., Cvikic, D., Damjanovic, M. and Zdravkovic, M. 2004. Srbija-late afila type of pea (*Pisum sativum* L.) in the Centre for Vegetable Crops. *Agroznanje Agroknowledge J.* **5(1):** 105–109.

Osborne, H.T. 1935. Incubation period of pea mosaic in the aphid, *Macrosiphum pisi. Phytopathology* **25:** 160–177.

Ovchinnikova, A.M and Andryukhina, R.M. 1984. Isolation and determination of the races of the causal agent of dark spot *Ascochyta* disease of pea. *Mikolo. -i-Fitopatol.* **18(6):** 493–497

Ovchinnikova, A.M. and Andrukhina, R.M. 1977. Study on the resistance of pca to *ascochytosis. Referativnyl Zhurnal* **12:**455–469

Oyarzun, P., Gerlagh, M. and Hoogland, A.E. 1993. Pathogenic fungi involved in root rot of peas in the Netherlands and their physiological specialization. *Neth. J. Pl. Pathol.* **99:** 23–33.

Oyarzun, P.J., Postma, J., Luttikholt, A.J.G. and Hoogland, A.E. 1994. Biological control of foot and root rot in pea caused by *Fusarium solani* with nonpathogenic *F. oxysporum* isolates. *Can. J. Bot.* **72(6):** 843–852.

Pal, A.B.; Brahamappa, Sohi, H.S. and Rawal, R.D. 1979. Studies on inheritance of resistance to rust (*Uromyces fabae* Pers. de Bary) on peas. *SABRAO J.* **11(2):** 101–103.

Pal, A.B.; Brahmappa, Rawal, R.D. and Ullasa, B.A. 1980. Field resistance of pea germplasm to powdery mildew (*Erysiphe polygoni*) and rust (*Uromyces fabae*). *Plant Dis.* **64:** 1085–1086.

Patel, M.K. 1933. India : diseases in the Bombay Presidency. *Int. Bull. Pl. Prot.* **7(11):** 246.

Paul, Y.S. and Rathour, R. 1998. New blight of pea from Himachal Pradesh. *Indian Phtopath.* **51:** 197–198.

Pavgi, M.S. and Singh, U.P. 1965. Parasitic fungi from North India –V. *Mycopath. Et Mycol. App.* **27:** 85–89.

Peters, D. 1982. *Pea enation mosaic virus. Descriptions of Plant Viruses.* No. 257. CMI, AAB, Kew Surrey, England.

Phal, R. and Choudhary, B. 1983. Fusarium wilt of garden pea: race situation. *Indian J. Agric. Sci.* **53(9):** 863–864.

Prasad, P. and Dwivedi, S.N. 2007. Fungicidal management of field pea (*Pisum sativum* L.) powdery mildew caused by *Erysiphe polygoni* DC. *Prog. Res.* **2(1/2):** 116–118.

Prasada, R. and Verma, U.N. 1948. Studies on lentil rust, *Uromyces fabae. Indian Phytopath.* **1:** 142–146.

Pretorius, J.C., Craven, P., Watt, E., Vander, der–Watt–E–van and Van–der–Watt, E. 2002. *In vivo* control of *Mycosphaerella pinodes* on pea leaves by a crude bulb extract of *Eucomis autumnalis*. *Ann. App. Biol.* **141:** 125–131.

Price, M. and Cishahayo, D. 1986. *Ascochyta* blight on major food crops in Rwanda, East Africa. *Plant Dis.* **70:** 1159.

Punithalingam, E. and Holliday, P. 1972. *Descriptions of the pathogenic fungi and bacteria No 340.* Commonwealth Mycological Research Institute, Kew, Surrey, England.

Ram, R. S. 1992. Inheritance of powdery mildew resistance in peas (*Pisum sativum* L.). *Exp. Genet.* **8(1–2):** 37–40.

Rana, D.P.S., Bhardwaj, P.K., Rao, M.V.B. and Chatterjee, D. 1991. Field evaluation of fenarimol and triadimefon for control of powdery mildew (*Erysiphe polygoni* D. C.) of pea. *Indian J. Pl. Prot.* **19(1):** 31–35.

Rana, J. C., Banyal, D. K., Sharma, K. D., Sharma, M. K., Gupta, S. K. and Yadav, S. K. 2013. Screening of pea germplasm for resistance to powdery mildew. *Euphytica* **189:** 271–282.

Rathi, A.S. and Tripathi, N.N. 1992. Variability in *Erysiphe polygoni* DC. on peas. 44th Annual Meeting of Indian Phytopath. Soc. Bhavanagar Univ., Bhavanagar, Gujarat, 30th Dec., 1991 to 2 nd Jan., 1992 pp. 71 (Abstr.).

Rich, A.E. 1987. 'Sure Crop', a new pea (*Pisum sativum* L.) cultivar resistant to Fusarium root rot. *Phytopathology* **77(1):** 121.

Roger, C. and Tivoli, B. 1996. Spatio–temporal development of pycnidia and perithecia and dissemination of spores of *Mycosphaerella pinodes* on pea (*Pisum sativum*). *Plant Pathol.* **45(3):** 518–528

Roger, C., Tivoli, B. and Huber, L. 1999. Effects of interrupted wet periods and different temperatures on the development of *Ascochyta* blight caused by *Mycosphaerella pinodes* on pea (*Pisum sativum*) seedlings. *Plant Pathol.* **48(1):** 10–18

Rossignol, Y. 1988. Pea seed treatments: new oxadixyl based mixtures for broad spectrum disease control. *Monog. Brit. Crop Prot. Coun. No.* **39:** 99–104.

Rudakov, O.L. and Lepikhova, R.M. 1974. Intra specific and morphological and physiological differentiation of *Ascochyta pisi* Lib. and *Ascochyta pinodes* L. K. Jones. *Mikol. i. Fitopatol.* **8:** 408–412

Sackett, W.G. 1916. A bacterial stem blight of field and garden peas. *Bull. Colo. Agric Exp. Stn. No.* 218.

Safarova, D., Navratil, M., Petrusova, J., Pokorny, R. and Piakova, Z. 2008. Genetic and biological diversity of the Pea seed–borne mosaic virus isolates occurring in Czech Republic. *Acta Virologica* **52(1):** 53–57.

Sagar, V. and Sugha, S.K. 1998. Effect of soil type and available nutrients on Fusarial population and severity of pea root rot. *J. Mycol. Pl. Pathol.* **28(3):** 294–299.

Sagar,V. and Bhardwaj C.L. 1997. Identity of 'Mega' form of *Phoma medicaginis* var. *pinodella* on pea from Bajaura, Himachal Pradesh. *J. Mycol. Pl. Pathol.* **27:** 195–198.

Saksirirat, W., Sirmung Kararat, S., Steinmetz, J. and Schoenbeck, F. 1994. Applicability of *Gliocladium roseum* Bain. for biological control of *Fusarium* wilt of pea. *Kaen Kaset–Khon Kaen Agri. J.* **22(1):** 37–42.

Sandhu, K.S. and Dhillon, G.S. 1984. Reaction of pea varietes to *Ascochyta* blight. *J. Res. Punjab Agric. Univ.* **21:**391–395.

Sangar, R.B.S. and Singh, V.K. 1994. Effect of sowing dates and pea varieties on the severity of rust, powdery mildew and yield. *Indian J. Pulse Res.* **7(1):** 88–89.

Sanssene, J., Boulos, G. and Gascoin, B. 1998. Efficiency of seed treatment on root necroses of peas (*Pisum sativum* L.). 3rd European conference on grain legumes. Opportunities for high quality, healthy and added–value crops to meet European demands. Valladolid, Spain, 14–19 November 1998. 263.

Sarkar, D., Muchlbeuer, F.J. and Kraft, J.M. 1982. Techniques of screening peas for resistance to *Phoma medicaginis* var. *pinodella*. *Crop Sci.* **22:**988–992.

Schroeder, W.T. and Previndeti, R. 1965. Breakdown of the er resistance to powdery mildew in *Pisum sativum*. *Phytopathology* **55:**1075.

Shailbala and Khulbe, R.K. 2009. Management of pea powdery mildew with organic inputs. *Pl. Dis. Res.* **24(2):** 199–200.

Sharma, A.K. 1985. Epidemiological studies on white rot (*Sclerotinia* rot) of pea. *Indian Phytopath.* **38(2):** 319–321.

Sharma, A.K. 1987. Evaluation of fungicides for the control of *Sclerotinia* rot of pea. *Indian Phytopath.* **40(3):** 399–401.

Sharma, A.K. 1992. Technique for creation of artificial epiphytotic condition of *Sclerotinia* rot (*Sclerotinia sclerotiorum*) of pea (*Pisum sativum*) in field and evaluation of germplasm. *Indian J. Agric. Sci.* **62(3):** 232–234.

Sharma, B.K. and Singh, B.M. 1990. Biological control of white rot of pea caused by *Sclerotinia sclerotiorum* (Lib.) de Bary. *J. Biol. Control* **4(2):** 132–134.

Sharma, B.K., Basandrai, A.K and Singh, A.K. 1999. Management of white rot of pea. *Pl. Dis. Res.* **14(2):** 161–164.

Sharma, B. K., Singh, B.M. and Sugha, S.K. 1992. Integrated effect of biological and chemical control on sclerotial viability of *Sclerotinia sclerotiorum* (Lib.) de Bary. *J. Biol. Control* **6(1):** 29–34.

Sharma, B.K. and Basandrai, A.K. 1997. Effect of biocontrol agents, fungicides and plant extracts on sclerotial viability of *Sclerotinia sclerotiorum. Indian J. Agric. Sci.* **67(3):** 132–133.

Sharma, B.L., Parashar, R.D. and Bohare Sudha. 1989. Studies on survey of *Fusarium* wilt of pea in Northern region of Madhya Pradesh. *Legume Res.* **12(3):** 151–152.

Sharma, C.L. and Kapoor, A.S. 1997. Some epidemiological aspects of white rot of pea. *Indian Phytopath.* **50(3):** 342–349.

Sharma, C.L. and Kapoor, A.S. 1999. *In vitro* evaluation of herbicides and plant extracts against *Sclerotinia sclerotiorum* causing white rot in peas. *Pl. Dis. Res.* **14(10):** 79–81.

Sharma, C.L. and Kapoor, A.S. 2001. Evaluation of resistance of pea germplasm against *Sclerotinia sclerotiorum. Himachal J. Agric. Res.* **25(1):** 90–92.

Sharma, R.C. and Gill, S.S. 1996. Cultural control of Sclerotinia sclerotiorum under irrigated rice farming system. Seed Res. **24(2):** 171–173.

Sharma, R.C. and Munshi, G.D. 1992. White mold of pea and its control. *Pl. Dis. Res.* **7(2):** 235–239.

Sharma, S., Yadav, O.P. and Singh, S. 1994. Effect of date of sowing, irrigation regimes and spacing on wilt incidence and yield of field pea. *Crop Res.* (Hisar) **8(3):** 582–584.

Sharma, S.K. and Sharma, A. 2003. Effect of time of planting, soil hydrothermal regimes and some meteorological parameters on the development of Fusarium wilt in autumn pea. *Pl. Dis. Res.* **18(2):** 127–130.

Sheridan, J.J. 1973. The survival of *Mycosphaerella pinodes* on pea haulm buried in soil. *Ann. App. Biol.* **75:** 195–203

Shukla, Pratibha and Amin, K.S. 1991. Integrated disease management of powdery mildew of field peas (*Pisum sativum* L.) caused by *Erysiphe polygoni* D. C. *Legume Res.* **14(2):** 59–63.

Shyam, K.R., Gupta, S.K. and Verma, A.K. 1995. Management of powdery mildew and leaf miner in pea through fungicide and insecticide applications, In: *Integrated disease management and plant health.* (Eds. V. K. Gupta and R. C. Sharma), Scientific Publishers (India), Jodhpur, pp. 119–122.

Sidenko, I.E. 1966. Specialization of *Uromyces pisi,* the causal agent of rust of *Lathyrus sativus. Nov. Sist. niz. Rast.* 188–194.

Sindhan, G.S. and Prashar, R.D. 1990. Effect of macro and micro-nutrients on the development of powdery mildew of pea. *Indian J. Mycol. Pl. Pathol.* **19(2):** 219–221.

Singh, A and Banyal, D. K. 2007. Integrated management of pea diseases in Lahaul valley. *Pl. Dis. Res.* **22(2):** 132–138.

Singh, A., Banyal, D.K. and Tyagi, P.D. 2012. Host range and perpetuation of *Erysiphe pisi,* the causal agent of powdery mildew of pea. *Indian Phytopath.* **65(1):** 102–104.

Singh, D. 1991. Biological control of *Sclerotinia sclerotiorum* (Lib.) de Bary by *Trichoderma harzianum. Trop. Pest Mgt.* **37(4):** 374–378.

Singh, D. and Singh, D. 1991. Development of white rot of pea (*Sclerotinia sclerotiorum*) in relation to meteorological factors. *Indian J. Mycol. Pl. Pathol.* **21(2):** 192–195.

Singh, D. and Singh, D. 1995. Effect of sowing date on white rot (*Sclerotinia sclerotiorum*) development in pea (*Pisum sativum*). *Indian J. Agric Sci.* **65(8):** 621–623.

Singh, D., Dhiman, J.S. and Saimbhi, M.S. 1989. Chemical control of stemfly-wilt complex of peas. *Trop. Pest Mgt.* **35(2):** 176–179.

Singh, D., Singh, A. and Thakur, H.L. 2002. White rot disease in pea. The Tribune, Chandigarh, March 25th, 2002.

Singh, H. and Dickinson, C.H. 1980. Metalaxyl for the control of downy mildew of pea caused by *Peronospora viciae*. *Plant Dis*. **64:** 1090–1092.

Singh, H.B. 1985. *In vitro*, glasshouse and field evaluation of some plant extracts and oil for the control of powdery mildew of pea. *Proc. of Seminar on Chemistry, biology, cultivation and utilization of medicinal and aromatic plants* p. 13.

Singh, H.B. and Singh, U.P. 1988. Powdery mildew of pea (*Pisum sativum* L.) *Int. J. Trop. Pl. Dis.* **6:** 1–18.

Singh, L., Bagri, R.K., Thakore, B.B.L. and Singh, J. 2011. Effect of different media on growth and enzyme production of *Fusarium oxysporm* f. sp. *pisi*. *Pl. Dis. Res.* **26(1):** 40–46.

Singh, P.P., Wadhwa, A. and Sandhu, K.S. 2005. Effect of epidemiological parameters on the development of Ascochyta blight of pea. *Indian Phytopath.* **58(2):** 200–205.

Singh, R.B. Singh, M.N., Singh, U.P. and Singh, R.M. 1983. Inhertance of resistance to powdery mildew in pea and its use in breeding. *Indian J. Agric. Sci.* **48:** 86–88.

Singh, R.B., Jain, A.K., Jat, R.R. and Mathur, S.S. 1988. Varietal screening of pea to powdery mildew. *Pl. Pathol. Newsl.* **6(1–2):** 20.

Singh, R.N. 1985. Genotypic resistance in peas to powdery mildew and variation in the pathogen. *Indian J. Plant Pathol.* **3**:201–202.

Singh, R.S. 1987. *Diseases of vegetable crops*. Oxford and IBH Publishing Co. Pvt. Ltd. New Delhi, 362 p.

Singh, S.J. and Sokhi, S.S. 1980. Pathogenic variability in *Uromyces viciae fabae*. *Pl. Dis. Reptr.* **64:** 671–672.

Singh, S.J., Sokhi, S.S. and Grewal, R.K. 1981. Relative efficacy of fungitoxicants against pea rust. *Indian Phytopath.* **34:** 272–275.

Singh, S.K. 1994. Foot rot of peas caused by *Phoma medicaginis* var. *pinodella* (Jones) Boerema: a new record from India. *Indian J. Mycol. Pl. Pathol.* **24:** 238.

Singh, S.K., Rahman, S.J., Gupta, B.R. and Kalha, C.S. 1992. An integrated approach to the management of the major diseases and insect pests of peas in India. *Trop. Pest Mgt.* **38:** 265–267.

Singh, U.P. and Singh, H.B. 1983a. Differential responses of *Erysiphe polygoni* conidia to living and non–living substrata. *Phytopath. Z.* **107:** 140–145.

Singh, U.P. and Singh, H.B. 1983b. Comparative efficacy of some commercial fungicides, plant extracts and oil for the control of powdery mildew (*Erysiphe polygoni* D. C.) of pea (*Pisum sativum* L.) *Aust. Pl. Pathol.* **12:** 22–24.

Singh, U.P., Prithiviraj, B., Mishra, G.D. and Singh, H.B. 2000. Modes of germination of *Erysiphe pisi* conidia on pea (*Pisum sativum*) leaves and glass slides. *Indian J. Plant Pathol.* **18 (1&2):** 57–59.

Singh, U.P., Singh, H.B. and Chauhan 1984. Effect of some fungicides, plant extracts and oil on inoculum density of different nodal leaves of pea infected by *Erysiphe polygoni*. *Z. Pflkrankh. Pflschutz.* **91**:20–26.

Smith, H.C. and Close, R.C. 1966. Elimination of bacterial blight from pea crops. *Canterbury Chamber of Commerce, Agril. Bull.* 440.

Snyder, W.C. 1933. A new vascular Fusariumn disease of peas. *Science* **77:** 327.

Snyder, W.C. and Hansen, H.N. 1940. The species concept in *Fusarium*. *Am. J. Bot.* **27:** 64–67.

Snyder, W.C. and Walker, J.C. 1935. *Fusarium* near–wilt of pea. *Zentralbl. Bakteriol. Parasitenk. Abst.* 11, **91:** 355–378.

Sohi, H.S., Sokhi, S.S. and Rawal, R.D. 1974. Varietal reaction of peas to powdery mildew (*Erysiphe polygoni*) and rust (*Uromyces fabae*). *Mysore J. Agric. Sci.* **8:** 529–532.

Sood, A.K. 1988. Survey of bacterial blight damage on pea in Kullu Valley of Himachal Pradesh. *Indian Phytopath.* **41:** 615–616.

Sontakke, N.A., Singh, H.N. and Singh, N. 2008. Chief features of physiographic rainfall variations across India during instrumental period (1813–2006). Pashan, India : Indian Institute of Tropical Meteorology.

Spencer, D.M. 1978. *The powdery mildews*. Academic Press, London.

Srivastava, L.S. and Gupta, D.K. 1990. Occurrence of powdery mildew, rust and *Ascochyta* blight of pea in Sikkim. *Pl. Dis. Res.* **5:** 86.

Srivastava, S.L. and Bisht, S. 1983. Studies on biological control of *Erysiphe polygoni* D. C. and pea. *Geobios* **10:**117–119

Stegmark, R. 1988. Downy mildew resistance of various pea genotypes. *Acta Agric. Scandinavica* **38:** 373–379.

Stegmark, R. 1990. Variation for virulence among Scandinavian isolates of *Peronospora viciae* f. sp. *pisi* (pea downy mildew) and response of pea genotypes. *Plant Pathol.* **39:** 118–124.

Stevenson, W.R. and Hagedorn, D.J. 1969. A new seed-borne virus of peas. *Phytopathology* **59:** 1051–1052.

Stone, R.E. 1912. The life history of some *Ascochyta* on some leguminous plants. *Ann. Mycol.* **10:** 564–592

Sugha, S.K. 1999. Efficacy of carbendazim granules against *Sclrotinia sclerotiorum* of pea (*Pisum sativum*), cabbage (*Brassica oleracea* var. *capitata*) and cauliflower (*Brassica oleracea* var. *botrytis*). *Indian J. Agric. Sci.* **69(4):** 302–303.

Sugha, S.K., Chauhan, R.S. and Singh, B.M. 1994. Sensitivity of aeciospores and urediospores of the pea rust pathogen to selected systemic fungicides. *Trop. Agric.* **71:** 27–30.

Sugha, S.K., Banyal, D.K. and Rama, S.K. 2008. Management of pea (*Pisum sativum*) rust (*Uromyces fabae*) with fungicides. *Indian J. Agric. Sci.* **78:** 269–271.

Sukapure, R.S., Bhide, V.P. and Patel, M.K. 1957. Fusarium wilt of garden peas in Bombay State. *Indian Phytopath.* **10(1):** 11–17.

Sulaiman, M. and Utikar, P.G. 1967. Existence of three physiological races in Fusarium wilt of peas (*Pisum sativum* L.). *Beitr. Trop. Subtrop. Landwirt. Tropenvot. Med.* **5(1):** 33–40.

Surwase, A.G., Badgire, D.R. and Suryawanshi, A.P. 2009. Management of pea powdery mildew by fungicides, botanicals and bio-agents. *Ann. Pl. Protec. Sci.* **17(2):** 384–388.

Susuri, L. 1976. Contribution to the study of some factors of *Ascolyta pisi* Lib. parasite of pea. *Zastata Bilja*. **27:** 69–87

Sydow, H. and Butler, E. J. 1906. Fungi Indiae Orientalis. *Pars–I Ann. Mycol.* **4:** 424–445.

Taylor, J.D. 1972. Races of *Pseudomonas pisi* and sources of resistance in field and garden peas. *N. Z. J. Agric. Res.* **15:** 441–447.

Taylor, J.D. and Dye, D.W. 1976. Evaluation of streptomycin seed treatments for the control of bacterial blight of peas (*Pseudomonas pisi* Sackett, 1916). *J. Agric. Res.* **19:** 91–95.

Taylor, J.D., Bevan, J.R., Crute, I.S. and Reader, S.L. 1989. Genetic relationship between races of *Pseudomonas syringae* pv. *pisi* and cultivars of *Pisum sativum. Plant Pathol.* **38:** 364–375.

Taylor, J.D., Mosek, M., Primrose, S.B. and Littlejohn, I.H. 1982. Genetics of resistance to *Pseudomonas pisi*. In: *32nd Annual Report for 1981. Natl. Veg. Res. Sta., Warwick, U. K.*

Taylor, J.D., Vivian, A., Malik, A. and Reader, S.L. 1985. Genetics of host specificity in races of *Pseudomonas syringae*. In: *35th Annual Report for 1984. Nat. Veg. Res. Sta., Wellesbourne, Warwick, U.K.* pp. 85–86.

Tiwari, A. and Narsinghani, V.G. 1986. Wilt resistant genotype of peas. *Indian J. Agric. Sci.* **56(2):** 145–146.

Tornos, T., Cebrian, M.C., Cordoba Selles, M.C., Alfaro Fernandez, A., Herrera Vasquez, J.A., Font, M.I. and Jorda, M.C. 2008. First report of Pea enation mosaic virus infecting pea and broad bean in Spain. *Plant Dis.* **92(10):** 1469.

Toros, S., Schotman, C.Y.L. and Peters, D. 1978. A new approach to measure the LP 50 of pea enation mosaic virus in the vector. *Virology* **90:** 235–240.

Trivoli, B., Beasses, C., Lemarchand, E. and Marsson, E. 1996. Effect of Ascochyta blight (*Mycosphaerella pinodes*) on yield components of single pea (*Pisum sativum*) plants under field conditions. *Ann. App. Biol.* **129:** 207–216.

Tsilosani, G.A. and Barbakadze, A.N. 1976. Some representatives of the genus *Pseudomonas* on legumes. *Trudy Nauchnoissledovatel, Kogo Instituta Zashchity Rastenii.* **26:** 112–119.

Tu, J.C. 1991. Etiology, biology and control of a soil borne rot complex of green peas. *Pl. Prot. Bull. Taipei* **33(1):** 17–35.

Upadhyay, A.L. and Gupta, R.P. 1994. Fungicidal evaluation against powdery mildew and rust of pea (*Pisum sativum* L.) *Ann. Agric. Res.* **15(1):** 114–116.

Uppal, B.N., Patel, M.N. and Kamat, M.N. 1935. Pea powdery mildew in Bombay. *Bull. Bombay Dept. Agric.* 177 12 p.

Utikar, P.G., Gadre, U.A. and More, B.B. 1978. Seed treatment to control Fusarium wilt of pea. *Pesticides* **12(11):** 29–30.

Van der Gaag, D.J., Frinking, H.D. and Van der Gagg, D.J. 1996. Homothallism in *Peronospora viciae* f. sp. *pisi* and the effect of temperature on oospore production. *Plant Pathol.* **45:** 990–996.

Van der Gaag, D.J. and Frinking, H.D. 1997. Survival characteristics of oospore populations of *Peronospora viciae* f. sp. *pisi* in soil. *Plant Pathol.* **46:** 978–988.

Vaughan, R.E. 1913. *Mycosphaerella pinodes*, the ascigerous stage of *Ascochyta pisi. Phytopathology*, **3:** 71.

Venkatakrishnaiah, N.S. 1951. The perithecial stage of *Erisiphe polygoni* DC. on pea (*Pisum sativum* Linn.). *Curr. Sci.* **20(5):** 137–138.

Verma S. and Dohroo, N.P. 2005a. Effect of sowing dates on development of Fusarium wilt of autumn pea caused by *Fusarium oxysporum* f. sp. *pisi. Pl. Dis. Res.* **20(2):** 177–179.

Verma S. and Dohroo, N.P. 2005b. Effect of spacing on the development of Fusarium wilt in autumn pea. *Pl. Dis. Res.* **20(2):** 180–183.

Verma, S. and Dohroo, N.P. 2003. Evaluation of botanicals *in vitro* against *Fusarium oxysporum* f. sp. *pisi* causing wilt of pea. *Pl. Dis. Res.* **18(2):** 131–134.

Verma, S. and Dohroo, N.P. 2003a. Morphological, cultural and pathogenic variability in *Fusarium oxysporum* isolates from pea. *Pl. Dis. Res.* **18(1):** 5–8.

Verma, S. and Dohroo, N.P. 2005. Novel approach for screening different antagonists against *Fusarium oxysporum* f. sp. *pisi* causing *Fusarium* wilt of autumn pea. *Pl. Dis. Res.* **20:** 58–61.

Vigier, B., Raghavan, G.G.V. and Drake, G. 1983. Effect of soil compaction and sterilization on root rot disease and yield of canning peas. *Can. J. Pl. Sci.* **63(1):** 327–331.

Vladimirtseva, L.V. 1990. Evaluation of pea breeding material for susceptibility to *Fusarium. Nauchno-Tekhniches-Kii Byulleten* 1.

Vladimirtseva, L.V., Guseva, N.N. and Ovchinnikova, A.M. 1990. Forms of pea resistant to *Ascochyta pisi*. *Sel. Semenovodstvo-Moskva. No.* 3, 30–31.

Wallen, V.R. 1965. Field evaluation and the importance of *Ascochyta* complex on peas. *Can. J. Pl. Sci.* **45:** 27–33

Wallen, V.R. 1974. Influence of three *Ascochyta* diseases of peas on the plant development and yield. *Can. Pl. Dis. Surv.* **54:** 86–90

Wang, D. and Maule, A.J. 1992. Early embryo invasion as a deteriminant in pea of the seed transmission of pea seed-borne mosaic virus. *J. Gen. Virol.* **73:** 1615–1620.

Warkentin, T.D., Rashid K.Y. and Xue, A.G. 1996. Fungicidal control of *Ascochyta* blight of field pea. *Can. J. Pl. Sci.* . **76:** 1, 67–71

Warkentin, T.D., Xue, A.G. and McAndrew, D.W. 2000. Effect of mancozeb on the control of *Mycosphaerella* blight of field pea. *Can. J. Pl. Sci.* **80:** 403–406

Watson, D.R.W. and Dye, D.W. 1971. Detection of bacterial disease in garden pea seed stocks. *Pl. Dis. Reptr*. **55:** 517–521.

Yarwood, C.E. 1957. Powdery mildews. *Bot. Rev.* **23:** 235–293.

Yarwood, C.E. 1978. History and taxonomy of powdery mildews. In: *The powdery mildews*, pages 1–37 (Ed. Spencer, D. M.) Academic Press, London.

Young, J.M. and Dye, D.W. 1970. Bacterial blight of peas caused by *Pseudomonas pisi* Sackett. in New Zealand. *N.Z.J. Agric. Res.* **13:** 315–324.

■■■■

Chapter 2. Disease Problems in Vegetable Production

Plate - 1

1a. Symptoms of Pre-emergence damping-off showing patchy appearance

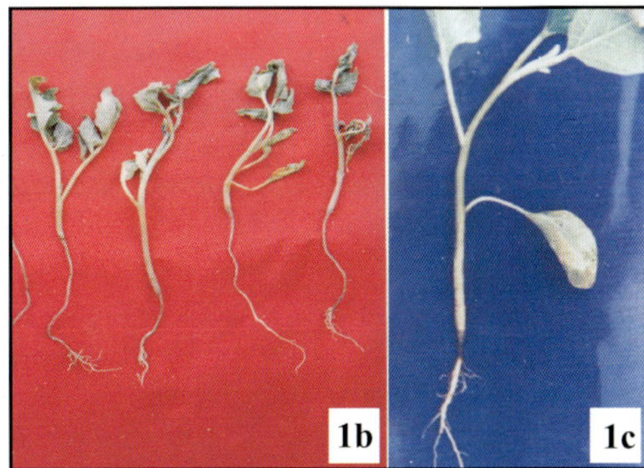

1b. Tomato 1c. Brinjal
Symptoms of post-emergence damping off

Chapter 3. Disease of Potato

1a. Upper surface of leaf

1b. Lower surface of leaf

1c. Tubers

1d. Crops

Plate 1. Symptoms of late blight of Potato

Plate 2. Symptoms of early blight on potato leaves

Plate 3. Symptoms of black scurf on potato tubers

4a. Warts on tuber

4b. Wart infected plant

Plate 4. Symptoms of potato wart (Courtesy CPRI, Shimla)

Plate 5. Symptoms of (a) pitted scab; (b) shallow scab

Chapter 3. Disease of Potato

Plate 6a. Wilted plant

Plate 6b. Symptoms of brown rot on tuber

Plate 6. Symptoms of bacterial wilt of Potato

Plate 7. Symptoms of (a) PVY and (b) PLRV

Chapter 4. Disease of Tomato

1a. Infected leaves

1b. Infected fruits

Plate 1. Symptoms of late blight of tomato

Chapter 4. Disease of Tomato

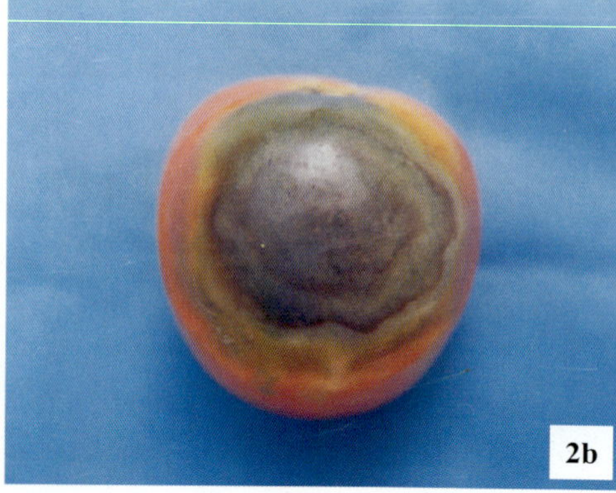

2a. Infected tomato plant

2b. Infected fruit

Plate 2. Symptoms of Buckeye rot of tomato

3a. Infected leaf

3b. Infected fruit

3c. Infected stem

Plate 3. Symptoms of early blight *Alternaria solani* of tomato

4a. Symptoms of *Alternaria alternata*

4b. Symptoms of *Alternaria alternata* f.sp. *lycopersici* on leaf

4c. Symptoms of *A.alternata* f.sp. *lycopersici* on stem

Plate 4. Symptoms of Alternaria leaf spots

Chapter 4. Disease of Tomato

5b

5a. Infected leaf

5b. Infected fruit

Plate 5. Symptoms of Septoria leaf spot

6a. Infected leaf

6b. Infected fruit

Plate 6. Symptoms of bacterial spots

Plate 7. Bacterial wilt infected plant

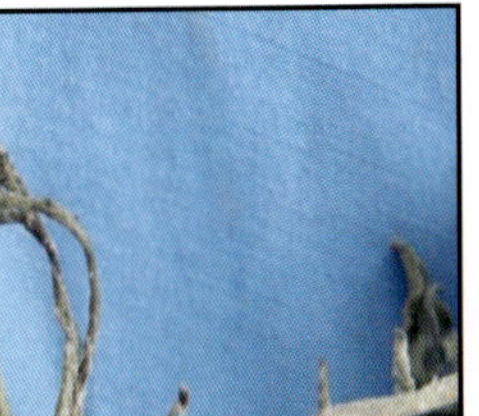

8a. Canker on stem

8b. Canker on fruits

Plate 8. Symptoms of bacterial canker on tomato

Plate 9. Symptoms of Tomato mosaic virus

Chapter 5. Disease of Cruciferous Vegetables

Plate 1a. On leaf

Plate 1b. On curd

Plate 2. Symptoms of Alternaria leaf spot

Plate 1. Symptoms of Downy mildew

3a. On petioles

3b. On stalk with sclerotia

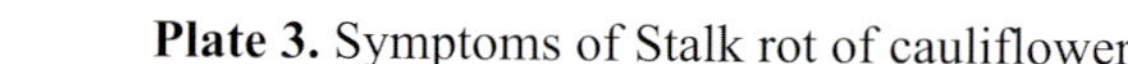

Plate 3. Symptoms of Stalk rot of cauliflower

Plate 4. Symptoms of white rust of crucifers

Plate 5. Symptoms of wire stem of cauliflower

Chapter 5. Disease of Cruciferous Vegetables

Plate 6. V' shape lesions of black rot of crucifers

Plate 7. Curd rot of cauliflower

Plate 8. Symptoms of cauliflower mosaic

Chapter 6. Disease of Cucurbits

1a. Upper leaf surface

1b. Lower leaf surface

1b. Field view

Plate 1. Symptoms of downy mildew of cucurbits

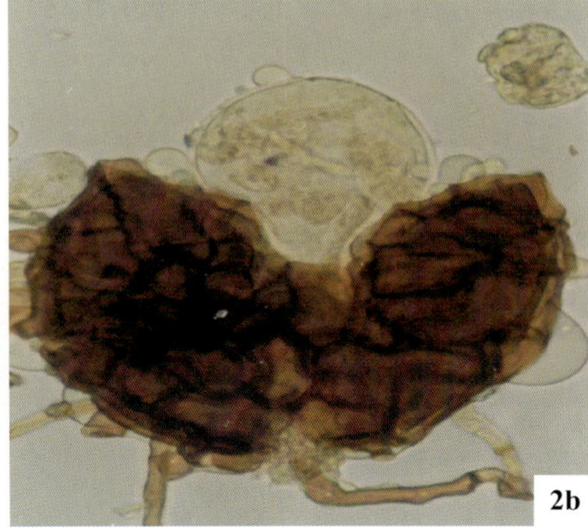

2a. On leaves **2b.** Cleistothecium of *Sphaerotheca fuliginea*

Plate 2. Symptoms of Powdery mildew of cucumber

3a. On leaves **3b.** On fruits

Plate 3. Symptoms of cucumber anthracnose

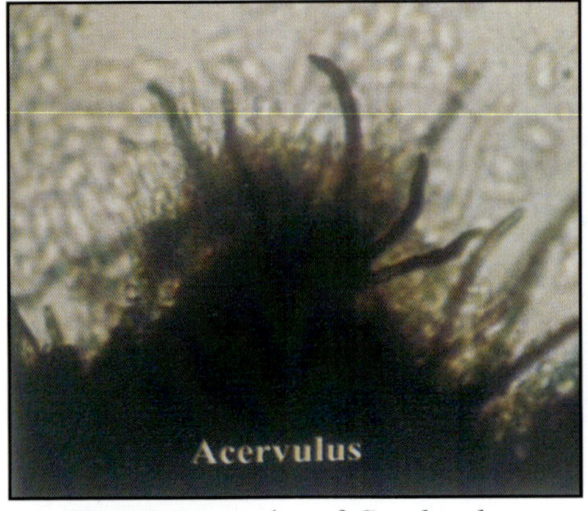

Plate 4. Acervulus of *C. orbiculare*

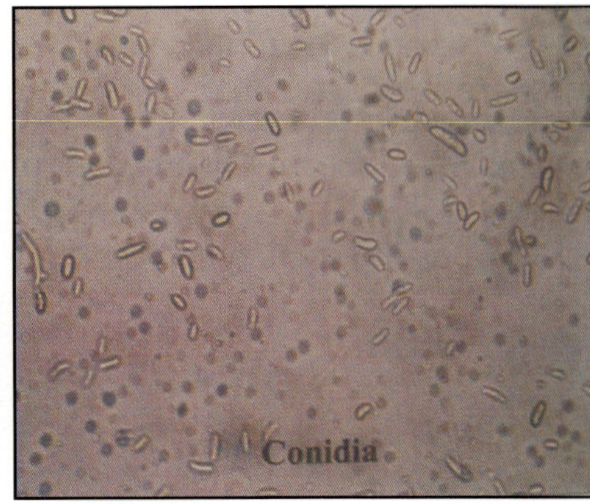

Plate 5. Conidia of *C. orbiculare*

Chapter 6. Disease of Cucurbits

Plate 6. Symptoms of Fusarium wilt of muskmelon

Plate 7. Symptoms of angular leaf spot

Plate 8. Symptoms of cucumber mosaic virus on summer squash

Chapter 7. Disease of Pea

Plate 1. Symptoms of powdery mildew of pea

Plate 2. Conidiophore and conidia of *Erysiphe pisi*

3a. Aerial stage

3b. Telial stage

Plate 3. Symptoms of Pea rust

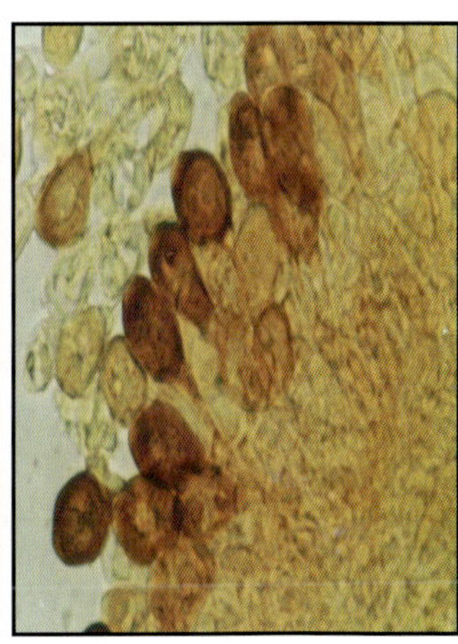

Plate 4. Teleutospers of *U. viciae fabae*

Chapter 7. Disease of Pea

5a. Blight

5b. Foot rot

5c. Pod rot

Plate 5. Symptoms of Aschochyta blight of Pea

Plate 6. Symptoms of Fusarium wilt

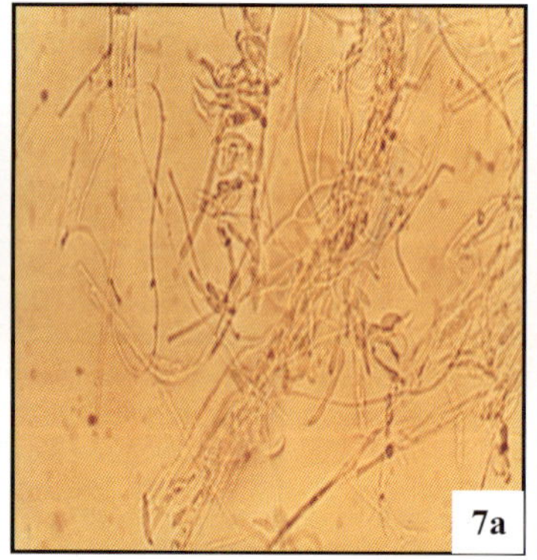

Plate 7a. Mycelium and spores of *F. oxysporim* f. sp. *pisi*

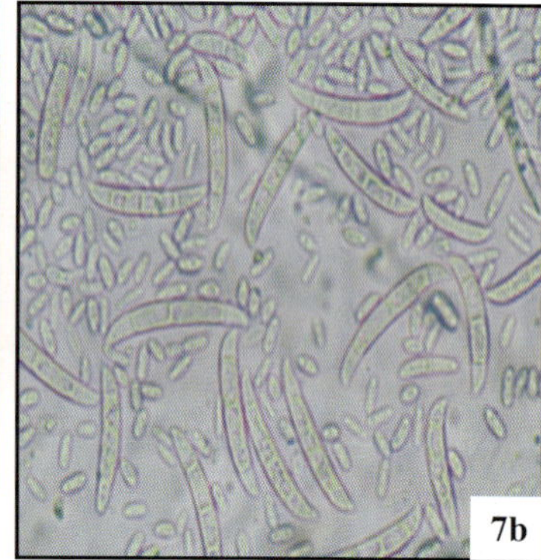

Plate 7b. Macro and Micro-conidia of *F. oxysporim* f. sp. *pisi*

Plate 8. Symptoms of white rot of pea (Courtesy A.S. Kapoor)

Plate 9. Symptoms of bacterial blight of pea

Chapter 8. Disease of French bean

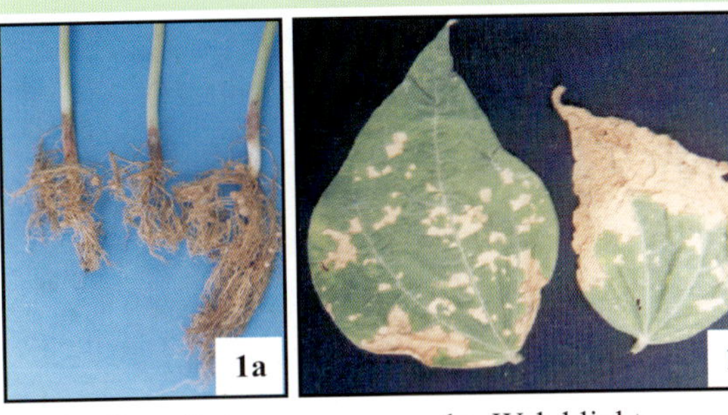

1a. Root Rot **1a.** Web blight **1c.** Pod blight

Plate 1. Symptoms of root rot and web blight

2a. On leaves

2b. On pods

Plate 2. Symptoms of angular leaf spot

Plate 3. Symptoms of Anthracnose on pods

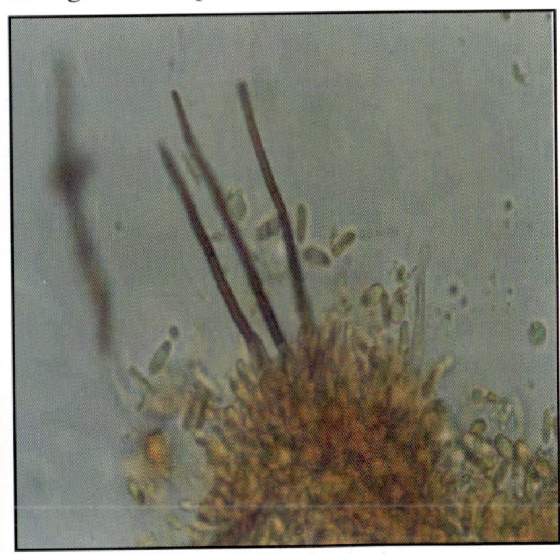

Plate 4. Acervulus of *Colletotrichum lindemuthianum*

Chapter 8. Disease of French bean

Plate 5. Symptoms of Floury leaf spot of french bean

Plate 6. Symptoms of bean rust

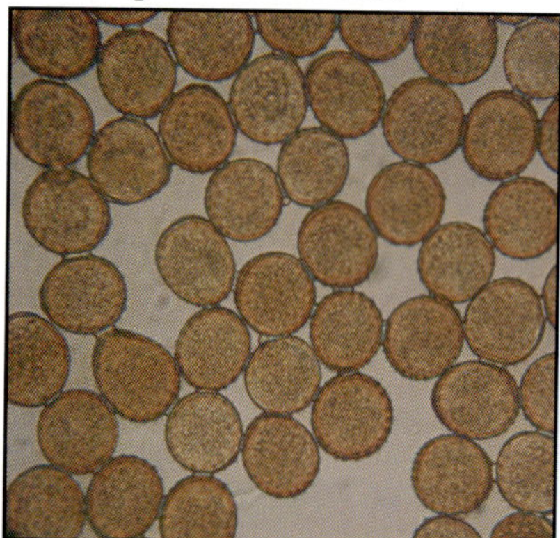

Plate 7. Urediniospores of *Uromyces appendiculatus*

Plate 8. Symptoms of bean common mosaic

Chapter 9. Disease of Chillies and Bell Pepper

1a. On Chillies

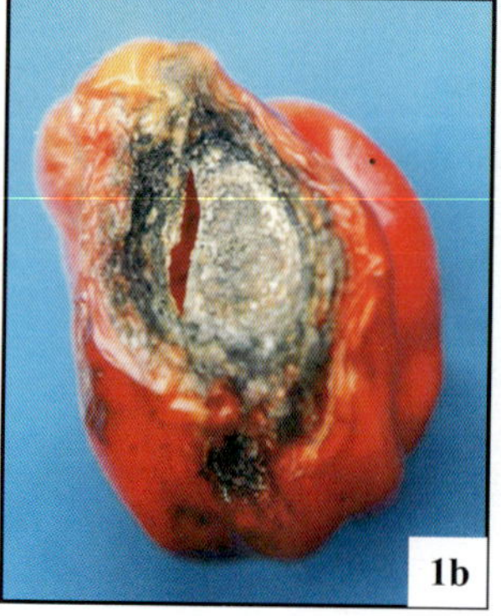

1b. On Bell pepper

Plate 1. Symptoms of anthracnose on ripe fruit of bell pepper

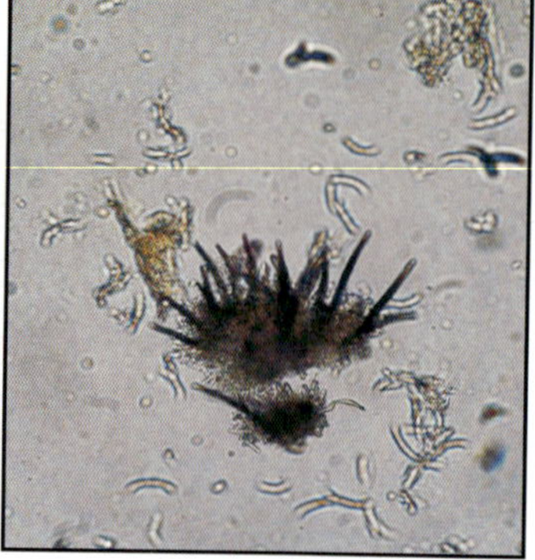

Plate 2. Acervulus of *Colletotrichum capsici*

3a. On leaf **3b.** On fruit

Plate 3. Symptoms of Phytophthora blight

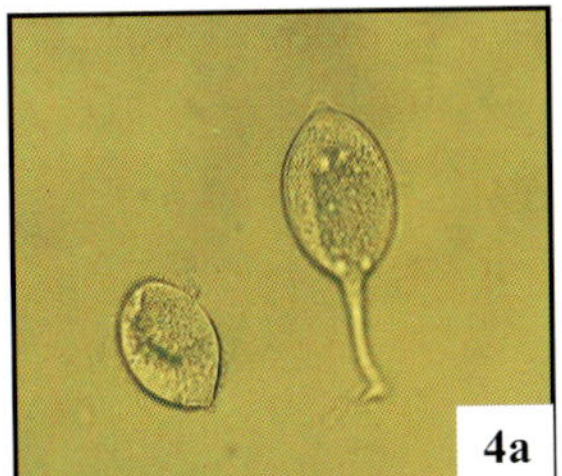

Plate 4a. Sperangia of *P. nicotianae* var. *nicotianae*

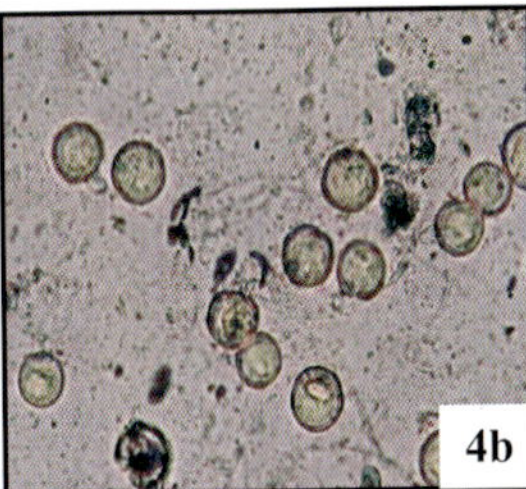

Plate 4b. Oosphores of *P. nicotianae* var. *nicotianae*

Plate 5. Symptoms of Cercospora leaf spot

6a. On leaves **6b.** On fruits

Plate 6. Symptoms of bacterial spot

Plate 7. Symptoms of leaf curl of chilli

Plate 8. Symptoms of Mosaic of chilli

Plate 1. Symptoms of downy mildew

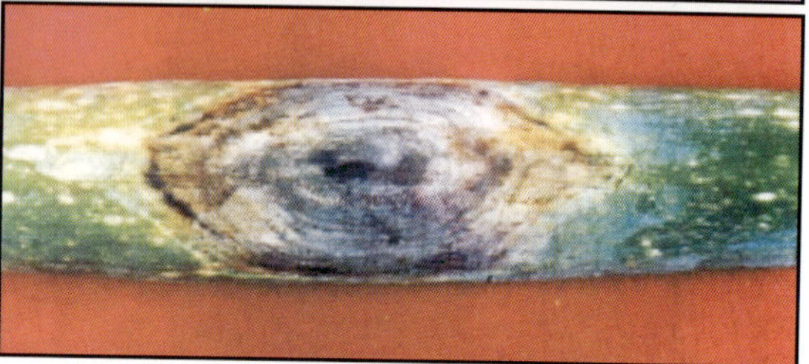

Plate 2. Symptoms of purple blotch of onion

Plate 3. Symptoms of Stemphylium blight

Plate 4. Symptoms of *Fusarium* basal rot

Chapter 11. Disease of Egg Plant

1a. Leaf **1b.** Fruit

Plate 1. Symptoms of Phomopsis blight

Chapter 13. Disease of Colocasia

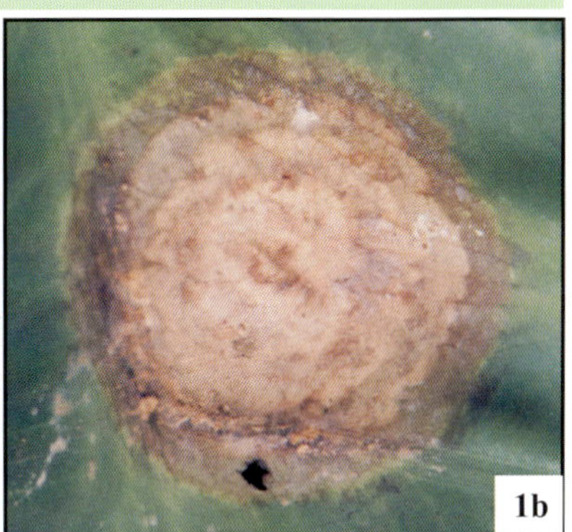

1a. On Leaf **1a.** Close up of lesion

Plate 1. Symptoms of Phytophthora blight

Chapter 14. Disease of Okra

Plate 1. Symptoms of Fusarium wilt of okra

Plate 2. Symptoms of Cercospora leaf spot on okra

Chapter 14. Disease of Okra

Plate 3. Symptoms of powdery mildew of okra

Plate 4. Symptoms of okra yellow vein mosaic

Chapter 15. Disease of Leafy Vegetables

Plate 1a. Symptoms of stem gall of coriander

Plate 1b. Symptoms of gall on flowers and fruits of coriander (Photo courtesy: N. Mehta)

2a. On Field view

2ab. On leaves closeup

2a. On Sem (courtesy: N. Mehta)

Plate 2. Symptoms of Powdery mildew

Chapter 15. Disease of Leafy Vegetables

Plate 3. Symptoms of Sclerotinia lettuce drop (Photo courtesy: Meenu Gupta)

Chapter 16. Disease of Ginger

1a. Field showing infection of rhizome rot

1b. Dry rot

1c. Wet Rot (Photo courtesy: V.B. Santhakumar)

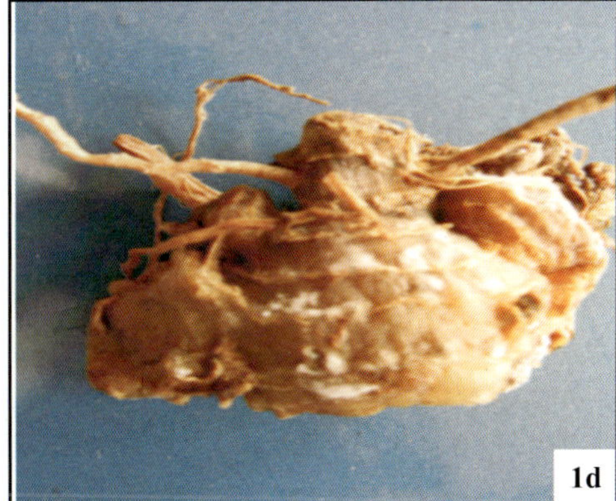

1d. Rotten Rhizome

Plate 1. Symptoms of Rhizome rot

Chapter 16. Disease of Ginger

Plate 2. Symptoms of Phyllosticta leaf spot

Chapter 18. Disease of Vegetables under Protected cultivation condition

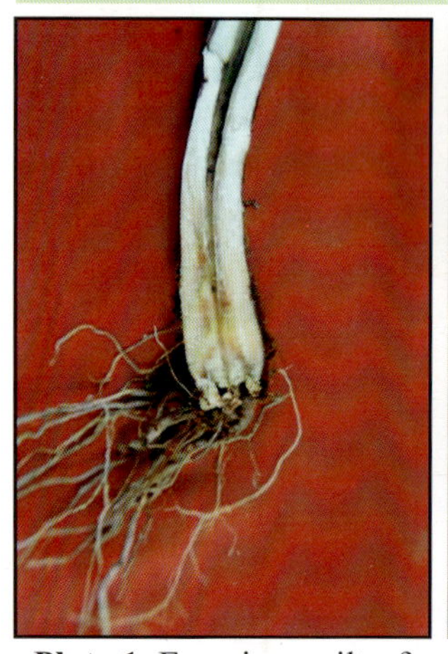

Plate 1. Fusarium wilt of capsicum

Plate 2. Sclerotium stem rot of capsicum

Plate 3. Phyllostica leaf spot of tomato

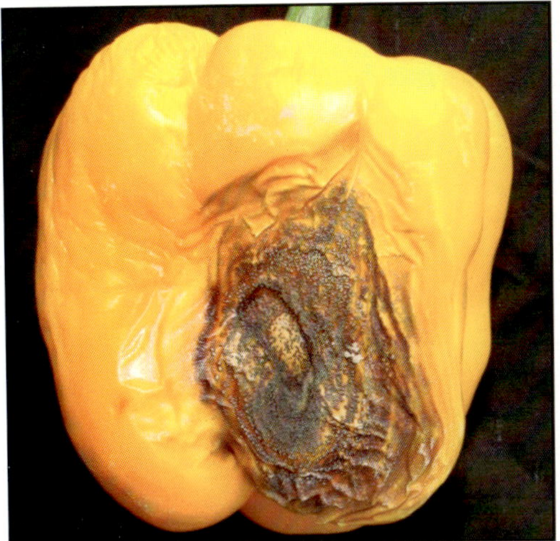

Plate 4. Anthracnose and ripe rot of capsicum

Plate 5. Symptoms of powdery mildew of capsicum

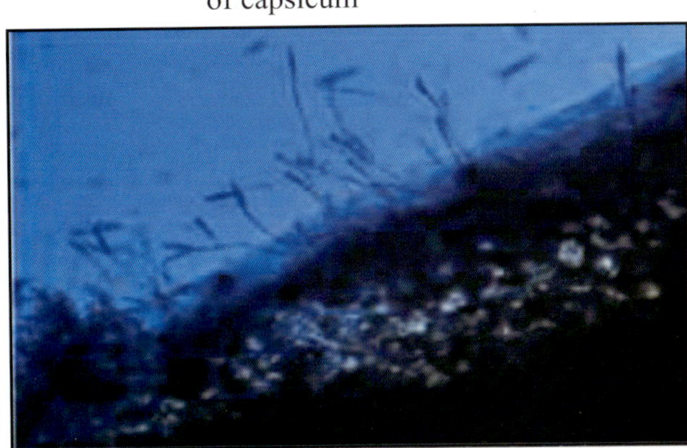

Plate 6. Conidia of *Leveillula taurica* borne on hyphal branches

Plate 7. Downy mildew of cucumber

Plate 8. Symptoms of tomato spotted wilt

Plate 1. Symptoms of root knot on plants

Plate 2. Symptoms of root knot on tomato roots

8

DISEASES OF FRENCH BEAN

Various types of beans are grown throughout the world belonging to different genera of the family Leguminosae. The important members of broad group termed as beans are common bean or French bean (*Phaseolus vulgaris* L.), lima bean (*P. lunatus* L.), scarlet bean (*P. multiflorus* Wild.), tepary bean (*Mucuna* sp.), field bean or lablab (*Dolichos lablab*), sword bean (*Canavalia ensiformis* DC.) and cluster bean (*Cyamopsis tetragonoloba* (L.) DC.). These are grown as grain crops, as vegetables for green pods and as green manure crops for incorporating into the soil.

French bean is cultivated throughout the world both for green pods and dry pulse. It has gained increasing popularity due to its quality proteins and nutritional balance besides certain medicinal properties. In India, the crop is largely grown in Himachal Pradesh, Jammu and Kashmir, Uttar Pradesh, North Eastern Hills, Darjeeling, South Plateau, South Indian Hills (Nilgiri and Palni hills) Mahabaleshwar, Ratnagiri (Maharastra) and Chickmangalore (Karnataka) having mild climate with humid environmental conditions. In India, French bean occupies an area of 137.54 thousand hectares with a production of 1370.21 thousand MT (Anonymous, 2014).

Successful cultivation of this crop is hindered due to the attack of various diseases of fungal, bacterial and viral in nature. Under favourable environmental conditions, epiphytotics of these diseases have often reduced the yields considerably. The information on various diseases of French bean is given in this chapter.

(A) FUNGAL DISEASES

1. ROOT ROT AND WEB BLIGHT

Root rot and web blight is a very serious disease of this crop throughout the world. In warm and wet conditions it may assume serious proportions and cause significant losses.

Geographical distribution

Root rot of French bean was first recorded during the year 1924 from New Jersey (Cook, 1924). Thereafter, it was reported from Egypt, England, Ceylon, Uganda, Tanganyika, Rhodesia, Brazil, Manitoba, Medellin Columbia, Iran, Malaysia, Japan, Venezuella, Spain, Cuba, North Minas Gerais, Turkey and France ((Temizel and Ertunc, 1992; Kouassi, 1993).

Web blight was first described as destructive wilt of beans and cowpeas in Puerto Rico (Matz, 1921). Subsequently, it was reported on beans from Sierre Leone, Florida, Texas, Brazil, Burma, Ceylon, Japan, Philippines, Mexico, Costa Rica, Central America and Trinidad (Rajnauth, 1987).

In India, root rot disease of French bean was recorded for the first time in 1980 from Bangalore (Sharma and Sohi, 1981). Later Srivastava and Gupta (1989) reported the aerial blight phase of this disease from Sikkim. Paul and Sharma (1990) reported the occurrence of root rot from the Solan area of Himachal Pradesh, while web blight was recorded during 1994 crop season (Mathew and Gupta, 1996) from the same locality.

Losses

Losses from root rot vary greatly in different localities and from year to year. Five to ten per cent loss is not uncommon and more has been noted in situations especially favourable to the fungus. Several workers have reported varying degree of losses due to web blight. Rajnauth (1987) reported 25–100 per cent yield losses in Trinidad. In Brazil, under favourable environmental conditions the bean crop has been destroyed within 48–72 hours (Sartorato, 1988b).

In India, Sharma and Sohi (1980) reported that yield losses of green pods at different stages of plant growth varied from 8.4 to 64.6 per cent. They also observed up to 15 per cent incidence of root rot in fields around Bangalore (Sharma and Sohi, 1981).

Symptoms

The disease occurs on roots, stem, leaves, pods and seed (Mathew and Gupta, 1996). Sanchez and Cardenaz (1988) reported that root rot affects plant populations from emergence to the first 30 days of crop growth. The distinguishing symptoms of the disease occur in two phases and are described below:

(a) Root rot: Symptoms appear on roots and stem above and below the surface of the soil as reddish brown cankers. The lesions enlarge rapidly and engirdle the stem at the collar region, extending longitudinally downward to the roots. Roots are affected in advanced stages of infection leading to partial or complete rotting of the root system (Plate 1a).

(b) Web blight: The symptoms on leaves appear as small, circular, water-soaked spots (Plate 1b). These spots increase rapidly in size and cover extensive areas of the leaf blade. Initially, the spots are much lighter in colour than the surrounding unaffected area. At later stages, they become tan brown and surrounded by dark borders. The spots coalesce to form larger areas on the leaf blade. Light tan hyphae develop on both surfaces of the infected leaf and at later stages are studded with minute sclerotia of the fungus. Defoliation occurs in advanced stages of infection. The pods are attacked at all stages of growth. On young pods, circular slimy water-soaked spots appear which are light tan in colour. On mature pods, the spots are dark brown, more or less circular, slightly zonate and definitely sunken (Plate 1c). Spots coalesce to cover the entire pod. The distal end of the pods is the most affected. In favourable weather, runner hyphae of the fungus are seen spreading from the point of infection to healthy parts, which later become studded with sclerotia. Tan brown to reddish brown discolouration is observed on infected seed, located immediately below the spots on the affected pods.

The Pathogen

The disease is caused by *Rhizoctonia solani* Kuhn. The fungus consists of septate mycelium. The colour of the hyphae varies from hyaline when young to brown on maturity. The mycelium is branched near the distal septum of the mother hyphal cells at right or acute angles and the branches are constricted at or near the point of origin of the septum. The size of hyphae varies from 117.7–173.4 x 6.19–9.29 μm. In culture, sclerotia develop within 5–6 days. The size of sclerotia ranges from 1 to 3 mm in diameter (Butler and Bracker, 1970).

The sclerotia consist of brown barrel shaped monilioid cells measuring 20–35 x 10–15 μm. The fungus could be seen growing as white crust on infected foliage and as white mycelial web at the distal ends of severely infected pods. In advanced stages, smooth sclerotia measuring 1.0 to 2.5 mm in diameter are seen adhering to infected plant parts. The basidial stage of the fungus (*Thanatephorus cucumeris*) also occurs but very rarely.

Variability in *R. solani* has been studied with respect to cultural, morphological, pathological, physiological, anastomosis grouping and isoenzyme analysis (Ogoshi, 1987). In Himachal Pradesh, Upmanyu *et al.* (2003) also carried out these studies with 18 isolates. The colony colour of all the isolates varied from white to brown and produced varied patterens of sclerotial formation except two (isolate I_7 and I_{18}), which did not produce sclerotia. All the isolates of *R. solani* exhibited slight variation in their physiological requirements. Most of the isolates preferred Czapek's dox agar as solid medium, while potato dextrose, broth supported the maximum growth of the isolates. All the isolates grew rapidly within a temperature range of 25 to 30°C and RH above 95 per cent. pH ranging between 5.0 and 6.0 supported the maximum growth of the isolates. However, isolates preferred slightly higher pH as compared to lower ones. Glucose among carbon and ammonium tartrate as well as potassium nitrate among nitrogen sources yielded maximum mycelial growth of most of the pathogen isolates (Upmanyu and Gupta, 2009). Isolates also varied significantly in aggressiveness within each cultivar while cultivars showed significant variation in horizontal resistance. Two isolates (I_8 and I_9) were the least sensitive to Bavistin while I_{10}, I_{13} and I_{17} exhibited maximum sensitivity. The isolates were placed in AG-1-1A, AG-1-1B and AG-4 anastomosis groups on the basis of their behaviour. Isolates produced different banding patterns for esterase as well as NADH dehydrogenase and were clustered into different groups on the basis of similarity coefficients. Banding pattern of esterase clearly indicated the relationship with anastomosis groups of *R. solani* isolates. Isolates of AG-4 varied very little while AG-1-1B and AG-1-1A were slightly more variable (Fig. 1a and 1b).

Sharma *et al.* (2005) also studied the morphological and molecular variability among 24 isolates of *R. solani* collected from different hosts and locations of north India and classified the isolates into four groups. However, French bean isolates were grouped into two separate clusters indicating the high genetic variability in the pathogen population prevalent in different epidemiological regions of Himachal Pradesh.

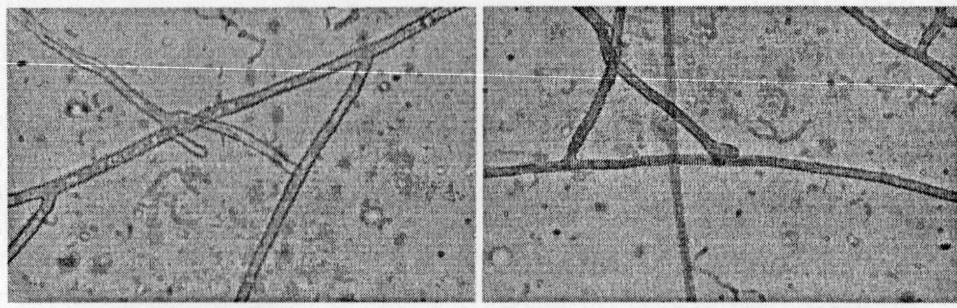

Fig. 1a. Type 2 reaction (AG-4 x I_{15}) **Fig. 1b.** Type 3 reaction (AG-1-IB x I_4)

Host range

The fungus is polyphagous in nature and has been recorded on a number of hosts from various parts of the world. It is most widely distributed and attacks 287 genera belonging to dicotyledonous and monocotyledonous families of Angiosperms and also three each of Gymnosperms and Pteridophytes (Ahuja, 1976).

Disease cycle

The pathogen perennates as mycelium or sclerotia in the soil on plant debris. The fungus is present in most of the soils and once established in a field remains there indefinitely. The germination of sclerotia gives rise to infective propagules, which are sufficient to cause infection. The fungus spreads with rain, irrigation or floodwater, tools and with infected or contaminated propagation material. The association of *R. solani* with bean seed has been reported (Nitshe and Cafati, 1985). Seed transmission upto 46 per cent in cultivars Kentucky Wonder and Contender (Gupta *et al*., 2000) have also been reported. Shama and Shama (1989) indicated that seed penetration of *R. solani* occurs directly through intact seed coat or through the hilum tracheids and that the pathogen is transferred from infected seed to the soil. The infected seeds after germination either rot in the soil itself or give rise to plants that later damped off. The soil borne inoculum is splashed by rain to the foliage that causes web blight. On some plants or plant parts the pathogen can enter only through wounds or through openings in the epidermis, while in others, it thrives in the soil or in mud splashed on to the plants until it can enter through healthy tissue. In favourable weather, runner hyphae of the pathogen are seen spreading from the point of infection to healthy parts, which later become studded with sclerotia (Fig. 2).

Epidemiology

Root rot is favoured by high soil temperature and soil moisture, while web blight is more prevalent in the wet season. *Rhizoctonia* diseases of roots are most severe under moist soil conditions (Tu *et al*., 1992). Bruggen *et al.* (1986) reported positive correlation between lesion size and soil moisture. If the moisture is sufficient enough or there is frequent drizzle, pronounced blight symptoms will appear. During humid periods fan shaped whitish hyphae spread from the point of infection and produce the conspicuous and destructive phase of the disease. Diaz Plaza *et al.* (1991) reported that disease incidence and severity increased after rain. Upmanyu and Gupta (2005) studied the role of abiotic factors on the development of this disease and reported that root rot development was favoured at high (80%) soil moisture and 25°C temperature, while 100 per cent relative humidity coupled with 25°C temperature favoured web blight. Continuous leaf wetness for atleast 6h was essential for disease initiation and increase in leaf wetness durations from 6 to 12 h showed a corresponding decrease in incubation period from 36 to 26 h whereas, further increase in leaf wetness did not exert any effect on incubation period. High rainfall and soil moisture coupled with high relative humidity and soil temperature (23–25°C) were found to be favourable for web blight (Mathew and Gupta, 1996). A high soil moisture (80%) and a temperature of 25°C were the most favourable for root rot development, while web blight development was optimum at >85% relative humidity coupled with 25°C temperature (Upmanyu and Gupta, 2005).

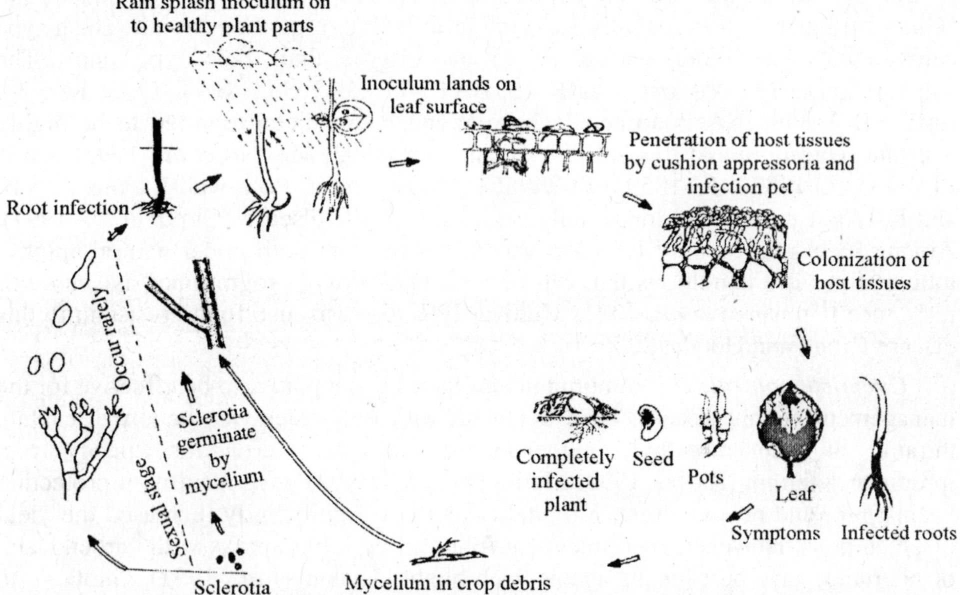

Fig. 2. Disease cycle of root rot and web blight of French bean

Management

The causal fungus persists in the soil, therefore, an integrated approach of disease management combining the use of cultural, biological and chemical methods is required.

(i) Cultural practices: Cultural practices *viz.,* planting disease free seed, mulching, adjustment of planting date, medium (30–60 x 15 cm) plant spacing, crop rotation and soil solarization have been recommended to keep the disease under check (Sartorato *et. al.*, 1988; Tu, 1992, Sharma and Gupta, 2003). Deep tillage (30 cm subsoiling) can reduce soil compaction and thereby, root rot severity in addition to increased yield (Tan and Tu, 1995). Voland and Epstein (1994) reported the efficacy of fresh manure over composted manure in reducing bean hypocotyl lesions, while soil amendment with cotton, mustard and neem cakes was found effective in reducing root rot incidence and increased yields (Upmanyu *et al.*, 2002). The effectiveness of these amendments may be due to enhanced populations of non-parasitic microorganisms exhibiting antagonistic effects on the pathogen, thereby inhibiting the mycelial growth and sclerotia production of the pathogen. Soil amendments with low soil C:N ratio increased incidence of *R. solani* whereas high C:N ratio amendment did not interfere with the incidence, while incidence was independent of soil moisture conditions (Fenille and Souzo 1999). Poultry manure, composted *Urtica* sp. and composted *Lantana camara* treatments were superior to other composts in suppressing Rhizoctonia root rot (Joshi *et al.*, 2009).

(ii) Host resistance: Prasad and Weigle (1976) observed that dark seeded PI accessions of French bean *viz.*, 109859, 163583, 174908, 226895 and 300665 were resistant to *R. solani,* while most of the white seeded cultivars were susceptible to both seed infection and pre-emergence damping off. However, Sharma and Sohi (1981) reported all the three French bean cultivars *viz.,* "Contender", "Premier" and "Giant Stringless" to be equally susceptible irrespective of their colour. Bush type cultivars/lines are more susceptible to the disease than pole type ones. The cultivars/lines *viz.,* Wisconsin RRR 46, BAT 477, BAT 332, BAT 1753, RIZ 30, EMP 81, A 300, ICA Pijao and Jackson Wonder have been reported to be highly resistant (Hagedorn and Rand, 1980; Tu and Park 1993; Muyolo *et al.*, 1993), while PLB-43, EC-18834, IC-18593, PI-249554, Phagu Chitra, Pole wall-I, Pole wall-II, and K-13522 were found moderately susceptible to the disease (Gupta *et al.*, 1997). Among 93 cvs./lines, only ET 8396 was found resistant both under natural epiphytotic and in vitro conditions that can be used in breeding programmes as source of resistance (Upmanyu *et al.*, 2004). Cultivar IPR 96-4 was also found resistant to this disease (Khati and Hooda, 2006)

(iii) Chemical control: Different fungicides have been reported to be effective for the management of this disease. Seed treatment with fungicides *viz.,* benomyl, captan, thiram, thiophanate-methyl, carbendazim, carboxin, terrachlor, pencycuron, iprodione + thiram (Raffat, 1992; Maringoni *et al.*, 1992) gave maximum protection against pre- and post-emergence mortality and also significantly increased the yield of green pods. However, seed treatment followed by foliar sprays with carbendazim or bitertanol gave best results against web blight (Thakur *et al.*, 1991). Gupta *et al.* (1999) reported good efficacy of Bavistin (0.2%), Celest (0.1%), Topsin-M (0.2%) and Raxil (0.2%) against pre-emergence damping off of beans. Good control of web blight phase of this disease was achieved by seed treatment with Celest (0.2%) or Raxil (0.2%) or Bavistin (0.2%) followed by foliar sprays of Bavistin (0.05%) or tebuconazole (0.05%) (Mathew and Gupta, 1996; Upmanyu *et al.*, 2002).

(iv) Biological control: Exploitation of the populations of bioagents may prove useful to combat the ravages of soil borne diseases. Biological treatment of soil under glass house conditions with *Trichoderma viride* showed results similar to chemical control (Dumitras and Sesan, 1990). Elad *et al.* (1980) found that in naturally infested soils, wheat bran preparations of *Trichoderma harzianum* significantly decreased the *Rhizoctonia* disease of beans. Tu and Vaartaja (1981) reported that *Gliocladium virens*, a hyperparasite of *R. solani*, inhibited production of sclerotia. Management of this disease by seed/soil treatment and foliar sprays with *T. harzianum, T. virens* and *T. viride* has also been reported (Mathew and Gupta, 1998; Hazarika and Das, 1998; Upmanyu *et al.*, 2002). Inhibition of mycelial growth and sclerotial germination of this pathogen by *T. harzianum* and *T. viride* culture filterate has also been observed (Hazarika and Das, 1998). Roberti *et al.* (1993) indicated that *T. virde* produced non-volatile compounds while *T. harzianum* produced coils and hook pincer shaped structures, which prevented the growth of *R. solani*. The efficacy of potential bacterial antagonists *viz.*, fluorescent pseudomonads and *Bacillus* spp. against root rot has also been reported (Sanchez *et al.*, 1994; Wolk and Sarkar, 1994, 1995). An endophytic *Pseudomonas florescens* carrying *Serratia marcescens* Chi A gene either on the plasmid or integrated into the chromosome controlled effectively the phytopathogenic fungus *Rhizoctonia solani* on bean seedlings under plant growth chamber conditions (Downing and Thomson, 2000). Coinoculation of seed with *Bacillus subtilis* MB1600 (Epic) and *Rhizobium tropici* significantly reduced root rot severity and enhanced yield (Estevez-de-Jensen *et al.*, 2002). Efficacy of seed treatment and soil application of *B. subtilis* in the reduction of pre- and post-emergence root rot have also been reported (Sharma and Gupta, 2003). The water extracts obtained from green parts of potato, tomato and rape and *Trichoderma* PBG-1 showed marked antagonistic activity against *R. solani* (Smolinska and Kowalska, 2006). Out of sixty isolates of *T. harzianum*, 7 isolates significantly reduced the incidence of root rot compared to the control. *T. harzianum* isolates Tr-34 and Tr-14 were the most effective, resulting in only 20.8 and 23.4 per cent root-rot incidence, respectively (Joshi *et al.*, 2008).

(v) Integrated disease management: Integration of seed treatment with bioagents and fungicides has been found to be more effective to manage soil borne pathogens. Tu and Zheng (1993) studied the compatibility of seed treatment with *Bacillus subtilis*, *Pseudomonas fluorescence* or *Gliocladium virens* and subsequently dressed with diazinon + captan + thiophanate-methyl. They observed that fungicide treated seed showed inhibition of bioagents, which was transient, and the growth and sporulation of the bioagents resumed as the effect of the fungicides diminished. They further suggested that bioagents such as *B. subtilis* and *P. fluorescence* are more mobile in the rhizosphere of seedlings grown from treated seed and hence, their incorporation together with fungicides can extend the effect of the chemical to give protection against root rot. Upmanyu *et al.* (2002) also reported the tolerance in *T. viride* to carboxin, tebuconazole and carbendazim. The combined seed treatment of French bean with Contaf (0.025%) and *Gliocladium virens* (0.1%) resulted in higher seed germination, lower incidence of the disease and higher grain yield (Mukherjee *et al.*, 2001). Combined treatment of seed (soaking in *T. harzianum*, 1 x 10^7 conidia/ml) and soil (20 and 100 kg/calcium) and micronized sulfur (40% micron sulfur), respectively before sowing reduced the root rot incidence and increased green pod yield (Ziedan and Mahamoud, 2002). Soil amendment with mustard cake alongwith carbendazim seed treatment and foliar sprays were found highly effective in reducing the pre-emergence and post-emergence root rot and web blight severity (Upmanyu *et al.*, 2002). A combination treatment including soil solarization (SS) for 50 days + soil amendment (mustard cake) + combination of *Allium sativum* ethanol extract and *Bacillus subtilis* was found effective in reducing pre- and post-emergence damping-off, while web blight was best contained by SS + soil amendment + *A. sativum* (ST + FS) and increased yields (Sharma and Gupta, 2003).

2. ANGULAR LEAF SPOT

It is an important disease of French bean particularly in mid hill conditions where moderate temperatures with high humidity conditions favour the development of this disease.

Geographical distribution

The disease has been reported from all the French bean growing areas of the world. For the first time, it was reported in the year 1878 from Italy (Saccardo, 1878). Since then, it has been reported from different continents of the world like U.S.A, Europe, Africa and Asia (Gupta and Shyam, 2003).

In India, the disease was reported for the first time from Nilgiri hills in South India (Srinivasan, 1953). Subsequently, the disease was recorded in Solan, Himachal Pradesh (Sohi *et al.*, 1963), Chaubattia, Uttarakhand (Bose and Sindhan, 1972), Bangalore, Karnataka (Ponnappa *et al.*, 1976) and Tadong, Sikkim (Srivastava and Gupta, 1994). The disease has been appearing regularly in varying intensities in all bean-growing areas.

Losses

The losses due to angular leaf spot are of two types. Firstly, there is reduction in photosynthetic capacity because of the destruction of foliage which results in subsequent loss of green pod and secondly, the green pod infection makes the pods unfit for market. Several workers have reported varying degree of losses in different parts of the world. Up to 50 per cent yield losses under epiphytotic conditions has been reported from USA whereas in Mexico, it has been considered as a principal fungal disease where up to 90 per cent incidence has been recorded causing significant losses (Campos and Fucikovsky, 1981). In Brazil, Sarotorato (1988a) also observed severe yield losses due to this disease.

In India, due to this disease the total loss including damaged and unmarketable pods has been estimated to be about 40–70 per cent (Singh and Saini, 1980) while Ponappa *et al.* (1976) reported 80 per cent yield loss and premature death of the plants from Bangalore area of Karnataka. In Sikkam, Srivastava and Gupta (1994) reported up to 20–25 per cent loss in grain yield every year due to this disease.

Symptoms

The pathogen infects all green parts of the plants viz., leaves, petioles, stems, pods and seed, producing varied symptoms. The detailed symptoms are described below:

(i) Leaves: On cotyledonary leaves, the symptoms appear as circular spots while on true leaves, 3–5 angled spots appear in between veins and veinlets (Plate 2a). The spots vary from 0.5 to 8.5 mm in size and are dark grayish on upper surface and light gray on the lower surface of the leaves, but with the passage of time, change to reddish brown and finally attain dark brown colour. On the lower surface, light reddish discolouration appears on the veins and veinlets, which delimit such spots. Gray moldy coating covers the spots on the lower surface of the leaves. On close observation, the spots reveal the presence of coremia bearing large number of spores. Many of such spots coalesce covering a large portion of the leaf giving it dirty brown appearance, whereas the remaining part turns yellowish. Such severely infected leaves show upward curling and defoliate prematurely.

(ii) Petioles and stems: Elongated dark brown spots appear both on petioles and stems. These are concentrated towards the point of leaf attachment on the petiole while on the stem, at the point of origin of branches. Under severe conditions, complete defoliation occurs resulting in pre-mature death of the plants. Coremia formation is rare on these parts under natural conditions.

(iii) Pods and seed: Pods are infected at all the developmental stages. Lesions on the pods are smooth, usually circular and rarely elongated (Plate 2b). At first, these spots are superficial having reddish brown centre with well-defined ashy black borders, but with the passage of time, deeper tissues are involved. In the central reddish brown portion, coremia and spores are produced. Severely infected pods either bear no seeds or produce only shrivelled seeds. In late infections *i.e.,* at maturity stage, the seeds underneath the pod lesions show yellowish brown discolouration which is more prominent on hilum region, but can occur any where on the seed surface.

The Pathogen

The disease is caused by fungus *Phaeoisariopsis griseola* (Sacc.) Ferr. On the host tissue, the fungus produces a stromatic structure in the substomatal cavity. Conidiophores are laterally joined together in the form of synnemata or coremia, which are darker at the base and lighter towards the tips, and either curved flexuous or septate. The coremia measure 128–680 μm in length and 28–69 μm in thickness. Conidia are light grey to hyaline, cylindrical to spindle shaped and straight to slightly curved. Conidia are borne singly on the tips of columnar hyphae, which are mostly 1–3 (rarely 4–6) septate, with dimensions ranging from 50–60 × 7–8μm. Karanja *et al.* (1994) also reported cylindrical to spindle shaped brown coloured conidia of average dimensions ranging from 17–95 × 7–24 μm and 2–3 septate. No sexual stage of the fungus has yet been observed.

Villegas (1959) working with 30 isolates of *P. griseola* collected from different bean production sites in Colombia reported the existence of 13 pathogenic races. In India, existence of a new strain of *P. griseola* from Himachal Pradesh, India on the basis of morphological characters *viz.,* length and width of synnemata, conidiophores, conidia and conidial septation has been reported (Sharma and Sohi, 1978). Similarly, Lacerda *et al.* (1994) identified five races (*viz.,* G1, G2, G3, G4 and G5) from Pennambuco state, Brazil of which G1 was the most common and G3 most virulent. Earlier Hocking (1967) recovered a Tanzanian isolate of *P. griseola*, which produced circular lesions and speculated the role of single mutation within natural isolates. Gupta and Shyam (2007) reported highly significant differences in the aggressiveness of 12 isolates collected from three districts of Himachal Pradesh.

Using isozyme analysis Correa-Victoria (1987) divided 55 isolates of *P. griseola* from Latin America, the United States and Africa into two groups. All African isolates displayed the same electrophoretic pattern and were associated with large seeded bean cultivars originating in the Andean region of South America, while a second electrophoretic pattern was associated with small seeded bean cultivars from Central and North America. Similarly, Guzman *et al.* (1995) using RAPD markers differentiated 62 isolates of *P. griseola* from Malawi, USA and Brazil into two major groups (Andean and Mesoamerican). Corrales and Jara (1995) further suggested that these races have evolved with large seeded Andean cultivars, while Mesoamerican races have evolved with small or medium seeded varieties from Central America.

Host range

The pathogen has a very narrow host range restricted only to the family leguminosae. In 1931, Abramoff reported *I. griseola* to cause angular brown spot on soyabean in Russia. Out of six species of legumes namely *Phaseolus lunatus* L., *Dolichos lablab* L., *D.* sp., *Indigofera sumatrama* Gaertn. *Cajanus* sp. and *Vigna sinensis* L., only *P. lunatus* L. was attacked by this pathogen (Cardona-Alvarez and Walker, 1956). In India, among thirteen leguminous plants tested, only *Phaseolus multiflorus, P. vulgaris* and *P. lunatus* were found susceptible (Verma, 1979).

Disease cycle

In areas where the crop is grown regularly year after year in the same field, the fungus overwinters in the infected plant debris. Seed borne nature of the pathogen has also been reported. In India, the pathogen has been reported to be both seed and soil borne (Sohi and Sharma, 1974; Sindhan and Bose, 1980). Infected seeds used for sowing serve as the source of primary infection. The infected seed after germination gives rise to a plant with circular brown spots on the cotyledonary leaves. These spots on the under surface bear numerous dot like structures known as coremia which bear condiophores and conidia of the fungus. The conidia may be disseminated from the debris by rain splashes or along with wind blown soil particles and from sporulating lesions by wind currents. The pathogen infects host tissue by entering through stomata and advances intercellularly in the mesophyll and palisade parenchyma. After 7–10 days of infection the fungus develops intracellularly throughout necrotic lesions, stromata develop in the substomatal cavity and sporulation may occur during periods (24 to 48 h) of continuous moisture. The conidia cause secondary infection and the cycle continues. In severe infections, the leaves turn yellow and fall down prematurely. Infection from pods also goes to the developing seed (Fig. 3).

Epidemiology

The pathogen (*P. griseola*) overwinters both on plant debris and in infected seed (Sohi and Sharma, 1974; Sindhan and Bose, 1980). The viability of conidia on plant debris is 6 and 8 months in laboratory and field, respectively (Sindhan and Bose, 1980), thus it serves as a source of primary inoculum. Other source of primary inoculum is infected seed. Though the infestation site has been identified in the hilum of seed but in some cultivars both hilum and seed coat seem to be infected (Saettler and Corea, 1988). While disease can initially result from the planting of infected seed causing spots on cotyledonary leaves, the primary route in most bean growing areas involves the rain splashed conidia from plant debris or wind blown conidia with soil particles attacking leaves, pods and stems. Infected leaves become inoculum source for secondary spread of the disease. However, the rate and degree to which secondary spread occurs is dependent on the interactions and influences of all factors involved with the disease and its spread such as climate, pathogen and the host. Moderate temperatures and moist conditions favour the arrival of viable inoculum and successful field infections. Optimum temperature for growth and sporulation, spore germination and development of the disease are 21–24°C, 18–24°C and 24°C (Sindhan and Bose, 1980; Sarotorato, 1988a), while optimum RH for sporulation, spore germination and development of symptoms has been reported as >90 per cent.

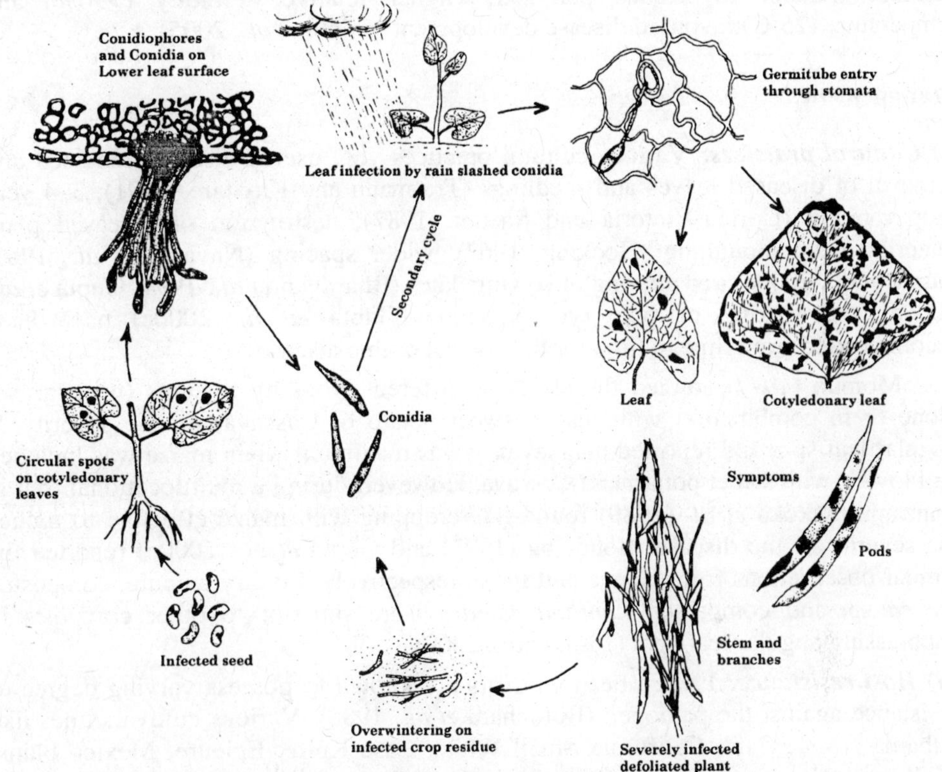

Fig. 3. Disease cycle of angular leaf spot of French bean

Mathew *et al.* (1998) studied the effect of leaf wetness durations at optimum temperature and reported that leaf wetness of at least 12 h duration coupled with moderate temperature (25°C) exhibited minimum incubation period of 9 days. Gupta *et al.* (2005) reported that increase in relative humidity from 85.7 to 89.9 per cent coupled with moderate temperatures of 25°C resulted in sharp increase in disease development. Diurnal periodicity of conidial release of *P. griseola* revealed a distinct rhythm at different times of the day with maximum conidial density being observed between mid night and early morning hours (Gupta *et al.*, 2005). Conidial formation and release was favoured by temperature (23–24°C), high relative humidity (>90%), prevalence of dew (20.00 to 8.00 h) and dark period.

These workers also studied the role of environmental factors on disease development and reported the consistent effect of rainfall, soil moisture and relative humidity on the disease development. Simple correlation coefficients between disease severity and rainfall (0.7596), soil moisture (0.7070) and relative humidity

(0.5253) were found to be highly significant and positive. But Chhetry *et al.* (1994) reported non-significant linear correlation and regression coefficients of disease incidence and severity with microclimatic factors. With inoculum concentration from $1x10^4$ to $5x10^4$ conidia of *P. griseola* and leaf wetness duration from 3 to 9 h, there was a progressive decline of incubation period of angular leaf spot on French bean. Intermittent wetness up to 3 cycles (12 h dry and 12 h wet) was found effective to increase number of lesions per leaf. Higher relative humidity (>90%) and temperature (25°C) favoured disease development (Gupta *et al.*, 2005).

Management

(i) Cultural practices: Various cultural practices viz., use of disease free seed and removal of diseased leaves and seedlings (Trutmann and Kayitare, 1991); 3–4 year crop rotation (Correa-Victoria and Saetler, 1987); destruction of diseased plant material; deep ploughing (Hocking, 1967); wider spacing (Navarro *et al.*, 1981; Gupta *et al.*, 2000e) and altering of sowing dates (Bhardwaj *et al.*, 1994, Gupta *et al.*, 2000e) and staking of pole type varieties (Gupta *et al.*, 2000e) have been recommended from time to time for the control of this disease.

Moreno (1977) studied the effect of different cropping systems (bean grown alone or in combination with maize, sweet potato or Cassava) on the severity of angular leaf spot and reported that severity was maximum when maize was included and lowest with sweet potato and Cassava. However, during a multilocational trial in Tanzania, Kikoka *et al.* (1989) found intercropping with maize effective to reduce the severity of the disease. Boudreau (1993) and Gupta *et al.* (2000e) reported the similar observations from Kenya and India, respectively. Poultry manure, composted *Urtica* sp. and composted *Lantana camara* were superior to other composts in suppressing angular leaf spot (Joshi *et al.*, 2009).

(ii) Host resistance: French bean cultivars are known to possess varying degree of resistance against the pathogen (Buruchara *et al.*, 1988). Various cultivars/lines like Albama No. 1, Café, California Small White, Case Knife, Epicure, Mexico blank, McKastan, Navy bean, Negro Costa Rica, Scotia Rojo Chico, Mexico 11, Mexico 12 and Cauca 27, Carpoata (Santos Filho *et al.*, 1978); Negro Tacana (Lopez *et al.*, 1997) and Manteiga Maravilha (Paula *et al.*, 1998) are reported as resistant. Araiyo *et al.* (1989) reported that non-black cv. Ouro and black coloured cvs. Rio Tibagi and Milionario 1732 possess resistance to *P. griseola*.

In India, cvs./lines *viz.,* EC 38921, EC 44621, PLB 148, Kentucky Wonder, Canadian Round, Canadian Long Red, IC 47651-6, EP 146 (BAT-482), HPR-54, HPR-63, HPR-92, HPR-93, HPR-111, HPR-232, HPR-260, HPR-299, HPR-332, HPR-326, Hans, Nagar Local, SVM-1, JK-8, Him-12, IIHR 42-2, IIHR 901, EC 97830, EC 44758 and NIC 14402 were reported to be resistant against this disease (Singh and Sharma, 1975; Bhardwaj *et al.*, 1992; Srivastava *et al.*, 1995; Mohan *et al.*, 1995; Gupta *et al.*, 2000d). Of the 68 French bean genotypes/lines screened for resistance to *P. griseola* during 2002 and 2003 at two locations namely, Shalimar and Wadoora in Kashmir, 9 genotypes/lines, *viz.,* SL-4-PL-1, HIM-1, Bountiful, JKRO-3, L-27, L-28, L-8 and Local Big Beans, exhibited highly resistant reaction on pods. On leaves, 6 genotypes, *viz.*, Uri-Red, EC-285559, Medur Tral, Local Red, Local Big

Beans and SAW/GH/TA/41A at Wadoora location, and 4 genotypes, *viz.* Gurez Local, Big Bean Black, EC39855 and Local Big Beans at Shalimar location showed highly resistant reaction, while the genotype Shalimar French Bean-1 gave highly susceptible response on leaves at both locations (Bashir *et al.*, 2006).

The effectiveness of supplementing local bean mixture with 25 per cent or more resistant lines (BAT 76 and A 285) was evaluated and it resulted in significant reduction of angular leaf spot severity in Africa (Pyndji and Trutmann, 1992; Trutman and Pyndji, 1994).

(iii) Chemical control: Various fungicides have been reported to give good control of this disease in India and other countries. However, during continuous rains which are essential both for good growth of bean plants and rapid development and spread of the disease, routine application of fungicides is sometimes delayed which often results in heavy build up of the disease. Proper timing of initial sprays is of prime importance to achieve desirable level of disease control.

Attempts have been made to control the disease by both seed treatment with mercurial and other fungicides and foliar sprays with various fungicides. Sprays of fungicides like wettable sulphur, mancozeb, Bordeaux mixture, zineb, carbendazim, benomyl, triforine, triademorph, (Fortugno, 1977; Sindhan, 1984; Issa *et al.*, 1982; Hidalgo and Araya, 1993; Singh *et al.*, 1995), chlorothalonil, TPTH, TPTA (Castro *et al.*, 1991; Vierra *et al.*, 1998) and mancozeb (Issa *et al.*, 1982; Castro *et al.*, 1991; Vierra *et al.*, 1998), have been reported effective to reduce disease intensity and increase yield. Oliveira *et al.* (1992) reported stannic triphenyl (fentin) acetate, bitertanol and tebuconazole as more effective when compared to Bavistin, while Srivastava and Gupta (1994) reported captan, thiophanate methyl, carbendazim, ziram and captafol to be approximately equally superior to mancozeb, copper oxychloride and tridemorph. Combi fungicide sprays like benomyl + captafol, chlorothalonil + copper oxychloride (Issa *et al.*, 1982) and thiophanate methyl + thiram (Rodrigues *et al.*, 1987) have also been found effective. Combination treatments of carbendazim (seed treatment + foliar spray) and bitertanol (seed treatment + foliar spray) were also found effective in reducing the severity and increasing the green pod yields than seed treatment or foliar spray alone (Mathew *et al.*, 1998).

Various EBI fungicides have longer post-infection and eradicant activity being systemic in nature. Mathew *et al.* (1998) studied the mode of action of EBI's like hexaconazole, penconazole and bitertanol along with mancozeb and carbendazim by providing artificial weather conditions for infection. Carbendazim provided hundred per cent control of the disease up to 120 h while mancozeb upto 96 h. EBI's exhibited poor protective activity only up to 72 h. On the other hand, EBI fungicides demonstrated post-infection activity upto 72 h, while it was only 24 h in case of carbendazim and mancozeb. When applied as pre-symptom spray, mancozeb had little effect and produced sporulating spots while EBI's and carbendazim produced reddish brown non-sporulating spots when applied on 7th and 8th day after inoculation. As post-symptom treatment carbendazim, hexaconazole, flusilazole and bitertanol had high level of antisporulant activity and their residue persisted for more than 10 days. These properties of EBI fungicides can be utilized in planning different spray programmes for effective control under different situations.

Of late, spray schedules involving the strobilurine (azoxystrobin), tria-zoles (tebuconazole, propiconazole) and benzimidazole chemical groups were found effective in controlling angular leaf spot (Oliviera and Oliveira, 2003).

(iv) Integrated disease management: Fungicidal treatments coupled with careful selection of varieties and sowing dates proved highly effective to reduce disease severity and increased the yield of the crop (Bhardwaj *et al*., 1994). Effect of two mulches (Pine needles and *Eucalyptus* leaves) alone and in combination with carbendazim (0.1%) sprays on mulch and foliage on the disease severity and pod yields has also been studied by Mathew *et al.* (1998). Sprays of carbendazim on mulches and foliage were highly effective in reducing the disease severity and increasing green pod yields. Sprays of carbendazim on mulches alone were also found quite effective and reduced the severity and increased green pod yields significantly with additional advantage of residue free pods.

3. ANTHRACNOSE

It is an important disease of common bean throughout the world but more severe in tropical and subtropical regions (Pastor-Corrales *et al.,* 1994). Scribner (1888) was the first to use the name anthracnose, which is now generally used. The term anthracnose is derived from the Greek word meaning ulcer and is appropriate because of the ulcer like depressed lesions on the pods.

Geographical distribution

The pathogen was actually collected by Mycologists as early as 1843 in France (Desmazieres, 1843), but the first description of bean anthracnose in 1875, was based on the specimens collected from Lindemuth, Bonn, Germany (Zaumeyer and Thomas, 1957). Subsequently, it was reported from England, Italy, France, Germany, North America, Japan, Taiwan, East and West Indies, Belgian Congo, Uganda, Brazil, Guatemala, Venezuela, Canada, Cuba, Australia, New Zealand and erstwhile USSR (Zaumeyer and Thomas, 1957; Plopper, 1983).

In India, its occurrence was noticed for the first time in Nilgiri hills in 1915 (Hutchinson and Ram Ayyar, 1915). Since then the disease has been recorded in the entire bean growing areas of the country, which have cool and moist weather during growing season.

Losses

Though this disease occurs worldover including tropical and subtropical regions but it causes greater losses in the temperate zones than it does in tropics. Frequent epidemics have occurred in Italy and Germany. It was at one time (between 1912–1920) considered as one of the most serious diseases of bean in the United States, east of Mississippi river (Zaumeyer and Thomas, 1957). Serious economic losses have also been reported from Latin American countries like Venezuela, Colombia and Brazil. Losses from this disease can approach to 100 per cent when highly contaminated seed is planted under favourable disease development conditions. Yield losses of 90–100 per cent have been reported in many countries throughout tropical America and Africa (Pastor-Corrales and Tu, 1989; Lenne, 1992). In epidemic years, the disease can cause considerable yield losses (Sharma *et al.*, 1994; Padder and Sharma, 2010)

Symptoms

The disease may appear on any above ground plant part depending upon time of infection and source of inoculum, but most striking symptoms appear on immature pods. Initial symptoms may, infact, appear on the cotyledonary leaves as small, dark brown to black lesions. Conidia and hyphae then may be disseminated by rain or dew to the developing hypocotyl where rust coloured specks develop. The specks gradually enlarge lengthwise along, and partially around the hypocotyl and young stems, forming a sunken lesion. On leaves, the infection may occur on both sides, but early sign of infection usually appear on the under leaf surface as blackened dead portions of the veins. These may extend to limited adjoining areas. Later, such spots also appear on the upper leaf surface. Lesions are also formed on petioles and stems. On pods, black sunken spots with lighter or grey central area are seen (Plate 3). The central portion of the spots shows pinkish masses of spores of the fungus, especially in wet weather. Later, the sides of these spots appear raised. The seeds obtained from heavily infected pods show brown to light chocolate coloured sunken cankers on the seed coats.

The Pathogen

The fungus, *Colletotrichum lindemuthianum* (Sacc. and Magn.) Briosi and Cav. is responsible for this disease. The name is still widely used in scientific literature despite the fact that the perfect stage (*Glomerella cingulata* (Stonem) Spauld & Shrenk) of the fungus has been described and occurs very rarely (Kimati and Galli, 1970). The mycelium is branched, septate and hyaline at first becoming dark coloured with age. On culture media, conidia may form singly at the ends of hyphal threads. When stromatoid masses form, acervuli are produced upon them (Plate 4) and on the host they form stroma below the cuticle, rupturing the latter if it is still present on the lesion. Conidia, which are borne acrogenously on short conidiophores, develop successively under favourable conditions until pinkish spore masses appear on the surface. Setae occur sparingly. Conidia (4.4–5.3 × 13 to 22 μm) are hyaline, cylindrical with rounded ends or somewhat pointed at one end; a clear vacuole like body often occurs near the centre. One to four, usually two germ tubes arise from a conidium. Appresoria are formed at the tips of germ tubes, especially when they come into contact with host.

The perfect stage of the fungus was originally called *Glomerella lindemuthiana* Shear (Shear and Wood, 1913), but now it has been renamed as *G. cingulata* (Kimati and Galli, 1970). The fungus produces perithecia with a diameter of 120 to 210 μm and rostrum 30 to 80 μm in length. Perithecia contain hyaline and filiform periphyses until 27 days of age and asci which measure 8 × 48–68 μm disappear after 27–30 days. Each ascus contains eight ascospores, which may be alantoid (605 × 20 μm) in shape. Ascospores are ejected from the tip of ascus.

Meyer *et al.* (2001) reported that during attack of this disease, CO_2 assimilation, stomatal conductance to water vapour, and photosynthetic electron transport rate decreased, whereas in dark respiration increased.

Barrus (1911) was the first to differentiate two distinct pathogenic races with in a morphologic species of *C. lindemuthianum* on the basis of reaction of host varieties and designated them as Alpha and Beta. Since then many pathotypes of this pathogen has been described in Europe (Fouilloux, 1979), South and Central America (Menezes and Dianese, 1988), Canada (Tu *et al.*, 1984) and Uganda (Pastor-Corrales and Tu, 1989) etc. Unfortunately, differential bean cultivars were not identical which made it difficult to compare results and to exchange data or resistant germplasm (Pastor-Corrales and Tu, 1989). To overcome this difficulty, a set of twelve bean differential cultivars has been defined with binary values (Pasto-Corrales, 1991). A number designates a race, which is the sum of the binary values assigned to differentials susceptible to the isolate (Pastor-Corrales, 1991).

Using this system, atleast 130 different *C. lindemuthianum* races have been identified. Race characterization has been carried out on isolates from South America (Rava *et al.*, 1994; Tamayo-Molano *et al.*, 1995; Sicard *et al.*, 1997), Central America (Balardin and Kelly, 1998; Gonzalez *et al.*, 1998), USA (Kelly *et al.*, 1994), Canada (Balardin *et al.*, 1997), Brazil (Carneiro, 1999), India (Sharma *et al.*, 1999) and Spain (Fernandez *et al.*, 2000). In many African countries, even though anthracnose is severe, little or no information exists on diversity of *C. lindemuthianum*. Only 56 isolates have been characterized using the binary system in Africa, and a total of 36 races have been identified up till now (Koch, 1996; Bigirimana *et al.*, 2000). Races 401 and 485 have been recorded for the first time in Burundi, Central Africa (Bigirimana *et al.*, 2000). Occurrence of Delta mutant (Race 7), a new race in North of Spain was recorded (Fernandez *et al.*, 2000).

In India, 85 isolates of *C. lindemuthianum* collected from different bean growing areas of Himachal Pradesh were characterized on the basis of their reaction types on International and CIAT differentials (Drijfhout and Davis, 1989) and 12 and 19 races were identified, respectively (Sharma *et al.*, 1999). International differentials, specifically Alpha-Brazil, Beta, Gamma and IND-I to IND-IX were characterized. The races designated as IND I to IND IX were different from those identified in Europe and USA, thus forming a new race group from the Indian subcontinent. On CIAT differential set, only races 65 and 73 resembled the North American races. Out of 90 isolates, 12 races resembled American and European virulences, whereas 17 races (101, 103, 115, 119, 131, 195, 537, 551, 581, 598, 613, 615, 631, 639, 707, 775 and 935) were identified as new, thereby enhancing the total race number to more than 140 globally. RAPD data analysis with nine 10-mer random primers producing polymorphic bands, grouped 90 isolates into two major clusters accommodating 36 and 54 isolates, respectively. There was no congruence between the RAPD pattern and virulence phenogram (Sharma *et al.*, 2007).

Padder *et al.* (2007) analyzed five *C. lindemuthianum* populations causing anthracnose in common bean to know about genetic diversity, migration and the probable rate of spread of races capable of overcoming resistance present in elite cultivars. Diversity within each population (HS) was high with values ranging from 0.26 to 0.31. Average differentiation among populations (GST) was 0.12 and populations were isolated by distance between the two populations. This indicates that less short distance gene flow occurs in Himachal Pradesh. Ribosomal DNA analysis of fifteen isolates representing 15 races of *C. lindemuthianum* restricted with

2 restriction enzymes (*Msp I* and *Hea* III), five haplotypes were observed in the dendrogram thereby revealing low diversity (Padder *et al.*, 2008).

Host range

Besides *Phaseolus vulgaris*, the host range of *C. lindemuthianum* includes *Vigna radiata, V. sinensis, P. multiflorus, P. lunatus, Macrotyloma uniflorum, Dolichos lablab, D. biflorus* and *Canavalia ensiformis.*

Disease cycle

The pathogen overwinters in infected seeds and plant debris. Ravi *et al.* (1999) reported that in seed, the pathogen was mostly present in seed coats and cotyledons, and rarely in embryonic axes. In infected seed, conidia can remain viable for many years (Tu, 1983) and infect young seedlings. The conidia may germinate in 6 to 9 h under favourable environmental conditions and produce one to four germtubes, which in turn form appressorium and get attached to the host cuticle by a gelatinous layer. The later produces infective hyphae, which penetrate the cuticle and epidermis by mechanical pressure, applied by appressoria. Infective hyphae initially produce biotrophic primary hyphae (100–130 μm) that are wide in diameter and entirely intracellular, followed by necrotrophic secondary hyphae (25–40 μm) that are narrow and either intercellular or intracellular (Perfect *et al.*, 2001). The hyphae enlarge and grow between the cell wall and protoplast for 2 to 4 days without apparent damage to the host cells. Several days later, the cell walls are degraded and the protoplast dies, leading to the appearance of water-soaked lesions. Mycelium then may aggregate within the lesion site and form an acervulus, which ruptures the host cuticle. The acervulus contains a layer of 3 to 50 conidiophores depending on the size of lesion. The conidia are dispersed by rain splashes to nearby plants and the cycle can be repeated several times during the growing season. The perfect stage of the fungus is very rarely encountered in nature. Perithecia are formed in the infected tissue and asci, and ascospores are produced within the perithecia. The role of ascospores in disease cycle has not been investigated because their occurrence is very rare (Fig. 4).

Epidemiology

Conidia and/or dormant mycelia in the infected seeds and/or infected plant debris germinate and infect the young seedlings. It can remain viable in infected seed for several years (Tu, 1983). When infected seed germinates, lesions appearing on cotyledons serve as the source of secondary inoculum. The spores are almost entirely water-borne. Primary leaves and the hypocotyl are foci of secondary infections. Borucka and Marcinkowska (2001) reported the importance of wet seasons at early growing period besides availability of a source of inoculum for the initiation and occurrence of this disease. The pathogen is very sensitive to changes in temperature and humidity. It develops most abundantly in cool, wet weather and largely disappears under hot and dry conditions. A relative humidity of 92 per cent and above is essential for infection, the optimum being close to 100 per cent. The fungus requires about 10 mm of rain to establish initial infection (Tu, 1981). The optimum temperature for disease development ranges from 18 to 27°C with maximum intensity at 21°C (Sindhan, 1983) and is markedly reduced at 13°C. Heavy and frequent rains with moderate temperatures (19–25ºC) and high relative humidity (>70%) favoured the progress of the disease in terms of vertical and horizontal spread of the disease (Kumar *et al.*, 1999). Moderate rainfall at frequent intervals are also essential for the local dissemination of conidia present in water-soluble gelatinous matrix and the development of severe anthracnose epidemics. The movement of insects, animals and man may spread conidia particularly when foliage is moist.

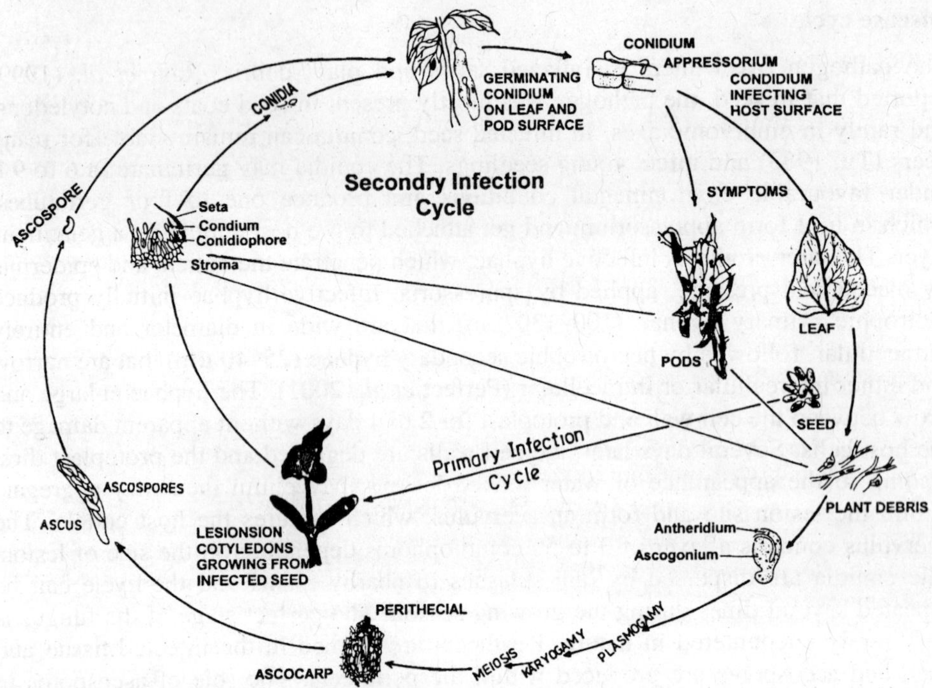

Fig. 4. Disease cycle of anthracnose of French bean

Management

Various management strategies have been attempted to reduce yield losses due to bean anthracnose, which are based on use of fungicides, resistant cultivars, cultural practices and biocontrol of seed borne infection. The most effective method of reducing the losses from this disease is to integrate as many of the available control measures as possible into a disease management programme.

(i) Cultural practices: Use of healthy seed, clean cultivation and three-year crop rotation has been recommended for the management of this disease (Sharma and Sohi, 1989). Disease free seed can be produced either by surface irrigation in semi-arid regions where conditions of high temperature and low humidity prevent infection of this fungus, or under a pedigree seed programme, in which seed plots were isolated and subjected to strict inspection for disease free seed. Best sowing time of French bean in hill regions is in between mid April to mid May for obtaining maximum yield and minimum anthracnose incidence.

(ii) Host resistance: The most promising strategy to control this disease is the use of resistant cultivars. However, the diversity and the high variability of the fungus with the continuous emergence of new races pose problems in breeding programme. Although cultivars with new resistance genes are resistant to majority of pathotypes, however, this resistance can be broken when such a plant is widely planted in fields (Lenne, 1992). Anderson *et al.* (1963) reported that the resistance in cv. Charlevoix against alpha and beta races of *C. lindemuthianum* is conferred by single independent dominant gene. A glycoprotein inhibitor in the cell wall of resistant varieties imparts resistance (Sharma *et al.,* 1986). Due to the presence of much pathogenic variability in the pathogen, non-race specific resistance has been used in breeding for anthracnose. Cultivar IAPAR-Rio Negro-8 is resistant to all races of the pathogen under field conditions (Alberini *et al.,* 1987), while Menezes and Dianese (1988) have identified varieties (Aroana 80, Ayso, Carioca 80, Iguaco, Moruna 80, Riuo Piqiuri, Rio Vermelho, Mulatao, Olho-de-Pombo and Vermelhao) which were resistant to atleast 7 of the 9 races of the pathogen and no race was virulent to all the 70 cvs. screened in Brazil. Chakrabarty and Shyam (1990) demonstrated resistance in cvs./lines VL 60, VL 63, Jawala, HPR 33, B 4 B 6 and P 48 against eight isolates of the pathogen. However, Jawala later exhibited highly susceptible reaction, which may be due to the development of new pathotypes. Kumar *et al* (1997) screened sixty cultivars/lines of kidney bean against ten races viz., beta, gamma, Ind I, Ind II, Ind II, Ind IV, Ind V, Ind VI (alpha-Brazil), Ind VII and Ind VIII of *C. lindemuthianum* under field and artificial epiphytotic conditions and found two accessions (AB-136 and G 2333) as highly resistant to all the above races prevalent in Himachal Pradesh, India, while cultivars Cornell 49-242, EC 43036, EC 57080, KRC 5, PI 207-262 and Widusa exhibited resistance to more than five races, whereas rest were susceptible and these can be utilized in breeding of race specific resistant varieties. Later Gonalves-Vidigal *et al.* (2001) reported that a single dominant gene controls the resistance of cv. AB 136 to both races 31 and 69 and the symbol Co-6 was assigned to the gene. In addition, linkage analysis using random amplified polymorphic DNA marker indicated that Co-6 also controls the resistance of this cultivar to other races of the pathogen, or that different genes are present in the same linkage block.

In the recent years few cvs./lines/land races viz., Negro Otomi (Acosta-Gallegos *et al.*, 2001), SRC 74, SRC 89A, SRC 90, SRC 95, SRC 201, PDR 14 (Sud and Sood, 2002), Morden 003 (Mundel *et al.,* 2004) and HUR-15,HUR-137, IPR 96-4, VL-63, PDR-14, HUR -385 (Khati and Hooda, 2006) have been reported resistant to this disease. Some exotic accessions like G 2333, Cornell 49242, PI 207262, Mexique 222, TO, Perry Marrow, Kaboon and Widusa were resistant to more than five Indian races, whereas two Indian accessions KRC-5 and Hans showed resistance to six and four races, respectively (Pathania *et al.*, 2006). Ninety nine common bean germplasm lines comprising 65 indigenous and 34 exotic accessions were screened against 4 races *viz.,* 03, 515, 529 and 598 of *C. lindemuthianum* under laboratory conditions (Sharma *et al.*, 2012). Among the indigenous collections 47, 34, 32 and 18 accessions showed resistance to race 3, 515, 598 and 529, respectively, whereas among exotic accessions 16, 22, 18 and 11 exhibited resistance to these races. Seven indigenous collections *viz.,* IC-328537, IC-328538, IC-448888, IC313294, IC-278723, IC339645, IC341862 and 3 exotic accessions *viz,* eC-169813, EC-398530 and EC-500226 were reistant to all the evaluated races.

Gupta *et al.* (2010) studied the role of oxidative enzymes and biochemical constituents in imparting resistance to French bean varieties against anthracnose and reported that higher amount of oxidative enzymes (catalase, polyphenol oxidase, peroxidase) was present in diseased leaves than healthy ones and resistant varieties (Kentucky Wonder and Pant Anupam) showed higher enzyme activity than their susceptible (Contender and Laxmi) ones. In case of biochemical constituents, higher amount of total, reducing and non-reducing sugars, phenol content were present in resistant cultivars than susceptible ones. Higher amount of total and ortho-dihydroxy phenols and protein content was observed in diseased pods as compared to healthy pods. Higher content of chlorophyll was observed in healthy leaves of resistant varieties as compared to diseased leaves.

(iii) Chemical control: Various systemic and non-systemic fungicides such as captan, thiram, carbendazim etc.have been used by various workers to reduce seed infections (Sindhan and Bose, 1981; Trutman *et al.*, 1992). Foliar sprays of carbendazim, benomyl, zineb, maneb and mancozeb have been recommended for its control (Chakrabarty and Shyam, 1988). Benlate and Bavistin persisted in the plant upto 15–20 days after seed treatment (Sindhan and Bose, 1981). EBI fungicides like triadimefon and triforine were effective against the pathogen under *in vitro* conditions, but were phytotoxic to *P. vulgaris* seedlings when used as seed treatment (Chakrabarty and Shyam, 1988). Carbendazim alone or in combination with thiram as seed treatment followed by two or three sprays of mancozeb at 45, 60 and 75 days after sowing effectively managed this malady (Sharma and Sugha, 1995). A combination of seed treatment with tebuconazole (0.1%) and foliar sprays of tricyclazole (0.03%) reduced the disease severity both on leaves and pods, while maximum seed yield was recorded in case of seed treatment and foliar sprays of carbendazim (Gupta *et al.*, 2000c).

Continuous use of systemic fungicides has been shown to be associated with the development of resistant biotypes (Tu and McNaughton, 1980). Maringoni *et al.* (2002) reported that all isolates of *C. lindemuthianum* collected from *P. vulgaris* fields located at Paranpanema Valley, Brazil were insensitive to benomyl, carbendazim and thiophanate methyl showing the occurrence of cross resistance to different benzimidazole fungicides, while all isolates were sensitive to chlorothalonil. Recently, Oliveira and Oliveira (2003) reported that strobilurin fungicides like azoxystrobin and trifloxystrobin efficiently controlled this disease, while chemicals like carbendazim and fluquinconazole applied without rotation were least efficient treatments. However, combination of systemic fungicides with non-systemics and rotation of fungicides can be an effective approach for management of this disease instead of constant use of one systemic fungicide.

(iv) Biocontrol: Although research in biocontrol has advanced considerably in many major plant diseases in recent years but this is not the case with bean anthracnose. Limited work has been done like *in vitro* evaluation of the antagonists against this pathogen (Gupta *et al.*, 1991) or the phenomenon of cross protection by inoculating a less virulent strain of the fungus to reduce the severity from severe strains (Sutton, 1979). Infected seed when soaked in 10 per cent culture filterate or treatred with 0.4 per cent talc formulation of *Trichoderma viride* recorded minimum seed infection and maximum seed germination (Ravi *et al.*, 1999). Treatment with *P. chlororaphis*

PCL1391 resulted in best biocontrol of anthracnose, while *P. fluorescens* WCS365 showed no significant difference compared to the positive control (Bardas *et al.*, 2009). Padder *et al.* (2010) evaluated three bioagents (*Trichoderma viride, T. harzianum* and *Gliocladum virens*) and five biopesticides (Achook, Neemgold, Wannis, Spictaf and Neemazal) under *in vitro* and *in vivo* conditions against *C. lindemuthianum*. All the three antagonistic fungi caused significant inhibition of mycelial growth, maximum being with *T. viride* (69.21%) followed by *T. harzianum* (64.20%). Among the biopesticides tested at four concentrations, Wanis applied @ 1000 μl/ml caused maximum inhibition of 82.12 per cent followed by Spictaf (52.85%). *T. viride* and Wanis @ 1000 μl/ml were most effective in reducing the seed borne infection.

(v) Plant extracts: Under greenhouse conditions, the extracts of *M. argyrophylla* and *O. vulgare* caused the maximum reductions (41.82% and 37.65%) in disease severity when a local effect assay was carried out (Pinto *et al.*, 2010). In this study, antimicrobial activity of cascalote phenolics was evaluated against phytopathogenic fungus *C. lindemuthianum.* The results indicated that cascalote phenolics have *in vitro* fungistatic activity against *C. lindemuthianum* races R-0 and R-1472. Moreover, these compounds provoked on the whole, *in vitro* inhibition of spores germination and cellulase and polygalacturonase activities in these fungi. *In vivo* assays with cascalote phenolics under greenhouse conditions using susceptible cv. PI 206272 of common bean resulted in a good protection against anthracnose and reduced disease severity especially as a preventive treatment (Garcia *et al.*, 2010).

4. FLOURY LEAF SPOT

Floury leaf spot also known as Farinose spot is an important disease of beans in relatively high altitudes in the tropics where moderate temperatures along with plentiful moisture in the form of rain or dew is prevalent.

Geographical distribution

The occurrence of this disease was first reported from Ecuador in 1950 (Petrak, 1950). Subsequently it was reported from Brazil, Colombia, Nicaragua, Zambia Uganda, Tanzania, Kenya, Ethiopia, North West India, Malaya, New Guinea, Venezuela, Honduras, Panama, Guatemala and the Dominican Republic.

In India, this disease appears in an epidemic form in the northwestern Himalayan region and its occurrence was first reported from Solan during the year 1964 (Sohi *et al.*, 1965). Subsequently, it was reported from Uttar Pradesh (Tripathi, 1985) and Sikkim (Srivastava and Gupta, 1989a).

Losses

Srivastava and Gupta (1989a) reported about 65 per cent infection of lower leaves, which were defoliated prematurely. The premature defoliation phase of this disease is most damaging which reduces yield considerably.

Symptoms

The disease appears on lower older leaves as yellowish spots and progresses upward. In severe infection, leaves show yellow mottling and in older infection, the spots become brown with yellow margins. These spots are generally 2–3 mm in diameter, but several spots merge together to affect most of the area. On the under surface, white fluffy growth of the fungus conidiophores can be seen which give to the leaf the appearance of having been sprinkled with coarse flour, hence the disease is named as floury leaf spot (Plate 5). Infected leaves turn brown and finally shed prematurely. Pre-mature defoliation causes heavy yield losses.

The Pathogen

The fungus responsible for this disease is *Mycovellosiella phaseoli* (Drummond) Deighton (Syn.: *Ramularia phaseolina* Petrak). On the under surface of host, colourless conidiophores emerge in tufts from the stomata and intertwine in clusters around the leaf hairs. Erect, more or less divergent, somewhat tapering towards the tip, simple or irregularly dichotomous, a few short, erect, single conidiophores are also found scattered over the leaf surface. Conidia are variable in shape and size, mostly ellipsoid or ovoid, tips acute or obtuse, base papillate at the point of attachment, straight hyaline, commonly aseptate, rarely one septate and measure 5.24–21.4 × 3.5–5.3 μm.

Disease cycle

Very little information is available in literature regarding the disease cycle of this pathogen. The pathogen cannot be cultured (Sharma and Kaushal, 1990) and no perfect stage is known. The fungus is not a seed borne and perpetuates in infected plant debris from one season to another (Sharma and Kaushal, 1990). Conidia loose germination with the passage of time and germinate poorly after one year of dry storage, but when supplemented with sugars (0.1%) and amino acids like DL-methionine, L-leucine, valine and DL-threonine, good germination of one-year-old conidia was obtained (Sharma *et al.*, 1998). This indicates the possibility that conidia can remain viable in crop debris. The initial occurrence of lesions on lower most leaves on the plants grown in disease prone areas indicates that soil or debris may act as source of primary inoculum. Overwintering conidia may be splashed by rain to lower leaves, which in the presence of free water germinate on host surface and cause infection. Conidia produced on primary spots are blown by wind and cause secondary infections.

Epidemiology

The disease is more common in altitudes of 1500 to 2000 m with heavy rainfall. Sharma *et al.* (1998) studied the effect of temperature and different humidity levels on the germination of *M. phaseoli* and reported that germination of conidia start within 6 h of incubation at 20 and 25°C.

Management

Since the pathogen overwinters in infected plant debris, the disease can be managed by using cultural, chemical and resistant varieties.

(i) Cultural practices: Cultural practices like collection and destruction of infected fallen leaves, two-year crop rotation with non-leguminous crops and wider plant spacings (Gupta and Mathew, 1999) alongwith weed free cultivation can be useful in reducing the disease.

(ii) Host resistance: French bean cultivars/lines *viz.*, PIB 326, EC 30020, Mantulan, 12-A, Sel.-P-2-3 and EC 10037 have been reported resistant (Singh *et al.*, 1974; Kapoor *et al.*, 1988) which can be successfully utilized in breeding programmes for developing resistant varieties. Sugha *et al.* (1998) reported that White Local developed minimum disease as compared to other genotypes and observed that all the genotypes irrespective of their being pole or bush types were equally susceptible to the disease which are contrary to the observations of Kapoor *et al.* (1988) who reported pole types as resistant and bush types as susceptible to this disease. Two hybrids of *Phaseolus vulgaris* (Contender) x *P. acutifolius* (G 40015 and G 40020) and five of *P. vulgaris* x *P. coccineus* (Pc1 and Pc2) were free of floury leaf spot lesions (Kumar *et al.*, 2005).

(iii) Chemical control: Sprays of various fungicides like benomyl, carbendazim and cercobin have been reported effective in controlling this disease (Singh and Sharma, 1976).

(iv) Integrated disease management: Two foliar sprays of *Pseudomonas fluorescens* strain RPB 14, 1st at seedling stage (10^{-7} cfu ml^{-1}) and 2nd at flowering (10^{6} cfu ml^{-1}), followed by single spray of carbendazim (0.05%) at pod developmental stage was found effective in the management of this disease (Mondal, 2004).

5. RUST

Rust being a polycyclic disease which can cause serious economic losses under environmental conditions favourable to its development by premature defoliation of bean plants, decrease in number of pods/plant and their reduced weight.

Geographical distribution

The disease has been reported from the entire bean growing areas of the world. This disease was described for the first time from Germany in 1795 (Persoon, 1795). It has been reported from USA, North America, Brazil, Portugal, Ethiopia, France, Uganda, Guatemala, Philippines, Bulgaria, Greece, Tanganyika, Chiles, Hawaii, Peru and Turkey.

In India, the disease was reported as early as in 1918 (Butler, 1918). Now the occurrence of this disease has been reported from Karnataka (Sharma, 1989), Tamil Nadu, Uttarakhand (Sharma, 1998), Meghalaya (Bhat *et al.*, 1999) and Sikkim (Bhat, 2002). Recently, the occurrence of this disease has been reported on pencil beans in Solan area of Himachal Pradesh (Gupta *et al.*, 2008).

Losses

Severe rust infection results in premature defoliation, stunted growth, effects on most physiological activities and subsequent reduction in yields, while infected pods may be rejected in the market due to the development of disfiguring lesions (Partridge, 1997). Various workers have reported varying degree of losses due to this disease in different parts of the world. Issa *et al.* (1980), while assessing yield losses in 5 cultivars of bean (*P. vulgaris*), encountered 29–76 per cent losses solely due to

Uromyces appendiculatus. Whereas, 36.7 per cent yield losses were observed by Singh and Musyimi (1981) in plants, which had 79 per cent infection. The losses in green pod yield have been reported from 4.7 to 69. 0 per cent depending on the cultivar, age of the crop at the time of infection and per cent disease intensity (Sharma, 1989). Gonzalez and Garcia (1996a) concluded from their studies that this disease could cause yield losses ranging from 28 to 50, 8 to 33 and 13 to 29 per cent in different bean cultivars like ICA-Pijao, Velasco Largo and Bonte 11, respectively. In Himachal Pradesh, severity of rust was observed to be 15–80 per cent in different areas of Solan district (Gupta *et al.,* 2008).

Symptoms

Leaves and pods are the principal plant parts attacked by the disease, but occasionally, tender parts of stem and branches are also infected. On leaves, the disease may appear on both the surfaces, but is more common on the abaxial surface. Pustules of the disease usually appear first as small and slightly raised spots that are almost white in colour. These pustules enlarge by forming reddish brown sori, up to 2 mm in diameter, containing the urediniospores (Plate 6). After coalescing, they may occupy larger areas. A ring of secondary sori may develop around the original infection on susceptible cultivars. Telial stage is dark brown to black coloured and linear. The affected leaves may turn yellow and dry or fall down prematurely.

The Pathogen

The disease is caused by *Uromyces phaseoli* var. *typica* Arth. and *U. appendiculatus.* Being an autoecious rust, all its spore stages are produced on bean plant, but pycnial and aecial stages are not commonly encountered in nature. The urediniospores are brownish, globoid or ellipsoid, one celled echinulate-measuring 23–33 × 16–23 μm (Plate 7). Spore wall is golden brown 1–1.5 um thick having two equatorial or superequatorial pores. Teliospores are one celled, pedicellate, globoid or broadly ellipsoid. These measure 24–32 × 20–26 μm and may be smooth or possessing a few verucose marks on the wall. The teliospores wall is chestnut brown in colour, 3–3.5 μm thick with a hyaline papilla over the pore. Telia arise from uredinia and appear as dark brown or black in colour. Pycnia appear as yellowish spots on upper surface of the leaf. Orange coloured aecia are cupulate 0.25–0.30 mm in diameter. Aeciospores are ellipsoid, one celled and measure 20–26 × 10–20 μm.

Variation amongst different isolates of *U. appendiculatus* had been observed as early as in 1939 (Dundas and Scott, 1939). Since then lot of work has been done in this direction and many races have been identified on different set(s) of differentials (Harter and Zaumeyer, 1941, De Menzes, 1952, Chirsten and Echandi, 1967, Zuniga de Rodriguez and Victoria, 1975, Stavely *et al.,* 1989 and Faleiro *et al.*, 1999). From 80 monopustular isolates, 53 races on a set of only eight differential hosts (Kentucky Wonder 814, Early Gallatin, 51051, NEP-2, Ecuador 299, Olathe, Mexico 309 and Redlands Pioneer) more distinguished, instead of 20 differentials. Nine and 13 different races have been identified from Tnazania and Minas Gerain, Brazil (Mmbaga and Stavely, 1988, Faleiro *et al.*, 1999).

Variability among bean rust pathogen isolates was also measured by virulence and the random amplified polymorphic DNA (RAPD) technique (Mohammed,

1994). Diversity of *U. appendiculatus* has been measured classically using host differential cultivars having different single or multiple rust resistance genes (Araya *et al.*, 2004; Souza *et al.*, 2007). The new international classification system and the binary nomenclature grouped the 47 single pustule isolates of *U. appendiculatus* into 9 different races. Hierarchical cluster analysis grouped the races into two major clusters depending on the virulence of the races on the host differential cultivars. The most resistant genes for pyramiding in Kenya were identified as Ur-5, Ur-11 and Ur-CNC (Arunga *et al.*, 2012).

Host range

Besides French bean (*P. vulgaris*), the host range of rust pathogen includes *P. lunatus, Dolichos lablab, D. biflorus, Vigna chinensis* etc.

Disease cycle

Being autoceious rust *i.e.,* it has its life cycle confined to a single host. The aecia are rare in nature, but have been observed in bean fields in Oregon (Eastman, 1943). The Urediniospores are produced in great numbers in sori on the upper and lower surfaces of the leaf. They spread the disease over great distances and cause severe epidemics. The urediniospores will germinate as soon as they are mature. One or two germ tubes developed from germinated urediniospores (Xin *et al.*, 2000). The penetration of the germtube occur mainly through the stomata, but occassionally penetrated directly through the epidermal cell. A well defined appressorium is formed after penetration by the pathogen. The fungus forms a sub-stomatal vesicle when it penetrates through the stomata and a round inflated body usually occurs. One or two primary hyphae then develop from the inflated body. Once within the bean leaf, the fungus grows intercellularly and the penetration to the leaf cell occurs after a haustorium mother cell is formed at the hyphal tip in contact with a host cell. The branches emerge from behind the site of the primary hyphae where a haustorium mother cell develops. The secondary hyphae spread intercellularly. Under favourable environmental conditions, urediniospores produce another generation of same kind of spores in about 10–15 days and produce several generations in a single season (Zaumeyer and Thomas, 1957). When climate becomes unfavourable, teliospores are produced in the same sorus. Teliospores, which require a rest period and retain their viability for many months, are largely instrumental in carrying over the organism from one season to another.

Epidemiology

Obligate parasitism and autoecious nature of rust pathogen helps it to complete entire life cycle on bean plants. It is not seed borne. In cooler regions, survival is ensured through plant debris in the form of teliospores, while in other areas through continuous growing of crop including other hosts. Cloudy and humid days, allowing dew deposition on leaves for some time in the morning favour germination of spores and subsequent infection. Harter *et al.* (1935) have reported optimum temperatures for urediniospores germination and infection as 14.5 and 17°C, respectively. Davison and Vaughan (1963) suggested that rust overwnters in or as urediniospores in trash and on trellis poles because 16 per cent urediniospores of *U. phaseoli* race 33 germinated even after storage at –18°C for >600 days and produced abundant infection on bean leaves. Daily mean temperature of 34°C prevents disease development. Sharma (1977) observed that irrigation given at 3 days intervals favoured disease development than with 18-day intervals. Disease intensity is not influenced by varying levels of N, P and K. Farm implements, insects, humus and animals help in local dissemination of the rust fungus (Sumatrini, 1998). Sharma (1998) analyzed the meteorological data for eight years (1986–1993) and showed that excessive precipitation was deleterious for disease development, which explains the occurrence of severe infection during periods of low rainfall.

Management

The disease can be managed by cultural, chemical, biological methods and host resistance.

(i) Cultural practices: Moreno and Mora (1984) have suggested minimum tillage or no tillage to prevent spread of the disease and intercropping of bean with maize, since, severity and rate of infection was more in monoculture. Mixed cropping reduced rust incidence level by 51 and 25 per cent with sole cropping and row intercropping, respectively (Fininsa, 1996). Plant density and disease incidence are directly related (Flores Revilla *et al.*, 1994). Mishra *et al.* (1998) evaluated seven dates of sowing under Orissa conditions and Arka Komal variety was sown in between 12 October and 23 November, 9th November was found as the optimum date of sowing because the rust incidence was minimum. Sharma and Sohi (1989) also recommended some cultural practices like long crop rotations, collection and destruction of plant debris, wider plant spacings and removal of weeds to permit aeration and drying which check rapid build-up of inoculum and spread of the disease under field conditions.

(ii) Host resistance: More than 35 physiologic races of the rust fungus (*U. appendiculatus*) are present in Australia, Brazil, Hawaii and the USA. Extensive cultivation of the new cultivars may make them susceptible to new virulence(s). Golden Gate Wax and Brown Kentucky Wonder 298 have been used in breeding for resistance to four races (Dundas, 1942). Cultivars/lines like Virginia 119, No.780 (Foster, 1947), Hawaiian Wonder (Frazier and Hendrix 1949), Seminole, Redlands Greenleaf and Redlands Pioneer (Anonymous, 1966), Jackson Wonder and Strikton (Macraw *et al.*, 1973), Brazil K-12973 (Sokol *et al.*, 1977), Catu and Carioca (Issa, 1985) Serva Negra, Ag497 and Agrorrico (Azevedo and Kushalppa, 1986), Gresham (Anonymous, 1989), PV 136, PV 88, PV 89 and PV 3 (Prakasam and Thumburaj, 1991), white seeded Enganador, Chevere and CC 25-9B, Negro Inifap (Villar-Sanchez and Lopez-Salinas, 1993), Negro Cotaxtle 91 (Becerra Leor *et al.*, 1994), IBRN-6, IBRN-11, Hawaiian Wonder, Bush Bean S-9, Bush Blue Lake and Contender (Ghimire *et al.*, 1995), lines 83-10 and 83-11 with stringless pods (Guu, 1995), Ouro Negro (Faleiro *et al.*, 1996), IIHR 220, stringless bean (Mohan *et al.*, 1997), Mimoso Rasteiro (Paula *et al.*, 1998), SVM-1, Hans (Sharma, 1998) and IC 47643, PI 201489 and PI 247761 (Bhat *et al.*, 1999) have been found to be resistant in different parts of the world. Cv. Ouro Negro showed immunity to 3 out of 4 races tested and PI 181395 was the best source of resistance to rust, as it was immune to all the four races of *U. appendiculatus* (Faleiro *et al.*, 1998). Inheritance studies indicated that resistance to rust was controlled by a single, dominant gene (Aghora *et al.*, 2007). Line 'Arka Anoop' developed by breeding programme is resistance to rust and has high pod yield and good pod quality. Germplasm lines/cvs. Like Alapa Trey, Sing Tamey and Local Collection were found resistant to this disease under Solan, Himachal Pradesh conditions (Sharma *et al.*, 2014).

(iii) Biological control: *Verticillium lecanii* has been suggested to be a potential biological agent for the management of rust as it penetrates hyphae and invades urediniospores of the fungus (Allen, 1982). Under field conditions *Verticillium lecanii* reduced the disease incidence and severity to 56 and 17 per cent as compared to 98 and 85 per cent, respectively, in agrochemical package while seed yield was slightly less in biocontrol than agrochemical package which may be due to the use of fertilizer in latter package (Carrion *et al.*, 1999). Different isolates of *Bacillus subtilis* like APPL-1 (Baker *et al.*, 1983 and Baker *et al.*, 1985) and FF-1, FF-5 and FF-6 (Mizubuti *et al.*, 1995) have been selected as the potential bioconrol agents. In culture medium *B. subtilis* isolates released a thermo stable substance(s) antagonistic to the rust fungus as pustule number was reduced by >95 per cent (Centurion *et al.*, 1994). Isolate B 206 showed the greatest efficacy, when cells of the antagonist were automized on to the plant.

(iv) Chemical control: Different workers suggested various chemicals from time to time for keeping the disease under check. Straib (1943) suggested the disinfection of stakes with formaline (0.1%) before reuse. Preventive application of zineb (5%) with a mist blower or as a spray (0.35%) proved effective. Systemic fungi-cides are more effective than the non-systemic owing to the difficulty of obtaining under leaf coverage with the latter. Efficacy of various systemic fungicides like oxycarboxin, carboxin, triforine, bitertanol, triadimefon, tridemorph, hexaconazole and HWG 1608 has been reported against this disease (Jaffer, 1971; Rolim *et al.*, 1981; Singh and Musymi, 1979; 1984; Prakasam and Thamburaj, 1991; Gonzalez and Garcia, 1996; Sumartini, 1998) and possibly eradicated the already established infection of *U. phaseoli* in bean plants after three sprays without any sign of phytotoxicity (Singh and Musymi, 1979; 1984; Gonzalez and Garcia, 1996b).

Seed treatment with the thiram and subsequent sprays with mancozeb was found best combination treatment for controlling rust (Mahanta and Dhal, 2000). Three sprays of hexaconazole (0.1%) at 15 days interval were found best in keeping the disease under check (Bhat, 2002). Sprays of Contaf (0.1%) alone and in combination with mancozeb (0.2%) were highly effective and reduced the disease incidence from 86.6 per cent to 24.1 per cent (Singh and Bhat, 2002). Sprays of propiconazole were found to be most effective in reducing the severity of rust to 10.93 per cent as compared to control (41.20%) and gave about 91.54 per cent disease control (Shukla and Sharma, 2009).

(v) Systemic induced resistance: Treatment of bean leaves with salicylic acid (SA) or 2, 6-dichloro-isonicotinic acid induces resistance against the rust fungus *U. fabae* resulting in reduced rust pustule density (Rauscher *et al.*, 1999). Saccharin application in broad bean (*Vicia faba*) induced resistance to the rust fungus *Uromyces viciae- fabae* (Boyle and Walters, 2005) and provided systemic protection to rust infection after fourteen days of application. Bion, was also used to induce resistance in bean cultivars like Borlotto Nano Lingua di Fuoco (BLF), Borlotto

Taylor, Cannellino, Cannellino Montalbano, Saxa and Top Crop, against *U. appendiculatus*. A single 0.3 mm BTH spray 7 days before inoculation was sufficient to fully control the disease in all the examined cultivars (Iriti and Faoro, 2003).

(B) BACTERIAL DISEASES

1. COMMON BLIGHT

Common blight of beans is a major constraint for their production particularly in tropics (Saettler, 1989). In favourable weather conditions, the disease cause significant yield reductions.

Geographical distribution

The disease was first reported from New York, U.S.A. (Beach, 1892) and subsequently, reported from Australia, Bulgaria, Canada, China, France, Hungary, Japan, Madagaskar, Norway, Phillipines, Russia, Spain, Turkey, South Africa, Uruguay, Yugoslavia, Argentina, Brazil, Colombia, Paraguay, Venezuela, Guatemala, Nicaragua, Honduras, Mexico, the Dominican Republic, Uganda, Burundi, Zaire, Rawanda and Zambia. In India, this disease was reported for the first time from Bombay by Uppal *et al.* (1946).

Losses

Common blight can cause significant yield losses if conditions are favourable for disease development. Warm and humid weather favours severe outbreaks of the disease, thus the disease is more serious in the tropics (Saettler, 1989). Epidemics of common blight occurred in Ontario, Canada in early 1960's and in 1962, when warm weather along with high humidity occurred (Sutton and Wallen, 1970). In Colombia, yield losses were estimated upto 22 and 45 per cent in natural and artificial infected fields, respectively (Yoshi *et al.*, 1976). It has also been estimated that common blight is responsible for an average 17 per cent yield losses in the Cauca Valley of Colombia (Saettler, 1989).

In India, upto 84 per cent yield losses due to this disease in French bean cvs. Contender and Arka Komal have been reported at different crop growth stages (Kishan, 1988).

Symptoms

The disease affects stem, branches, leaves, pods and seed. The symptoms appear as characteristic water soaked spots on the under side of the leaves that enlarge and coalesce together to form large blighted areas. Infected areas often become flaccid and turn brown giving it a sunscald appearance. If infection occurs through the vascular tissues, reddish discolouration of veins and midribs is observed. In advanced stages, tissues become necrotic and defoliation may also take place. Lesions on stems appear as sunken water soaked areas, which gradually enlarge to form reddish streaks extending longitudinally in the stem. In early stages of pod formation, characteristic stem girdle or joint rot frequently develops. Infection on pods may occur at any place as small water soaked spots, which are surrounded by distinct zones of narrow reddish brown or brick red band of tissue. On seeds, discolouration appears especially near the hilum region. Plants arising from infected seeds exhibit characteristic wilting (Saettler, 1989).

The Pathogen

The disease is caused by *Xanthomonas axonopodis* pv. *phaseoli* (Smith) Dye (XAP). The bacterium is a motile, non-capsulated, non-spore forming, gram-negative rod, measuring 1.9 x 0.8 μm in size with a single polar flagellum. The pathogen forms characteristic yellow coloured colonies due to a water insoluble eacotenoid and a mucoid growth on nutrient glucose agar medium. The characteristic yellow colour of the bacterial colonies is due to the production of brominated aryl polyenes, more frequently referred to as xanthomonadins, that have absorption peaks in between 443 and 446 nm (Starr, 1983). Some other useful properties of this bacterium include hydrolysis of starch, gelatin and casein, lipolysis of Tween 80, and lack of pectinase activity. Acid is produced as a metabolic byproduct when cells grow on media containing arabinose, glucose, mannose, galactose, trehlose or cellubiose. It also causes proteolysis of milk. The pathogen is a strict aerobe, and metabolism of glucose is through oxidative pathway.

Pazos *et al.* (1986) failed to differentiate in Cuban isolates of *X. c.* pv. *phaseoli* (XCP) isolated from bean and by using bacteriophages, 3 pathotypes were distinguished among 35 isolates. Highly significant differences among fifteen isolates of this pathogen were observed among varieties and isolates in relation to resistance and pathogenecity (Rava and Romeiro, 1990) and isolate Xp CNP 28, characterized by high degree of pathogenecity, was suggested as a new standard isolate to replace Xp CNF 15.

Report regarding serological variability among *X. campestris* pv. *phaseoli* also exists in literature and 43 isolates from Sao Paulo and Perana States of Brazil were classified into 3 serological groups of which 22 did not produce melanin *in vitro* and reacted only with the homologous antiserum AS-Feij I (serogroup I), 17 reacted only with antiserum Feij 20 (serogroup II) and four did not react with the antisera tested (serogroup III) by using the agar double diffusion test (Maringoni *et al.*, 1994). The strains of Xap were also grouped in 3 groups through profiles of total proteins (PAGE-SDS) (Maringoni and Bach, 1999).

On *Phaseolus acutifolius*, 30 isolates of XAP interacted differently and 8 physiologic races were defined suggesting an underlying gene- for-gene relationship (Opio *et al.*, 1996). They further suggested that levels of resistance already available within *P. vulgaris* are adequate, so that breeder incorporating new complete resistance from *P. acutifolius* could risk destabilizing the host-bacterium relationship by introducing race-specific resistance, that could be transient in agriculture. Genotypes useful for differentiation of XAP physiological races have been identified (Zapata, 1997). DNA probes were also used to differentiate *X. campestris* pv. *phaseoli* and *X. campestris* pv. *phaseoli* var. *fuscans* isolates originating from different geographical locations and RFLP analysis revealed DNA polymorphism within *X.campestris* pv. *phaseoli* isolates from the same geographical location (Otoya and Pastor-Corrales, 1992).

Host range

The host range of this pathogen includes both cultivated and weed host species *viz., Dolichos lablab, Emilia sonchifolia, Glycine max, Ipomoea crassicaulis, Lupinus polyphyllus, Phaseolus acutifolius, P. aureus, P. coccineus, Pisum sativum, Strophostyles helvola, Vigna mungo* and *V. radiata* var. *radiata,* (Saettler, 1989; Rodriguez *et al.,* 1991; Chand, 1992). The bacterium was also isolated from diseased leaves of peas, cowpeas, *Ipomoea purpure, Malva sysiaca* and *Vigna luteola* which may play significant role in the harbouring and spread of the pathogen in the neighbouring bean fields (Khlaif and Qadous, 1995).

Disease cycle

The bacterium overwinters in infected seed, crop debris and weed hosts (Irigoyen and Garbagnoli, 1998) of which contaminated seed has been observed to be the most efficient mode of survival and dispersal over long distances (Saettler, 1989). The pathogen survived for more than five months in the plant debris on the ground surface (Arnaud-Santana *et al.*, 1991), but the survival of the pathogen is influenced by tillage (Gilbertson *et al.*, 1990). Survival of the pathogen on the weed host *Parthenium hysterophorus* and subsequent transmission of the bacterium to *P. vulgaris* and *vice verse* has also been reported (Ovies and Larsinaga, 1988). Upon germination, the bacterium from infected seed infects the developing cotyledons and can then enter the vascular system of the plant. Alternatively the bacterium can be splashed to the plants from the plant debris. Once inside the plant, the bacteria invade and cause breakdown of the middle lamella. Bacterial multiplication takes place, and masses of bacteria are formed that expand into what macroscopically appears as water-soaked spot. The bacteria also exude to the surface of the tissue and can be disseminated in the field by rain and wind to initiate new infections. Bacteria enter the plants via wounds caused by leaf eating insects or natural openings such as stomata or hydathodes (Saettler, 1989). The bacteria can infect stem via stomata or via the vascular system. Most vascular infections probably initiate from seedborne inoculum. In some of these cases, the bacteria can cause plugging of xylem elements, causing wilting. The bacteria can infect seed via the vascular system, moving through the pod pedicel into the funiculus and through the raphe into the seed coat, where they remain beneath the seed coat until seed germination.

Epidemiology

The bacterium is seed borne in nature. Infection of the pedicels of flower buds and small pods resulted in transmission of the bacteria through the vascular tissues of the pod to the seeds, causing internal infection without any external symptoms on either pods or seeds (Aggour *et al.*, 1989). The minimum population of *X.c. pv. phaseoli* required to initiate infection was 10^2 c.f.u./seed while seed infection levels of 0.2 per cent resulted in serious disease incidence (Opio *et al.*, 1993). The bacterium also enters plants through wounds and/or natural openings and infects stems via stomata, infected cotyledons or vascular system. The incidence and severity of the diseases increase with the increase in rainfall and temperature (Mayek-Perez *et al.*, 1995). Warm and humid conditions favour disease development and the disease can develop at temperatures ranging from 15-40°C and RH > 30 per cent with optimum 24°C and

>50 per cent RH (Qadous and Khlaif, 1997). Wind splashed rains help to spread the pathogen from infected to healthy plants. Infected seed from resistant genotypes had no visible symptoms and such seeds may play an important role in the epidemiology of this disease because they are difficult to detect and may occur at low frequency in seed lots (Mabagala, 1997). The disease was more severe under short photoperiod and high temperature than longer photoperiod and lower temperature (Santana *et al.*, 1993). Akhavan *et al.* (2009) showed that the epiphytic population size and disease severity were higher on spray inoculated plants irrigated with overhead sprinkler system during pods filling.

Management

The common blight of beans is particularly a difficult disease to manage due to intricate association of the pathogen with the seed. An integrated approach is, therefore, useful to keep the disease under check.

(i) Cultural practices: Use of pathogen free seed is must because it eliminates an important source of primary inoculum for common blight (Irigoyen and Garbagnoli, 1998), which is an important component of effective disease management strategy for this disease. The disease free seed can be produced in arid zones with low rainfall, but its use cannot guarantee a clean crop. Because blight bacteria can colonize symptomless plants and the bacteria can systemically invade seed via vascular elements, certified pathogen-free seed may still be contaminated by blight bacteria (Weller and Saettler, 1980). The reduction of inoculum achieved by the use of the pathogen free seed greatly reduces the probability of common blight epidemics developing in the field.

Deep ploughing, 2–3 years crop rotation and destruction of crop debris/ weed hosts, avoidance of irrigation with sprinklers help to reduce the probability of epidemics (Saettler, 1989; Gilbertson *et al.*, 1990). Bacterial population in plant debris decline over time (Gilbertson *et al.*, 1990) and atleast in temperate regions, the bacteria probably cannot survive for more than one winter. In such situations, crop rotation will be an effective management strategy. But this may not be an effective approach in tropical region particularly in case of the farmers with small land holdings,. Moreover, the pathogen is having a vast host range; then again rotation will be less effective strategy.

Mixed intercropping of bean with maize was also found effective in reducing the incidence of this disease as compared to sole cropping and row intercropping, respectively (Fininsa, 1996). Delay in planting time to avoid frequent rains early in the season along with use of disease free seed has also been recommended to alleviate considerable yield loss and quality deterioration (Tu, 1999).

(ii) Host resistance: Since there is no satisfactory chemical control available to combat the ravages of this disease, the use of resistant varieties/cvs. is an important control measure (Rodrigues *et al.*, 1999). Various varieties/lines *viz.*, Fullgreen, PLS-63, 101, MS 9471, DL 2434 (Patel and Jindal, 1973); P 597 (*P. acutifolius*) Feijao 60 Diaz, PI 207262, Coll. 10 B, Mexico, 168, Mexico 240, Desconhecido Amarelo (Rava *et al.*, 1990), BR-40, Contista, Saginan, WW 55-0, Kuovskaya, Yushchoyasya (Ovchonnckova and Zaitsev, 1992); OAC 88-1 (Scott and Michaels, 1992); XR 235 (Wright and Wood, 1993); Great Northern Jules, GN Nebr1, Sel. 27, Pinto Nebr,

EP1, Walley, Xr1, Xr2 (Velich *et al.*, 1994); BAC-6, XAN-159 (Mohamed *et al.*, 1995); IIHR 220 (Mohan *et al.*, 1997); T-589, T-769; T-352, T-647 (Vizgarra, 1998); Cut 54, XAN 280, 9356-26 (Rodriguez *et al.*, 1999) which possess resistance to this disease has been reported from different parts of the world. Accession CF 800375 has also been reported to be resistant to this disease (Rava *et al.*, 2003).

Randomly amplified polymorphic DNA (RAPD) molecular markers have been effectively utilized in the identification of regions on the linkage map associated with resistance to common bacterial blight. The chromosomal region (a BCM interval in linkage group 5 was significantly associated with resistance to XAP strains DR-7 and EK-11 in leaves, pods and seeds. Moreover, regions in linkage group 1 were also significantly associated with resistance to XAP strains in more than one plant organ. The V locus of the flower colour gene (v^{lae}) was linked to a QTL (quantitative trait locus) with major effect on CBB resistance (Jung *et al.*, 1997). Five interspecific hybrids of *P. vulgaris* (Nim 1) x *P. coccineus* and cv. IPR96-4 were resistant to this disease (Kumar *et al.*, 2005, Khati and Hooda, 2006). Cultivars Bayo Madero, Negro 8025 and Pinto Villa were tolerant against common blight pathogen (Prudencio-Sains *et al.*, 2008). Plants of 17 bean genotypes were evaluated under greenhouse conditions for their reaction to *X. campestris* pv. *phaseoli* following leaf-spray inoculation with bacterial suspension (10^8 cfu/ml) and the cultivar Oreol was the most resistant (Todorovic *et al.*, 2008).

(iii) Chemical method: Different chemicals have been used both as seed treatment or foliage application in attempts to control this disease. Although some success has been achieved but complete control is still intractable. Some chemicals used by various workers both seed and/or foliage application are mentioned as below:

(a) Seed treatment : Hot water seed treatment at 50°C for ten minutes followed by diping in streptocycline (10 μg/ml) solution has been found effective in reducing the seed infection (Jindal, 1991).

(b) Foliar sprays: Several fungicides/antibiotics have been applied to foliage to control this disease, but complete control is still lacking. Application of copper fungicides to some extent checks this disease (Saettler, 1989). However, heavy use of copper based fungicides to control diseases caused by Xanthomonads can result in selection of copper-resistant strains of the bacteria (Stall *et al.,* 1986) hence need based use of copper fungicides is warranted. Moderate control of the disease coupled with higher yields has been achieved by three sprays of copper oxychloride + zineb or copper oxychloride + mancozeb (Maringoni, 1990).

The chemicals can form the effective part of integrated approach for this dreaded disease. Sometimes when the environment becomes abruptly favourable for the development/multiplication of bacteria, spot application of chemicals can keep the bacterial populations below damage threshold level.

(iv) Biological control: Bacterial isolates DFs093 offered complete control, DFs041 and DFs1297 offered 90 per cent and DFs490, DFs769, DFs831, DFs842 and DFs843 offered 80 per cent control of common blight pathogen in the cotyledonary leaves when bean seeds immersed in suspension of each one of these isolates were agitated for 5 hours at 10°C and showed in non-sterilized soil (Zanatta *et al.*, 2007). *Paenibacillus polymyxa* strain HKA-15 showed strong antagonism against *Xanthomonas campestris* pv. *phaseoli* strains M-5 and CP-1-1. In agar diffusion assay, the antibacterial metabolite from *P. polymyxa* HKA-15 showed a clear zone of inhibition against M-5 and CP-1-1. Under phytotron conditions, the biocontrol activity of *P. polymyxa* HKA-15 against bacterial common blight pathogen *X. campestris* pv. *phaseoli* M-5 was assessed. At four days after inoculation of M-5, lowest mean disease rate (MDR) (1.13) and per cent disease incidence (PDI) (28.25) were recorded in streptomycin sulphate at 100 ppm as positive chemical treatment. The application of crude metabolite from *P. polymyxa* HKA-15 at 100 ppm was at par with positive chemical control in suppression of bacterial common blight disease in French bean plants (Mageshwaran *et al.*, 2012).

2. HALO BLIGHT

Halo blight has been recognized as a serious disease of beans known to occur world over, causing considerable crop losses. The disease has been often referred to as the greasy spot disease due to the appearance of greasy water soaked lesions on affected organs.

Geographical distribution

Halo blight was first reported from New York, USA during the year 1926 (Burkholder, 1926). Subsequently, it was reported from the Netherlands, Switzerland, New Zealand, France, Tasmania, Australia, Denmark, Poland, Brazil, Germany, Russia, South Africa, Belgium, England, French Morocco, Spain, Canada, Uruguay, Kenya, Mauritius, Chile and Italy (Bazzi and Calzolari, 1988).

Losses

Halo blight is slightly less important as compared to common blight because it is more prevalent in those areas where temperature is low. It is a major constraint of dry bean production in South Africa and Poland especially in the moderate to cooler areas of the country (Marcinkowska, 1994; Fourie, 1998). In Italy, it is one of the most frequent diseases of this crop (Randhi and Parisi, 1997). Yield losses in the range of 23–43 per cent in Michigan fields have been reported (Saettler and Potter, 1970).

Symptoms

Symptoms appear as small isolated water-soaked spots on the under surface of the leaves which gradually enlarge and coalesce. Later, these spots are surrounded by a characteristic greenish yellow halo measuring 2–3 cm in diameter, thereby giving the disease its name. Systemic plant chlorosis with leaf yellowing and malformation may also develop without much external infection. Later, the pathogen has been isolated from such leaves. At 20ºC, the systemic chlorosis is more pronounced and uniform (Coyne and Schuster, 1974). During infection process, the pathogen produces a non-specific toxin "phaseolotoxin' which in turn produces systemic chlorosis and typical halo in the plant. Infections are initiated on the lateral veins which progress to the main vein often accompanied by a marked water-soaking and reddish discolouration of the interveinal regions. On young stems, sunken water-soaked spots gradually enlarge and appear as reddish streaks, extending longitudinally along the stem. In advanced stages of infection, the stem epidermis split open and are often accompanied by light cream to silver coloured bacterial exudations. Brown to red zonate lesions surround water-soaked spots on affected pods. The bacterium soon invades the tissue and progress to the developing seed often leading to shrivelling, discolouration and even rotting of seed.

The Pathogen

The disease is caused by *Pseudomonas savastanoi* pv. *phaseolicola* (Burkholder) Garden *et al.* (Young *et al.*, 1996). It is a strictly aerobic, gram-ve, rod shaped, motile bacterium occurring singly or in pairs. The organism exhibits a smooth to rough colony variation in culture. The presence of D-gluconate, L (+) arabinose, sucrose, succinate, DL-β-OH butyrate, transaconitase, L-serine, L-alanine and P-hydroxybenzoate favour growth of the bacterium in culture while it does not utilize glutarate, mesotartarate, DL-glycerate, isoascorbate, erythritol, sorbitol and meso-inositol. White to cream coloured colonies with a bluish hue and accompanied by a green fluorescent pigment appear on agar medium. Sensitivity of bacterial strains to bacteriophages 11P, 12P and 48P and serological tests have also been suggested for the confirmation of identity of *P.s.* pv. *phaseolicola* (Taylor, 1970) which were later successfully used in the identification of this pathogen (Mabagala and Saettler, 1992; Fourie, 1998).

Variation among the populations of this pathogen is well known (Mabagala and Saettler, 1992). Jenson and Livingstone (1944) were the first to demonstrate host specificity among 13 strains of this pathogen. Later, the occurrence of races 1 and 2 in the United States on the basis of the reaction of bean cultivar Red Mexican U.I.-3, which is resistant to race 1, but susceptible to race 2 (Patel and Walker, 1965). The occurrence of these two races has been reported from Tanzania, Colombia, New Zealand and Kenya (Taylor *et al.*, 1987). The existence of a third race in Nabraska was first suggested by Schuster *et al.* (1979) and Coyne *et al.* (1979), who recovered strains of halo blight bacteria that were virulent to the usually resistant Great Northen bean cultivars, such as Great Northern U.I.-59 and California Pink. Taylor *et al.* (1987) confirmed the existence of race 3 of *P.s.* pv. *phaseolicola* in South Africa. Race 3 causes a hypersensitive reaction in cv. Tendergreen, which is susceptible to race 1 and 2. Mabgala and Saettler (1992) also recorded the existence of races 1, 2 and 3 in northern Tanzania. Besides these, some strains of race 2, designated as 2P, produced a brown, diffusible pigment on agar media and it can grow at 34°C. Whereas the non-pigment producing strains of race 2 and of race 1 failed to grow at this temperature. Taylor and Teverson (1996) extended the range of differentials to eight cultivars/lines, and accordingly identified nine races of this pathogen. Recently, in South Africa, seven races (1, 2, 4, 6, 7, 8 and 9) were recorded among 1128 isolates, with race 8 the most prevalent (Fourie, 1998). Races 1, 2, 6 and 8 were widely distributed through the whole production area, while races 4, 7 and 9 were restricted to 1 or 2 localities.

Host range

The pathogen has a wide host range including *Neonotonia wightii, Parthenium hysterophorus, Phaseolus acutifolious, P. angularis, P. coccineus, P. lunatus, P.*

polyanthus, P. polystachyus, P. radiatus, Pueuaria hirsuta, P. thunbergia and *Glycine max.* (Mabagala and Saettler, 1992).

Disease cycle

The bacterium is an important seed-borne pathogen of this crop (Calzolari, 1997). Seed borne inoculum is carried on the surface of the seed or beneath the testa and in between the cotyledons. Pathogen also survives in bean debris and volunteer bean plants (Mabgala and Saettler, 1992). Within the same genotype, *P.s.*pv. *phaseolicola* tended to survive longer in the stems than in leaf debris. The bacterium enters directly through wounds or through stomata during periods of high relative humidity and free moisture, and start multiplying inside the host tissue. Symptoms of blight appear on the leaf surface.

Epidemiology

The bacterium is seed borne and also overwinters in infected plant debris. In the periods of high humidity and free moisture, the bacterium enters the host through stomata or wounds. Symptoms appear in 6–10 days at 24–28°C, but may be delayed for 2–3 days more at higher temperature (Zaumeyer and Thomas, 1957). Under high soil moisture conditions and low pH, survival of this bacterium did not exceed 2–3 months. The bacterium apparently cannot survive in colder regions in soil or plant debris.

Management

(i) Cultural practices: Cultural practices include use of disease free seed, a 2–3 year crop rotation, ploughing in plant debris immediately after harvest and avoiding irrigation with sprinklers (Calzolari, 1997).

(ii) Host resistance: Breeding for halo blight resistance is generally considered the best method of disease control (Mabgala and Saettler, 1992) and has been used successfully in several areas of the world (Schwartz, 1989). The development and deployment of resistant cultivars requires knowledge of the distribution of races of the pathogen in a particular area. Resistance against this disease has been reported in *Phaseolus vulgaris* hybrid/cvs. Ph-64-23-144, Ph-65-9-9, Ph-64-9-61 and Ph-64-9-56 (Radkov *et al.*, 1978), *P. radiatus*, *P. calcaratus*, *Vigna angularis*, PI 150414, cv. Edmund (Kiessling and Skiebe, 1987), V4141had higher tolerance to race1 and 2 of the pathogen. (Hanan and Khan, 1999). Accession CF 800375 has also been reported to be resistant to this disease (Rava *et al.*, 2003). Common bean 12 accessions had resistance to halo blight from the Iberian Peninsula (Monteagudo *et al.,* 2006). Cv. 'Chase' has been found resistant to halo blight pathogen (Singh *et al.,* 2007).

(iii) Chemical control

(i) Seed treatment : Hot air treatments (70°C for 2 h for moistened seed or 50°C for 3 h for dried seed) reduced 72–74 per cent seed infection. Hot water soaks at >48°C reduced the number of bacteria present in the seeds but tended to decrease germination and prolonged soaking rendered the seed unfit for sowing. Antibiotics, kasugamycin and streptomycin eradicated seed infection when used as a hot water soak (60 min) and at relatively higher concentration, while a successive tap water

rinse and a further 30 min soaking in sodium hypochlorite (0.5%) improved the effectiveness of the streptomycin used at low dosage and reduced its phytotoxic effects (Tamietti and Garibaldi, 1984).

(ii) Foliage treatment : Sprays of cupric hydroxide significantly reduced or limited *P. s.* pv. *phaseolicola* on bean foliage. Copper oxychloride (0.2–0.25%) sprays starting from seedling stage to flowering are also found effective in reducing pod infection. Application (3–4 times) of copper products after seedling emergence was also recommended (Calzolari, 1997). Though, streptomycin is effective in reducing foliage infection but continuous use can induce resistant mutants in the bacterial population.

(C) VIRAL DISEASES

1. BEAN COMMON MOSAIC

Bean Common Mosaic is one of the most commonly prevalent viral diseases of French bean, causing considerable losses in yield and quality.

Geographical distribution

Ever since its first record from Russia (Iwanowski, 1894) the disease has been reported from various countries *viz.*, U.S.A., Canada, Bermuda, Agrentina, Belgian Congo, Chile, Columbia, Czechoslovakia, Ecuador, France, Guatemala, Honduras, Italy, Japan, Java, The Netherlands, Peru and Uruguay, China, England, New Zealand, Australia, Brazil, Germany, Iran and Venezuella.

Yaraguntaiah and Nariani first reported occurence of the disease in India in 1963 from New Delhi and subsequently from Shimla, Himachal Pradesh, Bangalore, Karnataka, Pantnagar, Uttar Pradesh and Pune, Maharashtra.

Losses

Yield losses up to 26 to 84 per cent has been reported to be caused by the disease (Sastry *et al.*, 1981; Sharma and Dhar, 1994).

Symptoms

Typical symptoms consist of well-defined dark green areas of foliar tissue against a lighter green background of affected leaf lamina (Plate 8). During early stages, affected leaves appear crinkled, chlorotic, stiff and accompanied by downward curling. In advance stages of infection much reduced leaves with sinuous or filiform leaflets are produced, whereby plants become stunted and appear dwarfed and bushy. Seedlings from infected seed show characteristic mottling, blistering, downward cupping and chlorosis on the fruit and succeeding compound leaves. Such plants usually fail to set pods or develop undersized pods. The pods on severly-infected plants are usually undersized and contain fewer ovules than those produced on normal plants. They are occassionally covered with small, dark green spots and often mature very late. The seed may be aborted, smaller or malformed. The symptoms are temperature dependent and infection of this disease causes delayed seed maturity. The roots of affected plants turn dark and during periods of severe infection the tap root system turns almost black due to vascular necrosis giving rise to what are typically called Black root' symptoms. Plants affected by black root, initially exhibit slight wilting of the leaflets near the top of the plants, which subsequently loose their bright green colour and appear greyish, green. A permanent general wilting of all the leaves follows often resulting in death of the plants.

The Virus

At least 20 viruses belonging to 12 taxonomic groups are shown to infect French bean under natural conditions. Dhanju *et al.* (1995d) reported that the viruses associated with mosaics of French bean in Himachal Pradesh consisted mostly strains of Bean Common Mosaic and Bean Yellow Mosaic viruses of which the former is most predominant.

The *Bean Common Mosaic Virus* (BCMV) is an ubiquitous pathogen of worldwide occurence. Taxonomically the virus belongs to the *potyvirus* group. Earlier two serotypes of the virus namely A and B had been described to cause mosaic diseases. However, those once grouped as serotype A are now considered to be isolates of a seperate potyvirus species and several strains (Drijfhout *et al.*, 1978) once considered to be distinct have now been shown to be strains of this virus (McKern *et al.*, 1992). BCMV consists of filamentous flexuous particles measuring 730–750 × 12–15 nm in size. The physico-chemical properties relating to this virus has been summarized in Table 1.

Table 1. Physicochemical properties of Bean Common Mosaic Virus

Genome	ssRNA, Monopartite
Transmission	Sap, Seed, Insects
Properties of particles in sap	
TIP	60°C
LIV	1–4 days
DEP	10^{-3}–10^{-4}
Physical Properties	
Sedimentation Coefficient	154–158 S
Density	1.31–1.32 g cm^{-3} in Cscl

TIP = Thermal inactivation point; LIV =Longevity *in vitro;* DEP = Dilution end point

(Source: Brunt *et al.*, 1996)

Pathogenic variability studies in bean common mosaic virus (BCMV) infecting common bean (*Phaseolus vulgaris* L.) revealed the existence of two pathogroups PG-I and PG-II and four strains (NL-1, NL-1n, NL-7 and NL-7n) in Himachal Pradesh, a North-Western Himalayan state of India. Two strains, NL-1 and NL-7 were identical to the previously described NL-1 and NL-7 strains from Europe and USA. Whereas, the other two designated as NL-1n and NL-7n differed from earlicr identified strains with respect to their necrotic reaction on cultivar Jubila at high temperature (730°C). Reverse phase HPLC peptide profiling of tryptic digests of coat protein of these strains further confirmed that NL-1, NL-1n, NL-7 and NL-7n are distinct from each other (Fig. 5) (Kapil *et al*., 2011).

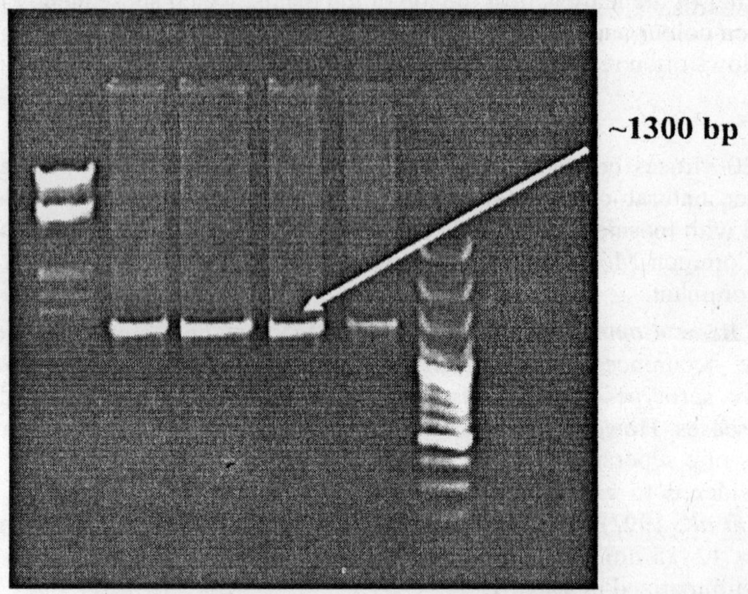

Fig. 5. RT-PCR amplification of four strains of BCMV using BCMV specific primers. Lane1: EcoRI /Hind III double digest ladder. Lane 2: Nl-1. Lane 3: NL-1n. Lane 4: NL-7. Lane 5: NL-7n and Lane 6: 100 bp ladder.

Four strains (NL-1, NL-1n, NL-7 and NL-7n) of bean common mosaic virus (BCMV) prevalent on common bean in Himachal Pradesh were compared at the 3' terminal region of the viral genome to elucidate variation and relationship among these strains. 3' terminal region comprising of partial nuclear inclusion b (NIb) (171–233 bp), complete coat protein (CP) (860 bp) and 3' untranslated region (UTR) (208–244 bp) shared 96–98% nucleotide and 96–99% amino acids identity among various strains of BCMV. Multiple alignment as well as cluster dendrograms of the 3' terminal region placed the test isolates in BCMV species of genus Potyvirus. Phylogenetic analysis of complete CP as well as 3' UTR also showed Indian strains to be distinct strains (Sharma *et al*., 2011).

Host range

The virus is known to infect various cultivated species as well as weed hosts *viz*., *Cajanus cajan, Cassia tora, Chenopodium amaranticolor, C. quinoa, Cicer arietinum, Crotolaria spectabilis, C. striata, Cyamopsis tetragonoloba, Glycine max, Gomphrena globosa, Lens esculenta, Lupinus albus, L. angustifolius, Macroptilium lathyroides, Melilotus alba, Nicotiana benthamiana, N. clevelandii, Phaseolus lathyroides, P. lunatus, P. vulgaris, Rhynchosia minima, Sesbania exaltata, Tetragonia expansa, Trifolium incarnatum, T. subterraneum, Vicia faba, V. sativa, V. villosa, Vigna angularis, V. radiata* and *V. unguiculata* (Zaumeyer and Goth, 1964; Drijfhout and Bos, 1977).

Transmission

The efficient transmissibility of the virus through bean seed is responsible for the worldwide distribution of the disease (Schmidt, 1992). Seed transmission rates ranging from 0 to 93 per cent have been reported in cultivars of different hosts (Edwardson and Christie, 1991). In nature, aphids like *Acyrthosiphon pisum, Aphis fabae* and *Myzus persicae* also transmit the virus in a non-persistant manner.

Epidemiology

Infected seed, plants of susceptible French bean cultivars and weed hosts serve as natural sources of primary inoculum (Robertson 1962). Aphid vectors are responsible for secondary field spread of BCMV. Gupta and Chowfla (1988) observed that nymphs and apterous aphids are more efficient as compared to winged adults. Air temperature and relative humidity significantly affected vector populations (Berg, 1984).

Management

No single method has proved effective to manage the disease. However, an integrated approach involving alteration of planting dates, manipulation of plant population density, use of insecticides to control insect vectors and breeding for disease resistance have been found to be effective (Jeyarajan and Doraisamy, 1986).

(i) Cultural practices: Use of virus free seed is essential because of efficient transmissibility of the virus through seed (Schmidt, 1992). The production of BCMV free seed is taken up in isolated regions away from other bean plantings and aphid vectors or under screenhouse/greenhouse conditions (Morales, 1989). Alteration of planting dates coinciding with periods of reduced vector activity, rouging of infected bean plants from the field or weed hosts from in and around the field help to eradicate natural reservoirs of the virus and planting of non host barrier crops especially maize, and closer spacing help in minimizing virus incidence by interfering with flight of the vector and thereby inactivating the virus (Dhanju *et al.*, 1995b).

(ii) Host resistance: *Phaseolus vulgaris* cvs./lines *viz.*, Bean 1 49-3, Bean 2 119-3, Bean Br II 42-1, Contender, HPK 69, Iran 431, Kentucky Wonder, Master, Pusa Parvati, Turkey 326, Turkey 434 (Gupta and Chowfla, 1990); ESAL 547, ESAL 550, Jalo (Pellicano *et al.*, 1993); Topcrop, Masterpiece (Devanathan *et al.*, 1994); Premier (Sharma and Dhar, 1994); PDR 14 (Bhagawati and Bhagabati, 1994); EC94469, EC 94929 (Dhanju *et al.*, 1995a); IC 41654, IC 41661, P414, EC 19505, EC 27518, EC 24939, EC 43636, EC 271562, EC 271523, EC 271491 and EC 241107 (Dhanju *et al.*, 1995c) have also been reported to possess resistance to BCMV. Of late, fourteen cvs./lines like Contender, EC-3, EC-29, EC-41, EC-43, IC-7, IC-11900, IC-16453, IC-17851, IC-18546, IIHR-220, IIHR-909, Pant Anupama and Premier were found resistant to this disease (Ramanjula *et al.*, 2004).

Sharma *et al.* (2006) evaluated ninety four bean accessions comprising of land races, exotic collections and recommended cultivars against three strain groups (I, IV, VI) of Bean Common Mosaic Virus (BCMV). Two exotic varieties Top Crop and Amanda were found highly resistant to all the three strain groups. The existence of two pathogroups PG-I and PG-II and four strains (NL-1, NL-1n, NL-7 and NL-7n) in Himachal Pradesh, a North-Western Himalayan state of India was reported by Kapil *et al.* (2011). Two strains, NL-1 and NL-7 were identical to the previously described NL-1 and NL-7 strains from Europe and USA, whereas the other two designated as NL-1n and NL-7n differed from earlier identified strains with respect to their necrotic reaction on cultivar Jubila at high temperature (>30°C). Reverse phase HPLC peptide profiling of tryptic digests of coat protein of these strains further confirmed that NL-1, NL-1n, NL-7 and NL-7n are distinct from each other. Bean accessions *viz.,* KR 7, KR 225, KR 295, KRC 4, KRC 7, KRC 11, KRC 12, KRC 13, KRC 16, KRC 22, Amanda, Black Turtle Soup, Contender, Hans, Great Northern UI 123, Improved Tender Green 40031, Jubila, Kentucky wonder, Monroe, Premier and Sanilac, were found resistant to NL-1n and NL-7n strains of bean common mosaic potyvirus strains (Sharma *et al*., 2008).

Chemical control: Soil application of granular insecticides (Thimet @ 1.5 kg a.i./ha) at sowing followed by 2–3 foliar sprays of either monocrotophos (0.05%) or dimethoate (0.05%) (Singh *et al.*, 1981) has been seen effective to manage aphid population resulting in reduced spread of the virus.

2. BEAN YELLOW MOSAIC

This disease can cause significant losses, if the favourable environment exists for its spread.

Geographical distribution

The occurrence of bean yellow mosaic virus (BYMV) was first reported in 1934 from USA (Bos, 1970). Susbsequently it was reported from East Africa, Germany, Canada, Holland, Italy and Switzerland.

Losses

The losses caused by BYMV are frequently very large. Since the symptoms of this virus and its strains and Common bean mosaic often overlap, it is almost impossible to assign any particular amount of loss caused by an individual virus.

Symptoms

Mosaic symptoms consist of contrasting dark and yellowish green areas, often accompanied by bright yellow spots, which intensify with the age of the plants. Some isolates cause only a mild and diffuse chlorotic mottle and limited plant stunting, while others cause a prominently coarse mosaic, rugosity, malformation and severe stunting. Some varieties may show necrotic spots, veinal and apical necrosis, wilting and premature death. Late infection generally causes less prominent foliar symptoms and if the infection coincides with pod formation, the pods eventually will exhibit a light green mottle and slight malformation. BYMV infected plants are usually stunted and bushy due to reduction of internode length and proliferation of lateral branches. Plant maturity is delayed and quality and quantity of the seed is affected. Certain strains of BYMV induce systemic necrosis that appear as a purplish discolouration at the base of the lower leaves, accompanied by veinal, stem and petiole necrosis. The necrosis at the terminal growing point consequently result in to death of the plant. Some strains incite local lesions and others produce malformed pods bearing reddish brown nrecrotic spots. However, type and severity of symptoms vary according to strain of virus and cultivar of French bean involved.

The Virus

Bean Yellow Mosaic Virus (BYMV) belongs to the *potyvirus* group. The virus is characterized by elongate, flexuous, filamentous particles about 750–810 × 11–15 nm. The virus has TIP of 65ºC, LIV of 2 to 4 days and DEP of 10^3 to 10^{-5} (Brunt *et al.*, 1996).

Host range

BYMV has a wide host range and can infect 233 species in 71 genera of 21 families (Edwardson and Christie, 1991). Besides French bean, its host range includes mung bean, lima bean, pigeon pea, chickpea, lentil, peas, broadbean, soybean, fenugreek, *Cucurbita pepo*, *Gladiolus* sp. and *Robinia pseudoacacia* (Raizada *et al.*, 1991).

Transmission

BYMV is readily transmitted through sap inoculation and by many species of aphids in non-persistent manner, but not through the seeds of French bean though it has been reported in broadbean and other leguminous hosts (Bos, 1970). BYMV has been reported to be experimentally transmitted by 47 aphid species in 25 different genera (Edwardson and Christie, 1991), but *Acyrthosiphum pisum, Macrosiphum euphorbiae, Myzus persicae, Aphis craccivora* and *A. fabae* are the important vectors of this virus under field conditions.

Epidemiology

BYMV is not a seed borne in French bean though its seed borne nature has been reported in other leguminous hosts. The virus has a wide host range and aphid vectors transmit it. Field spread by aphid vectors from infected French bean and other collateral hosts is primarily responsible for the epidemics of this disease.

Management

(i) Cultural practices: Destruction of collateral hosts of this virus from in and around the field of French bean, plantation of barrier crops like maize and crop rotation reduce the infection of this disease.

(ii) Host resistance: This is the most reliable method for the control of this disease. Sources of resistance to a BYMV strain including pod malformation have been identified in various Great Northern lines like GNUI No. 31, 59, 123 and 1140.

(iii) Chemical control: Application of various systemic insecticides like metasystox, dimethoate etc. reduces aphid population in French bean fields or other host crops and reduces the spread of virus.

REFERENCES

Acosta-Gallegos, J.A., *et al.* 2001. Registration of "Negro Otoni" shiny black bean. *Crop Sci.* **41:** 361.

Aghora, T.S., Mohan, N., Somkumar, R.G. and Grija Ganeshan 2007. Breeding French bean (*Phaseolus vulgaris*) for resistance to rust (*Uromyces phaseoli* Reben Wint). *J. Hort. Sci.* **2(2)**: 104–107.

Allen, D.J. 1982. *Verticillium lecanii* on the bean rust fungus, *Uromyces appendiculatus. Trans. British Myco. Soc.* **79(2)**: 362–364.

Aggour, A.R., Coyne, D.P., Vidaver, A.K. and Eskridge, K.M. 1989. Transmission of the common blight pathogen in bean seed. *J. Am. Soc. Hortic. Sci.* **114(6)**: 1002–1008.

Ahuja, S.C. 1976. Banded leaf and sheath blight of maize. Ph.D. Thesis, Post-graduate School, IARI, New Delhi. 153p.

Akhavan, A., Bahar, M., Saeidi, G. and Lak, M. 2009. Factors affecting epiphytic population of *Xanthomonas axonopodis* pv. *phaseoli* in epidemiology of bean common blight. *J. Sci. Technol. Agric. Nat. Res.* **13**: 265–277.

Alberini JL, Mohan SK Menezes JR de, Silva WR da, Oliari L. and Meyer RC 1987. IAPAR 8 – Rio Negro, a new French bean cultivar. *Pesquisa-Agropecuaria Brasileira* **22(9–10)**: 995–998.

Anderson AL, Adams MW and Whitford G. 1963. Charlevoix anthracnose resistant dark red kidney bean. *Mich. State Univ. Agric. Stn. Res. Rep.* 6.

Anonymous. 2014. http://www.nhb.gov.in

Anonymous. 1966. Division of Science Services, *Biol. Branch, Plant Pathol. Section. Rep. Dep. Agric. NSW,* 1964–65: 75–80.

Anonymous. 1989. French bean (*Phaseolus vulgaris*). *Pl. Varieties J.* **2(2)**: 15–16.

Anonymous. 2001. http//www.fao.org.

Araiyo, G.A.de A., Viera, C. and Chagas, J.M. 1989. Performance of bean (*Phaseolus vulgaris* L.) cultivars in the Zona da Mata area of Minas Gerais V. *Revista Ceres* **36**: 382–390.

Araya C M, Alleyne A T, Steadman J R, Eskridge K M and Cöyne D P. 2004. Phenotypic and genotypic characterization of *Uromyces appendiculatus* from *Phaseolus vulgaris* in the Americas. *Plant Dis.* **88**: 830–836.

Arnaud–Santana, E., Pena-Matos, E., Coyne, D.P. and Vidaver, A.K. 1991. Longevity of *Xanthomonas compestris* pv. *phaseoli* in naturally infested dry bean (*Phaseolus vulgaris*) debris. *Plant Dis.* **75(9)**: 952–953.

Arunga E E, Ochuodho J O, Kinyua M G and Owuoche J O. 2012. Characterization of *Uromyces appendiculatus* isolates collected from snap bean growing areas in Kenya. *African J. Agric. Res.* **7(42)**: 5685–5691.

Azevedo, L.A.S.de. and Kushalppa, A.C. 1986. Quantification of the resistance of *Phaseolus vulgaris* cultivars to *U. phaseoli typical. Fitopatologia Brasileira* **11(3)**: 599–608.

Baker, C.J., Stavley, J.R., and Mock, N. 1985. Biocontrol of bean rust by *Bacillus subtilis* under field conditions. *Plant Dis.* **69(9)**: 770–772.

Baker, C.J., Stavley, J.R., Thomas, C.A. Sasser, M. and Macfall, J.S. 1983. Inhibitory effect of *Bacillus subtilis* on *Uromyces phaseoli var. typica* and on development of rust pustules on bean leaves. *Phytopathology* **73(8)**: 1148–1152.

Balardin, R.S. and Kelly, J.D. 1998. Interaction between *Colletotrichum lindemuthianum* races and gene pool diversity in *Phaseolus vulgaris. J. Am. Soc. Hortic. Sci.* **123**: 1038–1047.

Balardin, R.S., Jarosz, A.M. and Kelly, J.D. 1997. Virulence and molecular diversity in *Colletotrichum lindemuthianum* from South, Central and North America. *Phytopathology* **87**: 1184–1191.

Bardas, G.A., Lagopodi, A.L., Kadoglidou, K. and Tzavella Klonari, K. 2009. Biological control of three *Colletotrichum lindemuthianum* races using *Pseudomonas chlororaphis* PCL1391 and *Pseudomonas fluorescens* WCS365. *Biol. Cont.* **49(2):** 139–145.

Barrus MF. 1911. Variation of varieties of bean in their susceptibility to anthracnose. *Phytopathology* **1:** 190–195.

Bashir, S., Mustaq, A, Vinay Sagar and Najib, M. 2006. Identification of sources of résistance in French bean to *Phaeoisariopsis griseola* causing angular leaf spot disease. *Indian J. Plant Protec.* **34:** 221–224.

Bazzi, C. and Calzolari, A. 1998. Bacterial diseases of bean: two years of observations in the Po valley. *Informatore Fitopatol.* **38(9):** 49–51.

Beach, S.A. 1892. Some bean diseases. *New York State Agric. Expt. Stn. Bull.* **48:** 305–333.

Becerra Leor, E.N., Lopezsalinas, E. and Acosta, G. J. A. 1994. Genetic resistance and chemical control of bean rust in the humid tropics of Mexico. *Revista Maxicana de Fitopatologia* **12(1):** 35–42.

Berg, G.M. 1984. Nonpersistent viruses: epidemiology and control. *Adv. Virus Res.* **31:** 387–429.

Bhagawati, R and Bhagabati, K.N. 1994. Incidence of bean common mosaic virus of French bean and its effect on grain yield. *Indian J. Virol.* **10:** 141–143.

Bhardwaj, C.L., Kalia, N.R. and Singh, L.N. 1992. Reaction of French bean germplasm against angular leaf spot. *Indian J. Mycol. Pl. Pathol.* **22 (3):** 270–271.

Bhardwaj, C.L., Nayital, S.C., Verma, S. and Kalia, N.R. 1994. Effect of sowing date, variety and management of angular leaf spot (*Phaeoisariopsis griseola*) on yield of French bean (*Phaseolus vulgaris*). *Indian J. Agric. Sci.* **64(5):** 336–338.

Bhat M N, Singh A K and Medhi R P. 1999. Reduction of French bean (*Phaseolus vulgaris*) cultivars/accession to rust pathogen *Uromyces appendiculatus*. *Indian J. Agric. Sci.* **69(12):** 849–850.

Bhat M N. 2002. Chemical control of rust and angular leaf spot of French bean in Sikkim. *Pl. Dis. Res.* **17(1):** 90–92.

Bigirimana, J., Fontaine, R. and Hofte, M. 2000. Bean anthracnose: virulence of *Colletotrichum lindemuthianum* isolates from Burundi, Central Africa. *Plant Dis.* **84:** 491.

Borucka, K. and Marcinkowska, J.Z. 2001. Occurrence of *Colletotrichum lindemuthianum* on *Phaseolus* beans in Poland. *Veg. Crops Res. Bull.* **54(1):** 147–151.

Bos, L. 1970. Bean yellow mosaic virus. In: *Description of Plant Viruses*. No. 73. CMI/ AAB, Kew, Surrey, England.

Bose, S.K. and Sindhan, G.S. 1972. Leaf spot of French bean caused by *Isariopsis griseola* Sacc. and its control. *Prog. Hortic.* **4:** 69–75.

Boudreau, M.A. 1993. Effect of intercropping beans with maize on the severity of angular leaf spot of beans in Kenya. *Plant. Pathol.* **42(1):** 16–25.

Boyle C and Walters D. 2005. Induction of systemic protection against rust infection in Broad bean by saccharin: effects on plant growth and development. *New Phytologist* **167**: 607–612.

Bruggen, Van, A.H.O., Whalen, C.H. and Arneson, P.A. 1986. Emergence, growth and development of dry beans seedlings in respect to temperature, soil moisture and *Rhizoctonia solani. Phytopathology* **76:** 568–572.

Brunt, A.A.; Crabtree, K; Dallwitz, M.J; Gibbs, A.J. and Watson, L. 1996. In *Viruses of Plants, Descriptions and lists from the VIDE Database*, CAB International, University Press, Cambridge, 1484p.

Burkholder, W.H. 1926. A new bacterial disease of the bean. *Phytopathology* **16:** 915–927.

Buruchara, R.A., Gathuru, E.M. and Mukunya, D.M. 1988. Disease progress of angular leaf spot caused by *Isariopsis griseola* Sacc. and its implications on resistance of some bean (*Phaseolus vulgaris* L.) cultivars. *Acta Hortic.* **218:** 321–328.

Butler E J. 1918. *Fungi and diseases in plants.* ThackerSprink and Co. Calcutta VI, 547 pp.

Butler, E.E. and Bracker, C.E., 1970. Morphology and cytology of *Rhizoctonia solani.* In: *Rhizoctonia solani, Biology and Pathology.* (Ed. Parmeter Jr. J.R.) University of California Press, Berkley, pp. 32–51.

Butler, E.J. 1918. *Fungi and Diseases in Plants.* Thacker Spink & Co., Calcutta, 547p.

Calzolari, A. 1997. Halo and common spot of beans. *Informatore Agrario* **53(25):** 66–67.

Campos, A.J. and Fucikovsky, Z.L. 1981. Study of some aspects of angular spot caused by *Isariopsis griseola* Sacc. on bean. *Fitopatologia* **16(1):**16–19.

Cardona–Alvarez, C. and Walker, J.C. 1956. Angular leaf spot of bean. *Phytopathology* **46:** 610–615.

Carneiro, S.M.de T.P.G. 1999. Physiological races of *Colletotrichum lindemuthianum* in Parana State – Brazil. *Summa Phytopathol.***25(3):** 275–278.

Carrion, G., Romero, A. and Rico–Gray, V. 1999. Use of *Verticillium lecanii* as a biocontrol agent against bean rust. *Association Latinoamericana de Fitopatologia (ALF) Fitopatologia* **34(4):** 214–219.

Castro, J.L. de, Ito, M.F., Dudienas, C., Bulisani, E.A., D'Artegnan–de–Almeida, L. and De–Castro, J.L. 1991. Fungicide activity on two bean cultivars in Capao Bonito, Sau Paulo State, Brazil. *Bragantia* **50(2):** 309–321.

Centurion MAP, D.A.C., Kimati, H. and Pereira, G.T. 1994. Machanism of action of antagonisits selected for the biological control of bean rust (*Uromyces phaseoli* (Reben.) Wint.). *Cientifica (Jaboticobal)* **22(2):** 163–175.

Chakrabarty PK and Shyam KR 1988. Evaluation of systemic fungitoxicants against *Colletorichum lindemuthianum*, the incitant of bean anthracnose. *Indian J. Plant Pathol.* **6(1):** 67–70.

Chakrabarty, P.K. and Shyam, K.R. 1990. Reaction of some promising lines of French bean to different isolates of *Colletotrichum lindemuthianum* in Himachal Pradesh. *Indian J. Plant Pathol.* **8(1&2):** 86–88.

Chand, R. 1992. Occurrence of *Xanthomonas compestris* pv. *phaseoli* (Smith) Dye on pea (*Pisum sativum* L.) in India. *Z. Pflanzenkr. Pflanzenschutz* **99(4):** 426–429.

Chhetry, G.K.N., Mishra, R.R. and Sharma, G.D. 1994. Effect of microclimatic factors on epidemiology of leaf spot disease caused by *Colletotrichum lindemuthianum* and *Phaeoisariopsis griseola* of French bean (*Phaseolus vulgaris*). *Indian J. Agric. Sci.* **64(9):** 667–670.

Christen, R.and Echandi, E. 1967. The more common races of rust *Uromyces appendiculatus* in Costa Rica and evaluation of the rust resistance of some French bean cultivars. *Turrialba* **17(1):** 7–10.

Cook, M.T. 1924. Common diseases of beans and peas. *Agric. Exp. Stn. Circ.* **142 :** 8.

Corrales, M.A. and Jara, C.E. 1995. Evolution of *Phaeoisariopsis griseola* with common bean in Latin America. *Fitopatol. Colombiana* **19(1):** 15–24.

Correa–Victoria, F.J. 1987. Pathogenic variation, production of toxic metabolites and isozyme analysis in *Phaeoisariopsis griseola* (Sacc) Ferr. Michingan, USA, Michigan State University, Ph.D. Thesis.

Correa–Victoria, F.J. and Saettler, A.W. 1987. Angular leaf spot of red kidney beans in Michigan. *Plant Dis.* **71(10):** 915–918.

Coyne, D.P. and Schuster, M.L. 1974. Breeding and genetic studies of tolerance to several bean (*Phaseolus vulgaris* L.) bacterial pathogens. *Euphytica* **23:** 651–656.

Coyne, D.P., Schuster, M.L. and Erwin, D.C. 1979. Reaction of *Phaseolus vulgaris* germplasm to a new virulent strain of *Pseudomonas phaseolicola. Ann. Rep. Bean Improv. Coop*. **22:** 20–21.

Davison, A.D. and Vaughan, E.K. 1963. Longevity of uredospores of race 33 of *Uromyces phaseoli* var. *typica* in storage. *Phytopathology* **53(6):** 736–737.

De Menezes, O.B. 1952. Identification of four physiologic races of bean rust. *Dusenia* **3(4):** 309–312.

Desmazieres, J.B.H.J. 1843. Dixieme Notice Sur Quelques Plantes Cryptogames La Plupart Inedites Recemment Decouvertes En France, Et Qui Vont Paraitre en Nature Dans La Collection Publiee Par L'Auteur. *Ann. des Sci. Nat. (Ser. 2)* **19:** 335–375.

Devanathan, M.; Muniyappa, V. and Peter, G.B. 1994. Screening of French bean (*Phaseolus vulgaris* L.) varieties for bean common mosaic virus (BCMV) resistance. *South Indian Hortic.* **42:** 53–54.

Dhanju, K.S.; Chowfla, S.C. and Garg, I.D. 1995d. Characterization of viruses causing French bean mosaic in Himachal Pradesh. *Int. J. Trop. Pl. Dis.* **13:** 33–41.

Dhanju, K.S.; Chowfla, S.C. and Handa, A.K. 1995a. Screening of French bean germplasm under field and glasshouse conditions against bean common mosaic virus. *Legume Res.* **18:** 53–55.

Dhanju, K.S.; Chowfla, S.C. and Handa, A.K. 1995b. Effect of barrier crops and spacings on the incidence of mosaic disease and yield of French bean. *Legume Res.* **18:** 113–116.

Dhanju, K.S.; Chowfla, S.C. and Handa, A.K. 1995c. Relative response of French bean germplasm against common mosaic under glass house conditions. *Legume Res.* **18:** 135–136.

Diaz–Plaza, R.; Teliz, O.D. and Munoz, D.A. 1991. Effect of diseases of bean grown under rainfed conditions at Mixteca peblana. *Revista–Mexicana–de–Fitopatol.* **9:** 21–30.

Downing, K.J. and Thomson, J.A. 2000. Introduction of the *Serratia marcescens* Chi A gene into an endophytic *Pseudomonas flourescens* for the biocontrol of phytopathogenic fungi. *Can. J. Microbiol.* **46(4):** 363–369.

Drijfhout E. and Davis J.H.C. 1989. Selection of a new set of homogeneously reacting bean (*Phaseolus vulgaris*) differentials to differentiate races of *Colletotrichum lindemuthianum. Plant Pathol.* **38:** 391–396.

Drijfhout, E. and Bos, L. 1977. The identification of two new strains of bean common mosaic virus. *Neth. J. Pl. Pathol.* **83:** 13–25.

Drijfhout, E., Silbernagel, M.J. and Burke, D.W. 1978. Differentiation of strains of bean common mosaic virus. *Neth. J. Pl. Pathol.* **84:** 13–26.

Dumitras, L. and Sesan, T. 1990. Feature of antagonism of *Trichoderma viride* Pers ex Fr. to *Rhizoctonia solani* Kuhn. *Biologie Vegelata.* **32:** 87–92.

Dundas, B. 1942. Breeding beans for resistance to powdery mildew and rust. *Phytopathology* **39(9):** 828.

Dundas, B. and Scott, G.W. 1939. Physiologic strains of bean rust. *Phytopathology* **29(9):** 820–821.

Eastman, J. 1944. Report of provincial plant pathologist. *British Colombia Deptt. Agric. Ann. Rept.* **1943:** R59–R63.

Edwardson, J.R. and Christie, R.G. 1991. *The potyvirus Group*. Vol. I–IV, *Fl. Agric. Exp. Stn. Monograph* 16.

Elad, Y.; Chet, I. and Katen, I. 1980. *Trichoderma harzianum* biocontrol agent effective against *Sclerotium rolfsii* and *Rhizoctonia solani. Phytopathology* **70:** 119–121.

Estevez-de-Jensen, C., Percich, J.A., Graham, P.H. and de-Jensen-C-Estevez. 2002. Integrated management strategies of bean root rot with *Bacillus subtilis* and *Rhizobium* in Minnesota. *Field Crop Res.* **74(2–3):** 107–115.

Faleiro, F.G., Paula Jr. T. J. De, Barros, E. G. De, Frietos, M. A. S. and Moreira, M. A. 1996. Common bean resistance to *Uromyces appendiculatus* from zona da Mata region of the state of Minas Gerais, Brazil. *Fitopatologia Brasileira* **21(1):** 123–125.

Faleiro, F.G., Vinhadelli, W.S., Ragagnin, V.A., Paula, T.J.Jr., Moreira, M.A. and Barros, E.G. 1998. Resistance of the common bean to four races of *Uromyces appendiculatus. Revista Ceres*. **263**: 11–18.

Faleiro, F.G., Vinhadelli, W.S., Ragagnin, V.A., Zambolin, L., Paula, T.J.Jr., Moreira, M.A. and Barros, E.G. 1999. Identification of physiological races of *Uromyces appendiculatus* in the state of Minas Gerais, Brazil. *Fitopatologia Brasileira* **24(2):** 166–169.

Fenille, R.C. and Souza, N.L. de. 1999. The role of the organic material amended and the soil moisture on the pathogenecity of *Rhizoctonia solani* Kuhn AG–4 HGI in snap bean. *Pesquisa Agropecuaria Brasileira* **34(10):** 1959–1967.

Fernandez, M.T., Fernandez, M., Casares, A., Rodriguez, R. and Fueyo, M. 2000. Bean germplasm evaluation for anthracnose resistance and characterization of agronomic traits: a new physiological strain of *Colletotrichum lindemuthianum* infecting *Phaseolus vulgaris* L in Spain. *Euphytica* **114(2):** 143–149.

Fininsa, C. 1996. Effect of intercropping bean with maize on bean common bacterial blight and rust diseases. *Int. J. Pest Mgt*. **42**(1): 51–54.

Flores Revilla, C. Neito Angel, D. and Torre Almaraz, R.de la. 1994. Effect of varieties, foliar nutrients and seedling rates on disease severity and yield in French bean (*Phaseolus vulgaris). Revista Mexicana de Fitopatologia* **12(1):** 69–74.

Fortugno, C. 1977. Angular leaf spot (*Isariopsis griseola* Sacc.) of bean. *IDIA* **317/320:** 26–33.

Foster, H.H. 1947. Reaction to rust under Misssippi conditions, of certain lines and vars. of pole snap beans. *Pl. Dis. Reptr*. **31(10):** 378–383.

Fouilloux, G. 1979. New races of bean anthracnose and consequences on our breeding programmes. Pp. 221–235 in: *Int. Symp. Dis. of Trop. Food Crops* (Eds. H. Maraite and J.A. Meyer). Universite Catolique de Louvain-la-Neuve, Belgium.

Fourie, D. 1998. Characterization of halo blight races on dry beans in South Africa. *Plant Dis.* **82(3):** 307–310.

Frazier, W.A. and Hendrix, J.W. 1949. Hawaiian Wonder, new rust resistant pole green bean. *Circ. Hawaii Agric. Exp. Sta*. **28**: 7.

Garcia, Veloz R., Marin, M.R., Veloz, R.R., RodriguezGuerra, R., Torres, P.I., Gonzalez, C.M.M., Anaya Lopez, J.L., Guevara Olvera, L., Feregrino Perez, A.A., Loarca Pina, G. and Guevara Gonzalez, R.G. 2010. Antimicrobial activities of cascalote (*Caesalpinia cacalaco*) phenolics-containing extract against fungus *Colletotrichum lindemuthianum. Indust. Crops Prod.* **31(1):** 134–138.

Ghimire, S.R., Pradhanang, P.M. and Lohar, D.P. 1995. Use of fungicides and identification of resistant varieties for the management of bean rust and anthracnose disease in common beans. *Working Paper Lumle Regional Agric. Res. Centre No.* **95/76**: 42p.

Gilbertson, R.L., Rand, R.E. and Hagedorn, D.J. 1990. Survival of *Xanthomonas compestris* pv. *phaseoli* and pectolytic strains of *Xanthomonas compestris* in bean debris. *Plant Dis.* **74(4):** 322–327.

Gonalves-Vidigal, M.C., Sakiyama, N.S., Vidigal-Filho, P.S., Amaral-Junior, A.T., Poletine, J.P.and Oliveira, V.R. 2001. Resistance of common bean cultivar AB 136 to races 31 and

69 of *Colletotrichum lindemuthianum*: the Co–6 locus. *Crop Breeding App. Biotech.* **1(2):** 99–104.

Gonzalez, M. and Garcia, C. 1996a. Evaluation of fungicides for the control of bean rust (*Uromyces appendiculatus*). *Agronomia Mesoamericana* **7(1):** 86–89.

Gonzalez, M. and Garcia, C. 1996b. Evaluation of losses due to the rust on bean (*Phaseolus vulgaris* L.) in four sowing times in Cuba. *Agronomia Mesoamericana* **7(1):** 95–98.

Gonzalez, M., Rodrigues, R., Zavala, M.E., Jacobo, J.L., Hernandez, F., Acosta, J., Martinez, O. and Simpson, J. 1998. Characterization of Mexican isolates of *Colletotrichum lindemuthianum* by using differential cultivars and molecular markers. *Phytopathology* **88:** 292–299.

Gupta S.K., Jarial K. and Kumar M. 2008. Occurrence of French bean rust in Himachal Pradesh. *J. Mycol. Plant Pathol.* **38(2):** 352–353.

Gupta, S.K. and Mathew, K.A. 1999. Fungal and bacterial diseases of beans. In: *Diseases of Horticultural crops– Vegetables, Ornamentals and Mushrooms* (Eds. L.R. Verma and R.C. Sharma). Indus Publishing Co., New Delhi pp. 301–335.

Gupta, S.K. and Shyam, K.R. 2003. Angular leaf spot of French bean: epidemiology and management. *Annu. Rev. Plant Pathol.* **2:** 221–237.

Gupta, S.K. and Shyam, K.R. 2007. Variation among *Phaeoisariopsis griseola* isolates, the incitant of angular leaf spot of French bean. *Pl. Dis. Res.* **22(1):** 1–7.

Gupta, S.K., Dohroo, N.P. and Shyam, K.R. 1991. Antagonistic studies on seed-borne mycoflora of French bean (*Phaseolus vulgaris* L.). *Indian J. Plant Pathol.* **9(1&2):** 62–63.

Gupta, S.K., Mathew, K.A and Shyam, K.R. 2000. Seed transmission of *Rhizoctonia solani* in French bean cultivars. *Pl. Dis. Res.* **15(1):** 83–84.

Gupta, S.K., Mathew, K.A. and Shyam, K.R. 2005. Influence of inoculum density and environmental factors on angular leaf spot of French bean. *Pl. Dis. Res.* **20(1):** 25–28.

Gupta, S.K., Mathew, K.A. and Shyam, K.R. 2000e. Management of angular leaf spot of French beam through cultural practices. In: Proceedings of Indian Phytopathological Society-Golden Jubilee International Conference on "Integrated Plant Disease Management for Sustainable Agriculture" Vol. II, Indian Phytopathological Society, IARI, New Delhi: 701–704.

Gupta, S.K., Mathew, K.A. and Shyam, K.R. 2005. Influence of inoculum density and environmental factors on angular leaf spot of French bean. *Pl. Dis. Res.* **20(1):** 25–28.

Gupta, S.K., Mathew, K.A., Shyam, K.R. and Gupta, D.K. 2000c Role of fungicides in management of anthracnose of French bean. In: Proceedings of Indian Phytopathological Society–Golden Jubilee International Conference on "*Integrated Plant Disease Management for sustainable Agriculture*" Vol. II, Indian Phytopathological Society, IARI, New Delhi: 705–707.

Gupta, S.K., Mathew, K.A., Shyam, K.R. and Sharma, A. 1999. Fungicidal management of root rot (*Rhizoctonia solani*) of French bean. *Pl. Dis. Res.* **14(1):** 20–24.

Gupta, S.K., Mathew, K.A., Shyam, K.R. and Sharma, A.K. 2000d. Evaluation of French bean germplasm against angular leaf spot. *Indian Phytopath.* **53(4):** 488–489.

Gupta, S.K.; Mathew, K.A., Shyam, K.R. and A.K. Joshi. 1997. Screening of French bean germplasm against root rot and web blight. *Legume Res.* **20(1):** 57–60.

Gupta, Sachin, Mallick, S.A and Gupta, Moni. 2010. Role of oxidative enzymes and biochemical constituents in imparting resistance to French bean varieties against bean anthracnose. *Indian Phytopath.* **63(10)**: 47–50.

Gupta, Y. and Chowfla, S.C. 1988. Susceptibility of French bean to Bean Common Mosaic virus in relation to plantage, populationage and age of *Myzus persicae Indian J. Mycol. Pl. Pathol.* **18:** 93–94.

Gupta, Y. and Chowfla, S.C. 1990. Screening of French bean germplasm for resistance to bean common mosaic virus. *Indian Phytopath.* **43:** 434–436.

Guu, Jin Wen. 1995. Development of rust resistant and stringless bean lines. Bull. Taichung District Agric. Imp. Stn. **46:** 11–18.

Guzman, P., Gilbertson, R.L., Nodari, R.l, Johnson, W.C., Temple, S.R.; Mendala, D., Mkandawire, A.B.C. and Gupta, P. 1995. Characterization of variability in the fungus *Phaeoisariopsis griseola* suggests coevolution with the common bean (*Phaseolus vulgaris*). *Phytopathology* **85(5):** 600–607.

Hagedorn, D.J. and Rand, R.E. 1980. Wisconsin RRR 46 Snap bean breeding line. *Hortic. Sci.* **15:** 529–530.

Hanan, F. and Khan, I. 1999. Resistance of common bean (*Phaseolus vulgaris*) cultivars to two races of halo blight (*Pseudomonas phaseolicola* Burkh.). *Sarhad J. Agri.*

Harter, I.L., Andrus, C.F. and Zaumeyer, W.J. 1935. Studies on bean rust caused by *Uromyces phaseoli* var. *typica. J .Agric. Res.* **1(9):** 737–759.

Harter, L.L. and Zaumeyer, W.J. 1941. Differentiation of physiologic races of *Uromyces phaseoli* var. *typica* on bean. *J. Agric. Res.* **6(12):** 717–731.

Hazarika, D.K. and Das, K.K. 1998. Biological management of root rot of French bean (*Phaseolus vulgaris* L.) caused by *Rhizoctonia solani. Pl. Dis. Res.* **13(2):** 101–105.

Hidalgo, R. and Araya, C.M. 1993. Optimum growth stage of common bean for chemical control of anthracnose (*Colletotrichum lindemuthianum*) and angular leaf spot *(Isariopsis griseola)* in San Carlos, Costa Rica. *Agronomic Costarricense* **17(1):** 75–80.

Hocking, D. 1967. A new virulent form of *Phaeoisariopsis griseola* causing circular leaf spot of French bean. *Pl. Dis. Reptr.* **51:** 276–278.

Hutchinson CM and Ram Ayyar CS. 1915. The Indian rice beer ferment. *Mem. Dept. Agric. Indian Bact. Ser.* **1:** 137–168.

Irigoyen, E.D. and Garbagnoli, C. 1998. Common bacteriosis in bean (*Xanthomonas compestris* pv. *phaseoli* (E.F. Smith) Dowson): detection, infection and transmission through seeds. *Fitopatologia* **32:** 166–172.

Iriti, M. and Faoro, F. 2003. Benzothiadiazole (BTH) induces cell death independent resistance in *Phaseolus vulgaris* against *Uromyces appendiculatus. J. Phytopathol.* **15(3):** 171–180.

Issa, E. 1985. Resistance to four diseases in cultivars of French bean, *Phaseolus vulgaris* L., bred in Sao Paulo. *Biologico* **51**: 113–117.

Issa, E., Cruz, B.P.B, Wattanabe, K. and Arruda, H.V. 1980. Damage done by bean (*Phaseolus vulgaris* L.) diseases. *Biologica* **46(7):** 135–140.

Issa, E., Sinigaglia, C. and Oliveira, D.A. 1982. Chemical control of angular leaf spot of bean (*Phaseolus vulgaris* L.) due to *Isariopsis griseola* Sacc. *Biologico* **48(12):** 299–303.

Iwanowski, d. 1894. Ueber die Mosaikkrankheit der tabaksflanze. *Acad. Imp. Sci. St. Petersbourg Bull.* **35(3):** 67–70.

Jaffer, A.A. 1971 Effect of fungicides on bean rust (*Uromyces appendiculatus*) (Pers.) *Lev. Misc. Rep. Pestic. Res. Inst.* **765**: 8.

Jensen, J.H. and Livingstone, J.E. 1944. Variation in symptoms produced by isolates of *Phytomonas medicaginis* var. *phaseolicola. Phytopathology* **34:** 471–480.

Jeyarajan, R. and Doraisamy, S. 1986. Virus diseases of pulses and their management. In *Pest* and *Disease Management Centre for Pl. Protect Studies*, TNAU, Coimbatore, 109–114.

Jindal, K.K. 1991. Physical and chemical agents for the eradication of *Xanthomonas compestris* pv. *phaseoli* from bean seeds. *Pl. Dis. Res.* **6(1):** 68–71.

Joshi, Deeksha, Hooda, K.S., Bhatt, J.C., Mina, B.L. and Gupta, H.S. 2009. Suppressive effects of composts on soil–borne and foliar diseases of French bean in the field in the western Indian Himalayas. *Crop Protec.***28(7):** 608–615.

Joshi, D., Hooda, K.S. and Bhatt, J.C. 2008. Biocontrolefficacy of indigenous *Trichoderma* isolates against root rot pathogens of French bean Uttrakhand hills. *J. Biol. Contr.* **21:** 119–126.

Jung, G.H., Scrotch, P.W., Coyne, D.P., Nienhuis, J., Arnaud–Santana, E., Ariyarathen, H.M., Kaeppler, S.M. and Bassett, M.J. 1997. Molecular marker based genetic analysis of tepary bean derived common bacterial blight resistance in different developmental stages of common bean. *J. Am. Soc. Hortic. Sci.* **122(3):** 329–337.

Kapil, R., Sharma, P., Sharma, S.K., Sharma, O.P., Sharma, O.P., Dhar, J.B.and Sharma, P. N. 2011. Pathogenic and molecular variability in *bean common mosaic virus* infecting common bean in India. *Archives Phytopathol. Pl. Protec.* **44(11):**1081–1092.

Kapoor, K.S., Gill, H.S. and Sharma, S.R. 1988. Screening of French bean against floury leaf spots. *Indian Phytopath.* **41:** 650–654.

Karanja, T.W., Mwang'ombe, A.W. and Mibey, K. 1994. The effect of media and light regimes on cultural and morphological characteristics and sporulation of *Phaeoisariopsis griseola* Deighton. *East African Agric. For. J.* **59(3):** 241–251.

Kelly, J.D., Afanador, L. and Cameron, L.S. 1994. New races of *Colletotrichum lindemuthianum* in Michigan and implication in dry bean resistance breeding. *Plant Dis.* **78:** 892–894.

Khati, Pinkey and Hooda, K.S. 2006. Multiple disease resistance in rajmash genotypes. *Indian Phytopath.* **59(4):** 521–522.

Khlaif, H.M. and Qadous, N.S. 1995. Common blight of beans in the Jordon Valley: sources of inoculum. *Dirasat. Series B, Pure App. Sci.* **22(3):** 717–728.

Kiessling, D. and Skiebe, K. 1987. Studies on the causes of variability in incomplete resistance of bean, *Phaseolus vulgaris* L. to *Pseudomonas syringae* pv. *phaseolicola* (Burkholder) Young *et al., Arch. Zichtungs–forschung* **17:** 191–200.

Kikoka, L.P., Katunzi, A.L. and Teri, J.M. 1989. Evaluation of effect of cropping systems on bean diseases and yield. In *Proc. Int. Pest Mgt. in Trop. Sub-trop. Crop. Syst.* **89(3):** 853–858.

Kimati H and Galli F 1970. *Glomerella cingulata* (Stonem.) Spauld. *et* V. Schrenk. f. sp. *phaseoli* n.f., faseascegena de agente causal de anthracnose do feijoeiro. *Annuals da Escola Superior* de *Agricultura "Luisde Queiroz"* **27:** 411–437.

Kishan, R. 1988. Loss in yield of French beans affected with common blight. Pp, 29–31. In: *Advances in Research on Plant Pathogenic Bacteria.* (Eds. S.S. Gnanamanickan and A. Mahadevan). Today and Tomorrow's Printers and Publishers, New Delhi.

Koch, S.H. 1996. *Colletotrichum* spp. on dry beans and lupins in South Africa. Ph.D. Thesis, University of Pretoria, South Africa.

Kouassi, A.S. 1993. Tracing *Rhizotonia solani. UNILET–Informations* **81:** 18–19.

Kumar, A., Kaushal, R.P. and Sharma, R. 2005. Transfer of resistance to floury leaf spot and bacterial blight in French bean from alien germplasm. *Indian Phytopath.* **58(2):** 136–142.

Kumar, A., Sharma, P.N., Sharma, O.P. and Tyagi, P.D. 1997. Resistance to *Colletotrichum lindemuthianum* in kidney bean accessions of diverse origin in Himachal Pradesh. *Indian Phytopath.* **50:** 59–64.

Kumar, A., Sharma, P.N., Sharma, O.P. and Tyagi, P.D. 1999. Epidemiology of bean anthracnose *Colletotrichum lindemuthianum* under sub–humid mid–hills zone of Himachal Pradesh. *Indian Phytopath.* **52(4):** 393–397.

Lecerda, J.T., Coetho, R.S.B., Mariano, R.L.R. and Menezes, M. 1994. Pathogenic variability of *Isariopsis griseola* on bean in the state of Pernambuco. *Summa Phytopathol.* **20:** 93–96.

Lenne, J.M. 1992. Colletotrichum diseases of legumes. Pp.134–166 In: *Colletotrichum: Biology, Pathology* and *Control*. (Eds. J.A. Bailey, and M.A Jeger), C.A.B. International, Wallingford, U.K.

Lopez, S.E., Acosta, G.J.A., Becerra, L.E.N., Frayre, V.G., Orozco, S.H. and Beebe, S.E. 1997. Registration of Negro Tacana common bean. *Crop. Sci.* **37:** 1022.

Mabgala, R.B. 1997. The effect of populations of *xanthomonas campestris* pv. *Phaseoli* in bean reproductive tissues on seed infection of resistant and susceptible bean genotypes. *European J. plant Pathol.* **103(2):** 175–181.

Mabgala, R.B. and Saettler, A.W. 1992. Races and survival of *Pseudomonas syringae* pv. *phaseolicola* in Northern Tanzania. *Plant Dis.* **76(7):** 678–682.

Macraw, W.M., Nasser, S.H., Sidky, S.T. and Faris, F.S. 1973. Producing a var. of beans, resistant to the rust disease. 1. Screening local and imported vars. against the disease. *Agric. Res. Rev.* **51(3):** 145–152.

Mageshwaran V., Mondal K.K., Kumar U. and Annapurna K. 2012. Role of antibiosis on suppression of bacterial common blight disease in French bean by *Paenibacillus polymyxa* strain HKA–15. *African J. Biotech.* **11(60):** 12389–12395.

Mahanta I C and Dhal A. 2000. Management of leaf spot and rust of rajmash (*Phaseolus vulgaris*) through seed treatment and sprays of chemicals. *Legume Res.* **23(4):** 271–272.

Marcinkowska, J. 1994. Diseases of bean recorded in Poland. *Biuletyn Instytutu Hodowli I Aklimatyzagi Roslin* **190:** 169–175.

Maringoni, A.C. 1990. Chemical control of the common bacterial blight of bean and its effect in the transmission of *Xanthomonas compestris* pv. *phaseoli* (Smith) Dye by seeds. *Pesquisa Agropecuaria Brasileira* **25(8):** 1151–1156.

Maringoni, A.C. and Bach, E.E. 1999. Characterization of *Xanthomonas axonopodis* pv. *phaseoli* strains through electrophoretic analysis of total proteins. *Summa Phytopathol.* **25(2):** 173–177.

Maringoni, A.C., Barros, E.M. de and de Barros, E.M. 2002. Occurrence of isolates of *Colletotrichum lindemuthianum* resistant to benzimidazole fungicides. *Summa Phytopathol.* **28(2):** 197–200.

Maringoni, A.C., Kimati, H. and Kurozawa, C. 1994. Serological variability among isolates of *Xanthomonas compestris* pv. *phaseoli. Summa Phytopathol.* **20(3/4):** 164–167.

Maringoni, A.C., Tofoli, J.G. and Fregonese, L.H. 1992. Fungicides action *in vitro* on *Rhizoctonia solani* and on control of dry bean damping off. *Summa Phytopathol.* **18:** 269–275.

Mathew, K. A. and Gupta, S.K. 1996. Studies on web blight of French bean caused by *Rhizoctonia solani* and its management. *Indian J. Mycol. Pl. Pathol.* **26:** 171–176.

Mathew, K. A., Gupta, S.K. and Shyam, K.R. 1998. New strategies in fungicidal management of angular leaf spot (*Phaeoisariopsis griseola*) of French bean. *J. Mycol Pl. Pathol.* **28(2):** 123–133.

Mathew, K.A and Gupta, S.K. 1998. Biological control of root rot of French bean caused by *Rhizoctonia solani. J. Mycol. Pl. Pathol.* **28(2):** 202–205.

Matz, J. 1921. Una enfermendad damina de la haleichuela. *Porto Rico Insular Stn. Cir.* **57:** 8.

Mayek–Perez, N., Pedroza–Flores, J.A., Villneal–Garcia, L.A. and Valdes–Lozano, C.G.S. 1995. Genetic and environmental factors related to temporal dynamics and diseases effects on dry beans (*Phaseolus vulgaris* L.) at Marin, Nuero, Mexico. *Revista Mexicana de Fitopatol.(RPP 98/ 415)*

McKern, N.M., Mink, G.I., Barnet, O.W., Mishra, A., Whittaker, L.A., Silbernagel, M.J., Ward, C.W. and Shukla, D.D. 1992. Isolates of bean common mosaic virus comprising two district potyviruses. *Phytopathology* **82(5):** 923–929.

Menezes, J.R. and Dianese, J.C. 1988. Race characterization of Brazilian isolates of *Colletotrichum lindemuthianum* and detection of resistance to anthracnose in *Phaseolus vulgaris. Phytopathology* **78:** 650–655.

Meyer, S., Saccardy–Adji, K., Rizza, F. and Genty, B. 2001. Inhibition of photosynthesis by *Colletotrichum lindemuthianum* in bean leaves determined by chlorophyll fluorescence imaging. *Plant Cell Environ.* **24(9):** 947–955.

Mishra, H.N., Mahapatra, A.K., Sahu, S. and Mahapatra, A.K.B. 1998. Performance of French bean under different dates of sowing in eastern ghat highland zone of Orissa. *Ann. Agric. Res.* **19(1):** 83–88.

Mizubuti, C.S.G., Maffia, L.A., Muchovej, J.J., Romeiro, R.S. and Batista, U.G. 1995. Selection of isolates of *Bacillus subtilis* with potential for the control of dry bean rust. *Fitopatologia Brasileira* **20**(4): 540–544.

Mmbaga M T and Stavely J R. 1988. Pathogenic variability in *Uromyces appendiculatus* from Tanzania and rust resistance in Tanzanian bean cultivars. *Plant Dis.* **72**: 259–262.

Mohamed, M.F., Coyne, D.P. and Read, P.E. 1995. A radiation–induced mutant with resistance to common bacterial blight disease in common beans. *Hortic. Sci.* **30(3):** 577–578.

Mohammed M. 1994. Bean rust pathogen variability measured by virulence and the random amplified polymorphic DNA (RAPD) technique. Ph.D thesis. The University of Nebraska – Lincoln 139 pp.

Mohan, N., Aghora, T.S. and Somkumar, R.G. 1997. Identification of French bean lines resistant to bacterial blight. *J. Maharashtra Agril. Univ.* **21(2):** 309–10.

Mohan, N., Aghora, T.S. and Somkuwar, R.G. 1995. Reaction of French bean germplasm to angular leaf spot (*Isariopsis griseola* Sacc.). *Legume Res.* **18:** 215–216.

Mohan, N., Aghora, T.S., Somkwar, R.G. and Girija Ganeshan. 1997. Sources of resistance to rust in French bean (*Phaseolus vulgaris* L.). *Indian J. Agric. Sci.* **67(2):** 85–86.

Mondal, K.K. 2004. Integrated management strategies for fuscous blight and floury leaf spot of rajmash. *Indian Phytopath.* **57(2):** 135–139.

Monteagndo, A.B., Rodino, A.P., Lema, M., Fuente, M de la, Santalla, M., Ron, A. M. de and Singh, S.P. 2006. Resistance to infection by fungal, bacterial and viral pathogens in a common bean core collection from the Iberian Peninsula. *Hort. Science* **41:** 319–322.

Moralles, F.J. 1989. Bean common Mosaic–Screening for Disease Resistance. *Centro Internacional de Agricultura Tropical (CIAT) Publ.* No. 33. Cali, Colombia. 28p.

Moreno, R.A. 1977. Efecto de diferentes sistemas de cultivo sobre la severidad de la mancha angular del frijol (*Phaseolus vulgaris* L.) caused for *Isariopsis griseola* Sacc. *Acta Agron. Palmira* **7:** 164–169.

Moreno, R.A. and Mora, L.E. 1984. Cropping pattern and soil management influence on plant disease 2. Bean rust epidemiology. *Turrialba* **34(1):** 41–45.

Mukherjee, S., Tripathi, H.S. and Rathi, Y.P.S. 2001. Integrated management of wilt complex in French bean (*Phaseolus vulgaris* L.). *J. Mycol. Pl. Pathol.* **31(2):** 213–215.

Mundel, H.H., Kiehn, F.A., Huang, H.C., Conner, R.L., Saindon, G. and Kemp, G. 2004. Morden 003 navy bean. *Can. J. Plant Sci.* **84:** 231–233.

Muyolo, N.G., Lipps, P.E. and Schmitthenner, A.F. 1993. Reactions of dry bean, lima bean, and soyabean cultivars to Rhizotonia root and hypocotyl rot and web blight. *Plant. Dis.* **77:** 234–238.

Navarro, A.R., Puerta–E, O.D. and Isaza, L. 1981. Combined effect of population density and chemical control of leaf diseases and yield of bean. *Fitopatol. Colombiana* **10(1/2):** 7–9.

Nitshe, M.J. and Cafati, K.C. 1985. Identification of internal fungi present in bean seeds. *Agricultura Tecnica* **45:** 227–234.

Ogoshi, A. 1987. Ecology and pathogenecity of anastomosis and intraspecific groups of *Rhizoctonia solani* Kuhn. *Annu. Rev. Phytopath.* **25:** 125–143.

Oliveira, S.H.F., Barros, B.C. and Castro, J.L. 1992. Effect of fungicides on the control of bean foliar diseases and seed quality. *Summa Phytopathol.* **18(2):** 178–184.

Oliviera, S.H.F. de and de Oliveira, S.H.F. 2003. New fungicides and spraying schedules in the control of anthracnose and angular leaf spot. *Summa Phytopathol.* **29:** 45–48.

Opio, A.F., Allen, D.J. and Teri, J.M. 1996. Pathogenic variation in *Xanthomonas compestris* pv. *phaseoli*, the causal agent of common bacterial blight in *Phaseolus* beans. *Plant Pathol.* **45(6):** 1126–1133.

Opio, A.F., Teri, J.M. and Allen, D.J. 1993. Studies on seed transmission of *Xanthomonas campestris* pv. *phaseoli* in common beans in Uganda. *Afr. Crop Sci. J.* **1(1):** 59–67.

Otoya, M.M. and Pastor–Corrales, M.A. 1992. DNA probes for studies of genetic diversity of pathogens causing common bacterial blight of beans. *Fitopatol. Columbiana* **16(1/2):** 149–157.

Ovichinnikova, A.M. and Zaitsev, V.N. 1992. Resistance of French bean to diseases in the south of the Non–chermozem zone. *Selektsiya I Semenovodstvo (Moskva)* **4–5:** 17–18.

Ovies, J. and Larrinaga, L. 1988. Transmission of *Xanthomonas compestris* pv. *phaseoli* by a wild host. *Ciencia y Tecnica en la Agricultura Proteccion de Plantas* **11(2):** 23–30.

Padder, B.A. and Sharma, P.N. 2010. Assessment of yield loss in common bean due to anthracnose (*Colletotrichum linemuthianum*) under glass house conditions. *Res. J. Agri. Sci.* **1:** 184–188.

Padder, B.A., Sharma, P.N., Kapil, R., Pathania, A., Sharma, O.P. 2010. Evaluation of bioagents and biopesticides against *Colletotrichum lindemuthienum* and its integrated management in common bean. *Nat. Sci. Biol.* **2:** 72–76.

Padder, B.A., Sharma, P.N. Sharma, O.P. and Kapoor, V. 2007. Genetic diversity and gene flow estimates among five populations of *Colletotrichum lindemuthianum* across Himachal Pradesh. *Physiol. Molec. Pl. Path.***70**: 8–12.

Padder, B.A., Sharma, P.N., Sharma, O.P. and Kapoor, V. 2008. Ribosomal DNA analysis of *Colletotrichum lindemuthianum* virulences from Himachal Pradesh (India). *Appl. Biol. Res.* **10**:6–10.

Pastor–Corrales, M.A. 1991. Estandarizacion de variedades differenciales y de designacion de razas de *Colletotrichum lindemuthianum*. *Phytopathology* **81:** 694. (Abstr.).

Pastor–Corrales, M.A. and Tu, J.C. 1989. Anthracnose. Pp. 77–104 in: *Bean Produc–tion Problems in Tropics*. 2nd edition (Eds. H.F. Schwartz, and M.A. Pastor–Corrales) CIAT, Cali, Colombia.

Pastor–Corrales, M.A., Erazo, O.A., Estrada, E.I. and Singh, S.P. 1994. Inheritance of anthracnose resistance in common bean accession G 2333. *Plant Dis.* **78:** 959–962.

Patel, P.N. and Jindal, J.K. 1973. Studies on resistance in crops to bacterial diseases in India. VI. Reaction of moth bean (*Phaseolus acutifolius*) and hycinth bean (*Dolichos lablab*) lines to bacterial blight and leaf spot diseases caused by *Xanthomonas phaseoli. Indian Phytopath*. **26:** 607–608.

Patel, P.N. and Walker, J.C. 1965. Resistance in *Phaseolus vulgaris* to halo blight. *Phytopathology* **55:** 889–894.

Pathania, A. Sharma, P.N., Sharma, O.P., Chahota, R.K., Ahmad, B. and Sharma, P. 2006. Evaluation of resistance sources and inheritance of resistance in kidney bean to Indian virulences of *Colletotrichum lindemuthianum.Euphytica* **149**: 97–103.

Partridge J E. 1997. Rust of dry beans. University of Nebraska–Lincoln, Department of Plant Pathology.

Paul, Y.S. and Sharma, R.C. 1990. Additions to the host range of *Rhizoctonia solani. Pl. Dis. Res.* **5:** 234.

Paula, T.J. Jr, Pinto, C.M.F., Silva, M.B. da, Nietsche, S., Carvalho, G.A. de and Falerio, F.G. 1998. Resistance of snap bean cultivars to anthracnose, angular leaf spot and rust. *Revista Ceres* **45(258):** 171–181.

Pazos, V., Lagomasino, I. and Martinez, R. 1986. Sensitivity of *Xanthomonas compestris* pv. *phaseoli* isolates to bacteriophages. *Revista de Proteccion Vegetal* **1(1):** 21–24.

Pellicano, I.J.; Ramalho, M.A.P.; Figueira, A. and Dos, R. 1993. Evaluation of bean (*Phaseolus vulgaris*) cultivars for resistance to bean common mosaic virus. *Giencia Pratica* **13:** 112–115.

Perfect, S.E., Green, J.R. and O'–Connell, R.J. 2001. Surface characteristics of necrotrophic secondary hyphae produced by the bean anthracnose fungus, *Colletotrichum lindemuthianum. Euro. J. Pl. Pathol.* **107(8):** 813–819.

Persoon C H. 1795. Observations Mycologicae. **In**: Usteri, P., *Annalen der Botanic* **3(15):** 1–43. Leipzig, Germany.

Petrak, F. 1950. Beitrage zur Pilsflora von Ekuador. *Sydowia* **4:** 450–587.

Pinto, J.M.A., Souza, E.A. and Oliveira, D.F. 2010. Use of plant extracts in the control of common bean anthracnose. *Crop Protec.***29(8):** 838–842.

Ploper, L.D. 1983. *Enfermedades del poroto en el Noroeste Argentino y su control,* Publication Miscelanea No. 74, Tucuman, Argentina.

Ponnappa, K.M. Hiremath, P.C. and Sulladmath, V.V. 1976. Occurrence of *Phaeoisariopsis griseola* (Sacc.) Ferr. on French bean. *Curr. Sci.* **45:** 836–837.

Prakasam, V. and Thamburaj, S. 1991. Efficacy of fungicides in control of rust disease of French bean caused by *Uromyces phaseoli* (Reben.) Wimb. *Indian J. Mycol. Plant Pathol.* **21(2):** 158–159.

Prasad, K. and Weigle, J.L. 1976. Association of seed coats factors with resistance to *Rhizoctonia solani* in *Phaseolus vulgaris*. *Phytopathology* **66:** 342–345.

Prudencio Sains, J.M., Navarrete Maya, R., Navarrete Maya, J. and Acosta Gallegos, J.A. 2008. Dynamics of the bean common and halo blights in the valley of Mexico. *Agricultura Tecnica en Mexico* **34(2):** 201–212.

Pyndji, M.M. and Trutmann, P. 1992. Managing angular leaf spot of common bean in Africa by supplementing farmer mixtures with resistant varieties. *Plant Dis*. **76(11):** 1144–1147.

Qadous, N. and Khlaif, H. 1997. Common blight of beans in Jordon valley: disease development and response of different host cultivars to the disease. *Dirasat. Agric. Sci.* **24(2):** 260–267.

Radkov P., Nedelchev NK and Dekov D. 1978. Study of the field resistance of some bean hybrids and cultivars to bacteriosis under non–irrigated and irrigated conditions. *Rastenievdni–Nauki* **15(9–10):** 149–153.

Raffat, F.M. 1992. Efficiency of certain fugicides in controlling *Rhizoctonia* damping off of bean. *Assuit J. Agric. Sci.* **23:** 37–47.

Raizada, R.K., Mohd Aslam QMR Haq and Singh, B.P. 1991. *Vicia birsuta*: a natural host of bean yellow mosaic virus. *Int. Conf. Virol. Trop*. Dec. 5, NBRI, Lucknow, pp.37.

Rajnauth, G. 1987. Web blight, an important disease of bean and pak–choi in Trinidad. *Trop. Agric.* **84:** 356–358.

Ramanjulu, V., Rajaram Reddy, D., Babu, G.P.and Gopal, K. 2004. Identification of sources of resistance to bean common mosaic virus in French bean (*Phaseolus vulgaris* L.). *Pl. Dis. Res.* **19(2):** 134–139.

Randhi, P. and Parisi, B. 1997. Viral and bacterial diseases of French beans. *Informatore Agrario* **53(32):** 56–57.

Rauscher M, Adam A L, Wirtz S, Guggenheim R, Mendgen K and Deising H B. 1999. PR-1 protein inhibits the differentiation of rust infection hyphae in leaves of acquired resistant broad bean. *Plant J.* **19(6):** 625–633.

Rava, C., Sartorato, A. and Romeiro, R.S. 1990. Evaluation of bean cultivars for resistance to *Xanthomonas compestris* pv. *phaseoli* in greenhouse and field conditions. *Summa Phytopathol.* **16(2):** 83–91.

Rava, C.A. and Romeiro, R.S. 1990. Variability among isolates of *Xanthomonas compestris* pv. *phaseoli* in relation to their pathogenecity on *Phaseolus vulgaris* cultivars. *Summa Phytopathol.* **16(3/4):** 225–232.

Rava, C.A., Costa, J.G.C.–da, Fonseca, J.R., Salgado, A.L. and da–Costa, J.G.C. 2003. Identification of sources of resistance to anthracnose, common bacterial blight and bacterial wilt in common bean. *Revista Ceres* **50:** 797–802.

Rava, C.A., Molina, J., Kauffman, M., and Briones, I. 1994. Determination de razas fisiologicas de *Colletotrichum lindemuthianum* en Nicaragua. *Fitopatol. Brasileira* **18:** 388–391.

Ravi, S., Doraiswamy, S., Valluvaparidasan, V. and Jeyalakshmi, C. 1999. Effect of biocontrol agents on seed– borne *Colletotrichum* in French bean. *Pl. Dis. Res.* **14(2):** 146–151.

Roberti, R., Ghisellini, L., Pisi, A., Flori, P. and Filippini, G. 1993. Efficiency of two species of *Trichoderma* as a biological control against *Rhizoctonia solani* Kuhn. isolated from string bean root rot in Italy. *Adv. Hortic. Sci.* **7:** 19–25.

Robertson, R.S. 1962. The role of seed transmission in epidemiology of bean mosaics in central Washington. *Pl. Dis. Reptr.* **56:** 71–72.

Rodriguez, A.M., Ovies, J. and Cruz, R. 1991. Two weed hosts of *Xanthomonas compestris* pv. *phaseoli*. *Proteccion de Plantas* **1(3/4):** 43–48.

Rodriguez, O., Faure, B., Benitez, R., Carballo, R.M. and Capote, I. 1999. Advances in the study of common bean resistance to bacterial diseases in Cuba. *Agronomia Mesoamericana* **10(1):** 55–58.

Rodriques, C.H., Zambolim, L. and Martins, H.C.P. 1987. Effectiveness of fungicides in the control of angular leaf spot (*Isariopsis griseola*) of bean (*Phaseolus vulgaris*) *Fitopatol. Brasileira* **12(1):** 40–45.

Rolim, P.R.R., Brignani Neto, F., Roston, A.J. and Oliveira, D.A. 1981. Chemical control of bean diseases.1. Rust (*Uromyces phaseoli* (Pers.) Wint. Var. *typica* Arth.). *Biologica* **47(7):** 201–205.

Saccardo, P.A. 1878. Fungi veneti novi criticivel Mycologiae venetae addendi *Isariopsis griseola* Sacc. *Michelia* **1:** 273 p.

Saettler, A.W. 1989. Common bacterial blight. In bean production problems in the tropics, 2nd ed. (Eds. H.F. Schwatz and M.A. Pasto-corrales), CIAT, Cali, Colombia, 726pp.

Saettler, A.W. and Corea, F.J. 1988. Transmission of *Phaeoisariopsis griseola* by bean seed. *J. Seed Tech.* **12:** 133–142.

Saettler, A.W. and Potter, H.S. 1970. Chemical control of halo bacterial blight in field beans. Pp. 1–8, *Michigan Agri. Exp. Stn. Res. Rep. 98.*

Sanchez, A., Echavez, B.R. and Schroder, E.C. 1994. *Psedudomonas cepacia*, a potential biofungicides for root rot pathogens of beans. *J. Agric. Univ. Puerto Rico* **78:** 55–57.

Sanchez, A.J.H. and Cardenas, A.M. 1988. Etiology and damage of root rot disease in beans *Phaseolus vulgaris* L. in the state of Durango. *Revista Chapingo* **12:** 43–49.

Santana, E.A., Coyne, D.P., Beaver, J.S. and Zaiter, H.Z. 1993. Effect of photoperiod and temperature on common bacterial blight disease of common beans (*Phaseolus vulgaris* L.). *Euphytica* **66(3):** 211–216.

Santos Filho, H.P., Ferraz S. and Sediyama, C.S. 1978. Effect of the time of inoculation of *Isariopsis griseola* Sacc. on three bean cultivars. *Fitopatol. Brasileira* **3(2):** 175–180.

Sarotorato, A. 1988a. Angular leaf spot. Piracicaba Brazil, *Asasociacao Brasileira para Pesquisa de Potassa e do Fosfato*, pp. 491–501.

Sartarato, A., Zimmerman, M.J., Rocha, M. and Yamada, T. 1988. Web blight. In *Cultura do Feijoero Factores que afetam a Productividade*, 503–520.

Sartorato, A. 1988b. Web blight. In: *Cultura do Feijoero Factores que afetam a Productividade,* 503–520.

Sastry, K.S.M., Singh, S.J. and Rajendran, R. 1981. Screening of different bean cultivars against bean common mosaic virus. *Pulse Crop Newsl.* **1:** 30.

Schmidt, H.E. 1992. Bean Mosaics. In: *Plant Diseases* of *International Importance Vol II. Diseases of Vegetables and Oilseed Crops.* (Eds. H.S. Chaube, J. Kumar, A.N. Mukhopadhyay and U.S. Singh). Prentice Hall Inc. 40–73p

Schuster, M.L., Coyne, D.P. and Smith, C. 1979. New strains of halo blight bacterium in Nebraska. *Annu. Rep. Bean Improv. Coop.* **22:** 19–20.

Schwartz, H.F. 1989. Halo blight. Pp. 285–301. in Bean Production Problems in the Tropics. 2nd ed. (Eds. H.F. Schwatz and M.A. Pastor–Corralles), CIAT, Cali, Colombia.

Scott, M.E. and Michaels, T.E. 1992. *Xanthomonas* resistance of *Phaseolus* inter–specific cross selections confirmed by field performance. *Hortic. Sci.* **27(4):** 348–50.

Scribner, F.L. 1888. Anthracnose of the bean *Gloesporium lindemuthianum. U.S. D.A. Rep.* (1887): 361.

Shama, Saad and Shama, S. 1989. Transmission of *Rhizoctonia solani* Kuhn. in seeds of bean (*Phaseolus vulgaris* L.). *Curr. Sci.* **58:** 972–974.

Sharma, A K. 1998. Epidemiology and management of rust disease of French bean. *Vegetable Science* **25(1):** 85–88.

Sharma, Neeraj, Gupta S K, Bharat N K and Sharma M. 2014. Evaluation of French bean germplasm against bean rust (*Uromyces appendiculatus*). *Pl. Dis. Res.* **29 (1):** 115–117.

Sharma P.N., Sharma O.P. and Tyagi P.D. 1994. Status and distribution of bean anthracnose in Himachal Pradesh. *Himachal J. Agric. Res.* **20 (1&2):** 91–96.

Sharma, K.D. and Kaushal, R.P. 1990. Studies on spore viability and perpetuation of *Mycovellosiella phaseoli* (Drummond) Deighton field. *Indian J. Plant Pathol.* **8(1 & 2):** 28–32.

Sharma, K.D., Kaushal, R.P. and Singh, B.M. 1998. Effect of bioregulators on spore germination of *Mycovellosiella phaseoli* (Drummond) Deighton, a floury leaf spot pathogen of French bean. *Indian Phytopath.* **51(1):** 43–46.

Sharma, M. and Gupta, S.K. 2003. Ecofriendly methods for the management of root rot and web blight (*Rhizoctonia solani*) of French bean. *J. Mycol. Pl. Pathol.* **33:** 345–361.

Sharma, M., Gupta, S.K. and Sharma, T.R. 2005. Characterization of variability in *Rhizoctonia solani* by using morphological and molecular markers. *J. Phytopathol.* **153:** 449–456.

Sharma, O.P. and Dhar, J. 1994. Studies on mosaic disease of french bean (*Phaseolus vulgaris* L.) in Himachal Pradesh. In: *Virology in the Tropics*. (Eds. N. Rishi; K.L. Ahuja and B.P. Singh), Malhotra Publishing House. New Delhi. pp. 319–325.

Sharma, P., Sharma, P.N., Kapil, R., Sharma, S.K. and Sharma, O.P. 2011. Analysis of 3'–terminal region of bean common mosaic virus strains infecting common bean in India. *Indian J. Virol.* **22(1)**:37–43.

Sharma, P.N. and Sugha, S.K. 1995. Management of bean anthracnose through chemicals. *Indian Phytopath.* **48(3)**: 304–307.

Sharma, P.N., Banyal, Kajal, Rana, J.C., Nag, Ruby, Sharma, S.K. and Pathania, Anju. 2012. Screening of common bean germplasm against *Colletotrichum lindemuthianum* causing bean anthracnose. *Indian Phytopath.* **65(1)**: 99–101.

Sharma, P.N., Kumar, A., Sharma, O.P., Sud, D. and Tyagi, P.D. 1999. Pathogenic variability in *Colletotrichum lindemuthianum* and evaluation of resistance in *Phaseolus vulgaris* in the North–Western Himalyan region of India. *J. Phytopath.* **147(1)**: 41–45.

Sharma, P.N., Padder, B.A., Sharma, O.P., Pathania, A. and Sharma, P. 2007. Pathological and molecular diversity in *Colletotrichum lindemuthianum* (bean anthracnose) across Himachal Pradesh, a north–western Himalayan state of India. *Aust. Pl. Pathol.* **36**: 191–197.

Sharma, P.N., Pathania A., Kapil, R., Sharma, P., Sharma, O.P., Patial, M. and Kapoor V. 2008. Resistance to bean common mosaic potyvirus strains and its inheritance in some Indian land races of common bean. *Euphytica* **164**: 173–180.

Sharma, R.C., Barthe J.P., Lafitte, C. and Touze, A. 1986. *In vitro* absorption of endopolygalacturonase secreted by *Colletotrichum lindemuthianum* on cell walls of monogenically resistant and susceptible cultivars to anthracnose. *Can. J. Bot.* **64:** 2903–2908.

Sharma, R.D. and Sohi, H.S. 1978. Evidence of a new strain of *Phaeoisariopsis griseola* Sacc. from India. *Curr. Sci.* **47:** 348–349.

Sharma, S.K., Sharma, O.P. and Sharma, P.N. 2006. Sources of resistance to some strains of Bean Common Mosaic Virus (BCMV) and patterns of inheritance of resistance in selected cultivars of kidney bean. *J. Genet. & Breed.* **60**: 1–4.

Sharma, S.R. 1977. Annual Report, IIHR, Bengalore, 1976.

Sharma, S.R. 1989. Yield loss caused by rust in French bean. *Pl. Dis. Res.* **4(1)**: 92– 94.

Sharma, S.R. and Sohi, H.S. 1989. Diseases of French bean in India and their control. *Indian Rev. Life Sci.* **9**:153–168.

Sharma, S.R. and Sohi, H.S. 1980. Assessment of losses in French bean caused by *Rhizoctonia solani* Kuhn. *Indian Phytopath.* **33:** 366–369.

Sharma, S.R. and Sohi, H.S. 1981. Effect of different fungicides against Rhizoctonia root rot of French beans (*Phaseolus vulgaris* L.). *Indian J. Mycol. Pl. Pathol.* **11:** 216–220.

Sharma, S.R. and Sohi, H.S. 1989. Diseases of French bean in India and their control. *Indian Rev. Life Sci.* **9:** 153–168.

Shear, C.L. and Wood, A.K. 1913. Studies of fungus parasites belonging to the genus *Glomerella. U.S.D.A., Bureau Pl. Ind. Bull.* 252.

Shukla A and Sharma H R. 2009. Fungicidal management of angular leaf spot and rust of bean (*Phaseolus vulgaris*). *J.Plant Dis. Sci.* **4(2)**: 222–223.

Sicard, D., Michalakis, Y., Dron, M. and Neema, C. 1997. Genetic diversity and pathogenic variation of *Colletotrichum lindemuthianum* in the three centers of diversity of its host, *Phaseolus vulgaris. Phytopathology* **87:** 807–813.

Sindhan G.S. 1983. Effect of temperature and relative humidity on the development of anthracnose of French bean. *Prog. Hortic.* **15:** 132–135.

Sindhan, G.S. 1984. Evaluation of fungicides against occurrence of angular leaf spot of French bean caused by *Phaeoisariopsis griseola. Indian J. Plant Pathol.* **2(1):** 39–42.

Sindhan, G.S. and Bose, S.K. 1980. Effect of temperature and relative humidity on the development of angular leaf spot of French bean. *Prog. Hortic.* **12:** 18–24.

Sindhan G.S. and Bose S.K. 1981. Evaluation of fungicides against anthracnose of French bean caused by *Colletotrichum lindemuthianum Indian Phytopath.* **34(3):** 325–329.

Singh A K and Bhat M N. 2002. Control of French bean rust through fungicides and a neem based formulation. *Indian Phytopath.* **55(2):** 241–243.

Singh, A.K. and Saini, S.S. 1980. Inheritance of resistance to angular leaf spot (*Isariopsis griseola* Sacc.) in French bean (*Phaseolus vulgaris* L.). *Euphytica* **29:** 175–176.

Singh, B. M. and Sharma, Y.R. 1976. Screening of fungicides to control angular and floury leaf spots of beans. *Indian J. Mycol. Pl. Pathol.* **6:** 148–151.

Singh, B.M. and Sharma, Y.R. 1975. Screening of bean lines for resistance to angular leaf spot caused by *Isariopsis griseola. Indian Phytopath.* **28:** 435–436.

Singh, B.M., Agarwal, V.D. and Sharma, Y.R. 1974. Screening of bean cultivars to floury leaf spot caused by *Ramularia phaseoli. Indian Phytopath.* **27:** 260–261.

Singh, J.P. and Musiyimi, A.B.K. 1984. Chemical control of bean rust in Kenya. *East African Agric. Forestry. J.* **45(3):** 207–209.

Singh, J.P. and Musyimi, A.B.K. 1981. Effect of rust on bean yield. *Indian Phytopath.* **34(3):** 378– 379.

Singh, J.P. and Musymi A.K.B. 1979. Control of rust and powdery mildews by a new systemic fungicide, Bayleton. *Pesticides* **13(4):** 51–53.

Singh, R.Y., Tripathi, S.P. and Yadav, G.R. 1995. Evaluation of different fungicides against various leaf spot diseases of beans. *Recent Hortic.* **2:** 111–112.

Singh, S.J.; Sastry, K.S. and Sastry, K.S.M. 1981. Field tests with insecticides and mineral oil for the protection of French beans from yellow mosaic virus disease. *Gartenbauwissenchaft* **46:** 88–91.

Singh, S.P., Teran, H., Lema, M., Webster, D.M., Strausbaugh, C.A., Miklas, P.N., Schwartz, H.F. and Brick, M.A. 2007. Seventy–five years of breeding dry bean of the Western USA. *Crop Sci.***47(3):** 981–989.

Smolinska, U. and Kowalska, B. 2006. The effectivity of plant extracts and antagonistic microorganisms on the growth inhibition of French bean pathogenic fungi. *Veg. Crops Res. Bull.* **64:** 67–76.

Sohi, H.S. and Sharma, R.D. 1974. Mode of survival of *Isariopsis griseola* Sacc. The causal agent of angular leaf spot of beans. *Indian J. Hortic.* **31:** 110–113.

Sohi, H.S., Sharma, S.L. and Verma, B.R. 1963. Annual Report of the section of Plant Pathology, Himachal Pradesh for the year 1962–63. 30 p.

Sohi, H.S., Sharma, S.L., Verma, B.R. and Sachdeva, K.B. 1965. Floury leaf spot of bean caused by *Ramularia phaseolina* Pertak. *Indian Phytopath.* **18:** 384.

Sokol, P.F., Gracia, R.M., Zimina, T.A. and Pivovarov, V.F. 1977. Some results of studies on the improvement of horticultural plants under Cuban conditions. *Inforem Cientifico–Tecnico Instituto de Investigaciones Fundamentales en Agricultura Tropical No.* **39:** 15–19.

Souza T, Alzate–Marin A L, Dessaune S N, Nunes S E, Quiroz V T, Moreira M A and Barros EG. 2007. Inheritance study and validation of SCAR molecular marker for rust resistance in common bean. *Crop Breeding and Applied Biotechnology* **7:**11–15.

Srinivasan, K.V. 1953. *Isariopsis griseola* Sacc. on *Phaseolus vulgaris* L. *Curr. Sci.* **22:** 20.

Srivastava, L.S. and Gupta, D.K. 1989. Aerial blight of French bean, groundnut, soyabean, black, green and horse gram – a new record from India. *Pl. Dis. Res.* **4:** 163–164.

Srivastava, L.S. and Gupta, D.K. 1994. Management of angular leaf spot of 'rajma' through fungicides in Sikkim. *Indian Phytopath.* **47(2):** 211–213.

Srivastava, L.S., Gupta, D.K., Dhiman, K.R. and Singh, G. 1995. Source of resistance in French bean (*Phaseolus vulgaris*) to angular leaf spot (*Phaeoisariopsis griseola*) in Sikkim. *Indian J. Agric. Sci.* **65(4):** 305–307.

Srivsastava, L.S. and Gupta, D.K. 1989a. Farinose spot of French bean–a new disease. *Curr. Sci.* **58(14):** 802.

Stall, R.E., Loschke, D.C. and Jones, J.B. 1986. Copper resistance and avirulence loci are linked to a self–transmissible plasmid in *Xanthomonas compestris. Phytopathology* **76:** 240–243.

Starr, M.P. 1983. The genus Xanthomonas, In the Prokaryotes: A handbook on habitats, isolation, and identification of bacteria, (Ed. Starr, M.P.), Springer–Verlag, New York, 742 pp.

Stavley, J.R., Steadman, J.R. and McMillan, R.T.Jr. 1989. New pathogenic variability in *Uromyces appendiculatus* in North America. *Plant Dis.* **73(5):**428–432.

Striab, W. 1943. Investigation on the biology and control of the bean rust (*Uromyces phaseoli* var. *typica* (Pers.) Wint.). *Gartenbavwirs* **17:** 397–445.

Sud, A.K. and Sood, V.K. 2002. Screening of indigenous and exotic collections of kidney bean against bean anthracnose under dry temperate conditions of Himachal Pradesh. *Crop Res.* Hisar **24(1):** 202–203.

Sugha, S.K., Sharma, O.P. and Sharma, P.N. 1998. Performance of 'vegetable type' French bean genotypes against floury leaf spot disease. *Pl. Dis. Res.* **13(1):** 85–86.

Sumartini 1998. Rust disease in French bean and its control. *J. Penelitian Pengembangan Pertanian* **17(4):** 149–153.

Sutton, D.C. 1979. Systemic cross protection in bean against *Colletotrichum lindemuthianum. Aust. Pl. Pathol.* **8:** 4.

Sutton, M.D. and Wallen, V.R. 1970. Epidemiological and ecological relations of *Xanthomonas phaseoli* var. *fuscans* in Southwestern Ontario, 1961–1968. *Can. J. Bot.* **48:** 1329–1334.

Tamayo–Molano, P.J., Otoya, M.M. and Pastor–Corrales, M.A. 1995. Diversidad de razas de *Colletotrichum lindemuthianum* el patogeno de la anthracnosis del frijol, en Rionegro, Antioquia. *Fitopatol. Colombiana* **19:** 1–6.

Tamietti G and Garibaldi A. 1984. Eradication of Pseudomonas syringae pv. phaseolicola in bean seeds in Italy. In: Phytopathology and plant bacterial diseases of quarentine significance, pp. 69–76.

Tan, C.S. and Tu, J.C. 1995. Tillage effect of root rot severity, growth and yield of beans. *Can. J. Pl. Sci.* **75:**183–186.

Taylor, J.D. 1970. Bacteriophage and serological methods for identification of *Pseudomonas phaseolicola* (Burkh.) Dowson. *Ann. App. Biol.* **66:** 387–395.

Taylor, J.D. and Teverson, D.M. 1996. Interactions between pseudomonads and *Phaseolus* beans and exploitation of genotype mixtures in subsistence agriculture. In: *Brighton Crop Prot. Conf. Pests Dis.–1996* Vol. 2.

Taylor, J.D., Teverson, D.M. and Davis, J.H.C. 1987. Diseases of *Phaseolus* beans–Biology, resistance and control. Pp. 63–64 in: *Annu. Rep. Inst. Hortic. Res., Wellsbourne, Warwick,* England.

Temizel, M.M. and Ertunc, F. 1992. Investigations on the detection of bean diseases of Van province. *J. Turk. Phytopathol.* **21:** 20–31.

Thakur, R.S., Sugha, S.K. and Singh, B.M. 1991. Evaluation of systemic fungicides for control of *Rhizoctonia solani* under glass house conditions. *Indian J. Agric. Sci.* **61:** 230–232.

Todorovic, B., Milijasevic, S., Rekanovic, E., Potocnik, I. and Stepanovic, M. 2008. Susceptibility of bean genotypes to *Xanthomonas campestris* pv. *phaseoli* in greenhouse conditions. *Pesticidi i Fitomedicina* **23(3):** 167–173.

Tripathi, K.C. 1985. Occurrence of floury leaf spot of French bean in Uttar Pradesh hills. *Indian Bot. Rep.* **4:** 175.

Trutmann O, Paul K.B. and Cishabayo 1992. Seed treatments increase yield of farmer varietal field bean mixtures in the Central African highlands through multiple disease and beanfly control. *Crop. Prot.* **11(5):** 458–464.

Trutmann, P. and Kayitare, E. 1991. Disease control and small multiplication plots improve seed quality and small farm dry bean yields in Central Africa. *J. App. Seed Prod.* **9:** 36–40.

Trutmann, P. and Pyndji, M.M. 1994. Partial replacement of local common bean mixtures by high yielding angular leaf spot resistant varieties to conserve local genetic diversity while increasing yield. *Ann. Appl. Biol.* **125(1):** 45–52.

Tu, J.C. 1981. Anthracnose (*Colletotrichum lindemuthianum*) on white bean (*Phaseolus vulgaris*) in Southern Ontario: Spread of the disease from an infection focus. *Plant Dis.* **5:** 477–480.

Tu, J.C. 1983. Epidemiology of anthracnose caused by *Colletotrichum lindemuth–ianum* on white bean (*Phaseolus vulgaris*) in Southern Ontario: Survival of the pathogen. *Plant Dis.* **67:** 402–404.

Tu, J.C. 1992. Management of root rot diseases of peas, beans and tomatoes. *Can. J. Pl. Path.* **14:** 92–99.

Tu, J.C. 1999. Relationship between disease severity and yield loss of navy bean caused by bacterial blight (*Xanthomonas campestris* pv. *phaseoli*). In: 51st International Symposium on crop protection, Gent, Belgium, 4th May, 1999, Part II. *Mededelingen–Faculteit Landbouwkundige en Toegepaste Biologische Wetenschappen, Universiteit Gent* **64(3b):** 637–642.

Tu, J.C. and McNaughton, M.E. 1980. Isolation and characterization of benomyl resistant biotypes of the delta race of *Colletotrichum lindemuthianum Can. J. Pl. Sci.* **60:** 585–592.

Tu, J.C. and Park, S.J. 1993. Root rot resistance in common bean. *Can. J. Pl. Sci.* **73:** 365–367.

Tu, J.C. and Varrtaja, O. 1981. The effect of the hyperparasite (*Gliocladium virens*) on *Rhizoctonia solani* and *Rhizoctonia* root rot of the white beans. *Can. J. Bot.* **59:** 22–27.

Tu, J.C. and Zheng, J. 1993. Compatability of biocontrol agents with DCT. *Mededelingen van de Faculteit Landbouwwetenschappen Univrsiteit Gent.* **58:** 1359–1364.

Tu, J.C., Sheppard, J.W. and Laidlaw, D.M. 1984. Occurrence and characterization of the epsilon race of bean anthracnose in Ontario. *Plant Dis.* **68:** 69–70.

Upmanyu S. and Gupta, S.K. 2009. Physiological variation among French bean isolates of *Rhizoctonia solani. J. Pl. Dis. Sci.***4(2):** 160–163.

Upmanyu, S, Gupta, S.K. and Shyam, K.R. 2002. Innovative approaches for the management of root rot and web blight (*Rhizoctonia solani*) of French bean. *J. Mycol. Pl. Pathol.* **32(3):** 317–331.

Upmanyu, S, Gupta, S.K. and Shyam, K.R. and Kaur, R. 2003. Patterns of variation among isolates of *Rhizoctoria Solani, Indian Phytopath.* **56:** 308 (Abstr.)

Upmanyu, S. and Gupta, S.K. 2005. Influence of environmental factors on the progress of root rot and web blight (*Rhizoctonia solani*) of French bean. *Indian Phytopath.* **58**(1): 79–83.

Upmanyu, S., Gupta, S.K. and Sharma, M.K. 2004. Evaluation of French bean (*Phaseolus vulgaris*) germplasm against web blight (*Rhizoctonia solani*). *Indian J. Agric. Sci.* 74: 448–450.

Uppal, B.N., Patel, M.K. and Nigam, B.G. 1946. Bacterial blight of French bean in Bombay province. *Proc. Natl. Inst. Sci.* 12: 351–359.

Velich, I., Szarka, J. and Neda, P. 1994. Biotic and abiotic stresses in bean: I. Genetic background of complex resistance to bacterial diseases. *Kerteszeti Tudomany* **26(1):** 49–54.

Verma, B.R. 1979. Studies on angular leaf spot of beans (*Phaseolus vulgaris* L.) caused by *Isariopsis griseola* Sacc. Ph.D Thesis, HPKVV, College of Agriculture, Solan. 221 p.

Vierra, R.F., Paula, T.J. Jr. and Berger, P.G. 1998. Effect of fungicides and cartap insecticides on the control of bean foliar diseases during the winter–spring season. *Summa Phytopathol.* **24(1):** 17–22.

Villar– Sanchez, B. and Lopez–Salinas, E. 1993. Negro Inifap, a new var. of French bean for Chiapas state and similar tropical regions. *Revista Fitotecnia Mexicana* **16(2):** 208–209.

Villegas, J.M. 1959. Variabilidad del *Isariopsis griseola* Sacc. agente caused de la mancha angular del frijol (*Phaseolus vulgaris* L.). Thesis Ing. Agr., Univesidad de Cldar, Mcnizales, Colombia 61 p.

Vizgarra, O.N. 1998. Carioca beans: beans of untra diti oval colour for Argentina. *Avance Agroindustrial* **19(75):** 24–26.

Voland, R.P. and Epstein, A.H. 1994. Development of suppressiveness to diseases caused by *Rhizoctonia solani* in soils amended with composted and non–composted manure. *Plant Dis.* **78:** 461–466.

Weller, D.M. and Saettler, A.W. 1980. Evaluation of seedborne *Xanthomonas phaseoli* and *Xanthomonas phaseoli* var. *fuscans* as primary inocula in bean blights. *Phytopathology* **70:** 148.

Wolk, M. and Sarkar, S. 1995. *In vivo* antagonism of fluorescent pseudomonads against *Rhizoctonia solani* and *Pythium aphanidermatum* on cucumbers and beans. *Anzeiger fur Schadlingskunde pflanzenschutz Umweltschutz* **68:** 3–8.

Wolk, M. and Sarkar, S.. 1994. Antagonism in vivo of Bacillus spp. against Rhizoctonia solani and Pythium spp. Anzeiger fur Schadlingskunde pflanzenschutz Umweltschutz **67:** 1–5.

Wright, D.G. and Wood, B.A. 1993. Testing three cultivars of French bean (*Phaseolus vulgaris*) from Florida for resistance to common blight caused by *Xanthomonas compestris* pv. *phaseoli. Aust. Pl. Pathol.* **22:** 58–61.

Xin, F.G., Gouren, Z., Baodong, L., Jing, H.H. and Jine, Z. 2000. Microscopic and ultramicroscopic investigation into the infection process of snap bean rust (*Uromyces appendiculatus*). *Acta Phytopathologica Sinica* **30(1):** 53–59.

Yaraguntaiah, R.C. and Nariani, T.K. 1963. Bean Mosaic virus in India. *Indian J. Microbiol.* **3:** 147–150.

Yoshii, K., Galvez, G.E. and Alvarez, G. 1976. Estimation of yield losses in beans caused by common blight. *Proc. Am. Phytopathol*. Soc. **3:** 298–299.

Young, J.M., Saddler, G.S., Takikawa, Y., De Boer, S.H. Vanterin, L., Grozdyak, R.UI. and stead, D.E. 1996. Normes of plant pathogenic bacteria 1864–1995. *Rev. Plant Pathol.* **75:** 721–763.

Zanatta, Z.G.C.N., Moura, A.B., Maia, L.C. and Santos, ASdos. 2007. Bioassay for selection of biocontroller bacteria against bean common blight (*Xanthomonas axonopodis* pv. *phaseoli*). *Brazilian J. Microbiol.* **38(3):** 511–515.

Zapota, M. 1997. Identification of *Xanthomonas campestris* pv. *phaseoli* races in *Phaseolus vulgaris* leaves. *Agronomia Mesoamericana* **8(1):** 44–52.

Zaumeyer, W.J. and Goth, R.W. 1964. A new severe symptom inducing strain of common bean mosaic virus. *Phytopathology* **54:** 1378–1385.

Zaumeyer, W.J. and Thomas, H.R. 1957. A monographic study of bean diseases and methods for their control. *U.S.D.A. Tech. Bull. No.* **868:** 255 pp.

Ziedan, E.H. and Mahamoud, S.Y.M. 2002. Calcium and sulfur soil treatment to improve biological control with *Trichoderma harzianum* for root rot disease control. *Assiut J. Agril. Sci.* **33(3):** 149–160.

Zuniga de Rodriguez, J.E. and Victoria, K.J.I. 1975. Determination of physiologic races of bean rust (*Uromyces phaseoli var. typica*) Arth in the Cauca Valley. *Acta Agronomica* **25**(1/2/3/4): 75–85.

■■■■

9

DISEASES OF CHILLIES AND BELL PEPPER

Chillies and bell pepper (*Capsicum annum* L.) popularly known as 'Lal Mirch' and 'Shimla Mirch', respectively, are important solanaceous crops grown throughout the world. The fruits are a good source of vitamin A and ascorbic acid. These are native of Mexico with secondary centre of origin in Guatemala (Smith and Heiser, 1957) and Portuguese explorers introduced it to Asia. In India, these are grown in most of the states of the country. During cultivation, these crop are affected by various fungal, bacterial and viral diseases, which not only reduce the potential yields drastically but also affect quality of the produce. A detailed account of these diseases has been given in this chapter.

(A) FUNGAL DISEASES

1. ANTHRACNOSE AND RIPE FRUIT ROT

This is a baffling disease of chilli and bell pepper seed crops in those areas where moisture is plentiful with moderate temperatures. In northern plains and hills, these crops are grown in rainy season and fruiting coincides with the monsoon rains. In the second fortnight of August when rains normally stop and dew deposition starts, which provides sufficient moisture for the germination of conidia and the disease assumes serious proportions.

Geographical distribution

The disease has been reported from various parts of the world. The samples were collected by Mc Rae in India in 1912, while H. Sydow first described the pathogen associated with this disease in 1913 (Sydow and McRae, 1929). Since then, it has been reported from China, Georgia, Philippines, Uganda, Ceylon, Tanganyika, United States, Florida, Iowa, Porto Rico, Argentina, Malaya, Pakistan, Bulgaria, Malaysia, Bangladesh, Copiah, Desoto and Oklibbeha countries in Mississippi (Mirdha and Siddique, 1989; Roy *et al*., 1997).

In India, Butler (1918) was the first to observe its incidence. Later, it was reported from Bihar and Assam by Dastur (1922) and Chowdhry (1957), respectively. Subsequently, it was reported from Himachal Pradesh. Now it is common in Kumaon hills of Uttranchal, Punjab, Haryana and Southern states of India (Saha and Singh, 1988; Thind and Jhooty, 1990). Most predominant pathogen associated with this disease was *Colletotrichum capsici* (Thind and Jhooty, 1985) but now the association of *C. gloesporioides* (*Glomerella cingulate*) with chilli has been reported from Maharashtra (Khodke and Gahukar, 1995) and on bell pepper from Himachal Pradesh (Gupta *et al.*, 2009).

Losses

Losses from this disease are very heavy and vary from season to season. These losses result from, (1) rotting and premature dropping of the severely infected fruits (2) reduction in the number of healthy seeds due to ripe fruit rot and (3) die-back of the infected twigs and branches. Several workers have reported varying degree of losses from this disease. The disease causes considerable damage, inflicting both qualitative and quantitative losses in India (Ananad *et al.*, 2010). Bansal and Grover (1969) reported crop loss from 10 – 35 per cent due to this disease in six districts of Punjab and Haryana. Saha and Singh (1988) reported that in suitable weather, it might cause up to 25 per cent damage to the crop. Roy *et al.* (1997) observed the occurrence of this disease on 3 to 15 per cent fruits in home gardens in Mississippi. Ripe fruit rot is more conspicuous as it caused 10-15 per cent loss to mature fruits during transit and storage (Bagri *et al.*, 2004).

Symptoms

The symptoms of the disease appear in two phases *i.e.,* die-back and anthracnose and ripe fruit rot.

(a) Die-back : Symptoms appear as necrosis of tender twigs from the tip downwards. The entire plant or branch may wither away. The twigs become straw coloured in advanced stages of the disease. Large numbers of black dots (acervuli of the fungus) are seen scattered all over the necrotic parts of the plants. Only the top or side branches may be killed or sometimes in case of severe attack, the entire plant may be withered. The fungus may also attack stem.

(b) Anthracnose and ripe fruit rot

(i) Colletotrichum capsici: The disease is noticed on red ripe fruits although green fruits are also sometimes infected. In bell pepper, both fully green and red fruits are infected, while in chillies, the infection is more common on red ripe fruits. The symptoms are characterized by appearance of small, black, circular slightly sunken spots on the skin of the fruits which spread in the direction of long axis. As the fruit matures, these spots enlarge, coalesce and form patches on the fruit covering larger areas (Plate 1a,b). During moist weather, these spots are covered by pinkish mass of fungal spores. With the advance of the disease, the spots spread, forming concentric markings with dark fructifications (acervuli of the fungus). The fruits with many lesions drop off pre-maturely resulting in heavy reduction in the yield. When the diseased fruit is cut open, the lower surface of the skin is found covered with minute elevated spherical, black stromatic masses of the fungus. After invading the outer hull of the fruit, the fungus travels into the seed cavity and infect the seed. Infected seeds are rusty in colour due to heavy infection in the seed cavity and shrivelled in appearance.

(ii) Colletotrichum gloeosporiodes: The characteristic symptoms of the disease appear as small circular or elongated, slightly sunken bleached lesions on the fruits which spread in long axis. In moist weather, these lesions bear many light black,

brown or orange coloured acervuli either in rings or in scattered fashion. On fruits, many acervuli appear which are subepidermal with light to dark brown setae disrupting outer epidermal cell walls of the host.

The Pathogen

The disease is caused by three species of *Colletotrichum* namely *C. capsici* (Syd.) Butler and Bisby, *C. gloeosporoides* Penz (*Colletotrichum* state of *Glomerella cingulata* (Stonem) Splang and Schrenk) and *C. piperatum* (E. and E.) E. and H. Among these, *C. capsici* is more prevalent and damaging (Grover and Bansal, 1970; Thind and Jhooty, 1990).

Colletotrichum capsici

The mycelium of *C. capsici* is septate, intercellular as well as intra-cellular and aerial mycelium appears light to dark grey in colour. Acervuli are round and elongated in shape (Fig. 1). Setae are scattered, brown, 1–5 septate, rigid and swollen at base and acute at apex. Conidiophores are short, hyaline to faintly brown, cylindrical, septate or aseptate. Conidia are falcate, fusiform with acute apices and narrow truncate base (Plate 2). They are one celled, hyaline and uninucleate. Conidia germinate to form small germ tubes and when these come in contact with any hard surface they form appresoria. The appresoria are medium brown, clavate to circular.

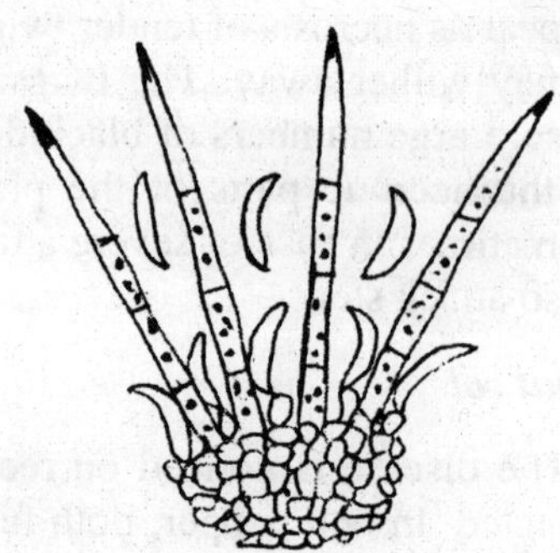

Fig. 1. Acervulus of *C. capsici*

Several workers have reported existence of variants in genus *Colletotrichum*. Kumar and Mahmood (1984) reported cultural and nutritional variability among five isolates of *C. capsici* collected from chilli seeds and plants from different Indian locatlities on the basis of their virulence, growth characters and nutritional behaviour, and were assigned different physiological races. A marked variation in symptoms, lesion type and morphological characters in *C. annuum* infected by *C. capsici* and *C. gloeosporoides* was observed (Thind and Jhooty, 1990). They further grouped >150 isolates of these species of *Colletotrichum* into 8 distinct categories (5 for *C. capsici* and 3 for *C. gloeosporoides*) and all the five categories of *C. capsici* were widely prevalent in Punjab, India, but their individual incidence varied in different areas while *C. gloeosporoides* was less prevalent. Variation in response to mancozeb and captafol was more pronounced among isolate categories of *C. capsici,* while isolates of *C. gloeosporoides* did not show much difference. Kolte and Sapkal (1994) also studied the variation in 10 isolates of *C. capsici* and virulence pattern of these isolates showed a clear relationship with N content of mycelium. The isolates with highest infectivity (96%) had the highest N (6.44%), while infectivity was least (76%) by isolates with lowest N (3.36%). The most virulent isolates produced symptoms of dieback in inoculated seedlings in 15 days compared with up to 30 days for the less virulent isolates. Five isolates of *C. capsici* collected from five different conventional chilli growing areas of southern Tamil Nadu revealed differences in morphology, physiology and virulence. Isolate I_1 was most virulent as determined by fruit rot incidence and intensity (Jeyalakshmi and Seetharaman, 1999). 67 isolates of *C. capsici* were studied by Mohan *et al.* (2006), which were further grouped into 4 broad groups (C1, C2, C3 and C4) based upon morphological and pathological variations. Wherein isolates of group C1 were having dirty white fluffy colonies with sparse or low sporulation, Isolates of group 2 exhibited grayish white colonies with moderate sporulation, dark and jet black colonies with appressed and high sporulation were characterstic in isolate 3 and 4. Christopher *et al.* (2013) studied morphological and molecular variability among *C. capsici* isolates and reported that molecular polymorphism generated by RAPD confirmed the variation of the different isolates and they were grouped in three clusters. However, morphological and RAPD grouping of isolates suggested no correlation among the isolates. Kaur *et al.* (2016) categorized *C. capsici* isolates into 5 groups on the basis of colony appearance and arrangement of accervuli. Studies on molecular variability among these isolates using 27 microsatellite markers resulted in the division of isolates in three major clusters.

Colletotrichum gloeosporioides

The mycelium of *C. gloeosporioides* is septate and in culture forms brown to orange acervuli. Acervuli bearing setae are rarely observed in the culture, however, on fruits the acervuli were subepidermal with light to dark brown setae disrupting outer epidermal cell walls of the host (Fig. 2). The conidiophores are closely packed on the surface of the acervulus and are hyaline. The conidia of the fungus are oblong to cylindrical with obtuse ends having one to three oil globules. The conidia measure 4.12–24.72 × 1.03–8.24 μm in size.

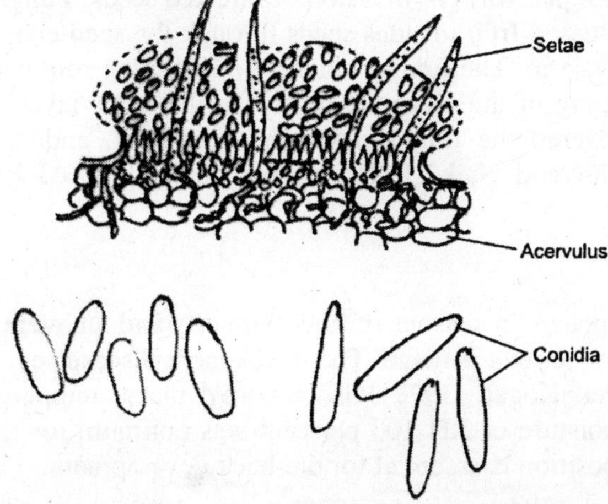

Fig. 2. Acervulus and conidia of *C. gloeosporoides*

Virulence spectrum of 14 *C. gloeosporioides* isolates determined on a host differential set comprised of three *Capsicum* sp. (*C. baccatum, C. chinense* and *C. annuum*) revealed the presence of six pathotypes *viz.*, CgA-1, CgA-2, CgA-3, CgA-4, CgA-5 and CgA-6 (Sharma and Katoch, 2014). They further reported that race groups CgA-1 and CgA-5 were most virulent infecting 8 of 11 differential lines. Genetic variability of these isolates determined through ISSR, ERIC- and BOX-PCR markers exhibited wider diversity. ISSR-PCR markers exhibited maximum polymorphism using 22 alleles, clustering the 14 isolates into four sub clusters at 0.70 coefficients. A combined dendrogram generated using 33 alleles amplified by ISSR, ERIC and BOX primers identified the isolates into two major groups having early divergence from each other.

Host range

Colletotrichum capsici: This pathogen has a very wide host range both in cultivated and wild plants. Besides, *C. annuum*, it can infect *Alsotonia scholaris, Tinospora cordifolia, Amaranthus spinosus, Adhatoda vasica, Bougainvillea spectabilis, B. glabra, Cassia occidentalis, Croton roxburghii, Evolus nummularis, Sida cordifolia, Cicer arietinum*, cotton, eggplant, jute, tomato and turmeric (Kamal and Singh, 1978; Kumar and Singh, 1979; Singh, 1987).

Colletotrichum gloeosporioides: The host range of this pathogen includes *Ardisia solanacea, Barringtonia acutangula, C. annuum, Clerodendrum viscosum, Croton roxburghii, Ficus religiosa, Hemigraphis hirta, Holarrhena antidysentrica, Ipomoea fistulosa, Jasminum flexile, J. sambac, Justicia betonica, Mallotus philippensis, Streblus asper* and *Syzygium heyneanum* (Kamal and Singh, 1978).

Disease cycle

The pathogen survives both in infected plant debris (Chowdhury, 1957) and in the infected seed (Jindal *et al.*, 1994). The fungus can survive in plant debris in the soil for atleast nine months, which serves as source of primary infection, whereas secondary infection takes place through wind borne conidia. Histopathological studies indicated two pathways of invasion of infected seeds. Fungal mycelium in the inner surface of diseased fruit invades seeds through the seed coat and colonized the outer layer of endosperm. The fungus could also infect the ruminent of the placenta, then enter the opening of the placenta, or invade the outer layer of the endosperm. Basak (1998) recovered the fungus from the seed coat, endosperm and embryo. However, Padaganur and Naik (1991) reported that *C. capsici* is externally seed borne.

Epidemiology

The disease appears in seasons of high humidity and an average RH of 78.5 per cent is necessary for serious damage. There was more disease severity when weather was completely wet. Kogan (1977) also reported that a temperature of 26ºC and presence of free moisture or RH 100 per cent was optimum for this disease. Heavy prolonged dew deposition is essential for die-back development.

Thirupathiah and Subramanian (1979) reported that the extracts from fruits of *C. annuum* contain endo-polygalacturonate transeliminase, which appears to increase the permeability of the cells, and this effect was greater in red ripened fruits, which makes them more susceptible to *C. capsici* infection than in green fruits. The amount of wax also plays a significant role in the infection of fruits. Prakasam and Jeyarajan (1989) studied the amount of wax at different ages of fruits and reported that it was higher in 5-day-old resistant green fruits than in susceptible ripe fruits.

Management

The disease can be kept under check by using host resistance, cultural practices, biological method and chemicals.

(i) Cultural practices: Since the pathogen overwinters in infected plant debris and seed, cultural practices like procurement of seeds from healthy and spotless fruits, collection and destruction of infested crop debris after harvest (Saha and Singh, 1988), crop rotation for at least one year with non-solanaceous crops should be followed. Solanaceous weed hosts should not be allowed to grow in the vicinity of the field.

(ii) Host resistance: Use of resistant varieties is one of the important methods to reduce the initial inoculum of the pathogen population in the field. Due to wide variation in the pathogen *C. capsici*, it is very difficult to develop a variety, which is equally resistant for different areas. Ullasa *et al.* (1981) evaluated 298 sweet pepper genotypes to *C. capsici* and reported California Wonder, Canapes F_1, Hungarian Wax, Mogidas Cruzes-2, Pazardiska, Kapija, Soroksari, Merrimack Wonder and *C. microcarpus* Linn. as resistant. Several cvs./lines/hybrids like NP Hybrid, Kashmir Yellow and Kalyanpur No. 1 (Singh and Thakur, 1979), lines 1951 and 2158 (Sankaralingam and Suthanthira-pandian, 1983), Kumchung No.2, Bulamhouse, Pakistan and Hongilpum (Chang and Chung, 1985), Pant- C 1 and B 7-9 Self (Perane and Joi, 1986), S 20-1, H-1, H-4, H-6, S 34-13, *C. pendulum*, BG-1, Lorai and Perennial (Singh and Thind, 1980; Singh *et al.*, 1990), CO 2220 (AVRDC, 1990) were reported resistant against this disease. Two cultivars KDC-1 and Bhagyalaxmi were observed as moderately resistant against this disease in Northern Karnataka region (Patil and Naik, 2000). Perennial has been used as a resistant donor to breed chilli varieties like Punjab Lal, CH-1, CH-2 and Ch-3 in Punjab. Cvs./lines like LCA 301, LCA 324, K-1 and Byadgi Kaddi were found resistant (Hegde and Anahosur, 2001). Sharma and Katoch (2014) evaluated 195 *Capsicum* genotypes comprising local land races, commercial varieties and exotic accessions against this disease (*C. gloeosporioides)* and reported that exotic *C. baccatum* (EC 631750) was resistant to five races followed by *C. chinensis* (EC631751) and two commercial genotypes Surajmukhi and California Wonder.

Bhullar *et al.* (1972) studied the levels of phenols in resistant and susceptible cultivars of chillies and reported that after inoculation of 90 days old plants with *C. capsici*, phenol level rose in both resistant and susceptible plants, but to a greater extent in stems, leaves and fruits of resistant plant whereas, a slight decrease was noted in the fruits of susceptible cultivars. Azad (1991) also studied the role of chemical constituents in resistant, moderately susceptible and susceptible cultivars of chilli fruits during infection with *C. capsici* and reported that sulfur, capsicin and ascorbic acid were maximum in Rukuni (resistant) and minimum in Clustarol (susceptible), while it was *vice versa* for sugars (reducing, non-reducing and total).

Hegde and Anahosur (2001) also reported that resistant genotypes contained higher capsaicin, ascorbic acid and lower total sugars.

(iii) Chemical methods: Fungicides like ziram, captan, zineb, mancozeb, copper oxychloride, carbendazim and iprobenphos have been reported to control this disease effectively.

Seed dressing with various systemic and non-systemic fungicides like thiram, bisdithane, organic mercurials, captafol, captan, carbendazim, benomyl, streptocycline and cycloheximide have been found effective (Grover and Bansal, 1970; Kumar and Mahmood, 1986; Perane and Joi, 1988; Mirdha and Chowdhury, 1990). The treated seeds recorded high percentage of seed germination in comparison to untreated ones. However, Chakravarti and Anilkumar (1975) reported seed soaking with thiram (0.2%) gave best control. Drenching with Bavistin (0.1%) in nursery bed has also been reported effective (Biswas, 1992).

Spray of various systemic and non-systemic fungicides like Bavistin (0.1%), Dithane M-45 (0.25%), Dithane Z-78 (0.25%), Fenapanil (0.4%), Difolatan (0.2%), captan (0,25%), combination of Bavistin with Agrimycin, Brestan, Cuman-L, Miltox and Kitazin-P have also been reported effective against this disease under field conditions (Thind and Jhooty, 1987; Raju and Rao, 1989; Datar *et al*., 1990 and Biswas, 1992). Three sprays of mancozeb, thiophanate-methyl, captafol, ziram and carbendazim at 15 days interval significantly reduced the disease intensity and mancozeb (0.25%) gave highest net profit of Rs. 11896/ha, followed by thiophanate methyl, captafol, ziram and carbendazim (Kumawat, 1997). Efficacy of prochloraz (0.125%), difenoconazole (0.05%), tebuconazole (0.05 to 0.1%) and strobilurin fungicides (azoxystrobin, trifloxystrobin and pyraclostrobin, 0.1%) against this disease has also been reported (Alexander and Waldenmaier, 2002; Lewis and Miller, 2003; Ekbote, 2003; Rathore, 2004; Mohan *et al*., 2004).

(iv) Biological methods: Several workers attempted the biological control of *C. capsici* using bacteria and actinomycetes. Gupta and Dixit (1982) reported *Streptomyces griseus* as an antagonist of *C. capsici* and correlated its population on the surface of chilli with intensity of the disease. The population of *S. griseus* were high in September under Mathura, India conditions and fruit rot (*C. capsici*) did not develop in the field, but from October onwards, *S. griseus* populations started decreasing and maximum disease intensity was observed in December when the populations of *S. griseus* was minimum. Seedling dip in the suspension of *Azospirillum* sp. gave best control of shoot dieback followed by *Azotobacter* as compared to untreated seedlings (Kolte *et al*., 1994). However, yields of green and red chillies were highest when the plants were dipped in the mixture of *Azospirillum* and *Azotobacter* suspension. Among various antagonists, *Saccharomyces cerevisiae* exhibited the maximum reduction of mycelial growth, followed by *Trichoderma viride,* while *Bacillus subtilis* showed maximum growth reduction followed by *Pseudomonas fluorescens* under *in vitro* conditions. Whereas in pot experiments *S. cerevisae* recorded the maximum reduction of fruit rot and dieback incidence, followed by *B. subtilis* when sprayed 105 and 120 days after sowing (Jeyalakshmi *et al.,* 1998). *Pseudomonas fluorescens* showed antagonistic activity against *Colletotrichum capsici,* while Nimbicidin (0.3%) inhibited the growth upto 43.18 per cent under *in vitro* conditions ((Hegde *et al*., 2002). Under greenhouse conditions, significantly less mortality was obtained on *P. fluorescens* (11.12%) and Nimbicidin (17.26%) sprayed plants compared to the mortality in the control plants. Rani *et al.* (2008) studied the antagonistic potential of *Trichoderma* species against *Colletotrichum capsici* and reported that seed treatment with pure culture of *Pseudomonas fluorescens* and *Trichoderma viride* gave maximum control in addition to higher fruit yield.

(v) Plant extracts: Root extracts of *Abrus precatorius* and *Rawolfia tetraphylla* showed significant inhibitory effects on both the conidia germination and radial growth of *C. capsici* (Kumaran *et al.*, 2003). Among different plant extracts evaluated, neem seed kernel extract and nimbicidine caused considerable inhibition of growth of *C. capsici (*Naik *et al.*, 2008). Raj *et al.* (2008) studied *in vitro* efficacy of plant extracts like *Allium sativum* (20%), *Datura metel* (60%), *Eucalyputs globulus* (60%) and *Prosopis julifora* (60%) on the production of Cellulolytic and Pectinolytic enzymes and maximum inhibition of the activity of Cellulolytic enzymes such as Cellulase C_1 and Cellulase C_x and Pectinolytic enzymes like Polygalacturonase (PG), Pectin transeliminase (PTE) and Polygalacturonase-Trans-eliminase (PTGE) was recorded with extract of *Allium sativum.*

2. PHYTOPHTHORA LEAF BLIGHT AND FRUIT ROT

It is an important disease in those areas where fruiting coincides with the onset of monsoon rains. Due to favourable climatic conditions the disease is so severe that sometimes it causes complete defoliation of the plants.

Geographical distribution

The disease is prevalent world over where warm-humid conditions exist during the growing season. Leonian (1922) was the first to report its occurrence from Mexico. Subsequently, it was reported from Puerto Rico, Argentina, Florida, Greece, United States, Russia, Venezuela, Japan, France, Pakistan and Brazil.

In India, the disease caused by *Phytopthora capsici* Leon. was reported for the first time from Solan area of Himachal Pradesh by Sohi *et al.* (1971). Later on Sharma and Bhardwaj (1976) reported the disease caused by *Phytophthora nicotianae* var. *nicotianae* Waterhouse in Himachal Pradesh. In hot pepper, during June to January cropping season of 2011 and 2012, severe foliar blight (*Phytophthora boehmeriae*) epidemics were observed in Karnataka and Tamilnadu (Chowdappa *et al.*, 2014).

Losses

Losses from leaf blight and fruit rot disease are greater than any other disease of bell pepper. These result from (i) seedling mortality in unprotected field nursery and mortality of young plants at post transplanting stage, (ii) premature defoliation of leaves and fruit rot and (iii) reduction in fruit production due to flower blight.

Warm, humid conditions in rainy season (July–September) may result in heavy blight development on leaves, rotting of fruits and necrotic lesions on stems and branches, thereby, resulting in reduction of potential yields, drastically.

The reduction in yield is mainly due to pre-mature defoliation of plants and rotting of fruits. The yield loss of infected plants has been reported to range from 43–

100 per cent (Liang *et al.*, 1992). In India, Shyam (1969) reported the incidence of fruit rot varying between 58 and 100 per cent in Himachal Pradesh, while Bhardwaj (1983) reported that leaf blight severity and fruit rot incidence during crop season varied between 35 to 62 and 30 to 68 per cent on leaves and fruits, respectively.

Symptoms

The pathogen have been reported to attack all the aerial plant parts such as leaves, stem, branches, growing shoots, collar region and fruits as described below:

Leaves: The first symptoms appear as water soaked to dull or faded green areas differing slightly from the colour of the normal healthy leaf tissue (Plate 3a). Under humid conditions, the diseased areas rapidly enlarge covering entire leaf, resulting in premature leaf fall. However, with the decrease in relative humidity and intervening clear sky, further development of the disease is restricted and the spots get bleached to light tan colour with brown coloured margins.

Stem and shoots : Symptoms on stems appear first as dark green water-soaked areas, irregular in shape, turning dark brown to black with the passage of time giving a blighted appearance. The infection of growing shoots results in the rapid killing of the growing tips as well as flower buds progressing downwards towards the base. At such a stage, complete defoliation is often observed.

Collar region: Symptoms on collar region of adult plants appear as water-soaked areas with whitish growth of mycelium engirdling the collar region and the point of contact of the soil line. The rot often progresses downwards to the roots in the affected plants and there is sudden drooping of leaves of the whole plant giving the appearence of sudden wilt.

Fruits: The fruits are vulnerable to infection irrespective of their size and age except the fruits nearing maturity with the change of fruit colour from light green to red. The disease first appears in the form of dark green water-soaked spots in contrast to light green colour of the healthy fruit tissue. The spots enlarge rapidly, ultimately covering the entire fruit (Plate 3b). The fruit showing initial symptoms, when cut open exhibits profuse mycelial growth inside the fruit cavity even covering the normal surface of the pericarp as well as immature seeds. Completely rotten fruits may fall down on the ground often invaded by saprophytic fungi and bacteria causing soft rots.

The Pathogen(s)

Two species of *Phytopthora* have been found associated with this disease. *P. capsici* Leon. has been reported from various parts of the world. Besides, *P. capsici* (Sohi and Shyam, 1976), *P. nicotianae* (Breda de Hann) var. *nicotianae* Waterhouse (Sharma and Bhardwaj, 1976) has also been found associated with this disease from Himachal Pradesh. Association of *P. boelmeriae* with this disease in South India has also been reported (Choudappa *et al.,* 2014). Morphological characters of both species are described below:

(i) Phytophthora capsici

The fungus grown on oatmeal agar medium has fluffy and cottony white colonies. Mycelium is hyaline, branched 3.3–8.6 μm wide and non-septate, but few septa are found in case of old hyphae. Typically, hyphal branches arise at right angles often variously swollen and tuberous and of abnormal diameter. Sporangia are rarely produced or almost absent in culture, but are abundantly produced in tap water or 0.1 M KNO_3 solution. The sporangia are hyaline, ovoid to pyriform or sometimes round to lemon shaped, measuring 29.7–59.8 × 33.0–72.5 μm, non-pedicillate with a predominant, hemispherical papilla at the apex. The mean size of the sporangia is 38.4 × 45.5 μm. Sporangia germinate readily in tap water at room temperature (17–23°C) by giving rise to 15–20 zoospores. The zoospores are reniform to oval, biflagellate and remain motile for 20–30 seconds then shed flagella and become encysted. The encysted zoospores germinate by means of germtubes. Oospores are formed abundantly in aerial as well as submerged mycelium in the medium. The oogonia are hyaline, circular to spherical. The oogonial stalk is encompassed by persistent amphygynous antheridium. The oospores are circular to spherical, 16.5 to 26.4 μm, averaging 21.28 μm. Single oosporic cultures of the pathogen exhibited the phenomenon of homothalism. The oospores are found to germinate either by germ tubes or by stalked or sessile germ sporangia. The fungus does not produce chlamydospores (Fig. 3).

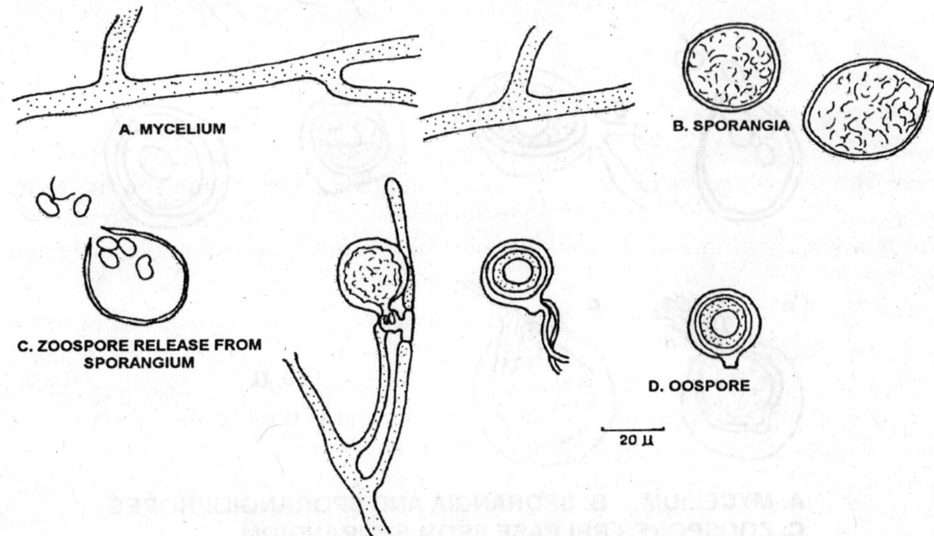

Fig. 3. Morphological structures of *Phytophthora capsici* (Courtsey K.R. Shyam)

(ii) Phytophthora nicotianae var. *nicotianae*

The colonies of the fungus on oatmeal agar are cottony white and fluffy in appearance. The hyaline branched and non-septate mycelium is 3.5–8.5 μm in width with branches at right angles. The sporangia are not produced in culture, but are abundantly produced in tap water or 0.1 M KNO_3 solution. The sporangia are hyaline, non-pedicellate, globose to subglobose with a prominent beak like papilla at the apex(Plate 4a) and measured 15.0–42.0 x 17.0–47.0 μm. The length : breadth ratio is generally 1.2 : 1, however, in certain cases it rises to 1.4 : 1. The sporangia germinate by production of 15 to 20 zoospores. Oospores are formed abundantly on aerial and submerged mycelium in culture. The oogonia are hyaline, globose and measured 23.0 x 21.0 μm. The oogonial stalk is encompassed by a persistent amphigynous antheri-dium. The oospores are thick walled golden brown in colour (Plate 4b) and measured 19.0–37.0 μm (Fig. 4).

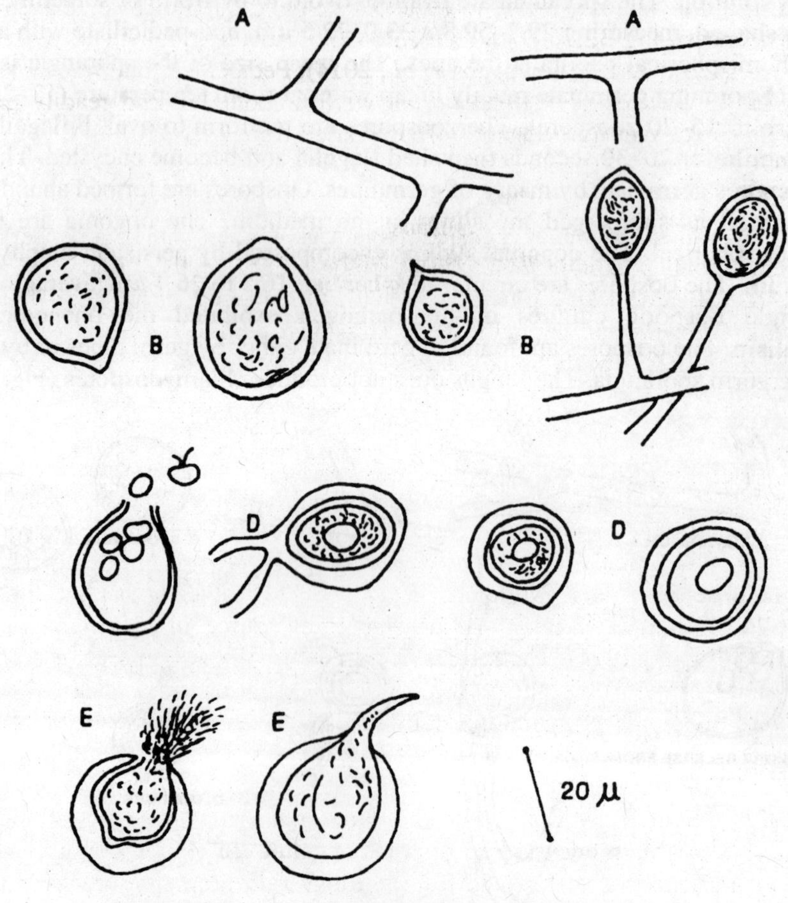

A. MYCELIUM B. SPORANGIA AND SPORANGIOPHORES
C. ZOOSPORES RELEASE FROM SPORANGIUM
D. OOSPORES E. GERMINATING OOSPORES

Fig. 4. Morphological structures of *Phytophthora nicotianae* var. *nicotianae* (Courtesy S.S. Bhardwaj)

Gupta *et al.* (2012) studied variability among twenty three isolates of *P. nicotianae* var. *nicotianae* at morphological/genetic level collected from different locations of North India. Variation among isolates was observed in mycelial colour, appearance, size, and shape of sporangia. Oospore production with variable size was recorded only in ten isolates. Genetic variation was also studied with twenty four 10-mer oligonucleotide RAPD primers, out of which 14 primers yielded informative strong and reproducible DNA amplicons of *P. nicotianae* var. *nicotianae*. The dendrogram produced from computerized cluster analysis of DNA finger prints revealed a great deal of heterogeneity among isolates.

(iii) Phytophthora boehmeriae

Sporangia are caduceus and papillate, and borne on simple sporangiophores. Most of the sporangia are globose, ovoid to ellipsoid or nearly spherical or obturpinate to obpyriform measuring 32.76–38.80 x 18.21–28.57 μm with length to breadth ratio varied from 1.22 to 1.96 (Chowdappa *et al.*, 2014). Pedicel length of the isolates was <5 μm. All 43 isolates of *P. boehmeriae* were homothallic and readily produced oospores on carrot agar figs. Size of oospore varied from 21.9–25.1 μm.

Host Range

The host range of *P. capsici* as reported by various workers includes pepper, brinjal, tomato, cucumber, pumpkin, watermelon, bottle gourd, squash melon, fig, tobacco and petunia (Shyam, 1969; Bhardwaj, 1978). Ornamental plants like *Peperomia, Saintpaulia, Gloxinia* and carnation have also been reported to be infected by this fungus (Rattink, 1981) while Shukla *et al.* (1981) reported it on *Hibiscus*.

Disease cycle

In areas, where the crop is grown regularly in the same field, the fungus overwinters in the form of oospores in soil (Shyam, 1969) which remain viable during unfavourable weather conditions in plant debris or in infected seeds from dropped rotten fruits in the fields. *P. capsici* could survive in moist soil for at least 15 months even in the absence of host, while oospores remained viable for at least 7–8 months in the soil. Infected seed constitute an important source of dissemination of the fungus (Jindal *et al.*, 1994). The fungus is both externally and internally seed borne. With the onset of monsoon rains in the second fortnight of June, when soil moisture is high with moderate temperatures (24–26°C), the oospores germinate either by germtubes or by the rupture of oosporic walls or by stalked or sessile germ sporangia. The sporangia in the presence of free water produce biflagellate zoospores, which after swimming for some time in a film of water lose flagella and become encysted. Such encysted zoospores constitute the primary inoculum, which bring about infection on leaves and fruits through rain drop splashes from the soil. The zoospores of *P. capsici* encyst on the surface of Capsicum leaves and penetrate either directly or by a germ tube or by an infection peg from an appressorium into an epidermal cell, occasionally through stomata. Penetration is complete in 4 h. The fungus invades intracellularly, downward as well as radially. The hyphae penetrate deep into the tissue. The symptoms develop on different parts after 24–72 h. The mycelium in the epidermal layer of the affected portion of the fruit agglomerates, pushes through the epidermis and cuticle by rupturing it and grows on the fruit surface under humid conditions (RH >95%), producing a crop of sporangia on simple branched sporangiophores. No sporangia are produced on the infected leaf tissue. These sporangia from fruits spread rapidly during the rainy season and cause secondary infection. When the conditions become unfavourable, the fungus goes in sexual reproduction and produces oospores (Fig. 5).

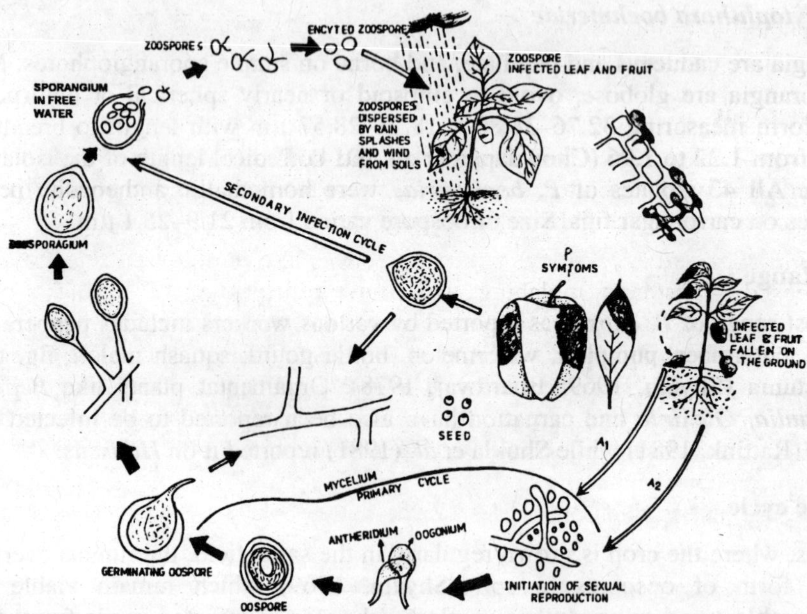

Fig. 5. Disease cycle of leaf blight and fruit rot of bell pepper

Epidemiology

Both pathogens survive in the form of oospores in the soil as well as in infected seed (Shyam, 1969; Bhardwaj, 1983). Both fungi remain viable for more than one year under ordinary field conditions. Both species were found pathogenic on leaves, fruits, stems and roots with shortest incubation period on leaves followed by fruits (Gupta and Jarial., 2014). *P. nicotianae* var *nicotianae* had shorter incubation periods on leaves (26 h) and fruits (28 h) than *P. capsici,* while the reverse was true in *P. capsici* having shorter incubation period on roots (42 h) and stems (52 h). The fruits situated on the lower branches are first to be infected. Presence of abundant rainfall, high RH and warm weather are essential for initiation of this disease, while it is completely absent in dry weather. The inoculum surviving in soil multiplies rapidly when soil moisture is quite high during frequent rains and the standing water in the field helps in abundant sporangial formation and their germination by zoospores, which are splashed from the soil with spattering rains. After infection, sporangia are produced on the fruit surface and are carried by wind to the adjoining fruits and foliage.

Various workers have reported warm temperature ranging between 25 to 30ºC in combination with high relative humidity (>70%) as optimum conditions for maximum disease development on fruits and leaves (Shyam, 1969 and Jia, 1992). Bhardwaj *et al.* (1985) also reported a positive correlation between disease development and high relative humidity, rainfall and temperature. The distrubution of the disease in the field has been reported to follow field drainage pattern. Splashing rain and run off water have been observed to be the main source of secondary spread of the disease.

Zoospores penetrate the host either through stomata or directly through epidermal cells within 60–120 minutes and fruits of all ages except the mature ones are susceptible to this disease. The incidence of *P. capsici* is more in clay loam soil in comparison to sandy loam soil (Yang *et al.*, 1991).

Management

Since, the causal fungi persist either in the soil or in infected seed, therefore, integration of various cultural, biological, host resistance and chemical methods is expected to help substantially in the management of this disease.

(i) Cultural practices : Various cultural practices like careful seed selection, roguing of diseased plants/parts, mulching with straw, transplanting of the plants on top of furrow increased to 30 cm (Fereyra *et al.*, 1984), soil solarization with transparent polythene sheet alongwith application of fumigants (Cartia *et al.*, 1989), amendment of soil with high C/N ratio material (Kim *et al.*, 1990) and cultivation of Capsicum with non host plants like groundnut, sesame, onion, garlic, pea and spinach (Lee *et al.*, 1993) have been reported effective in reducing the severity of the disease. Different mulches and soil amendments have been evaluated for the management of this disease and application of pine needle mulch before the onset of monsoon rains was found effective against this disease (Gupta and Jarial, 2014).

(ii) Host resistance: Several workers evaluated the available cultivars and germplasm lines to identify sources of resistance against *P. capsici*. Cultivars/lines like PI-188376, PI-202132, PI-201234 (Kimble and Grogan, 1960), Waxy Globe (Saini and Sharma, 1978), PI- 123469, PI-163192, PI-201232, PI-201234, PI-224445, P 51, AC 2258, Riogrande and Gosung (Byung, 1988), Sweet peppers like Smith 5, Sweet pepper 469, Stal Cristal and L 5279 (Fernandez, 1988), Kiraseong (Choi *et al.*, 1990), Kingkun (Hwang and Kim, 1990), Calafyuco INTA (Galmarini *et al.*, 1991), Serrano Criolla de Morelos (Tamietti *et al.*, 1992), Szechwan (Hartman and Hwang, 1992), PM-702 (Ustun, 1993), Santaka Pimento Naciional (Saborio and Pacheco, 1993), PI-59236 (Boiteux *et al.*, 1993), Tian Jiao-2 (Niu *et al.*, 1994), PBC-408 (Riley and Bosland, 1995), PI-201232 and SCM-334 (Ortega *et al.*, 1995). However, information on sources of resistance against *P.n.* var. *nicotianae* is almost lacking except PBG-631 (Kohli, 1997) and Kandaghat Selection (Verma *et al.*, 2001). PBG-631 is horticulturally superior as fruits are non pungent and large in size (15–16 x 3–4 cm), with nearly appropriate shape.

Out of 28 cvs./lines of bell pepper evaluated against *P. nicotianae* var. *nicotianae* both under field and laboratory conditions, only one line i.e., DG (Dark Green) was found resistant, while rest were moderately resistant to highly susceptible (Gupta *et al.*, 2007). Higher degree of horizontal resistance was present in the moderately resistant lines "HC 201" and "Feroz" as reflected by slowest and moderate apparent infection rate, respectively, besides comparatively longer incubation periods under laboratory conditions. Out of seven bell pepper lines screened against 23 isolates of *P. nicotianae* var. *nicotianae* (Gupta *et al.*, 2012), only one genotype *i.e.,* Dark Green was found tolerant to 16 isolates, while Feroz and Danish Wonder were tolerant to 14 isolates. Molecular markers *viz.,* RAPD revealed genetic variation among fruit rot resistant and susceptible lines of bell pepper. Fruit rot tolerant lines *viz.,* Feroz, Danish Wonder and Dark Green clustered in one group. Such lines can be used in developing fruit rot resistant varieties/hybrids to broad range of isolates.

The sources of resistance against this disease mostly belong to pungent peppers, however, a few sweet pepper genotypes were also reported to be resistant (Fernandez, 1988) which are of tremendous use because many generations of backcrossing are needed to remove undesirable traits of pungent peppers introduced together with resistance. Removal of these undesirable traits *viz.,* pungency and improper sized fruits is the most combersome because both the traits are controlled by polygenes (Greenleaf, 1986). The inheritance of resistance to *P. capsici* has been reported to be monogenic dominant (Saini and Sharma, 1978), digenic (Reifshneider *et al.*, 1992), trigenic (Ortega *et al.*, 1992), polygenic (Palloix *et al.*, 1990) and epistatic (Barutal *et al.*, 1994).

While studying the inheritance of resistance, Sharma and Saini (1979) concluded that resistance in Waxy Globe appeared to be controlled by one or several major dominant genes with some modifiers. In the same year, Yamakawa *et al.* (1979) reported that resistance in line LS 279 was conferred by a single incompletely domonant gene. Matsuoka *et al.* (1984) identified resistance in pungent types of *C. annuum* and the resistance has been observed to be controlled by two recessive genes and a dominant modifier. When one of the resistance genes is present in the heterozygous conditions, resistance occurs if a modifier is present. When both resistance genes are present in heterozygous conditions, the modifier must be homozygous.

(iii) Chemical Control: Various systemic and non-systemic fungicides have been reported to be quite efficacious in the management of this disease under both laboratory and field conditions. However, the frequent rains during the months of July–August, which are essential for good growth of bell pepper plants, also promote rapid multiplication of inoculum of the pathogen and spread of the disease and sometimes, when routine application of fungicides is delayed, often results in heavy disease outbreak. Proper selection of fungicide and timing of initial spray is, therefore, of prime importance to keep the disease under check.

Fungicides like captafol, prothiocarb, metalaxyl, fosetyl-Al, copper sulphate and captan are reported effective under *in vitro* conditions against *P. capsici* (Papavizas and Bower, 1981; Abdellaoui - Maane *et al.*, 1988 and Zhang *et al.*, 1992), while dichlone, captafol, pyroxychlor, copper oxychloride, zineb, propineb and ziram have been found effective against *P. nicotianae* var. *nicotianae* (McIntyre and Lukens, 1977; Bhardwaj, 1983). Efficacy of potassium phosphonate and monoethyl phosphonate against isolates of *Phytophthora* species has also been reported (Quinette and Coffey, 1989). Bordeaux mixture, Kirti copper, Zincop, zineb and captafol have been reported effective in controlling the disease on bell pepper caused by *P. capsici* under field conditions (Sohi *et al.*, 1973, McIntyre and Lukens, 1977 and Papavizas and Bower, 1981). Bhardwaj and Sharma (1985) reported Blitox -50 (0.3%) to be most efective in checking the infection of *P. nicotianae* var. *nicotianae* and increased fruit yields, followed by ziram (0.2) and Bordeaux mixture (0.8%). Kim (1989) observed that soil drenching with metalaxyl was the most effective chemical treatment and gave 40 to 60 per cent control of the disease while Hussain *et al.*(1990) recommended seed treatment with captan and Ridomil and spraying with metalaxyl for the control of this disease. Matheron and Matejka (1995) studied the activites of sodium tetrathiocarbonate (STTC) and metalaxyl on the viability of *P. capsici* within colonized *C. annuum* tissue and found that pathogen was not recovered from colonized tissue buried in soil that was drenched with 490 μg/ml of STTC, while recovery from metalaxyl (10 μg/ml) drenched stem tissue did not differ from that obtained from water drench only. Akomine, a non-systemic fungicide was reported most effective with 100 per cent inhibition of mycelial growth at 50 ppm concentration as compared to 83.15 per cent inhibition at 300 ppm with systemic fungicide Curzate against *P.n.* var. *nicotianae* when evaluated *in vitro* (Verma *et al.*, 2006). Gupta and Jarial (2010) evaluated different systemic and non-systemic fungicidesagainst this disease both by detached leaf technique and in protective spray programme. Combi-fungicides like metalaxyl + mancozeb or cymoxanil + mancozeb were more effective in reducing the disease and increasing the fruit yield as compared to non-systemics yet strobilurin fungicides like azoxy-strobin and kresoxim methyl though provided slightly less control of the disease, but being ecofriendly with least residual toxicity can be applied at critical stages of the crop even after fruit formation for obtaining produce with minimum residues.

(iv) Biological Control : Several fungal and bacterial antagonists *viz., Trichoderma* sp., *Pseudomonas cepacia, Bacillus polymyxa* and *Streptomyces parvullus* (Jee and Kim, 1988; Lee *et al.*, 1990 and Puri *et al.*, 1994) inhibited the growth of *P. capsici* and *P. n.* var. *nicotianae* under *in vitro* conditions. Immersion of seed or dipping of roots and lower stems of seedlings into the homogenate of C/2-9 (new race of *Streptomyces ochraceiscleroticus*) before planting in natural, inoculated soil gave 73.3 per cent protection of *C. annuum* against *P. capsici* (Turhan, 1981). Promising results were also obtained when antagonists like *Pseudomonas cepacia, Bacillus polymyxa* and other *Bacillus* sp. were applied with suitable organic soil amendments and calcium fertilizers (Cho, 1987). Ahn and Hwang (1992) reported the complete control of *Phytophthora* blight in Capsicum plants by application of cultural filtrates of actinomycetes. *Streptomyces californicus* (MTCC No. 8771), an actinomycete which is having siderophore production and phosphate solublizing activity besides antagonistic activity to *P. nicotianae* var. *nicotianae* when applied in field both as soil and foliar application reduced fruit rot incidence to 11.7 as compared to 48.3 per cent in control and increased fruit yield by > 100 per cent (Thakur *et al.*, 2009).

(v) Integrated disease management: Use of integrated disease management practices like cultural methods (drainage with high ridges, crop rotation, intercropping and mixed cropping), resistant cultivars along with application of appropriate fungicides (mancozeb and metalaxyl) have been suggested by Hwang and Kim (1995) for the effective management of this disease. Application of pine needle mulch on the field floor alone and in combination with the sprays of metalaxyl + mancozeb (0.25%) on mulch and foliage resulted in significantly lower disease levels followed by sprays of copper oxychloride (0.3%) both on mulch and foliage and increased fruit yield (Gupta and Jarial, 2008)

3. CERCOSPORA LEAF SPOT

Cercospora leaf spot or frogeye leaf spot is prevalent in those areas of the world where warm moist weather is prevalent during the pepper cultivation. This disease is of minor importance, but if favourable climatic conditions prevail for a longer period, then the disease may assume serious proportions and lead to heavy losses.

Geographical distribution

Cercospora leaf spot occurs in all *C. annuum* growing areas of the world. This disease was first time reported from Georgia in 1923 (Higgins, 1923). Subsequently, it was reported from Porto Rico, Spain, Florida, Bermuda, Ceylon, Morroco, Nicaragua, Trinidad, Peru, Ivory Coast, Cuba, Korea and Brazil (Suang *et al.*, 1984; Matsuoka *et al.*, 1996).

In India the disease was reported long back in 1929 (Sydow and McRae, 1929) and subsequently, it was reported from Jabalpur, M.P. (Nema and Agarwal, 1960), Bangalore (Rangaswami *et al.*, 1970) and in Kumaon Hills (Muneem *et al.*, 1995).

Losses

There are no exact estimates of losses due to this disease available in literature but in favourable weather, disease can cause significant losses. Meon (1990) reported that the attack by this disease appreciably reduced the photosynthetic activity of the infected plants, which lead to significant yield losses. Premature defoliation of the plants is the most devastating phase of this disease.

Symptoms

Symptoms of this disease appear as scattered or confluent, circular to ellipsoid, brownish to black diffused spots, which are 5–10 mm in diameter. Small necrotic lesions on the leaf surface later coalesce giving an irregular and blighted appearance. The spots have brown borders with light coloured, faded or greyish central part (Plate 5). When many spots appear on the leaves, they turn yellow and fall down prematurely. Symptoms also appear on the stems and petioles as elongated or irregular spots. In severe cases, the fruits are also attacked and fruits get decayed. Defoliation is the most damaging phase of the disease. Spots on leaves resemble the frogeye, so the disease is also called as frogeye leaf spot.

The Pathogen

The fungus responsible for this disease is *Cercospora capsici*. The fungus is characterized by dark brown stromata, 34–39 x 28.5–45 μm in size, from which arise fasciculate, straight or slightly curved, continuous, simple yellowish brown conidiophores, 31.2–37.5 x 4.5–5.5 μm in size, producing polymorphous, cylindrical, fusoid, 1 to 3 septate, acrogeneous yellow to brown conidia, 34.2–77.1 x 3.5–5.5 μm in size, broad and tapering towards the apex (Fig. 6).

Disease cycle

The fungus mainly perennates through the infected plant debris in the form of mycelium or stromata. In areas where fruit infection occurs, the fungus can also overwinter in the form of infected seed. The conidia develop on the overwintering stromata which serve as the source of primary infection. These conidia are disseminated by wind and rain splashes to the leaf surface where they germinate and cause infection. The spots are produced on the leaves on which conidia are formed which serve as source of secondary infection. When weather becomes unfavourable, the fungus survives in the infected plant debris or in the infected seed if the fruits are attacked and the cycle is repeated next year.

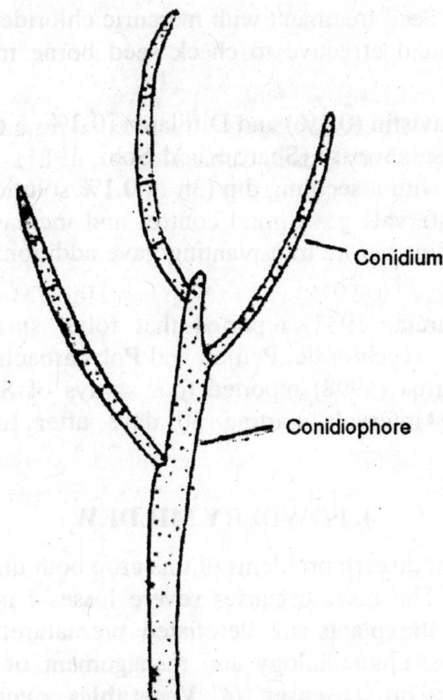

Fig. 6. Conidia and conidiophore of *Cercospora capsici*

Epidemiology

Disease development is favoured by high humidity and temperature (Suang *et al.*, 1984). The disease progressed fast when temperature was around 20–25°C with RH > 95 per cent (Kawagoe, 1998). The pathogen sporulates abundantly at 20–30°C, poorly at 10°C and there is no sporulation at 40°C (Meon, 1990). Sporulation is maximum when moisture is at saturation level and declines with decreasing moisture. Kawagoe (1997) reported 17–18°C in the light and 25°C in the dark as optimum temperature for sporulation in this pathogen. He further observed that in plastic house, spore release was sharply increased at 10 A.M., when the plastic house was opened and reached at its maximum around 12 noon and spore dispersal rapidly decreased after closing the plastic house in the evening. This clearly indicated that spore release and dispersal in a plastic house increased with an increase in air currents and temperature and a decrease in relative humidity.

Management

(i) Cultural practices: Cultural practices *viz.*, collection and burning of infected crop residue, use of disease free seed, and 3–4 years crop rotation have been found effective to reduce initial inoculum levels.

(ii) Host resistance : Out of 150 chilli collections, IC 99910, IC 99912, IC 92150, P-1718, P-1939, N-1015, EC 362901, EC 362910, EC 362913, EC 362925, EC 362934, EC 362935 and EC Eec 3637 were found free from *Cercospora* leaf spot, while rest were susceptible (Ullasa *et al.*, 1981).

(iii) Chemical control: Seed treatment with mercuric chloride (1 : 2000) solution for 2 minutes has been found effective to check seed borne inoculum and protected seedlings in nursery.

Foliar sprays of Bavistin (0.1%) and Difolatan (0.3%) 5 to 6 times at fortnightly intervals reduced disease severity (Sharma and Sohi, 1981). Bavistin seed dressing (0.2%) in combination with a seedling dip (3h in 0.1% solution) together with foliar sprays at fortnightly intervals gave good control and increased yield (Sharma and Sohi, 1984). Seedling dips before transplanting gave additional protection for 30–40 days.

Gonzalez and Garcia (1987) reported that foliar sprays with mancozeb + metallic copper, copper oxychloride, Pelijon and Polycarbacin effectively controlled the disease, while Sharma (1998) reported four sprays of Score (0.05%) or Roko (0.05%) at fortnightly intervals starting 30 days after transplanting as highly effective.

4. POWDERY MILDEW

It is one of the important disease problems of this crop both under open and protected cultivation conditions. The disease causes severe losses particularly in warm and humid climates when the plants are defoliated prematurely. Disease symptoms, pathogen, disease cycle, epidemiology and management of this disease has been described in **Chapter on Diseases of Vegetables grown under Protected Cultivation conditions.**

5. WET ROT OR CHOANEPHORA BLIGHT

It is a minor disease of chilli. It does occur occasionally especially in warm and humid weather conditions.

Geographical distribution

The disease is reported to occur in tropical and sub-tropical countries including India and USA. In India, the occurrence of this disease has been reported from Haryana, Punjab, Andhra Pradesh and West Bengal (Thind and Jhooty, 1985; Prabhavathy and Reddy, 1995; Panja, 1999).

Losses

The disease causes maximum damage when the crop is grown in rainy season especially in tropical and subtropical environment due to blossom blight. A severe outbreak of this disease occurred in Florida during 2003 crop season where almost 100 per cent plants were affected by this disease.

Symptoms

The symptoms of the disease mostly appear on buds and flowers. Injured leaves, side branches and green immature fruits are affected by the pathogen. The fungus first gets established on the fallen flower petals and buds and then spread to stem and leaves producing a wet rot and finally die-back. Affected tissues become water-soaked, dark green and soft. The fungus develops fast on these tissues and appears as a mass of tiny black-headed pins in a pincushion. Infected flowers and buds finally wilt and young plants may die.

The Pathogen

The disease is caused by *Choanephora cucurbitarum* (Berk. & Ravenel) Thaxter. The mycelium is generally coenocytic. Reproduction of this fungus is by modified sporangia called sporangiola. The sporangiola (some times called conidia) result from the evolutionary reduction of large sporangia that contain hundreds of spores to a small sporangium *i.e.,* monosporus. The spore and sporangial wall appear fused, hence serving as a single asexual reproductive unit similar to a conidium. Masses of these sporangiola give the fungus its characteristic pincushion appearance. The size of sporangiola varies from 8–30 x 5–10 μm. Since the pathogen is heterothallic, the zygospore formation is only possible if both mating types are present in a field. The mycelium develops intercalary chlamydospores, which are formed in chains.

Disease cycle

The fungus may survive in soil as chlamydospores and produce mycelium with sporangiola for infection. The sporangiola spread by wind and wind-driven rain to the crop. Fallen flowers catch initial infection and spread the disease to injured leaves, shoots and fruits.

Epidemiology

C. cucurbitarum is especially aggressive under hot and wet conditions. Sporangiola cause infection of flowers and succulent parts when temperatures and relative humidity are high and temperatures are abundant (Dougherty, 1979).

Management

(i) Cultural practices: The disease once appeared in the field is very difficult to control but it can be kept under check by following some cultural practices like wider plant spacing which will allow better air movement and more rapid drying of plant surfaces, avoidance of overhead irrigation and excessive fertilization especially nitrogenous and manipulation of planting dates to avoid excessive hot and wet weather.

(ii) Chemical control: Sprays of fungicides like zineb, mancozeb, captan, ziram and copper oxychloride have been reported effective against this disease and increased pepper yields (Chahal and Grover, 1974; Panja, 1999). Efficacy of carboxin in the management of this disease has also been reported (Meah and Mian, 1981).

(B) BACTERIAL DISEASES

1. BACTERIAL SPOT

This is a very important and common disease of Capsicum in those areas of the world where significant precipitation and moderate temperature occur during growth and fruiting stages and cause potential yield reductions due to premature defoliation. Same disease also occurs on tomato.

Geographical distribution

The disease was first reported in 1918 from Florida, USA (Sherbakoff, 1918). Subsequently, it has been reported from various tropical and subtropical areas of the world including India, which experience warm and humid weather conditions.

In India, the disease was recorded for the first time at Agricultural College Farm, Poona in 1948 (Patel *et al.*, 1950). Since then it has been recorded in Rajasthan, Kerala and Himachal Pradesh (Shekhawat and Chakravarti, 1977a).

Losses

Most devastating phase of this disease is defoliation. Premature defoliation affects the yields drastically. Besides, spotting of fruits make them unfit for market. In Rajasthan, the disease cause up to 16 per cent losses in chillies (Shekhawat and Chakravarti, 1977a).

Symptoms

The symptoms of the disease appear on all the above ground parts like leaves, stems and fruits of the plants. Initially, the lesions are water-soaked, circular or irregular that becomes necrotic with brown centres and thin chlorotic borders (Plate 6a). Lesions may enlarge to 10 mm in diameter. These lesions are generally sunken on the top surface of the leaf and slightly raised on the bottom. In favourable environmental conditions these spots on leaves coalesce and have a blighted appearance (Zapata, 1995). Such leaves turn yellow and fall down prematurely. Stem lesions are narrow, elongated and raised expanding upto 6 mm. The lesions become light brown and rough in appearance. Fruit lesions initially begin as green spots, which enlarge and become brown in colour (Plate 6b). These spots are raised with a cracked, roughened wart like appearance and are up to 5 mm in diameter. During periods of high humidity, fruit around the lesions may start rotting. Similar spots may appear on floral parts, cotyledons and petioles.

The Pathogen

The bacterium responsible for this disease is *Xanthomonas vesicatoria* (ex Doidge) Vauterin *et al.* The bacteria are motile, strictly aerobic, Gram-negative, rod shaped with round ends, non-spore forming and possess a single pollar flagellum. Each cell measures 1–1.8 x 0.5–0.7 μm. Bacteria occur singly but occasionally in pairs and rarely in short chains. Its colonies on nutrient agar or on specific medium i.e. yeast dextrose agar (YDA) are lemon yellow, slightly convex, wet and shining. Recently Ramyashreedevi *et al.* (2017) collected 31 isolates of Xanthomonads isolated from leaf and fruit samples of tomato, chilli and bell pepper from 5 different states of India viz., Himachal Pradesh, Karnatka, Uttarakhand, Tamil nadu and Haryana and were characterized as translucent yellow, raised colonies on YGCA medium, Gram-negative, rod shaped, positive in starch hydrolysis, variation in pectin utilization. These isolates showed symptoms on chilli cv. Pusa Jwala and tomato cv. Pusa Ruby and they were grouped in to pepper - tomato group (XCVPT), belonged to *Xanthomonas euvesicatoria* which was further confirmed by specific primers i.e. Bs-XeF and Bs-XeR and Seu 2.4 and Seu 2.5 and amplified DNA of all the isolates at 173 bP and 208 bP, respectively.

Eleven pepper races have been identified based on differential reactions to four resistance genes (Minsavage *et al.*, 1990). Strains are pathogenic either on pepper only or tomato only or on both pepper and tomato.

Disease cycle

The pathogen is seed borne in nature and present within or on the seed surface. The bacterium is also able to survive in infected plant debris (Peterson, 1963). Dissemination of the bacterium in or on the seed and transplants is an important factor in disease cycle. The bacterium penetrates directly through the injuries created by wind-driven sand, insect punctures or mechanical means, hydathodes and stomata and lesions develop more frequently on the wounded fruits than unwounded ones (Pohronezny *et al.*, 1993). After entry, the bacterium produces slimy masses of cells within the stomatal cavities and adjoining intercellular spaces. The adjoining cells swell and break the epidermis producing lesions. In infected fruits, the seed may be infected.

Epidemiology

The bacterium persists in infected seed and plant debris. Moderate temperatures along with high precipitation and relative humidity favour disease development. Low relative humidity is unfavourable to the disease and can greatly inhibit disease development following infection. Infection of the plants can take place at a wide range of temperature *i.e.,* 15 to 35°C. The optimum temperature lies between 22 to 34°C and maximum disease develops in between July and September (Shekhawat and Chakravarti, 1977b). Mean temperature (25±1°C) coupled with high humidity (60–80%) and low sunshine hours with intermittent rainfall favoured disease development (Prashad and Gupta, 2012).

Plant stresses like repeated instability of water table, sudden increase in soluble salts, lack of adequate nutrients, use of tank mixtures of pesticides which have high electrical conductivity also play important role in disease development and spread (Simons, 1987). The bacterium is disseminated within a field by wind driven rain droplets, clipping of transplants and aerosols (McInnes *et al.*, 1988).

Management

Integration of cultural practices with host resistance and chemical methods can be useful in the management of this disease.

(i) Cultural practices: Since the disease requires high precipitation and high relative humidity, fields should be well drained free of low-lying areas to minimize water-logged conditions. The field should be rotated with non-solanaceous crops and pepper should not be rotated with tomato. The field should be kept free from weeds and other volunteer plants. Use of disease free seed also reduces the immediate availability of primary inoculum and ensures disease free transplants. Sprinkler or overhead irrigation of the field should be limited to keep the disease under check.

(ii) Host resistance: Cultivars Medalla–de ovo, True Heart 27 A, True Heart 2 B and Florida XVR 3-25 have shown tolerance/resistance to leaf spot disease (Cook, 1984; Marrero and Gonzalez, 1985). The resistance has been reported to be controlled by two recessive genes (Stall and Lozano, 1982). However, three genes BS_1, Bs_2 and

Bs_3 have been reported for resistance (Kim and Hartman, 1985). None of the genes result in resistance to all strains of the pathogen.

(iii) Seed treatment: Since the pathogen is seed borne in nature and if seed treatment reduces the bacterial populations it may be an effective method for reducing disease incidence in the field. Hot water treatment may be used in which seeds are submerged in water heated at 50°C for 25 minutes. The shortcoming of this method is that it may reduce the seed germination and is not completely effective although incidence is reduced very much. The other method involves disinfecting the seeds for 1 minute in 1.3 per cent sodium hypochlorite solution and drying in shade.

(iv) Chemical control: The disease is difficult to control with the conventional bactericides. With repeated sprays of antibiotic like streptomycin, resistant strains develop fast and disease control is reduced (Thayer and Stall, 1961). Copper fungicides can be used for the control of this disease provided copper sensitive strains are present. Copper tolerant strains are effectively controlled by the combination of copper fungicides and mancozeb (Stall *et al.*, 1986). A combination of Agrimycin 100 with copper oxychloride provided effective control of this disease and increased yields (Mandge *et al.*, 1992). Prashad *et al.* (2011) reported the efficacy of Streptocycline alone and in combination with Blitox-50 and antibiotic Bacterinashak in the management of this disease.

2. BACTERIAL WILT

Bacterial wilt is a limiting disease problem in tropical, subtropical and some warm temperate climates. This disease also occurs under protected cultivation conditions.

Geographical distribution

Kelman (1953) was the first to review the literature and indicate the worldwide economic importance of bacterial wilt on its many hosts. Since then the disease is occurring through out the world where ever favourable climate for the growth and spread of the bacterium are available. In India, the disease has been reported from Kerala, Uttaranchal, Orissa, West Bengal, and Andeman.

Losses

Losses from this disease are significant particularly during warm humid weather. Reports of complete crop loss in the field are not uncommon. Under protected conditions, this is a common disease and some times plants wilt prematurely.

Symptoms

The initial symptoms of the plants include sudden wilting of foliage on the youngest leaves, which may recover during night. With the progress of the disease, the initial wilt is succeeded by a sudden and permanent wilt of entire plants, with a slight or no leaf yellowing. Infected young plants die immediately while older plants may first show leaf drooping and discolouration only on one side or part of the plant and eventually the whole plant wilt and dies. The vascular system in affected plants is discoloured from pale yellow to brown in colour. A simple diagnostic test is to place cross section of the infected tissues suspended in water that yields milky white exudates of bacteria.

The Pathogen

The disease is caused by the bacterium *Ralstonia solanacearum* (Smith) Yabuuchi *et al.* The bacterium is Gram-negative rod, 0.5 x 1.5 μm, aerobic and motile by one polar flagella.

Strains of *R. solanacearum* differ in host range, geographical distribution, pathogenecity and physiological properties. Strains are subdivided informally into five races based upon host range or differentiated into five biovars based upon the utilization and/or oxidation of three disaccharides and three sugar alcohols as carbon sources. In India, race 1, 3 and 4, and biovar II, III, and IV (Mathew *et al.*, 2000; Sunaina and Tomar, 2003) are prevalent.

Host Range

Bacterial wilt causes disease on more than 200 plant species belonging to over 40 families. Important hosts of *R. solanacearum* include potato, pepper, tobacco, tomato, eggplant and banana. Besides these, *R. solanacearum* could survive in the rhizosphere, on the rhizoplane and inside the roots of a large number of weed plants like *Solanum nigrum, S. anguivi, S. torvum, S. xanthocarpum, Croton bonplandianum, Euphorbia hirta, Colocasia esculenta, Portulaca oleracea, Phyllanthus niruri* and *Parthenium hysterophorus* in West Bengal (Samaddar *et al.*, 1998).

Disease cycle

The bacterium has a wide host range and can also survive in moist soil for longer periods in the absence of plant debris. Diseased plant debris also helps in the overseasoning of the bacterium. Entry of the pathogen in the host is via roots through mechanical/nematode/insect injury or growth cracks. The bacterium colonizes the xylem vessels and spreads throughout the plant. The incubation period varies, depending upon age and susceptibility of the plant involved. The bacteria invade the intercellular spaces of the parenchyma cells in the cortex and pith and disintegrate the cell walls resulting in pockets filled with bacterial cells and polysaccharide and cellular debris. Bacteria are released back into the soil from infected roots and decaying plant material (Bruehl, 1987).

Epidemiology

The pathogen perenates in soil and it is having a wide host range. The pathogen generally does not survive and reproduce below 10ºC (van Elsas *et al.,* 2000) although it can survive over a wide range of temperature *i.e.,* 15 to 37ºC and optimum being 35 to 37ºC. The disease usually develops when mean temperature is above 20ºC and more severe wilting is seen at higher temperatures of 30ºC or higher and high soil moisture. Though the pathogen can grow over a wide range of pH i.e., from 6.2 to 7.4 but alkaline soils are generally unfavourable for disease development. Dissemination of the bacteria occurs with irrigation or rainwater, on infected plants, infested seeds of tomato and eggplant and the soil (Goto, 1990).

Management

No chemical has been found effective against this disease. However, nematodes in the soil should be eradicated with use of some nematicides because in the presence of nematodes, the incidence of the disease increases. The disease can be kept under control by using following cultural practices and resistant sources.

(i) Cultural practices: Cultural practices like avoidance of planting in the infested soil, a long crop rotation, use of disease free seed, weed free cultivation, avoiding movement of soil from infected fields to disease free fields through equipments and workers and disinfestations of the equipment should be followed to keep the disease under check. Nursery should be grown in disease free soil and should be disinfested before sowing either with soil solarization or steam sterilized or treated with chemicals like Formaldehyde.

(ii) Host resistance: Wilt diseases are very difficult to manage and can be kept under check with the use of resistant cultivars. Various chilli cultivars/lines like Kandhari (Rahim and Samraj, 1974), CA 219, CA 33 (Gopalakrishnan and Peter, 1991), Manjeri (Jyothi *et al.*, 1993), Ujwala (Gopalakrishnan, 1996), Kulai (Rahman and Abdulah, 1997), PP-9656-06, PP-977127, PP-977195-1 and PP-977635 (Kishan *et al.*, 2001) have been recommended for cultivation in disease prone areas. Sweet pepper cultivar Cholo has been identified as tolerant with only 10 per cent disease incidence (Jimnez *et al.*, 1988). It was further reported that, the restriction of pathogen spread associated with the resistance in *C. annuum* to *R. solanacearum* is mainly attributed to some induced, morphological and physical barriers, which reduce the rate of multiplication of the pathogen in resistant host (Rahman and Abdullah, 1997; Rahman *et al.*, 1999).

(iii) Biologicl Control: Effectiveness of seven organic inputs was evaluated against *R. solanacearum* under protected cultivation. Biosol, matka khad, fermented buttermilk, panchgavya, biodynamic compost tea, cow urine and vermiwash were used as organic inputs and tetrazolium chloride medium (TZC) was amended at different concentrations (0.5–4.0%) with both sterilized and unsterilized organic inputs. Results revealed that unsterilized biodynamic compost tea showed 91.30 and 95.60 per cent inhibition at 2.0 and 4.0 per cent concentration in the bacterial growth, respectively, whereas biosol (unsterilized) was also found effective in inhibiting bacterium with 92.30 per cent. Drench with mixture of biosol, fermented butter milk, cow urine, vermiwash before and after transplanting along with seedling dip in biosol for 10 min. before transplanting provided maximum (85%) disease control (Ashlesha and Paul, 2012).

(C) VIRAL DISEASES

1. LEAF CURL

It is an important disease of this crop and can cause significant yield reductions if the vector is present in the area.

Geographical distribution

There are various reports of leaf curl disease affecting peppers in Southeast and East Asia, particularly India (Vasudeva and Samraj, 1948; Muniyappa and Veeresh, 1984) and Sri Lanka (Sugiura *et al.*, 1975). In India, this disease is prevalent throughout the country, although severity is less in southern India.

Losses

The disease causes considerable yield losses in Chillies (Mishra *et al.*, 1963). However, losses up to 80 per cent have been reported in many parts of the northern India (Singh *et al.*, 1979).

Symptoms

The symptoms of the disease appear as curling of the leaves followed by reduction in their size which later turn pale yellow in colour (Plate 7). The older leaves become leathery and brittle. The affected plants become stunted. Fruits that are formed are small and deformed.

The Virus

The disease is caused by *Tobacco Leaf Curl Virus* (TLCV), which belongs to *bigeminivirus* group. The virus consists of geminate isometric particles measuring 18 to 20 nm in diameter.

Host range

The virus has a wide host range. Besides tobacco and C*apsicum annuum*, the virus can be experimentally transmitted to 34 other plant species, including *Lycopersicon esculentum, Beta vulgaris, Carica papaya, Cyamopsis tetragonoloba, Sesamum indicum, Phaseolus vulgaris* and *Petunia hybrida* (Valand and Muniyappa, 1992).

Transmission

Tobacco Leaf Curl Virus (TLCV) is transmitted by whitefly (*Bemisia tabaci*) vector.

Epidemiology

The virus survives from one season to another on various annual and perennial hosts and transmitted to pepper by the vector *Bemisia tabaci*. The epidemic of the disease depends on the availability of the vector, sources of inoculum, vector activity and their population and prevailing environmental conditions. During wet weather, the spread of the disease becomes slow due to reduced activity of vector (Shivanathan, 1982).

Management

(i) Host resistance: Several cultivars/lines like Perennial, BG-1, Lorai, Punjab Lal (Thakur *et al.*, 1987; Singh and Kaur, 1990), JCA 248, JCA 218, Pant C-1, NP 46A, Pusa Jwala, JCA 196 (Sangar *et al.*, 1988), Punjab Surkh (Hundal *et al.*, 1996), DC-18 (Roy *et al.*, 1997), Pusa Sadabahar, Pant C-2, Jawahar Mirch-2 (Singh *et al.*, 1998), Surya Mukhi and Japani Loungi (Kumar *et al.*, 1999) were reported as resistant/tolerant to this virus. Resistance in different chilli cultivars has been reported to be controlled by monogenic recessive genes (Bal *et al.*, 1995).

(ii) Chemical control: Since, whitefly transmits this disease in a persistent manner, so the timely application of various insecticides at regular intervals could be effective in reducing the activity and population of the vectors. Sprays of systemic insecticides like Metasystox at 10–14 days interval causes reduction in disease. Maximum disease control can be obtained by applying cypermethrin, dimeththoate and endosulfan (Valdez and Wolfenberger, 1995). However, combined use of yellow traps followed by cypermethrin spray reduced the population of *B. tabaci* by 78 per cent. Some recently available insecticides like imidaclorpid, acephate and acetamaprid can be effectively used for the control of whiteflies, but continuous use of one insecticide may not prove effective in long run. It is always better to use these insecticides in rotation or alternate them with some neem-based insecticides like Nimbecidine (1ml/ l) or NeemAzal (1.5 ml/l). Alternate spraying of trizophos 40 EC (1.5 ml/ l) and Neemark (5ml/ l) both in nursery and main field at 15 days interval reduced the leaf curl incidence to 8.31 per cent (Venkatesh *et al.*, 1998).

2. MOSAIC

Potyvirus and *Cucumovirus* groups mainly cause mosaic in chillies. The viruses like pepper severe mosaic virus, pepper mild mosaic, pepper veinal mosaic, potato virus Y and cucumber mosaic virus are associated with the mosaic complex. Sometimes *Tobamoviruses* like tobacco mosaic virus (TMV), tomato mosaic virus (ToMV) and pepper mild mottle virus (PMMoV) are also found associated with this disease world over.

Geographical distribution

The mosaic diseases have worldwide distribution. In India, the natural occurrence of Cucumber mosaic virus (CMV) in chilli was recorded for the first time by Anjaneyulu and Appa Rao (1967). Since then these have been reported from all parts of the country including Himachal Pradesh (Kapoor, 2012).

Losses

The mosaic diseases have been reported to cause considerable yield losses particularly in susceptible cultivars. The incidence of CMV alone ranged from 20 to 50 per cent in infected pepper fields (Benner *et al.*, 1985), while 66.7 to 74 per cent yield reductions in bell pepper due to mixed infection of potato virus Y (PVY) and CMV was recorded (George *et al.*, 1995).

Symptoms

(a) *Cucumovirus* genus: Plants infected early in the season have more severe symptoms and reduced fruit set and quality. Stunting is a common symptom of the infected plants and is frequently accompanied by a lighter than normal shade of green and overall thickening of the leaves (Plate 8). Leaves may become elongated and narrow which become filiform. Kapoor (2012) reported mosaic, leaf deformation, ring spots on leaves and fruits of bell pepper.

(b) *Potyvirus* genus: *Potato virus Y* (PVY) and its strains generally induce mosaic or mottle and are frequently accompanied by dark green vein bending followed by necrosis of the veins, petioles and stems of the chillies. Pepper veinal mottle virus causes mosaic mottling, distortion and filiformy of the leaves with bushy appearance of the infected plants. Symptoms developed by pepper mottle virus include systemic mottling and deformities of the infected leaves. The virus also causes systemic necrosis of the leaves resulting in their death, while pepper severe mosaic virus produces mosaic mottle, vein clearing and deformation of infected leaves.

(c) *Tobamovirus* genus: A general mosaic or mottle of dark and light green or yellow is observed on the leaves of plants infected with these viruses especially TMV and ToMV. PMMoV usually induces less change in foliar colour although it frequently causes leaves to pucker. Infected plants remain stunted. In susceptible cultivars some strains of TMV or ToMV strains cause more severe symptoms like defoliation, premature flower drop, wilting stem death and necrosis of the tissue inside the fruit wall. High temperature and low light intensity may increase symptom severity and disease development.

The Virus(s)

Cucumber mosaic virus (CMV) is the type member of the *Cucumovirus* genus of RNA viruses. CMV forms icosahedral particles of approximately 29 nm in diamter.

Potato virus Y (PVY), *Pepper mottle virus* (PepMov) and *Tobacco etch virus* (TEV) are the type member of the *Potyvirus* genus of RNA viruses. All potyviruses form flexuous, rod-shaped particles ranging from 680–900 nm in length.

TMV, ToMV and PMMoV are the type members of the *Tobamovirus* genus of RNA viruses. All tobamoviruses form rigid, rod-shaped particles approximately 300 nm in length.

Host Range

The host range of *Cucumovirus* group like cucumber mosaic virus (CMV) is distributed in 100 families containing over 1200 plant species. Besides pepper, major crops susceptible to CMV are tomato, cucumber, melon and squash as well as numerous weed species.

Potyvirus like TEV has a fairly wide host range infecting both solanaceous plants and members of at least nine other plant families. The virus occurs on many weeds including *Solanum* sp., *Cirsium vulgare, Chenopodium album, Nicandra physalodes* and *Datura stramonium*. PVY and PepMoV are primarily restricted to solanaceous plants.

Tobamoviruses like TMV has a wide host range, while ToMV and PMMoV have more limited host ranges. The Solanaceae family contains many species susceptible to infection by all the three viruses.

Transmission

Cucumo and *Potyviruses* are sap transmissible and also transmitted by more than 75 species of aphids in a non-persistent manner. The most efficient vectors are *Aphis gossypii, Myzus persicae* and green peach aphid and *M. persicae* of *Cucumo* and *Potyviruses*, respectively (Lopez-Moya and Garcia, 1999). Tobamo-viruses are readily transmitted by mechanical means including worker's hands, clothing or tools (Gibbs, 1977). No insect vectors are required for their transmission.

Epidemiology

The severity of mosaic (*Cucumo* and *Potyviruses*) disease depends on the activity of the vectors and the sources of primary inoculum (infected weeds, solanaceous and other hosts) available in and around the field, and environmental conditions.

The plant debris can harbour the infectious particles of tobamoviruses for many years. These are also carried on infected seed. Secondary spread within a field or glasshouse occurs by contact between plants, which is frequently mediated by workers handling plants or using virus-contaminated tools during staking.

Temperature affects both vector population as well as virus multiplication. Maximum population of *M. persicae* and *Aphis craccivora* in pepper fields was recorded at a temperature range of 26 to 30ºC along with corresponding increase in the incidence of mosaic disease (Sharma *et al.*, 1991).

Management

(i) Cultural practices: Cultural practices like destruction of previous year's plant debris, removal of weeds that can harbour viruses, use of disease free seed and distancing of pepper plantings from potato, tomato and other solanaceous crops are the best ways to limit viral infection.

The aphids transmit the viruses in non-persistent manner in nature. These viruses can be kept under check by using the practices which affect the vector activity and transmission capacity. Barrier crops, which are taller than the pepper, can impede aphid entry into the fields and proved quite useful in checking CMV and PVY in peppers. Significant reduction in the primary spread of mosaic disease in chillies and bell pepper has been observed by planting sunflower, snap bean, sesame, sorghum, maize and pearlmillet as a barrier crop (Deol and Rataul, 1978; Chowfla and Sharma, 1990; Ragupathi and Veeraragavathatham, 2002). Growing of marigold with pepper has also been found useful in reducing the incidence of viruses infecting pepper in Mexico (Chew-Medinaveitia *et al.*, 1995). If these barrier crops are weekly sprayed with insecticide like Metasystox, the aphid population is significantly reduced (Anandam *et al.*, 2002). Intercropping of chilli with onion resulted in the lowest aphid population and disease incidence (Basavarajappa and Rajasekhar, 2001a). Mulching of the field with aluminium foil has also been found effective in reducing the aphid population as well as mosaic incidence in chilli (Basavarajappa and Rajasekhar, 2001b). Similarly Kapoor (2012) also evaluated different mulches and reported that amongst the different coloured mulches tested, silver (polyester sheet) followed by white polythene mulch, showed minimum disease incidence and maximum average yield/plot thus proved better than the yellow, blue, black and green coloured mulches used. Planting of maize as barrier crop was also found effective in reducing virus disease incidence and increasing the yield of bell pepper. Minimum disease incidence was observed when maize was planted as border row plus as an intercrop in the middle of the plot.

(ii) Host resistance: Cultivars/lines like Punjab Lal, Gauhati Black, Perennial (Thakur *et al.*, 1985; Chowfla and Sharma, 1990), HC 1-1, HC 15, HC 22, HC 28, HC 69, HC 226, Pusa Sadabahar and Virus Free-1 (Dhawan *et al.*, 1996), Pusa Sadabahar, RHRC-clustering erect, Punjab Lal (Acharyya and Acharyya, 1999), VR 42, VR-55, VR-42 x VR- 55 (Prasad *et al.*, 2001) have been identified to possess resistance against various viruses causing mosaic complex in peppers.

(iii) Plant extracts : *Datura stramonium* leaf extract caused complete inhibition of chilli mosaic virus infection (Singh and Sharma, 1999). Spraying of Neem seed kernel extract (5%) reduced the incidence of chilli mosaic to 9.92 per cent and increased the yield (Ragupathi and Veeraragavathatham, 2002).

(iv) Chemical control: Application of insecticides like quinalphos, malathion, methyl parathion, chlorpyriphos, monocrotophos, phosphamidon, cypermethrin, dimethioate and fenpropathrin decreased transmission percentage with increase in aphid mortality and ultimately reduced the mosaic incidence (Devi *et al.*, 1995).

REFERENCES

Abdellaouni-Maane, S., Seng, J.M. Saindrenan, P. and Bompeix, G. 1988. Fosetyl-Al is effective against mutants of *Phytophthora capsici* resistant to metalaxyl. *Cryptogamie Mycologie* **9:** 47–56.

Acharyya, P. and Acharyya, P. 1999. Screening of chilli germplasm against strains of cucumber mosaic virus. *Environ. Ecol.* **17:** 484–487.

Ahn, S.J. and Hwang, B.K. 1992. Isolation of antagonistic-producing actinomycetes antagonistic to *Phytophthora capsici* from pepper growing soils. *Korean J. Mycol.* **20(3):** 259–268.

Alenxander, S.A. and Waldenmaier, C.M. 2002. Management of anthracnose in bell pepper, 2001. *Fungic. Nematicide Tests* (online) *Rep.* **57:** 1094.

Anand, M., Naik, M.K., Ramegowda, G. and Devika Rani, G.S. 2010. Biocontrol and plant growth promotion activity of indigenous *Pseudomonas fluorescens* isolate. *J. Mycopathol. Res.* **48:** 45–50.

Anandam, R.J., Sabitha Doraiswamy and Doraiswamy, S. 2002. Role of barrier crops in reducing the incidence of mosaic disease of chilli. *Z. Pflanzenkr. Pflanzenschutz* **109:** 109–112.

Anjaneyulu, A. and Appa Rao, A. 1967. Natural occurrence of cucumber mosaic virus of chilli in India. *Indian Phytopath.* **20:** 380–381.

Ashlesha and Paul, Y.S. 2012. Evaluation of organic inputs against *Ralstonia solanacearum* – A bacterial wilt pathogen of capsicum under protected cultivation. *Pl. Dis. Res.* **27(1):** 66–70.

AVRDC, 1990. Screening for anthracnose resistance. In: *Progress Report, 1988*: 71–72.

Azad, P. 1991. Fate and role of chemical constituents of chilli fruit during infection with *Colletotrichum capsici. Indian Phytopath.* **44(1):** 129–131.

Bagri, R.K., Choudhary, S.L. and and Rai, P.K. 2004. Management of fruit rot of chilli with different plant products. *Indian Phytopath.* **57:** 107–109.

Bal, S.S., Singh, J. and Dhanju, K.C. 1995. Genetics of resistance to mosaic and leaf curl viruses in chilli (*Capsicum annuum* L.). *Indian J. Virol.* **11(1):** 77–79.

Bansal, R.D. and Grover, R.K. 1969. Reaction of chilli (*Capsicum frutescens*) varieties to *Colletotrichum capsici. J. Res. Punjab Agric. Univ.* **6:** 345–348.

Bartual, R., Lacasa, A., Marsal, J.I. and Tello, J.C. 1994. Epistasis in the resistance of pepper to Phytophthora stem blight (*Phytophthora capsici* L.) and its significance in the prediction of double cross performances. *Euphytica* **72(1/2):** 149–152.

Basak, A.B. 1998. Studies on the location of *Colletotrichum capsici* (Syd.) Butler & Bisby in the infected chilli seeds. *Seed Res.* **26(1):** 101–104.

Basavarajappa, M.P. and Rajasekhar, D.W. 2001a. Effect of intercropping in the management of chilli mosaic transmitted by insect vector. *Insect Environ.* **7:** 63–64.

Basavarajappa, M.P. and Rajasekhar, D.W. 2001b. Effect of mulching on chilli mosaic transmitted by insect vector. *Insect Environ.* **7:** 64–65.

Benner, C.P., Kuhn, C.W., Demski, J.W., Doboson, J.W., Colditz, P. and Nutter, F.W. Jr. 1985. Identification and incidence of pepper viruses in northeastern Georgia. *Plant Dis.* **69:** 999–1001.

Bhardwaj, S.S. 1978. Pathogenic variability in *Phytophthora nicotianae* var. *nicotianae* causing fruit rot of bell pepper (*Capsicum annuum* L.). *Pantnagar J. Res.* **3:** 140–141.

Bhardwaj, S.S. 1983. Studies on fruit rot and leaf blight of bell pepper (*Capsicum annuum* L.) caused by *Phytophthora nicotianae* B. de Haan var. *nicotianae* Waterhouse. Ph.D. thesis, HPKVV, Solan 114 pp.

Bhardwaj, S.S. and Sharma, S.L. 1985. Chemical control of fruit rot and leaf blight of bell pepper in India. *Agric. Ecos. Environ.* **14:** 151–154.

Bhardwaj, S.S., Sharma, S.L. and Dohroo, N.P. 1985. Role of meteorological factors in development of bell pepper fruit rot (*Phytophthora nicotianae* var. *nicotianae* Waterhouse). *Indian J. Mycol. Pl. Pathol.* **15:** 328–329.

Bhullar, B.S., Bajaj, K.L. and Bhatia, I.S. 1972. Studies on the phenols resistant and susceptible varieties of chillies in relation to anthracnose disease. *Phytopathol. Z.* **75:** 236–240.

Biswas, A. 1992. Efficacy of fungicides in the control of anthracnose disease of chilli in Sunderban region of West Bengal. *J. Mycopathol. Res.* **30(1):** 31–35.

Boiteux, L.S., Cupertino, F.P. and Reifschneider, F.J.B. 1993. *Capsicum chinensis* PI 159236: a source of resistance to *Phytophthora capsici* and tomato spotted wilt virus (TSWV). *Capsicum Eggplant Newsl.* **12:** 76.

Bruehl, G.W. 1987. *Soilborne Plant Pathogens*. Mcmillan Publishing Co., 866 Third Avenue, New York, USA.

Butler, E.J. 1918. *Fungi and Diseases in Plants*. Thaker Spink and Co., Calcutta, India, 300pp.

Byung, S.K. 1988. Characteristics of bacterial spot resistant *Phytophthora* blight resistant lines of *Capsicum* pepper. *J. Korean Soc. Hortic. Sci.* **29(4):** 247–252.

Cartia, G., Cipriano, T. and Greco, N. 1989. Effect of solarization and fumigants on soil borne pathogens of pepper in green house. *Acta Hortic.* **255:** 111–116.

Chahal, A.S. and Grover, R.K. 1974. Chemical control of soft fruit rot of chillies caused by *Choanephora cucurbitarum. Haryana J. Hortic. Sci.* **3:** 190–194.

Chakravarti, B.P. and Anilkumar, T.B. 1975. Control of seed borne infection of *Colletotrichum capsici* in chillies. *Curr. Res.* **4(10):** 172.

Chang, S.H. and Chung, B.K. 1985. Studies on the varietal resistance and effects of nutrients for fungal growth of pepper anthracnose caused by *Colletotrichum dematium* f.sp. *capsici. Korean J. Mycol.* **13(4):** 227–233.

Chew-Medinaveitia, Y.I, Zavaleia-Mejia, E., Delgadillo-Sanchez, F., Valfivis-Alcala, R., Pena-Martinez, M.R. and Cardenas-Suniano, E. 1995. Evaluation of control strategies for virus disease of pepper (*Capsicum annuum* L.). *Fitopatologia* **30:** 74–78.

Cho, E.K., 1987. Strategies for biological control of soil borne disease in economic crops in Korea. *Korean J. Pl. Pathol.* **3(4):** 313–317.

Choi, J.K., Park, D.Y., Woo, Y.H. and Sung, J. M. 1990. Studies on the resistance of red pepper varieties to *Phytophthora* blight and anthracnose. *Res. Rep. Rural Dev. Admn. Hortic.* **32(2):** 1–9.

Chowdappa, P., Madhura, S., Nirmal Kumar B.J., Mohan Kumar, S.P. and Hema, K.R. 2014. *Phytophthora boehemerie* as dominant pathogen responsible for severe foliar blight of *Capsicum annuum* L. in South India. *Plant Dis.* **98:** 90–98.

Chowdhury, S. 1957. Studies on the development and control of fruit rot of chillies. *Indian Phytopath.* **10(1):** 55–62.

Chowfla, S.C. and Sharma, P.N. 1990. Management of bell pepper mosaic disease complex in Himchal Pradesh. *Indian Phytopath.* **43:** 349–351.

Christopher, D.J., Raj, T.S. and Kumar, S.R. 2013. Morphological and molecular variability in *Colletotrichum capsici* causing anthracnose of chilli in TamilNadu. *Pl. Dis. Res.* **28(2):** 121–127.

Cook, A.A. 1984. Florida XVR 3–25 bell pepper. *Hortic. Sci.* **19:** 735.

Dastur, J.F. 1922. Dieback of chillies (*Capsicum* spp.) in Bihar. *Mem. Dept. Agric. India.* **11:** 129–144.

Datar, V.V., Sontakke, M.B., Purandhari, N.D. and Shinde, N.N. 1990. Fungicidal control of anthracnose of chillies. *Indian J. Mycol. Pl. Pathol.* **20(2):** 156–158.

Deol, G.S. and Rataul, H.S. 1978. Role of various crops in reducing the incidence of cucumber mosaic virus in chilli (*Capsicum annuum* L.). *Indian J. Entomol.* **40:** 261–264.

Devi, P.S., Reddy, H.R. and Devi P.S. 1995. Effect of insecticides on aphid transmission on pepper vein banding virus and cucumber mosaic virus of chilli (*Capsicum annuum* L.). *Mysore J. Agric. Sci.* **29(2):** 141–148.

Dhawan, P., Dang, J.K., Sangwan, M.S. and Arora, S.K. 1996. Screening of chilli cultivars and accessions for resistance to cucumber mosaic virus and potato virus Y. *Capsicum Eggplant Newsl.* **15:** 55–57.

Dougherty, D.E. 1979. Bud rot of pepper. *Proc. Fla. State Hortic. Soc.* **92:** 103–106.

Ekbote, S.D. 2003. Efficacy of prochloraz 45 EC against fruit rot and die-back of chilli. *Indian J. Pl. Prot.* **31:** 139–140.

Fernandez, M.C. 1988. Evaluation of sweet and chilli pepper (*Capsicum annuum*) genotypes for resistance to *Phytophthora capsici* Leonian. *Agric. Tecnica* **48(4):** 359–362.

Ferreyra, E.R., Tosso, T.J. and Fernandez, M.C. 1984. Effect of irrigation water management on *Phytophthora capsici* Leonian, causal agent of *Capsicum annuum* blight. *Agric. Tecnica* **44:** 319–324.

Fulton, H.D. 1908. Diseases of pepper and beans. *Lousiana Agric. Exp. Sta. Bull.* **293:** 119–152.

Galmarini, A., Sentiner, A. and Galmarini, H. 1991. Breeding pepper (*Capsicum annuum* L.) for resistance to *Phytophthora capsici* in Argentina: 'Calfayuco INTA' a new cultivar. *Capsicum Newsl.* **10:** 61.

George, T.E., Anand, N. and Singh, S.J. 1995. Influence of potato virus Y (PVY) and cucumber mosaic virus (CMV) on growth and yield of bell pepper (*Capsicum annuum* L.). *Indian Cocoa, Arecanut Spices J.* **19:** 1–4.

Gibbs, A.J. 1977. *Tobamovirus* group. CMI/AAB Descriptions of plant viruses No. 184.

Gonzalez, M. and Garcia, E. 1987. Chemical control of *Cercospora* disease of *Capsicum. Argentina de Cuba* **19(2):** 25–30.

Gopalakrishnan, T.R. and Peter, K.V. 1991. Screening and selection of bacterial wilt resistance in chilli. *Indian J. Genet. Pl. Breed.* **51:** 332–334.

Gopalkrishnan, T.R. 1996. Ujwala, a bacterial wilt resistant chilli variety. *ICAR News Sci. Tech. Newsl.* **2(4):** 23.

Goto, M. 1990. *Fundamentals of Bacterial Plant Pathology*. Academic Press, Inc. 1250 Sixth Avenue, San Diego, CA, USA.

Greenleaf, W.H. 1986. Pepper breeding. In: *Breeding Vegetable Crops*, (Ed. M.J. Basett), Avi Publishing Co. pp. 67–134.

Grover, R.K. and Bansal, R.D. 1970. Seed borne nature of *Colletotrichum capsici* in chilli seeds and its control by seed dressing fungicides. *Indian Phytopath* **23(4):** 664–668.

Gupta, J.S. and Dixit, R.B. 1982. *Streptomyces griseus* in relation to *Colletotrichum capsici* on the surface of chilli. *Geobios* **9(5/6)**: 249–250.

Gupta, S.K. and Jarial, Kumud 2008. Management of fruit rot (*Phytophthora nicotianae* var. *nicotianae*) of bell pepper (*Capsicum annuum*) with mulch and fungicide application. *J. Mycol. Pl. Pathol*. **38(3)**: 689–680.

Gupta, S.K. and Jarial, Kumud 2010. Comparative efficacy of two strobilurins in management of leaf blight and fruit rot (*Phytophthora nicotianae* var. *nicotianae*) of bell pepper. *Indian Phytopatho.* **63:** 216–218.

Gupta, S.K. and Jarial, Kumud 2014. Epidemiology and management of leaf blight and fruit rot of bell pepper. *Rev. Plant Pathol*. **6:** 473–499.

Gupta, S.K., Jarial, Kumud and Kansal, S. 2009. *Colletotrichum gloeosporioides* causing anthracnose in bell pepper seed crop – a new record from Himachal Pradesh. *J. Pl. Dis. Sci.* **4(1)**: 126–127.

Gupta, S.K., Jarial, Kumud, Kansal, S., Thakur, M.C., Vikram, Amit and Mehta, Seema. 2007. Evaluation of bell pepper (*Capsicum annuum*) germplasm for resistance to leaf blight and fruit rot caused by *Phytophthora nicotianae* var. *nicotianae. Indian J. Agric. Sci*. **77(4)**: 267–269.

Gupta, S.K., Jarial, Kumud, Saharan, M.S., Kaur, R. and Saxena, Bhawana. 2012. Morphological and genetic variation among *Phytophthora nicotianae* isolates causing fruit rot of bell pepper and their correlation with genetic variability among host genotypes. *Indian phytopath*. **65(1)**:38–44.

Halsted, B.D. 1890. Rep. Bot. Dept., N.J. Agric. Exp. Stn. Rep. **11:** 358–360.

Hartman, G.L. and Hawang, T.C. 1992. *Phytophthora* blight of pepper: screening for disease resistance. *Trop. Pest Mgt*. **38(3)**: 319–322.

Hegde, G.M. and Anahosur, K.H. 2001. Biochemical basis of resistance to fruit rot (*Colletotrichum capsici*) in chilli genotypes. *Karnataka J. Agric. Sci.* **14:** 686–690.

Hegde, G.M., Anahosur, K.H. and Kulkarni, S. 2002. Biological control of *Colletotrichum capsici* causing fruit rot of chilli. *Pl. Pathol. Newsl*. **20:** 4–5.

Higgins, B.B. 1923. The diseases of pepper. *Georgia Agric. Exp. Stn. Bull*. **141:** 48–75.

Hundal, J.S., Khurana, D.S. and Kaur, S. 1996. Punjab Surakh-a new high yielding variety of chilli. *J. Res. Punjab Agric Univ*. **32:** 240.

Hussain, A., Ahmad, M.N. and Akhtar, A.S. 1990. Protect chillies crop from *Phytophthora*-a new disease in Pakistan. *Capsicum Newsl.* **8–9:** 59.

Hwang, B.K. and Kim, C.H. 1995. Phytophthora blight of pepper and its control in Korea. *Plant Dis*. **79:** 221–227.

Hwang, B.K. and Kim, Y.J. 1990. Capsidiol production in pepper plants associated with age-related resistance to *Phytophthora capsici. Korean J. Pl. Pathol.* **6(2)**: 193–200.

Jee, H.J. and Kim, C.H. 1988. Studies on biological control of *Phytophthora* blight of red pepper, isolation of antagonists and evaluation of antagonistic activity *in vitro* and in green house. *Korean J. Pl. Pathol*. **4(4)**: 305–312.

Jeyalakshmi, C. and Seetharaman, K. 1999. Studies on the variability of isolates of *Colletotrichum capsici* (Syd.) Butler and Bisby causing chilli fruit rot. *Crop Res. (Hisar)* **17(1):** 94–99.

Jeyalakshmi, C., Durairaj, P., Seetharaman, K. and Sivaprakasam, K. 1998. Biocontrol of fruit rot and dieback of chilli using antagonistic microorganisms. *Indian Phytopath*. **51(2):** 180–183.

Jia, J.S. 1992. On chilly pepper *Phytophthora* blight and its control in Xinjiang. *Acta Phytopathol. Sin*. **22:** 257–263.

Jimnez, J.M, Bustamante, E., Bermudez, W. and Gamboa, A. 1988. Response of four cultivars of sweet pepper to bacterial wilt in Costa Rica. *Manejo Tantegrado de Plagas* **7:** 19–28.

Jindal, K.K., Gupta, S.K. and Shyam, K.R. 1994. Studies on the germination, vigour and microflora of bell pepper seeds from healthy and diseased fruits. *Indian J. Mycol. Pl. Pathol.* **24(3):** 227–228.

Jyothi, A.R., Koshy, A., Mathew, J. and Abraham, J. 1993. Bacterial wilt of chillies in Kerala and reaction of certain chilli accession to *Pseudomonas solanacearum* Smith. *Bacterial Wilt Newsl.* **6:** 5.

Kamal and Singh, R.P. 1978. Fungi of Gorkhpur VI. Some new host records from India. *Acta Botanica Indica* **6(Suppl.):** 188–189.

Kapoor, P. 2012. Detection of cucumber mosaic virus in bell pepper crop from Himachal Pradesh. *Pl. Dis. Res.* **27 (1):** 71–80.

Kapoor, Prpmil. 2012. Effect of polythene mulches and barrier crops on the virus disease incidence and yield of bell pepper. *Indian Phytopath*. **65(4):** 391–394.

Kaur, Sandeep, Rattan, G.S., Kaur, Jaspal and Pathak, D. 2016. Morphological and molecular variability among different isolates of *Colletotrichum capsici. Pl. Dis. Res.* **31(1):** 29–34.

Kawagoe, H. 1997. Formation and dispersal of *Cercospora capsici* spores on sweet pepper leaves in aplastic house. *Ann. Phytopathol. Soc. Jpn.* **63(3):** 188–192.

Kawagoe, H. 1998. Influence of temperature and relative humidity on disease occurrence of sweet pepper caused by *Cercospora capsici* in a plastic house. *Ann. Phytopathol. Soc. Jpn.* **64(2):** 137–138.

Kelman, A. 1953. The bacterial wilt caused by Pseudomonas solanacearum. *North Carolina Agric. Exp. Stn. Tech. Bull.* **99:** 194pp.

Khodke, S.W. and Gahukar, K.B. 1995. Changes in biochemical constituents of chilli fruits infected with *Colletotrichum gloeosporiodes* Penz. *J. Maharashtra Agri. Univ.* **20:** 142–143.

Kim, B.S. and Hartman, R.W. 1985. Inheritance of a gene (Bs_3) conferring hypersensitive resistance to *Xanthomonas campestris* pv. *vesicatoria* in pepper (*Capsicum annuum*). *Plant Dis.* **69:** 233–235.

Kim, C.H. 1989. Soil borne diseases: *Phytophthora* blight and other diseases of red pepper in Korea. *Extension Bull. -ASPAC, Food Ferti. Tech. Centre* **302:** 10–17.

Kim, K.Y., Park, S.K., Shin, Y.A. and Lee, E.J. 1990. Survey on the actual condition of cultural methods and continuous cultivation injury of red pepper in field. *Res. Rep. Rural Dev. Adm., Hortic.* **32(1):** 1–10.

Kimble, K.A. and Grogan, R.G. 1960. Resistance to *Phytophthora* root rot of pepper. *Pl. Dis. Reptr.* **44:** 872–873.

Kishan, S., Suryanarayana, M.A., Aṭtri, B.L. and Swaroop, K. 2001. Evaluation of AVRDC chillies for yield and bacterial wilt resistance under Andeman conditions. *Hortic. J.* **14(3):** 56–60.

Kogan, E.D. 1977. The pathogen of anthracnose of fruits of solanaceous vegetable crops in Maldavia. *Referatavnyi Zhurnal* **8:** 1317.

Kohli, U.K. 1997. Breeding for resistance against *Phytophthora* blight of bell pepper. In: *Breeding for resistance to biotic and abiotic stresses in vegetable crops*, CAS Horticulture (Vegetables), Dr. Y.S. Parmar University of Horticulture and Forestry, Solan, pp. 3–8.

Kolte, S.O. and Sapkal, K.N. 1994. Variation in *Colletotrichum capsici* isolates causing fruit rot and dieback of chilli (*Capsicum annuum*). *J. Soil Crops* **4(1):** 88.

Kolte, S.O., Korekar, V.B. and Lokhande, N.M. 1994. Biocontrol of dieback and fruit rot of chillies (*Capsicum annuum*) caused by *Colletotrichum capsici* under field conditions. *Orissa J. Agril. Res.* **7(3/4):** 80–84.

Kumar, K.P. and Singh, R.P. 1979. New host records of *Colletotrichum* species. *Indian Phytopath.* **32(3):** 461–462.

Kumar, R., Rai, N. and Lakpale, N. 1999. Field reaction of some chilli genotypes for leaf curl virus in Chhatisgarh region of India. *Orissa J. Hortic.* **27:** 100–102.

Kumar, S. and Mahmood, M. 1984. Cultural and nutritional variabilities among five isolates of *Colletotrichum capsici. Sci & Cult.* **50(10):** 291–294.

Kumar, S. and Mahmood, M. 1986. Evaluation of fungicides against *Colletotrichum capsici,* the incitant of dieback and fruit rot of chilli. *Pesticides* **20(4):** 28–29.

Kumaran, R.S. Gomathi, V. and Kannabiran, B. 2003. Fungitoxic effects of root extracts of certain plant species on *Colletotrichum capsici* causing antracnose in *Capsicum annuum. Indian Phytopath.* **56:** 114–116.

Kumawat, G.L. 1997. Field evaluation of fungicides for control of anthracnose of chilli, *Capsicum annuum* Linn. *Indian Cacao, Arecanut and Spices J.* **21(3):** 71–73.

Lee, H.U., Kim, C.D., Lee, Y.S. and Cho, C.T. 1993. Effects of soil chemical properties and soil microflora by anti plants on *Phytophthora* blight incidence of red pepper. *RDAJ Agril. Sci., Soil Ferti.* **35(2):** 305–310.

Lee, H.U., Kim, C.D., Lee, Y.S. and Yoo, I.D. 1990. Antifungal antibiotic against fruit rot disease of red pepper from *Streptomyces parvulus. Korean J. App.Mycol. Biotech.* **18(2):** 142–147.

Leonian, L.H. 1922. Stem and fruit blight of pepper caused by *Phytophthora capsici* sp. nov. *Phytopathology* **12:** 401–408.

Lewis, I.M.L. and Miller, S.A. 2003.Evaluation of fungicides and biocontrol agents for the control of anthracnose on green pepper fruit. *Nematicide Test Rep.* **58:** 62–63.

Liang, Y., Lu, J.D. and Ji, R.Z. 1992. Study on the occurrence and yield loss of pepper due to *Phytophthora* blight in Shaanxi. *Plant Prot.* **1:** 14–15.

Lopez-Moya, J.J. and Garcia, J.A. 1999. Potyviruses (Potyviridae). In: *Encyclopedia of Virology*" (Eds. A. Granoff and R.G. Webster). San Diego, Academic press pp. 1369–1375.

Malenson, G. and Delecluse, R. 1957. Pathogenic fungi observed in Morroco. *Bull. Soc. Sci. Natl. Morocco* **17(2):** 132–144.

Mandge, A.S., Datar, V.V. and Sontakke, M.B. 1992. Chemical management of bacterial leaf spots and thrips of chilli. *J. Maharashtra Agril. Univ.* **17(2):** 280–281.

Marrero, T.A. and Gonzalez, B.J. 1985. Response of sweet pepper varieties to *Xanthomonas vesicatoria. Cienc. Agric.* **23:** 14–19.

Matheron, M.E. and Matejka, J.C. 1995. Comparative studies of sodium tetrathiocarbonate and metalaxyl on *Phytophthora capsici* and root and crown rot on chilli pepper. *Plant Dis.* **79:** 56–59.

Mathew, S.K. Girija, D., Devi, S.N., Sadhankumar, P.G. Rajan, S. 2000. Variability in isolates of *Ralstonia solanacearum* affecting solanaceous vegetables in Kerala. *Veg. Sci.* **27(2):** 189–191.

Matsuoka, K., Casali, V. W. D. and Saraiva, T.R. C.B. 1984. Sources of resistance to *Phytophthora capsici* in *Capsicum annuum. Fitopatol. Brasileira.* **9:** 193–201.

Matsuoka, K., Vanetti, C.A., Costa, H. and Pinto, C.M.F. 1996. Diseases caused by fungi in Capsicum and pepper. *Informe Agropecuario (Belo Horizonte)* **18(184):** 64–66.

McInnes, T.B., Gitaitis, R.D., McCarter, S.M., Jaworski, C.A. and Phatak, S.C. 1988. Airborne dispersal of bacteria in tomato and pepper transplant fields. *Plant Dis.* **72:** 575–579.

McIntyre, J.C. and Lukens, R.J. 1977. Screening chemicals to control *Phytophthora parasitica* var. *nicotianae*. *Pl. Dis. Reptr.* **61:** 366–370.

Meah, M.B. and Mian, M.A.W. 1981. Pathogenicity of *Choanephora cucurbitarum* on chilli and possibility of its chemical control. *Acta Phytopathol. Acad. Sci. Hung.* **16:** 1–7.

Meon, S. 1990. Infection of chilli by *Cercospora capsici. Pertanika* **13(3):** 321–325.

Minsavage, G.V., Dahlbeck, D, Whalen, M.C., Kearney, B., Bonas, U., Staskawicz, B. and Stall, R.E. 1990. Gene-for-gene relationships specifying disease resistance to *Xanthomonas campestris* pv. *vesicatoria*–pepper interactions. *Mol. Pl.-Microbe Interact.* **3:** 41–47.

Mirdha, M.A.U. and Chowdhury, M.A.H. 1990. Efficacy of some selected fungicides against seed borne infections of chilli fruit rot. *Seed Res.* **18(1):** 98–99.

Mirdha, M.A.U and Siddiqui, A.B.M. 1989. Fruit rot disease of chilli in relation to seed infection. *Seed Res.* **17(2):** 174–177.

Mishra, M.D., Raychaudhuri, S.P. and Jha, A. 1963. Virus causing leaf curl of chilli (*Capsicum annuum*). *Indian J. Microbiol.* **3:** 73–76.

Mohan, C., Bhawna and Thind, T.S. 2006. Morphological and pathological variations in isolates of *Colletotrichum capsici* causing fruit rot and die back of chillies. *Pl. Dis. Res.* **21 (2):** 101–105.

Mohan, C., Thind, T.S., Raj, P. and Arora, J.K. 2004. Promising activity of triazoles and other fungicides against fruit rot of chilli and Stemphyllium blight of onion. *Pl. Dis. Res.* **19(2):** 200–204.

Muneem, K.C., Verma, S.K. and Pant, K.C. 1995. Performance of some chillies against leaf spot, powdery mildew and fruit rot in Kumaon hills. *Indian Phytopath.* **48(2):** 206.

Muniyappa, V. and Veeresh, G.K. 1984. Plant virus diseases transmitted by whiteflies in Karnataka. *Proc. Indian Acad. Sci. (Anim. Sci.)* **93:** 397–406.

Naik, M.K., Angadi, H.D., Devika Rani, G.S. and Patil, M.G. 2008. Evaluation of chilli (*Capsicum annuum* L.) genotypes, fungicides and plant products for management of anthracnose disease. *J. Mycopathol. Res.* **46:** 213–219.

Nema, K.G. and Agarwal, G.P. 1960. Fungi causing plant diseases at Jabalpur M.P. IV. *Proc. Natl. Acad. Sci., India.* **30:** 55–68.

Niu, E., Shuquin, Q., Niu, E.Z. and Qu, E.Z. 1994. New released pungent *Capsicum* pepper cultivar Tian Jiao 2. *China Veg.* **6:** 6–7.

Ortega, R.G., Espanol, C.P. and Zueco, J.C. 1995. Interactions in the pepper-*Phytophthora capsici* system. *Plant. Breed.* **114:** 74–77.

Ortega, R.G., Palazon, E.C. and Cuartero, Z.T. 1992. Genetic relationships among four pepper genotypes resistant to *Phytophthora capsici. Plant Breed.* **108(2):** 118–125.

Padaganur, G.M. and Naik, K.S. 1991. Mycoflora of chilli seeds from fruit rot affected and healthy fruit. *Curr. Res. Univ. Agric. Sci., Bangalore* **23(9):** 183–184.

Palloix, A., Daubeze, A.M., Phaly, T. and Pochard, E. 1990. Breeding transgressibe lines of pepper for resistance to *Phytophthora capsici* in a recurrent selection system. *Euphytica* **51(2):** 141–150.

Panja, B.N. 1999. Relative efficacy of different fungicides *in vitro* and in field for controlling twig blight disease of chilli caused by *Choanephora cucurbitarum* (Berk and Rav.) Thaxter economically. *Pl. Prot. Bull., Faridabad* **51:** 17–19.

Papavizas, C.C. and Bower, J.H. 1981. Comparative fungitoxicity of captafol and metalaxyl to *Phytophthora capsici. Phytopathology* **71:** 123–128.

Patel, M.K., Kulkarni, Y.S. and Dhande, G.W. 1950. Bacterial leaf spot of chillies. *Indian Phytopath.* **3:** 95–97.

Patil, M.G. and Naik, M.K. 2000. Identification resistance cultivar against anthracnose of chilli. *Package of Practices*, University of Agricultural Sciences, Dharwad.

Perane, R.R. and Joi, M.B. 1986. Reaction of chilli cultivars to fruit rot and dieback of chilli incited by *Colletotrichum capsici*. *Curr. Res. Reptr.* **2(1):** 52–54.

Perane, R.R. and Joi, M.B. 1988. Studies on seed borne infection of fruit rot and dieback of chillies. *J. Maharashtra Agric. Univ.* **13(2):** 231–232.

Peterson, G.H. 1963. Survival of *Xanthomonas vesicatoria* in soil and diseased tomato plants. *Phytopathology* **53:** 765–767.

Pohronezny, K., Hewitt, M. and Infante, J. 1993. Susceptibility of pepper fruit to infection by *Xanthomonas campestris* pv. *vesicatoria*. *Can. J. Pl. Pathol.* **15:** 107–109.

Prabhavathy, K.G. and Reddy, S.R. 1995. Post harvest fungal diseases of chilli (*Capsicum annuum*) from Andhra Pradesh. *Indian Phytopath.* **48(4):** 492.

Prakasam, V. and Jeyarajan, R. 1989. Role of surface wax of chilli fruits in inducing resistance to fruit rot pathogens *Colletotrichum capsici* and *Helminthosporium rostratum*. *Curr. Sci.* **58(17):** 976–977.

Prasad, B.C.N., Reddy, K.M., and Sadashiva, A.T. 2001. Development of F1 hybrids with resistance to Cucumber mosaic virus (CMV) in chilli (*Capsicum annuum* L.). *Capsicum Eggplant Newsl.* **20:** 78–81.

Prashad, D. and Gupta, S.K. 2012. Role of epidemiological factors on the development of bacterial spot (*Xanthomonas campestris* pv. *vesicatoria*) in bell pepper. *Indian Phytopath.* **65(4):** 412–414.

Prashad, D., Gupta, S.K. and Shukla, Arti 2011. Management of bacterial spot of bell pepper (*Capsicum annuum*) through chemicals and antibiotic. *Indian Phytopath.* **64(2):** 204–205.

Puri, A., Agarwal, A. and Mehrotra, R.S. 1994. Antagonism of *Trichoderma* against fungal pathogens and their enzyme production. *Adv. Pl. Sci.* **7(1):** 147–153.

Quinette, D.G. and Coffey, M.D. 1989. Comparative antifungal activity of four phosphonate compounds against isolates of nine *Phytophthora* species. *Phytopathology* **79:** 761–767.

Ragupathi, N. and Veeraragavathatham, D. 2002. Management of chilli mosaic virus disease using insecticides, botanicals and barrier crops. *South Indian Hortic.* **50:** 273–275.

Rahim, A. Na Samraj, 1974. Comparative resistance to certain varieties of chillies to the bacterial wilt caused by *Pseudomonas solanacearum* Smith. *Agric. Res. J. Kerala* **12:** 105.

Rahman, M.A. and Abdullah, H. 1997. Susceptibility of Capsicum species and cultivars to *Ralstonia solanacearum*: anatomical differences and bacterial multiplication in resistant and susceptible cultivars. *Pertanika J. Trop. Agril Sci.* **20(1):** 1–11.

Rahman, M.A., Abdullah, H. and Vanhaecke, M. 1999. Histopathology of susceptible and resistant *Capsicum annuum* cultivars infected with *Ralstonia solanacearum*. *J. Phytopathol.* **147(3):** 129–140.

Raj, T.S., Christopher, D.J. and Rani, U. 2008. Effect of plant products on the *in vitro* production of cellulolytic and pectinolytic enzymes by*Colletotrichum capsici*. *Pl. Dis. Res.* **23 (1):** 24–27.

Raju, K.S. and Rao, G.S. 1989. Studies on the bioefficacy of a new systemic fungicide fenapanil against chilli fruit rot. *Madras Agril. J.* **66(12):** 21–22.

Rangaswami, G. Seshadri, V.S. and Lucy Channamma, K.A. 1970. *Fungi of South India*. Univ. Agril. Sci., Bangalore, 193pp.

Ramyashreedevi, G.S., Singh, D., Srivastava, Arpita, Biswas, K.S. and Gupta, A.K. 2017. Characterization of *Xanthomonas* species causing bacterial leaf spot disease of pepper (*Capsicum annuum*) in India. *Indian J. Agric. Sci.* **87(12):** 1679–1686.

Rani, S.U., Kumar, R.U. and Pratheema, K. 2008. Effect of bioagents against *Colletotrichum capsici* on seed quality of chilli. *Pl. Dis. Res.* **23 (1):** 16–19.

Rathore, B.S. 2004. Performance of some new fungicides in controlling dieback and fruit rot of red pepper. *Pl. Dis. Res.* **19(2):** 196–199.

Rattink, H. 1981. Characteristics and pathogenecity of six isolates of *Phytophthora capsici* from pot plants. *Neth. J. Pl. Pathol.* **87:** 83–80.

Reifschneider, F.J.B., Boiteux, L.S., Vechia, della, P.T., Poulos, J.M., Kuroda, N. and Della Vechia, P.T. 1992. Inheritance of adult plant resistance to *Phytophthora capsici* in pepper. *Euphytica* **62(1):** 45–49.

Riley, M.K. and Bosland, P.W. 1995. Additional sources of resistance to *Verticillium* wilt and *Phytophthora* root rot of *Capsicum. Capsicum Eggplant Newsl.* **14:** 65–67.

Roy, A., Bordoloi, D.K. and Paul, S.R. 1997. Screening of resistant genotypes of red pepper (*Capsicum annuum* L.) against leaf curl infection. *PKV Res. J.* **21:** 175–176.

Roy, K.W., Killebrew, J.F. and Ratnayake, S. 1997. First report of *Colletotrichum capsici* on bell pepper in Mississippi. *Plant Dis.* **81(6):** 693.

Russel, T.A. 1934. Report of the Plant Pathologist. *Rep. Agric. Bermuda, 1934*, pp. 24–32.

Saborio, M.M. and Pacheco, L.H. 1993. Reactions of progenitors and hybrids of capsicum to artificial inoculation with *Phytophthora capsici* under field conditions. *Manejo Integradode Plages* **30:** 11–14.

Saha, L.R. and Singh, H.B. 1988. Diseases of chilli and their management. *Int. J. Trop. Pl. Dis.* **6:** 135–143.

Saini, S.S. and Sharma, P.P. 1978. Inheritance of resistance to fruit rot and induction of resistance in bell pepper. *Euphytica* **27:** 721–723.

Samaddar, K.R., Chakraborty, M. and Kanjilal, S. 1998. Identification of the races of *Pseudomonas solanacearum* causing wilt of solanaceous vegetables in West Bengal and its survival. *J. Mycol. Res.* **36:** 51–58.

Sangar, R.B.S., Katwale, T.R., Saraf, R.K. and Parihar, M.S. 1988. Field screening of chilli varieties to viral diseases in Madhya Pradesh. *Farm Sci J.* **3:** 69–71.

Sankaralingam, A. and Suthanthirapandian, I.R. 1983. Reaction of chilli genotypes to fruit rot. *Proc. Nat. Semi. Prod. Tech. Tomato and Chillies,* March 22–23.

Schifriss, C., Pilowsky, M. and Zacks, J.M. 1992. Resistance to *Leveillula taurica* mildew in *Capsicum annuum. Phytoparasitica* **20:** 279–283.

Sharma, P.N. and Katoch, A. 2014. Population structure of *Colletotrichum gloeosporioides* associated with capsicum anthracnose in north western Himalayas and evaluation of disease resistance. *J. Mycol. Plant Pathol.* **44(4):** 412–416.

Sharma, P.N., Chowfla, S.C. and Andotra, P.S. 1991. Epidemiology of mosaic disease complex of bell pepper in Himachal Pradesh. *Indian Phytopath.* **44:** 112–114.

Sharma, P.P. and Saini, S.S. 1979. Inheritance of resistance to *Phytophthora capsici* Leon. in *Capsicum annuum* L. *Haryana J. Hortic. Sci.* **8:** 51–57.

Sharma, S.K. 1998. Efficacy of different fungitoxicants against frogeye leaf spot of bell pepper (*Capsicum annuum* L.). *Pl. Dis. Res.* **13(1):** 62–63.

Sharma, S.L. and Bhardwaj, S.S. 1976. A fruit rot and leaf blight of bell pepper caused by *Phytophthora nicotianae* var. *nicotianae. Indian Phytopath.* 29: 93–94.

Sharma, S.R. and Sohi, H.S. 1981. Studies on non-chemical control of *Cercospora* leaf spot of chilli. *Indian J. Mycol. Pl. Pathol.* **11(1):** 30–34.

Sharma, S.R. and Sohi, H.S. 1984. Control of *Cercospora* leaf spot of chillies with Bavistin. *Pesticides* **18(4):** 21–23.

Shekhawat, P.S. and Chakravarti, B.P. 1977a. Assessment of loss, symptomatology and occurrence of bacterial leaf spot of chillies caused by *Xanthomonas vesicatoria* in Rajasthan. *Indian J.Mycol. Pl. Pathol.* **7:** 11–14.

Shekhawat, P.S. and Chakravarti, B.P. 1977b. Factors affecting development of bacterial leaf spot of chillies caused by *Xanthomonas vesicatoria. Indian Phytopath.* **29:** 392–397.

Sherbakoff, C.D. 1918. Report of the Associate Plant Pathologist. *Fla. Agri. Exp. Stn. Rep. 1916–1917*: 66–86.

Shivanathan, P. 1982. Epidemiology of chilli leaf curl disease in Sri Lanka. *Trop. Hortic.* **138:** 99–109.

Shukla, R.S., Sattar, A., Kumar, S., Singh, K.P. and Hussain, A. 1981. *Phytophthora* leaf blight of *Hibiscus moschatus* – new record from India. *Indian Phytopath*. **34:** 539.

Shyam, K.R. 1969. The fruit rot of bell pepper (*Capsicum annuum* L.) caused by *Phytophthora capsici* Leon. M.Sc. Thesis, Punjab University, Chandigarh, 80pp.

Simons, J.N. 1987. Observations on double cropping bell pepper and control of bacterial leaf spot in Palm Beach County. *Proc. Fla. State Hortic. Soc.* **100:** 226–232.

Singh, H. Jarnail, S. and Kaur, S. 1990. Resistant sources to fruit rot *Colletotrichum capsici* in pepper (*Capsicum annuum* L.). *J. Res. Punjab Agric. Univ*. **27(3):** 419–420.

Singh, J. and Kaur, S. 1990. Development of multiple resistance in chilli pepper. *Proc. 3rd Int. Conf. 20–23, March, Kualalumpur, Malaysia*.

Singh, J.S. and Thind, T.S. 1980. Search for anthracnose and die-back resistant varieties of chilli (*Capsicum* spp.) in Punjab. *The Punjab Veg.Grow.* **XV-XVI:** 37–39.

Singh, R.S. 1987. *Diseases of vegetable crops*. Oxford & IBH Publishing Co. Pvt. Ltd., New Delhi, 362pp.

Singh, S.A. and Thakur, D.P. 1979. Reaction of chilli (*Capsicum fruitesencs*) varieties to *Colletotrichum capsici* (Syd.) Butler & Bisby. *Curr. Sci.* **48(11)**: 512–513.

Singh, S.J., Sastry, K.S. and Sastry, K.S.M. 1979. Efficacy of different insecticides and oil in the control of leaf curl disease of chillies. *J. Pl. Dis. Prot*. **86:** 253–256.

Singh, S.S. and Sharma, R.K. 1999. Control trial on chilli mosaic virus after inhibiting their infectivity by leaf juices of some angiospermic plants. *J. Living World* **6:** 18–21.

Singh, U.C., Reeti Singh, Nagaich, K.N. and Singh, R. 1998. Reaction of promising chilli varieties against major insect pests and leaf curl disease. *Indian J. Entomol*. **60:** 181–183.

Smith, P.G. and Heiser, C.B. 1957. Breeding behaviour of cultivated peppers. *Proc. Am. Soc. Hortic. Sci*. **10:** 286–290.

Sohi, H.S. and Shyam, K.R. 1976. Fruit rot of bell pepper caused by *Phytophthora capsici* Leonian. *Indian J. Hortic.* **33 (3&4):** 266–273.

Sohi, H.S., Shyam, K.R. and Sharma, S.L. 1971. Fruit rot of *Capsicum annuum* caused by *Phytophthora capsici* Leon. *Int. Symp. Pl. Pathol., Banggalore* pp. 31 (Abstr.).

Sohi, H.S., Shyam, K.R. and Sharma, S.L. 1973. Chemical control of fruit rot of bell pepper caused by *Phytophthora capsici* Leon. *Pesticides* **7:** 30–32.

Stall, R.E. and Lozano, J.C. 1982. Selection for component of horizontal resistance to bacterial spot of pepper. *Proc. Fifth Int. Conf. Pl. Pathogenic Bacteria*. Pp. 511–517.

Stall, R.E., Loschke, D.D. and Jones, J.B. 1986. Linkage of copper resistance and avirulence loci on a self-transmissible plasmid in *Xanthomonas campestris* pv. *vesicatoria. Phytopathology* **76:** 240–243.

Suang, J.M., Cho, D.J. and Kand, S.W. 1984. Undescribed fungal leaf spot disease of pepper caused by *Cercospora capsici* in Korea. *Korean J. Mycol*. **12(2):** 75–77.

Sugiura, M., Bandaranayake, C.M. and Hemashandra, G.H. 1975. Chilli virus diseases in Sri Lanka. *Trop. Agric. Res. Cent. Ministry Agric. Forestry, Jpn. Tech. Bull.* **8:** 62.

Sunaina, V. and Tomar, D.K. 2003. Distribution of bacterial wilt of potato and its races/ biovars in Uttaranchal hills. *Bionotes* **5:** 13–14.

Sydow, H. and McRae, W. 1929. Hyphomyces Indiae Orientali. *Ann. Crypt. Exot*. **2:** 262–271.

Tamietti, G., Nervo, G. and Restaino, F. 1992. Screening of Capsicum genotypes for resistance to *Phytophthora capsici* Leon. and *Verticillium dahliae* Kleb. *Capsicum Newsletter, Spl. Issue on EUCARPIA meeting, Rome*, 194–195.

Thakur, M.C., Gupta, S.K., Kansal, S. and Vikram, A. 2009. Developing superior varieties/ hybrids of bell pepper resistant to fruit rots and their epidemiology and management. Final Progress Report of the ICAR Adhoc research project, YSPUHF, Solan, 113pp.

Thakur, M.R., Singh, J., Singh, H. and Sooch, B.S. 1987. Punjab Lal -a new multiple disease resistant chilli variety. *Prog. Fmg.* **23(9):** 11–16.

Thakur, P.D., Chowfla, S.C. and Sharma, O.P. 1985. Reaction of pepper varieties to bell pepper leaf distortion mosaic caused by potato virus Y (PVY$^{o\text{-}stop}$ strain). *Veg. Sci.* **12:** 120–122.

Thayer, P.L. and Stall, R.E. 1961. A survey of *Xanthomonas vesicatoria* resistance to Streptomycin. *Proc. Fla. State Hortic. Soc.* **75:** 163–165.

Thind, T.S. and Jhooty, J.S. 1985. Relative prevalence of fungal diseases of chilli fruits in Punjab. *Indian J. Mycol. Pl. Pathol.* **15:** 305–307.

Thind, T.S. and Jhooty, J.S. 1987. Relative performance of some fungicides in controlling anthracnose and black rot of chillies. *Indian Phytopath.* **40:** 543–545.

Thind, T.S. and Jhooty, J.S. 1990. Studies on variability in two *Colletotrichum* species causing anthracnose and fruit rot of chillies in Punjab. *Indian Phytopath.* **43(1):** 53–58.

Thirupathaiah, V. and Subramanian, D. 1979. Leakage of electrolytes in chilli fruits infected with *Colletotrichum capsici. Proc. Indian Acad. Sci., B.* **88(II,2):** 109–113.

Tucker, C.M. 1928. Report of Plant Pathologists. *P.R. Agric. Exp. Stn. Rep.* pp. 25–27.

Turhan, G. 1981. A new race of *Streptomyces ochraceiscleroticus* in the biological control of some soil borne plant pathogens: *in vitro* studies on the possibilities of using C/2–9 against some important diseases. *Z. Pflanzenkr. Pflanzenschutz* **88(7):** 422–434.

Ullasa, B.A., Rawal, R.D., Sohi, H.S., Singh, D.P. and Joshi, M.C. 1981. Reaction of sweet pepper genotypes to anthracnose, Cercospora leaf spot and powdery mildew. *Plant Dis.* **65(7):** 600–601.

Ustun, A.S. 1993. Effect of inoculum concentrations on root rot (*Phytophthora capsici* Leon.) in peppers. *Doga, Turk Tarim ve Ormancilik Dergisi.* **17(3):** 683–693.

Valand, G.B. and Muniyappa, V. 1992. Epidemiology of tobacco leaf curl virus in India. *Ann. App. Biol.* **120:** 257–267.

Valdez, J.A. and Wolfenberger, D.A. 1995. Yellow traps and insecticides for control of a strain of sweet potato whitefly and associated virus incidence on pepper. *J. Ento. Sci.* **30:** 342–348.

van Elsas, J.D., Kastelein, P., van Bekkum, P., van der Wolf, J. M., de Vries, P.M. and van, Overbeek, L.S. 2000. Survival of *Ralstonia solanacearum* biovar 2, the causative agent of potato brown rot, in field and microcosm soils in temperate climates. *Phytopathology* **90:** 1358–1366.

Vasudeva, R.S. and Samraj, J. 1948. A leaf curl disease of tomato. *Phytopathology* **38:** 364–369.

Venkatesh, H.M., Muniyappa, V., Ravi, K.S., Krishnaprasad, P.R., Reddy, P.P., Kumar, N.K.K. and Verghese, A. 1998. Advances in IPM in horticultural crops. *Proceedings of First National Symposium on Pest Mangement in Horticultural Crops: environmental implications and thrusts* (ed. P.P. Reddy and N.K.K. Kumar), Bangalore, 15–17 Oct., 1997.

Verma, S., Shyam, K.R. and Gupta, S.K. 2006. Fungicidal management of Phytophthora leaf blight and fruit rot of bell pepper (*Capsicum annuum* L.). *Pl. Dis. Res.* **21 (2):** 128–131.

Verma, S., Shyam, K.R. Gupta, S.K. and Sharma, H.R. 2001. Evaluation of bell pepper (*Capsicum annuum*) germplasm for resistance against leaf blight and fruit rot (*Phytophthora nicotianae* var. *nicotianae*). *Indian J. Agric. Sci.* **71(3):** 219–221.

Yamakawa, K., Mochizuk, T. and Yasui, H. 1979. Screening cultivated and wild red peppers for *Phytophthora capsici* resistance and its inheritance. *Bull. Veg. Ornamental Crop Res. Sta.* **6:** 29–37.

Yang, S.S., Kim, C.H., Cho, E.K. and Lee, E.J. 1991. Distribution and characteristics of suppressive soil to *Phytophthora* blight of red pepper in Korea. *Res. Rep. Rural Dev. Adm., Crop Prot.* **33(3):** 18–22.

Zapata, M. 1995. Bacterial spot on pepper caused *Xanthomonas campestris* pv. *vesicatoria* race 2, from south east of Puerto Rico. *J. Agric. Univ. Puerto Rico* **79:** 1–2.

Zhang, P.R., Chen, H.S. and Gu, H.F. 1992. On the control effects of fourteen fungicides against bell pepper Phytophthora rot fungus. *Plant Prot.* **18:** 25–26.

■■■■

10

DISEASES OF ONION AND GARLIC

Onion (*Allium cepa* L.) and garlic (*Allium sativum* L.) are two most important bulbous vegetable crops cultivated in many countries of the world. In India, it is grown in an area of 0.29 million ha with annual production of 3.14 million tones production of bulbs for local consumption as well as for export. The main onion growing states are Maharashtra, Gujarat, Tamil Nadu, Orissa, Madhya Pradesh, Uttar Pradesh, Haryana, Punjab and Andhra Pradesh. Among these, Maharashtra is leading with about 26 per cent of country's production (Pathak, 1994). The onion and garlic crops are susceptible to several diseases of fungal, bacterial and viral nature, which cause substantial production losses. A detailed review of various diseases aff-ecting these crops alongwith management strategies is given in this chapter.

(A) FUNGAL DISEASES

1. DOWNY MILDEW

The disease causes widespread damage on the worldwide basis wherever this crop is grown in cool and humid conditions.

Geographical distribution

The disease was first reported from England in 1841 by Berkeley (Yarwood, 1943). Subsequently, it was reported from Bermuda Islands, USA, Canada, Venezuella, Brazil, Iraq, Holland, Australia, U.S.S.R., China, Japan, Sweden, Norway, Germany, Israel and Poland (Palti *et al.*, 1958; Jones and Mann, 1963; Rondomanski, 1967).

In India, the disease was first reported from Kashmir valley during 1974–75 and 1975–76 seasons (Mirakhur *et al.*, 1978). Since then it has been reported from Kurukshetra area of Haryana (Gupta *et al.*, 1984) and Himachal Pradesh (Singh *et al.*, 1987).

Losses

The pathogen attacks all types of onions, but is more destructive to common onion (*A. cepa* L.) where it has been reported to cause losses as high as 60- 75 per cent of the crop (Cook, 1932). Severely affected leaves cause retarded development of bulbs and as a result, a large number of "bottle necked" onion bulbs develop (Rondomanski, 1967). In India, the losses from this disease ranges from 12-75 per cent in crop yield depending on the time of disease outbreak and disease severity (Sugha and Singh, 1991).

Symptoms

Symptoms differ with the type of infection. Infection develops from three different sources like systemically infected perennating leaves, plants grown from infected onion bulbs or from local lesions resulting from air borne inoculum. Systemic infection occurs when the plants are raised from diseased bulbs or infected seedlings used for planting. Plants raised from such bulbs remain stunted, become distorted and light green in colour. In humid weather conditions, sporulation develops on the leaves and covers them with felty whitish to grayish fungal growth (Fig. 1). In secondary and local infections, spots with greenish and chlorotic zones appear and at times at the point of infection, the plants fall off. In humid weather the fungus develops as white to purplish downy growth on these spots (Plate 1). In such cases undersized bulbs are produced.

Fig. 1. Symptoms of Downy mildew of Onion

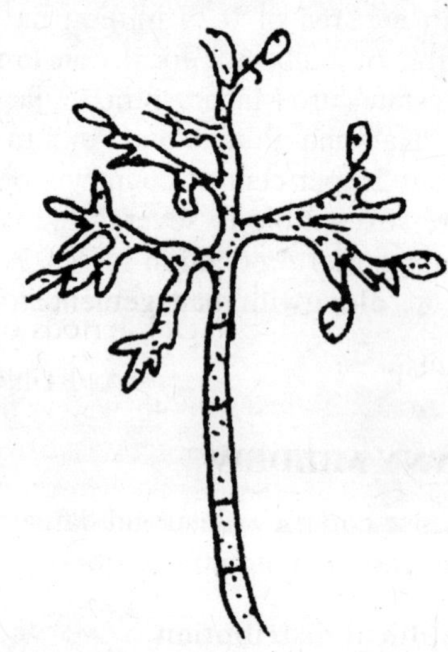

Fig. 2. Sporangiophore and sporangica of *Pseudoperonospora destructor*

The Pathogen

The disease is caused by *Peronospora destructor* (Berk.) Casp. The sporangiophores are aseptate, hyaline swollen at the base, 160–485 x 9–14 μm in diamension and 3 to 4 times dichotomously branched (Fig. 2). The sterigmata are acute to subacute, 9–14 x 2–4 μm and bear pyriform to fusiform sporangia, which measure 36–69 x 18–25 μm in size. Oospores develop late in the season and measure 40–50 μm in diameter. They germinate by germtube.

Host range

Besides common onion, *Allium cepa*, the pathogen also attacks other species of *Allium* like *A. fistulosum*, *A. ascalonicum*, *A. nigrum*, *A. porrum*, *A. ursinum*, *A. oleraceum*, *A. sativum*, *A. cepa* var. *multiplicans*, *A. cepa* var. *bulbifera* and *A. schoenoprasum* (Yarwood, 1943).

Disease cycle

The carry over of the fungus from one season to another may occur as mycelium in the bulbs, seed or soil, as oospores on or in the seed or in plant debris (Yarwood, 1943; Popkova *et al.*, 1980) and systemically infected onion bulbs are another important source of primary infection. But role of oospores has not been established in induction of disease even under controlled conditions (Viranyi, 1981). The onion seedlings from disease prone areas play significant role in introduction of disease in new localities. Secondary infection is caused by wind borne sporangia disseminated from the plants with primary infection.

Epidemiology

The pathogen perpetuates as mycelium in infected plant debris and the systemically infected bulbs. The sporulation and infection by *P. destructor* usually takes place overnight while dissemination of the newly formed conidia occurs during morning hours (Yarwood, 1943; Rondomanski, 1967). The pathogen requires cool and moist nights to initiate infection. The temperature of 13ºC and >95 per cent relative humidity are optimum for the development of the disease. Presence of dew or raindrops on the leaf surface further enhances the chances of infection. The disease is worse in damp conditions (Devlash and Sugha, 1997b). Germination of sporangia occurs at 7–16ºC temperatures, optimum being 10 to 13ºC. The infection cycle is characterized by long latent periods of 9 to 16 days and 1 to 2 day periods of sporulation, dispersal and infection. Closely spaced and densely planted crops developed more disease compared with widely spaced ones (Devlash and Sugha, 1997a), while an increase in level of nitrogen fertilization increased disease severity proportionately and potassium decreased its severity.

Battilani *et al.* (1996) developed a preliminary model, ONIMIL to forecast primary infection, which is able to determine for each day the probability of *P. destructor* establishing an infection on onion and its infectivity level. Recently, MILIONCAST, a model was developed based on the data from the controlled environment studies to predict the rate of sporulation in relation to temperature and relative humidity (Gilles *et al.*, 2004).

Management

Various practical measures, to preclude sources of infection and to prevent spread of the disease, are of great importance in controlling this disease.

i. Cultural practices: Cultural practices like use of healthy bulbs for planting, collection and destruction of infected crop debris (Jorgensen and Nistrup, 1987), cultivation on well drained soils, crop rotation avoiding the *Allium* crops in rotation, application of balanced doses of fertilizers alongwith potassium as the increase in level of potassium decrease disease severity (Devlash and Sugha, 1997a) or avoidance of frequrent irrigations (Ahmad and Karimullah, 1998) help in the reduction of the initial inoculum and spread of the disease in the field. Widely spaced, inter and mixed cropping of onion with Indian mustard (*Brassica juncea*) alone or in combination with coriander (*Coriandrum sativum*) developed less disease compared with pure crops of seed and bulb onions (Devlash and Sugha, 1997a). Soil solarization for six weeks during July–August retarded symptom development of downy mildew for 20 days and increased the bulb yields by >97 per cent over that of shaded plots (El-Shanawany *et al.*, 2003).

ii. Host resistance: White skinned varieties were generally more susceptible to the disease than red and lilac skinned onion varieties (Szurke and Szurke, 1982). Resistance in onion varieties against this disease was correlated to flat leaves and xeromorphic tissue structure (small cell, thick wall, high lignification). Susceptible species were characterized by fistular leaves and swollen flower stalk, cells were larger and thin walled and lignification less marked (Furst, 1976). Talieva (1976) correlated the resistance to chemical nature of cuticular wax and reported those species as resistant which had acid content in the wax. Cvs/lines like IC 48045, IC-32149, IC-33617, IC-49371 and DOP-11 have been reported as good sources of resistance and can be utilized in breeding programme (Sharma, 1997a).

iii. Chemical control: Various systemic and non-systemic fungicides like mancozeb or zineb, copper oxychloride, propineb, metalaxyl + thiram, metalaxyl + mancozeb, oxadixyl + mancozeb have been reported to check onion downy mildew effectively (Mir *et al.*, 1987; Jorgensen and Nistrup, 1987; Khokhar and Jaffrey, 2000; Khalid *et al.*, 2002). The efficacy of different fungicides increased with the addition of adjuvants and Acrobat MZ with Synertrol oil was found effective in both pre- and post infection treatments against this disease (Mac-Manus *et al.*, 2002). Combining fortnightly mancozeb sprays with an alert-based metalaxyl application reduced the total number of fungicide applications by 8 whilst maintaining effective disease control (Wright *et al.*, 2003). Iqbal *et al.* (2009) reported the efficacy of Melody Duo, Ridomil Gold and Curzate M-8 which not only reduced the disease severity by 90 per cent but also increased the yield by > 130 per cent.

2. PURPLE BLOTCH

Purple blotch is another important disease of onion and garlic growing areas of the world. The disease affects both bulb and seed crop through out the country (Dhiman *et al.*, 1986) and it is more serious in *kharif* crop than *rabi* season crop (Quadri *et al.*, 1982).

Geographical distribution

The disease was first recorded from Bombay (Ajrekar, 1921). Since then it has been reported from many states of the country like Maharashtra, Gujarat, Tamilnadu, Rajasthan, Haryana, Uttar Pradesh and Punjab (Pandotra, 1964; Utikar and Pandule, 1981).

Losses

The disease causes 20–25 per cent losses to the onion seed crop (Pandotra, 1964). The disease is further aggravated by earlier damage done to the leaves by thrips (Thind and Jhooty, 1982) and *Botrytis* (Gupta and Srivastava, 1981).

Symptoms

The characteristic symptoms of the disease appear as small, light coloured, sunken lesions, which become zonated with purplish centre. These lesions rapidly enlarge and eventually girdle the leaf or inflorescence stalk (Plate 2). In moist weather the surface of the spot is covered with the brown or almost black sporulation of the fungus. Usually, the affected leaf or stem falls down and dies within 3 or 4 weeks under favourable environmental conditions. In garlic, the disease appears on leaves with the similar symptoms.

The Pathogen

The fungus responsible for this disease is *Alternaria porri* (Ellis) Cif. Conidiophores arise singly or in groups and are straight or flexuous, often geniculate, septate, pale to mild brown in colour measuring up to 120 μm in length and 5–10 μm breadth. Conidia are muriform and are usually solitary, straight or curved, obclavate and taper to a beak that is commonly about the same length or slightly larger than the body of the conidium (Fig. 3). The conidia are mid golden brown in colour, smooth or minutely verucose and measure 100 - 300 μm in length with the broadest part 15–20 μm thick. Each conidium has 8 to 12 transverse and zero to several longitudinal or oblique septa. The beak is flexuous, pale, 2–4 μm thick and tapering. Each cell of the conidiophore is capable of germination by germ tube. Existence of chlamydospores in the fungus is also reported (Basu, 1971). Sexual stage of the fungus is unknown.

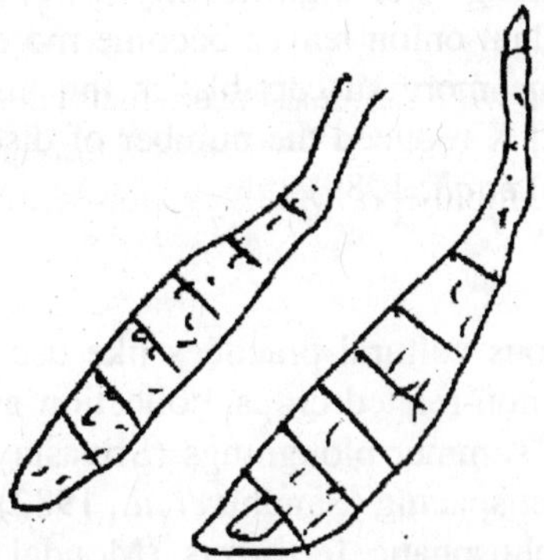

Fig. 3. Conidia of *Alternaria porri*

Host range

Besides onion and garlic, the host range of the pathogen includes Egyptian onion (*Allium proliferum*), Welsh onion (*A. fistulosum*), leek (*A. ampeloprasum*) and also on some other species of *Allium*.

Disease cycle

The fungus survives from one season to other in infected plant debris as dormant mycelium. The fungus can also survive in diseased onion leaves and seed stalk debris for 12 months buried at 5–7.5 cm depth (Gupta and Pathak, 1988). Wherever the chlamydospores of the fungus are formed they can also serve as source of perennation. Invasion takes place either through stomata or directly through the cuticle, to form an intercellular mycelium, which in turn forms conidiophores and conidia. Conidia produced on the primary infection lesions serve as the source of secondary inoculum.

Epidemiology

The main source of perennation is dormant mycelium in the infected plant debris. Persistent dew for extended periods or high relative humidity (>85%) are essential for infection and sporulation (Everts and Lacy, 1990). Length of dew period was significantly positively correlated with lesion numbers, but not with lesion size (Everts and Lacy, 1996). Khare and Nema (1981) recorded maximum spore population at 8.00 A.M. and minimum at 12 noon. It had a bimodal peak with a slight increase at 4.00 P.M. Maximum sporulation took place on days immediately after rains. Sporulation took place in night and the spores discharged at early morning. Temperature of 22°C and RH of 90 per cent were most congenial for disease development. Spore production was maximum in 88 days old plants (395/cm^2 leaf) and minimum in 45 days old plants (120/cm^2 leaf). The conidia were readily dispersed by air currents (Chawda and Rajasab, 1992). Lakra (1999) reported that numerous purple spots/blotches were observed on older leaves and stalks when fortnightly dewfall was >1.0 mm, mean maximum RH >75 per cent and mean maximum temperature 20–30ºC with >18 h favourable temperature (10–30ºC) duration. Exposure of leaf and/or scape to wetness for 8 h was a pre-requisite. Miller (1983) studied the relationship between onion leaf age and susceptibility to *A. porri* at weekly intervals. Leaf damage was significantly lower on younger than on older leaves in each week. Individual onion leaves become more susceptible as they aged and emerging leaves become more susceptible as the bulbs approached maturity. Increased doses of N, P, and K reduced the number of diseased inflorescence stalks and infected umbels (Mondal *et al.*, 1989).

Management

i. Cultural practices: Various cultural practices like use of healthy seed/ planting material, crop rotation with non-related crops, collection and destruction of infected plant debris, good drainage, summer ploughings (Srivastava *et al.*, 1996), alteration of date of sowing, wider plant spacing (Sandhu *et al.*, 1982) and use of recommended doses of nitrogenous and phosphatic fertilizers (Mondal *et al.*, 1989) reduce the disease incidence. Hot water (50°C for 20 min) soak was found to be highly effective in reducing the seed borne inoculum, although the percentage of germination and emergence of onion seeds were slightly reduced (Aveling *et al.*, 1993).

ii. Host resistance: Cvs./lines/hybrids viz., Pusa Red (Raju and Mehta, 1981), hybrid P x 76 (Alves *et al.*, 1982), IIHR-56-1 (Pathak *et al.*, 1986), hybrid PVM-7 (Perez-Moreno and Chavez-Hernandez, 1992), IC 48059, IC 48179, IC 39887, IC 48025, ALR (Sharma, 1997b), Ro-1 (Sharma and Sain, 2003) have been reported as resistant to this disease. Higher phenolic compounds and higher oxidase activity are responsible for the resistance in cv. Taliani Red against this disease (Michail and Salem, 1981).

iii. Chemical control: Many workers have evaluated several fungicides for the control of this disease. Pandotra (1965) found Bordeaux mixture (4:4:50) the best among 11 fungicides tested against *A. porri*. Aujla *et al.* (2013) evaluated the efficacy of nine new fungicides against this disease for their effectiveness under *in vitro* conditions and reported that all the tested fungicides significantly inhibited mycelial growth of the pathogen, Score and Nativo being the most effective as both completely inhibited the mycelial growth at 0.1 per cent. Sprays of other fungicides like copper oxychloride (Suhag and Duhan, 1985), metalaxyl + mancozeb (Georgy *et al.*, 1983; Sugha and Tyagi, 1994) mancozeb (Gupta *et al.*, 1986), captafol (Thind and Jhooty, 1984), fosetyl-al, iprodione (Perez-Moreno and Chavez-Hernandez, 1992), difenoconazole (Sastrahidayat, 1994) were reported effective against this disease. Adding stickers like Triton can increase efficacy of these fungicides (Sastrahidayat, 1994). Thind and Jhooty (1984) reported that five sprays of Dithane M-45 (0.3%) or Difolatan (0.2%) mixed with malathion (0.2%) and Triton (0.2%) gave effective control of this disease on onion seed crop. Spraying of mancozeb with malathion or monocrotophos (0.05%) or cypermethrin (0.01%) was found effective in the management of this disease and thrips attack (Thind and Jhooty, 1984; Memane *et al.*, 2001). Foliar application of a mixture of hexaconazole (0.05%) + mancozeb (0.3%) gave the maximum disease control of 79.58 per cent. (Deshmukh *et al.*, 2007). Among new fungicides evaluated for their efficacy against the disease under pot conditions and Folicur and Tilt were found as the promising alternatives over the conventional fungicides for efficient management of this disease on seed crop (Aujla *et al.*, 2010).

iv. Plant extracts: Plant extracts from *Pithecellobium dulce* and *Prosopis juliflora* leaves and plant oils like *Palmarosa* and neem were evaluated against this disease under field conditions and *Palmrosa* oil (0.1%) sprayed at the first appearance of symptoms and after 15 days reduced the disease significantly (Mohan *et al.*, 2001).

v. Biological control: ***Cladosporium herbarum*** inhibited the mycelial growth of *A. porri* under *in vitro* conditions, while the infection under field conditions was reduced by 66.6 per cent (Tyagi *et al.*, 1990). *Bacillus thermophillus* inhibited the conidial germination and mycelial growth of *A. porri* at 2.2 x 10^9 and 2.2 x 10^8 cfu/ml, respectively, under *in vitro* conditions, however, its population declined after 7 days of spray on onion leaves (Mandhare and Suryawanshi, 2003). *B. thermophillus* produced a thermo-labile metabolite toxic to *A. porri* which restricted the fungal growth and invasion in host plants and caused morphological changes in the germ tubes or mycelium.

vi. Integrated disease management: Heat treatment of onion bulbs at 35°C for 8 h before sowing followed by single prophylactic spray of metalaxyl + mancozeb (0.25%) at the bolting stage or no heat treatment and 3 sprays of metalaxyl + mancozeb (0.3%) at 15 day intervals from the appearance of the disease gave the most effective control of the disease, while heat treatment of bulbs at 40 and 45°C reduced crop growth (Sugha *et al.*, 1993). The most effective treatment combination for controlling this disease includes pendimethaline (3.5 l/ha), nitrogen (100 kg/ha) and mancozeb (0.25%) application (Srivastava *et al.*, 1999)

3. STEMPHYLIUM BLIGHT

This is an important disease of these crops and causes significant yield losses under favourable environmental conditions.

Geographical distribution

The disease is prevalent in almost all the onion and garlic growing countries of the world including India. The disease has been reported from United States, South

Africa, Venezuella (Miller *et al.*, 1978; Aveling and Naude, 1992; Cova and Rodriguez, 2003). In India the occurrence of this disease on onion was first reported from Uttar Pradesh (Rao and Pavgi, 1973) while on garlic it was reported from Kullu, Himachal Pradesh during 1973 (Singh and Sharma, 1977). Now the disease is appearing in all the onion and garlic growing areas of the country particularly in northern parts (Thind *et al.*, 1985).

Losses

The disease attains severity during prolonged wet weather and causes significant losses due to the blightening of the foliage. However, exact data on losses due to this disease have not been reported. The disease caused significant damage during 1976 in South Texas (Miller *et al.*, 1978). In Himachal Pradesh, there are reports on occurrence of this disease in epidemic form (Sugha and Kumar, 2005)

Symptoms

The main symptoms of the disease appear in the middle of the leaf as small, yellow to pale orange flecks or streaks which later develop into elongated spindle shaped diffused spots surrounded by characteristic pinkish margin (Plate 3). The spots turn grey at the centre and later become brown to dark brown with the appearance of conidiophores and conidia of the fungus. Gradually the entire foliage is blighted. Similar symptoms also develop on inflorescence stalks.

The Pathogen

The disease is caused by *Stemphylium vesicarium* (Wallr.) E. Simmons (Tel. *Pleospora allii* (Rabenh) Ces. & de Not). The conidia of the fungus are oblong or broadly oval and sometimes inequilateral measuring 12–22 x 25–42 μm in size. They have one to six transverse septa and one to three longitudinal septa. Conidia are constricted at the major transverse septa. Conidia are light or medium golden brown to olive brown, and each has a conspicuous basal scar like zone. Conidiophores are straight to variously curved, 1–4 septate measuring 5–8 x 33–47 μm in size with frequent nodular swellings. Perithecia mature within 3 to 6 months. Asci are cylindrical to clavate shape, which measure 35 x 170 μm at maturity. Young ascospores are ellipsoidal and the upper half is narrowly tapered. Ascospores have seven transverse septa and measure 18 x 38 μm in size.

Disease cycle

The pathogen survives in infected plant debris. In favourable weather, spores from the debris are splashed to the lower leaves and cause infection. Spores produced on the primary infections cause secondary infections.

Epidemiology

The disease attains severity during extended wet periods. The disease is also more severe in irrigated fields (Singh and Sharma, 1977). High humidity (>90%) and hailstorms during spring season also help in the establishment and development of this disease (Gupta and Pandey, 1986). Jakhar *et al.* (1996) reported that moderate temperature and high humidity coupled with precipitation favour disease initiation and development. Kumar and Katoch (2017) studied the effect of leaf wetness on incubation period and disease development and reported that minimum 24 h leaf wetness is necessary for infection to occur. Similarly, temperature ranging from 20–30ºC was found favourable for infection but 25ºC was most congenial for disease development as the symptoms of the disease appeared in 4 days only.

Management

i. Cultural practices: Use of healthy seed/planting material is of utmost importance. Two to three years crop rotation with non-related crops, summer ploughings, field sanitation and destruction of infected plant material are helpful in reducing the initial inoculum level of the pathogen in the field.

ii. Host resistance: Sources of resistance against this disease are very limited. Only one cultivar of garlic i.e. HG 1 had the lowest incidence of this disease and also performed well for yield, number of cloves/bulb and their weight (Pandey *et al.*, 1989) while 40 accessions of onion have been reported as tolerant (Bisht *et al.*, 1990).

iii. Chemical control: Fungicides like copper oxychloride, mancozeb, carbendazim, thiram, propineb, tebuconazole, chlorothalonil and difenoconazole have been reported effective in the management of this disease (Gupta and Srivastava, 1988; Mohan *et al.*, 2004). The most economical and effective control of the disease was reported by 4 sprays of mancozeb (0.25%) alongwith the sticker Triton/Sandovit in the onion bulb crop (Gupta and Srivastava, 1988, 1993; Gupta *et al.*, 1996) and 6 sprays in onion seed crop (Gupta *et al.*, 1989). However, in garlic the disease can be kept under check by 4 sprays of mancozeb (0.25%) or ziram (0.3%) alongwith sticker Triton/Sandovit @ 0.06% (Srivastava *et al.*, 1992). Khosla *et al.* (2007) gave the spray schedule comprising of first spray of Companion @ 0.25% (carbendazim 12% + mancozeb 63%) followed by second spray of Score @ 0.015% (difenoconazole 25 EC) after 15 days and third spray of Indofil M-45 0.25% (mancozeb 75 WP) again after 15 days with 91.6% disease control and cost benefit ratio of 1:3.12. Efficacy of different fungicides in the management of this disease have also been reported by Devlash (2011).

iv. Plant extracts: Prasad and Barnwal (2004) evaluated some plant extracts against this disease both *in vitro* and *in vivo*. Disease severity was minimum when the crop was sprayed with 20 per cent leaf extract of *Azadirachta indica* followed by *Datura metal* and increased the bulb yields.

4. BASAL ROT

Basal rot is an important disease of these crops particularly of onion and causes substantial yield losses by rotting the basal portions of the plants. Besides basal rot, *Fusarium* sp. also cause damping off, wilting, root rot and seed infection.

Geographical distribution

This disease is prevalent in almost all parts of the world where onions are grown (Everts *et al.,* 1985; Lopez and Cramer, 2004). In India, the occurrence of this disease was first reported from Rajasthan (Mathur and Sankhala, 1963).

Losses

Basal rot is a wide spread disease causing enormous losses and a severe outbreak of the disease occurred in Hokkaido, Japan (Kodama, 1983).

Symptoms

The main symptoms appear as wilting and rapid dying back of leaves from the tips of the plants near maturity (Plate 4). Watery decay on cutting open the soft bulbs are evident. Infected garlic shows reddish or reddish purple discolouration on stems and bulbs early in the season with some discolouration on bulb sheaths at harvest. Infected stem plates may show a brown discolouration. Infected bulbs may rot during storage.

The Pathogen

The disease is caused by *Fusarium oxysporum* Schlechtend ex Fr. f.sp. *cepae* (Hans.) Snyd. & Hans. Other *Fusarium* sp. viz., *F. moniliformae* var. *subglutinans*, *F. oxysporum* f.sp. *conglutinans*, *F. bulbigenum*, *F. solani* and *F. equiseti* are also associated in the basal rot. The fungus produces microconidia, macroconidia and chlamydospores.

Disease cycle

The fungus overwinters in the infected onion sets, garlic cloves and in soil. In soil the fungus perennates as chlamydospores. The pathogen may be disseminated widely through infected onion sets and garlic cloves.

Epidemiology

The pathogen is soil inhabitant and survives for longer periods once introduced. High temperature and low level of soil water was found favourable for high disease incidence (Kodama, 1983). The disease rarely occurs below 15ºC and becomes more prevalent when soil temperature and moisture are high. Garlic infection is more after periods of wet weather before harvest. Higashida and Ohasaki (1982) reported that the disease was more severe in gravel soils, which dried frequently and were poor in nutrients and microflora. Stadnik and Dhingra (1997) have studied root infection in onion by *F. oxysporum* f.sp. *cepae* at different stages. They also worked out the relationship of root infection with the development of onion basal rot.

Management

i. Cultural practices: The disease can be kept under check by following cultural practices like long crop rotation, mixed cropping with tobbaco and Jowar, green manuring to increase antagonistic microbial population in soil, good drainage, deep ploughing during summer months and avoiding injury to roots during intercultural operations.

ii. Host resistance: Onion cvs. like HF-555, Autumn Keeper, Autumn Splender, Autumn Ridge and Gial Spagnola were found tolerant under natural conditions, while hybrid onions Spanish and Mustang were resistant to bulb rotting after harvest. Cvs. Atena (Apaza and Mattos, 2000) and Akgun 12 (Ozer *et al.*, 2004) were reported as resistant. Cv. IIHR Yellow, breeding line Sl. 29 and Hybrid 1 have also shown some level of resistance (Somkuwar *et al.*, 1996).

iii. Chemical control: Various chemicals like mancozeb, dichloran, thiram, captafol, carbendazim and benomyl have shown good efficacy against basal rot (Naik and Burden, 1981; Quadri *et al.*, 1982 and Gupta *et al.*, 1983). Root and bulb rots caused by *F. oxysporum* are best controlled with those compounds that contain benomyl and vinclozolin (Ozer and Omeroglu, 1995). Pre-harvest sprays of carbendazim (0.1%) reduced decay of stored onion even after 5 months of storage (Srivastava and Tiwari, 1997).

5. RUST

This is a sporadic disease, but under favourable weather conditions may appear in severe form.

Geographical distribution

The disease is widely distributed in temperate regions of the world. The occurrence of the disease has been reported from different parts of Europe, USA and Asia (Walker, 1952; Cruger and Hommes, 1990). In India, the prevalence of this disease on garlic has been reported from Himachal Pradesh (Singh and Sharma, 1977) and Ludhiana (Punjab) in 1987 (Sandhu and Kang, 1988) and since then the disease is appearing sporadically. In Himachal Pradesh, after a gap of 33 years this disease was again recorded in 2009–2010 crop season in Shimla district (Bharat and Gupta, 2011) and since then it is appearing regularly.

Losses

In favourable weather conditions, the disease causes substantial yield reduction by causing significant loss of foliage and subsequent reduction in bulb size and quality.

Symptoms

Early symptoms appear as of small yellow to white flecks, streaks and spots on leaves. As these small areas expand, the leaf tissue covering them breaks and the orange spores (urediniospores) of the fungus become visible as pustules. Severely infected leaves can be almost entirely covered with pustules, resulting in complete yellowing, wilting and drying of the leaf. As the disease progresses, teliospores may occur on the same leaves, resulting in black pustules. On infected onions and chives, symptoms consist of small (<1/8 inch in dia.), white to tan spots. The orange pustules often form concentric groups on the spot periphery. Disease severity on onion and chives is significantly less severe than on garlic.

The Pathogen

The pathogen responsible for this disease is *Puccinia porri* (Sow.) Wint. (Syn. *Puccinia allii*). The urediniospores are small, 24–31 x 21–24 μm in size, orange yellow in colour and were ellipsoid to globose with thin echinulations. The teleutosori are small, scattered and contain teleutospores of brown colour. The teleutospores measuring 30–40 x 18–24 μm are unicellular or two celled, round or truncate at the top, pedicellate and are constricted at the septum.

Host Range

The fungus can infect chives, garlic, Japanese bunching onion, leek, onion, rakkyo and shallot.

Disease cycle

The fungus probably overwinters in plant debris and also in the soil as teliospores. Urediniospores are repeative spores and cause secondary infection.

Epidemiology

The fungus survives as either urediniospores or teliospores, which cause infection and subsequent disease development under cool temperature and high humidity. Urediniospores are the repeating stage of the fungus and are disseminated by wind and spread to long distances and cause fresh infections in healthy crops.

Management

i. Cultural practices: Since the pathogen overwinters in the infected plant debris and also in soil as teleutospores, crop rotation and sanitation help in reducing the initial inoculum level in the soil. Eradication of alternate hosts from nearby area to avoid contamination is useful.

ii. Chemical control: Various systemic and non-systemics fungicides like mancozeb, maneb, copper oxychloride, propiconazole, triadimefon, triadimenol etc. have been reported effective against this disease (Marchetti, 1977; Blumn and Gabrado, 1993; Garcia *et al.*, 1994).

6. ONION SMUDGE

It is a historical disease and in this pathosystem, the role of biochemical constituents (phenols) in host resistance was elucidated for the first time.

Geographical distribution

The disease is wide spread in temperate regions of the world and was first described in England. Since then it has been reported from many countries of the world like USA, Japan, Argentina and European countries. The disease is of common occurrence in India also.

Losses

This is a common disease in storage particularly on white varieties and affects both quality and quantity of the affected bulbs.

Symptoms

The disease is characterized as small dark green to black dots that appear on the outer scales. These small dots are scattered over the surface of the bulbs or grouped together in concentric rings giving a smudge appearance to the white onions. During moist conditions, acervuli of the fungus develop on the infected areas and spore masses and the setae can be easily recognized with the help of hand lens. The disease may spread to the underlying fleshy scales resulting in yellow depressions on the bulb.

The Pathogen

The fungus associated with this disease is *Colletotrichum circinans* (Berk.) Volino. The mycelium is septate, hyaline and becomes darker and thicker with age. The thick-walled hyphae intertwine and form dark green to black stromata. Acervuli are formed on stromata just beneath the cuticle by formation of a palisade layer of short, hyaline conidiophores which later rupture the cuticle. Conidiophores are formed in the acervuli along with scattered, thick walled, dark coloured setae, which are 1 to 3 septate measuring 80 to 315 μm long and arise from the stroma. The conidia are fusiform, falcate, hyaline measuring 14–30 x 3–6 μm in size.

Disease cycle

The fungus overwinters in the infected onion scales in soil or in bulbs and sets in storage. In field the disease appears shortly before harvest. Under favourable conditions, the fungus attacks the outer scales, and spores that are borne on the infected areas are carried by drops of water to other scales, where they germinate and cause infection.

Epidemiology

The fungus persists in soil as a free saprophyte for several years in the absence of host. The disease can develop at any temperature from 10 to 32°C, but the optimum is around 25°C alongwith abundance of rainfall. Moist conditions are essential for the production of conidia.

Management

i. Cultural practices: Cultural practices like crop rotation, good drainage, use of healthy seed, drying of bulbs immediately after harvest check the disease under storage conditions (Walker, 1952, Schwartz and Mohan, 1995). Protection of the harvested crop from rains, rapid and thorough curing and provision of well-ventilated storage also reduce the incidence of the disease.

ii. Host resistance: White onions are more susceptible than coloured onions. In disease prone areas, coloured varieties should be raised.

iii. Chemical control: Pre-harvest sprays either of mancozeb or carbendazim reduce the infection of this disease under storage conditions. Bulb treatment either with captan or thiram before storage also reduces this disease.

7. PINK ROOT

This is widely distributed disease in those areas where onion, garlic and shallots are grown.

Geographical distribution

The disease was first reported from Texas in 1917. Since then it has been reported from many parts of the world.

Losses

Under favourable weather conditions, the disease causes substantial yield losses. Heavy losses in onion due to this disease in Sudan have been reported (Yassin *et al.*, 1982).

Symptoms

Characteristic symptoms of the disease are the pink colouration of the roots. Affected roots initially turn yellow, shrivel and die, meanwhile taking on a distinctly pink colour. The new roots, which grow from the diseased plants, get immediately infected and become functionless. The affected plants are commonly not killed but the development of bulb is retarded as number and size of leaves are reduced at maturity. The disease is confined to the roots only.

The Pathogen

The disease is caused by *Pyrenochaeta terrestris* (Hansen) Gorenz *et al.*

Disease cycle and Epidemiology

The pathogen persists indefinitely in the soil and is distributed on onion sets and seedlings (Horst, 1979). The disease is disseminated through agricultural tools or by natural agencies like surface drainage water.

Management

Crop rotation of 4–5 years gives good control of pink root. Soil solarization during summer months of the fields having disease history also helps in reducing the disease. Early sown crops suffer more than late sowing. The incidence of pink root was decreased with the application of benomyl and procymidone (Korien *et al.*, 1991).

(B) VIRAL DISEASES

1. ONION YELLOW DWARF

Symptoms

The characteristic symptoms of the disease are yellow streaks at the base of the leaves with yellowing, crinkling and flattening of new leaves. Leaves and flower stalks are destroyed, twisted and shortened. Yield is considerably reduced. Bulbs of infected plants are malformed and difficult to store, being more susceptible to storage fungi.

The Virus

The disease is caused by *Onion Yellow Dwarf Virus* (OYDV). Mixed viral infection in onion seed crop has been detected using serological techniques.

Host Range

Primary hosts of the virus are onion, garlic and leek. Ornamental species of *Allium* are also susceptible to this virus. Strains of the virus isolated from each species affected are not transmitted to other species within the genus *Allium.*

Transmission

In nature, the virus is transmitted in a non-persistent manner by *Myzus persicae*. The virus is maintained by vegetative propagation of infected plants. OYDV is experimentally transmissable by mechanical inoculation. It is neither transmitted by contact between plants, nor by seed, or by pollen.

Management

There are no means of managing this disease other than the use of meristem tip culture form the production of virus-free plants. Rogueing of diseased plants and use of resistant varieties are recommended. Phytosanitary measures also assist in controlling the disease (Khaisin, 1965). Insecticidal sprays may give better control of the disease by checking the vector.

2. LEEK YELLOW STRIPE

This is also a common disease of this crop and in favourable weather conditions cause extensive damages.

Symptoms

Due to systemic nature of the disease, generally all plant parts become affected soon after infection, which may result from virus introduction by aphids. Plants of garlic derived from vegetative propagation are subjected to systemic infection from the onset of growth. In onion, symptoms are an irregular yellow striping of the whole lamina, but especially at its base. Whole leaves may become more or less evenly yellow. Affected plants are smaller and weigh less than normal. They are less juicy (with their leaves slightly flaccid), the white lower plant part is lustreless, and the keeping quality of the harvested product is impaired. Infected plants suffer more from frost and may be killed prematurely in winter. In garlic, chlorotic to yellow striping or streaking of leaves is similar to that in onions. These symptoms differ considerably in severity according to virus isolate and cultivar infected. Plant growth may be significantly reduced and yield severely impaired.

The Virus

The disease is caused by *Leek Yellow Stripe Virus* (LYSV).

Host Range

In nature, leek yellow stripe virus is almost limited to leek. Onions and shallots are highly resistant, but not immune to the virus. The garlic-specific strain of the virus, alone or in combination with the garlic-specific strain of onion yellow dwarf potyvirus, often occurs in complex and is a major cause of garlic mosaic, usually affecting all garlic plants wherever grown.

Transmission

Transmission through contact and by plant handling is unlikely because the leaves are covered with wax and experimental transmission in sap is often difficult. The virus is transmissible through aphids (*Aphis fabae* and *Myzus persicae*) in the non-persistent manner, but is undoubtedly spread in nature by many migrant aphid species that do not settle on *Allium* species, but briefly probe on them.

Management

1. Crops grown from seed or from vegetative planting material should be grown in isolation from infested crops of the same species.
2. The removal of diseased plants may reduce secondary spread within the crop.
3. Insect control with insecticides is ineffective against non-persistently aphid-transmitted viruses, and spraying with oil is too expensive and not permitted for crops grown for human consumption.

3. GARLIC MOSAIC

This viral disease is prevalent in different garlic growing areas. Ahlawat (1974) reported the disease from Laimpong town of Darjeeling (West Bengal) India for the first time.

Symptoms

The infected plants show typical mosaic symptoms and yield is also recorded comparatively less than in healthy plants.

The Virus

The disease is caused by *Garlic Mosaic Virus* (GMV). Ghosh and Ahlawat (1997) have detected filamentous viruses associated with mosaic disease of garlic in India. Likewise, four garlic viruses have been identified by reverse transcription polymerase chain reaction and their distribution has been determined in various regions of Japan (Takaichi *et al*., 1998).

Transmission

The virus is easily transmissible by two species of aphids viz., *Myzus persicae* Sulz. and *Aphis gossypii* Grover to garlic only and not to any other crop.

Management

Crop rotation and field sanitation help in reducing the disease. Szyndei *et al*. (1994) advocated thermotherapy and meristem tip culture for getting virus free plants of garlic.

REFERENCES

Ahlawat, Y.S. 1974. A mosaic disease of garlic in Darjeeling hills. *Sci and Cult*. **40:** 466–467.

Ahmad, S. and Karimullah. 1998. Relevance of management practices to downy mildew in onion. *Sarhad J. Agric.* **14:** 161–162.

Ajrekar, S.L. 1921. *Annu. Rept. Bombay Dept. Agric.* 1920–1921, pp. 102–104.

Alves, M.L.B., Paiva, W.O. and Assis, L.A.G. 1982. Incidence of purple blotch (*Alternaria porri* Ell. Cif.) on onion (*Allium cepa* L.) cultivars and hybrids in Manaus, Amazonia. *Acta Amazionica* **12:** 673–676.

Apaza, W.E. and Mattos, L. 2000. Reaction of onion cultivars to basal plate rot caused by *Fusarium oxysporum* f.sp. *cepae. Fitopatologia* **35:** 231–236.

Aujla, I.S., Goswami, S., Thind, T.S., Raheja, S. and Kumar, P. 2010. Relative effectiveness of fungicides in controlling purple blotch of onion seed crop. *Pl. Dis. Res.* **25(1):** 73–74.

Aujla, I.S., Amrate, P.K., Kumar, P. and Thind, T.S. 2013. Efficacy of some new fungicides in controlling purple blotch of onion under Punjab conditions. *Pl. Dis. Res.* **28(2):** 171–173.

Aveling, T.A.S. and Naude, S.P. 1992. First report of *Stemphylium vesicarium* on garlic in South Africa. *Plant Dis.* **76:** 426.

Aveling, T.A.S., Snyman, H.G. and Naude, S.P. 1993. Evaluation of seed treatments for reducing *Alternaria porri* and *Stemphylium vesicarium* on onion seed. *Plant Dis.* **77:** 1009–1011.

Basu, P.K. 1971. Existence of chlamydospores of *Alternaria porri* f.sp. *solani* as overwintering propagules in soil. *Phytopathology* **61:** 1347.

Battilani, P., Rossi, V., Racca, P. and Giosue, S. 1996. ONIMIL– a forecaster for primary infection of downy mildew of onion. *Bull. EPPO* **26:** 567–576.

Bharat, N.K. and Gupta, S.K. 2011. Outbreak of garlic rust in Shimla district of Himachal Pradesh. *J. Mycol. Pl. Pathol.* **41(1):** 132–133.

Bisht, I.S., Kamala, V. and Thomas, T.A. 1990. Evaluation of onion germplasm for resistance against Stemphylium blight. *Onion Newsl. Trop.* **2:** 36–37.

Blumn, L.E.B. and Gabrado, H. 1993. Chemical control of garlic rust in Curibanos. *Fitopatolo. Brasileira* **18(2):** 230–232.

Chawda, H.T. and Rajasab, A.H. 1992. Epidemiological investigations on anthracnose and purple blotch of onion (*Allium cepa* L.). *Onion Newsl. Trop.* **4:** 65–66.

Cook, H.T. 1932. Studies on the downy mildew of onions, the causal organism *Peronospora destructor* Casp. *N.Y. Cornell Agri. Exp. Sta. Mem.* **143:** 1–40.

Cova, J. and Rodriguez, D. 2003. Fungi associated with leaf blight of onion (*Allium cepa* L.) in Lara State, Venezuela. *Bioagro* **15(3):** 157–163.

Cruger, G. and Hommes, M. 1990. Pests and diseases of leek. *Gemuse Munch.* **26(2):** 130–135.

Deshmukh, V.S., Dhruj, I.U. and Chavan, R.V. 2007. Chemical control of purple blotch (*Alternaria porri*) (Ellis) Cif. of onion. *Pl. Dis. Res.* **22(1):** 34–36.

Devlash, R. 2011. Management of Stemphylium blight of garlic through fungicides. *Indian J. Plant Prot.* **39:** 157–159.

Devlash, R.K. and Sugha, S.K. 1997a. Factors affecting development of downy mildew (*Peronospora destructor*) of onion (*Allium cepa*). *Indian J. Agric. Sci.* **67:** 71–74.

Devlash, R.K. and Sugha, S.K. 1997b. Incidence of onion downy mildew and its impact on yield. *Indian Phytopath.* **50:** 127–129.

Dhiman, J.S., Chadha, M.L. and Sidhu, A.S. 1986. Studies on the reaction of onion genotypes against purple blotch. *Veg. Sci.* **13:** 304–309.

El–Shanawany, A.A., El–Sheikh, H.H. and Bashandy, A.A. 2003. Effect of solar heating of soil on downy mildew disease, growth, yield and quality of Egyptian onion. *African J. Mycol. Biotech.* **11(3):** 25–38.

Everts, K.L. and Lacy, M.L. 1990. The influence of dew duration, relative humidity and leaf senescence on conidial formation and infection of onion by *Alternaria porri. Phytopathology* **80:** 1203–1207.

Everts, K.L. and Lacy, M.L. 1996. Factors influencing infection of onion leaves by *Alternaria porri* and subsequent lesion expansion. *Plant Dis.* **80:** 276–280.

Everts, K.L., Schwartz, H.F., Epsky, N.D. and Capinerea, J.K. 1985. Effect of maggots and wounding on occurrence of Fusarium basal rot of onion in Calorado. *Plant Dis.* **69:** 878.

Furst, G. 1976. Anatomical and histochemical characters of resistance in onion downy mildew. *Referativery Zhurnal* **12:** 14–17.

Garcia, D.C., Barni, V. and Detimann, L.A. 1994. Fungicides for control of garlic rust. *Ciencia Rural* **24:** 275–283.

Georgy, N.I, Radwan, I.A., Mohamed, H.A. and Shahab, A.E. 1983. Chemical control of downy mildew and purple leaf blotch of onion in Egypt. *Agric Res. Rev.* **61(2):** 25–41.

Ghosh, D.K. and Ahlawat, Y.S. 1997. Filamentous viruses associated with mosaic disease of garlic in India. *Indian Phytopath.* **50:** 266–276.

Gilles, T., Phelps, K., Clarkson, J.P. and Kennedy, R. 2004. Development of MILIONCAST, an improved model for predicting downy mildew sporulation on onions. *Plant Dis.* **88(7):** 695–702.

Gupta, R.B.L. and Pathak, V.N. 1988. Survival of *Alternaria porri*, inducer of purple blotch of onion. *Indian Phytopath.* **41:** 406–409.

Gupta, R.P. and Pandey, V.B. 1986. Stemphylium blight of onion. *Indian Hortic.* 31: 13.

Gupta, R.P. and Srivastava, P.K. 1988. Control of Stemphylium disease of onion bulb crop. *Indian Phytopath.* **41:** 495–496.

Gupta, R.P. and Srivastava, P.K. 1993. Studies on interval and quality of mancozeb solution for the control of Stemphylium blight disease of onion. *AADF Newsl.* **13(1):** 9–11.

Gupta, R.P. and Srivastava, R.K. 1981. Botrytis leaf blight of onion–a new record from India. *Indian Phytopath.* **34:** 88–89.

Gupta, R.P., Srivastava, P.K. and Pandey, U.B. 1983. Efficacy of fungicides against *Fusarium oxysporum* f.sp. *cepae* incitant of basal rot of onion. *Pesticides* **17:** 16.

Gupta, R.P., Srivastava, P.K. and Pandey, U.B. 1984. Downy mildew of onion (*Allium cepa* L.): new disease from plains of India. *Indian J. Plant Pathol.* **2(1):** 77.

Gupta, R.P., Srivastava, P.K. and Sharma, R.C. 1996. Effect of foliar spray of different fungicides on the control of Stemphylium blight disease and yield of onion bulbs. *NHRDF Newsl.* **16(1):** 13–15.

Gupta, R.P., Srivastava, P.K. and Singh, J.B. 1989. Effect of number of sprays of mixed pesticides on the control of Stemphylium blight, purple blotch and thrips in onion seed crop. *Int. Cong. Recent Adv. in Medicinal, Aromatic and Spice crops*, 28–30 Jan., 1987. (Abstr.).

Gupta, R.P., Srivastava, V.K. and Pandy, U.B. 1986. Control of purple blotch disease of onion seed crop. *Indian Phytopath.* **39:** 303–304.

Higashida, S. and Ohasaki, I. 1982. The factual survey of Fusarium basal rot of onion and soil in some fields of Kitami district. *Bull. Hokkaido Perfactural Agric. Exp. Stn.* **47:** 31–38.

Horst, R.K. 1979. *Westocott's Plant Disease Book.* 4th Edn. Van Nostrand Reinhold Company, New York pp. 803.

Iqbal, M., Sahi, G.M., Tahir, H.A.S., Sahi, S.T. and Atif, M. 2009. Evaluation of different fungicides against downy mildew of onion. *Pakistan J. Phytopath.* 104–107

Jakhar, S.S., Duhan, J.C. and Suhag, L.S. 1996. Studies on the epidemiology and survival of *Stemphylium vesicarium* (Wallr.) Simmons in debris and seeds of onion. *Seed Res.* **24:** 135–140.

Jones, H. A. and Mann, L. K. 1963. Diseases and their control. In: *Onion and their allies.* Pp. 180–194.

Jorgensen, L.N. and Nistrup, J.L. 1987. Fungicide for controlling onion downy mildew. *Gartner Tidende* **103:** 432–433.

Khaisin, M.F. 1965. Susceptibility of common onion varieties to yellow dwarf virus. *Trudy Maldav Mauchno, Issled. Inst. Orosh Zemled. Ovoshch* **7:** 145–152.

Khalid, H., Gill, M.A. and Javed, I. 2002. Chemical control of downy mildew of onion. *Pakistan J. Phytopath.* **14:** 91–94.

Khare, V.K. and Nema, K.G. 1981. Studies on purple blotch of onion–sporulation on host and dispersal of conidia. *Indian Phytopath.* **34:** 214–215.

Khokhar, L.K. and Jaffrey, A.H. 2000. Efficacy of fungicides against downy mildew and yield of onion. *Pakistan J. Agric. Res.* **16:** 43–44.

Khosla, K., Thakur, B.S. and Bhardwaj, S.S. 2007. Chemical management of Stemphylium blight of garlic. *Pl. Dis. Res.* **22(1):** 47–51.

Kodama, F. 1983. Studies on basal rot of onion caused by *Fusarium oxysporum* f.sp. *cepae* and its control. *Rep. Hokkaido Perfactural Agric. Exp. Stn* **39:** 65.

Korien, S.O., Hussein, F.N. and Metwally, A.H. 1991. Chemical control of pink root rot, basal rot and neck rot diseases of onion produced by sets. *Assiut J. Agric. Sci.* **22 :** 81–97.

Kumar, Suman and Katoch, Shabnam 2017. Role of weather variables in the development of Stemphylium blight of garlic. *Pl. Dis. Res.* **32(1):** 73–76.

Lakra, B.S. 1999. Devlopment of purple blotch incited by *Alternaria porri* and its losses in seed crop of onion (*Allium cepa). Indian J. Agric. Sci.* **69:** 144–146.

Lopez, J.A. and Cramer, C.S. 2004. Screening short day onion varieties for resistance to Fusarium basal rot. *Acta Hortic.* **637:** 169–173.

Mac–Manus, G.P.V., Galea, V.J., O'Brien, R.G. and Napier, T. 2002. Oil–based adjuvants improve fungicide activity against downy mildew in onions. *Proc. Onions 2002 Conf. Nat. Veg. Ind. Centre Yanco Agril Inst., Australia 3–7 June, 2002*, pp. 53–57.

Mandhare, V.K. and Suryawanshi, A.V. 2003. Antagonistic effect of *Bacillus thermophillus* on some pathogens. *J. Maharashtra Agric. Univ.* **28:** 274–277.

Marchetti, L.L. 1977. Onion rust. *Informatore Fitopathol.* **27:**33.

Mathur, B.L. and Sankhala, H.C. 1963. Basal rot of onion. *Curr. Sci.* **31:** 40.

Memane, S.A., Khetmalas, M.B., Pawar, D.B. and Warade, S.D. 2001. Management of leaf blight and thrips of onion during rainy season. *J. Maharashtra Agric. Univ.* **26:** 347–349.

Michail, S.H. and Salem, M.A. 1981. Reaction of onion cultivars to scald disease incited by *Alternaria porri. Acta Phytopath.* **16:** 299–305.

Miller, M.E. 1983. Relationship between onion leaf age and susceptibility to *Alternaria porri. Plant Dis.* **67:** 284–286.

Miller, M.E., Taber, R.A. and Amador, J.M. 1978. Stemphylium blight of onion in South Texas. *Pl. Dis. Reptr.* **62:** 851–853.

Mir, N.M., Dhar, A.K., Khan, M.A., Dar, G.H. and Zargar, M.Y. 1987. Screening of fungicides for field control of downy mildew (*Peronospora destructor*) on onion. *Indian J. Mycol. Pl. Pathol.* **17:** 321–322.

Mirakhur, R.K., Dhar, A.K. and Kaw, M.P. 1978. Downy mildew of *Allium cepa* L. and its control with fungicide in Kashmir Valley. *Indian Phytopath.* **30:** 576–577.

Mohan, C., Thind, T.S., Raj, P. and Arora, J.K. 2004. Promising activity of triazoles and other fungicides against fruit rot of chilli and stemphylium blight of onion. *Pl. Dis. Res.* **19:** 200–203.

Mohan, K., Ebenezer, E.G. and Seetharaman, K. 2001. Management of leaf blight disease of onion caused by *Alternaria porri* by plant extracts and biocontrol agents. *Newsl. Nat. Hort. Res. Dev. Foun. 2001–2002*, **21–22(3–4):** 11–14.

Mondal, M.F., Hossain, I., Bakshi, A.I. and Nasiruddin, K.M. 1989. Effect of nitrogen and phosphorus on purple blotch of onion seed crop. *Bangladesh J. Plant Pathol.* **5(1–2):** 37–39.

Naik, D.M. and Burden, O.J. 1981. Chemical control of basal rot of onion in Zambia. *Trop. Pest Mgt.* **27:** 455–460.

Ozer, M. and Omeroglu, M. 1995. Chemical control and determination of fungal causal agents of wilt disease of onion in Terikdag province. *J. Turk. Phytopathol.* **24(2):** 47–55.

Ozer, N., Koycu, N.D. Chilosi, G. and Magro, P. 2004. Resistance to Fusarium basal rot of onion in greenhouse and field and associated expression of antifungal compounds. *Phytoparasitica* **32(4):** 388–394.

Palti, J., Chorin, M. and Nitizani, F. 1958. Survey of leaf and flower stalk diseases of onions in Israel. *Ktavim.* **9:** 167–175.

Pandey, U.C., Singh, J. and Singh, J. 1989. Performance of new varieties of garlic (*Allium sativum* L.). *Haryana Agric. Univ. J. Res.*

Pandotra, V.R. 1964. Purple blotch disease of onion in Punjab.I. Its occurrence, pathogenecity and host range. *Proc. Indian Acad. Sci Sec. B*, **60:** 336–340.

Pandotra, V.R. 1965. Purple blotch disease of onion in Punjab. II. Studies on the life history, viability and infectivity of the causal organism *Alternaria porri. Proc. Indian Acad. Sci. Sec. B.* **61:** 326–330.

Pathak, C.S. 1994. *Allium* crop situation in Asia. *Acta Hortic.* **433:** 53–74.

Pathak, C.S., Singh, D.P., Deshpande, A.A. and Sreedhar, T.S. 1986. Sources of resistance to purple blotch in onion. *Veg. Sci.* **13:** 300–303.

Perez–Moreno, L. and Chavez–Hernandez, L.F. 1992. Genotypes and fungicide evaluation for control of purple spot (*Alternaria porri*) and downy mildew (*Peronospora destructor* (Berk) Caps) of onion (*Allium cepa* L.) in Irapuato, GTO. *Revista Mexicana de Fitopatologia* **10:** 29–34.

Popkova, K.V., Palilov, N.A. and Kir'–yanova, EV. 1980. The sources of infection of onion downy mildew. *Mikologiya I Fitopatologiya* **14:** 435–440.

Prasad, S.M. and Barnwal, M.K. 2004. Evaluation of plant extracts in management of Stemphylium blight of onion. *Indian Phytopath.* **57(1):** 110–111.

Quadri, S.M.H., Srivastava, K.S., Bhonde, S.R., Pandey, U.B. and Bhagchandani, P.M. 1982. Fungicidal bioassay against some important pathogens of onion. *Pesticides* **16:** 11–16.

Raju, K.S. and Mehta, B.K. 1981. Note on the Varietal reaction of onion to purple blotch disease. *Indian Cocoa, Arecanut Spices J.* **4(4):** 117–118.

Rao, N.N. and Pavgi, M.S. 1973. *Pleospora allii* on onion from Varanasi. *Curr. Sci.* **42:** 734.

Rondomanski, W. 1967. Final Tech. Rep. (1962–67). *Res. Inst. Veg. Crops Skterniewice,* 59pp.

Sandhu, K.S. and Kang, M.S. 1988. Occurrence of rust on garlic in Punjab. *Pl. Dis. Res.* **3:** 50.

Sandhu, K.S., Gill, S.S. and Singh, H. 1982. Effect of cultural practices on purple blotch disease in onion seed crop. *J. Res, India* **19(2):** 118–120.

Sastrahidayat, I.R. 1994. A study on the effect of the fungicide difenoconazole in controlling purple blotch (*Alternaria porri*) of onion. *Agrivita* **17(2):** 97–101.

Schwartz, H.F. and Mohan, S.K. 1995. Compendium of Onion and Garlic Diseases. American Phytopathological Society, St. Paul, Minnesota, USA, 54pp.

Sharma, I.M. 1997a. Reaction of onion lines to downy mildew caused by *Peronospora destructor. J. Mycol Pl. Pathol.* **27:** 324–325.

Sharma, I.M. 1997b. Screening of onion varieties/lines against purple blotch caused by *Alternaria porri* under field conditions. *Pl. Dis. Res.* **12:** 60–61.

Sharma, S.N. and Sain, R.S. 2003. RO–1 an improved onion variety for the warmer areas of Rajasthan. *Indian J. Gen. Plant Breed.* **63:** 281–282.

Singh, B.M. and Sharma, Y.R. 1977. Occurrence of leaf blight of garlic caused by *Stemphylium botryosum* in India. *Indian Phytopath.* **30:** 272–273.

Singh, B.M., Singh, D. and Paul, Y.S. 1987. Out break of downy mildew of onion in Himachal Pradesh. *Pl. Dis. Res.* **2:** 124.

Somkuwar, R.G., Gowda, R.V., Singh, T.H. and Pathak, C.S. 1996. Screening of onion for resistance to onion basal rot. *Madras Agric. J.* **83:** 273–275.

Srivastava, A.K., Borse, V.A., Gupta, R.P. and Srivastava, P.K. 1996. Mode of perpetuation and role of nature in spread of *Stemphylium blight* and purple blotch diseases of onion. A short note. *Newsl. Nat. Hort. Res. Dev. Foun.* **16(4):** 9–11.

Srivastava, K.J., Bhardwaj, B.S. and Gupta, R.P. 1992. Management of garlic diseases in India. *NHRDF Newsl.* **15(3):** 6–9.

Srivastava, P.K. and Tiwari, B.K. 1997. Effect of pre–harvest fungicidal spray on the control of storage diseases of onion. *Newsl. Natl. Hortic. Res. Dev. Found.* **17:** 4–6.

Srivastava, P.K., Tiwari, B.K. and Srivastava, K.J. 1999. Studies on integrated disease management of onion. *Newslett. Nat. Hort. Res. Dev. Foun.* **19(4):** 7–9

Stadnik, M.J. and Dhingra, O.D. 1997. Root infection by *Fusarium oxysporum* f.sp. *cepae* at different stages and its relation to the development of onion basal rot. *Phytopathol. Mediter.* **36:** 8–11.

Sugha, S.K. and Kumar, Suman 2005. Outbreak of Stemphylium blight of garlic in Himachal Pradesh. *Pl. Dis. Res.* **20:** 190–191.

Sugha, S.K. and Singh, B.M. 1991. Status of downy mildew of onion in Kangra district of Himachal Pradesh and its effect on crop yield. *Pl. Dis. Res.* **6:** 35–38.

Sugha, S.K. and Tyagi, P.D. 1994. Effectiveness of metalaxyl + mancozeb against purple blotch (*Alternaria porri*) of onion (*Allium cepa*) in Himachal Pradesh. *Indian J. Agric. Sci.* **64:** 135–136.

Sugha, S.K., Tyagi, P.D. and Devlash, R.K. 1993. Effect of heat treatment of bulb and spray of fungicides in management of purple blotch (*Alternaria porri*) of onion (*Allium cepa*). *Indian J. Agric. Sci.* **63:** 303–305.

Suhag, L.S. and Duhan, J.C. 1985. Parasitism and field control of *Alternaria porri* causing purple blotch in onion. *Indian J. Plant Pathol.* **3:** 74–79.

Szurke, J. and Szurke, E. 1982. Rate and dynamics of downy mildew (*Peronospora destructor*) infection in spring sown Hungarian and foreign onion cultivars and selections in field and provocation trials, 1975 to 1980. *Fajtakiserletezes* **29:** 277–296.

Szyndei, M.S., Kozera, B. and Misterek, E. 1994. The elimination of virus from garlic (*Allium sativum* L.) plants by thermotherapy and meristem tip culture. *Acta Agrobotanica* **47:** 83–88.

Takaichi, M., Nagakubo, T. and Oeda, K. 1998. Four garlic viruses identified by reverse transcription–polymerase chain reaction and their regional distribution in northern Japan. *Plant Dis.* **82(6):** 694.

Talieva, M. 1976. Role of cultivar wax in the resistance of onion to Peronosporosis. *Referativnyl Zhurnal* **12:** 84–87.

Thind, T.S. and Jhooty, J.S. 1982. Association of thrips with purple blotch infection on onion plants caused by *Alternaria porri*. *Indian Phytopath*. **35:** 696–698.

Thind, T.S. and Jhooty, J.S. 1984. Use of malathion and stickers for increasing the efficacy of fungicides for controlling purple blotch of onion. *Indian J. Agric. Sci.* **54**(4): 299–302.

Thind, T.S., Sharma, R.C. and Jhooty, J.S. 1985. Occurrence of two fungal diseases of onion in Punjab. *Indian J. Mycol. Pl. Pathol*. **15:** 90.

Tyagi, S., Dube, V.P. and Charaya, M.U. 1990. Biological control of the purple blotch on onion caused by *Alternaria porri* (Ellis) Ciferri. *Trop. Pest Mgt*. **36:** 384–386.

Utikar, P.G. and Pandule, D.N. 1981. A virulent species of *Alternaria* causing leaf blight of onion. *Indian Phytopath*. **33:** 335–336.

Viranyi, F. 1981. Downy mildew of onion. In: *Downy Mildews* (Ed. Spencer, D.M.). Academic Press, New York, pp. 461–472.

Walker, J.C. 1952. *Diseases of Vegetable Crops*. McGraw Hill Rock Co. Inc. New York pp. 529.

Wright, P.J., Chynoweth, R.W., Bresford, R.M. and Henshall, W.R. 2003. Comparison of strategies for control of onion downy mildew. *Grower* **58**(2): 36–37.

Yarwood, C.E. 1943. Onion downy mildew. *Hilgardia* **14:** 595–691.

Yassin, A.N., Bubikar, A.G.T. and Ahmed, M.K. 1982. First report of pink root of onion in Sudan (*Pyrenochaeta terrestris*). *Plant Dis*. **66:** 741.

■■■■

11

DISEASES OF EGG PLANT

The egg plant (*Solanum melongena* L.), also known as aubergine, guinea squash or brinjal, is one of the important vegetable crops grown almost worldwide. It is native to India, where it has been cultivated since remote antiquity for its fleshy fruits. Eggplant fruits were a common food in China as long as 600 BC, when it was called Malayan purple melon. Now it is extensively grown in eastern and southern Asia, including India, as well as in USA and other countries. The raw vegetable contains only 15 calories/100 g, but its caloric value rises sharply when it is fried. Africans following folk medicine have long used brinjals to treat epilepsy and convulsions. In Southeast Asia it is still used to treat measles and stomach cancer. The brinjal crop is affected by several diseases of fungal, bacterial and phytoplasma nature, which inflict heavy losses in its production. A detailed account of various diseases affecting this crop and their management is given in this chapter.

(A) FUNGAL DISEASES

1. PHOMOPSIS BLIGHT AND FRUIT ROT

This is a serious disease of eggplant and can result into heavy losses under favourable weather conditions. It occurs on leaves as well as fruits, but the most destructive phase of the disease is fruit rot, which causes substantial damage both in field and transit.

Geographical distribution

The disease is almost worldwide in its occurrence. It was first reported from Lousiana in the year 1921 (Edgerton and Moreland, 1921). Subsequently, it was reported from Bermuda, Porto Rico, New Jersey, China, Philipine, Jamica, Canada, Brazil and Mauritius.

In India, the disease was first reported from Belgaon district of Bombay (Pawar and Patel, 1957). Since then it has been reported from almost all the states wherever eggplant is grown.

Losses

The disease has been reported to cause huge losses by way of damaging the fruits partly or completely both in the field as well as during transit. The affected fruits are smaller in size, which leads to reduced yield. In advanced stages of the disease, the seed quality is also impaired and the affected seed become discoloured and viability of the seed is also reduced. Edgerton and Moreland (1921) reported about one half-yield reduction of eggplants by *Phomopsis vexans* infection in Lousiana. Palo (1936) from Philippines reported that *P. vexans* killed 64 per cent fruits of the affected plants under field conditions and destroyed 10 and 20 per cent of the first and second harvest in transit, respectively. This disease caused such devastating losses of eggplant crop in Brazil that it rendered the control measures impracticable (De Figueiredo and Pereira, 1944).

In India, Panwar *et al.* (1970) reported 10 to 20 per cent yield losses due to fruit rot under Punjab conditions, while average yield reduction of 20 per cent by this disease has also been reported (Chowdhary and Hasija, 1979).

Symptoms

The pathogen causes damping-off of seedlings, stem canker, leaf blight and fruit rot in brinjal. In nursery bed, it causes damping off, which results from the infection of the seedlings at the collar portion just above the soil line. In early stages, the disease is more prominent on the leaves in the form of small spots, which are circular brown to deep brown with light coloured centres and dark margins (Plate 1a). In late stages, the lighter portion is studded with numerous black dots like pycnidia. The affected leaves turn yellow and die. The lesions on petioles or the lower part of the midrib also result in the death of the entire leaf. Affected leaves turn yellow and fall down pre-maturely.

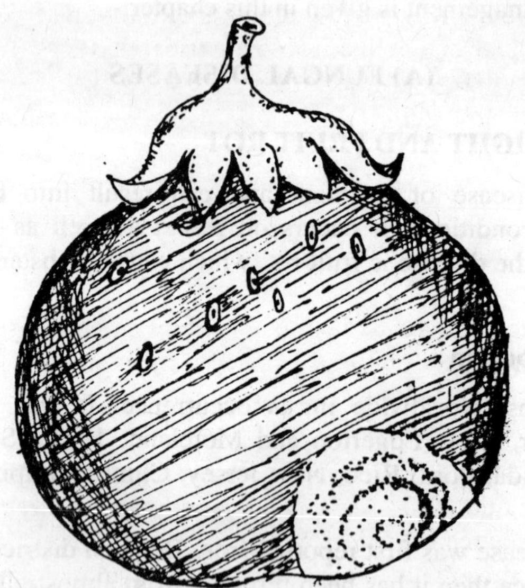

Fig. 1. Symptoms of Phomopsis blight on fruits

The disease also appears on the stem in the form of elongated, blackish-brown lesions. The affected plants bear smaller leaves and the axillary buds are often killed. When there is girdling of the stem at the base due to infection, the shoots above the infected area show wilting with pale yellow leaves and soon dry up. In strong wind, such plants topple down due to breaking of the main stem.

The fruits are affected while they are on the plant. On the fruit, disease manifests as pale dull sunken spots, which later merge and enlarge to form rotten areas and produce concentric rings of yellowish and brown zones (Fig. 1), and cover the entire fruit (Plate 1b). These spots are also marked by the presence of many black pycnidia, which may be scattered or arranged, in concentric rings. If infection enters the fruit through the calyx, the whole fruit becomes mummified due to dry rot. Fruits between 5–15 days are more susceptible with optimum temperature of 25ºC and relative humidity of 90 per cent (Kumar, 2012).

The Pathogen

The fungus responsible for this disease is *Phomopsis vexans* (Sacc. and Syd.) Harter. *Diaporthe vexans* Gratz is the perfect stage. However, this stage does not occur in nature, but is found in culture only (Pawar and Patel, 1957). The mycelium is hyaline and septate. The hyphae measure 2.7 to 3.9 μm in diameter. The pycnidium may or may not contain beak. The conidiophores (phialides) in the pycnidium are hyaline, simple or branched, sometimes septate, 10 to 16 μm long and arise from the innermost layer of cells lining the pycnidial cavity. The pathogen is reported to produce two types of conidia viz., **alpha** and **beta** in its pycnidium. Kumar and Sugha (1999) reported that formation of conidia in pycnidia of *P. vexans* is temperature dependent. At 10–16°C temperature the pathogen produces beta conidia and at 25–28°C, the alpha conidia. These two forms of conidia get inter-converted when subjected to specific temperature and further reported that alpha and beta are two forms of the same conidium. *P. vexans* produces only one type of conidia in its pycnidia which are hyaline, one celled, sub cylindrical and 5–9 x 2–2.8 μm in size during summer months which gradually changed into beta form. Beta form of conidia, the stylospores are filiform, curved, hyaline and septate. These spores measures 20–30 x 0.5–1.0 μm in size and normally do not germinate. But inoculation of host with beta conidia caused interveinal necrosis, which progressed towards leaf base and resulted in premature defoliation, indicating their role in pathogenesis. Isolations from such leaves produced pycnidia with alpha conidia at 25ºC and beta at 16ºC.

Gratz (1942) reported that the perithecia in culture are usually in clustures, 130–350 μm in diameter, beaked, carbonaceous, sinuate, irregular ostiole. The beak is 80–500 μm long. Asci are clavate, sessile and contain eight ascospores, each measuring 28–44 x 5–12 μm in size. Ascospores are hyaline, narrowly ellipsoid to bluntly fusoid, one-septate and constricted at the septum. They measure 9–12 x 3–4.4 μm.

Edgerton and Moreland (1921) reported distinct types of strains of *P. vexans* in Louisiana. One strain produced severe damping off of seedlings, but only produced moderate infection on leaves. Panwar and Chand (1968) studied the pathogenecity of three isolates *i.e.,* A, C and I of the fungus on eight brinjal cultivars and found isolate C to be most virulent on all the cultivars except N.S.L. The isolate A was more virulent to Pb. 8, No. 11, 15 and 34 while isolate I was least virulent. Islam *et al.* (1990) studied the variability of 16 isolates of *P. vexans* collected from different parts of India. Isolates varied in colony diameter (69-90 mm), dry weight (436.66 to 642.66 mg) and relative abundance of pycnidia production. Variations were also observed in colour, texture, zonation and arrangement of pycnidia in colonies.

Pathogenicity also varied with isolate. Intraspecific diversity in this pathogen has also been recorded (Jameel, 2006).

Host Range

The fungus *P. vexans* has its host range limited to the host *Solanum melongena* L. only. Other genera, even of family Solanaceae have not been found to be the host of the fungus. However, Chowdhury and Hasija (1979) reported that the pathogen infected injured fruits of tomato, but not of *Capsicum annuum*.

Disease cycle

The pathogen overwinters in the infected seed (Puttoo and Sohi, 1975) in the form of mycelium, which is present on the seed coat (Pan and Acharya, 1995), between the seed coat and endosperm as well as embryo region of the seed (Vishnuvat and Kumar, 1993, 1994) and in cotyledons (Pan and Acharya, 1995). The pycnidia of the fungus have also been observed in the seed coat, between the seed coat and endosperm, and in the endosperm tissue (Vishunavat and Kumar, 1994). Thus, infected seed may serve as a substrate for the pathogen survival (Pan and Acharya, 1995). The fungus also perennates both as mycelium and spores in the diseased plant refuse and soil. It remains viable for about 14 months in plant debris in soil and for 7 months in dried cultures exposed to different environmental conditions in the field and in the laboratory, respectively (Panwar *et al.*, 1970). If the infected seed is sown or soil of the nursery bed is already infested, the pathogen is carried into the field in the form of diseased seedlings. Long tendrils of conidia ooze out from the mature pycnidia and are locally disseminated by splashing rains and strong winds, on the feet of insects and on tools while doing cultural operations (Fig. 2).

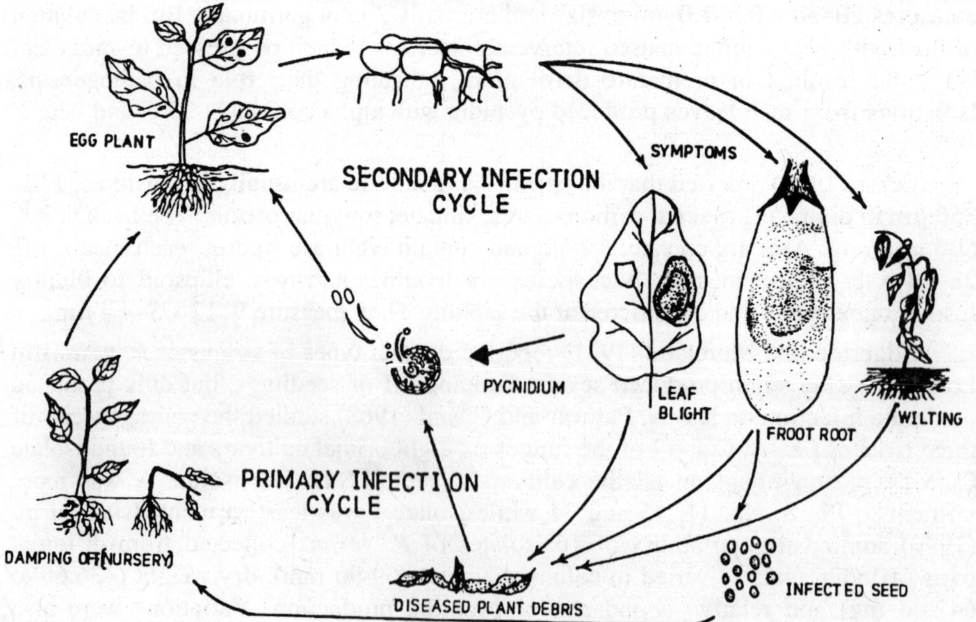

Fig. 2. Disease cycle of Phomopsis blight of brinjal

Epidemiology

The pathogen requires warm and wet weather conditions for infection and subsequent disease development. Pawar and Patel (1957) reported 7–11, 25–28 and 35–40°C as minimum, optimum and maximum temperature ranges, respectively, for the growth of *P. vexans* under *in vitro* conditions whereas Vasudeva (1960) observed optimum growth and sporulation of the fungus at 17–30 and 20–25°C, respectively. Chowdhury and Hasija (1979) found that the optimum relative humidity for disease development is 55 per cent or more alongwith 25°C temperature. Islam *et al.* (1992) studied the influence of environmental factors on leaf blight severity and reported that disease severity was significantly and positively correlated with maximum and minimum temperature, minimum relative humidity, total rainfall and number of rainy days and was negatively correlated with maximum relative humidity which clearly indicates that free water is required for the disease development. Storage rot is maximum at 25°C. Temperatures of 5, 10 and 40°C are unfavourable for disease development (Singh, 1987). The temperature of 25°C, 90 per cent RH, fruits of younger age (5–10 days old), higher inoculum load (>120 spores/ml) and inoculum of younger age (14 day old) were conducive for the development and progress of the disease (Sugha *et al.*, 2002). A single unit rise in micro temperature of crop canopy increased the severity of the disease to four times. (Kumar and Sugha, 2006).

Management

i. Cultural practices : Various cultural practices have been recommended from time to time to reduce the initial inoculum of the pathogen under field conditions as the fungus lives over from year to year readily on fragments of old plants left in the field and infested seed. Collection and burning of the old plants as soon as the crop is over (Edgerton and Moreland, 1921) and deep summer ploughing (Vishunavat, 1992) are some of the recommended cultural practices. Several workers recommended at least 3 years crop rotation to reduce the overwintering inoculum, as the pathogen is host specific, but the fungus can live as saprophyte on sterile vegetative structures of a number of field and garden crops like cauliflower petioles, beet and carrot roots, some of which may help the pathogen to perpetuate indefinitely. In such case, the efficacy of crop rotation in reducing the initial inoculum which may not be of much help. Use of the seed from healthy plants has been advocated for raising the healthy nursery. Only vigorous and completely healthy seedlings should be selected for transplanting in the field. Though the increase in the rate of N fertilizer resulted in the increase in the severity of leaf blight and fruit rot but N as ammonium sulfate and increased rates of K (30 to 60 kg/ha) reduced the disease (Kumar and Sugha, 2003).

ii. Physical treatment : Several workers recommended hot water seed treatment for reducing the seed borne inoculum. *P. vexans* was completely eliminated by 15 minutes immersion of the seed in hot water maintained at 55°C or 30 minutes immersion at 50°C.

iii. Host resistance : The earlier workers found no scope of resistance in brinjal varieties against Phomopsis blight and fruit rot. However, Edgerton and Moreland (1921) for the first time observed that eggplant varieties did not show equal susceptibility. Cvs./lines/hybrids like Florida Hih Bush, Excelsior, Pegan, Bengan, Florida Market, Florida Beauty, N.S.L., Annamalai, Pusa Dwarf, 11a, 264, 238,

P.Br.-125-5, 11a-12-2-1, Cica and Chu Qie Za 1 have been reported resistant against this disease from different parts of the world (Panwar, 1964; Kalda *et al.*, 1976; Datar and Ashtaputre, 1988; Reifschneider *et al.*, 1993), while wild species of *Solanum* like *S. gilo, S. integrifolium, S. xanthocarpum, S. indicum, S. khasianum, S. nigrum, S. torvum, S. sisymbrifolium* were immune (Pawar and Patel, 1957; Datar and Ashtaputre, 1988, Sokhi, 1994) and can be exploited as donor parents for resistance to *P.vexans*. Cvs. like ornamental brinjal and BPL-1, Ramnagar Giant and KS-233 were reported as resistant to moderately resistant at different stages of disease development (Sugha *et al.*, 2002a; Pandey *et al.*, 2002).

Kalda *et al.* (1977) studied the mode of inheritance to Phomopsis blight and fruit rot using three resistant parents viz., 11a-12-2-1, 264-1-1 and 238-4-10-3 and susceptible parents Pusa Purple Long, Pusa Kranti and Pusa Purple Round and reported that resistance was controlled by recessive genes of polygenic nature. Dominant gene effects were found more pronounced than the additive gene effects in most of the crosses. However, resistance is controlled by a pair of dominant genes in resistant *S. melongena* line 83-02 (Liu, 1998a).

iv. Chemical control: Mohanty *et al.* (1994) evaluated five fungicides and found that carbendazim caused hundred percent reductions at all concentrations. Kaushal and Sugha (1995) reported that seed treatment with carbendazim alone or carbendazim + thiram provided better control of pre- and post- emergence damping off.

Since the pathogen also subsists in infected plant debris, chemicals should be sprayed in the seedbeds and under field conditions to get the disease free crop. Sprays of various systemic and non-systemic fungicides like Bourdeaux mixture, copper oxychloride, captan, metiram, maneb, carbendazim, carboxin, mancozeb, captafol, tebuconazole etc. have been found effective for the control of this disease under field conditions (Teo, 1982; Grewal and Jhooty, 1987; Islam *et al.*, 1989, 1993, Jadeja, 2003; Manna *et al.*, 2004), but best benefit/cost ratio of 3.95 was obtained with three sprays of carbendazim. The calyx was the most infective part of the plant between the last 10 days in June to the last 10 days of July and fungicide application during the last 10 days of June resulted in a 54.9 per cent increase in fruit yield (Liu 1998). Various strains of *P. vexans* collected from Himachal Pradesh showed maximum sensitivity to prochloraz followed by carbendazim and mancozeb (Kumar and Sugha, 2004).

The development of resistance in *P. vexans* to benomyl has been reported by Teo (1984) and Jacqua and Gerion (1988) and observed its reduced efficacy in the management of this disease. Rotation of fungicides or application of systemic fungicides in combination of non-systemics is the best strategy. Sprays of copper oxychloride in combination with benomyl, mixture of captafol and carbendazim (Ceretal) and maneb plus thiophanate-methyl (Peltar) have been suggested to combat the problem of resistance and effective control of the disease (Teo, 1984; Jacqua and Gerion, 1988).

v. Plant extracts : Mohanty *et al.* (1995) studied the allelopathic control of *P. vexans* by aqueous leaf extracts of five plants and maximum fungal growth inhibition was caused by *Allemanda cathertica* (93.75%) followed by *Aegle marmelos, Catharanthus roseus, Polyalhia longifolia, Azadirachta indica,* while *Ocimum sanctum* was least effective. Similar results with *A. cathartica* have also been reported (Panda *et al.*, 1996). Kuprashvili (1996) used extracts of garlic and celery for treatment of seeds of aubergine infected with *P. vexans* and reported that the plant extracts disinfected seeds and increased yields. Spraying of neem aqueous extract and dhotra acetone extract (5%) controlled this disease on brinjal (Dhakate *et al.*, 2008).

vi. Biological control: All the isolates of *Beauveria bassiana* inhibited the growth and sporulation of *P. vexans* under *in vitro* conditions while minimum disease severity was observed when Bb-III (5 x 10^6 cfu/ml) applied seven days prior to inoculation (Chani and Bedi, 2004). The antagonist acted directly against *P. vexans* and not as an entomopathogenic fungus since insects do not have any role in the entry of this fungus. Further, Bala and Bedi (2010) evaluated the efficacy of *Beauveria bassiana* against *Phomopsis vexans* and reported increase in seed germination upto 29.2 per cent.

vii. Homoeopathic drugs : Mohanty *et al.* (1995a) evaluated five homoeopathic drugs against *P. vexans* under *in vitro* conditions and reported that Q potencies of Thuja and Tursicum at 2 per cent concentration provided hundred per cent inhibition, while Thuja was effective (73.9%) at 1 per cent concentration. They further observed that Q potency of the drugs was more effective than other potencies tested. Thuja and Turscium each at 2 per cent concentration were as effective as the fungicide Bavistin in inhibiting the growth of *P. vexans* (Panda *et al.*, 1996).

2. SCLEROTINIA WILT

It is a serious disease of eggplant. The disease is more severe in those areas where the plentiful moisture and cool weather are available. The damage to the plants is mainly through partial or complete wilting of the affected plants.

Geographical distribution

The pathogen, *Sclerotinia sclerotiorum* is geographically universally distributed and attacks a wide variety of hosts. The pathogen was first recorded in England in the year 1837 on vegetables (de Bary, 1886). Since then it has been reported from all parts of the world. In India, the fungus was reported for the first time on safflower (Joshi, 1924). In India, it has been reported on eggplant from Tarai region of Uttarakhand (Singh, 1987) and Katrain area of Himachal Pradesh (Kapoor, 1988).

Losses

About 1.5 per cent average annual crop loss due to this disease has been reported from USA alone (Purdy, 1979). In a seed crop, more than 40 per cent of the plants were infected in Katrain area of Himachal Pradesh due to this disease (Kapoor, 1988). The 1000 seed weight of seed extracted from diseased fruits was only 500 mg and germination 0 per cent compared to 3200 mg and 100 per cent from healthy fruits.

Symptoms

The initial symptoms of the disease appear as circular to elongated water-soaked lesions closer to the inflorescence followed by watery soft rot. At the point of infection, a dry, discoloured spot develops. As a result of tissue necrosis, the portion of the plant beyond the point of infection wilts. If the infection is at the base of the main stem, the entire plant wilts. However, if only few branches are attacked then partial wilting of the plant may take place. In the advanced stages under cool humid conditions, the mycelium emerges out and creamy coloured compact sclerotial initials as well as matured black sclerotia of varying sizes are evident on above ground parts (Kapoor, 1988). The affected plants show characteristic whitish growth of the fungus all over the infected surface under high moisture conditions. Fruits are also attacked. The rotting of the flesh takes place and in rotting tissue large number of sclerotia are seen. On splitting open the infected stems, a large number of sclerotia are seen embedded in the tissues.

The Pathogen

The fungus responsible for this disease is *Sclerotinia sclerotiorum* (Lib.) deBary. Mycelium in culture as well as on host surface is hyaline, cottony, branched, consisting of closely septate hyphae, which are 9.0 to 18.0 μm broad and filled with dense granular protoplasm. Microconidia (spermatia) are produced on short lateral branches of the vegetative mycelium in chains and measure 1.5 to 3.5 μm. When the food supply is exhausted and the vegetative growth ceases, the hyphae with granular protoplasm collect in small dense masses and form sclerotia. These sclerotia are first white in colour and then turn black. These are formed at the periphery of the colony and are round to irregular in shape measuring 2.5 to 15 mm. The sclerotia in culture are morphologically similar to those produced on the host.

The sclerotia also germinate by producing stalked apothecia, which range from one to five per sclerotium when proper humidity and light conditions are provided to overwintered sclerotia. The best appothecial development takes place at 7–11°C (Kruger, 1976) and their production ceases at 16–28°C with low soil moisture. Length of the stipe of the apothecia ranges from 6–10 mm, whereas diameter of the apothecial discs ranges from 2–9 mm. Apothecia are brown in colour and are round or lobate type. Asci are cylindrical measuring 108–153 x 4.5–10 μm in size. Each ascus contains eight ascospores, which are released in clouds. Ascospores are elliptical and ranged from 7–16 x 3.6–10 μm in size.

Host Range

S. sclerotiorum is a cosmopolitan pathogen and occurs worldwide. It is known to encompass 64 plant families, 225 genera and 383 species (Purdy, 1979). In Solanaceae family alone, it infects 19 species belonging to 10 genera including *Solanum melongena*.

Disease cycle

In the areas where brinjal crop is grown regularly year after year in the same field, the fungus overwinters in soil in the form of sclerotia in the plant debris (Zhong, 1987). In the presence of proper humidity and light conditions, sclerotia germinate by forming apothecium which in turn form asci and ascospores. If the spores, upon escaping from the ascus, lodge on a susceptible host, a new infection may originate. Mycelium from sclerotia is also capable of infecting eggplant. After infection, the fungus produces oxalic acid, which kills the cells in advance. During cool and moist weather fluffy mycelium appears on the host surface and when food is exhausted, the mycelium coagulates and starts forming sclerotia, which are initially white in colour, but later turns black (Fig. 3).

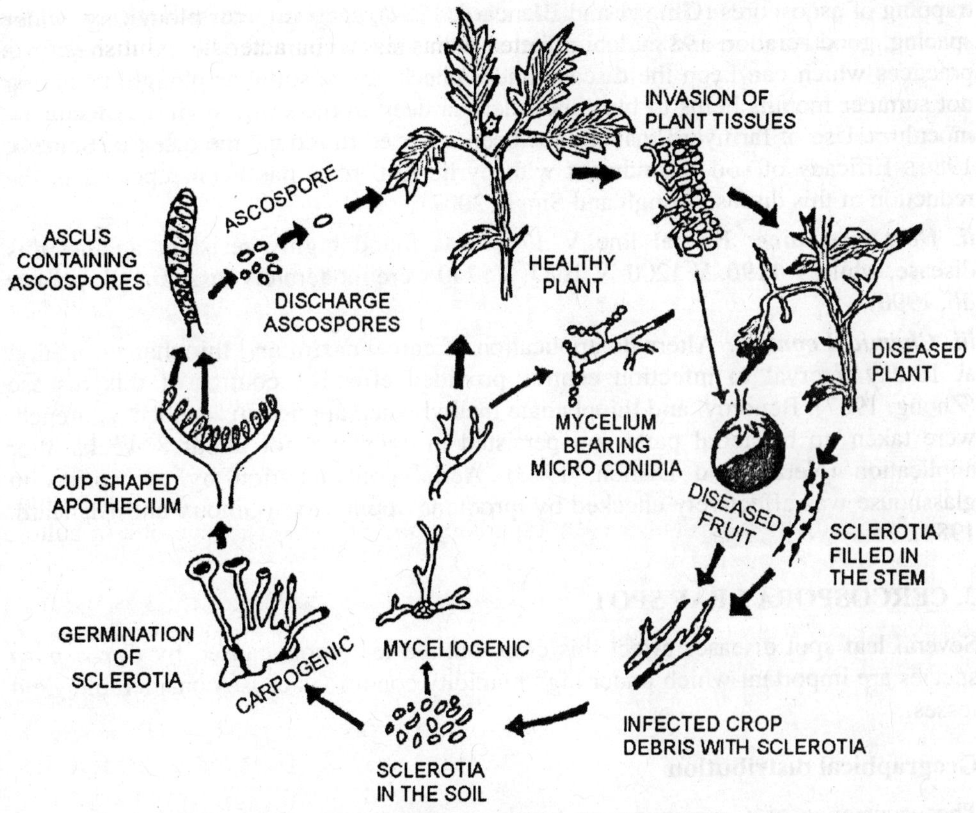

Fig. 3. Disease cycle of sclerotinia wilt of brinjal caused by *Sclerotinia sclerotiorum*

Epidemiology

The fungus responsible for this disease is soil borne in the form of sclerotia, which can infect the susceptible host over a wide range of temperatures i.e., from 0 to 25ºC with an optimum at 15 to 20ºC. High moisture and low temperature conditions are necessary for infection. The fungus can tolerate wide pH range, but is best adapted to an acidic substrate. Flowers are the most susceptible plant organ, which mark the end of the incubation period and the beginning of symptom expression (Tores and Moreno, 1991). Analysis of the tendencies of the number of infections and number of infected plants revealed a sigmoidal curve that was a function of the accumulated hours of which RH was >80 per cent. Application of nitrogenous fertilizers significantly enhances Sclerotinia diseases (Sandhu, 1986). The pathogen can attack a number of vegetable crops like pea, cauliflower, tomato, beans etc. that help in the build-up of inoculum.

Management

i. Cultural practices: Cultural practices like destruction of infected plant debris, suitable crop rotation, e.g., with paddy or maize, polyethylene mulching of irrigation channels to inhibit sclerotial germination, disturbed apothecial formation and trapping of ascospores (Ginoux and Blancard, 1984), deep summer ploughing, wider spacing, good aeration and mulching either with grass or pine needle are some of the practices which can keep the disease under check. Deep summer ploughing during hot summer months helps to bury the sclerotia deep in the soil and thus reduces the inoculum. Use of farmyard manure has been reported to reduce the disease (Sandhu, 1986). Efficacy of soil amendment with pyrite (2-0 t/ha) has been reported in the reduction of this disease (Singh and Singh, 2007).

ii. Host resistance: Brinjal line V 1755 was found highly resistant against this disease, while V 1196, V 1200, V 1687, V1740 were moderately resistant (Kapoor *et al*., 1990).

iii. Chemical control: Alternate application of carbendazim and thiophanate methyl at 10-day interval to infection centres provided effective control of this disease (Zhong, 1987). Benomyl and thiophanate methyl when applied in seedbed as drench, were taken up by aerial parts and persisted in eggplants for about 8 weeks after application (Netzer and Dishon, 1973). Aerial contamination by ascospores in glasshouse was effectively checked by iprodione application (Ginoux and Blancard, 1984).

3. CERCOSPORA LEAF SPOT

Several leaf spot diseases affect this crop, but the leaf spots caused by *Cercospora* species are important which under high humidity conditions cause considerable yield losses.

Geographical distribution

The occurrence of *Cercospora* sp. has been reported from almost all the tropical countries including Japan, South Africa, Sierra Leone and India (Bhartiya *et al*., 2000).

Losses

Occasionally, species of *Cercospora* can cause serious losses, especially when the crop is grown in rainy season and when the flower pedicel is infected. The loss due to this disease ranges between 60 to 80 per cent if the crop is not sprayed (Ullasa and Sohi, 1987).

Symptoms

Several species of *Cercospora* have been reported to be associated with this disease in India (Vasudeva, 1963), which can be distinguished on eggplant according to their symptoms produced by different species.

i. Cercospora melongenae : The symptoms of the disease appear as circular to irregular leaf spots, which are usually large and brown to grayish brown in colour.

Later spots coalesce together and affected leaves fall down prematurely. Sometimes fruits may rot.

ii. Cercospora egenulae: The occurrence of this species has been reported from Karnataka (Ullasa and Sohi, 1987). Early symptoms start as minute water-soaked specks, which gradually enlarge to become circular to angular leaf spots of 2–6 mm diameter. Early infection noticed on the lower leaves gradually spreads upwards with high humidity and rainfall. These spots often coalesce to form bigger spots having a papery pale yellow necrotic region in the centre. The necrotic region in the centre disintegrates bearing a shot hole. Infected leaves turn yellow, droop and fall prematurely resulting in quick defoliation of the entire plant.

iii. Cercospora solanigena: This leaf spot was reported from North Eastern Uttar Pradesh (Bhartiya *et al*., 2000). Leaf spots are amphigenous circular to irregular later coalescing, necrotic, dark brown in centre and light brown in margin, 0.5–2 mm wide.

The Pathogen

The disease is caused by several species of *Cercospora* like *C. melongenae* Wells. *C. solenigena* Bartiya, Dubey and Singh and *C. egenulae* (Syd.) Chupp and Doidge.

(a) C. melongenae: Conidiophores are simple, dark septate and 30–60 x 4–6 μm in size. Conidia are clavate, tapering to a blunt distal tip, subhyaline, 3–12 septate and 38–119 x 4–8 μm in size.

(b) C. egenulae : Fruiting both epi- and hypophyllous, stromata distinct dark brown, globose to irregular, 20 to 80 μm diameter, fascicles dense to olivacious brown, paler and narrower towards the tip, a septate, not branched, non-geniculate, undulate with rounded apex, 2.5–4.5 x 3–31 μm in diameter. Conidia are hyaline to subhyaline, shorter ones distinctly cylindric and very long ones obclavate or almost acicular, straight, 1–5 septate, base subtruncate to rounded, tip obtuse, 3–6 x 20–85 μm in diameter.

(c) C. solanigena: Mycelium internal, hyphae branched, septate, dark olivaceous. Stromata well developed, pseudoparenchymatous, dark, olivaceous, 10–30 μm in diameter. Conidiophores arising in fascicles (1–4) from stromata, macrone-matous to fasciculate, erect, straight to flexuous, unbranched, smooth, 1–6 transversely septate, geniculate, light olivaceous to dark brown, 16–100 x 3–5 μm. Conidiogenous cells integrated, terminal, polyblastic, cicatrized, scars conspicuously thickened paler in colour. Conidia acropleurogenous, holoblastic, dry, solitary, unbranched, 1–5 transversely septate, smooth walled, hyaline, acicular, straight to curved, base truncate, apex subacute to sub obtuse, hilum thickened, 15–84 x 2–5 μm.

Disease cycle

The fungus survives from one season to another in infected plant debris in soil and also in infected seed wherever fruit rot occurs. Diseased crop debris is the main source of primary inoculum. Dispersal of spores under field conditions takes place through air, water and insects.

Epidemiology

The pathogen survives in the infected plant debris in soil. Warm days and cool nights are ideal for infection. The disease is favoured by high humidity (Ullasa and Sohi, 1987). Water droplets must be present for spore germination and germ-tube penetration. Heavy persistent dews are more favourable. Moist wind, irrigation water and insects help in local transmission of the pathogen. Patel (2009) recorded higher concentration of spores of *Cercospora* sp. during August to October due to harvesting activity and light rain with high humidity.

Management

i. Cultural practices: Cultural practices like destruction of crop debris, crop rotation, use of disease free seed and wider plant spacing should be followed to reduce the primary inoculum of the pathogen in field.

ii. Host resistance: Cultivars like Manjari Gota, Junagadh Selection–1, Black Round (Ullasa and Sohi, 1987) Mullukanta, Aruna, MDU-1 and Arka Nidhi (Thiruvelvan *et al.*, 1999) showed resistant to moderately resistant reactions under field conditions, while Udupi Gulla, 86(M) GO-3, Composite-1 and *Solanum macrocarpon* showed immunity by reacting hypersensitively (Madalageri *et al.*, 1988).

iii. Chemical control: Sprays of various systemic and non-systemic fungicides like zineb, carbendazim and thiophanate methyl have been reported effective for the control of this disease (Ullasa and Sohi, 1987; Yadav and Singh, 1995; Patel *et al.*, 1999).

4. ALTERNARIA LEAF SPOT

This is minor leaf spot disease of eggplant and occasionally under favourable weather conditions may cause yield losses by premature defoliation.

Symptoms

The spots on leaves are brown coloured with concentric rings. Several spots coalesce to form large necrotic areas. Disease normally starts from the margins of the leaves. Severely infected leaves may fall off. Necrotic spots are also formed on fruits. The infected fruits turn yellow and sometimes drop prematurely.

The Pathogen

The fungus responsible for this disease is *Alternaria melongenae* Rang and Samb. No sexual stage of the fungus is known so far. The mycelium consists of septate, branched, light brown hyphae, which become darker with age. The size of conidia ranges from 31.5 – 89.5 x 10.10 – 5.21.0 μm .

Disease cycle

The fungus overwinters in the infected plant debris as mycelium. In favourable weather, the fungus starts producing conidia which come in contact with lower most leaves and cause primary infection. The conidia produced on primary spots are blown by moist wind and cause secondary infection.

Epidemiology

The fungus perennates in infected plant debris. The disease initiates in the month of late October to early November. However, it is more severe during November to December. Frequent rains or heavy dew favours the disease.

Management

i. Cultural practices: The pathogen survives as mycelium in the infected plant debris, it will be better if the infected material is collected and destructed. Follow suitable crop rotation and use healthy seed.

ii. Chemical control: Treat the seed with thiram or captan @ 0.3 per cent. With the initiation of the disease, spray the crop with mancozeb (0.25%) or copper oxychloride (0.3%) and repeat at 14 days interval.

(B) BACTERIAL DISEASES

1. BACTERIAL WILT

Bacterial wilt is a limiting factor in successful cultivation of eggplant crop particularly in tropical, subtropical and some warm temperate climates. In hills, the incidence of this disease was high at intermediate elevations i.e., between 800 to 1400 m.a.s.l. and decreased with the increase in altitude (Sunaina and Tomar, 2003).

Geographical distribution

Kelman (1953) was the first to review the literature and indicated the worldwide economic importance of bacterial wilt on its many hosts. Since then, the occurrence of the disease has been reported from Brazil, Bangladesh, Guadeloupe, Venezuela, Thailand, Tanzania, Poland, Pakistan and India. In India, the disease has been reported from almost all the states wherever this crop is grown and climate is favourable.

Losses

Losses from this disease are tremendous particularly during warm humid weather. Reports of complete crop loss in the field are not uncommon.

Symptoms

The initial symptoms of the disease on brinjal plants include sudden wilting of foliage of the youngest leaves, which may recover during night. With the progress of the disease, the initial wilt is succeeded by a sudden and permanent wilting of entire plant, with a slight or no leaf yellowing. Infected young plants die immediately, while older plants may first show leaf drooping and discolouration only on one side or part of the plant and eventually the whole plant wilts and dies. The vascular system in affected plants is discoloured from pale yellow to brown in colour. A simple diagnostic test is to place cross section of the infected tissues suspended in water that yields milky white exudates of the bacterial cells.

The Pathogen

The disease is caused by the bacterium *Ralstonia solanacearum* (Smith) Yabuuchi *et al.* The bacterium is Gram-negative rod, 0.5 x 1.5 μm, aerobic and motile by one polar flagella.

Strains of *R. solanacearum* differ in host range, geographical distribution, pathogenicity and physiological properties. Strains are subdivided informally in to five races based upon host range or differentiated in to five biovars based upon the utilization and/or oxidation of three disaccharides and three sugar alcohols as carbon sources. In India, race 1, 3 and 4, and biovar II, III, and IV (Mathew *et al.*, 2000; Sunaina and Tomar, 2003) are prevalent.

Host Range

Bacterial wilt affects more than 200 plant species belonging to over 40 families. Important hosts include potato, pepper, tobacco, tomato, eggplant, banana, *Datura metal, Momordica charantia, Cucurbita* sp. *Phaseolus vulgaris, Lablab purpureus, Vigna unguiculata* var. *sesquipedalis* (Bulbul *et al.*, 2001), *Melanthera discoides* (Coelho Netto *et al.*, 2001) and *Coleus barbatus* (Netto and Assis, 2002). Besides these, *R. solanacearum* could survive in the rhizosphere, on the rhizoplane and inside the roots of a large number of weed plants like *Solanum nigrum, S. anguivi, S. torvum, S. xanthocarpum, Croton bonplandianum, Euphorbia hirta, Colocasia esculenta, Portulaca oleracea, Phyllanthus niruri* and *Parthenium hysterophorus* in West Bengal (Samaddar *et al.*, 1998).

Disease cycle

The bacterium has a wide host range and can also survive in moist soil for longer periods in the absence of plant debris. Diseased plant debris also helps in the overseasoning of the bacterium. Seed borne nature of the pathogen in eggplant has also been reported (Sumithra *et al.*, 2000). Moreover, the bacteria can survive for long periods on surface water also (Mazzucchi *et al.*, 2000). Entry of the pathogen in the host is via root through mechanical/nematode/insect injury or growth cracks. Ghosh *et al.*, (2016) reported that prior infection of *Meloidogyne javanica* aggravated the disease severity (60–75%) than *R. solanacearum* alone (35.0%). Enhancement in the development of symptoms in the presence of root knot nematodes has also been observed (Verma *et al.*, 1997). The bacterium colonizes the xylem vessels and spreads throughout the plant. The incubation period varies depending upon age and susceptibility of the plant involved. The bacteria invade the intercellular spaces of the parenchyma cells in the cortex and pith and disintegrate the cell walls resulting in pockets filled with bacterial cells as well as polysaccharide and cellular debris. Bacteria are released back into the soil from infected roots and decaying plant material (Bruehl, 1987).

Epidemiology

The bacterium survives from one season to another in infected plant debris in soil, in seed, on surface water besides wide host range. The pathogen generally does not survive and reproduce below 10ºC (van Elsas *et al.*, 2000) although it can survive over a wide range of temperature i.e., 15 to 37ºC and optimum being 35 to 37ºC. The disease usually develops when mean temperature is above 20ºC and more severe wilting is seen at higher temperatures of 30ºC or more alongwith high soil moisture. Though the pathogen can grow over a wide range of pH i.e., from 6.2 to 7.4, but alkaline soils are generally unfavourable for disease development. Dissemination of the bacteria occurs with irrigation or rainwater, on infected plants, infested seeds of tomato and eggplant and the soil (Goto, 1990).

Management

No chemical has been found effective against this disease. However, nematodes in the soil should be eradicated with use of some nematicides because in the presence of nematodes, the symptoms are expressed early than alone and incidence of the disease increases. The disease can be kept under check by using following.

i. Cultural practices: Cultural practices like use of healthy seed, suitable crop rotation, control of weeds and volunteer plants, avoidance of surface water for irrigation (Baginska and Kordyla-Bronka, 1998), avoidance of movement of soil from infected fields to disease free fields through equipments and workers, and disinfestation of the equipment should be followed to keep the disease under check. Nursery should be grown in disease free soil and should be disinfested before sowing either with soil solarization or steam sterilized or treated with chemicals like formaldehyde.

ii. Host resistance: Cultivars/lines like Slim Jim and M701F1 (Chen *et al.*, 1997), Surya Swetha, Annapurna, BB 7, BB 13-1, BB 44 (Singh and Gopalakrishnan, 1998), SM-141, SM-6-6-C, DPL-B-4, Arka Keshav (Dalal *et al.*, 1999), 95-17, BB-46, CHES 243 (Kishan *et al.*, 2000), Arka Neelkanth, Arka Needhi (Chaudhary and Sharma, 2000), IHR-12, IHR-21, IHR-54, B-7, DPL-B-1, SM-6-6, BB-60-C (Fugro, 2001), EG-191, TS-3 (Sadashiva *et al.*, 2001) and Dingaras Multiple Purple (Tanimoto and Nakasone, 2002) have been reported resistant to this disease. Few F 1 hybrids like Surya x SM-116, Arka Keshav x SM-71 (Gopalakrishnan *et al.*, 2000) and Xiangqie 4 (Yao *et al.*, 2000) also showed resistant reaction to the disease under field conditions.

Bacterial wilt resistance was found to be recessive and polygenic in nature (Tanimoto and Nakasone, 2002) or is controlled by a number of minor, major and cytoplasmic genes (Feng *et al.*, 2003). Cv. Singnath was the most resistant nearly immune with 0.93 per cent of wilt, while cv. BB-40, BB-64 and Green Round were resistant under Tripura conditions (Biswas and Singh, 2007). Ramesh *et al.* (2016) identified a wild relative, *Solanum torvum* as a resistant one to this disease and used as root stock for grafting with cultivated local brinjal types provided complete protection from the disease when evaluated under natural epiphytotic conditions.

iii. Biological control: Root inoculation with Hrp-mutant strain of *Ralstonia solanacearum* induced protection against subsequent inoculation of pathogenic strain (Trigalet *et al.*, 1998). It was further observed that more invasive the Hrp mutant strain, the more protection it gives. Such isolates of *R. solanacearum* can be utilized in the management of this disease. Fluorescent Pseudomonads strains like FPP 8, FPP 9, FPP 13, FPP 5 and FPP 15 showed antagonism to *R. solanacearum* under *in vitro* conditions, while FPP 5 was the most promising strain for disease control (Chao *et al.*, 1997).

iv. Plant extracts: Seedlings dipped in the mixture of Asafoetida, turmeric powder and water (1g : 5g : 1 litre) before transplanting were protected from bacterial wilt (Bhattacharya *et al.*, 1994). The degree of protection increased with the length of time for which seedlings are dipped. Soil drenches of Asafoetida, turmeric powder and water at 20, 50 and 80 days after transplanting were also found effective for disease control (Pun and Das, 1997). Plant extracts viz., Asafoetida + turmeric, onion and garlic increase the yield of the crop (Sharma and Kumar, 2009).

v. Integrated Disease Management: Combination of seed treatment with streptomycin sulphate (1 g/kg seed), soil application of bleaching powder (20kg/ha), root dip treatment (streptomycin sulphate (1 g) + copper sulphate (1.75 g)/l of water) and cv. Arka Keshav recorded the highest fruit yield with minimum disease incidence (Arasumallaiah and Rangaswamy, 2008). Integration module consisting of green manuring with dhaincha leaf (25 q/ha) + bleaching powder (20 kg/ha) + lime (500 kg/ha) + consortium of fluorescent pseudomonads (PF-2 and PF-3) (Seedling treatment and soil spot application) was found best both in terms of disease reduction (72.25%) and enhancing yield (11.8 t/ha) over the control (Ghosh *et al.,* 2015).

(C) PHYTOPASMA DISEASES

1. LITTLE LEAF

This is an important disease particularly in Southeast Asian countries where the activity of the vector of the disease is more.

Geographical distribution

This disease was first reported from India in 1938 (Singh, 1987). Since then it has been reported from India, Sri Lanka and Bangladesh (Singh, 1987; Siddique *et a*l., 2001). In India this disease has been reported from almost all states wherever this crop is grown.

Losses

There are variable reports about its incidence from one state to another. Incidence of the disease depends on the susceptibility of the host varieties planted and availability of vector and primary inoculum in the area. However, the incidence of the disease was assessed in Warangal district of Andhra Pradesh and it was reported to be less than 8 per cent (Neerja *et al.*, 1997). The number, size and weight of fruits of the affected plants is reduced and seed production is almost negligible as compared to healthy plants (Joshi and Bose, 1983). Yield and growth of the plants were highly affected when plants were infected during early growth period i.e., before 35 and 55 days after sowing (Madhavi *et al*., 2008). The reduction in the primary productivity of little leaf-infected plants is reported to be 64.2 per cent (Mall and Sheikh, 1977).

Symptoms

The disease is characterized by reduction in the size of shoots of the affected plants. As the name indicates, the leaves become very small, thin and pale green. The foliage of the affected plants is clustered together and malformed. Internodal distance is reduced in the affected plants. Newly formed leaves are much more shorter.

Auxillary buds are stimulated and plants attain bushy appearance. The disease assumes serious proportions at flowering stage and reduces the yield. Flower formation is incomplete and virescence is observed (Patel and Shah, 1972). The flower apex after sepal and petal formation is transformed into a vegetative one. Flowering and fruiting is rare in affected plants. The growth of the fruits after infection is restricted or stopped and the fruits become hard and fail to mature (Anjaneyulu and Ramakrishnan, 1972).

The Pathogen

The disease is caused by a *Phytoplasma*. Phytoplasma bodies (230 – 770 nm) were found associated with the phloem cells of roots of the eggplant affected with little leaf (Verma *et al*., 1975). Each phytoplasma body contained ribosomes and strands of nuclear material surrounded by 16.5 nm wide triple layered unit membrane. Electron micrographs revealed pleomorphic phytoplasma bodies in the sieve elements of the affected plants (Kothari *et al*., 1982). Symptom remission occurred when affected plants were treated with achromycin (tetracycline hydrochloride), aureomycin, lederomycin and terramycin (Verma *et al*., 1975).

Host Range

The host range of this phytoplasma is very wide. Anjaneyulu and Ramakrishnan (1973) studied the host range of this disease both under artificial inoculation and natural conditions on 75 plant species belonging to 18 genera and six families and reported that the phytoplasma infected 60 species belonging to 13 genera and 5 families. In nature, its occurrence on weed hosts like *Datura* sp. and *Vinca rosea* have also been reported (Singh, 1987). It has also been successfully transmitted to tomato and tobacco.

The Vector

The vector of this disease is a jassid, *Hishimonus phycitis* (Dist.). Phytoplasma bodies were also found in the haemolymph and fat bodies of the vector, *H. phycitis* (Shantha and Lakshmanan, 1984).

Epidemiology

In nature, the phytoplasma perennates on several weed hosts like *Datura* sp. and *Vinca rosea* and several others. The vector *H. phycitis* transmits the disease from weed hosts to healthy plants. Early sown crops (June sown) escape the disease than July and August due to differences in vector populations (Kesharwal and Khare, 1986). Continuous cropping of the susceptible varieties increased disease incidence, as did the presence of weeds, sweet potato or tomato crops nearby which helped to increase vector populations. However, no appreciable effect of NPK application on the incidence of the disease was observed (Srinivasan and Chelliah, 1981).

Management

i. Cultural practices: Cultural practices like roguing of affected plants, destruction of weeds like *Datura* sp. and *Vinca rosea* from in and around the field, avoid planting of crops like sweet potato and tomato near by the eggplant field which helped to increase the population of vector and early sowing of the crop i.e., third week of June help in reducing the disease incidence (Kesharwal and Khare, 1986).

*ii. **Host resistance:*** Several cultivars/lines viz., Pusa Purple Cluster (Gupta, 1979), Chaklasi Doli, Doli-5 (Jyani *et al.*, 1995), CHBR 3, DBSR 44, DBSR 91-2, NDBH 8, HOE 414, JB 64-2, H 9, PBS 12 (Das and Mitra, 1999) were reported resistant, while ICI 11323, ICI 12938, ICI 13004 and ICI 13010 had longer incubation period and symptoms developed after 9–10 weeks. However, variety 'Hisar Shyamal' has been reported as tolerant (Kalloo *et al.*, 1991). Wild species of *Solanum* like *S. integrifolium* and *S. gilo* showed a hypersensitive response to the pathogen (Chakrabarti and Choudhary, 1975; Das and Mitra, 1999).

*iii. **Chemical control:*** Pre-transplant seedling dip in carbofuran (0.2%) for 24 h was reported to be effective against this disease (Mitra and Mazumdar, 1977). Application of phorate (1 kg/ha) to the seed bed or dipping of seedlings in aqueous solution of 0.05 per cent tetracycline + monocrotophos or soil application of phorate (1kg/ ha) 10 days after transplanting or spray application of monocrotophos in nursery and in field at 10 days interval for at least 10 times with the appearance of vector were found effective in reducing the incidence of the disease (Joi and Sonone, 1977). Application of Thiodan, Dipel (a formulation of *Bacillus thuringiensis* var. *alesti*) + Ekalux also reduced the incidence of this disease (Narayanasamy *et al.*, 1979).

Application of tetracycline to seedlings by girdling (via a strip of cotton tied round the stem) delayed the symptoms development by 4 months whereas control plants developed symptoms after 14 days (Verma and Dubey, 1976). The disease can also be kept under check by dipping the seedlings in tetracycline solution before transplanting followed by soil application of phorate (Maheashwari *et al.*, 1988).

REFERENCES

Anjaneyulu, A. and Ramakrishnan, K. 1972. Symptomatology of little leaf disease of eggplant. *Mysore J. Agric. Sci.* **6:** 147–154.

Anjaneyulu, A. and Ramakrishnan, K. 1973. Host range of eggplant little leaf disease. *Mysore J. Agric. Sci.* **7:** 568–579.

Arasumallaiah, L. and Rangaswamy, S.D. 2008. Effect of chemicals in the management of bacterial wilt on different brinjal varieties in coastal Karnataka. *Environ and Ecol.* **26:** 1445–1446.

Baginska, H. and Kordyla–Bronka, M. 1998. *Pseudomonas solanacearum* E.F. Smith (Syn. *Ralstonia solanacearum* (Smith Yabuuchi *et al.*) – the potential threat to crops in Poland. *Prog. Pl. Prot.* **38:** 480–483.

Bala, A. and Bedi, J.S. 2010. Biocontrol of seedling blight of brinjal caused by *Phomopsis vexans* using *Beauveria bassiana. Pl. Dis. Res.* **25(1):** 61–62.

Bhartiya, H.D., Ranjana Dubey and Singh, S.K. 2000. New *Cercospora* spp. associated with vegetable crops in North Eastern Uttar Pradesh. *Indian Phytopath.* **53(2):** 149–152.

Bhattacharya, S.K., Bora, L.C. and Saikia, B.K. 1994. An unconventional method for management of bacterial wilt of brinjal. *J. Agric. Sci. Soc. North East India* **7:** 229–231.

Biswas, Subrata and Singh, N.P. 2007. Effect of host genotypes and cultural practices for the management of bacterial wilt in brinjal (*Solanum melongena* L.) *Indian Phytopath.* **60 (4):** 438–441.

Bruehl, G.W. 1987. *Soilborne Plant Pathogens*. Mcmillan Publishing Co., 866 Third Avenue, New York, USA.

Bulbul, M.A., Mian, I.H. and Rashid, M.A. 2001. Screening of three important crops against *Ralstonia solanacearum* and evaluation of some plant species as hosts of the bacterial pathogen. *Bangladesh J. Pl. Pathol.* **17:** 63–67.

Chakrabarti, A.K. and Choudhary, B. 1975. Breeding brinjal resistant to little leaf disease. *Proc. Indian Natl. Sci. Acad.–B.* **41:** 379–385.

Chani, A.B. and Bedi, J.S. 2004. Evaluation of *Beauveria bassiana* as a biofungicide to control Phomopsis blight of brinjal. *Pl. Dis. Res.* **19(2):** 155–158.

Chao, Y.C., Chen, Y.L., Chao, Y.C. and Chen, Y.L. 1997. Influence of fluorescent pseudomonads isolated from eggplant roots on the growth and the disease development of bacterial wilt of eggplant. *Bull. Natl. Pingtung Polytech. Inst.* **6:** 101–112.

Chaudhary, D.R. and Sharma, S.D. 2000. Screening of some brinjal cultivars against bacterial wilt and fruit borer. *Agril. Sci. Dig*. **20:** 129–130.

Chen, N.C., Li, H.M., Wang, J.F., Chen, N.C., Li, H.M. and Wang, J.F. 1997. Bacterial wilt resistance sources in eggplant, *Solanum melongena. Bacterial Wilt Newsl.* **14:** 4–5.

Chowdhury, S.R. and Hasija, S.K. 1979. Pathological studies on *Phomopsis vexans* causing soft rot of brinjal fruits. *Indian Phytopath.* **32(3) :** 495–496.

Coelho Netto, R.A., Noda, H. and Boher, B. 2001. *Melanthera discoides*: a new *Ralstonia solanacearum* host. *Fitopatol. Brasileira* **26:** 781.

Dalal, N.R., Dalal, S.R., Golliwar, V.G. and Khobragade, R.I. 1999. Studies on grading and pre–packaging of some bacterial wilt resistant brinjal (*Solanum melongena* L.) varieties. *J. Soils and Crops* **9:** 223–226.

Das, A.K. and Mitra, D.K. 1999. Screening of brinjal varieties, lines and wild *Solanum* spp. for resistance to little leaf disease. *Indian J. Agric. Sci.* **69:** 726–728.

Datar, V.V. and Ashtaputre, J.U. 1988. Studies on resistance to Phomopsis fruit rot in eggplant. *Indian Phytopath.* **41 :** 637–638.

de Bary, A. 1886. Uebereinige Sclerotinien und Sclerotienkrankheiten. *Bot. Ztg*. **44:** 377–474.

De Figueiredo, E.R. and Pereira, H.F. 1944. A serious eggplant disease caused by *Phomopsis vexans. Biologico* **10(11)** : 349–352.

Dhakate, S.R., Ingle, Y.V., Bhongle, S.A. andGoramnagar, H.B. 2008. Efficacy of botanicals against *Phomopsis* blight of brinjal. *Plant Arch.* **8:** 585–588.

Edgerton, C.W. and Moreland, C.C. 1921. Eggplant blight Louisiana *Agric. Exp. Stn. Bull.* **178:** 1–44.

Feng, L.L., Qu, D. Y., Jin, L.P., Lian, Y., Feng, L.L., Qu, D.Y., Jin, L.P. and Lian, Y. 2003. Genetic analysis of resistance to bacterial wilt (*Ralstonia solanacearum*) in eggplant (*Solanum melongena* L.). *Acta Hortic. Sinica* **30:** 163–166.

Fugro, P.A. 2001. Sources of resistance to bacterial wilt in eggplant. *Capsicum and Eggplant Newsl.* **20:** 113–116.

Ghosh, P.P., Dutta, S. and Chattopadhy, A. 2015. Integration of organic and inorganic amendments with native bioagents for bio-intensive management of vascular bacterial wilt on eggplant (*Solanum melongena*). *Indian Phytopath.* **68:** 32–38.

Ghosh, P.P., Dutta, S., Khan, M.R. and Chattopadhyay, A. 2016. Role of *Meloidogyne javanica* on severity of vascular bacterial wilt of eggplant. *Indian Phytopath*. **69(3):** 237–241.

Ginoux, G. and Blancard, D. 1984. Sclerotinia disease of aubergine. Contribution to the setting up of a strategy of control. *Revue Horticola* **245:** 45–50.

Gopalakrishnan, T.R., Kutty, C.N. and Singh, P.K. 2000. Two new wilt resistant F 1 hybrids in brinjal (*Solanum melongena* L.). *J. Trop. Agric.* **38:** 5–8.

Goto, M. 1990. *Fundamentals of Bacterial Plant Pathology*. Academic Press, Inc. 1250 Sixth Avenue, San Diego, CA, USA.

Gratz, L.O. 1942. The perfect stage of *Phomopsis vexans*. *Phytopathology* **32(6)** : 540–542.

Grewal, R.K. and Jhooty, J.S. 1987. Control of Phomopsis fruit rot of egg plant with different fungicides. *Indian Phytopath.* **40(2)** : 263–264.

Gupta, D.K. 1979. Reaction of brinjal lines/ varieties to little leaf at Udaipur under natural conditions. *Indian Phytopath*. **32:** 99–100.

Islam, S.J., Pan, S. and Pan, S. 1989. Chemical control of leaf blight and fruit rot of brinjal caused by *Phomopsis vexans. Indian J. Mycol. Res.* **27(2)** : 159–163.

Islam, S.J., Pan, S. and Pan, S. 1990. Variabilities among isolates of *Phomopsis vexans. Env. and Ecol.* **8(1B)** : 315–319.

Islam, S.J., Pan, S. and Pan, S. 1992. Functional relationship of Phomopsis leaf blight of eggplant with environmental factors and a linear model for predicting the disease. *Indian Phytopath.* **45(2)** : 199–202.

Islam, S.J., Pan, S. and Pan, S. 1993. Economics of fungicidal control of Phomopsis leaf blight of eggplant. *Indian Phytopath.* **46(4)** : 383–388.

Jacqua, G. and Gerion, A.L. 1988. *Corynespora cassicola* and *Phomopsis vexans* in aubergine crops. *Bull. Agronomique Antilles Guyane* **7** : 39–43.

Jadeja, K.B. 2003. Evaluation of different herbicides, fungicides, Phytoextracts and bioagents against *Phomopsis vexans* causing stem and branch blight of brinjal. *J. Mycol. Pl. Pathol.* **33**: 446–450.

Jameel, A. 2006. Intraspecific diversity in *Phomopsis vexans*, inciting Phomopsis blight of brinjal. *J. Res. Birsa Agril. Univ.* **18:** 81–87.

Joi, M.B. and Sonone, H.N. 1977. Effect of different insecticides and antibiotics for the control of little leaf disease of brinjal. *J. Maharashtra Agric Univ*. **2:** 74–76.

Joshi, R.D. and Bose, K. 1983. Yield loss studies on little leaf disease of eggplant. *Indian Phytopath.* **36:** 604–607.

Joshi, S.D. 1924. Wilt disease of safflower. *Mem. Dep. Agric. Indian Bot. Sci.* **13:** 39–46.

Jyani, D.B., Patel, N.C., Ratanpara, H.C., Patel, J.R. and Borad, P.K. 1995. Varietal resistance in brinjal to insect pests and diseases. *Gujarat Agril. Univ. Res. J.* **21:** 59–63.

Kalda, T.S., Swarup, V. and Chaudhary, B. 1976. Studies on resistance to Phomopsis blight in eggplant (*Solanum melongena* L.). *Veg. Sci.* **3** : 65–70.

Kalda, T.S., Swarup, V. and Chaudhary, B. 1977. Resistance to Phomopsis blight in eggplant. *Veg. Sci.* **4** : 90–101.

Kalloo, G., Dixit, J. and Baswana, K.S. 1991. Hisar Shyamal is superior to BR 112 brinjal. *Indian Hortic*. **36:** 29–34.

Kapoor, K.S. 1988. *Sclerotinia sclerotiorum* – a threat to seed production of eggplant. *Capsicum Newsl*. **7:** 92.

Kapoor, K.S., Sharma, S.R. and Gangopadhyay, S. 1990. Assessment of resistance in eggplant against Sclerotinia wilt with a new screening technique. *Capsicum Newsl*. **8–9:** 70.

Kaushal, N. and Sugha, S.K. 1995. Role of *Phomopsis vexans* in damping off of seedlings in eggplant and its control. *Indian J. Mycol. Pl. Pathol.* **25** : 189–191.

Kelman, A. 1953. The bacterial wilt caused by Pseudomonas solanacearum. North Carolina Agric. Exp. Stn. Tech. Bull. **99:** 194pp.

Kesharwal, R.L. and Khare, M.N. 1986. Note on occurrence and spread of little leaf disease of brinjal. *Indian J. Mycol. Pl. Pathol.* **16(2):** 237–240.

Kishan, S., Suryanarayan, M.A. and Swaroop, K. 2000. Performance of bacterial wilt resistant brinjal varieties/lines under two locations during rainy season in Andamans. *Madras Agric. J.* **87:** 356–358.

Kothari, I.L., Dube, H.C. and Doshi, S.K. 1982. Ultrastructural changes in phloem cells of "little leaf" diseased brinjal. *Indian Bot. Reptr.* **1:** 91–95.

Kruger, W. 1976. Investigations on the effect of temperature on apothecial develop–ment of *Sclerotinia sclerotiorum* (Lib.) de Bary. *Nacherichtenbl. Pflanzen–schutz.* **28(9):** 129–135.

Kumar, S. 2012. Factors affecting the development of fruit rots of brinjal. *Pl. Dis. Res.* **27(1):** 39–44.

Kumar, S. and Sugha, S.K. 1999. Role of alpha and beta conidia in pathogenesis of *Phomopsis vexans. J. Mycol. Pl. Pathol.* **29(2):** 166–171.

Kumar, S. and Sugha, S.K. 2003. Effect of fertilizers on Phomopsis disease (*Phomopsis vexans*) of brinjal (*Solanum melongena*). *Indian J. Agric. Sci.* **73:** 32–34.

Kumar, S. and Sugha, S.K. 2004. Sensitivity of different strains of *Phomopsis vexans* to selected fungicides. *J. Mycol. Pl. Pathol.* **34:** 100–101.

Kumar, S. and Sugha, S.K. 2006. Role of weather variables in the development of Phomosis blight of Brinjal. *Pl. Dis. Res.* **21 (1):** 30–32.

Kuprashvili, T.D. 1996. The use of Phytoncides for seed treatment. *Zashchita i Karantin Rastenii* **5 :** 31.

Liu, X.M. 1998. Study on the infective law and chemical control of eggplant *Phomopsis* rot. *J. Jilin Agric. Univ.* **20(1):** 1–5.

Liu, X.M. 1998a. Studies on resistance inheritance of *Solanum melongena* L. to *Phomopsis* rot. *J. Jilin Agric. Univ.* **20(4):** 1–7.

Madalageri, B.B., Dharmatti, P.R., Madalageri, M.B. and Padaganur, G.M. 1988. Reaction of eggplant genotypes to *Cercospora solani* and *Leucinodes orbònalis. Pl. Pathol. Newl.* **6(1–2):** 26–27.

Madhavi, P., Bajpai, R.K. and Mall, T.P. 2008. Effect of little leaf disease (MLO's) infection on growth and yield loss of brinjal. *Indian Phytoptah.* **61:** 372–378.

Maheshwari, S.K., Gupta, P.C. and Suhag, L.S. 1988. Studies on chemical control of little leaf disease of brinjal. *Indian J. Mycol Pl. Pathol.* **18:** 182.

Mall. L.P. and Sheikh, A.S. 1977. The influence of mycoplasmal infection on primary productivity of brinjal var. Pusa Purple Long. *J. Indian Bot. Soc.* **56:** 38–40.

Manna, D.K., Jash, S., Srikanta, D, Das, S.N. and Das, S. 2004. Effects of environmental factors on Phomopsis blight of brinjal and its management. *Ann. Pl. Prot. Sci.* **12(1):** 229–231.

Mathew, S.K. Girija, D., Devi, S.N., Sadhankumar, P.G. Rajan, S. 2000. Variability in isolates of *Ralstonia solanacearum* affecting solanaceous vegetables in Kerala. *Veg. Sci.* **27(2):** 189–191.

Mazzucchi, U., Manzo, A., Forte, L. and Mainolfi, P. 2000. Control of *Pseudomonas solanacearum* obligatory in Italy too. *Informatore Agrario* **56:** 97–101.

Mitra, D.K. and Majumdar, M. 1977. Control of *Hishimonus phycitis* and *Amrasca biguttula* Ishida little leaf disease of brinjal, by pre–transplant seedling soak with carbofuran. *Pesticides* **11:** 25–26.

Mohanty, A.K., Kar, A.K. and Sethi, P.N. 1995. Efficacy of crude leaf extracts of some selected plants in controlling brinjal blight and fruit rot pathogen, *Phomopsis vexans. Crop Res. Hisar* **9(3) :** 447–448.

Mohanty, A.K., Panda, R.N., Kar, A.K. and Sethi, P.N. 1995a. Effect of certain homeopathic drugs on the growth of *Phomopsis vexans* causing fruit rot of brinjal. *Crop Res. Hisar* **9(1) :** 165–167.

Mohanty, A.K., Panda, R.N., Sethi, P.N. and Kar, A.K. 1994. Efficacy of certain fungicides in controlling *Phomopsis vexans* causing fruit rot of brinjal. *Orissa J. Agric. Res.* **7(suppl.) :** 85–86.

Narayanasamy, P., Ragunathan, V. and Bhaskaran, P. 1979. Incidence of brinjal mosaic virus and brinjal little leaf diseases in Dipel– organic insecticide combination treatment. *Madras Agric. J.* **66:** 170–173.

Neeraja, G., Rachel, E.K., Reddy, S.R., Reddy, S.M., Srivastava, H.P., Purohit, D.K. and Ram Reddy, S. 1997. A survey of little leaf disease of brinjal (*Solanum melongena* L.) in Warangal (Dist.) Andhra Pradesh. *Microbial Biotechnol.* 160–163.

Netto, R.A.C. and Assis, L.A.G. 2002. *Coleus barbatus*: a new *Ralstonia solanacearum* host. *Fitopatol. Brasileira* **27:** 226.

Netzer , D. and Dishon, I. 1973. Persistence of benomyl and thiophanate compounds in soil and various plants following soil application. *Phytoparasitica* **1:** 33–37.

Palo, M.A. 1936. The Phomopsis disease of eggplants and its control. *Philipp.J. Agric.* **7(3) :** 289–315.

Pan, S. and Acharya, S. 1995. Studies on the seed borne nature of *Phomopsis vexans* (Sacc. & Syd.) Harter. *Indian Agricultuurist* 39(3) : 193–198.

Panda, R.N., Tripathy, S.K., Kar, J. and Mohanty, A.K. 1996. Antifungal efficacy of homeopathy drugs and leaf extracts in brinjal. *Env. and Ecol.* **14(2)** : 292–294.

Pandey, K.K., Pandey, P.K., Kalloo, G. and Chaurasia, S.N.G. 2002. Phomopsis blight in brinjal and sources of resistance. *Indian Phytopath.* **55:** 507–509.

Panwar, N.S. 1964. Studies on the fruit rot of brinjal (*Solanum melongena* L.) caused by *Phomopsis* species in Punjab. M.Sc. Thesis, Punjab Agricultural University, Ludhiana, India.

Panwar, N.S. and Chand, J.N. 1968. Cultural characters and pathogenecity of three isolates of *Phomopsis vexans* (Sacc. & Syd.) Harter. *Indian J. Microbiol.* **8 :** 203–205.

Panwar, N.S., Chand, J.N., Singh, H. and Paracer, C.S. 1970. Phomopsis fruit rot of brinjal (*Solanum melongena* L.) in the Punjab. I. Viability of the fungus and role of seeds in disease development. *J. Res., PAU.* **7 :** 641–643.

Patel, J.D. and Shah, J.J. 1972. Pathological anatomy of brinjal affected by "little leaf virus". *J. Indian Bot. Soc.* **51:** 18–22.

Patel, K.K, Yadav, F.D. and Shivani. 1999. Efficacy of fungicidal spraying on control of leaf spot disease of brinjal var. Muktakeshi in Faizabad. *J. Liv. World.* **6:** 16–18.

Patel, S.L. 2009. Observation on the air borne *Cercospora* spores causing leaf spot disease on brinjal at Nashik, India. *Flora and Fauna (Jhansi)* **15(1):** 65–67.

Pawar, V.H. and Patel, M.K. 1957. Phomopsis blight and fruit rot of Brinjal. *Indian Phytopath.* **10(2) :** 115–120.

Porter, R.P. 1943. Seed–borne inoculum of *Phomopsis vexans* – its extent and effects. *Pl. Dis. Reptr.* **27 :** 167–169.

Pun, K.B. and Das, G.R. 1997. Management of bacterial wilt of brinjal using asafoetida and turmeric. *Bacterial Wilt Newsl.* **14:** 6.

Purdy, L.H. 1979. *Sclerotinia sclerotiorum*, history, diseases and symptomatology, host range, geographic distribution and impact. *Phytopathology* 69: 875–880.

Puttoo, B.L. and Sohi, H.S. 1975. Seed mycoflora of eggplant. *Indian J. Mycol. Pl. Pathol.* **5 :** 111–113.

Ramesh, R., Achari, G, Asolkar, T., Dsouza, M. and Singh, N.P. 2016. Management of bacterial wilt of brinjal using wild brinjal (*Solanum torvum*) as root stock. *Indian Phytopath.* **69(3):** 260–265.

Reifschneider, F.J.B., Madeira, M.C.B. and Silva, C. 1993. Cica : a new aubergine hybrid resistant to antharacnose and Phomopsis rot. *Hortic. Brasileira* **11(1)** : 57.

Sadashiva, A.T., Reddy, K.M., Girija Ganeshan and Prasad, B.C.N. 2001. Evaluation of eggplant (*Solanum melongena* L.) lines for bacterial wilt resistance. *Capsicum Eggplant Newsl.* **20:** 117–119.

Samaddar, K.R., Chakraborty, M. and Kanjilal, S. 1998. Identification of the races of *Pseudomonas solanacearum* causing wilt of solanaceous vegetables in West Bengal and its survival. *J. Mycol. Res.* **36:** 51–58.

Sandhu, K.S. 1986. *Sclerotinia sclerotiorum* a destructive pathogen of cauliflower seed crop in Punjab. *Punjab J. Res.* **23(1):** 78–81.

Shantha, P.R. and Lakshmanan, M. 1984. Vector borne MLOs of brinjal little leaf. *Curr. Sci.* **53:** 265–267.

Sharma, J.P. and Kumar, S. 2009. Management of *Ralstonia* wilt of tomato through microbes, plant extract and combination of cake and chemicals. *Indian Phytopath.* **62(4):** 417–423.

Siddique, A.B.M., Agarwal, G.K., Alam, N. and Krishna Reddy, M. 2001. Electron microscopy and molecular characterization of phytoplasma associated with little leaf disease of brinjal (*Solanum melongena* L.) and periwinkle (*Catharanthus roseus*) in Bangladesh. *J. Phytopathol.* **149:** 237–244.

Singh, P.K. and Gopalakrishnan, T.R. 1998. Evaluation of brinjal genotypes for yield and bacterial wilt resistance. *J. App. Biol.* **8:** 70–74.

Singh, R. and Singh, L.B. 2007. Evaluation of different soil amendments against Sclerotinia blight of brinjal. *Ann. Plant Protec. Sci.* **15(1):** 265–266.

Singh, R.S. 1987. *Diseases of Vegetable Crops*. Oxford & IBH Publishing Co. Pvt. Ltd., New Delhi. 363pp.

Sokhi, S.S. 1994. Integrated approaches in the management of vegetable diseases in India. *Indian Phytopath.* **47** : 371–376.

Srinivasan, K. and Chelliah, S. 1981. Effect of different levels of fertilizers on the incidence of little leaf disease of brinjal (*Solanum melongena* L.). *Agric. Res. J. Kerala* **19:** 121–122.

Sugha, S.K., Kaushal, N. and Kumar, S. 2002. Factors affecting development of Phomopsis fruit rot of brinjal. *Indian Phytopath* **55:** 26–29.

Sugha, S.K., Kumar, S., Pathania, N.K.and Kumar, S. 2002a. Evaluation of brinjal germplasm against Phomopsis disease. *Capsicum Eggplant Newsl.* **21:** 106–109.

Sumithra, K.U, Krishnappa, M., Vasanth, T.K., Shetty, H.S., Mortensen, C.N. and Mathur, S.B. 2000. Seed borne nature of *Ralstonia solanacearum* in eggplant (*Solanum melongena* L.) cultivars in India. *Seed Sci. Tech.* **28:** 291–299.

Sunaina, V. and Tomar, D.K. 2003. Distribution of bacterial wilt of potato and its races/biovars in Uttranchal hills. *Bionotes* **5:** 13–14.

Tanimoto, H. and Nakasone, W. 2002. Gene analysis of the bacterial wilt resistance in "Dingaras Multiple Purple" (*Solanum melongena* L.). *Hortic. Res. Jpn.* **1:** 103–106.

Teo, C.H. 1982. Plant Pathology– Brinjal (*Solanum melongena*). In *Annual Report of the Research Branch, Dept. of Agriculture, Sarawak, for the year 1980*, Kuching, Sarawak, Ministry of Agriculture and Community Development, 240pp.

Teo, C.H. 1984. Plant Pathology– Brinjal. In *Annual Report of the Research Branch, Dept. of Agriculture, Sarawak for the year 1982*, Ministry of Agriculture and Community Development, 280pp.

Thiruvelvan, P., Thamburaj, S., Natarajan, S., Veeraragavathatham, D. and Prakasam, V. 1999. Reaction of brinjal genotypes, varieties and hybrids to *Cercospora* and *Alternaria* leaf spot diseases. *South Indian Hortic.* **47:** 308–309.

Tores, J.A. and Moreno, R. 1991. *Sclerotinia sclerotiorum*: epidemiological factors affecting infection of greenhouse aubergine crops. *J. Phytopathol.* **132:** 65–74.

Trigalet, A., Trigalet, D.D. and Feuillade, R. 1998. Aggressiveness of French isolates of *Ralstonia solanacearum* and their potential use in biocontrol. *Bull. OEPP* **28:** 101–107.

Ullasa, B.A. and Sohi, H.S. 1987. Studies on *Cercospora* leaf spot disease of Brinjal in Karnataka. *Veg. Sci.* **14(1):** 92–97.

van Elsas, J.D., Kastelein, P., van Bekkum, P., van der Wolf, J. M., de Vries, P.M. and van, Overbeek, L.S. 2000. Survival of *Ralstonia solanacearum* biovar 2, the causative agent of potato brown rot, in field and microcosm soils in temperate climates. *Phytopathology* **90:** 1358–1366.

Vasudeva, R.S. 1960. Report of the Division of Mycology and Plant Pathology. *Sci. Rep. Agric. Res. Inst. New Delhi,* 1957–58, pp. 111–130.

Vasudeva, R.S. 1963. *Indian Cercosporae*, ICAR, New Delhi, pp. 295.

Verma, A., Raychaudhuri, S.P., Chenulu, V.V., Singh, S., Ghosh, S.K. and Prakash, N. 1975. Yellows type of diseases in India: eggplant little leaf. *Proc. Indian Natl. Sci. Acad.–B.* **41:** 355–361.

Verma, R.K. and Dubey, G.S. 1976. Control of little leaf disease of brinjal: chemotherapy and screening for resistance. *Acta Botanica Indica* **4:** 144–150.

Verma, S.K., Gupta, N.N. and Phogat, K.P.S. 1997. Studies on the interaction between root–knot nematodes, *Melodogyne* sp., fungi, *Fusarium oxysporum* and *F. solani* and bacterium *Pseudomonas solanacearum* (*Ralstonia solanacearum*) on brinjal, *Solanum melongena* L. under valley conditions of Garhwal hills. *Prog. Hortic.* **29:** 188–193.

Vishunavat, K. 1992. Phomopsis blight and fruit rot of eggplant. In : *Plant Diseases of International Importance Vol II. Diseases of vegetables and oil seed crops.* (Eds. Chaube, H.S., Kumar, J., Mukhopadhyay, N. and Singh, U.S.). Prentice Hall, Inc., A Simon & Schuster Co. Engelwood Cliffs, New Jersey.

Vishunavat, K. and Kumar, S. 1993. Detection and transmission of seed borne inoculum of *Phomopsis vexans* (Sacc. & Syd.) Harter and the effect of infection on seed quality in eggplant (*Solanum melongena* L.). *Seed Res.* **21** : 66–71.

Vishunavat, K. and Kumar, S. 1994. Location of infection of *Phomopsis vexans* in brinjal seed. *Indian J. Mycol. Pl. Pathol.* **24** : 226.

Yadav, F.D. and Singh, H.B. 1995. Management of *Cercospora* leaf spot disease of brinjal by fungicides in Faizabad (U.P.). *J. Liv. World.* **2:** 23–26.

Yao, Y.G., Yang, J.G, Yao, Y.G. and Yang, J.G. 2000. A new eggplant F1 hybrid–Xiangqie No. 4. *China Veg.* **2:** 28–29.

Zhong, S.T. 1987. Preliminary observations on eggplant Sclerotinia blight. *Pl. Prot.* **13:** 25.

■■■■

12

DISEASES OF CARROT AND SUGARBEET

CARROT

The carrot (*Daucus carota* L.), a member of family Umbelliferae, is grown throughout the world. Total world production is about 13 million tonnes, the top ten producer countries being China, USSR, USA, Poland, Japan, UK, France, Italy, Germany and Canada. Carrots are also grown in the tropics including India at high altitudes or during the cool season. This crop is attacked by several diseases, which reduce the potential yields drastically. The present chapter deals with the commonly occurring fungal and bacterial diseases of carrot and various methods for their management.

(A) FUNGAL DISEASES

1. ALTERNARIA BLIGHT

This is an important disease of carrot foliage in temperate and mediterranean regions of the world and causes serious yield losses.

Geographical distribution

The disease was first reported in 1855 from Germany (Chupp and Sherf, 1960). Subsequently, it has been reported from many countries of temperate and mediterranean regions of the world like, Belgium, Holland, USA, Denmark, Brazil, Israel, France, India, etc. (Souza *et al.*, 2001; Farrar *et al.*, 2004; Jensen *et al.*, 2004).

Symptoms

On the foliage, symptoms appear first as small, dark-brown to black spots along the leaf margin and tend to be more elongated, irregular in shape and lack definite margin (Fig. 1). As the spots increase in number, the inter-veinal tissue dies until the entire leaf-let is killed. In moist weather, the disease spreads rapidly. Sometimes, leaf petioles are also infected and show elongated dark spots and the entire leaf dies without spots on the foliage. It also causes damping-off of seedlings, blight of seed stalks and sometimes a black decay of roots.

The Pathogen

The disease is caused by the fungus *Alternaria dauci* (Kuhn) Groves and Skolko. The conidiophores are short and dark. Conidia are dark coloured, 30–100 μm long and have multiseptate character typical of *Alternaria* sp. The broader end of conidium is attached to the conidiophore.

Fig. 1. Symptoms of Alternaria blight of carrot

Host range

Besides carrot, *A. dauci* can infect parsley (*Petroselinum crispum*), celeriac (*Apium graveolens* var. *rapaceum*) and weed hosts like *Daucus maximus, Ridolfia segetum* and *Caucalis tenella.*

Disease cycle

The fungus is seed borne, mycelium being found both on the seed surface and as a compressed layer between the pericarp and testa (Vala and Kapoor, 1985). It also remains viable for longer period in infected plant debris as well as in volunteer carrot plants growing in fallow land (Pryor *et al.*, 2002). The spores are splashed by rain or come in contact with the plants through running water.

Epidemiology

The fungus is both seed and soil borne. It requires free moisture in the form of prolonged dew or rains for infection, which is maximum at 27.8°C. The spread of fungal spores takes place by running water, splashing rains or in contaminated soil. Infection takes place rather slowly except in the most favourable weather. When wounds are present, the incubation period is reduced appreciably. Lower N doses enhance the disease severity (Westerveld *et al.*, 2002).

Management

(i) Cultural practices: The disease can be kept under check by collection and destruction of infected plant debris, using healthy seed produced in disease free areas, well-drained soils, long crop rotations and deep cultivation to place infected debris below 20 cm depth. Seed borne inoculum can be reduced by treating the infected seed by hot water at 50°C for 20 to 30 minutes without significant losses in germination (Nega *et al.*, 2003). Severity of this disease was also decreased with the increasing N doses (Westerveld *et al.*, 2002).

(ii) Host resistance: Muniz *et al.* (1988) screened 7 cvs. against this disease in Brazil and found Kuronan and Brasilia highly resistant with 3 and 5 per cent infection and root yields of 20.5 and 23.5 t/ha, respectively. Shin Kuroda was resistant and Nova Kuroda and Tropical were moderately resistant. Lines IHR 230, IHR 231, IHR 232, IHR 244, IHR 248, IHR 253 and IHR 264 were found resistant to this disease in Bangalore, India (Gowda *et al.*, 2000), while cvs. Brasilia, Alvorado, Tropical, Carandai and Kuronan were reported as resistant in Brazil (Lopes *et al.*, 2000).

(iii) Chemical control: Since the pathogen is seed borne, consequently seed treatment is of utmost importance. Seed treatment with fungicides like Vitavax (carboxin) + thiram and Voronit @ 200g/100kg seed (Mirkove, 1979), Fundazole (benomyl), Rovral (iprodione), Sumisclex (procymidone) and Polycar-bacin (metiram) (Markova and Golyshin, 1988; Nowicki, 2002) have been reported quite effective. It improved seed germination, yield and storage life of carrots.

The disease can be managed in field conditions by spraying Blitox (0.25%) or Dithane M-45 (0.25%). Difenoconazole, an EBI fungicide, @ 125g/ha when applied at 21 days interval provided excellent control of this disease in comparison to mancozeb (Follas *et al.*, 1992). Sprays of iprodione and hexaconazole have been recommended for the control of this disease in France (Poissonnier *et al.*, 1995). Recently available fungicides like azoxystrobin and difenoconazole provided better control of this disease as compared to traditional fungicides like copper oxychloride and sulfur (Ben-Noon *et al.*, 2001; Siviero *et al.*, 2004; Alkseeva, 2009). However, higher efficacy of azoxystrobin (Amistar 250 SC) on the already established disease under field conditions has been reported (Mazur *et al.*, 2005). Efficacy of pyraclostrobin + metiram (Cabrio Top) and pyraclostroin against this disease have also been reported which not only reduced the disease severity but also increased the root weight (Maringoni *et al.*, 2012).

(iv) Biological control: Highly infected carrot seed with *A. dauci* when bioprimed with *Clonostachys rosea* strain IK 726, reduced incidence of the pathogen to 4.8 per cent with a lower infection rate (Jensen *et al.*, 2004).

2. STEMPHYLIUM ROOT ROT

The disease occurs widely in different carrot growing areas of the world and causes huge losses.

Geographical distribution

The disease is probably occurring ever since the carrots have been stored in large quantities. The occurrence of the disease has been reported from temperate and mediterranean areas of the world like Australia and Poland (Nowicki, 2002; Coles and Wicks, 2003).

Losses

Up to 88 per cent plants in some areas were infected and caused huge losses, while up to 70 per cent of carrots with crown and shoulder infection were reported unmarketable (Coles and Wicks, 2003).

Symptoms

The fungus attacks older leaves and the growing roots in the field. Symptoms produced on leaves and petioles are similar to those of *A. dauci,* but with more widely distributed lesions. Lesions on the roots are circular to irregular, shallow and slightly depressed, with sharp margins and have greenish black to black mealy tissues. Decay at the crown extends deep into the core. Black mould produced by the fungus may or may not be present on the surface of the decayed areas.

The Pathogen

The disease is caused by the fungus *Stemphylium radicinum* (Meier, Drechsler and Eddy) Neergaard. Dark blackish-brown to black colonies contain conidiophores which arise singly or in groups. Conidiophores are simple and unbranched, upto 200 μm long and 3–9 μm thick with one to several conidial scars. Conidia are solitary or borne in chains of two or three; they are of variable shape from elliptical to pear-shaped, with the broadest end attached to the conidiophores. The size of conidia ranges from 27–57 × 9–27 μm, having 3 to 7 transverse septa and one to several longitudinal septa, possibly with a constriction at each septum.

Host range

Beside carrots, *S. radicinum* is also pathogenic to parsley (*Petroselinum crispum*), parsnip (*Pastinaca sativa*) and dill (*Anethum graveolens*).

Disease cycle

The fungus responsible for this disease is seed borne, but it may also overwinter in soil on dead plant material. The fungus has longer longevity in mineral soils *i.e.,* upto 8 years (Dixon, 1981). The fungus may infect roots directly from the soil, or infection may occur from the infected tops. The prevalence of black rot lesions on the sides of the roots show that the crown is not necessarily the pathway for infection from diseased leaves. Most carrots at harvest probably carry spores of this fungus on their surface. Under favourable conditions these spores may infect the carrots mostly through broken root tips.

Epidemiology

The fungus is both seed and soil borne. The fungus can infect carrots at temperature range of 5 to 33.8°C with the optimum of 27.7°C. Black rot is also a serious storage problem because the fungus can grow at low temperatures.

Management

(i) Cultural practices : Care should be taken during harvesting and handling to avoid mechanical injuries. Remove field heat immediately after harvest and store carrots at temperature near 1–2°C and 80–85 per cent RH (Gor-Kovenko, 1991). Use of healthy seed can also be helpful in reducing the initial inoculum of the disease in the field.

(ii) Chemical control: Seed treatment with fungicides as suggested for blight (*A. dauci*) is also applicable for the control of this disease. Seed treatment with captan, carbendazim and thiophanate methyl reduced seedling mortality as compared to untreated seeds where mortality reached up to 85.2 per cent after 10 days (Murtaza *et al.*, 1988). Fungicidal sprays may be used to control field infections using materials such as thiram or zineb (Dixon, 1981). Root application of Rovral (iprodione) reduced root rot and increased yield (Timchenko *et al.*, 1989).

(iii) Biological control: *Trichoderma viride* showed strong antagonism to *S.* (*A.*) *radicina* on stored carrots (Sesan, 1990). Hentschel (1991) applied *Bacillus subtilis* str. Tqq by soaking carrot seed with seed borne *S.* (*Alternaria*) *radicina,* which increased seed germination and seedling health. Application of small amount of metiram and iprodione enhanced the effectiveness of this antagonist in glasshouse experiments.

3. CERCOSPORA LEAF SPOT

This is a wide spread disease of this crop throughout the world and is rated as an important foliar disease of this crop.

Geographical distribution

This disease was first described in Italy during 1889 (Chupp and Sherf, 1960). Subsequently, it has been reported from almost all countries of the world wherever carrots are grown intensively but it is common in the north temperate regions like Washington, USA (Inglis *et al.*, 2001) than in other parts of the world.

Symptoms

Symptoms of this disease appear on all the above ground parts of the plant. In contrast to *A. dauci*, the pathogen infects younger leaves rather than older ones. On the leaf, sub-circular, grey to brown or black spots occur at first on the lobes. As the spots increase in number, the intervening parts die and the foliage shrivels. The spots on the petioles and stems are elliptic with a pale centre and dark margin. The fungus also attacks floral parts and late infections permit the fungus to enter the seed.

The Pathogen

The disease is caused by *Cercospora carotae* (Passerini) Solhein. The mycelium of the fungus accumulates in sub-stomatal spaces in the leaf as intertwining mass of hyphae. From these stromatic structures fascicles of conidiophores develop and emerge through the leaf stomata. The conidiophores are indeterminate in growth and show scars where conidia were attached. The conidia are slightly obclavate, hyaline, several celled and measure 40–110 × 2.2–2.5 μm.

Host range

The pathogen attacks both cultivated and wild forms of *Daucus*.

Disease cycle

The fungus perennates on seed or in infected plant debris. Conidiophores are produced in minute black stromata present in the plant debris, which in turn produce conidia. When they are wind blown to wet leaves, they germinate and enter stomata and cause infection. The younger leaves are more susceptible than older ones. Conidia produced on primary lesions, cause secondary infections.

Epidemiology

The fungus overwinters in the infected seed and plant debris. Wind, water and tools disseminate conidia produced on the stromata. Penetration takes place through stomata with 28°C as optimum temperature for germination and infection. Secondary infection is through wind-borne conidia.

Management

(i) Cultural practices: Since the fungus survives in infected plant debris and seed, field sanitation and use of long crop rotations will be helpful in reducing the disease. Use of healthy seed is also helpful in reducing the disease.

(ii) Host resistance: Cultivars such as Brasilia, Alvorado, Tropical, Carandai and Kuronan were found as resistant to this disease in Brazil (Lopes *et al.*, 2000).

(iii) Chemical control: The seed treatment with fungicides recommended for *A. dauci* is also applicable to this disease. Mancozeb or copper oxychloride is effective in the management of this disease as foliar sprays. Follas *et al.* (1992) recommended sprays of difenoconazole @ 125 g/ha at 14 days interval which is better than mancozeb.

4. POWDERY MILDEW

This is another important disease of carrot and most of the umbelliferous crops are susceptible. Losses upto 80 per cent in seed yield are reported for late season crop (Dixon, 1981).

Geographical distribution

The occurrence of the disease has been reported from Mediterranean regions like Isreal, Italy, France, Jersey and East Anglia (U.K.) (Dixon, 1981).

Symptoms

Symptoms include development of a white mycelial cobweb on the upper leaf surface. Younger and inner leaves are spotted first with weft like mats of mycelium that soon coalesce forming larger areas and finally covering the entire leaf as it matured. Slight necrosis may occur when plants are slightly affected. When young plants are attacked they are either killed or their vigour is greatly reduced.

The Pathogen

The disease is caused by *Erysiphe heraclei* D.C. Fungal growth appears as typical white, dense, persistent powdery mildew colonies on leaves, inflorescence, and stems. Hyphae are 6–10 μm wide. Conidiophores are 64–154 μm long with straight foot cells 24–42 μm long and 6–10 μm wide. Conidia develop either singly or in chains, are cylindrical to ovate, and range in length from 34–46 μm and width from 14–20 μm. Germ tubes are simple and formed from the ends of conidia. The cleistothecia are initially globular in shape but then become flattened, produced in a scattered or thinly grouped manner measuring 85–120 μm in diameter. Appendages are narrow to rather coarse, mycelium like and are often interlaced with the mycelium. In each cleistothecium 3–10 asci develop which are 55–70 x 30–45μm in size. Each ascus contains 3–6 ascospores.

Host range

The host range of *Erysiphe heraclei* covers at least 85 umbelliferous genera. Besides carrot, it also infects angelica (*Angelica archangclica*), celery (*Apium graveolens*), coriander (*Coriandrum sativa*), dill (*Anethum graveolens*), fennel (*Foeniculum vulgare*), parsnip (*Pastinaca sativa*) and parsley (*Petroselinum crispum*).

Disease cycle

Wherever cleistothecia of the fungus are formed, they provide an idea regarding the mode of perpetuation. Otherwise the pathogen survives from one season to other in conidial state either on the perennial weeds or on other related crops grown in the area. Ganeshan (1987) recorded the perfect stage in Karnataka on some new cultivars. Seed transmission has also been reported.

Epidemiology

The disease is more serious on winter-sown crops. Ganeshan (1987) reported the occurrence of this disease on January planted crop during June–July prior to flowering.

Management

(i) Host resistance: Lines IHR 230, IHR 231, IHR 232, IHR 244, IHR 248, IHR 253 and IHR 264 were found resistant to powdery mildew of carrot in Bangalore, India (Gowda *et al.*, 2000). Cv. "Stefano" also has high degree of resistance to this disease (Anonymous, 2015).

(ii) Chemical control: The disease can be managed effectively by applying the conventional fungicides like wettable sulphur, Karathane and Bavistin. Recently available EBI fungicides like hexaconazole (Anvil @ 10 and 20 g a.i./ha), pyrifenox (Dorado @ 50 g a.i./ha) and penconazole (Topas 10 EC @ 25 g a.i./ha) when applied on carrots cv. Nantes at 14 days interval provided best control of this disease, while sulfex was least effective owing to the long period between 1st and 2nd spray (Cvjetkovic *et al.*, 1993). The level of fungicides residues in the carrot root determined 14 and 30 days after the first treatment was high for EBI fungicides except hexaconazole. Watson (2016) observed that this disease can be managed, but not eliminated by fungicide application and reported the efficacy of fungicides like

Amistar, Amistar Top, Folicur, Cabrio, Fontelis and sulphur in managing this disease.

5. WATERY SOFT ROT

The disease is very serious in those areas, which have cool and moist conditions and may result into considerable losses at harvest. The occurrence of watery soft rot after harvest is largely dependent upon the presence of the active disease at harvest.

Geographical distribution

Watery soft rot was first recorded in England in the year 1837 on vegetables (de Bary, 1886). Subsequently, it was reported from many parts of the world like England, Canada (Kora *et al.*, 2003). In India, the disease was recorded during 1999 crop season in Varanasi (Pandey *et al.*, 2003).

Symptoms

The main symptoms are water soaking and softening of decaying tissues. An important diagnostic symptom in the advanced stages of decay is the presence of white fluffy to appressed mould of the fungus. In later stages of decay, sclerotia of the causal organism may be formed in the mouldy growth. Sclerotia are initially white, then bownish and finally black.

The Pathogen

The fungus responsible for this disease is *Sclerotinia sclerotiorum* (Lib.) de Bary. The fungus has been described in detail under the disease stalk rot of crucifers.

Host range

The pathogen is polyphagous in nature and has a wide host range of over 350 species in 60 families including plants of uneconomical importance (Gupta and Dohroo, 1993). Khan (1976) reported that all the vegetable hosts *viz.,* knol khol (*Brassica oleracea* var. *caulorape*), lettuce (*Lactuca sativa*), cabbage (*Brassica oleracea* var *capitata*), Brussel's sprout (*B. oleracea* var. *gemmifer*), spinach (*Spinacia oleracea*), fennel (*Foeniculum vulgare*), tomato (*Lycopersicon esculentum*), pea (*Pisum sativum*), French bean (*Phaseolus vulgaris*), capsicum (*Capsicum annuum*), mustard (*Brassica oleracea* var. *sarson*), brinjal (*Solanum melongena*), turnip (*Brassica rapa*), beans (*Phaseolus coccinens*), carrot (*Daucus carota*), celery (*Apium graveolens* var. *dulca*) were found susceptible to the pathogen.

Disease cycle

The fungus is soil borne in nature and survives in the form of sclerotia. The sclerotia in favourable weather conditions germinate either myceliogenically or carpogenically and cause infection. Ascospores are blown by moist wind and cause infection.

Epidemiology

The disease is more severe if the weather at harvest is cool and rainy, making conditions favourable for the development of the fungus. Stored carrots may develop rot during a period of 3 to 5 months. Low temperature and high humidity favour this disease.

Management

(i) Cultural practices: Cultural practices like burning of crop residues, deep ploughing during summer months, flooding of the infected fields etc. should be followed to keep this disease under check. Carrots should be stored at 0°C and 95 per cent relative humidity. Proper measures should be taken to avoid the condensation of moisture on carrots in storage. Ozone concentration (30 or 60 fl/l) caused 50 per cent reduction in daily growth of *S. sclerotiorum* in storage, indicating that ozone was fungistatic (Liew and Prange, 1994).

(ii) Chemical control: Pre-harvest root application of carbendazim or thiophanate-methyl under field conditions can reduce this disease during storage. Timchenko *et al.* (1989) reported that root application with Rovral (iprodione) or Sumisclex (procymidone) controlled this disease effectively and increased yields.

(iii) Biological control: Sesan (1990) reported the antagonistic activity of *Trichoderma viride* against this pathogen. Soil application of wheat bran: saw dust preparation of *T. viride* under field conditions may be of use in checking the ravages of this disease. Efficacy of *Bacillus circulans* strains 1, 7 and 8 and *B. coagulans* strain 6 against *S. sclerotiorum* has also been reported (Czaczyk *et al.*, 2000).

6. CAVITY SPOT

Cavity spot of carrot can result in serious economic loss both in the fresh market and processing industries (Groom and Perry, 1985).

Geographical distribution

The disease is of common occurrence in U.K., Norway, Israel, U.S.A., Canada (British Columbia) and France.

Symptoms

Lesions may be small (1–2 mm) and almost circular to elongated, upto 1.0 cm, transversely to form lens-shaped cavities (Green and Makin, 1985). The skin over the larger lesions tends to rupture, leaving a ragged fringe and the interior of the cavity is grayish in colour. Even after peeling the roots for processing, small lesions may still be apparent as dark blemishes with raised borders. The spot develops in growing crop and is evident at harvest time.

The Pathogen

Breton and Rouxel (1994) reported that *Pythium violae* Chesters and Hickman is the most frequent causal agent of this disease followed by *P. sulcatum, P. sylvaticum* and *P. intermedium*. In addition to these, three new species of *Pythium* (*P. acantticum, P. paroecandrum* and *P. mamillatum*) have also been found associated with this disease in British Coloumbia (Benard and Punja, 1995).

Disease development

The causes are complex and several hypotheses have been advanced including calcium deficiency (Maynard *et al*., 1963), mineral nutrition (Jacobsohn *et al*., 1984) and infection by bacteria (Soroker *et al*., 1984). In general, incidence of this disease tends to be very high during wet season and poorly drained soils. High soil moisture (near field capacity) or flooding and temperature of 15°C were found optimum for disease development (Benard and Punja, 1995).

Management

(i) Cultural practices: The incidence of cavity spot can be substantially reduced by following crop rotation (Soroker *et al*., 1984) and growing crop on narrow ridges rather than on wide flat beds (Jacobsohn *et al*., 1984). Eitarabily *et al*. (1997) studied the effect of lime and gypsum amendment on the development of this disease and found that lime application significantly reduced the disease incidence, while gypsum had no effect on disease incidence. They further suggested that the decrease in incidence was not due to calcium concentration in soil, but it was due to increase in soil pH associated with application of lime.

(ii) Host resistance: Benard and Punja (1995) reported cvs. Panther, Caropride, Fannia and Navajo resistant to the cavity spot.

(iii) Chemical control: Application of phosphonate @25 kg/ha, 12 weeks after sowing reduced the incidence of cavity spot (Walker, 1991). Seed dusting with metalaxyl (1.5–6.0 g/kg seed) and sprays effectively controlled the disease (Shlevin *et al*., 1987; Walker, 1991), but varying degrees of sensitivity of *P. violae* and *P. sulcatum* may jeopardize the continued effectiveness of chemical control (Breton and Rouxel, 1994). The cavity spot tends to be less prevalent in alkaline soils at pH 8 and above, fungicide treatment may be unnecessary or a reduced rate of application may be sufficient (White, 1988). Martinez *et al*. (2005) evaluated five different fungicides against 14 isolates of *Pythium* sp. and reported that most of the isolates were highly sensitive to both metalaxyl and zoxamide, but insensitive to fludioxonil, fosetyl-Al and chlorothalonil, however, under greenhouse assays, only zoxamide provided significant and consistent disease control. To avoid the risk of development of resistance to fungicides, fungicides should be used in rotation.

(B) BACTERIAL DISEASES

7. SOFT ROT

This is an important disease of carrots in field (Chand and Kishun, 1989) and storage. In some breeding lines, mortality upto 70 per cent due to this disease has been reported. While it is not that important in cold stored roots, but it can cause heavy losses at ordinary temperatures.

Geographical distribution

The occurrence of the disease has been reported from U.S.A., erstwhile USSR, New Zealand, Italy, Romania, Venezuela, Egypt, Poland, New Zealand and India (Rodriguez *et al*., 2002).

Symptoms

The disease initially appears on the lower part of roots as small, water-soaked, brown spots. Later, these spots develop rapidly and spread into the core of the root from where they progress from inside to outside. In the advanced stage of the disease, infected part becomes soft due to maceration of tissues, which emit foul smell. Sometimes plants show wilting symptoms.

The Pathogen

The disease is caused by *Erwinia carotovora* sub.sp. *carotovora* (Jones) Bergey *et al.* It is a weak parasite and facultative anaerobe. The bacterium is Gram-negative rod, with peritrichous flagella. It produces grayish white colonies on nutrient agar medium and cloudy growth in nutrient broth. It produces gas in the nutrient medium utilizing several carbohydrates as carbon source.

Disease cycle

Bacteria survive on plant debris in the soil and may infect the growing crop especially after heavy rains or in badly drained fields. Roots may be contaminated at the time of harvesting or during storage.

Management

(i) Cultural practices: Improvement in drainage can reduce the field infection. Contamination of harvested roots can be reduced by washing in chlorinated water that should be renewed frequently and checked regularly for its disinfectant capacity. If the roots are subsequently to be stored at ambient temperature, dry the surface of roots thoroughly because the bacterium is sensitive to desiccation. If roots are stored at temperatures close to 0°C then soft rot may be controlled without recourse to chlorination or surface drying. Joe *et al.* (2017) evaluated the antibacterial activity of spent green tea (SGT) and reported that SGT treated carrots recorded low degree of maceration, relative electrolyte leakage values and also maintained high β-carotene content, phenolic content and total antioxidant percentage which could be utilized to protect carrots against this disease under post-harvest storage conditions.

(ii) Host resistance: Cultivar Rola was found to be least susceptible to the disease (Kozak, 1990).

(iii) Biological control: Liquid formulation of *Bacillus subtilis* IIHR BS-2 (1 x 10^8 cfu/ml) was evaluated both as seed treatment (10 ml/kg seed) and soil application (after vermicompost enrichment) individually and in combination, and found seed treatment along with soil application of *B. subtilis* (5–1/2 tons/ha) effective as it reduced the disease incidence by 70.2 per cent and increased carrot yields (Rao *et al.*, 2017).

SUGAR BEET

Beet (*Beta vulgaris* L.) includes red beet, sugarbeet, fodder beet and Swiss chard. During cultivation, several diseases of fungal, bacterial and viral nature attack these crops, which not only reduce the quantity of the produce but also the quality.

(A) FUNGAL DISEASES

1. SEEDLING DISEASES

Seedling diseases such as damping-off, seedling blight, collar rot, root rot etc. are caused by pathogens such as *Sclerotium rolfsii, Rhizoctonia solani, R. bataticola* and species of *Pythium* and *Fusarium* in India. Of these *S. rolfsii, R. solani* and *Pythium* sp. cause acute mortality. *Pythium* sp. produce pre and post-emergence damping-off, while others mostly produce post-emergence damping-off. High temperature alongwith high moisture conditions favour these diseases.

Management

The pathogens associated with this disease are soil and seed borne in nature. Use of cultural practices, chemical methods and biological control can provide effective management of these diseases.

(i) Cultural practices: Cultural practices like collection and destruction of infected plant debris, deep ploughing during summer months, long crop rotations and use of healthy seeds are important and effective methods for the management of these diseases. Dimov *et al.* (1988) reported that when beet crop was rotated with alfalfa, the occurrence of *Trichoderma, Fusarium, Phytophthora, Pythium, Rhizoctonia* and *Alternaria* varied with the rotation and buildup of pathogenic fungi did not occur in such soils.

(ii) Chemical control: Plant base application of Pencycuron (0.2%) decreased root rot (*R. solani*) considerably (Uchino *et al.*, 1987). Tolclofos methyl + hydroxy isoxazole (hymexol) when mixed with nursery soil and also applied on the soil surface, effectively controlled damping-off (*R. solani*) better than PCNB and hymexazol (Fuji *et al.*, 1987). While seed treatment with hymexazole alone provided protection against seedling disease caused by *Pythium* sp. and *Aphanomyces cochlioides* (Payne and Williams, 1990). Thiabendazole effectively controlled pre-emergence damping-off and root rot (*Epicoccum nigrum, Fusarium solani, Pythium* sp. and *R. solani*) under *in vivo* conditions when applied as seed treatment (Abdel-Rehim *et al.*, 1992). Srivastava and Tripathi (1998) have reported effective management of sugarbeet seedling disease complex by combination of fungicides. The effectiveness of five combinations of 4 compatible fungicides, PCNB (Brassicol), TMTD (Thiram), carboxin (Vitavax) and carbendazim (Bavistin) @ 2.5 g/kg seed (mixed in equal proportion) incorporated in seed pellets of sugarbeet was evaluated against seedling diseases complex caused by 4 soil borne pathogens, namely, *Pythium aphanidermatum, Rhizoctonia solani, R. bataticola* and *Sclerotium rolfsii*. Seed pelleted with all combinations of fungicides provided better protection to seedlings as against steeping of seeds in aqueous suspension of a combination of fungicides. Of various combinations used, a mixture of Bavistin + Thiram was most efficacious in reducing seedling mortality as compared to other treatments. Azoxystrobin application in both treated and natural infection trials provided consistent level of control (Kiewick *et al.*, 2001).

(iii) Biological control: The causal fungi responsible for damping-off of seedlings are soil borne in nature. The use of resident antagonists in reducing this disease can be an effective approach. Seed pelleting of sugar beet seeds with oospores of

Pythium oligandrum reduced damping-off of sugarbeet in soils naturally infested with *Aphanomyces cochlioides* and *Pythium* sp. and control was equivalent to that achieved with hymexazol seed coating (McQuilken *et al.*, 1990). *P. oligandrum* oospores produced by liquid fermentation in cane mollases medium and pelleted on sugarbeet seeds resulted in significant control of damping-off caused by *P. ultimum* (Whipps *et al.*, 1993).

2. CERCOSPORA LEAF SPOT

This is a disease of considerable importance in warm temperate areas, Cercospora leaf spot is the most devastating foliar disease of beet crop causing significant yield reductions.

Geographical distribution

This is a common disease of Central and Southern Europe, Italy, Poland, Belgium, Greece, Japan, Russia and certain parts of North America (Walczak *et al.*, 2002). In India, it was first observed in the tarai region of Uttarakhand (Mukhopadhyay, 1968). Now the disease is widely prevalent in Kashmir valley, Kalpa (Himachal Pradesh), Sriganganagar (Rajasthan) and Punjab.

Losses

In 1998 crop season, Cercospora leaf spot appeared in severe form in Minnesota and North Dakota where sugar beet industry ($ 3 billion) is an economic activity and growers lost $ 45 million (Khan, 2009).

Symptoms

The symptoms of the disease appear mostly on the leaves and rarely on the petioles. The characteristic symptoms include discrete circular lesions, 3–5 mm in diameter, with a necrotic centre and reddish to dark brown margin during damp and cool weather. In case of severe infection, the petioles are also infected. The spots are scattered at first, but in case of severe attack they coalesce and cover the entire leaf blade and affect the quality and yield of seeds. In seed crop, all above ground parts including seed clusters are affected.

The Pathogen

The disease is caused by *Cercospora beticola* Sacc. The mycelium of the fungus is septate, dark coloured and intracellular. Small sclerotial masses are formed on the host tissue from which dark coloured conidiophores arise in clusters. Conidiophores are 4–6 μm in diameter at the apex and 60–19 μm long. The conidia are borne on the tip of the conidiophores. These are hyaline, elongated, filiform, multiseptate, broadly rounded at the point of attachment to the conidiophore, measuring 78–228 μm in length, being 4.4–6.3 μm wide at the base and 1.6–3.2 μm in size and tapering slightly toward the opposite end.

Host range

The pathogen infects only beet plants and its close relatives like mangel wurzel, Swiss chard, spinach and certain closely related weeds.

Disease cycle

The fungus chiefly overwinters in infected plant debris as mycelium or on the seed. The infected seeds result in diseased seedlings. In the spring season, overwintering mycelium in plant refuse starts producing conidia which are disseminated by wind currents, rain splashes and insects to the leaves.

Epidemiology

The disease mostly appears in the month of November–December, but severe attacks occur in the month of March–April in the plains. In the hills, it appears during summer months. High relative humidity is an essential pre-requisite for sporulation. The conidial germination is best in the presence of free water and moderate temperatures *i.e.,* in between 15–32°C.

Management

(i) Cultural practices: The inoculum for the annual recurrence of the disease comes from infected leaf debris, cultural controls are therefore, aimed at the destruction of infected debris by burning and deep ploughing and use of long rotations to prevent an accumulation of the soil borne inoculum. Sowing of cereals (sorghum, wheat) and lucerne before sugarbeet crop alongwith weed free cultivation reduced the disease (Rajpurohit and Singh, 1996).

(ii) Host resistance: The resistance to Cercospora leaf spot has been found to be correlated with the 3-hydroxytyramine content of the leaves (Rautela and Payne, 1970). Rajpurohit and Singh (1992) screened 35 varieties of sugarbeet and none was found free from the disease but the variety Desperzpoly RC showed low incidence (3.2%) and also gave highest root yield as well as sucrose content.

(iii) Chemical control: The disease can be kept under check by elaborate chemical spray programme including both systemic and non-systemic fungicides. Three sprays of fentin hydroxide @ 0.75 kg/ha were found best followed by mancozeb, zineb, captafol and copper oxychloride @ 2 kg/ha (Rajpurohit and Singh, 1993). Systemic fungicides, such as benomyl, carbendazim and thiophanate methyl derivatives were found superior to non-systemics (Mukhopadhyay and Upadhyay, 1977). Rajpurohit and Singh (1993) reported that two sprays of carbendazim, benomyl or thiophanate methyl @ 0.5 kg/ha at 20 days interval were found to be best for controlling this disease and increasing the root and foliage yield as well as sucrose content. Good efficacy of recently available strobilurin fungicide (pyraclostrobin) has been reported against this pathogen (Manaresi *et al.*, 2002). This compound has a favourable toxicological and environmental profile, and it is safe to users. Application of difenoconazole (0.1%) was also found effective in reducing the severity of this disease (Haller *et al.*, 2013).

The exclusive use of benomyl or carbendazim should be restricted as the pathogen may develop resistance (Dixon, 1981). In those areas where this disease is of common occurrence and the resistance has not been noticed, these fungicides should be used carefully and in combination with protectant fungicides like mancozeb. This strategy can prolong the effectiveness of benzimidazole fungicides.

3. POWDERY MILDEW

This is a widely distributed disease of beet crop and sometimes appears in severe dimensions.

Geographical distribution

The disease is widely distributed in several countries of the globe wherever beet is grown with the possible exception of South America, Australia and Eastern Asia (Dixon, 1981). In India this disease was first recorded in Pantnagar (Mukhopadhyay, 1969) and later it was observed extensively in Phaltan and Ahmednagar (Maharashtra) areas under warm and dry weather.

Symptoms

The disease initially appears on the lower leaves and gradually spreads towards the top. The formation of first white and later grey-tan mildew areas on both the sides of the leaf characterize the disease. In general, infection is more on the upper surface of the leaf. In advanced stages of the disease development, mildew patches enlarge and coalesce and the leaf appears as if dusted with wheat flour. Severely affected leaves turn pale and ultimately dry up. Under favourable climatic conditions, cleistothecia develop as small dark round structures on the infected surface of the leaf.

The Pathogen

The disease is caused by *Erysiphe betae* (Vanha) Weltzien. Mycelium is superficial and persistent. Conidiophores are unbranched and erect. Conidia arise singly, are hyaline, ovoid, 30–50 × 15–20 μm in size. Cleistothecia are globose, dark brown to black, 80–120 μm in diameter with 4–8 asci per cleistothecium. There are generally 2–3 ascospores per ascus.

Host range

The pathogen is specialized to *Beta* sp., and infects only *B. cicla, B. diffusa, Boerhavia maritime, B. patellaris, Beta patula, B. repens, B. trigyma* and *Beta vulgaris*.

Disease cycle

Wherever cleistothecia of the fungus are formed these serve as the possible mode of perpetuation. In normal situation, the pathogen survives from one season to other in conidial state either on the perennial weeds or on other related crops grown in the area.

Management

The disease can be managed by dusting sulphur or spraying wettable sulphur, benomyl, carbendazim, thiophanate-methyl, tridemorph, thiabendazole lactate (Asher, 1986). Ergosterol biosynthesis inhibitors like triadimefon, hexaconazole, penconazole, propiconazole, flusilazole, cyproconazole, carbendazim + flusilazole etc. can be effectively used in reducing the intensity of this disease (Asher, 2000). Efficacy of strobilurin fungicide (pyraclostrobin) has been reported against this pathogen (Manaresi *et al*., 2002). Gado (2013) evaluated the impact of different plant extracts and fungicides against this disease and reported the efficacy of fungicide Eminent was better compared to plant extracts.

4. SCLEROTIUM ROOT ROT

This is a very common and destructive disease of sugarbeet and under favourable weather conditions reduces yields up to 55 per cent in certain varieties (Mukhopadhyay and Thakur, 1971).

Symptoms

The disease is characterized by sudden rotting of mature roots below the soil surface. The affected root is usually covered with white mycelium that contains numerous brown sclerotia which is the most conspicuous sign of this disease. The fungal growth and sclerotia can also be seen in the soil around such roots. Later, when enough damage has been done to the roots, the leaves show yellowing and wilting, and such plants can be easily pulled out.

The Pathogen

The pathogen responsible for this disease is *Sclerotium rolfsii* Sacc. The basidial stage of this fungus falls in the species of *Athelia*. Initially, the mycelium of the fungus in pure culture is at first silky white, but gradually it loses its luster and becomes somewhat dull in appearance. The hyphae are hyaline, thin walled, sparsely septate when young. The cells are 60–350 μm long and 2–8 μm wide. The broader hyphae show clamp connections, which are absent in thin hyphae. The number of nuclei/cell is highly variable. Mostly there are 2 nuclei in cells of secondary and tertiary branches. Sclerotial initials are formed from hyphal strands that consist of 3–12 hyphae lying parallel. A spherical shape is soon assumed even if it is only a loose mass of hyphae. With further hardening, differentiation takes place and the sclerotium shows an outer layer of polyhedral cells, 3–4 μm in diameter, surrounding the compacted hyphae. Mature sclerotia are dark brown, but variation from lighter brown to darker colour may be found. These are small, about the size of radish seed, hard and usually round.

The basidial stage grows as a spreading white hymenium on the host surface or on the surface of the culture medium. Although the hymenia may be pure white in culture, on the host they may be grey, yellow or buff coloured. The basidia are obvoid, 7–9 μm long and 4–5 μm wide. Each basidium bears 2–4 parallel or divergent sterigma, which are 2.5–4 μm long. Basidiospores are unicellular, elliptical to obvate, sometimes rounded or pyriform, smooth walled, hyaline and epiculate at the base measuring 6–7 × 3.5–5 μm in size.

Host range

The pathogen is polyphagous in nature and has a very wide host range. *Sclerotium rolfsii* infects more than 500 species of plants belonging to 100 families and causes economically important diseases in warm moist climates throughout the world (Aycock, 1966). Roy (1977) also observed this fungus in Assam, on vegetables like chilli (*Capsicum annuum* L.), French bean (*Phaseolus vulgaris* L.), carrot (*Daucus carota* L. cv. *Nantes*), Pea (*Pisum sativum* L.), cauliflower (*Brassica oleracea* var. *botrytis* L.), cabbage (*Brassica oleracea* var. *capitata*), potato (*Solanum tuberosum*) and arum (*Colocasia* sp.).

Disease cycle

The fungus is soil borne in nature and can survive as saprophyte on crop debris. The fungus also produces sclerotia that are left in the field. These germinate under favourable weather conditions and cause infections.

Epidemiology

The pathogen is soil borne in the form of sclerotia. High temperature and humidity favour the disease development. In the plains of India, sugarbeet is usually grown on ridges and this practice is likely to stimulate disease incidence because the lower leaves become covered with soil resulting in a 'bridge' of dead tissue which furnishes an ideal medium for initiating pathogenesis.

Management

(i) Cultural practices: Since the fungus survives in infected plant debris as a saprophyte or by forming sclerotia, destruction of infected plant debris, crop rotation, deep summer ploughing, flooding, solarization during summer months etc. can be helpful in reducing the initial inoculum load in the field. Reduction in the incidence of root rot has also been recorded through the use of nitrogenous fertilizers including calcium nitrate, anhydrous ammonia, urea, ammonium sulphate and calcium ammonium nitrate (Thakur and Mukhopadhyay, 1972).

(ii) Chemical control: Carboxin and chlorneb significantly reduced fungal growth *in vitro* at low concentrations and completely inhibited sclerotia formation. Under field conditions, ridge soil drenching with carboxin and chlorneb at 2 and 15 kg/3000 l water/ha, respectively, significantly reduced root rot and increased yields (Mukhopadhyay and Thakur, 1971).

Das and Raj (2004) suggested the control of this disease through seed soaking with non-conventional chemicals like 2,3,5-trichloro-phenoxy acetic acid (10^{-4} M), chitosan and zinc chloride (10^{-4} M). The treatment not only reduced the disease but also increased root yield, foliage yield and sugar yield as compared to other tested chemicals.

(iii) Plant extracts: The methanol extract of leaves of *Aegle marmelos* decreased the growth and sclerotia production by *S. rolfsii* and no sclerotia were formed in the presence of extract at 150 ppm even after 21 days of incubation (Prithiviraj *et al.*, 1996).

(iv) Biological control: Application of *Trichoderma harzianum* inoculum was found effective in reducing the disease caused by *S. rolfsii* in sugarbeet (Ciccarese *et al.*, 1992). Correa *et al.* (1995) tested a metabolite trichorzianines A and B obtained from *T. harzianum* on the mycelium of *S. rolfsii* and produced a change in the morphogenetic pattern of mycelia. Papavizas and Collins (1990) showed the association of *Gliocladium virens* with sclerotia of *S. rolfsii* in naturally infested soil. When sclerotia were incubated in soil infested with *G. virens*, sclerotial viability and capability for infection of host tissue was reduced. *G. virens* was found to colonize, penetrate and sporulate inside the sclerotia of *S. rolfsii*.

(B) VIRAL DISEASES

5. RHIZOMANIA

Rhizomania is an economically important disease of sugar beet and causes severe yield losses. From the infected crop, only 2–3 t sugar/ha is obtained instead of 8–10 t in the healthy crop (Putz *et al*., 1990).

Geographical distribution

Rhizomania has been reported from most of the sugar beet growing areas of Europe, Asia and the U.S.A. (Putz *et al*., 1990).

Symptoms

Leaves may show flabbiness, wilting and mild yellowing on the infected sugarbeet plants in the field. The most characteristic symptoms found in beet leaves are a pattern of yellowing areas along the veins. However, the virus mostly remains confined to roots and rarely moves to the foliage. Symptoms on roots are characterized by root stunting and a proliferation of lateral rootlets on the main tap root, giving it bearded appearance.

The Virus

The virus responsible for this disease is *Beet Necrosis Yellow Vein Virus* (BNYVV) which is a probable member of the furovirus (fungus borne rod shaped virus) group (Shirako and Brakke, 1984). Dilution end point of this virus ranged in between 10^{-4} and 10^{-5}. The infectivity of BNYVV is usually abolished when sap is heated for 10 min. to 65–70°C or frozen to –20°C. Ultrathin sections of BNYVV-infected tissue show the virus particles scattered in the cytoplasm and in angled layer aggregates. The virus is not seen in large aggregates, but shows an erratic distribution in the mesophyll cells of systemic plants (Russo *et al*., 1981).

Host range

This virus infects all cultivars of sugarbeet, fodder beet, red beet, swiss chard (*Beta vulgaris* L. var. *cycla*), spinach (*Spinacia oleracea* L.), and several other species of Chenopodiaceae serving as hosts of the Plasmodiophoromycete fungus *Polymyxa betae*. It can also be mechanically inoculated to other herbaceous species, such as *Tetragonia expansa* Murr. (Aizoaceae), *Gomphrena globosa* L. (Amaranthaceae) and *Nicotiana clevelandii* Gray (Solanaceae).

Alopecurus myosuroides, Lolium multiflorum, Sorghum halepense, S. vulgare, Convolvulus arvensis, Matricaria inodora, Stellaria media, Calystegia sepium, Capsella bursa-pastoris, Centaurea cyanus and *Galinosorga parviflora* have reported to act as alternate hosts of BNYVV (Mouhanna, 2000).

Vector

Polymyxa betae has been considered to be the vector of BNYVV and was also shown to transmit soil borne virus (Ivanovic, 1983). This fungus has worldwide distribution. The abnormal proliferation of the root lets, characteristic of rhizomania was found in the presence of non-viruliferous *P. betae,* but zoospores transmit rhizomania symptoms to whole sugarbeet plants, in the absence of detectable virus in the infected plants, indicating that BNYVV is not requisite to the etiology of the disease. The infectivity of BNYVV was retained with *P. betae* in dry soil for atleast 15 years. ELISA detected BNYVV in extracts of resting spore clusters of a viruliferous *P. betae* isolate (Abe and Tamada, 1986).

Management

Management of rhizomania disease is presently based on the development of varieties resistant to the vector since field trials with fungicides failed to show any yield benefits (Blunt and Asher, 1989).

(i) Host resistance: Sugar beet varieties Laetitia and Nagano have been reported to possess resistance to this disease in Belgium (Anonymous, 2003).

(ii) Chemical control: Soil disinfection with dichloropropene was recommended for the control of fungal vector *P. betae* (Hess and Schlosser, 1984). Pre-plant application of fumigants like dichloropropene, telone 11 (1,3- dichloropropene), Vorled (1,3-dichloropropene + methyl isothiocyanate), Vorlex 201 (1,3- dichloropropene + chloropicrin + methyl isothiocyanate) and Pichlor 60 reduced disease incidence and significantly increased yields (Martin and Whitney, 1990).

(iii) Biological control: Some studies have also been carried out on the biological control of vector *P. betae* with *Trichoderma* sp. Soil inoculation with *T. harzianum* reduced infection by vector, *P. betae* upto 25 per cent and the *Trichoderma* population introduced remained high throughout experimental period (D'Ambra *et al.*, 1987).

REFERENCES

Abdel-Rehim, M.A., Aziza, K.D., Tarabeih, A.M. and Hassan, A.A.M. 1992. Damping-off and root rot of okra and table beet with reference to chemical control. *Assiut. J. Agric. Sci.* **23(4):** 19–36.

Abe, H. and Tamade, T. 1986. Association of best necrotic yellow vein virus with isolates of *Polymyxa betae* Keskin. *Ann. Phytopathol. Soc. Jpn.* **52:** 235–247.

Alkseeva, K. L. 2009. Score for control of *Alternaria* on carrot. *Zashchita I Karantin Rastenii* **7:** 26.

Anonymous, 2003. The choice of varieties for 2003. *Betteravier–Bruxelles* **37:** 391.

Anonymous. 2015. http://www.spssales.com.au/

Asher, M. 2000. Control of foliar diseases. *Brit. Sugarbeet Rev.* **68(2):** 32–33.

Asher, M.J.C. 1986. The occurrence and control of sugarbeet powdery mildew. *Aspects App. Biol.* **13:** 131–137.

Aycock, R. 1966. Stem rot and other diseases caused by *Sclerotium rolfsii. NC Agric. Exp. Stn. Tech. Bull.* 174. 202 pp.

Benard, D. and Punja, Z.K. 1995. Role of *Pythium* species in cavity spot development on carrots in British Columbia. *Can. J. Pl. Path.* **17:** 31–45.

Ben-Noon, E. Shtienberg, D., Shlevin, E., Vintal, H. and Dinoor, A. 2001. Optimization of chemical suppression of *Alternaria dauci*, the causal agent of Alternaria leaf blight in carrots.*Plant Dis.* **85:** 1149–1156.

Blunt, S.J. and Asher, M. 1989. *Polymyxa* – the rhizomania vector. *Brit. Sugarbeet Rev.* **57(2):** 31–33.

Breton, D. and Rouxel, F. 1994. Recent findings relevant to cavity spot of carrots in France. *Acta Hortic.* **354:** 159–170.

Chand, R. and Kishun, R. 1989. Studies on soft rot of carrot. *Indian Phytopath.* **42:** 137–139.

Chupp, C. and Sherf, A. 1960. *Vegetable diseases and their control.* John Wiley and Sons. Inc., New York, 693 pp.

Ciccarese, F., Frisullo, S., Amenduni, M. and Cirulli, M. 1992. Biological control of *Sclerotium rolfsii* root rot of sugar beet with *Trichoderma harzianum.* In : *Biological control of plant diseases: Progress and Challenges for the Future* (eds. E.C. Tjamos, G.C. Papavizas and R.J. Cook), Plenum Press, New York, pp. 243–248.

Coles, R.B. and Wicks, T.J. 2003. The incidence of *Alternaria radicina* on carrot seeds, seedlings and roots in South Australia. *Aust. Pl. Pathol.* **32:** 99–104.

Correa, A., Rebuffat, S., Bodo, B., Roquebert, M.F., Dupont, J. and Bettucci, L. 1995. *In vitro* inhibition activity of trichorzianines on *Sclerotium rolfsii* Sacc. *Cryptogamie Mycologie* **16(3):** 185–190.

Cvjetkovic, B., Hrlec, G. and Isakovic, L. 1993. Effect of fungicides for the control of powdery mildew (*Erysiphe heraclei* D.C. sp. *dauci* Jacz.) on carrot and the problem of residue in roots. *Difesa delle Plante* **16:** 21–26.

Czaczyk, K., Stachowiak, Trojanowska, K. and Gulewicz, K. 2000. Antifungal activity of *Bacillus* sp. isolated from compost. *Folia Microbiol.* **45(6):** 552–554.

D'Ambra, V., Mutto, S. and Causin, R. 1987. Activity of *Trichoderma harzianum* against *Polymyxa betae* in glasshouse trials. *Rivista di Patol. Vegetale* IV **23(3):** 100–107.

Das, S. and Raj, S.K. 2004. Management of *Sclerotium* root rot of sugarbeet through seed soaking with non-conventional chemicals. *Indian Phytopath.* **57:** 112–113.

de Bary, A. 1886. Ubereinige Sclerotinien und Sclerotienkrankheiten. *Bot. Ztg.* **44:** 377–474.

Dimov, T., Lapotishkina, A. and Vasileva, I. 1988. Effect of different crop rotations with sugarbeet on the number of soil microscopic fungi. *Pochvoznaniel Agrokhimiya* **23(5):** 54–58.

Dixon, G.R. 1981. *Vegetable Crop Diseases.* AVI Pub.Co. Inc. Westport, Connecticut, 404pp.

Eitarabily, K.A., Hardy, G.E.S.J., Sivasithamparam, K. and McKay, A.G. 1997. Amendment of soil with lime or gypsum and its effect on cavity spot disease of carrots *(Daucus carota* L.) caused by *Pythium coloratum. Aust. J. Expl. Agric.* **37(2):** 265–270

Farrar, J.J., Pryor, B.A. and Davis, R.M. 2004. Alternaria diseases of carrot. *Plant Dis.* **88:** 776–784.

Follas, G.B., Beetz, H.M. and Popay, A.J. 1992. Control of leaf spots in carrots and early blight in potatoes and tomatoes with difenoconazole. *Proc. 45 N Z Pl. Prot. Conf., Wellington*, pp. 46–49.

Fuji, K. Tokumoto, H., Niizuma, S., Narita, M., Sugawara, H. and Tomiyama, N. 1987. Controlling black leg on paper pot seedling of sugarbeet by tolclofos methyl and hydroxyisoxazol. *Proc. Res. Assoc., Jpn.* **29:** 50–62.

Gado, E.A.M .2013. Impact of treatment with some plant extracts and fungicides on sugar beet powdery mildew and yield components. *Aust. J. Basic Appl. Sci.* **7(1):** 468–472

Ganeshan, G. 1987. Occurrence of *Erysiphe heraclei* with perfect stage on new cultivars of *Daucus carota. Indian J. Mycol. Pl. Pathol.* **17(1):** 81.

Gor–Kovenka, V.S. 1991. Carrot rots during storage. *Zaschita Rastenii Moskva* **11:** 29–30.

Gowda, R.V., Pathak, C.S. and Ganeshan, G. 2000. Resistance source for powdery mildew and Alternaria leaf blight diseases in carrot. *J. Trop. Agric.* **38(1–2):** 84–86.

Green, C.D. and Makin, T. 1985. Soil borne transmission of cavity spot of carrots grown in Lincolnshire for processing. *Crop Prot.* **4:** 351–353.

Groom, M.R. and Perry, D.A. 1985. Induction of 'cavity spot like' lesions on roots of *Daucus carota* by *Pythium violae*. *Trans. Brit. Mycol. Soc.* **84(4):** 755–757.

Gupta, S.K. and Dohroo, N.P. 1993. *Sclerotinia* rot of cauliflower. *Int. J. Trop. Pl. Dis.* **11:** 131–138.

Haller, H., Jadesha, G. and Prakasam, V. 2013. Efficacy of fungicides and botanicals against *Cercospora beticola* Sacc. in sugarbeet. *Bioinfolet.* **10:** 473 – 479.

Hentschel, K.D. 1991. Biocontrol of seed borne *Alternaria radicina* on carrots by antagonistic *Bacillus subtilis. Bull. SROP* **14(8):** 73–76.

Hess, W. and Schlosser, E. 1984. Rhizomania VI. Infection yield loss relation and control with dichloropropene. *Mededelingen van de Faculteit Landbouwwe–tenschappen Rijksuniversiteit Gent* **49(2b):** 473–480.

Inglis, D.A., Derie, M.L. and Volker, K.C. 2001. Evidence that *Cercospora carotae* causes leaf spot on carrot in western Washington. *Plant Dis.* **85:** 559.

Ivanovic, M. 1983. Viruses of sugarbeet associated with *Polymyxa betae. Ann. Rep. Rothamsted Expt. Stn 1982*, pp. 189–190.

Jackobsohn, R. Zutra, D., Dan, H. and Kelman, Y. 1984. Studies on carrot cavity spot. *J. Hortic. Sci.* **59:** 529–535.

Jensen, B., Knudsen, I.M.B., Madsen, M. and Jensen, D.F. 2004. Biopriming of infected carrot seed with an antagonist, *Clonostachys rosea*, selected for control of seedborne *Alternaria* spp. *Phytopathology* **94:** 551–560.

Joe, M.M., Chanbhasa, M., Henry, A.J and Benson, A. 2017. Treatment of carrots with spent-green tea extract reduces the soft rot of *Pectobacterium* spp. *Arch. Phytopathol. Plant Protec*. **50:** 491–503.

Khan, M.A. 1976. Studies on stalk rot of cauliflower caused by *Sclerotinia sclerotiorum*. M. Sc. Thesis, HPU, College of Agriculture, Solan, 80 p.

Khan, M.F.R. 2009. Successfully controlling Cercospora leaf spot on sugar beet with fungicides in Minnesota and North Dakota, USA. *Int. Sugar J.* **111:** 646–649.

Kiewnick, S., Jacobsen, B.J., Braun, A., Kiewnick, Eckhoff, J.L.A. and Bergman, J. 2001. Integrated control of Rhizoctonia crown and root rot of sugar beet with fungicides and antagonistic bacteria. *Plant Dis.* **85:** 718–722.

Kora, C., McDonald, M.R. and Boland, G.J. 2003. Sclerotinia rot of carrot: an example of phonological adaptation and bicyclic development by *Sclerotinia sclerotiorum. Plant Dis.* **87:** 456–470.

Kozak, 1990. Susceptibility of carrot cultivars to the bacterial soft rot induced by *Erwinia carotovora. Genet. Polonica* **31:** 205–212.

Liew, C.L. and Prange, R.K. 1994. Effectof ozone and storage temperature on post–harvest diseases and physiology of carrots (*Daucus carota* L.). *J. Am. Soc. Hortic. Sci.* **119(3):** 563–567.

Lopes, C.A., Ritschel, P.S., Vieira, J.V. and Lima, D.B. 2000. Behaviour of summer carrot genotypes in fields with different etiologies of the leaf blight complex. *Hortic. Brasileira* **18(2):** 119–122.

Manaresi, M., Coatti, M., Brunelli, A. and Canova, A. 2002. F 500 (pyraclostrobin): a newest broad spectrum strobilurin fungicide. *Atti–Giornate–fitopatologiche,–Baselga–di–pine, Trento, Italy,* April 7–11, 2002. **2:** 119–124.

Markova, G.F. and Golyshin, N.M. 1988. Testing of fungicides for carrot seed treatment. *Agrokhimiya* **1:** 101–103.

Mazur, R.N., Kadow, K.J. and Anderson, H.W. 2005. Damping-off control; an evaluation of seed and soil treatments. *Can J Res* **439**: 291–348.

Maringoni, AC., Baldini, L.F.G., Demant, L.A.R., Evangelista, R.M. and Fujita, E. 2012. Fungicide action in the control of leaf blight and physical-chemical characteristics of the carrot cv. Nantes. *Revista–Brasileira–de–Produtos–Agroindustriais* **14(2):** 161–165.

Martinez, C., Levesque, C.A., Belanger, R.R. and Tweddell, R.J. 2005. Evaluation of fungicides for the control of carrot cavity spot. *Pest Mngt. Sci.* **61(8):** 767–771.

Martin, E.N. and Whitney, E.D. 1990. In bed fumigation for control of rhizomania of sugarbeet. *Plant Dis.* **74(1):** 31–35.

Maynard, D.N., Gersten, B., Young, R.E. and Vernell, H.F. 1963. The influence of plant maturity and calcium level on the occurrence of carrot cavity spot. *Proc. Am. Soc. Hortic. Sci.* **83:** 506–510.

McQuilken, M.P., Whipps, J.M. and Cooke, R.C. 1990. Control of damping-off in cress and sugarbeet by commercial seed coating with *Pythium oligandrum. Plant Pathol.* **39(3):** 452–462.

Mirkove, E. 1979. Carrot seed treatment against some fungal pathogens. *Gradinarrkai Lozarska Nauka* **16(5–6):** 92–96.

Mouhanna, A.M. 2000. I Rhizomania: investigations on epidemiology. II Rhizomania: investigations on systemic acquired resistance (SAR) of sugar beet. III Molecular phyto-pathological studies on the reaction of sugar beet plants to rhizomania and the resistance inducer BION (R). *Rhizomania:-Untersuchungen-zur-epidemiologie-und-Systemische-Aktivierten-Resistenz-SAR-bei-der Zuckerrube 2000,* 239pp.

Mukhopadhyay, A.N. 1968. Protect your sugarbeet from foliage diseases. *Indian Farmers Dig.* **1(10):** 41–42.

Mukhopadhyay, A.N. 1969. Powdery mildew of sugarbeet, a new record for India. *Sci and Cult.* **35:** 276–277.

Mukhopadhyay, A.N. 1971. *Sclerotium* rot of sugarbeet in India. *Mycopath et. Mycol. Appl.* **44:** 265–270.

Mukhopadhyay, A.N. and Thakur, R.P. 1971. Control of *Sclerotium* root rot of sugarbeet with systemic fungicides. *Pl. Dis. Reptr.* **55:** 630–634.

Mukhopadyay, A.N. and Upadhyay, J.P. 1977. Relative efficacy of carbendazim and mancozeb against Cercospora leaf spot of sugarbeet. *Pesticides* **11:** 15–18.

Muniz, J.O., De, L. and Ponte, J.J. Da. 1988. Resistance of carrot cultivars to leaf blight in the Baturite mountains, State of Ceara (Brazil). *Fitopatol. Brasileira* **13(1):** 48–51.

Murtaza, M., Wahid, A., Ali, S. and Nadeem, A. 1988. Seedling blight of carrot caused by *Stemphyllium radicinum. Pakistan J. Agric. Res.* **9(4):** 601–603.

Nega, E., Ulrich, R., Werner, S. and Jahn, M. 2003. Hot water treatment of vegetable seed – an alternative seed treatment method to control seed-borne pathogens in organic farming. *Z. Pflanzenkr. Pflanzenschutz* **110:** 220–234.

Nowicki, B. 2002. Carrot protection against complex of pathogens causing damping-off. *Veg. Crop. Res. Bull.* **57:** 83–94.

Pandey, K.K., Pandey, P.K. and Kapoor, K.S. 2003. Collar rot of carrot caused by *Sclerotinia sclerotiorum. Indian Phytopath.* **56:** 122.

Papavizas, G.C. and Collins, D.J. 1990. Influence of *Gliocladium virens* on germination and infectivity of sclerotia of *Sclerotium rolfsii. Phytopathology* **80:** 627–630.

Payne, P.A. and Williams, G.E. 1990. Hymexazol treatment of sugarbeet seed to control seedling diseases caused by *Pythium* spp. and *Aphanomyces cochlioides. Crop Prot.* **9(5):** 371–377.

Poissonnier, J., Reulet, P., Guery, B. and Tison, P. 1995. *Alternaria* blight in carrot. *INFOS (Paris)* **112:** 42–44.

Prithiviraj, B., Khiste, S., Ram, D. and Singh, U.P. 1996. Effect of methanol extract of *Aegel marmelos* leaves on mycelial growth and sclerotium formation in *Sclerotium rolfsii. Int. J. Pharmacognosy* **34(2):** 148–150.

Pryor, B.M., Strandberg, J.O., Davis, R.M., Nunez, J.J. and Gilbertson, R.L. 2002. Survival and persistence of *Alternaria dauci* in carrot cropping systems. *Plant Dis.* **86:** 1115–1122.

Putz, C., Mcrdinoglu, D., Lemaire, O., Stocky, G. Valentin, P. and Sabine Wiedemann 1990. Beet necrotic yellow vein virus, causal agent of sugarbeet rhizomania. *Annu.Rev. Pl. Pathol.* **69:** 247–254.

Rajpurohit, T.S. and Singh, R.D. 1992. Field evaluation of sugarbeet varieties against Cercospora leaf spot. *Indian J. Mycol. Pl. Pathol.* **22(3):** 294–296.

Rajpurohit, T.S. and Singh, R.D. 1993. Chemical control of Cercospora leaf spot of sugarbeet. *Indian J. Mycol. Pl. Pathol.* **23(1):** 42–45.

Rajpurohit, T.S. and Singh, R.D. 1996. Effect of date of sowing, previous crop, irrigation and weeds on incidence of Cercospora leaf spot of sugarbeet. *Indian J. Mycol. Pl. Pathol.* **26(1):** 130 (Abstr.).

Rao, M.S., Kamalnath, M., Umamaheswari, R., Rajinikanth, R., Prabu, P., Priti, K., Grace, G.N., Chaya, V.C. and Gopalakrishnan, C. 2017. *Bacillus subtilis* IIHR BS-2 enriched vermin-compost controls root knot nematode and soft rot disease complex in carrot. *Scientia Hortic.* **218:** 56–62.

Rautella, G.S. and Payne, M.G. 1970. The relationship of peroxidase and ortho-diphenol oxidase to resistance of sugarbeet to Cercospora leaf spot. *Phytopathology* **60:** 238–245.

Rodriguez, S.A., Albarracin, N.S. de, Arjona, A.E., Albarracin, M.; Hernandez, Y.A. and de Albarracin, N.S. 2002. Post-harvest diagnostics of bacterial diseases in Zanahoria. *Fitopatol. Venezolana* **15(1):** 17–20.

Roy, A.K. 1977. Attack of *Sclerotium rolfsii* on some plants and capability of the fungus to cause storage rot. *Indian Phytopath.* **30:** 425–426.

Russo, M. Martelli, G.P. and Di Franco, A. 1981.The fine structure of local lesions of beet necrotic yellow vein virus in *Chenopodium amaranticolor. Physiol. Pl. Pathol.* **19:** 237–242.

Sesan, T. 1990. Relationships between main fungal pathogens of stored carrots and saprophytic fungi. *Biologie Vegetala* **42:** 149–152.

Shirako, Y. and Brakke, M.K. 1984. Two purified RNAs of soil borne wheat mosaic are needed for infection. *J. Gen. Virol.* **65:** 119–127.

Shlevin, E., Ben Nun, R., Tzror, L., Nachmias, A. and Ohali, I. 1987. Reduction of cavity spots in carrots by Ridomil applications. *HASSADEH* **67(11):** 2272–2274.

Siviero, P., D'–Ascenzo, D, Prencipe, N. and Politi, A. 2004. Introduction of new agricultural plant protection chemicals against *Alternaria* in carrot. *Informatore Agrario* **60(7):** 69–71.

Soroker, E., Bashan, Y. and Okoni, Y. 1984. Reproducible induction of cavity spot in carrots and physiological and microbial changes occurring during cavity formation. *Soil Biol. Biochem.* **16**: 541–548.

Souza, R.T., Forcelini, C.A., Reis, E.M. and Calvete, E.O. 2001. Frequency of *Alternaria dauci* and *Cercospora carotae* as causal agents of carrot leaf blights in Passo Fundo. *Fitopato. Brasileira* **26(3):** 614–618.

Srivastava, S.N. and Tripathi, R.C. 1998. Management of sugarbeet seedling disease complex by combination of fungicides. *Indian Phytopath.* **51(1):** 75–77.

Thakur, R.P. and Mukhopadhyay, A.N. 1972. Nitrogen fertilization of sugarbeet in relation to Sclerotium root rot caused by *Sclerotium rolfsii* Sacc. *Indian J. Agric. Sci.* **42:** 614–617.

Timchenko, V.I., Soladatenko, O.M. and Yarovoi, G.I. 1989. Diseases of carrot seed plants. *Zaschita Rastenii Moskva* **6:** 31–32.

Uchino, H., Kanzawa, K., Sugimoto, T., Naito, S. and Sayama, H. 1987. Studies on the control of Rhizoctonia root rot of sugarbeets. Effect of drenching pencycuron to sugarbeet seedlings in pepper rot transplanting system. *Proc. Sugarbeet Res. Assoc., Jpn.* **29:** 50–57.

Vala, D.G. and Kapoor, J.N. 1985. Detection of *Alternaria* species on seeds of carrot and cabbage. *Indian Phytopath.* **38:** 760.

Walczak, F., Grendowicz, L., Jakubowska, M., Skorupska, A., Strugala, N., Tratwal, A. and Wojtowicz, A. 2001. Harmfulness of some agrophages of cultivated plants in Poland in 2000 and infestation of plants by weeds. *Prog. Pl. Prot.* **41:** 330–349.

Walker, S.E. 1991. Effect of metalaxyl and phosphonate on incidence of cavity spot in carrots. *Aust. Pl. Pathol.* **20:** 21–26.

Watson A. 2016. Managing carrot powdery mildew. *Aust. Pl. Pathol.* **45:**29–35.

Westerveld, S.M., McDonald, M.R., Scott-Dupree, C.D. and Mckeown, A.W. 2002. The effect of nitrogen on insect and disease pests of onions, carrots and cabbage. *J. Veg. Crop Prod.* **8(2):** 87–99.

Whipps, J.M., McQuilken, M.P. and Budge, S.P. 1993. Use of fungal antagonists for biocontrol of damping-off and Sclerotinia diseases. *Pesticide Sci.* **37(4):** 309–313.

White, J.G. 1988. Studies on the biology and control of cavity spot of carrots. *Ann. Appl. Biol.* **113:** 259–268.

■■■■

13

DISEASES OF COLOCASIA

Colocasia (*Colocasia esculenta* L., Schott), commonly known as taro, is grown throughout the world including India either as staple or subsistence crop. It belongs to the family Araceae and is a tuberous perennial herb with cordate peltate leaves. The crop is not only used as vegetable but also has medicinal and industrial values (Ghosh and Pan, 1989). It is rich in proteins, minerals, fibre, phosphorus, iron etc. The starch, mucilage and tissues of this crop possess good industrial value both in textile, paper and alcohol industries. Like other crops, colocasia also suffers from the attack of a large number of fungal and bacterial diseases, both in field and storage. A comprehensive account of fungal dand viral iseases including their management has been given in this chapter.

1. PHYTOPHTHORA BLIGHT

Phytophthora blight is known to be the most serious disease of colocasia and causes considerable yield losses.

Geographical distribution

Raciborski first reported this disease in 1900 from Java. Subsequently, it was reported from India, Formosa, Philippines, China, Sri Lanka, Langawl Island and Malaya, Marinas, Carolines, Burma, British Solomons, Fiji, N. Borneo, Hong Kong, Japan, Indonesia, Malaysia, Sarawak, Africa and Caribbeans, Hawaiin and other Pacific Islands (Hunter *et al.*, 1998). In India, the occurrence of the disease was first recorded by Butler and Kulkarni in 1913 (Hegde *et al*, 2011).

Losses

Losses from this disease are greater than any other disease of Colocasia. These results from (i) premature defoliation of the leaves due to severe blightening, (ii) reduced corm production due to reduction in net photosynthetic area, (iii) rotting of the corms both under field and storage conditions and (iv) increased expenses due to spraying operations. The losses in yield have been estimated to range between 25–50 per cent (Kay, 1973). Jackson and Gollifer (1975) found that infected leaves collapse within 20 days of unfurling as compared to 40 days in healthy leaves. They have also reported 30–40 per cent loss in corm yield when the crop was attacked within 40–70 days after planting. Vasquez (1990) tested four colocasia accessions with varying degrees of resistance to disease and reported yield loss of 3–5 per cent in case of resistant and 24.4–36.5 per cent in moderately resistant to susceptible accessions.

Serious losses in taro production were also reported from Samoa due to this disease (Rogers *et al.*, 2000). The disease assumes serious proportions in those areas having high relative humidity with frequent rainfall and causes 25–50 per cent losses in crop yield (Gurung, 2001, Mishra *et al.*, 2011).

Symptoms

The symptoms of the disease start appearing in the later part of the rainy season. The earliest symptoms include the appearance of small, dark, round spots on the leaves (Plate 1a). These spots rapidly increase in size by centrifugal growth and become circular or assume an oval or irregular shape (Plate 1b). Several such spots may appear on the leaf and the whole leaf lamina may get infected. Clear, yellow liquid drops ooze out from the surface of the spots in the initial stages, later the central portions become dry and drop off and thus forming shot holes in the leaf. The periphery of the spots has zonations of different shades of brown, green and yellow and during moist conditions may contain a delicate white haze of sporangial growth. In severe attacks, the petioles are also infected and the infection seems to travel down the underground portions. Badly infected leaves fall down pre-maturely on the soil surface. The corms are also attacked by the sporangia from the fallen leaves and begin to rot. Sometimes, the corms may rot completely during the unusually wet weather in the field itself but the rot develops more rapidly later on in the storage.

The Pathogen

The fungus responsible for this disease is *Phytophthora colocasiae* Racib. The mycelium is hyaline, coenocytic, intercellular in freshly attacked tissues and intracellular when rotting has advanced. The haustoria are slender, long and unbranched. Sporangiophores are formed on the surface of the leaf and are upto 50 µm long, unbranched, very slender and extremely narrow at the tip. They bear single pear-shaped or elongated sporangia (Fig. 1a), which measure 38–46 x 18–26 µm, germinating directly or indirectly depending on the weather conditions. When they germinate indirectly as many as 12 reniform, bi-flagellate zoospores are formed in each sporangium. Chalamydospores are also formed (Fig. 1b). These are thick walled, round and hyaline structures. Being a heterothallic fungus, whereever, A1 and A2 mating types are present in nature, the pathogen goes in sexual reproduction and produce anthredium and oogonium. The oogonium is spherical and yellowish in colour. The amphigynous antheridium persists at the base of the oogonium that measures 99 µm in diameter. The oospores are spherical, 20–28 µm in diameter and lie free in the oogonium (Fig. 1c). Padmaja *et al.* (2017) studied variation among the size of oospores in four isolates of *P. colocasiae* and reported that the mean diameter of oospore ranging from 19.20 to 27.80 µm.

Host range

The fungus has a very limited host range. Besides *Colocasia esculenta*, it also attacks *Alocasia, C. antiquorum var. esculenta* (L.) Schott., *Amorphophallus companulatus, Solanum tuberosum, Citrus aurantium, Artocarpus integrifolia, Cocus mucifera, Areca catechu, Theobroma cacao, Lycopersicon esculentum, Hevea braziliensis, Malus sylvestris, Ananas squamosa, Phaseolus vulgaris* and *Piper betle* (Butler and Bisby, 1960; Paharia and Mathur, 1961).

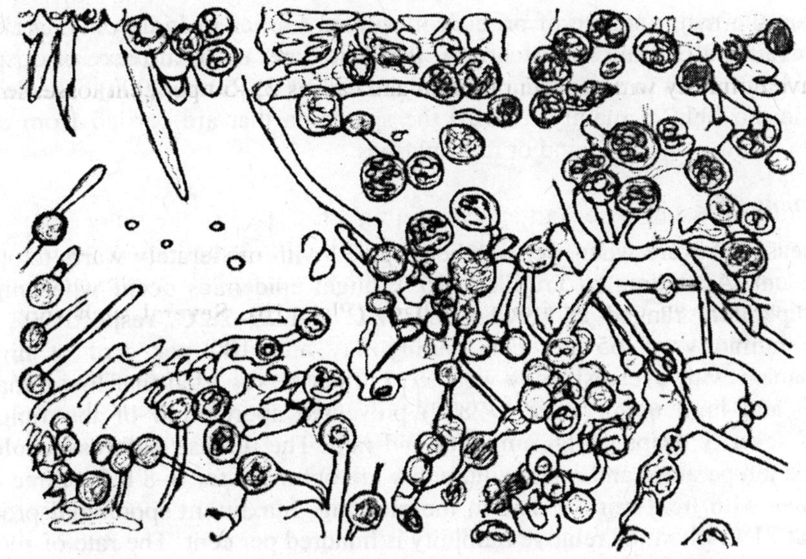

(b) *Phytophthora colocasiae* Racib (Chlamydospores)

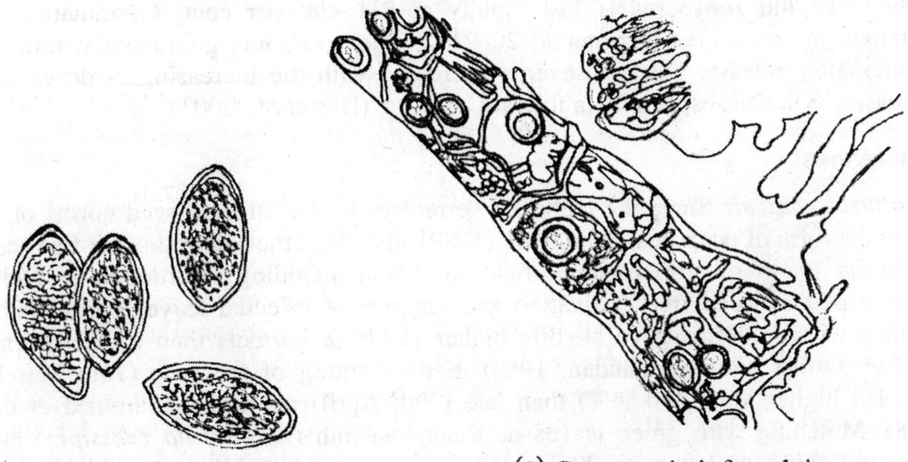

(a) Semipapillate sporangia (c) Oospores in infected tissues

Fig. 1. Microscopic structures of *P. colocasiae*

Disease cycle

The pathogen perpetuates either through the mycelium and oospores present in the crop debris, corms lying in the soil or infected corms used for planting (Gomez, 1925, Rama, 2006). Narula and Mehrotra (1984) could isolate the fungus from a field with or without *Colocasia* crop only during rainy season. Gollifer *et al.* (1980) observed that inoculum on detached leaf lesions or in soil remained viable for only a few days and that on petiole bases for 2 days if stored dry, but for 14 days if planted in the field immediately. In artificially infected soil, the fungus survived for 51 days but infected leaves buried in soil did not yield the fungus after five days. The fungus has poor competitive ability and cannot survive in soil in free state. Thus infected corms are the major source of primary inoculum. Earlier in India only 'A1' mating type is available (Narula and Mehrotra, 1980) but later on occurrence of oospores in infected leaf tissues in soil were observed (Atlar, 2003, Rama, 2006). The secondary spread in the field is mainly through the sporangia that are carried from diseased leaves to healthy ones by wind or rain splashes.

Epidemiology

The disease is severe when wet weather coupled with moderately warm temperature is prevalent. According to Trujillo (1965), blight epidemics occur when night and day temperature ranged in between 20–22ºC and 23–28ºC, respectively, with a relative humidity of 65 per cent during day and 100 per cent at night and accompanied with overcast rainy weather. Further he reported that low temperature (<20ºC) and high temperature (>28ºC) prevented sporulation of the fungus and reduced severity despite high humidity and rain. The disease outbreak could occur when the temperature and relative humidity are optimum for 6–8 h for three consecutive days with light rain or dew in the morning. Maximum sporangial production occurs at 21ºC when the relative humidity is hundred per cent. The rate of sporangial production was higher in susceptible varieties as compared to resistant ones (Mishra and Sri Ram, 2002). Relative humidity below 90 per cent inhibits sporulation. Viability of the zoospores is lost rapidly at RH <90 per cent. Germination of sporangia by zoospores is faster at 20–21ºC. The zoospores germinate within 30 minutes after release. Disease severity increased with the increasing N doses and decreasing K levels, especially in tolerant cultivars (Das *et al.*, 2003).

Management

i. Cultural control: Since the pathogen perrenates in the infected seed corms or in soil in the form of oospores, Sahu *et al.* (1989) advocated the use of disease free seed corms for planting, crop rotation, field sanitation including destruction of badly affected plants, weed free cultivation and removal of infected leaves periodically. Planting of side corms gave slightly higher yields of cormels than mother corms (Mohan Kumar and Sadanandan, 1988). Early planting of the crop (15th March) provided higher yield (40.56%) than late (29th April) planting (Ahammed *et al.*, 1988). Mulching with green leaves or water hyacinth (*Eichhornia crassipes*) and black polythene mulch gave 20 and 13 per cent more yields, respectively, over unmulched plots (Mohan Kumar and Sadanandan, 1988; Ahammed *et al.*, 1988). Planting of the crop in shady locations should be avoided. Ensure proper drainage of the field. Balraj *et al.* (1998) observed four inter crops in the Eucalyptus plantation and indicated that colocasia was the most profitable crop grown in open as well as in shade conditions with minimum incidence of blight. Closer plant spacing should be avoided as it enhances early infection and defoliation (Mishra and Sri Ram, 2002). Disease appearance was delayed by 5–6 days when Eupatorium (*Chromolaena odorata*) was used as mulch resulted in significant increase in yield and decrease in disease severity. (Rana and Sugha, 2007).

ii. Host resistance: Since the net gains from the crop are less hence fungicidal control may not be that much economical. Resistant varieties can provide better control of the disease. Deshmukh and Chibber (1960) in their study of field resistance colocasia varieties among reported only one variety 'Ahina' as fairly resistant. Paharia and Mathur (1964) screened varieties collected from various regions of India and found Poonam Pat as immune followed by Sakin-V from Ranchi as resistant and another seven varieties viz., Ahina, Bhadia, Kachu and Naga Kachu from Assam, Pusa from Bihar, Sakin-II from Ranchi and Simla from Himachal Pradesh as moderately resistant to blight. Among 11 cultivars screened against this disease under natural epiphytotic conditions, Burdwan local was found the best for commercial cultivation (Ghosh *et al.* 1991). A high degree of resistance to this disease was also observed in four cultivars viz., Jankhari, Muktakeshi, Topi and Manasaru (Mishra and Naskar, 1995). Genotypes like Kadma local, Muktakeshi or Nadia local, JCC 24 and JCC 25 were also reported resistant against colocasia blight (Alam *et al*., 1996; Kumar *et al*., 1996). Varieties like AN 65, 17, 50, 32, 46, 21, 12, AN 20, Gilin, Kugfel, Oglang, Pwetepwet, Thailand, Sushi and Pastora were found resistant against taro leaf blight (Okpul *et al*., 1997; Wall and Wiecko, 1999). The resistance in 'Bangkok' cultivar is controlled by at least three genes. A significant deviation from single-gene ratio and continuous variation in the level of resistance in the progeny indicated that resistance in this cultivar is polygenic and likely to be durable (Singh *et al*., 2001).

iii. Chemical control: The disease has been claimed to be successfully controlled by application of copper fungicides in many places like Fiji and Solomon Islands (Bergquist, 1974; Jackson, 1980). Various conventional fungicides like Bordeaux mixture (2:2:50), Blitox-50 or Dithane M-45 can be applied for the control of this disease at 10–15 days interval (Singh, 1987). Aggarwal and Mehrotra (1988) observed that Apron 35 FW, Blitox, Blimix, Cuman-L, Demosan 65 W, Dexon, Difolatan, Fytolan, Hexaferb, Kitazin, Milton, Ridomil 25WP and Syllit gave complete inhibition of mycelial growth and reduced sporangial formation at varying concentrations, while maximum inhibition of pectolytic and cellulolytic enzymes was caused by Ridomil 25 WP followed by Apron 35 WF. These two fungicides were also highly effective when tested under field conditions. Sahu *et al.* (1989) recommended four sprays of zineb at 15 days interval starting from the end of July for reducing the incidence of this disease and increased yields, but maximum financial return was obtained by spraying Ridomil MZ at 15 d interval (Ghosh *et al*., 1991; Singh and Singh, 2003). Das (1997) also recommended three sprays of metalaxyl + mancozeb or mancozeb alone starting at the onset of the disease and repeating at fortnightly interval for effective management of this disease. Ridomil MZ and mancozeb were also found inhibitory to the pathogen by Mishra and Sri Ram (2002). Sugha (2007) evaluated different fungicides under *in-vitro* conditions and reported 'True' non-systemic fungicides and systemic fungicide 'Krilaxyl' as most effective against the pathogen.

Ridomil MZ when sprayed on lamina and petiole gave best control with 35.50 to 85.76 per cent disease control with 38 to 192 per cent increase in yield over control showing direct positive relationship between sprays of Ridomil MZ and per cent disease control and that of corm yield (Sugha, 2010).

iv. Homoeopathic drugs: Aggarwal *et al.* (1993) studied the effect of 3–200 potencies of four homoeopathic drugs on mycelial growth, sporangial production, pectolytic and cellulolytic enzyme production and control of *P. colocasiae* under glass house conditions. Kali iodide and *Arsenicum album* caused maximum inhibition of mycelial growth while sporulation was poor in 30 and 200 potencies of Kali iodide and *Arsenicum album*, respectively and 200 potencies of *Thuja occidentalies* and *Blatta orientalis*. The maximum decrease of pectolytic and cellulolytic activity was shown by Kali iodide followed by *Arsenicum, Thuja* and *Blatta*, respectively. Effective potencies of these drugs when applied on leaves before inoculation with *P. colocasiae* did not allow rotting to develop.

v. Biological control : Although information on biological control of this disease is very limited, but Narula and Mehrotra (1987) made an attempt to control *P. colocasiae* with phylloplane microflora with different degrees of success. Two bacteria, three actinomycetes and four fungi showed antagonistic potential under *in vitro* conditions while *in vivo* only *Streptomyces albidoflavus* and *S. diastaticus* reduced per cent infection by 90–93 and 76 per cent, respectively, as compared to control. Antagonistic activity of *Trichoderma* sp. to *P. colocasiae* has also been reported (Sitansu *et al.*, 1997).

Integrated Disease Management

Integration of corm dip (1 h) in Indofil M-45, mulching with Eupatorium leaves and three fortnightly sprays of Ridomil MZ (0.25%) resulted in 76.9 and 131,3 per cent disease control and increase in crop yield, respectively (Rana *et al.*, 2007)

2. PYTHIUM ROOT AND CORM ROT

Many species of *Pythium* have been reported to be associated with this disease. In warm and humid weather, this disease causes significant losses both under field and storage conditions. It can cause yield reductions ranging from 10 to 100 per cent (Parris, 1941).

Geographical distribution

The disease was first reported from Gold Coast in 1930 to be caused by *P. aphanidermatum* associated with rotten corms (Wright, 1930). Subsequently, its occurrence has been reported from other colocasia growing countries like Hawaii, Islands of Fernanedo, Gulf of Guinea, South East Asia and Pacific region, Puerto Rico, Fiji, Western Samoa and France (Anonymous, 1985).

Symptoms

a. Root rot : The symptoms of *Pythium* root rot usually originate at the base of the corm which turns light yellow to grey blue or dark purple and may assume a parallel sided bottle necked shape in contrast to the normal ovoid or oblong. The roots infected by *P. ultimum* become mucilaginous and decay rapidly, rhizomes become soft and cheese like, petiole bases appear water soaked and the leaves collapse eventually (Alvarez-Garica *et al.*, 1971).

b. Corm rot: Corm rot is characterized by softness of the corms that exude a watery substance and are deficient in starch content. Suckers from *Pythium* rotted plants that put out new leaves and develop individual root systems are also liable to produce similar symptoms.

The Pathogen

Pythium species like *P. aphanidermatum, P. ultimum, P. myriotylum, P. gracile* and *P. irregulare* (Reddy, 1970; Kerz-Moehlendick *et al.*, 1983) have been reported to be associated with this disease. The mycelium of *Pythium* sp. is hyaline and coenocytic. Chlamydospores occasionally develop in old cultures grown on oatmeal and potato dextrose agar. Sporangia are formed within 48–60 h at 28–30ºC measuring 16–41.6 x 12.8–33.6 μm and are borne terminally on short side branches of main hyphal strand. These are cut off from mother hyphae by a septum. Free-swimming zoospores are produced within 20–30 minutes. A thin-walled vesicle was extruded into which the protoplasm passed before the swarm spores become differentiated but no sexual bodies were observed (Parris, 1941).

Disease cycle

Pythium is a soil borne fungus and perennates in the form of chlamydospores/ oospores. In favourable weather, they germinate and start producing sporangiophores which inturn produce sporangia and zoospores. Zoospores after swimming for some time in a film of water become encysted and cause infection. The pathogen causing soft rot gains entry through the root particularly in the region of root hair development.

Epidemiology

Pythium species require warm and humid weather for infection and spread. The temperature of 28–30ºC has been reported as optimum for the production of sporangia (Parris, 1941).

Management

Cultural practices like deep ploughing followed by drying of land, use of healthy seed corms, improved drainage and crop rotation have been reported as the means of control against this disease (Alvarez-Garica *et al.*, 1971). Resistant varieties can be successfully used for the control of this disease.

3. FUSARIUM ROT

Several species of *Fusarium* are reported to be associated with this disease and cause losses both in field and storage. The losses from the disease may be enormous because the corms are rotten in the field itself with the attack of these pathogens. However, no exact estimates of losses have been reported.

Geographical distribution

Fusarium oxysporum has been recorded in British Solomon islands (Gollifer and Brown, 1973), while *F. solani* has been reported from Southern USA, Egypt and India (Michail *et al.*, 1969) and *F. proliferatum* from Guam (Wall and Cruz, 1991).

Symptoms

F. oxysporum causes a whitish grey spongy soft rot with a distinct brown margin. Decay usually begins at the base or side of the corms that may be partially or completely destroyed (Gollifer and Brown, 1973). *F. solani* causes a dry corky or powdery rot with a definite margin. White mould may also develop on the surface and the diseased corm becomes lighter in weight than the healthy ones. The rotten tuber yielded mushy odour (Sharma and Upadhyay, 1977). *F. proliferatum* produces yellowish spongy rot, sometimes found on the bottom of the freshly harvested corms (Wall and Cruz, 1991).

The pathogens

Several species of *Fusarium* like *F. oxysporum*, *F. solani* and *F. proliferatum* are responsible for this disease.

Host range

Fusarium sp. being a soil borne pathogen has a wide host range and infect many plant species. Potato tuber has been reported as the collateral host for *F. solani* (Michail *et al.*, 1969).

Disease cycle

Fusarium sp. responsible for this disease are soil borne in nature and can survive in the form of thick-walled resting spores (Chlamydospores). *Fusarium* sp. produce abundant asexual spores (conidia), which can cause infection of the corms usually through wounds sustained during harvesting.

Epidemiology

Although *F. solani* is primarily a wound parasite but it is highly aggressive in causing the dry rot disease in tubers with a wide range of temperature adaptability. Warm temperature (30ºC) and humid (90–100% RH) conditions of storage, continuous darkness and near optimum pH favoured conidial germination (Mishra and Rath, 1986).

Management

i. Cultural control: Cultural practices like field sanitation, long crop rotation, corm selection before storage and planting, destruction of collateral host like potato (Michail *et al.*, 1969) and avoiding injury to the corms, while harvesting and storage may be helpful in containing this disease. Cold storage is also effective in reducing this disease.

ii. Chemical control: Treatment with fungicides resulted in lower disease incidence as compared to control and carbendazim treatment resulted in the lowest disease incidence and crop weight loss (Bhaskar *et al.*, 2003). Application of Benlate to the crop, 15 days prior to harvesting, consistently decreased severity of decay of corms (Quevedo *et al.*, 1991). Dipping of corms in hot water (50ºC) with 200-ppm benomyl suspension for 5 minutes has also been found effective against Fusarium rot of Colocasia (Quevedo *et al.*, 1991).

VIRAL DISEASES

Dasheen Mosaic

It is an important viral disease of colocasia and widely distributed throughout the world wherever this crop is grown.

Geographical Distribution

The occurrence of this disease has been reported from Hawaii, Egypt, India, Japan, Papua New Guinia, Puerto Rico,Trindad, USA and Venezuela (Lebot, 2009, Reddy, 2010).

Losses

It is an important viral disease and in infected field up to 90 per cent plants can be affected by this virus and upto 60 per cent losses due to this disease can occur (Lebot, 2009).

Symptoms

Mild light and dark mottled symptoms appear on leaves and a feather pattern of light yellow tissue may develop alongwith veins. Partial or complete yellowing of the leaf lamina along the veins and veinlets are the characteristic symptoms of the disease. In severely infected plants, petioles remain short and the leaves remain dwarfed and fail to open properly.

The Virus

The disease is caused by *Dasheen Mosaic Virus* (DsMV) which is a *Potyvirus* having flexuoux rod of 750 nm (Zettler *et al*., 1970). The virus is a stylet borne and is transmitted by several species of aphids in non-persistent manner

Disease Cycle

Infected corms and cormels which are used for seed purpose may serve as a source of primary inoculum. The virus is also carried in infected planting stocks as this crop is exclusively vegetatively propagated. Secondary spread of the virus in the field is by aphids.

Management

The disease can be kept in check by using healthy planting material. Secondary spread of the disease can be checked by application of systemic insecticides.

REFERENCES

Aggarwal, A. and Mehrotra, R.S. 1988. Effect of various fungicides on mycelial growth, sporangial production, enzyme activity and control of *Phytophthora* leaf blight of *Colocasia esculenta* L. *Acta Phytopatho. et Entomol. Hungaricae* **23(3-4)**: 401–414.

Aggarwal, A., Kamlesh and Mehrotra, R.S. 1993. Control of Taro blight and corm rot caused by *Phytophthora colocasiae* using homoeopathic drugs. *Pl. Dis. Res.* **8(2)**: 94–101.

Ahammed, A., Siddique, M.A. and Rabbani, M.G. 1988. Effect of date of planting and mulching on the growth and yield of mukhikachu. *Bangladesh J. Agril. Res.* 13(1): 52-56.

Alam, S., Gogoi, R., Narzary, B.D. and Goswami, R.K. 1996. Screening of colocasia germplasm of Northeast India against Phytophthora blight. pp. 391–394. In: *Tropical tuber crops: Prolems, prospects and future strategies* (Eds. Kurup, G.T. , Palawiswami, M.S., Potty, V.P., Padmaja, G., Kabeerathumma, S. and Pillai, S.V.) Birsa Agricultural University, Ranchi, Bihar, India.

Alvarez-Garica, L.A., Cortes-Menllas, A. and Currtace. 1971. A *Pythium* soft rot of *Xanthosoma* and *Colocasia* spp. in Puerto Rico. *J. Agric. Univ. Puerto Rico* **55(1):** 78–84.

Anonymous, 1985. New Records, Tuvalu. Taro Disease. *Quarterly Newsl. F.A.O. Asia and Pacific Pl. Prot. Comm.* (1985), **28(3):** 29.

Atlar, B.S. 2003. Studies on leaf blight of colocasia caused by *Phytophthora colocasiae* Racib. M.Sc. Thesis, YSP University of Horticulture & Forestry, Solan, H.P., 82 pp.

Balraj, S., Vijay, S., Singh, R.P. and Srivastava, B.K. 1998. Economic prospects of vegetable intercropping in young Eucalyptus plantation. *Ann. Agric. Res.* **19(4):** 470–474.

Bergquist, R.C. 1974. Effect of fungicide rate, spray interval, timing of spray application and precipitation in relation to control of taro blight. *Ann. Bot.* **38:** 213–221.

Bhaskar, A.V., Rahman, M.A., Rao, S.C, Ali, S. and Rao, K.C. 2003. Efficiency of certain fungicides on the dry. Corm rot disease *(Fusarium solani)* in colocasia. *Indian J. Plant Protec.* **31:** 137–138.

Butler, E.J. and Bisby, G.R. 1960. *The fungi of India*. ICAR, New Delhi, 552 pp.

Das, S., Sen, H. and Basu, A. 2003. Effect of soil applied nitrogen and potash fertilizers on severity of Phytophthora blight and yield of taro (*Colocasia esculenta* var. *esculenta*) under field conditions. *Indian J. Agric. Sci.* **73:** 652–655.

Das, S.R. 1997. Field efficacy of fungicides for the control of leaf blight disease of taro. *J. Mycol. Pl. Pathol.* **27:** 337–338.

Deshmukh, M.J. and Chhibber, K.N. 1960. Field resistance to blight (*Phytophthora colocasiae*) in *Colocasia antiquorum*. *Curr. Sci.* **29:** 320–321.

Gomez, E.T. 1925. Leaf blight of Gabi. *Phillippines Agriculturists.* **14:** 429–440.

Ghosh, S.K., Sitansu, P. and Pan, S. 1991. Control of leaf blight of taro (*Colocasia esculenta* L.) Shott. caused by *Phytophthora colocasiae* Racib. through fungicides and selection of variety. *J. Mycopath. Res.* **29(2):** 133–140.

Ghosh, S.K.and Pan, S. 1989. A comprehensive account of the fungal diseases of *Colocasia esculenta* (L.) Schott. *Indian J. Mycol. Res.* **27:** 107–119.

Gollifer, D.E. and Brown, J.E. 1973. Corm rot of *Colocasia esculenta* caused by *Fusarium oxysporum. Plant Dis. Reptr.* **57:** 701–703.

Gollifer, D.E., Jackson, G.V.H. and Newhook, F.J. 1980. Survival of inoculum of leaf blight fungus *Phytophthora colocasiae* infecting Taro (*Colocasia esculenta*) in the Solomon Island. *Ann. App. Biol.* **94:** 379-390.

Gurung, K. 2001. Management and yield loss assessment of Colocasia blight . M.Sc. Thesis submitted to CSK HPKV, Palampur.

Hegde, V., Mukeshkumar, T. and Jeeva, M.L. 2011. Development of diagnostics for diseases of tropical tuber crops. In: *Production of disease free quality planting material propagated through tubers and rhizomes* (Eds. Singh, B.P., Dua, V.K. and Brajesh Singh), CPRI, Shimla (H.P.) India pp. 131–145.

Hunter, O., Pouno, K.and Semisi, S. 1998. The impact of taro leaf blight in the Pacific Island with special reference to Samoa. *J. South Pacific Agric.* **5(2):** 44–46.

Jackson, G.V.H. 1980. Diseases and pests of Taro Nonnea. New Caledonia, South Pacific Commission 52 pp.

Jackson, G.V.H. and Gollifer, D.E. 1975. Disease and pest problems of Taro (*Colocasia esculenta* L. Schott.) in British Solomon Islands. *PANS* **21:** 45–53.

Kay, D.E. 1973. Root crops. Tropical Prod. Inst. *Crop Prod. Dig.* **2:** 71–75, 168–179.

Kerz-Moehilendick, F.H., Fliege, F.H., Scholoesser, E. 1983. Morphology and pathogenicity of four *Pythium* spp. isolated with root and corm rot of taro, *Colocasia esculenta* (L) Schott. in Western *Somoa. Alafua Agril. Bull.* **8(3):** 53–62.

Kumar, Rajesh, Dubey, S.C. and Kumar, R. 1996. Screening of colocasia genotypes for resistance to Phytophthora leaf blight. pp. 388–390. In: *Tropical tuber crops: Problems, prospects and future strategies* (Eds. Kurup, G.T. Palawiswami, M.S., Potty, V.P., Padmaja, G., Kabeerathumma, S. and Pillai, S.V.) Birsa Agricultural University, Ranchi, Bihar, India.

Lebot, V. 2009. Tropical Root and Tuber Crops: Cassava, Sweet Potato, Yams and Aroids. *Crop Production Science in Horticulture Series*; 17. CABI Pub., UK 413 pp.

Michail, S.H., Elarosi, H. and Khairy, E.A. 1969. Dry rot of *Colocasia antiquorum* in U.S.R. (Egypt). *J. Phytopathol. WAR* **1:** 23–26.

Mishra, D. and Rath, G.C. 1986. Some pathological aspects of *Fusarium solani* (Mart.) Sacc. Causing post-harvest decay of colocasia and yam. *Indian J. Mycol. Pl. Pathol.* **16:** 323–325.

Mishra, R.S., Mishra, A.K., Sharma, K., Jeeva, M.L. and Hegde, V. 2011. Characterization of *Phytophthora colocasiae* isolates associated with leaf blight of taro in India. *Arch. Phytopathol. Plant Protec.* **44:** 581–591.

Mishra, R.S. and Naskar, S.K. 1995. Resistance in Taro against Phytophthora leaf blight. *Indian J. Mycol. Pl. Pathol.* **25(1 & 2):** 61.

Mishra, R.S. and Sri Ram, S. 2002. Phytonet Report. Bhuvneshwar, Orissa, CTCRI: pp 66–77.

MohanKumar, C.R. and Sadanandan, N. 1988. Effect of sources of planting material and mulching on the growth and yield of taro (*Colocasia esculenta* L.). *J. Root Crops* **14(2):** 55–57.

Narula, K.L. and Mehrotra, R.S. 1980. Occurrence of A' mating type of *Phytophthora colocasiae. Indian Phytopath.* **33:** 603–604.

Narula, K.L. and Mehrotra, R.S. 1984. The saprophytic survival of *Phytophthora colocasiae* in soil. *Indian Phytopath.* **37:** 256–261.

Narula, K.L. and Mehrotra, R.S. 1987. Biocontrol potential of Phytophthora leaf blight of colocasia by phylopane mycoflora. *Indian Phytopath.* **40(3):** 384–389.

Okpul, T., Ivancic, A. and Simin, A. 1997. Evaluation of leaf blight resistant taro (*Colocasia esculenta*) varieties for Bubia, Morobe Province, papua New Guinea. *Papua New Guinea J. Agric. Fty. Fisheries.* **40(1-2):** 13–18.

Padmaja, G., Devi, U., Mahalakshmi, B.K. and Sridevi, D. 2017. Characterization of isolates of *Phytophthora colocasiae* collected from Andhra Pradesh and Telangana causing leaf blight of taro. *Int. J. Curr. Microbial App. Sci.* **6(10):** 1901–1912.

Paharia, K.D. and Mathur, P.N. 1961. New host plant of Colocasia blight (*Phytophthora colocasiae* Racib.). *Curr. Sci.* **9:** 354.

Paharia, K.D. and Mathur, P.N. 1964. Screening of Colocasia varieties for resistance to colocasia blight (*Phytophthora colocasiae* Racib.). *Sci. & Cult.* **30:** 44.

Parris, G.K. 1941. Diseases of Taro in Hawaii. *Circ. Hawaii Agric. Exp. Sta.* **18:** 29pp.

Quevedo, M.A., Sanico, R.T. and Baliad, M.E. 1991. The control of post harvest diseases of taro corms. *Trop. Sci.* **31:** 359–364.

Raciborski, M. 1900. Parasitic algae and fungi. *Java Batavia Bulletin, New York State Mus.* **19:** 189.

Rana, U. 2006. Epidemiology and integrated management of Colocasia blight. Ph.D. Thesis, CSK, HPKV, Palampur, India, 90 pp.

Rana, U. and Sugha, S.K. 2007. Role of cultural practices in the management of colocasia blight. *Pl. Dis. Res.* **22(1):** 30–33.

Rana, U., Sugha, S.K and Rana, S.K. 2007. Integrated management of of Colocasia (*Colocasia esculenta*) blight. *Indian Phytopath.* **60(4):** 457–461.

Reddy, D.B. 1970. Quaterly Newsletter of the plant Protection for South East Asia and Pacific region. *FAO Publ. Bangkok, Thailand* **13(1):** 14pp.

Reddy, P.P. 2010. *Diseases of Horticultural crops: Nematode problems and their management.* Scientific Publishers India, 288 p.

Rogers, S., Tosefa, T., Hoponoa, T., Hazelman, S. and Hunter, O. 2000. Farmers in innovations in the South Pacific. Ileta Newsletter. Apia, American Samoa: The University of South Pacific **16(2):** 7–8.

Sahu, M.P., Singh, K.P. and Singh, J.R.P. 1989. Control of blight disease of taro. *Indian Fmg.* **39(2):** 22–23.

Sharma, K.D. and Upadhyay, S.K. 1977. Dry rot of *Colocasia esculenta* var. *antiquorum. Indian Phytopath.* **30:** 419–420.

Singh, D., Okpul, T., Gunua, T. and Hunter, D. 2001. Inheritance of resistance to leaf blight in taro cultivar 'Bangkok'. *J. South Pac. Agric.* **8(1):** 22–25.

Singh, H.K. and Singh, P.K. 2003. Efficacy of fungicides for the control of Phytophthora leaf blight of taro (*Colocasia esculenta* var. *esculenta*). *Prog. Hortic.* **35:** 239–240.

Singh, R.S. 1987. *Diseases of vegetable crops*. 3rd Edn. Oxford and IBH Publ Co. 406 p.

Sitansu, P., Ghosh, S.K. and Pan, S. 1997. Antagonistic potential of some fungi on *Phytophthora colocasiae* Racib. *J. Mycopathol. Res.* **35(1):** 41–46.

Sugha, S.K. 2007. *In vitro* evaluation of fungicides against *Phytophthora colocasiae* Racib. - The incitant of leaf blight of taro (*Colocasia esculenta* L.). *Pl. Dis. Res.* **22(2):** 147–149.

Sugha, S.K. 2010. Management of leaf blight (*Phytophthora colocasie* Racib.) of taro (*Colocasia esculanta* L.) with fungicides. *Pl. Dis. Res.* **25(1):** 19–25.

Trujillo, E.E. 1965. Effect of humidity and temperature on zoosporangia production and germination of *Phytophthora colocasiae. Phytopathology* **55:** 126.

Vasquez, E.A. 1990. Yield loss in taro due to *Phytophthora* leaf blight. *J. Root Crops* **61(1):** 48–59.

Wall, G.C. and Cruz, F.J. 1991. *Lasidiplodia theobromae* and *Fusarium proliferatum* causing storage rots of taro in Guam. *Plant Dis.* **75:** 1286.

Wall, G.C. and Wiecko, A.T. 1999. Screening of 29 cultivars (*Colocasia esculenta*) propagated *in vitro* for resistance to taro leaf blight. *J. South Pacific Agri.* **5(2):** 9–12.

Wright, J. 1930. Root rot of Coco-yams. Coast Department of Agriculture Year Book 1930 (Bull. 23), pp. 184–197.

Zettler, F.W., Foxe, M.J., Hartman, R.D., Edwardson, J.R. and Christe, R.C. 1970. Filamentous virus infecting dasheen and other araceous plants. *Phytopathology* **60:** 893–897.

■■■■

14

DISEASES OF OKRA

Okra (*Abelmoschus esculentus* (L.) Moench), popularly known as 'bhindi' or lady's finger, is one of the important vegetables grown throughout the country both under open and protected structures. During cultivation, the crop is severely affected by various diseases of fungal, bacterial and viral nature, which not only reduce the quantity but also affect the quality of the fruits. The present chapter deals with the fungal and viral diseases affecting this crop and their management.

(A) FUNGAL DISEASES

1. FUSARIUM WILT

Fusarium wilt is an important disease of okra wherever this crop is grown intensively.

Geographical distribution

Fusarium wilt was first described by Atkinson (1892), while he was studying the wilt of cotton in Albama, USA. Subsequently, the disease was recorded from North Carolina and Connecticut (Clinton, 1906; Stevens and Wilson, 1912; Wilson, 1913). Now the disease is prevalent in all those countries wherever the crop is grown intensively in warmer climates.

The occurrence of the disease in India was recorded in summer season of 1965 from various districts of Punjab and Haryana affecting both seedling and adult plant stages of okra (Grover and Singh, 1970).

Losses

The disease can appear at any stage of plant growth and the crops planted during summer months suffer more than the crops planted during February–March. May–June sown crop was more vulnerable to wilt and suffered 20–35 per cent loss due to this disease (Grover and Singh, 1970). The losses in dry weight of the seed due to this disease ranged from 20.3 to 24.2 per cent (Gangopadhyay and Kapoor, 1977).

Symptoms

All stages of plant growth are prone to the disease. The characteristic symptoms of the disease appear as yellowing and stunting of the plants followed by wilting and rolling of the leaves (Plate 1). Finally, the plant dies. Before the appearance of typical wilting symptoms, the leaves hang down during daytime and recover again in the night, but ultimately wilt and die. Vascular bundles of the affected plants appear as dark streaks. The whole stem is blackened in case of severe attack.

The Pathogen

Fusarium oxysporum f.sp. *vasinfectum* (Atkinson) Snyder and Hansen is the causal fungus responsible for causing wilt of okra. The mycelium is hyaline and intracellular in the host. Macroconidia are mostly 3 septate, but 0 to 1 septate microconidia and 4 to 5 septate macroconidia are also produced. Macroconidia are formed on sporodochia and pinnotes. In mass these conidia appear buff to salmon orange in colour. They are fusiform-falcate in shape and curved inward at both ends. The microconidia are 5–12 x 2–3.5 μm in size while macroconidia are 40–50 x 3–4.5 μm in size. Both intercalary and terminal chlamydospores are formed which are broadly ovate and measure 7–13 μm in diameter.

Host range

Besides okra, the fungus can infect cotton, Burly tobacco, *Cassia* sp. (Armstrong *et al.*, 1942), pea, cowpea and chillies (Grover and Singh, 1970).

Disease cycle

The fungus is mainly soil borne in the form of chlamydospores. However, Gangopadhyay and Kapoor (1977) reported the seed borne nature of this fungus. When in contact with the host roots, chlamydospores or conidia germinate and penetrate the root tissues. The pathogen can infect wounded as well as intact roots. Penetration through the cuticle is slow and the pathogen proliferates extensively in the corticle tissues, extending up and down. After some growth in root cortex, the pathogen reaches the xylem vessels where it multiplies very rapidly. The vessels are turned brown and clogged at further places.

Epidemiology

The fungus survives from one season to the other in the form of chlaymydospores and infected seed (Gangopadhyay and Kapoor, 1977). The fungus can grow at temperatures of 12.5 to 37.5°C but the optimum temperature for growth is 25°C and for disease development 22-28°C (Singh, 1987). Moisture is not that much important for infection.

Management

(i) Cultural practices : Fusarium is such a fungus that once it gets introduced in a field, it remains there indefinitely. Cultural practices like long rotations, exposing the soil to the sun during summer months by deep ploughing, soil solarization by tarping the soil with transparent polythene sheet (25 μm thickness) during summer months and destruction of diseased roots are some of the practices which can reduce the disease. The incidence of wilt can also reduced when mixture of Zn @10 or 15 ppm + P @ 60 ppm was applied to the potted plants (Randhawa *et al.*, 1977).

(ii) Host resistance: Some cvs. of okra like Pusa Sawani, Pusa Makhmali, Okra I.S. 9273, 9857 (Grover and Singh, 1970), C.S. 3232, 8899, I.S. 6653 and 7194 (Singh, 1987) have been reported as resistant to this disease. An introduction of *A. manihot*, a wild relative of okra, was found as highly resistant to *F. oxysporum* (McLeod *et al.*, 1983) and it can be used in breeding programmes for incorporating resistance to Fusarium wilt.

(iii) Chemical control: Seed treatment with Ziram (0.3% w/v) at 45 ± 1°C for half an hour has been reported effective for the control of seed borne inoculum (Gangopadhyay and Kapoor, 1977) or give dry seed treatment with carbendazim (0.2%) for the management of seed borne inoculum.

2. CERCOSPORA LEAF SPOT

Leaf spots and blight caused by *Cercospora* species in okra are very important particularly in moderate temperatures and high humidity conditions.

Geographical distribution

The disease has been reported from many countries like Mexico, Puerto Rico, India, China and Uganda (Chupp and Sherf, 1960). In India, the disease has been reported from almost all the states wherever this crop is grown (Jhooty *et al.*, 1977; Dhancholia and Singh, 1992; Srivastava *et al.*, 1992; Singh *et al.*, 2001).

Losses

Losses from this disease vary with disease intensity and sometimes the infection is so severe that the leaves drop down prematurely (Singh *et al.*, 2001).

Symptoms

Three different species of *Cercospora* are associated with this disease. The symptoms of each species are described below:

Cercospora abelmoschi: The fungus causes no definite leaf spot, but grows as a sooty mould mostly on the lower surface of the leaves. Severely affected leaves roll, wilt and fall down to the ground (Plate 2).

Cercospora hibiscina: The fungus produces dark olivaceous patches of mould on the lower surface of the leaves.

Cercospora malayensis: The spots caused by this specie are definite lesions on the leaves with grey centres and red to purple borders.

The Pathogens

The disease is caused by three species of *Cercospora* like *C. abelmoschi* Ell. & Ev., *C. hibiscina* Ell. & Ev. and *C. malayensis* Stevens. These species differ in the size of their conidiophores and conidia. In *C. abelmoschi,* the conidiophores are long, brown and bear pale olivaceous, slightly tapered conidia. The conidiophores of *C. hibiscina* are extremely long sometimes up to 1000 μm in size, narrow and bear conidia that are sometimes hyaline and appreciably more narrow than those of *C. abelmoschi.*

The conidiophores of *C. malayensis* are borne in clusters of 5–20 and bear conidia, which are colourless, narrow, long, and tapering from the blunt base to the sharp tip.

Host range

All the three species of *Cercospora* attack many species of *Abelmoschus* and are found wherever the genus is grown generally.

Disease cycle

The fungus overwinters in the infected plant debris as conidia or stromata in soil and also on the wild species of *Abelmoschus*. The conidia in favourable weather germinate and cause infection of the host through stomata. The spores produced on the primary spots are blown by wind and cause secondary infections.

Epidemiology

These fungi perennate in the infected plant debris and on the wild hosts. The three species cause infection in moderate temperatures (25–29ºC) and high humidity. Dhancholia and Singh (1992) reported that the disease (*C. abelmoschi*) was more serious when the mean minimum and maximum temperatures were 17.9 and 25ºC, respectively, along with 80 per cent RH and heavy rainfall (330 mm).

Management

(i) Cultural practices: Cultural practices like collection and destruction of infected plant debris, crop rotation and destruction of wild hosts from in and around the field should be followed to keep the disease under check.

(ii) Host resistance: Cvs./lines like IC 18960, IC 15055, Round Selection, Walgaon, EC 41292, IC 1542 (Jhooty ***et al.***, 1977), IC 3769, IC 6316, IC 4685, IC 6653, IC 9830, BH-9, Bhindi Gambo, Long Green, Sonepat Strain (Chauhan ***et al.***, 1980) have been reported as resistant.

(iii) Chemical control: Sprays of various systemic and non-systemic fungicides like copper oxychloride, carbendazim, benomyl, mancozeb, Bordeaux mixture, chlorothalonil and Topsin M have been found quite effective in controlling Cercospora leaf spots (Sokhi ***et al.***, 1979; Srivastava ***et al.***, 1992; Tripathi, 1994, Singh ***et al.***, 2001, Beaura ***et al.***, 2007, Farraq, 2011). Sprays of these fungicides not only decreased the severity of the disease but also increased fruit yield.

(iv) Plant Extract: Farraq (2011) evaluated different plant extracts in the management of this disease and reported the efficacy of lemongrass oil in the management of this disease.

3. POWDERY MILDEW

Powdery mildew is an important disease of this crop and in favourable weather causes significant yield reductions particularly if the infection takes place in early stages of plant growth.

Geographical distribution

The disease is of common occurrence wherever this crop is grown. The occurrence of the disease has been reported from Mexico (Diaz-Franco, 1999). In India the disease has been reported to occur in Delhi (Prabhu *et al.*, 1971), Karnataka (Sohi and Sokhi, 1974), Himachal Pradesh (Raj *et al.*, 1992) and Maharashtra (Jambhale and Nerkar, 1983).

Losses

Up to 43 per cent disease severity of this disease has been reported from Himachal Pradesh (Raj *et al.,* 1992). However, yield losses up to 86 per cent have been reported from Bangalore due to this disease.

Symptoms

Symptoms first appear as minute discoloured patches with thin fine meshwork of white mycelium arising at many places on the upper surface of lower leaves (Plate 3). These white patches soon join together to form larger white greyish powdery coating discernible on the severely affected leaves. In late stages, the affected leaves turn yellow and finally drop.

The Pathogen

Although the disease has been known to be caused by three pathogens like *Leveilllula taurica* (Bremer *et al.*, 1947), *Erysiphe cichracearum* (Allison, 1953) and *Sphaerotheca fuliginea* (Tarr, 1954), but among these *Erysiphe cichoracearum* D.C. is more common particularly in northern India (Raj *et al.*, 1992). The mycelium of *E. cichoracearum* is ectophytic, conidiophores unbranched and septate with cylindrical foot cell. Conidia are single celled, hyaline, oval to roundish (Fig. 1a) or barrel-shaped and in long chains and germinate by simple germtube (Fig. 1b). The conidial dimensions vary with the physiologic race and the host. They measure 22.50–45 × 9–27 μm (Pathania and Gupta, 2016). Cleistothecia are globose, dark, with hyaline to dark brown mycelioid appendages, 80 to 180 μm in diameter. They contain 8 to 18 asci. The asci are pedicellate, ovate to broadly ovate or ellipsoid, and measure 36–90 × 22.5–50. The number of ascospores per ascus is usually two, rarely three. The ascospores are one celled, hyaline oval to sub-cylindrical and measure 18–30 × 12–18 μm.

Disease cycle

The primary infection of the leaves is caused by wind blown ascospores (wherever cleistothecia are present) or by conidia formed on the earlier sown crop/indoor cultivation in the neighbouring areas. In the process of infection, conidia germinate and after penetration, colonization of the entire leaf takes place. The conidia formed abundantly on primary infections are blown by wind and air currents, and cause secondary infections and the cycle is repeated.

Epidemiology

Powdery mildews are generally favoured by moderate temperatures, reduced light, fertile dry soil and succulent plant growth (Yarwood, 1957), but two environmental factors *viz.*, temperature and humidity are closely inter-related. *E. cichoracearum* requires 60–80 percent relative humidity for the development of the disease and dry conditions were favourable for the growth and sporulation of the fungus. Temperature 25°C and 70.4 per cent RH were found optimum for maximum conidial germination (Pathania and Gupta, 2016). They further reported that moderate air temperature of 25±1°C coupled with slightly less relative humidity (<70%) and less cumulative rainfall helped significantly in disease development. The disease develops better in shade than in full light. Thus, it is more severe in fields with close plantings.

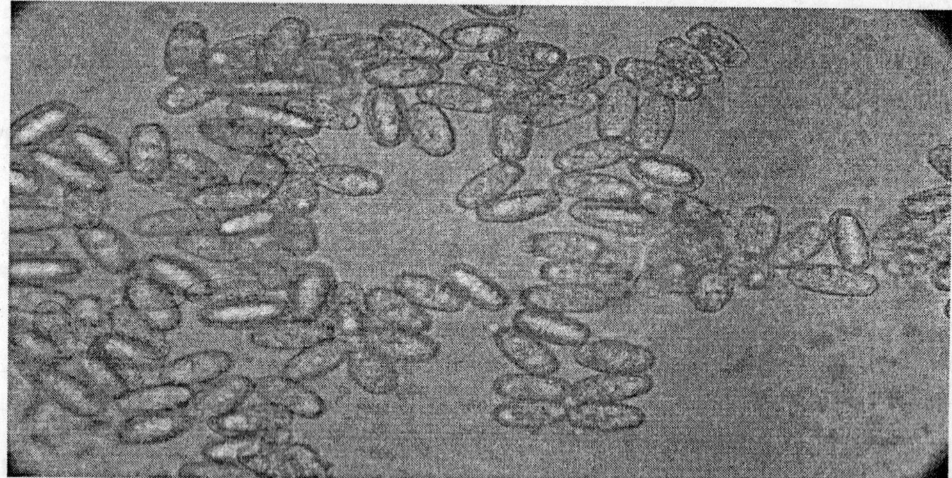

Fig. 1a. Oval to roundish or barrel shaped conidia

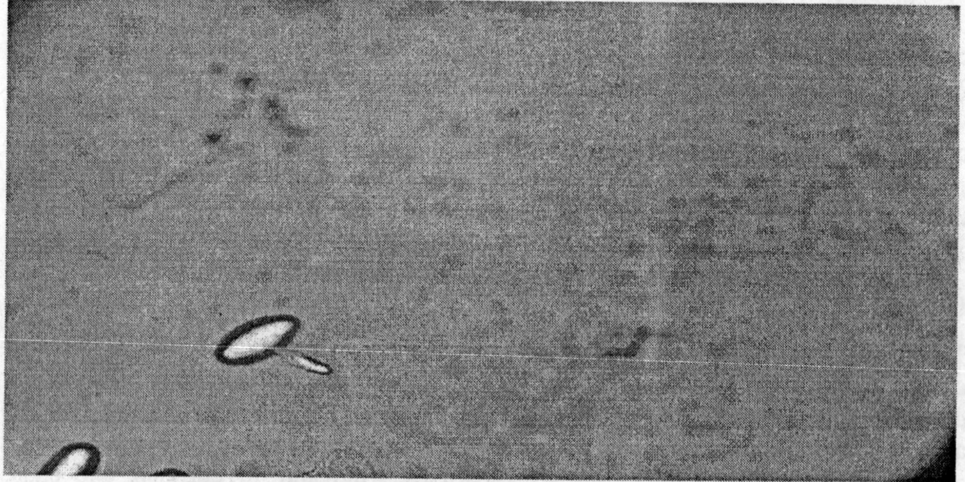

Fig. 1b. Conidial germination showing simple germ tube

Management

(i) Host resistance: Cv. Nigeria (Sohi and Sokhi, 1974), biotypes drawn from *Abelmoschus tetraphyllus, A. manihot, A. manihot* subsp. *manihot* and *A. moschatus*

(Jambhale and Narkar, 1992) were reported to be resistant to this disease, while cvs. like AE-37, AE-52, AE-79, BH-102, Clemson Spineless, Early Long Green, Long Green, Red Bhindi and Vaishali Vadhu were moderately resistant (Joi and Shende, 1979). The resistance due to this disease is controlled by a single, incompletely dominant gene (Jambhale and Nerkar, 1992).

(ii) Chemical control: Fungicides like wettable sulphur, thiophanate methyl, dinocap, carbendazim, benomyl, dimethirimol and tridemorph have been reported to provide effective control of powdery mildew of okra (Prabhu *et al.*, 1971; Choudhary, 1975; Sokhi and Sohi, 1975; Angelov and Gadzhonov, 1979; Joi and Shende, 1979, Singh *et al.*, 1998), but the highest cost benefit ratio was recorded in sulphur spray (Singh *et al.*, 1998). EBI fungicides like diniconazole, triforine, penconazole, cyproconazole, (Rossi *et al.*, 1989; Ragupathy *et al.*, 1998; Naik and Nagaraja, 2000) were found effective in reducing the disease severity and in increasing the yields. Topas (penconazole, 0.1%) provided complete control of the disease and was comparable with carbendazim (0.1%) and increased fruit yield (Naik and Nagaraja, 2000). Recently, Pathania and Gupta (2016) reported the efficacy of difenoconazol, hexaconazole and carbendazim in reducing the disease severity and increasing the fruit yield.

(iii) Plant extracts: Sprays of neem oil and neem seed kernel extract resulted in a lower incidence of powdery mildew of okra (Ragupathi *et al.*, 1994).

(iv) Biological control: A hyperparasitic fungus *Aphanocladium album* was found on powdery mildew affected okra plants grown under glasshouse conditions (Mitov and Ibrakhim, 1977). The fungus developed white aerial mycelium with numerous conidia, but no sexual stage. The optimum conditions for the development of the fungus were 28ºC and RH 80–100 per cent. This hyperparasitic fungus can be utilized for the control of this disease.

(B) VIRAL DISEASES

1. YELLOW VEIN MOSAIC (YVM)

This is a very destructive disease of this crop and if the infection takes place in the early stages of plant growth then the crop suffers very badly.

Geographical distribution

The occurrence of the disease was first reported from Bombay by Kulkarni (1924). Since then it has been reported from all parts of the country wherever it is grown intensively.

Losses

Due to infection by YVM virus, plant growth, fruit numbers and seed yield/plant are adversely affected, the effects being more severe if the symptoms appear early (Sinha and Chakrabarti, 1978). However, the losses from this disease range from 49.3 to 93.8 per cent depending on the stage of crop growth at which the infection occurs (Sastry and Singh, 1974) and it varies from location to location as well as season to season (Prakasha *et al.*, 2010).

Symptoms

The characteristic symptoms are clearing of small veins and veinal chlorosis of leaves. The yellow network of the veins is very conspicuous and veins, and veinlets are thickened (Plate 4). The chlorosis extends to the interveinal areas and in case of severe infection, the whole leaf turns yellow. The fruits are dwarfed, malformed and yellowish green in colour. The infected fruits fetch poor price in the market.

The Virus

The virus responsible for this disease is *Yellow Vein Mosaic Virus* (YVMV). It belongs to the bigeminivirus group. The size of the virus is 18 x 30 nm and has been reported to have close relationship with Indian Cassava Mosaic bigeminivirus (ICMV) in ELISA tests using polyclonal antiserum (Handa and Gupta, 1993). Virus particles are spherical, isometric, measuring 28–30 nm in diameter along with X-bodies as observed under electron microscope (Kothari *et al.*, 1982). Electron microscopy of YVMV infected *A. esculentus* plants revealed the presence of small, spherical or geminate virus like particles in the form of aggregates in the phloem sieve nuclei (Singh, 1990). Association of a DNA beta satellite with a begomovirus has been reported from India where begomovirus alone could produce mild leaf curling in okra and when it is coinoculated with DNA beta satellite typical symptoms of this disease appeared (Jose and Ramkrishna, 2003).

Transmission

The virus is not sap transmissible, but under artificial conditions it can be transmitted by grafting (Singh, 1987). In nature the virus is transmitted by insect vector, white fly (*Bemisia tabaci* Genn.).

Host range

Host range of this virus includes *Croton sparsiflora, Malvastrum tricuspidatum, Abelmoschus manihot, Althaea rosea, Hibiscus tetraphyllus* and *Ageratum* sp.

Epidemiology

This disease is not seed borne and overwinters on the wild hosts. Presence of wild hosts and the prevalent environmental conditions that favour the buildup of vector population are most important factors that determine the incidence of this disease in a particular area. Dry hot weather with little or no rainfall have been reported conducive for disease development and also for the multiplication of the vector population (Singh, 1990). He further reported that cooler weather along with high humidity and rainfall was detrimental to whiltefly multiplication and spread.

Management

(i) Cultural practices: Various cultural practices like destruction of wild hosts from in and around the field, avoidance of mixed cropping of pumpkin (Khan and Mukhopadhyay, 1985), adjustment of date of sowing to avoid the period of maximum population of the whiteflies (Lal *et al.*, 2001, Magar and Nirmal, 2010) and regular removal of affected plants upto 55 days (Sinha and Chakrabarti, 1982) are found to check the disease effectively. Application of yellow coloured polythene mulch significantly delayed the appearance of yellow vein mosaic symptoms to 69 days from the date of sowing compared with only 28 days in unmulched crop (Khan and Mukhopadhyay, 1985). Yellow polythene mulch also delayed the appearance of yellow vein mosaic symptoms under field conditions (Magar and Nirman, 2008). Highest marketable yield was obtained when 45 cm row to row distance was maintained in June sown crop (Kalita and Dhawan, 2006).

(ii) Host resistance: Several cvs./hybrids like KbS 312 (Singh and Singh, 1986), Parbhani Kranti (Jambhale and Nerkar, 1986), Okra Vishal (Nath and Saikia, 1992), Punjab-7 (Sharma *et al.*, 1993), Punjab Padmini, Punjab-8 (Arora, 1995), Hissar Unnat (Dhankhar *et al.,* 1996) AM-4-5 (Sarabani *et al.*, 2003), HIHBO 090, DVR-1, DVR-2 (Singh *et al.*, 2003), Azad Bhindi-1 (Yadav *et al.*, 2004),Azad Bhindi-2 (Yadav *et al.*, 2006) and ZOH-3002 (Biswas *et al.,* 2008) have been reported resistant to this disease from different parts of the country. Studies on the breeding for resistance to this disease were carried out in Punjab by using *A. esculentus* variety EC31830 from Ghana which has a symptomless carrier type of resistance to yellow vein mosaic and Punjab Padmini, tolerant to this disease have been selected (Sharma and Sharma, 1984). Limited inheritance studies revealed that tolerance is probably controlled by 2 dominant complementary genes or is under polygenic control. However, Rattan *et al.* (2000) indicated that resistance to this disease is monogenic and dominant. Presence of major gene(s) along with minor genes for resistance to Yellow Vein Mosaic Virus have also been reported (Arora *et al.*, 2008).

(iii) Chemical control: Systemic granular insecticides like aldicarb and carbofuran (1.0 kg a.i./ ha) applied to the soil at sowing time had no adverse effect on seed germination of okra and minimized the occurrence of yellow vein mosaic by controlling the vector *Bemisia tabaci* (Palaniswamy *et al.*, 1973). Four sprays each of parathion, oxydemeton-methyl or dimethioate at 10-day intervals from germination or one application of phorate 10-G (15 kg/ha) at sowing restricted the spread of the virus in okra (Sastry and Singh, 1973). Spraying of the crop immediately after germination is necessary other wise it will be difficult to check this disease. Soil application of phosphoro-dithioate (Furatox-10G) @ 15 kg/ha followed by four foliar sprays of Metasystox (demeton-S-methyl) at 15 days intervals from the sowing date reduced the incidence of the disease to 23.26 per cent (Khan and Mukhopadhyay, 1985). Soil application of carbofuran (1 kg a.i./ha) at sowing and 20 DAS followed by one spray of oxydemeton methyl 45 DAS significantly reduced the vector population resulting in a reduction of disease incidence and increased yields (Bhagat *et al.*, 1997). Four spraying of Metasystox (demeton-S-methyl) at 15, 30, 45 and 60 DAS gave the highest yield advantage with comparatively low incidence of the disease (Debnath and Nath, 2002).

(iv) Plant extracts: Alcoholic extracts of *Callistemon, Datura, Agave* and ginger provided good degree of suppression of symptoms on okra when sprayed under field conditions and there was a lower rate of dissemination of the disease in treated plants than in the control sprayed with water only (Chowdhury *et al.*, 1992). Leaf extracts of *Prosopis chilensis, Bougainvillea spectabilis*, neem oil (3%) and neem seed kernel extract (5%) were highly effective in reducing the infection of this disease to 83.3, 87.7, 88.3 and 86.7 per cent, respectively, over control (Pun *et al.*, 1999; 2000).

(v) Integrated disease management: Bhagat *et al.* (2001) recommended the sowing of cv. Parbhani Kranti during Kharif season and the susceptible stage (35–50 DAS) must be supplemented with systemic insecticide to reduce the whitefly population. Integration of Bajra as barrier crop and spray applications of imidaclorpid at 15 days interval starting from 15 days after germination of okra provided an excellent control of Yellow vein mosaic and Okra leaf curl virus besides increasing fruit yield (Kalita and Dhawan, 2010).

REFERENCES

Allison, J.L. 1953. Diseases of economic plants in Iraq. *FAO Pl. Prot. Bull.* **1:** 9–11.

Angelov, D. and Gadzhonov, I. 1979. Trials with chemicals for powdery mildew control on okra grown for seed. *Gradinarska l Lozarska Nauka* **16:** 75–79.

Armstrong, G.M. Hawkins, B.S. and Bennett, G.S. 1942. Cross inoculations with isolates of fusaria from cotton, tobacco and certain other plants subject to wilt. *Phytopathology* **32:** 685–698.

Arora, D., Jindal, S.K. and Singh, K. 2008. Genetics of resistance to yellow vein mosaic virus in inter-varietalcrosses of okra (*Abelmoschus esculentus* L. Monech). *SABRO J. Breed. Genet.* **40:** 93–103.

Arora, S.K. 1995. PAU is ahead in evolving okra virus resistant varieties. *Punjab Veg. Grow.* **30:** 66–68.

Atkinson, G.F. 1892. Some diseases of cotton. *Bull. 4, Ala. Agric. Exp. Stn.* Pp. 19–29.

Beaura, A.P., Mohapatra, K.B., Paul, P.K and Nandi, A. 2007. Chemical management of Cercospora blight of okra in Orissa. *J. Plant Protec. Environ.* **4:**134–137.

Bhagat, A.P., Yadav, B.I. and Prasad, Y. 1997. Management of bhindi yellow vein mosaic virus disease by insecticides. *J. Mycol. Pl. Pathol.* **27:** 215–216.

Bhagat, A.P., Yadav, B.P. and Prasad, Y. 2001. Rate of dissemination of okra yellow vein mosaic virus disease in three cultivars of okra. *Indian Phytopath.* **54:** 488–489.

Biswas, N.K., Naik, P.S., Das, S. and De, B.K. 2008. Management of yellow vein mosaic virus disease of okra (*Abelmoschus esculentus* (L.) Monech) through different plant oils. *Res. On Crops.* **9:** 345–347.

Bremer, H., Ismen, H., Karel, G., Ozkan, H. and Ozkan, M. 1947.Contribution to the knowledge of the parasitic fungi of Turkey. *Rev. Faculty Sci Univ. Istanbul* **13:** 122–172.

Chauhan, M.S., Dhankhar, B.S. and Duhan, J.C. 1980. Varietal resistance of okra to root rot and Cercospora leaf spot. *Macco Agril. Dig.* **5(1):** 17–18.

Choudhary, S.L. 1975. Fungicidal control of powdery mildew of bhindi (*Abelmoschus esculentus*). *Indian J. Mycol Pl. Pathol.* **5:** 42.

Chowdhury, A.K., Biswas, B. and Saha, A.K. 1992. Inhibition of bhendi (okra) yellow vein mosaic virus (BYVMV) by different plant extracts. *J. Mycopathol. Res.* **30:** 97–102.

Chupp, C. and Sherf, A.F. 1960. *Vegetable diseases and their control.* The Ronald Press Co., New York, 693pp.

Clinton, G.P. 1906. Notes on the fungus diseases. *Rep. 29, Conn. Agric. Exp. Stn.* Pp. 263–303.

Debnath, S., and Nath, P.S. 2002. Management of yellow vein mosaic disease of okra through insecticides, plant products and suitable varieties. *Ann. Pl. Prot. Sci.* **10:** 340–342.

Dhancholia, S. and Singh, D. 1992. Outbreak of Cercospora blight (*Cercospora abelmoschi* Ell. & Ev.) in okra in Himachal Pradesh. *Himachal J. Agric. Res.* **18:** 60.

Dhankhar, S.K., Dhankhar, B.S. and Saharan, B.S. 1996. Screening of okra genotypes for resistance to yellow vein mosaic disease. *Ann. Biol.* **12:** 90–92.

Diaz-Franco, A. 1999. Okra (*Abelmoschus esculentus*) powdery mildew in Mexico. *Revista Mexicana de Fitopatol.* **17:** 44–45.

Farraq, E.S.H. 2011. First record of Cercospora leaf spot disease on okra plants and its control in Egypt. *Plant Pathol J.* **10:** 175–180.

Gangopadhyay, S. and Kapoor, K.S. 1977. Control of Fusarium wilt of okra with seed treatment. *Indian J. Mycol. Pl. Pathol.* **7(2):** 147–149.

Grover, R.K. and Singh, G. 1970. Pathology of wilt of okra (*Abelmoschus esculentus* (L.) Moench) caused by *Fusarium oxysporum* f.sp. *vasinfectum* (ATK.) Snyder & Hansen, its host range and histopathology. *Indian J. Agric. Sci.* **40:** 989–996.

Handa, A. and Gupta, M.D. 1993. Characterization of yellow vein mosaic virus of bhindi (*Abelmoschus esculentus* (L.) Moench.) *Int. J. Trop. Pl. Dis.* **11:** 117–128.

Jambhale, N.D. and Nerkar, Y.S. 1983. Inheritance of resistance to powdery mildew in okra. Nat. Sem. Breeding crop plants for resistance to pests and diseases, May 25-27, 1983. Coimbatore, Tamil Nadu, pp.13.

Jambhale, N.D. and Nerkar, Y.S. 1986. Performance of yellow vein mosaic-resistant 'Parbhani Kranti' okra. *Indian J. Agric. Sci.* **56:** 677–680.

Jambahale, N.D. and Nerkar, Y.S. 1992. Screening of okra cultivars, related species and interspecific hybrid derivatives for resistance to powdery mildew. *J. Maharastra Agril. Univ.* **17:** 53–54.

Jhooty, J.S., Sokhi, S.S., Bains, S.S. and Rewal, H.S. 1977. Evaluation of germplasm lines of okra (*Abelmoschus esculentus*) against powdery mildew and Cercospora blight. *Veg. Hot. Humid Trop.* **2:** 30–32.

Joi, M.B. and Shende, R.V. 1979. Fungicidal control of powdery mildew of okra. *Pesticides* **13:** 32–34.

Joi, M.B. and Shende, R.V. 1979. Varietal reaction to powdery mildew (*Erysiphe cichoracearum* D.C.) in okra. *Veg. Sci.* **6:** 126–129.

Jose, J. and Ramakrishnan, U. 2003. Bhindi yellow vein mosaic disease in India is caused by association of DNA beta satellite with a begomovirus. *Virology* **305:** 310–317.

Kalita, M.K. and Dhawan, P. 2006. Management of yellow vein mosaic and leaf curl diseases of okra by adjusting date of sowing and row to row spacing . Indian *J. Agric Sci.* **76:** 762–765.

Kalita, M.K. and Dhawan, P. 2010. Ecofriendly management of okra yellow vein mosaic and leaf curl virus diseases by combination of barrier crop and insecticidal sprays.. *Adv. Plant Sci.* **23:** 289–291.

Khan, M.A. and Mukhopadhyay, S. 1985. Effect of different pesticide combinations on the incidence of yellow vein mosaic virus disease of okra (*Abelmoschus esculentus*) and its whitefly vector *Bemisia tabaci* Genn. *Indian J. Virol.* **1:** 147–151.

Khan, M.A. and Mukhopadhyay, S. 1985. Studies on the effect of some alternative cultural methods on the incidence of yellow vein mosaic (YVMV) disease of okra (*Abelmoschus esculentus* (L.) Moench.). *Indian J. Virol.* **1:** 69–72.

Kothari, I.L., Dube, H.C. and Joshi, S.K. 1982. Isolation and characterization of hibiscus virus I causing yellow vein mosaic of bhindi (*Abelmoschus esculentus* (L.) Moench.). *Curr. Sci.* **51:** 571–572.

Kulkarni, G.S. 1924. Mosaic and other related diseases of crops in Bombay Presidency. *Poona Agric. College Mag.* **16:** 6–12.

Lal, G., Singh, D.K., Jain, S.K. and Lal, G. 2001. Response of okra (*Abelmoschus esculentus* (L.) Moench.) cultivars to varying sowing dates under trai foot hills of Himalayas. *Adv. Hort. Forestry* **8:** 129–137.

Magar, S.J. and Nirmal, D.D. 2008. Effect of different mulching on the incidence of yellow vein mosaic disease of okra (*Abelmoschus esculentus*). *Indian J. Agric. Sci.* **78:** 726–729.

Magar, S.J. and Nirmal, D.D. 2010. Influence and sowing time on the increase of yellow vein mosaic disease, white fly population and yield of okra (*Abelmoschus esculentus*). *Indian Phytopath.* **63:** 112–113.

McLeod, J.M., Witcher, W., Epps, W.M. and Robbins, M.L. 1983. Resistance of okra plants to root knot nematode and Fusarium wilt. *Hort. Sci.* **18:** 249–250.

Mitov, N. and Ibrakhim, I. 1977. A new hyperparasitic fungus on the pathogen of powdery mildew of okra (*Erysiphe cichoracearum* DC). *Gradinarska I Lozarska Nauka* **14(4):** 93–98.

Naik, K.S. and Nagaraja, A. 2000. Chemical control of powdery mildew of okra. *Indian J. Pl. Prot.* **28:** 41–42.

Nath, P. and Saikia, A.K. 1992. Relative resistance of okra cultivars to yellow vein mosaic virus. *New Agriculturist* **3:** 199–102.

Palaniswamy, P. Thirumurthi, S. and Subramaniam, T.R. 1973. Effects of systemic granular insecticides on the incidence of the yellow mosaic disease of okra *Abelmoschus esculentus* (L.). *South Indian Hortic.* **21:** 104–106.

Pathania, Abhishek and Gupta, S.K. 2016. Morphological characterization and effect of meteorological factors on development of okra powdery mildew. *Pl. Dis. Res.* **31:** 23–28.

Pathania, Abhishek and Gupta, S.K. 2016. Management of okra powdery mildew through fungicides. *Pl. Dis. Res.* **31:** 125–128.

Prabhu, A.S., Phatak, K.D. and Singh, R.P. 1971. Powdery mildew of 'bhindi' (*Abelmoschus esculentus* (L.) Moench) in Delhi State. *Indian J. Hortic.* **28:** 310–312.

Prakasha, T.L. , Patil, M.S. and Benagi, V.I. 2010. Survey for bhendi yellow vein mosaic disease in parts of Karnataka. *Karnataka J. Agri. Sci.* **23:** 658–659.

Pun, K.B., Doraiswamy, S. and Jeyarajan, R. 1999. Screening of plant species for the presence of antiviral principles against okra yellow vein mosaic virus. *Indian Phytopath.* **52:** 221–223.

Pun, K.B., Doraiswamy, S. and Jeyarajan, R. 2000. Screening of virus inhibitory chemicals and neem products against okra yellow vein mosaic virus. *Indian Phytopath.* **53:** 95–96.

Ragupathi, N., Thamburaj, S. and Jeyarajan, R. 1994. Management of powdery mildew (*Erysiphe cichoracearum* DC) of bhendi. *South Indian Hortic* **42:** 278–280.

Ragupathi, N., Thiruvudainambi, S. and Thamburaj, S. 1998. Chemical control of powdery mildew disease of bhindi. *South Indian Hortic.* **46:** 102–103.

Raj, H., Bhardwaj, M.L., Sharma, I.M. and Sharma, N.K. 1992. Performance of commercial okra (*Hibiscus esculentus*) varieties in relation to diseases and insect-pests. *Indian J. Agric. Sci.* **63:** 747–748.

Randhawa, K.S., Bhandari, N.R. and Singh, D. 1977. Influence of phosphorus-zinc application on okra (*Abelmoschus esculentus* (L.) Moench). *Prog. Hortic.* **9:** 31–38.

Rattan, R.S., Bindal, A. and Bindal, A. 2000. Development of okra hybrids resistant to yellow vein mosaic virus. *Veg. Sci.* **27:** 121–125.

Rossi, C.E., Bacchi, L.M.A., Occhiena, E.M. and Watanabe, A. 1989. Chemical control of powdery mildew (*Erysiphe cichoracearum*) on okra. *Ecossistema* **14:** 110–113.

Sarabani, D., Nath, P.S. and Debnath, S. 2003. Performance of okra varieties in relation to yield and tolerance to YVMV. *Ann. Pl. Prot. Sci.* **11(2):** 400–401.

Sastry, K.S.M. and Singh, S.J. 1973. Restriction of yellow vein mosaic virus spread in okra through the control of vector, whitefly *Bemisia tabaci*. *Indian J. Mycol Pl. Pathol.* **3:** 76–80.

Sastry, K.S.M. and Singh, S.J. 1974. Effect of yellow-vein mosaic virus infection on growth and yield of okra crop. *Indian Phytopath.* **27:** 294–297.

Sharma, B.R. and Sharma, O.P. 1984. Breeding for resistance to yellow vein mosaic virus in okra. *Indian J. Agric. Sci.* **54:** 917–920.

Sharma, B.R., Arora, S.K, Dhanju, K.C. and Ghai, T.R. 1993. Performance of okra cultivars in relation to yellow vein mosaic virus and yield. *Indian J. Virol.* **9:** 139–142.

Singh, D., Maheshwari, V.K. and Gupta, Anuja. 2001. Fungicidal control of Cercospora leaf spot in seed crop of okra (*Abelmoschus esculentus* (L.) Moench.). *Seed Res.* **29(2):** 254–256.

Singh, D.K., Singh, S.K. and Jain, S.K. 2003. Evaluation of okra hybrids for growth, yield and yellow vein mosaic virus. *Scientific Hortic.* **8:** 129–133.

Singh, N.B., Sharma, H.K. and Srivastava, K.J. 1998. Chemical control of powdery mildew in okra seed crop. *Newsl. Natl. Hortic. Res. Dev. Found.* **18(4):** 1–4.

Singh, R.P. and Singh, L.N. 1986. Studies on the varietal resistance of okra (*Abelmoschus esculentus* (L.) Moench.) to yellow vein mosaic virus under field conditions. *J. Indian Bot. Soc.* **65**(Suppl.): 23.

Singh, R.S. 1987. *Diseases of Vegetable Crops.* Oxford & IBH Publishing Co. Pvt. Ltd., New Delhi. 362 pp.

Singh, S.J. 1990. Etiology and epidemiology of whitefly–transmitted virus diseases of okra in India. *Pl. Dis. Res.* **5(1):** 64–70.

Sinha, S.N. and Chakrabarti, A.K. 1978. Effects of yellow vein mosaic virus infection on okra seed production. *Seed Res.* **6:** 67–70.

Sinha, S.N. and Chakrabarti, A.K. 1982. Studies on the control of yellow vein mosaic in okra seed crop. *Veg. Sci.* **9:** 64–69.

Sohi, H.S. and Sokhi, S.S. 1974. Behaviour of okra varieties to damping off, powdery mildew and Cercospora blight. *Indian Phytopath.* **27:** 91–92.

Sokhi, S.S. and Sohi, H.S. 1975. Powdery mildew of okra in Karnataka State and its control. *Indian J. Mycol Pl. Pathol.* **5:** 69–73.

Sokhi, S.S., Jhooty, J.S., Bains, S.S. and Rewal, H.S. 1979. Chemical control of Cercospora blight of okra. *Indian J. Mycol. Pl. Pathol.* **9:** 247–248.

Srivastava, L.S., Gupta, D.K. and Varma, R.N. 1992. Chemical control of Cercospora blight of okra in Sikkim. *Indian J. Mycol. Pl. Pathol.* **22:** 73–74.

Stevens, F.L. and Wilson, G.W. 1912. Okra wilt (Fusariose), ***Fusarium vasinfectum***, and clover Rhizoctoniose. *34th Rep. (1910), N. Carol. Exp. Stn.* **11:** 70–73.

Tarr, S.A.J. 1954. Plant Pathology. Report of the Research Division, Ministry of Agriculture, Sudan, pp 46–55.

Tripathi, K.C. 1994. A note on the evaluation of fungicides for the control of angular leaf spot of okra caused by *Cercospora abelmoschi* Ell. & Ev. *Prog. Hortic.* **26:** 82–84.

Wilson, G.W. 1913. *Fusarium* or *Verticillium* on okra in North Carolina. *Phytopathology* **3:** 180–183.

Yadav, J.R., Singh, B., Kumar, R. and Srivastava, J.P. 2006. Azad Bhindi-2 a disease resistant variety of Bhindi (*Abelmoschus esculentus* (L.) Moench). *Plant Arch.* **6:** 395–396.

Yadav, J.R., Srivastava, J.P., Singh, B. and Kumar, R. 2004. Azad Bhindi (Azad Ganga) a disease resistant variety of Bhindi (*Abelmoschus esculentus* (L.) Moench.). *Pl. Arch.* **4 (1):** 205–207.

Yarwood, C.E. 1957. Powdery mildews. *Bot. Rev.* **23(4):** 235–307.

■■■■

15

DISEASES OF LEAFY VEGETABLES

Leafy vegetables such as spinach, coriander, lettuce etc. form an important constituent of our daily diet and fulfill the much needed component of fibre, minerals and elements in man's nutrition apart from adding flavour to the dishes like curries and soups. However, these vegetables suffer from several diseases caused by fungi, bacteria and viruses that affect their normal yield output and quality aspects. Important diseases of coriander, spinach and lettuce are discussed below.

A. CORIANDER

Coriander (*Coriandrum sativum* L) is an important spice crop grown for its leaves and fruits. It is mostly used as common flavouring substance. The stems, leaves and fruits have a pleasant aroma. Its leaves are used for flavouring curries, sauces and soups. Dry fruits are extensively used in preparation of curry powder, pickling spices, sauces and seasoning. During cultivation, it is affected by several diseases, which cause significant yield reductions. Important diseases affecting this crop are described in detail along with their management practices.

1. SEED ROT AND SEEDLING MORTALITY

Many fungi viz., *Alternaria alternata, Aspergillus candidus, A. flavus, Curvularia lunata, C. pallescens, Cladosporium oxysporum, C. herbarum, Fusarium* sp., *Penicillium* sp. and *Stachybotrys* sp. from seeds of coriander and one soil inhabitant *Pythium irregulare* (Prasad, 1982; Paul, 1992) have been reported to induce seed and seedling mortality. The pre-harvest rains increased the frequency of endophytic seed mycoflora (species of *Alternaria, Cladosporium* and *Curvularia*) and reduced the germination of seeds (Prasad, 1983). Seed treatment with mancozeb, captan, carbendazim + captafol and captafol proved effective to some extent to achieve better crop stand (Prasad, 1982; Paul, 1992).

2. CORIANDER WILT

It is another important soil borne disease of coriander reported from certain parts of India and other countries like Argentina.

Geographical distribution

The occurrence of the disease in India has been reported from Madhya Bharat as early as in 1952 (Anonymous, 1952). Since then it has been reported from Madhya Pradesh (Mall, 1968), Rajasthan (Srivastava, 1971), Tamil Nadu (Prakasam *et al.*,

1987). Occurrence of this disease has also been reported from Argentina and California (Madia *et al*., 1999, Koike and Gordon, 2005).

Losses

The pathogen associated with this disease is soil borne in nature and in favourable weather conditions causes significant yield reductions (Anonymous, 1952).

Symptoms

The symptom of the disease appear at various growth stages of the plants and are recognized as pre-emergence, post-emergence and adult stage symptoms. Initial infection occurs on radicles causing brown lesions, rotting of tissues and finally pre-emergence mortality. In post-emergence, underground portion of hypocotyls initially develops lesions, which slowly coalesced and girdled the region leading to softening and maceration of cells, finally plants collapsed and die due to spread of infection. Adult stage symptoms are typical and characteristic of *Fusarium* wilt. Plants, which are infected at early stage but could survive, develop these symptoms at later stage of growth. The wilt symptoms are initiated by yellowing and curling of lower leaves. Gradually these symptoms get extended up to the apical portion, and during this period lower leaves become dry and plants droop. Death is delayed and plants are able to survive with these symptoms and often reach the adult stage. Affected plants generally remain stunted and bear less flowers and fruits.

The Pathogen

The fungus responsible for this disease is *Fusarium oxysporum* Schl. emend. Snyder and Hansen f.sp. *coriandri* Kulkarni *et al.* The mycelium is abundant, fluffy and cottony white to light pink. Both micro- and macroconidia are produced. Microconidia are oval to elliptical, while macro-conidia are septate and sickle-shaped measuring 10.06–7.90 x 2.7–1.6 to 70.62–99.4 x 2.97–1.98 μm in size. Later in the season chlamydospores (14.04–9.9 x 9.9–11.5 μm) are also produced. Two culture types in this *Fusarium* have been reported (Mall, 1968). Type I has fluffy aerial mycelium and abundant microconidia and Type II with appressed growth and conidia in pionnotes. They further observed that members of Type II were variants of Type I which were initially isolated from diseased roots.

Disease cycle

The fungus is soil borne and overwinters in the form of chlamydospores. In favourable weather conditions, chlamydospores germinate and cause infection. The fungus enters in to root either through germinating seeds or wounds caused by insects, intercultural operations and nematodes. Microconidia and macroconidia are produced which cause secondary infections.

Epidemiology

The fungus is soil borne in the form of chlamydospores. Optimum temperature and pH for the disease is 28°C and 5.8 to 6.9, respectively (Srivastava, 1971). Infection increased with the increasing soil moisture content, but the population of the pathogen was highest at 52 per cent moisture holding capacity of the soil.

Management

(i) Cultural practices: Once the fungus is introduced in a field it is very difficult to eradicate. Follow long crop rotations, collect and burn the infected plant debris, plough deep the plant debris and solarize soil during summer months. Addition of oil cakes in the soil and raising pH of soil to 8.2 is has also been found effective in reducing the disease incidence (Srivastava, 1972).

(ii) Host resistance: Coriander varieties/lines like MP 5365 (Srivastava, 1972) and UD 373 (Prakasam *et al.*, 1987) showed least infection to the disease and can be used in breeding programme for developing resistant varieties.

(iii) Chemical control: Application of systemic fungicides like benomyl or carbendazim as seed treatment and foliar sprays were found effective in reducing the incidence of this disease (Jain and Agrawat, 1978).

(iv) Biological control: Seed treatment with *Trichoderma viride* (4g/kg seed) was found effective in reducing the incidence of the disease to 5.3 per cent as against 28.3 per cent in control (Muthulakshmi *et al.*, 2002).

(v) Integrated Disease Management: Seed treatment with *Pseudomonas fluorescens* (10g/kg seed) and soil application of neem cake (150 kg/ha) was found effective in keeping this disease in check. Soil application of bioagents like *T. harzianum* and *P. fluorescens* and field application of carbendazim (0.1%) and mancozeb (0.25%) at 45 DAS was also found effective (Singh 2009).

3. STEM GALL

It is the most important disease of this crop causing considerable yield losses and quality of essential oil.

Geographical distribution

The occurrence of the disease was first reported by Unger (1833) from Great Britain and now its presence has been reported from all coriander growing areas of the world. In India, the occurrence of the disease was first time reported from Pusa (Bihar) by Sydow and Butler (1911). Later it has been reported from northern parts of Madhya Pradesh and adjoining districts of Rajasthan and Uttar Pradesh (Gupta and Sinha, 1973), Himachal Pradesh (Paul, 1992), Arunachal Pradesh (Bag, 1998) and Haryana (Lakra, 1999).

Losses

The economic losses due to stem gall are direct and are threatening the farmers as infected fruits lose their value for consumption or as propagules. Stem gall causes severe yield losses and deteriorate the quality of seed (Singh *et al.*, 1984). In Uttar Pradesh, the disease reduced the yield by 9.6 to 33.8 per cent, while in Madhya Pradesh losses ranged from 2.2 to 11.9 per cent (Singh *et al.*, 1984; Tripathi *et al.*, 2003). A positive and significant correlation coefficient was recorded between stem gall intensity and percentage seed yield loss, which indicated that loss in seed yield increased with the corresponding increase in stem gall intensity (Tripathi *et al.*, 2003). Besides yield reductions, the disease also reduces height of the plants, number of branches/plant, number of umbles/branch, number of umblets/umbel, number of seeds/umbel (Lakra, 1999). A survey of coriander fields in 3 districts of Himachal Pradesh was conducted by Paul (1992) and recorded the disease intensity in between 10.76 to 60 per cent, 0.32 to 7.3 per cent and 0.2 to 2.35 per cent in the districts of Kangra, Mandi and Kullu, respectively.

Symptoms

The infection of stem gall disease takes place at an early stage of seed germination but the symptoms appear much later. The mycelium invades the host tissues intercellularly and concentrates mainly around the vascular bundles. They are altered morphologically and hypertrophy, and hyperplasia of infected plant host tissues result in the formation of distinct galls or swellings on stems, leaves, flowers, fruits and other aerial parts of plants. The disease appears in the form of tumour like swellings of leaf veins, leaf stalks, peduncles, stems and on fruits. The swellings on the veins give a swollen hanging appearance to the leaves. These tumours are at first glossy, but later rupture and become rough. The swellings are mostly elongated and their size varies as rule according to the size of the infected part, those on the stem being usually larger than those on the flower stalk (Plate 1a). Systemic infection provides greater distortion to the plant and plant parts. The inflorescence may show outgrowth on the surface and uniform invasion of the fruit makes it abnormally large, but partial invasion may lead to distortion.

The galls are generally elongated and rarely spherical. The galls on stem arise mainly from the cortical and phloem cells by hypertrophy and hyperplasia. The galls are composed of enlarged cortical cells and highly multiplied phloem in the early stage. The gall epidermis is continuous with the host epidermis and keeps pace with the developing gall. In advanced stages, the gall tissues possess vascular elements derived from the stealer region. The ground tissue is enlarged by the activity of the parasite. Infected parts of the flower become slightly massive. Pollen sacs are very much compressed, but on the whole the size of the anther is increased. Vascular tissue is profuse. Parasite could reach all the parts except the seed in a fruit. The fruit wall increases in its size due to hypertrophy and hyperplasia of the tissue (Plate 1b). The size of septum also increases. With the production of the abnormal tissue, the locules are slightly dislocated from their normal position. In all the infected parts, the oil canals were found very much affected and collapsed. This disease is usually apparent in early spring when the greatly distorted leaves first appear. On veins, mid-ribs and petioles, the gall appears as translucent yellow-white swellings that are often elongated and blister-like.

The Pathogen

The fungus responsible for this disease is *Protomyces macrosporus* Unger. Hyphae are intercellular, closely septate and broad. Branching is irregular. Scattered cells in the hyphae swell, form ellipsoidal or globose bodies, which later develop in to chlamydospores. As the chlamydospores mature, a thick, hyaline and three-layered wall measuring 50 to 60 μm in diameter surrounds them. The mycelium of the fungus is only found in the tumours although the resting spores of the fungus cause systemic infection.

Use of PCR-based molecular markers provide an efficient reproducible and reliable approach to characterize plant pathogen population particularly for the assessment of genetic diversity and phylogenetic relationships. Internal Transcribed Spacer (ITS) sequence data have been used to study the phylogenetic relationships of many *Protomyces* sp. (Nishida *et al.*, 1993). Molecular tools used in phylogenetic studies of Ascomycota (Sugiyama *et al.*, 2006) have included large and small subunits of ribosomal RNA genes and sequence analysis of the Internal Transcribed Spacer (ITS) regions of the rRNA genes (Bacigalova *et al.*, 2003).

Disease cycle

The disease is seed and soil borne in nature (Pavgi and Mukhopadhyay, 1969; 1972). In soil the fungus overwinters in the form of chlamydospores. The viability of chlamydospores decreases with age *i.e.* 10 per cent after 6 years and zero after 8 years (Gupta, 1973). The chalamydospores germinate in the presence of water by rupturing the outer wall. The inner wall is pushed out to form vesicle, which appears in continuation with the mouth of the crack. The protoplasm passes into this vesicle and the nucleus divides several times forming 100 to 200 daughter nuclei. The protoplasm also divides by cleavage into several uninucleate masses. On maturity these spores separate and collect in the centre of the vesicle. The latter bursts and the spores are set free. These spores further multiply by budding in yeast like fashion and cause infection of the host (Fig. 1).

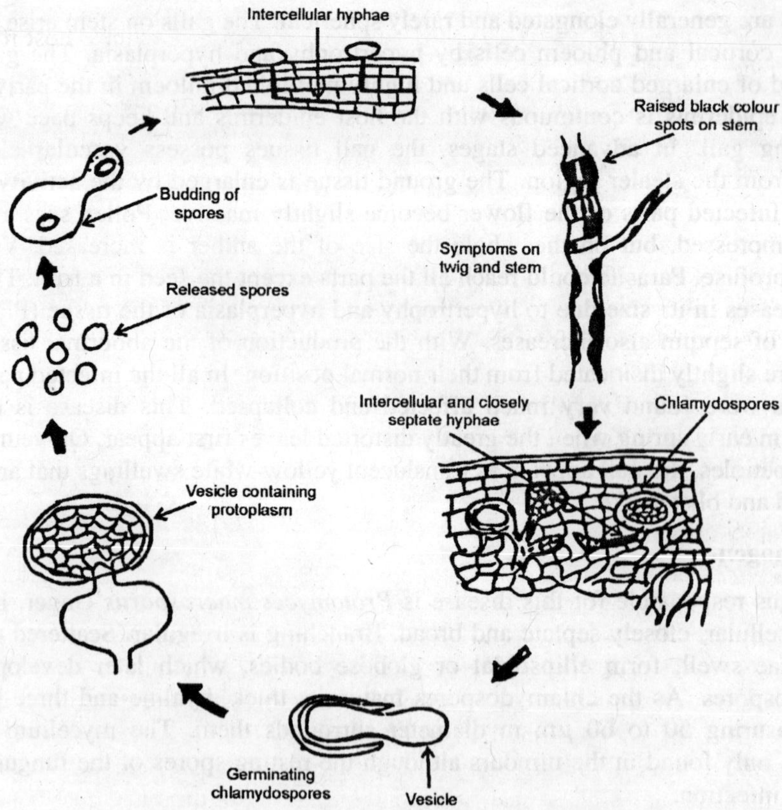

Fig. 1. Disease cycle of stem gall of coriander

Epidemiology

The disease is both seed and soil borne. Minimum/maximum atmospheric temperature and relative humidity play an important role in the initiation and development of this disease (Tripathi, 2003). The disease appeared on plants sown in the third week of October when the minimum and maximum temperatures were 13.2 and 30.9ºC, respectively, and the relative humidity was 57.2 per cent while maximum disease intensity was recorded in plants sown in the last week of January when the minimum and maximum temperature was 8.1 and 22.6ºC, respectively, and the relative humidity was 65.8 per cent. Rain had a positive but non-significant correlation with the disease intensity in various dates of sowing. Potassium and nitrogen fertilizers reduced stem gall incidence while phosphorus fertilizers increased it (Tripathi *et al.*, 2001). The highest soil moisture (IW: CPE 1.33) resulted in 46 percent disease while IW : CPE 0.7 recorded 22 per cent disease (Lakra, 2000). A pH of 7.5 was most suitable for infection, while minimum infection occured at pH 5.5 (Tripathi *et al.*, 2002a). However, under pH 5.5 the growth parameters and yield of coriander were drastically affected. Malhotra *et al.* (2016) observed that presence of excess soil moisture, low sunshine hours due to cloudy weather in winter season, fuels the spread of stem gall and damage the crop up to 100 per cent.

Management

The disease can be kept under check by following cultural practices, host resistance and chemical control.

(i) Cultural Practices: The disease affects less, the early and late sown crops. Tripathi *et al.* (2002c) and Tripathi (2003) advocated the sowing time around 16 October and 16 November to minimize losses due to this disease. To keep the disease under check use balanced doses of fertilizers. The application of K had no impact on the disease intensity, while P increased it in the absence of N (Tripathi *et al.*, 2001). They further suggested the application of 70 : 50 : 25 kg NPK/ha to obtain good seed yield. Besides these, use of clean and healthy seed, suitable crop rotation and field sanitation also help in keeping the initial inoculum level below economic threshold level.

(ii) Host resistance: Cvs./lines like JD 1, G-5365–91, Pant Haritma, UD 20, Rcr 41, Pant-1, CIMAP 2053, PH-7, Pant Haritma, COR-17 and COR-2 (Tripathi, 2001; Tripathi *et al.*, 2002b, Kalra *et al.*, 2003, Singh *et al.*, 2003) have been reported as resistant to this disease, while G-C-88–8, G-5365–91, Pant-1 and U.D. 20 were moderately resistant (Tripathi *et al.*, 1998).

(iii) Chemical control: Several systemic and non-systemic fungicides have been evaluated against this disease both as seed treatment and foliar sprays, but seed dressing with fungicides gave better results compared to seed dressing + foliar spray treatments (Tripathi and Bartaria, 2002). Out of various fungicides and combinations tested as seed dressers, captan, mancozeb, carbendazim + captafol (1 : 1) and captafol gave maximum inhibition percentage of *P. macrosporus* (Paul, 1992). Thiram applied as seed dressing or seed dressing + foliar spray resulted in the lowest disease intensity and highest seed yield, additional net returns and cost benefit ratio (Lakra, 2000; Tripathi and Bartaria, 2002).

Application of a variety of fungicides has lowered the disease incidence but has not given satisfactory control. Seed treatment with thiram, carbendazim, captan, thiram and Vitavax (carboxin) has been found effective by various workers. Better control is possible by combining seed and soil treatments with thiram, ceresan, omnamycin, captafol, carbendazim, carboxin, copper oxychloride, mancozeb, streptomycin and tridemorph and captan. Shukla *et al.* (2006) found that seed treatment with carbendazium (1g/kg) + foliar spray with bayleton (0.04%) was found most effective and reduced disease severity (57.66%) followed by seed treatment with carbendazim + foliar spray with Karathane (0.04%), foliar spray with Bayleton (0.4%). Kumar *et al.* (2014) reported that minimum disease incidence of 44.16 per cent was recorded by treatment of seeds with hexaconazole (0.2%) and foliar spray of hexaconazole (0.2%) after 40, 60 and 75 days of sowing followed by propiconazole.

Tripathi (2005) recorded the minimum disease intensity in seed treatment with thiram. Seed treatment with thiram reduces the disease up to 65 per cent over control. Minimum diseases intensity was recorded in soil treated with neem cake, which reduces the disease (by 50%) and increased yield (by 36%). Seed treatment with Thiram (3g/kg), soil treatment with neem cake (1.5 t/ha) and two spray of thiram (0.2%) gave better control of stem gall diseases (90%). Neem oil also gave a significant reduction in disease up to 27 per cent.

(iv) Integrated Disease Management: Malhotra *et al.* (2006) studied the influence of *Azospirillium* sp., nitrogen fertilizer, farmyard manure and their combinations on seed yield and incidence of stem gall of coriander. They observed that *Azosprillium* sp., 50 per cent nitrogen and 5 tonnes of FYM were the most economical treatment in increasing the seed yield and reducing the incidence of stem gall disease.

4. POWDERY MILDEW

Powdery mildew of coriander is an important disease of this crop and in favourable dry weather conditions can cause significant yield reductions.

Geographical distribution

The disease has been reported from all coriander growing areas of the world. In India, it is widely present and its occurrence has been reported from Rajasthan (Srivastava *et al.*, 1971), Madhya Pradesh (Ali *et al.*, 1999), Karnataka (Kalra *et al.*, 2002) and Uttar Pradesh (Kalra *et al.*, 2000).

Losses

The disease causes heavy yield reductions if infection starts at flowering stage. Infection at seed formation stage, however, does not cause much loss. In Rajasthan, the losses from the disease ranged from 15–20 per cent (Srivastava *et al.*, 1971).

Symptoms

The disease first appears as white floury patches on both sides of the leaves as well as on the stem inflorescence and seeds in case of severely infected plants (Plate 2a). These patches originate as minute discoloured specks from which a powdery mass radiates on all sides. Large areas on the aerial parts of the host may be covered with white floury patches (Plate 2b,c). The superficial mass consists of the mycelium and spores of the fungus. In advanced stages of attack, infected peduncles dry prematurely and render the flowers sterile or lead to poor seed setting.

The Pathogen

The disease is caused by *Erysiphe polygoni* DC which is an obligate parasitic fungus. The mycelium is generally fine, persistent, rarely thick. The conidiophores arise vertically from the superficial hyphae on the host surface. Conidia are formed singly or in short chains. They are ellipsoid or ovate with vacuolated cytoplasm. The ripe conidia fall off quickly and are dispersed by wind. Sometimes late in the season, but not always on coriander leaves in the field, the cleistothecia appear as sharp, black minute bodies scattered in the mycelial web. The cleistothecium is provided with a number of myceloid appendages. Usually four to eight asci are formed in each cleistothecium. Each ascus contains three to five ascospores.

Disease cycle

Wherever the cleistothecia develop on dead plant debris, these serve as the source of primary inoculum for the next season. Otherwise the fungus overwinters in conidial stage on different hosts in the area, which cause primary infection. Wind blown conidia cause secondary infections on other plants.

Epidemiology

The fungus has wide host range and overwinters in conidial stage. Conidial germination is best in between 20–24°C. A fairly dry soil and heavy application of nitrogenous fertilizers tend to increase the incidence of the disease.

Management

(i) Cultural practices: Cultural practices like collection and destruction of infected plant debris, application of balanced doses of fertilizers and early planting (Kalra *et al.*, 2000) reduce the overall disease severity.

(ii) Host resistance: Accessions CIMAP 2053 and CIMAP 2096 showed a high degree of tolerance to powdery mildew (Kalra *et al.*, 2003).

(iii) Chemical control: Efficacy of various fungicides like Cosan, Sultaf, Karathane WD, Elosal, Thiovit, Morocide, carbendazim and thiophanate methyl (Srivastava *et al.*, 1971; Ali *et al.*, 1999; Kalra *et al.*, 2002) has been reported against this disease. Four and three application of dinocap provided satisfactory control of the disease in the late and early maturing cultivars, respectively (Kalra, 2002). However, these could be further reduced to three and two if the sprays are timed at flowering, seed formation and full fruiting stages in the late maturing cultivars, and at seed formation and fruit formation stages in the early maturing cultivars. Recently available EBI's and strobilurin fungicides can keep this disease in check at very low doses and with minimum number of sprays.

B. SPINACH

Spinach (*Spinacia oleracea* L.) belongs to family Chenopodiaceae and is one of the important leafy vegetable crops grown in many countries for fresh and processed (frozen and canned) markets. Spinach apparently is native to central Asia and has been cultivated for more than 1300 years (Yamaguchi, 1983). During cultivation diseases impose significant production constraints affecting both yield and quality of spinach. Important diseases affecting this crop along with their management practices are described hereunder.

(A) FUNGAL DISEASES

1. ROOT ROT AND DAMPING-OFF

The crop is infected by many root pathogens which are known to induce pre- and post-emergence damping-off.

Symptoms

The symptoms of the disease appear in two phases *i.e.* pre-emergence and post-emergence damping-off. In the former case, there is a failure of seedling emergence from the soil either due to seed rots or killing of young seedlings before their emergence from the soil. In case of post-emergence damping-off, the disease outbreak is characterized by toppling over of infected seedlings any time after their emergence from the soil. The infected tissue initially appears to be water-soaked and soft and subsequently, the stem at the infection point get constricted resulting in toppling over and mortality of the seedlings.

The Pathogen

The fungi generally associated with this disease are different species of *Pythium* namely *P. aphanidermatum, P. ultimum, P. spinosum* (Akashi *et al.*, 1986), *Phytophthora cryptogea* (Krober and Beckmann, 1973) and *Rhizoctonia solani* (Akashi *et al.*, 1986). Among these fungi, species of *Pythium* are more common. *P. ultimum* and *P. spinosum* were pathogenic at cotyledonary stage, while *P. aphanidermatum* up to 5–leaf stage. The mycelium of *Pythium* is intracellular in the host. It consists of slender, coenocytic and much branched hyphae. Asexual reproduction is mostly by means of zoospores, which are always differentiated in a vesicle extruded by the zoosporangium. The sporangium forms a beak, the tip of which softens and out of which flows the protoplasmic mass of zoosporangium to form a spherical vesicle. In the vesicle, protoplasm cleaves to form zoospores that are kidney shaped and biflagellate. They rupture the wall of the vesicle and escape in the surrounding film of water. After swimming for a while, the zoospores encyst and germinate by germ tube. In sexual reproduction, the oogonium is plurinucleate, arises terminally, and less often intercalary, on long or short unmodified hyphae with homogeneous protoplasm and only one nucleus migrates into the periplasm. The antheridium is also plurinucleate at first but only one nucleus functions in the fertilization. The branch bearing the terminal antheridium may arise at a distance from the oogonium on a separate branch or from the same branch. Through a conjugation tube, one male nucleus passes into the egg, which then forms a thick wall and becomes oospore.

Disease cycle and Epidemiology

The pathogen overwinters either in soil or in infected seed or both. Species of *Pythium* attack the plants when they are in juvenile stage. Soil moisture near saturation stimulates mycelial growth as well as asexual reproduction. Young plants are more sensitive than older ones. High soil moisture and relatively higher soil temperature favour disease development.

Management

The disease can be effectively checked by following crop rotation, destruction of infected plant debris and sanitation. Regular application of humus can check the infection of *Phytophthora cryptogea* (Krober and Beckmann, 1973). Seed treatment with metalaxyl, terrazole, agrosan, seedex and ceresan also reduce the disease incidence under field conditions. For the control of *Rhizoctonia solani* root rot, seed treatment and soil drenches with tolclofos methyl, flutolanil, iprodione and pencycuron gave the best disease control (Liu and Liu, 1994).

2. DOWNY MILDEW

It is probably the most widespread and potentially destructive disease of this crop worldwide.

Geographical distribution

The occurrence of this disease has been reported from U.S.A, Japan, Iran, Netherlands, Italy etc. (Inaba *et al.*, 1983; Correll *et al.*, 1994; Lorenzini and Nali, 1994; Sadravi *et al.*, 2000).

Losses

Though exact figures of losses are not available but the disease under favourable weather conditions is potentially destructive and causes significant losses. Under conditions of prolonged leaf wetness and cool temperatures, epidemics can progress very rapidly and entire crop can be lost in a short period of time (Correll *et al.*, 1994).

Symptoms

The symptoms of the disease appear as slightly yellow, irregular, chlorotic lesions on leaves. Lesions frequently expand and coalesce and may become necrotic. Heavily infected leaves can appear curled and distorted. Under wet conditions and/or high relative humidity, blue-greyish sporangia and sporangiophores are produced and can be seen in a mass on the underside of the leaf. Severely infected plants either remain stunted or die prematurely.

The Pathogen

The disease is caused by *Peronospora farinose* (Fr : Fr) Fr. f.sp. *spinaciae* Byford (=*P. effusa* (Grev.) Ces). The fungus is a heterothallic obligate parasite. It produces asexual sporangiophores, which emerge from stomata either singly or in groups of two or three. Sporangia are hyaline, oval and germinate directly by germ tube. Sexual oospores are formed within the host tissues during cool, moist conditions. They germinate either by germ tube or by production of zoospores. Seven distinct physiological races of *P. farinose* f.sp. *spinaciae* such as race 1, 2, 3, 4, 5, 6 and 7 have been reported from USA and Europe (Brandenberger *et al.*, 1991; Irish *et al.*, 2003; Irish *et al.*, 2004). The occurrence of race 4 and race 7 of the pathogen outside USA was first time recorded in Italy and Netherlands, respectively (Lorenzini and Nali, 1994, Irish *et al.*, 2003).

Host range

Besides genus *Spinacia*, the fungus also infects several species of *Chenopodium* (Byford, 1981), but not lambsquarters. The downy mildew observed on lambasquarters does not infect spinach.

Disease cycle

The pathogen overwinters in the infected seed as oospores and mycelium (Inaba *et al.*, 1983) and the infected seeds have been shown to give rise to infected seedlings. Oospores can also survive in soil and may represent an important source of primary inoculum (Wright and Yerkes, 1950). Windborne sporangia from surrounding diseased fields may represent an important source of primary inoculum. The oospores/sporangia germinate and cause primary infection. The sporangia produced on primary pustules further cause secondary infections.

Epidemiology

The pathogen survives as oospores/sporangia from one season to another. The sporangia germinate directly on leaf surfaces with in 2–6 h under cool and wet conditions (Richards, 1939). Germination of sporangia and germ tube elongation can occur between 2 and 25ºC with an optimum at 10ºC (Sadravi *et al.*, 2000). Lesion development is favoured by temperatures of 15–25ºC with latent period of 6–12 days. Sporangia are readily dispersed by wind and rain splashes, but can rapidly lose viability when desiccated or exposed to sunlight (Richards, 1939; Wright and Yerkes, 1950). Flood or furrow irrigation help in the spread of oospores/sporangia within the field and in surroundings.

Management

(i) Cultural practices: Practices like use of healthy seed and destruction of plant debris and *Chenopodium* sp. from in and around the field, which can harbour the pathogen, should be followed to keep the disease under check. Follow crop rotation to reduce primary inoculum of the pathogen and should not be sown in the same row or bed year after year.

(ii) Host resistance: Planting cultivars with single-gene resistance to a prevalent race controls the disease (Brandenberger *et al.*, 1991; Jones and Dainello, 1982). Varieties Solomon, Samson and G-1 possessed good resistance to race 1 or race 2 (Shimazaki and Uchiyama, 1985) while Dynamo, Nores, Vital, Califlay and Medania to race 1, 2 and 3 (Hubbeling and Ester, 1978). Lines such as Bossanova, Bolero and RS 1250 had < 2 per cent diseased area (Koike *et al.*, 1992). Of late, cv. Byford (resistant to races 1, 2 and 3) was found susceptible to a new race '4' and the application of commercial formulations of metalaxyl, copper oxychloride and chlorothalonil were ineffective (Lorenzini and Nali, 1994). However, when new races of the pathogen appear, as it happened in California and Italy (Brandenberger *et al.*, 1991; 1992; Lorenzini and Nali, 1994), it may take several years to develop new commercial cultivars with single-gene resistance before these become available. Cvs. *viz.*, Nores, Duet, Early Hybrid 7, Dixie Market, Hybrid 424 Melody, Resistoflay against Race 1 and 2; Polka, Shasta, Woster, Debut, St. Helens, Baker, Jade, Olympia, Rainier against race 1,2 and 3 and Meridian against race 1 and 3 (Irish *et al.*, 2007) have been reported resistant to different races of the pathogen.

(iii) Chemical control: Several systemic and non-systemic fungicides like maneb, fosetyl, metalaxyl and chlorothalonil can be used to combat the ravages of this disease (Koike *et al.*, 1992).

3. WHITE RUST

It is also an important disease of this crop particularly in USA and causes considerable yield reductions.

Geographical distribution

The disease is widespread in USA. It was first recorded in Virginia in 1907 on spinach, in Texas in 1937 and in the Arkansas River Valley in 1945 (Wiant, 1939; Raabe and Pound, 1952). By 1970, this disease was considered as the most important problem of the South Texas (Thomas, 1970). Now it is an important foliar disease in all spinach production areas (Cornell *et al.*, 1994). The disease is restricted to some parts of United States only.

Losses

In susceptible cultivars, the disease causes significant yield reductions. Yield reductions exceeded 50 per cent in 32 of 35 spinach cultivars and breeding lines evaluated (Dainello *et al.*, 1981). Besides yield reductions, it also affects the quality of both fresh and processed spinach.

Symptoms

The symptoms of the disease manifest on leaf surface as small chlorotic lesions. With the progress of the disease small, glossy white pustules are produced frequently in concentric rings on the underside of the leaves and occasionally on the upper leaf surface. Lesions often coalesce, and white rust pustules can cover the entire leaf surface. As lesions mature, dark oospores are frequently produced, giving the lesion a grainy appearance prior to necrosis.

The Pathogen

The disease is caused by *Albugo occidentalis* G.W. Wils. The fungus is obligate parasite. The dry sporangia are hyaline and discoid and measure 10 x 14 μm in size. When hydrated, sporangia become spherical to ellipsoid and measure 10–19 x 20–22 μm. Oospores are finely reticulated and 44 –62 μm in size (Raabe and Pound, 1952).

Disease cycle

Oospores survive in infected plant debris for a year or more. Oospores probably function as over-wintering structures to initiate primary infection. They germinate directly or produce a sporangium to initiate infection (Vanterpool, 1959; Verma and Petrie, 1975). Sporangia typically germinate indirectly but occasionally directly (Raabe and Pound, 1952) and produce 6–9 biflagellate motile zoospores that encyst and produce a germ tube which directly penetrates the spinach leaf. Oospores are produced abundantly in mature lesions.

Epidemiology

The fungus overwinters in infected plant debris as oospores, which under favourable weather conditions germinate directly or produce sporangia. Severe epidemics of white rust are favoured by cool nights with heavy dew alternating with warm, dry sunny days (Raabe and Pound, 1952). The optimum temperature for disease development ranged from 12 to 18°C, while 6–12 h leaf wetness is required for disease development (Sullivan *et al.*, 2002). The sporangia can germinate at 2–25ºC with an optimum temperature of 12–16ºC. Approximately 22ºC is optimum for sporulation and higher temperature accelerates lesion development. Sporangia are produced in the leaf pustules or sori and are released when the epidermal tissue covering the sori rupture. Higher temperatures favour production of oospores more than the production of sporangia.

Management

(i) Cultural practices: The disease can be kept under check by following atleast two years crop rotation and destruction of infected plant debris.

(ii) Host resistance: Cultivars resistant to this disease can be developed by following recurrent selection breeding strategy, which have field or horizontal resistance. With horizontal resistance, latent periods are prolonged and both lesion development and sporulation are reduced (Dainello *et al.*, 1984; Brandenberger *et al.*, 1994). The flat or smooth leaf cultivars are generally more white rust resistant than the savoy or crinkled leaf cultivars (Dainello *et al.*, 1981). Medium to high degree of resistance in cvs. Ozarka and Green valley have been reported to this disease (Bowers and Goode, 1980).

(iii) Chemical control: The disease can be kept under check by application of metalaxyl in-furrows as granules or to the soil surface as bed spray and this practice controlled this disease for 48–60 days after planting, while seed treatment with metalaxyl controlled the disease for 30 days (Dainello and Jones, 1986). Other fungicides like chlorothalonil, metalaxyl, manccozeb can be applied as foliar sprays for the control of this disease (Jones and Dainello, 1983). Efficacy of acibenzolar-S-methyl (ACIB) and strobilurin derivatives alone and in combination in the management of white rust severity without causing leaf chlorosis has been reported (Leskovar and Kolenda, 2002). The surfactants like Naiad and sodium dodecyl sulfate (SDS) were found highly effective and comparable to fungicides in reducing white rust severity (Irish *et al.*, 2002).

4. CERCOSPORA LEAF SPOT

A number of pathogens have been reported to cause leaf spot diseases on spinach the worldover, but in India, Cercospora leaf spot is quite important and can cause considerable foliage damage under favourable weather conditions.

Geographical distribution

In India, the occurrence of this disease was first reported from Kanpur (Singh and Gupta, 1951). Since then it has been reported from Rajasthan and Punjab (Sehgal and Mathur, 1959; Singh *et al.*, 1987).

Symptoms

Symptoms of the disease appear as circular to angular grayish brown spots with reddish brown margin on the older leaves. They may coalesce and extend up to petioles. In the centre of these spots small black stroma of the fungus develop and produce clusters of conidiophores, which bear conidia.

The Pathogen

The disease is caused by *Cercospora beticola* Sacc. The mycelim of the fungus is septate, dark coloured and intracellular. Small sclerotial masses are formed in the host tissue from which dark coloured conidiophores arise in clusters. The conidia are borne on the tip of the conidiophores. These are hyaline, elongated, filiform, multiseptate, broadly rounded at the point of attachment to the conidiophore and tapering slightly toward the opposite end.

Host range

The pathogen has a wide host range (Verma *et al.*, 1968). Besides spinach and beet, it also infects *Datura fastuosa, Trifolium alexandrinum, Chenopodium murale, C. album, Amaranthus polygamous, Spinacia tetrandra, S. oleracea, Atriplex hortansis, Portulaca oleracea, Amaranthus tricolor, Beta vulgaris* var. *cicla* and certain closely related weeds (Verma *et al.*, 1968; Soylu *et al.*, 2003).

Disease cycle

The fungus chiefly survives on dried infected leaves in the soil, that acts as primary source of inoculum, or on susceptible wild plants (Verma *et al.*, 1968) or on the seed. The fungus remains viable for six months in infected leaves in soil (Verma *et al.*, 1975). The infected seeds give rise to infected seedlings. In favourable weather conditions, the overwintering mycelium in plant refuse starts producing conidia, which are disseminated by wind currents, rain splashes and insects to the healthy leaves.

Epidemiology

The pathogen overwinters as mycelium in the infected plant debris. The disease severity increases when the crop is sown by broadcasting than when sown in rows. The average number of spots per leaf decreased gradually with the increase in row to row distance (Singh *et al.*, 1987). The best growth of the fungus occurs at 22–30ºC temperatures and is supported with addition of calcium, potassium nitrate, glucose and pH 6.1 (Dange and Patel, 1968). Optimum temperature for infection and progress for the disease is 19–27°C at which the incubation period is short but increased as the temperature decreased (Verma *et al.*, 1968). High relative humidity is an essential pre-requisite to sporulation.

Management

Cultural practices like destruction of infected crop debris by burning and deep ploughing and use of long rotations to prevent an accumulation of the soil borne inoculum should be followed. Avoid overhead irrigation, especially during humid, cloudy weather. Bury crop residues deep immediately after harvest to reduce the overwintering inoculum. Eliminate volunteer spinach plants and weeds from in and around the field that can serve as the source of inoculum. Among several fungicides tested Bordeaux mixture (5:5: 50) followed by Blitox and Fungimar controlled the disease effectively (Mathur and Sehgal., 1964). Systemic fungicides like carbendazim and EBI fungicides can also be used to combat the ravages of this disease.

(B) BACTERIAL DISEASES

5. SOFT ROT

The disease is described in detail under lettuce diseases in this chapter.

(C) VIRAL DISEASES

Several viral diseases have been reported to occur in different parts of the world and cause heavy losses. In India, spinach is affected by several diseases of which mosaic and spinach leaf curl are important.

1. MOSAIC

The disease was first time recorded at Aligarh during 1971 (Naqvi and Mahmood, 1975). Since then it has been observed in many parts of the country and is known to cause considerable losses.

Symptoms

The characteristic symptoms of the disease include mosaic and mottling of the plants, which is characterized by light and green patches. The size of the leaves is reduced, distorted and they invariably show large chlorotic sectors, which are almost white. The infected plants remain stunted.

The Virus

The disease is caused by *Spinach Mosaic Virus* (SMV). The virus was found to be different from previous known virus diseases.

Host range

The host range of the virus includes *Achyranthes aspera, Amaranthus caudatus, A. leucocarpus, Lagenaria leucantha, Datura srtamonium, Lycopersicon glandulosum,*

Nicotiana glutinosa, N. rustica, N. tabacum, Petunia hybrida and *Solanum melongena* (Naqvi and Mahmood, 1975).

Transmission

The aphid *Myzus persicae* transmits this virus. The virus particles remain infective for 72 h at 10ºC and for 48 h at room temperature. The dilution and thermal inactivation end points were 10^{-4} and 70ºC, respectively.

Management

Collect and destroy all infected plants to reduce the initial inoculum in the field. Collect seed from disease free plants. Spray the crop with suitable insecticide or neem based botanicals for the control of aphid vectors.

2. SPINACH LEAF CURL

This is another important virus disease on spinach reported from India. The disease was first observed in Delhi during 1978 and 1979 seasons (Viswanath and Nariani, 1981). About 5 per cent plants were found affected due to this disease.

Symptoms

Characteristic symptoms of the disease include severe curling or rolling of leaves with vein thickening or presence of long leaf enations on the veins of lower leaf surface. The leaves become thick and leathery. Infected plants often remain stunted.

The Virus

Tobacco Leaf Curl Virus (TLCV) measuring 15–20 × 25–30 nm is responsible for causing this disease.

Transmission

The virus is not sap or mechanically transmitted. It is transmitted through white fly, *Bemissia tabaci,* (Viswanath and Nariani, 1981) in persistent manner. Only 30 minute acquisition access period and 60 minute access period is essential for whitefly to transmit this virus (Osaki *et al.*, 1979)

Management

Collect and destroy the infected plants immediately. Do not plant cotton or tomato and/or other crops susceptible to white fly. Effective management of white fly also provides an effective control of this disease.

C. LETTUCE

Lettuce (*Lactuca sativa* L.) is an important vegetable crop which is grown for salad, the world over. In temperate regions, mostly it is grown under glass and plastic house conditions, but sometimes it is grown as outdoor crop also. During cultivation, it is affected by a number of fungal, bacterial and viral diseases which cause significant yield reductions. Important diseases of lettuce along with their management practices are described below.

(A) FUNGAL DISEASES

1. ANTHRACNOSE

The disease is also known as shot hole, leaf perforation, ring spot and rust, but now anthracnose is the accepted name of the disease. Sometimes the disease causes considerable yield losses.

Geographical distribution

The disease has a worldwide distribution. The occurrence of the disease has been reported from Australia (Galea and Price, 1988), California, USA (Ochoa *et al.*, 1987), Serbia (Krnjaja and Ivanovic, 1996), Yugoslavia (Krnjaja *et al.*, 1999) and Japan (Horita *et al.*, 2001).

Losses

In some areas where cool and moist weather prevails, the disease causes significant yield reductions. Heavy losses due to this disease in Australia and Italy have been reported (Marte and Cappelli, 1979).

Symptoms

The symptoms appear as small, tan coloured, water-soaked spots on the lower leaves of infected plants. These spots later enlarge and turn straw coloured. The localized necrotic tissue may drop out resulting in shot hole effect. Infections are usually most prevalent along the midrib on the lower leaf surface.

The Pathogen

The fungus responsible for this disease is *Microdochium panattonianum* (Berl.) Sutton, Galea & Price. The sexual stage of the fungus has not been identified. Mycelium of the fungus is composed of hyaline, slender, septate hyphae that grow inter- and intracellularly within the host tissue. Acervuli consisting of densely packed, upright conidiophores arise from the mycelium. Conidia are typically two celled, hyaline, and slightly curved and may contain one to two large oil droplets. They are usually constricted at the septa and measure 4 x 17 μm in length. Microsclerotia may form in the infected tissue and serve as source of primary inoculum in next season (Parman and Price, 1991). The microsclerotia are 35–65 μm in diameter and composed of pseudoparenchymatous tissue of textura angularis and globulosa type. In California, USA, five races of the pathogen *i.e.* Race A, B, C, D and E have been identified out of 14 isolates on the basis of virulence phenotype (Ochoa *et al.*, 1987).

Host range

Besides *Lactuca sativa*, the pathogen also infects wild lettuce (*L. serriola*) and endives (*Cichorium endivia*) (Krnjaja and Ivanovic, 1996).

Disease cycle

The disease is not seed-borne (Galea and Price, 1988). The pathogen survives in the infected plant debris or as microsclerotia in the soil which serve as the source of primary inoculum (Patterson and Grogan, 1985a; 1991). After germination, microsclerotia produce hyphae and/or conidiophores, which in turn produce spores. Saturated atmosphere or overhead irrigation is required for conidia production, germination and infection. The fungus may enter either directly through the leaf epidermis or through the stomatal openings. Splashing water splash microsclerotia and conidia from soil on to leaves resulting in infection. Conidia are disseminated by rain splashes but these may also spread by mechanical means when the foliage is wet.

Epidemiology

The disease is favoured by cool and wet weather. Saturated atmosphere is essential for the conidial formation, germination and infection. Optimum conditions for germination and infection are 6 h of continuous leaf wetness at 20–22ºC temperature (Patterson and Grogan, 1991). Conidia are disseminated by rain splashes, but these may also spread by mechanical means when the foliage is wet.

Management

(i) Cultural practices: Cultural practices like crop rotation, ploughing deep to bury the plant debris and microsclerotia and destruction of the plants of wild lettuce from in and around the field should be followed to reduce the initial inoculum. The disease can be contained or minimized by planting lettuce during dry periods and avoiding sprinkler irrigation (Patterson *et al.*, 1986). Wider row to row and plant to plant spacing should be maintained to minimize the wetness of leaves and to avoid physical contact with the plants when the canopy is wet.

(ii) Host resistance: All cvs. *of L. sativa* were found susceptible to this disease while chicory, endive, *L. saligna* accessions were immune (symptomless) and the *L. serriola* accession was highly resistant (Galea and Price, 1988). These studies clearly indicate that *L. saligna, L. serriola,* endive and chicory are possible sources of genes for resistance to *M. panattonianum. L. saligna* line UC83US1 was resistant to all 5 races and can also be used in breeding programmes (Ochoa *et al.*, 1987).

(iii) Chemical control: Several fungicides like mancozeb, thiram, captafol, copper hydroxide, prochloraz, chlorothalonil, propiconazole, flutriafol, kresoxim-methyl, benomyl are reported effective against this disease (Parman *et al.*, 1991; Wicks *et al.*, 1994a; Kranjaja *et al.*, 1999). The overall most effective fungicide was prochloraz, which exhibited protectant and translaminar activity, decreased field disease expression and increased yields (Parman *et al.*, 1991).

2. POWDERY MILDEW

This disease is generally considered a minor problem but under certain conditions and particularly in protected environment it may cause substantial reductions in yield and quality of lettuce.

Geographical distribution

The occurrence of this disease has been reported from USA, Canada, Israel, France, Greece, Switzerland, Russia, Czechoslovakia, Germany and United Kingdom (Crute and Burns, 1983; Lebeda, 1985, Klemn, 1985; Dhanvantri and Jarvis, 1985).

Losses

The disease generally causes considerable yield losses when it appears in severe form (Klemn, 1985) and reduce marketable produce.

Symptoms

The symptoms of the disease appear as white lesions on seedlings, mature leaves and stems, which are initially discrete but gradually enlarge and coalesce. Under conditions of heavy infection the undersurface of the leaf will also be colonized. Affected foliage loses luster, curls, becomes chlorotic, then brown and dies.

The Pathogen

The fungus responsible for this disease is *Erysiphe cichoracearum* DC. The mycelium is well-developed, with conidia borne in long chains which are elliptical to barrel-shaped and measure 25–45 x 14–26 μm in size. In autumn during favourable weather conditions, large cleistothecia (90–135 μm in diam.) are formed with myceloid appendages. Asci are pedicellate, ovate to broadly ovate or ellipsoid and borne on a short stalk. The ascospores are one celled, hyaline oval to sub-cylindrical.

Disease cycle

Wherever cleistothecia are formed, the ascospores released from asci form the primary inoculum for infection. But where the cleistothecia are not formed, the fungus overwinters on the wild hosts in conidial stage. *Lactuca serriola* is the only other natural host found in the field (Schnathorst *et al.*, 1958). Wind borne conidia also cause primary infection. Such conidia are easily dislodged during periods of low humidity and are disseminated by wind over long distances.

Epidemiology

Wherever cleistothecia are produced, they explain the disease cycle of the pathogen. Otherwise the wind blown conidia from *L. serriola* cause primary infections. Most spores are disseminated between noon and 4 p.m. Minimum temperature for conidial germination is 6–10ºC and the maximum is 27ºC, while optimum is 18ºC. Highest percentage of conidial germination takes place with a relative humidity of 95–98 per cent, but a totally saturated atmosphere inhibits germination.

Management

(i) Cultural practices: Cultural practices like crop rotation, collection and destruction of infected plant material and a general hygiene should be followed to keep the disease under check. Periodic removal of older leaves help in reducing the initial inoculum of the pathogen.

(ii) Host resistance: Wild *Lactuca* species like *L. serriola* (PI 255665), *L. saligna* (LSA/92/1 and LSA/92/2), *L. virosa* (LVIR/26 and LVIR/57/1) and *L. dentate* (PI 234204) were completely free from powdery mildew infection and these can be used for breeding resistant cultivars (Lebeda, 1985). *L. saligna* and *L. virosa* can also be utilized as sources of resistance against this disease (Labeda and Mieslerova, 2011).

(iii) Chemical control: Several systemic and non-systemic fungicides like wettable sulphur, dinocap, carbendazim, hexaconazole, penconazole, difenconazole etc. can be applied for the effective control of this disease. Recently available strobilurin fungicides like azoxystrobin alone or in combination with difenoconazole can also be applied for effective control of this disease.

3. DOWNY MILDEW

This is a major disease of glasshouse and field grown lettuce crops the world over where there is adequate moisture and moderate to low temperature is prevalent.

Geographical distribution

Downy mildew of lettuce is known to occur in all parts of the world including Western Europe, USA, Israel and South Africa (Grogan *et al.*, 1955; Crute, 1987; Achar, 1996).

Losses

The disease can cause considerable damage to lettuce planted during periods of prolonged wetness from rain, dew, fog or sprinkler irrigation. Severely infected plants may require excessive trimming during packing, which increases production costs. Invasion of downy mildew lesions by secondary soft rot organisms is another major source of crop loss in the field or in transit and storage (Grogan *et al.*, 1955). The disease may cause significant post harvest losses by causing decay in transit and stores thus ultimately reducing the market value (Crute and Dixon, 1981).

Symptoms

Symptoms of the disease appear as pale green to yellow spots on the upper side of the infected leaf, which may coalesce to affect large areas. The downy growth of the fungus appears on the lower surface of the leaf under high humidity and low temperature conditions. Older lesions often become necrotic or translucent after invasion of infected tissue by secondary saprophytes. Sometimes the early infection of seedlings may lead to systemic infection that results in dark brown discolouration of vascular stem tissues (Davis *et al.*, 1997). Sometimes the disease may appear in the seedling stage and in favourable weather, cotyledons and primary leaves are completely covered with sporangiophores and sporangia of the fungus and such seedlings are killed immediately.

The Pathogen

The disease is caused by ***Bremia lactucae*** Regel. The fungus is an obligate parasite. Mycelium is coenocytic. One to three sporangiophores are formed from the mycelium in the sub-stomatal cavities. Branching of the sporangiophores is dichotomous or, rarely, trichotomous. At the tip of each branch sterigmata are formed, each with one sporangium. Sporangiophore length varies with environmental humidity from 200 to 1200 μm. The sporangia are hyaline, ovoid-ellipsoidal to globose and are 12–31 μm long and 11–27.5 μm wide. The fungus is mostly heterothallic *i.e.*, two mating types (B1 and B2) are needed for oospore production, although homothallic isolates have also been found (Davis *et al.*, 1997). However, oospores have rarely been observed in field-grown lettuce, lettuce debris or wild hosts. This may be due to the presence of only one of the two mating types. Oospores are abundantly produced in cotyledons or leaves artificially inoculated with two mating types. Oogonia and antheridia are produced about 4 days after inoculation, and oospores mature 10 days later. Oospores are spherical (27–30 μm in dia.) and are surrounded by a thick wall. Germination of oospores has only occasionally been achieved under controlled conditions after 1–6 months of storage in water and rotting debris.

A number of races in *B. lactucae* have been reported. There is considerable variation for specific virulence in *B. lactucae* populations and the use of new combinations for R-factors resulted in the increased frequency of pathotypes carrying matching virulences. Based on reaction of differential cultivars to various isolates of *B. lactucae*, 46 different virulence pathotypes possessing 2–12 (mostly 5–6) virulence factors have been reported (Thinggard, 1985). Since *B. lactucae* is predominantly heterothallic, it is likely that sexual recombinations are largely responsible for variations. Fifteen pathotypes of the fungus have been described in Europe (Giovannantonio *et al.*, 1998).

Host Range

The pathogen can infect chicory and numerous other compositae family plants besides lettuce (Garibaldi *et al.*, 2003). *L. serriole* is the most common wild host that may harbour *B. lactucae*.

Disease cycle

The pathogen can overwinter either as oospores in plant debris or in seed or wild *Lactuca* spp. Since both mating types are present in many parts of the world, oospores produced in lettuce debris may serve as the source of primary inoculum. Seed borne inoculum also forms an important part of primary inoculum. This disease is more severe on cotyledons and the initial infection may lead to systemic infection of the plants. *L. serriola* is the most common wild host that may harbour *B. lactucae*. Sporangia from the infected continuous and overlapping lettuce crops from nearby fields can be another source of primary inoculum. Secondary spread of the pathogen is through air borne sporangia.

Epidemiology

The pathogen overwinters either in the plant debris as oospores or the inoculum from wild hosts cause primary infections. Sporulation is affected by temperature with an optimum at 15ºC and by relative humidity, with sporulation increasing markedly at RH ≥ 90 per cent (Su *et al.*, 2004). Both RH and wind speed affected the number of plants with sporulation and the number of sporangia produced/plant. They further reported that sporulation may be affected by stomatal aperture in response to RH, as more closed stomata and correspondingly fewer sporangiophores are present at low RH. Dispersal of sporangia usually starts at sunrise with peaks between 10.00 a.m. and noon. Sporangia survived much longer at 23ºC (>12 h) than at 31ºC (2 to 5 h), regardless of RH (35 to 76%) (Wu *et al.*, 2000). Germination percentage was significantly reduced after exposure to 50 and 100 per cent sunlight. UVB significantly reduced sporangium viability, while fluorescent light and UVA had no effect relative to incubation in the dark. Solar radiation is the dominant factor determining survival of *B. lactucae* sporangia. The infection can be initiated within 3 h under optimum temperatures (10–22°C) in the presence of free water or relative humidity near saturation. The minimum and maximum temperatures for infection are –5°C and close to 30°C, respectively. Irrigation practices also play an important role in the development of epidemics of this disease. Generally, more disease is observed under sprinkler and under furrow irrigations than subsurface drip irrigation (Scherm and van Bruggen, 1995) because of the longer leaf wetness duration, increased canopy air humidity, and/or enhanced dispersal of inoculum.

Kushalappa (2001) developed a forecasting system "BREMCAST" to predict risk levels of downy mildew of lettuce. The criteria for the forecasting system were derived from the relationships of weather variables to sporulation, dissemination and infection processes of *B. lactucae*. In addition, the presence of inoculum source, or the disease, in the field was taken into consideration. The system predicted the occurrence of downy mildew on 84 and 80 per cent of the days under commercial and field plot conditions, respectively.

Management

(i) Cultural practices: The disease can be kept under check by destruction of infected plant debris, by using healthy seed and removal of wild *Lactuca* spp. from in and around the fields. Subsurface drip irrigation results in less conducive environment for downy mildew than furrow or sprinkler irrigation (Scherm and van Bruggen, 1995).

(ii) Host resistance: Lettuce downy mildew can be best managed by the use of resistant varieties. Resistant varieties have either specific host resistance genes (single dominant gene) or moderate level of field resistance. Accessions like 490999, 491204, 491205, 491207 and 401208 of *L. saligna* and 491178 of *L. serriola* (Gustafsson, 1989) have high resistance. High resistance was also observed in lettuce cultivars like Saffier and Mariska. Sixteen commercial lettuce cultivars resistant to all known races of *B. lactucae* were evaluated under greenhouse condition in Italy for resistance to pathotypes present in the area, all the cultivars were found resistant to the prevalent pathotypes (Giovannantonio *et al.*, 1998). However, small necrotic spots caused by the fungus were observed on the leaves of cv. Ninja and Bra 1939, while no symptoms were observed on the other 14 cultivars.

(iii) Chemical control: Before the introduction of systemic fungicides towards the end of 1970, copper-based fungicides and dithiacarbamates provided effective control of this disease, but they had certain limitations. These are surface-acting protectant fungicides and often fail to provide adequate control because it is difficult to achieve a continuous and sufficiently uniform presence of the chemical on all leaf surfaces exposed to spore deposition. Newly produced foliage of a rapidly growing crop such as lettuce is often left unprotected between applications and substantial amounts of chemical can be lost in irrigation or rain. Besides this they have some advantage that insensitive forms of the pathogen have not emerged to negate their effectiveness. Three distinct groups of chemical fungicides like phenylamide group (metalaxyl, benalaxyl, cyprofuram, oxadixyl and ofurace), fosetyl-Al and propamocarb having genuine systemic activity against *B. lactucae* were introduced during the 1970's. Out of these, metalaxyl provided an excellent control of this disease and this fungicide can be applied to foliage or via roots. These fungicides also have good curative properties and the capability of eradicating established infections. However, with the continuous use of these fungicides, pathotypes of *B. lactucae* have emerged which are insensitive to metalaxyl and fosetyl –Al and have resulted in control failure in many countries of the world (Maisonneuve *et al.*, 1989; Schettini *et al.*, 1991; Wicks *et al.*, 1994b; Cobelli *et al.*, 1998; Brown *et al.*, 2004). The metalaxyl resistant strains can be managed by mixtures of either dimethomorph or phosphonic acid with mancozeb effectively (Wicks *et al.*, 1994b). Recently available fungicides like azoxystrobin, Previcur N, and cymoxanil have also been found effective in the management of this disease (Paaske, 2000; Minuto *et al.*, 2000).

Use of lettuce resistant cultivars supplemented with sprays of either dimethomorph (0.75 g.a.i./l) or phosphoric acid (2.4 g.a.i/l) not only prolong the life of cultivars but also extend the effectiveness of fungicides (Wicks *et al.*, 1994a). Disease forecasting systems if followed can also help in keeping the disease in check.

4. SCLEROTINIA LETTUCE DROP

This is one of the most widespread and destructive diseases of this crop (Patterson *et al.*, 1986) particularly in those areas that have cool, moist environmental conditions.

Geographical distribution

Stevens and Hall (1911) first reported the occurrence of this disease in 1890 on greenhouse grown lettuce in Massachusetts. Later on, the occurrence of the disease has been reported from cool, moist regions of the world like USA, UK, Australia, Canada, Netherlands, Brazil, China, Israel, New Zealand, Venezuela and other parts of the world. However from India, the occurrence of this disease on lettuce was first recorded in 2014 from Himachal Pradesh (Bharat *et al.*, 2014).

Losses

The disease is responsible for rapid wilting and dying of infected plants. The incidence of the disease varies from season to season and ranges from few infected plants to >70 per cent in some fields depending on conditions such as cropping history, irrigation and tillage methods and other environmental and cultural characteristics of the field, however, in some cases the entire field may be lost (Davis *et al.*, 1997).

Symptoms

Lettuce drop infection usually occurs in two phases. The first phase of infection occurs 3–4 weeks after seedling emergence. As the chance of contact between sclerotia and lettuce at the seedling stage is low, only a very low percentage of plants become infected during this first phase. The second, economically important infection phase occurs near crop maturity and a much higher percentage of plants become infected. Hyphae that emerge from soil borne sclerotia initiate the infection on the outermost layers of leaves contacting the soil surface and subsequently, the infected leaves show initial symptom of wilting and giving the plant a stressed appearance. As the infection progresses, the mycelium of the pathogen grow inward from outermost leaf layers to other leaf layers and thus the entire plant wilts (Plate 3).

The layers of collapsed leaves lay flat on the soil surface and turn yellowish in colour. A soft watery decay develops on both above and below ground plant parts which differentiates lettuce drop infection from lettuce-root aphid infestation (Shrestha *et al.*, 2015). Under moist conditions, the fungus produces a snowy white mycelium across the infected tissue and black coloured sclerotia on infected plant parts. Usually sclerotia are produced on the lower surface of the leaves touching the soil, around the crown and throughout the root. The size of the sclerotia produced on the infected plant parts vary depending on the fungal species and temperature during the sclerotial formation.

The Pathogen(s)

Two different species of *Sclerotinia i.e., S. minor* Jagger and *S. sclerotiorum* (Lib.) de Bary are known to cause lettuce drop worldwide, however, *Sclerotinia nivalis* has also been found to be associated with lettuce crop from central China. Both species are widely distributed wherever lettuce is grown, but only one species or the other causes losses in most locations. In India, *S. sclerotiorum* is the pathogen involved with lettuce drop disease (Bharat *et al.*, 2014). Mycelium is hyaline, much branched, consisting of large, closely septate hyphae which are intercellular and intracellular and invade all the tissues of the host. The hyphae are 9–18 µm broad and filled with dense protoplasm. Sclerotia vary in size and shape according to environment and location and measure 2.36±1.07 mm in width, 3.33±1.11 mm in length, 2–10 mm in diameter and 6.8±3.7 mg in weight. These sclerotia after germination give rise to several columnar structures called stipes with 2.5–24 mm in length. The apothecial cups are well expanded and measure 1.5–10 mm in diameter. Ascospores are discharged in abundance from these cups. The ascospores are always eight in each ascus, which has an apical pore through which spore discharge occurs with violence. The asci varied from 81.0 to 199.4 µm in length and 4.3 to 12.4 µm in width and the ascospores from 6.0–17.0 µm in length by 2.0 to 8.4 µm in width which are non-septate, uniseriate, hyaline and elliptical.

Microconidia (spermatia) are produced on sclerotia, on the discs of mature over apothecia and in culture. These spermatia are formed in chains at the tips of short lateral branches of the vegetative mycelium. The spermatia germinate very sparsely in water and culture media and apparently do not serve as source of infection or dissemination of the fungus.

In *S. minor*, the sclerotia are small (0.5–3.0 mm), mostly circular and uniform in diameter. Carpogenic germination and production of apothecia are extremely rare in *S. minor* rather sclerotia germinate eruptively.

Host range

Both species of *Sclerotinia* have a wide host range and infect herbaceous and woody plant species in 64 families and 225 genera (Patterson *et al.*, 1986).

Disease cycle

Both species of *Sclerotinia* overwinter in the form of sclerotia in soil. The sclerotium of *S. minor* directly infect lettuce roots, crowns and leaves on or near the soil surface by eruptive myceliogenic germination (a mass of hyphae emerging from the sclerotium) and a senescent food base is not required for infection (Patterson *et al.*, 1986). In case of *S. sclerotiorum*, sclerotia 2–3 cm below soil surface germinate carpogenically when soil moisture is maintained near saturation for two or more weeks. The small tan, cup-shaped apothecia produce and release millions of airborne ascospores that are dispersed by wind throughout the field. The ascospores germinate in free water from dew, fog, rain or sprinkler irrigation. Infection is initiated on senescent or injured tissues, usually on lower leaves beginning of senescence near harvesting time or leaves injured mechanically or by insects. Mycelial infection from eruptively germinating *S. sclerotiorum* is rare in nature (Davis *et al.*, 1997).

Epidemiology

The disease is favoured by high moisture content and presence of sclerotial inoculum in the soils (Adams and Tate, 1975). Significant difference exists in the epidemiology of both species of *Sclerotinia* (Patterson and Grogan, 1985b). The optimum temperature for carpogenic germination of sclerotia is 15–20°C (Clarkson *et al.*, 2004). Lettuce drop normally develops after plants had begun to head and when periods of rainfall of 7–46 mm resulted in moist soil conditions that coincided with 4–10 days of daily minimum and maximum air temperature of 5–18 and 14–25°C, respectively. High temperatures, such as 37°C, for extended periods of time reduce sclerotial survival and wet soil combined with high temperatures severely reduces viability. Sclerotia buried at a depth of 20 cm for over 18 months survived, while surface sclerotia only survived for 6 months. Mila and Yang (2008) tested temperature fluctuations of 4, 8, 12, and 16°C around a median of 20°C and a constant of 20°C. Daily temperature fluctuations of 8°C resulted in highest levels of sclerotial germination and apothecial production.

Management

(i) Cultural practices: Cultural practices like crop rotation, deep ploughing during summer months (Subbarao *et al.*, 1996), collection and destruction of infected plant material etc. should be followed to keep the initial inoculum of the pathogen low. Incidence of the disease was less in those fields where alternate paddy and dry land cropping was followed (Li and Wang, 1994). Flooding of the field before raising lettuce is also recommended. Soil amendment with stable manure, fowl manure and lucerne hay were also found effective in reducing the infection of this disease (Asirifi *et al.*, 1994). Subsurface drip irrigation should be followed to keep the disease under check. Irrigation of the planting surface should be avoided because wet soil probably favours germination of sclerotia on or near the surface (Patterson *et al.*, 1986).

Host Resistance: Among the methods of disease control, host resistance generally offers the most predictable, convenient and least expensive option for managing disease. However, there are currently no commercial lettuce cultivars with resistance to *Sclerotinia* spp. Many genotypes exhibited partial resistance against *S. minor* but none exhibited complete resistance (Grube and Ryder, 2004).

Biological control: Some fungi and bacteria colonize and decay sclerotia in soil and a few of these organisms show potential as biological control agents. *Coniothyrium minitans* B 91 reduced sclerotial viability, their germination and formation of apothecia under *in vitro* and *in vivo* conditions (Cassiolato, 1998; Jones and Whips,

2002). *Trichoderma harzianum* has also been reported to be the potential antagonist of *S. sclerotiorum* (Inbar *et al.*, 1996). Dense coils of hyphae of *T. harzianum* around *S. sclerotiorum* hyphae caused partial degradation of *Sclerotinia* cell wall. Coating of lettuce seeds with *T. harzianum* conidia and seedling treatment with peat-bran preparation reduced pre-and post-emergence mortality and the disease under green house and field conditions. Smith *et al.* (2013) collected twenty-one isolates of *Trichoderma* sp. and out of which *T. asperellum, T. atroviride* and *T. harzianum* prevented germination of more than 70 per cent of sclerotia of *S. sclerotiorum* in bioassay tests.

Chemical control: Until the mid-1970s, only one fungicide, 2,6–dichloro-4–nitroaniline (DCNA) was used for the control of lettuce drop. Several systemic and non-systemic fungicides like ciprodinil + fludioxanil, procymidone + thiram (Minuto *et al.*, 1999), fluazinam, fludioxanil, vinclozolin, boscalid (Matheron and Porchas, 2004) have been reported to combat the ravages of this disease. Foliar fungicide sprays during the rosette stage of growth are required to protect senescing lower leaves from infection by ascospores from outside (Patterson *et al.*, 1986). Foliar application of the mixture of procymidone + thiram significantly controlled this disease (Minuto *et al.*, 1999a). Later, other fungicides such as benomyl, iprodione and vinclozolin sprayed over the bed and at the base of plants immediately after lettuce thinning, dramatically improved lettuce drop control. Drenching with carbendazim is effective in management of this disease as compared to foliar sprays.

(B) BACTERIAL DISEASES

1. SOFT ROT

It is an important disease of this crop in field, transit and market where it causes considerable losses.

Geographical distribution

The bacterium is omnipresent and causes soft rot of vegetables including lettuce throughout the world. The occurrence of this disease has been reported from USA, Hawaii, Jordan etc. (Cho, 1983; Rajeh and Hamed, 2000).

Losses

This is one of the serious transit, storage and market diseases of lettuce but heavy losses may also occur in the field, where it causes rapid wilt and collapse of crisp head and other lettuce types. In Hawaii, losses from this disease are as high as > 90 per cent in high rainfall regions (Cho, 1983).

Symptoms

The initial symptom of the disease appears as rapid wilting of outer wrapper leaves under field conditions but the plants at maturity are most susceptible. Wilting is caused by a collapse of vascular tissues, which develop a pinkish to brown discolouration. As the disease progresses, the pith of the stem becomes water-soaked, macerated and greenish. Eventually the entire lettuce head may become slimy.

The Pathogen

The bacterium responsible for this disease is *Erwinia carotovora* subsp. *carotovora* (Jones) Bergey *et al.* It is single celled, Gram-negative, straight rod measuring 0.5–1.0 x 1.0–3.0 μm and is motile by peritrichous flagella. It is facultatively anaerobic, oxidase negative and catalase positive; it reduces nitrate to nitrite, produces hydrogen sulfide from sodium thiosulfate and produces ß-galactosidase, but not urease.

Host Range

The bacterium has very wide host range which includes cauliflower, cabbage, Swiss chard, common mallow, gladiolus, onion, pepper, potato, spinach, squash, sweet melon, tomato etc.

Disease cycle

The bacterium is able to survive in soil, in aquatic environments and on crop residues. They are easily disseminated by insects, splashing water and contaminated equipments. The bacterium generally enters into the plant tissues through wounds or natural openings.

Epidemiology

Bacterium is capable of survival in soil or aquatic environments and on crop residues. The soft rotting bacteria are capable of rapid multiplication under favourable temperature conditions (25–30°C) and latent periods may be as short as 24 h if inoculum concentrations are high (Davis *et al.*, 1997). The disease is often associated with very moist to saturated soil conditions and warm temperatures. Lettuce at or near maturity appears to be much more susceptible than lettuce at an early growth stage, presumably because of excessive moisture levels remaining in the canopy for longer periods of time. Post-harvest infections are frequently related to improper handling or storage conditions. Excessive wounding during harvest or transport, contaminated water used in cooling or washing or higher than recommended storage temperatures are the factors commonly associated with disease outbreak in transit or storage.

Management

Cultural practices like crop rotation, use of furrow or drip irrigation than sprinkler irrigation, good soil drainage and application of balanced doses of fertilizers to maintain good vigour of the plants should be followed. Rapid removal of field heat through vaccum cooling, transit temperature below 5°C and extensive use of refrigeration by wholesellers and retailers have been proved successful in reducing the disease (Cho and Telede, 1976). Efforts should be made to avoid injury during harvest and transit. Protective sprays of copper fungicides also reduce infection of this disease. Tolerant/ resistant cultivars of lettuce if available should be used to reduce the infection of this disease.

(C) VIRAL DISEASES

1. LETTUCE MOSAIC

It is one of the most common and destructive diseases of lettuce the world over which causes substantial yield losses.

Geographical distribution

The disease was first described in 1921 in Florida, USA (Jagger, 1921). Subsequently it has been reported from Austria, Denmark, England, France, Germany, Holland, Hungry, Italy, Portugal, Switzerland, Mexico, Argentina, Brazil, Uruguay, Bermuda, Ghana, Zimbabwe, Sierra Leone, Tanzania, China, Japan, Syria, Australia, Tasmania, New Zealand, Egypt, Israel, Jordan, Iraq, Iran, Yemen, Finland, Spain, Yugoslavia, Czechoslovakia, Malaysia and Afghanistan (Dinant and Lot, 1992).

Losses

This is potentially a highly destructive disease of lettuce (Grogan, 1980). About 25 per cent infected plants due to this disease in green house have been recorded in Poland (Fiedorow and Andrzejczak, 2002). The disease causes severe losses mainly in field crops but the virus may be significant in early greenhouse winter crops when seedlings are not raised under insect proof conditions and are exposed to infection (Dinant and Lot, 1992). Symptoms like growth reduction, failure to 'heart', sometimes necrosis and/or yellowing are responsible for crop losses due to the large number of unmarketable plants. Due to mosaic infection, number of seeds/ plant and growth of seed stems is decreased (Fiedorow and Andrzejczak, 2002).

Symptoms

The symptoms of this disease on young plants appear as a slight inward rolling along the axis of the leaf. The first true leaves of infected plants are often irregular and slightly lobed. These symptoms are accompanied or followed by light green to yellow mottling. In the young rosette stage of growth, infected plants may show vein clearing and bronzing. Mottling symptoms are usually indistinct in older plants. Half grown to mature plants are severely stunted and dull green to light yellow. The tips of the outer leaves roll downward and the plant appears wilted.

The Virus

The virus responsible for this disease is *Lettuce Mosaic Potyvirus* (LMV), a typical member of the Potyviridae family of plant viruses. The virus particles are flexuous rods that measure approximately 750 x 13 nm. The genome of LMV is composed of one large, positive-sense, single stranded RNA. LMV isolates can be divided into four distinct pathotypes based on their ability to infect a series of differential lettuce cultivars (Davis *et al.*, 1997). Pathotypes I and II are unable to overcome the *mo* and *g* LMV resistance genes while III and IV can infect lettuce cultivars possessing the *mo* and *g* resistance genes. Pathotypes III and IV have been described from Europe while Pathotype II is prevalent in USA.

Host range

The virus has very wide host range including 20 genera in ten families. Lettuce is the main cultivated species getting infected. Other cultivated hosts getting infected to different levels are endive, escarole, spinach, pea, chickpea and safflower. Some natural hosts include common weeds such as *Senecia vulgaris, Sonchus asper, Lactuca* sp., *Helminthia echioides, Capsella bursa-pastoris, Lamium amplexicaule, Stellaria media* and *Anagallis arvensis.*

Transmission

The LMV is seed-borne in lettuce. The rate of seed-borne infection depends on the time of infection of the mother plant, variety and the environmental conditions (Dinant and Lot, 1992). It was shown that the rate might vary from 1 to 12 per cent in susceptible genotypes. Average seed transmission rate of AF 199 was 16.5 and 11.9 per cent in susceptible and tolerant cultivars (Jadao *et al.*, 2002). LMV is also transmitted in a non-persistent manner by many species of aphids, including *Myzus persicae* and *Aphis gossipii*. The virus is also readily sap-transmissible.

Epidemiology

Primary inoculum sources are infected seed and alternate hosts. LMV epidemics are usually associated with the planting of infected seed. Infected seedlings planted in the field may serve as source of primary inoculum. Within field the virus is transmitted by aphids in non-persistent manner i.e. only a few seconds are needed for the vector to acquire the virus from infected plants and to transmit to healthy plants.

Management

The disease is generally controlled by the use of virus-indexed seed and/or resistant cultivars. The state of Florida and two counties in California have adopted LMV-free seed programmes. All lettuce seed to be planted in these areas must pass an assay for LMV before being distributed for sale. If one seed in 30,000 is infected with LMV, the seed lot can not be sold or grown in the areas that have adopted the virus-free seed programme. Treatment of seeds with hot air at 80°C for 3 days (Howles, 1978) and incubation of seeds at 40°C for 6–10 days in polyethylene glycol, inactivates the virus (Walkey and Dance, 1979). Aphid transmission of the virus can be checked by using a suitable insecticide. Eradication of insect carriers, removal of potential weed protected structures hosts of the virus and ploughing of fields immediately after harvest are the important tools for the management of this disease. Resistant crisp-head lettuce varieties like Salinas 88 and Vanguard 75 have been developed in the USA and numerous resistant crisp-head, leaf, and romaine lettuce varieties have been developed in Europe, which can be used in the management of this disease.

REFERENCES

Achar, P.N. 1996. First report of downy mildew diseases in lettuce caused by *Bremia lactucae* in Natal, Southern Africa. *Plant Dis.* **80:** 464.

Adams, P.B. and Tate, C.J. 1975. Factors affecting lettuce drop caused by *Sclerotinia sclerotiorum. Pl. Dis. Reptr.* **59:** 140–143.

Akashi, K., Maeda, K. and Abe, H. 1986. Studies on root diseases of spinach and soil scientific research on the occurrence II: Pathogens of damping-off occurring in fields around Sapporo city. *Bull. Hokkaido Perfectu. Agric. Exp. Stn.* **54:** 1–8.

Ali, S.A., Saraf, R.K. and Pathak, R.K. 1999. Efficacy of fungicides in controlling powdery mildew of coriander (*Coriandrum sativum* L.). *J. Soils and Crops* **9:** 266–267.

Anonymous, 1952. Studies in the wilt disease of coriander in the Madhya Bharat. *Indian Council of Agricultural Research, New Delhi.*

Asirifi, K.N., Morgan, W.C. and Parbery, D.G. 1994. Suppression of Sclerotinia soft rot of lettuce with organic soil amendments. *Aust. J. Exp. Agric.* **34:** 131–136.

Bag, T.K. 1998. Stem gall of coriander – A new record from Arunachal Pradesh. *Indian J. Hill Fmg.* **11(1–2):** 119.

Bacigalova K, Lopandic, K, Rodrigues, MG, Fonseca, A, Herzberg, M, Pinsker, W and Prillinger, H. 2003. Phenotypic and genotypic identification and phylogenetic characterization of *Taphrina* fungi on alder. *Mycological Progress* **2:**179–196.

Bharat, NK, Thakur M, Bhardwaj, RK and Kumar, S. 2014. Occurrence of Sclerotinia rot and Botrytis gray mould on lettuce under protected conditions in Himachal Pradesh. *Int. J. Farm Sci.* **4(2):** 86–91.

Bowers, J.L. and Goode, M.J. 1980. Ozarka and Greenvalley: new disease resistant spinach cultivars. *Arkansas Farm Res.* **29:** 2, 6.

Bradenberger, L.P., Correll, J.C. and Morelock, T.E. 1991. Identification and cultivar reactions to a new race (race 4) of *Peronospora farinose* f.sp. *spinaciae* on spinach in the United States. *Plant Dis.* **75:** 630–634.

Bradenberger, L.P., Morelock, T.E. and Correll, J.C. 1992. Evaluation of spinach germplasm for resistance to a new (race 4) of *Peronospora farinose* f.sp. *spinaciae. Hortic. Sci.* **27:** 1118–1119.

Brandenberger, L.P., Correll, J.C. Morelock, T.E. and McNew, R.W. 1994. Characterization of resistance of spinach to white rust (*Albugo occidentalis*) and downy mildew (*Peronospora farinose* f.sp. spinaciae) *Phytopathology* **84:** 431–437.

Brown, S., Koike, S.T., Ochoa, O.E., Laemmlen, F. and Michelmore, R.W. 2004. Insensitivity to the fungicide fosetyl-aluminium in California isolates of the lettuce downy mildew pathogen, *Bremia lactucae. Plant Dis.* **88:** 502–508.

Byford, W.J. 1981. Downy mildews of beet and spinach. Pages 531–543. In: *The Downy Mildews.* (ed. D.M. Spencer), Academic Press, New York.

Cassiolato, A.M.R. 1998. Effect of *Coniothyrium minitans* on carpogenic germination and viability of *Sclerotinia sclerotiorum* sclerotia. *Fitopatol. Brasil.* **23:** 474–476.

Cho, J.J. 1983. Soft rot of crisphead lettuce incited by *Erwinia carotovora* subsp. *carotovora* in Hawaii. *Phytopathology* **73:** 1206–1209.

Cho, J.J. and Telede, K.Y. 1976. Evaluation of lettuce varieties for resistance to bacterial soft rot. *Hort. Digest* **35:** 4–5.

Clakson,, J.P., Phelps, K., Whipps, J.M., Young, C.S., Smith, J.A. and Watling, M. 2004. Forecasting Sclerotinia disease of lettuce: toward developing a prediction model for carpogenic germination of sclerotia. *Phytopathology* **94:** 268–279.

Cobelli, L., Collina, M. and Brunelli, A. 1998. Occurrence in Italy and characteristics of lettuce downy mildew (*Bremia lactucae*) resistant to phenylamide fungicides. *European J. Pl. Pathol.* **104:** 449–455.

Correll, J.C., Morelock, T.E., Black, M.C., Koike, S.T., Brandenberger, L.P. and Dainello, F.J. 1994. Economically important diseases of Spinach. *Plant Dis.* **78:** 653– 660.

Crute, I.R. 1987. The occurrence, characteristics, distribution, genetics and control of a metalaxyl-resistant pathotypes of *Bremia lactucae* in the United Kingdom. *Plant Dis.* **71:** 763–767.

Crute, I.R. and Burns, I.G. 1983. Powdery mildew of lettuce (*Lactuca sativa*). *Plant Pathol.* **32:** 455–457.

Crute, I.R. and Dixon, R. 1981. Downy mildew diseases caused by the genus *Bremia lactucae* Regel. In: *The Downy Mildews* (ed. Spencer, D.M.). Academic Press, London. Pp. 422–460.

Dainello, F.J. and Jones, R.K. 1986. Evaluation of use-patteren alternatives with metalaxyl to control foliar diseases of spinach. *Plant Dis.* **70:** 240–242.

Dainello, F.J., Heineman, R.R. and Black, M.C. 1984. Evaluation of spinach varieties and breeding lines for horticultural characteristics and resistance to white rust disease (*Albugo occidentalis*). *Texas Agric. Exp. Stn. Prog. Rep.* 4252.

Dainello, F.J., Heineman, R.R. and Jones, R.K. 1981. Relative white rust resistance and adaptability of spinach varieties in southwest Texas. *Publ. Texas Agric. Exp. Stn. No.* 3878, 7pp.

Dange, S.R.S. and Patel, P.N. 1968. Influence of nutrition and pH on growth and sporulation of *Cercospora beticola* Sacc. from spinach-beet (*Beta vulgaris* L.). *Indian Phytopath.* **21:** 434–439.

Davis, R.M., Subbarao, K.V., Raid, R.N. and Kurtz, E.A. 1997. *Compendium of Lettuce Diseases*. APS Press, St. Paul Minnesota. 79pp.

Dhanvantri, B.N. and Jarvis, W.R. 1985. Powdery mildew (*Erysiphe cichoracearum*) of greenhouse lettuce in Ontario. *Plant Dis.* **69:** 177.

Dinant, S. and Lot, H. 1992. Lettuce mosaic virus. *Plant Pathol.* **41:** 528–542.

Fiedorow, Z. and Andrzejczak, A. 2002. Susceptibility of lettuce cultivars to lettuce mosaic virus and cucumber mosaic virus and transmission of the viruses by lettuce seeds. *Prog. Pl. Prot.* **42:** 900–902.

Galea, V.J. and Price, T.V. 1988. Survival of the lettuce anthracnose fungus (*Microdochium panattonianum*) in Victoria. *Plant Pathol.* **37:** 54–63.

Garibaldi, A., Minuto, A., Gilardi, G. and Gullino, M.L. 2003. First report of *Bremia lactucae* causing downy mildew on *Helichrysum bracteatum* in Italy. *Plant Dis.* **87:** 315.

Giovannantonio, C. di, Ansanelli, C., Vecchio, M., Infantino, A and Di-Giovannan-tonio, C. 1998. Control of downy mildew of lettuce using resistant cultivars. *Coltura-Protette* **27(2):** 91–94.

Grogan, R.G. 1980. Control of lettuce mosaic with virus free seed. *Plant Dis.* **64:** 446–449.

Grogan, R.G., Snyder, W.C. and Bardin, R. 1955. Diseases of lettuce. *Univ. Calif. Div. Agric. Sci. Circ.* 448.

Grube R and Ryder EJ. 2004. Identification of lettuce (*Lactuca sativa* L.) germplasm with genetic resistance to drop caused by *Sclerotinia minor*. *J. Am. Soc. Hortic. Sci.* **129(1):** 70–76.

Gupta, R.N. 1973. Longevity of chlamydospores of coriander gall fungus. *Indian Phytopath.* **26:** 581–582.

Gupta, R.N. and Sinha, S. 1973. Varietal field trials in the control of stem gall disease of coriander. *Indian Phytopath.* **25:** 337–340.

Gustafsson, I. 1989. Potential sources of resistance to lettuce downy mildew (*Bremia lactucae*) in different *Lactuca* species. *Euphytica* **40:** 227–232.

Horita, H., Yasuoka, S., Ozawa, S. and Sato, M. 2001. Anthracnose of lettuce caused by *Microdochiam panattoninum* in Hokkaido. *52nd Annu. Rep. Soc. Pl. Prot. North Jpn.* 42–44.

Howles, R. 1978. Inactivation of lettuce mosaic virus by dry heat treatment. *J. Aust. Inst. Agric. Sci.* **44:** 131–132.

Hubbeling, N. and Ester, A. 1978. Testing spinach for resistance to downy mildew. *Zaadbelangen* **32(2):** 36–38.

Inaba, T., Takahashi, K. and Morinaka, T. 1983. Seed transmission of spinach downy mildew. *Plant Dis.* **67:** 1139–1141.

Inbar, J., Menendez, A. and Chet, I. 1996. Hyphal interaction between *Trichoderma harzianum* and *Sclerotinia sclerotiorum* and its role in biological control. *Soil Biol. Biochem.* **28:** 757–763.

Irish, B.M., Correll, J.C. and Morelock, T.E. 2002. The effect of synthetic surfactants on disease severity of white rust on spinach. *Plant Dis.* **86:** 791–796.

Irish, B.M., Correll, J.C., Koike, S.T., Schafer, J. and Morelock, T.E. 2003. Identifi-cation and cultivar reaction to three new races of the spinach downy mildew pathogen from the United States and Europe. *Plant Dis.* **87:** 567–572.

Irish, B.M., Correll, J.C., Raid, R.N. and Morelock, T.E. 2004. First report of *Peronospora farinose* f.sp. *spinaciae* race 5 causing downy mildew on spinach in Florida. *Plant Dis.* **88:** 84.

Irish, B.M., Correll, J.C,, Koike, S.T. and Morelock, T.E. 2007. Three new races of the spinach downy mildew pathogen identified by a modified set of spinach differentials. *Plant Dis.* **91:** 1392–1396.

Jadao, A.S., Pavan, M.A., Silva, N. da and Zerbini, F.M. 2002. Seed transmission of lettuce mosaic virus (LMV) pathotype II and IV in different lettuce genotypes. *Summa Phytopathol.* **28:** 58–61.

Jagger, I.C. 1921. A transmissible mosaic disease of lettuce. *J. Agric. Res.* **20:** 737–741.

Jain, K.L. and Agrawat, J.M., 1978. Control of Fusarium wilt of coriander by systemic fungicides. *Indian J. Mycol. Pl. Pathol.* **8**:30.

Jones, E.E. and Whipps, J.M. 2002. Effect of inoculum rates and sources of *Coniothyrium minitans* on control of *Sclerotinia sclerotiorum* disease in glasshouse lettuce. *European J. Pl. Pathol.* **108:** 527–538.

Jones, R.K. and Dainello, F.J. 1982. Occurrence of race 3 of *Peronospora effusa* on spinach in Texas and identification of sources of resistance. *Plant Dis.* **66:** 1078–1079.

Jones, R.K. and Dainello, F.J. 1983. Efficacy of metalaxyl and metalaxyl tank mixes in controlling *Albugo occidentalis* and *Peronospora effusa* on spinach (*Spinacea oleracea*). *Plant Dis.* **67:** 405–407.

Kalra, A., Gupta, A.K., Katiyar, N., Srivastava, R.K. and Kumar, S. 2003. Screening of *Coriandrum sativum* accessions for seed and essential oil yield and early maturity. *Pl. Genet. Resour. Newsl.* **133:** 19–21.

Kalra, A., Parmesameswaran, T.N. Ravindra, N.S. and Singh, H.B. 2002. Effect of number of timing of application of dinocap on control of powdery mildew and yield of coriander cultivars. *Trop. Agri.* **79:** 7–11.

Kalra, A., Parmesameswaran, T.N. Ravindra, N.S., Rao, M.G. and Kumar, S. 2000. Effects of planting date and dinocap applications on the control of powdery mildew and yields of seed and seed oil in coriander. *J. Agric. Sci.* **135(2):** 193–197.

Klemn, H. 1985 A notable injurious occurrence of powdery mildew (*Erysiphe cichoracearum* DC. Ex Merat) on lettuce (*Lactuca sativa* L.) in the GDR. *Arch. fur. Phytopathol. Und Pflanzenschutz* **21:** 287–295.

Koike, S.T. and Gordon, T.R. 2005. First report of Fusarium wilt of cilantro caused by *Fusarium oxysporum* in California. *Plant Dis.* **89:** 130.

Koike, S.T., Smith, R.F. and Schulbach, K.F. 1992. Resistant cultivars, fungicides combat downy mildew of spinach. *Calif. Agric.* **46:** 29–31.

Krnjaja, V. and Ivanovic, M. 1996. *Microdochium panattonianum* a pathogen of lettuce in Serbia. *Zastita Bilja* **47:** 219–227.

Krnjaja, V., Ivanovic, M. and Vuksa, P. 1999. Efficacy of fungicides against *Microdochium panattonianum* which causes lettuce anthracnose. *Pesticidi* **14:** 177–188.

Krober, H. and Beckmann, E, O. 1973 *Phytophthora cryptogea* as the cause of a root rot of spinach and its dependence on the nature of the soil and the weather conditions. *Phytopathol. Zeitsch.* **78:** 160–169.

Kumar, G., Yadav, S.K., Patel, J.S., Sarkar, A. and Awasthi, L.P. 2014. Management of stem gall disease in coriander using *Pseudomonas* and *Trichoderma* (bioagents) and fungicides. *J. Pure Appl. Microbiol.* **8:** 1–4.

Kushalappa, A.C. 2001. BREMCAST: development of a system to forecast risk levels of downy mildew on lettuce (*Bremia lactucae*). *Int. J. Pest Mgt.* **47:** 1–5.

Lakra, B.S. 1999. Assessment of losses due to stem gall in coriander. *Pl. Dis. Res.* **14:** 85–87.

Lakra, B.S. 2000. Management of stem gall of coriander (*Coriandrum sativum*) incited by *Protomyces macrosporus. Indian J. Agric. Sci.* **70:** 338–340.

Lebeda, A. 1985. Differences in resistance of wild *Lactuca* species to natural infection of lettuce powdery mildew (*Erysiphe cichoracearum*). *Euphytica* **34:** 521–523.

Lebeda, A. 1985. Occurrence of natural powdery mildew (*Erysiphe cichoracearum*) infection in the genus Lactuca in Czechoslovakia. *Acta Phytopathol. Acad. Scientia. Hung.* **20:** 149–162.

Lebeda, A. and Mieslerova, B. 2011. Taxonomy, distribution and biology of lettuce powdery mildew (*Erysiphe cichoracearum*). *Euphytica* **34:** 521–523.

Leskovar, D.I. and Kolenda, K. 2002. Strobilurin + acibenzolar-s-methyl controls white rust without inducing leaf chlorosis in spinach. *Ann. Appl. Biol.* **140:** 171–175.

Li, G.Q. and Wang, D.B. 1994. Preliminary study on occurrence, damage, and infection cycle of *Sclerotinia minor* Jagg on lettuce. *Pl. Prot.* **20(6):** 8–10.

Liu, H.L. and Liu, H.L.1994. Comparison of seed treatments and soil drenches with fungicides for the control of Rhizoctonia diseases of Spinach. *Bull. Taichung Distt. Agric. Imp. Stn.* **43:** 17–24.

Lorenzini, G. and Nali, C. 1994. A new race (race 4) of spinach downy mildew in Italy. *Plant Dis.* **78:** 208.

Madia, M., Gaetan, S. and Reyna, S. 1999. Wilt and crown rot of coriander caused by a complex of *Fusarium* species in Argentina. *Fitopatologia* **34(3):** 155–159.

Maisonneuve, B., Leroux, P. and Bellec, Y. 1989. Resistance to phenylamide fungicides in *Bremia lactucae* (lettuce downy mildew) in France. *ISPP Chem. Cont. Newsl.* **12:** 36.

Malhotra, SK, Kakani, RK, Sharma, YK and Singh, DK. 2016. Ajmer coriander- 1(NRCSS, ACR-1) resistant to stem gal disease- an innovative farming technology. *Indian J. Arecanut, Spices, Medicinal Plants* **18**:1–7.

Malhotra, SK, Vashishtha, BB and Apparao, VV. 2006. Influence of nitrogen, *Azospirillum* spp. and farmyard manure on growth, yield and incidence of stem gall disease of coriander (*Coriandrum sativum* L.). *J. Spices Aromatic Crops* **15:**115–117.

Mall, O.P. 1968. Studies on wilt disease of coriander with special reference to variability of the pathogen. *Indian Phytopath.* **21:** 379–387.

Marte, M. and Cappelli, C. 1979. Attacks of *Marssonina panattoniana* (Berl.) Magn. on lettuce in Perugia province. *Rivista di Patologia Vegetale-IV.* **15:** 117–126.

Matheron, M.E. and Porchas, M. 2004. Activity of boscald, fenhexamid, fluzinam, fludioxonil and vinclozolin on growth of *Sclerotinia minor* and *S. sclerotiorum* and development of lettuce drop. *Plant Dis*. **88:** 665–668.

Mathur, R.L. and Sehgal, S.P. 1964. Control of leaf spot disease of spinach caused by *Cercospora beticola. Indian Phytopath*. **17:** 75–77.

Mila, AL and Yang, XB. 2008. Effect of fluctuating soil temperature and water potential on sclerotia germination and apothecial production of *Sclerotinia sclerotiorum. Plant Dis.* **92:** 78–82.

Minuto, A., Garibaldi, A. Mocioni, M. and Gullino, M.L. 1999a. Chemical control of bottom rot of lettuce caused by Sclerotinia sclerotiorum. *Informat. Fitopatol*. **49(7–8):** 50–54.

Minuto, A., Gilardi, G., Garibaldi, A. and Gullino, M.L. 1999. New fungicides for the control of *Sclerotinia sclerotiorum* and *Botrytis cinerea* on lettuce. *Informat. Fitopatol*. **49(12):** 33–36.

Minuto, A., Gullino, M.L., Bertetti, D. and Garibaldi, A. 2000. Chemical control of downy mildew of lettuce. *Atti Gionnate fitopatologiche, Perugia*, 16–20 April, 2000, pp. 307–312.

Muthulakshmi, P., Chezhiyan, N., Muthukrishnan, K. and Doraisamy, S. 2002. Management of coriander wilt using biocontrol agents. *J. Spices Aromatic crops* **11:** 138–140.

Naqvi, A.Q. and Mahmood, K. 1975. Mosaic disease of spinach-a new record for India. *Indian Phytopath*. **28:** 268–271.

Nishida, H, Blanz, PA and Sugiyama, J. 1993. The higher fungus *Protomyces inouyei* has two group I introns in the 18S rRNA gene. *J. Mol. Evol.* **37**:25–28.

Ochoa, O., Delp, B. and Michelmore, R.W. 1987. Resistance in *Lactuca* spp. to *Microdochium panattoniana* (lettuce anthracnose). *Euphytica* **36:** 609–614.

Osaki, H., Kobatake and Inouye, T. 1979. Yellow vein mosaic of honeysuckle (*Lonicera japonica* Thumb.) a disease caused by Tobacco leaf *curl virus* in Japan. *Ann. Phytopathol. Soc. Japan* **45:** 62–69.

Paaske, K. 2000. Chemical control of lettuce downy mildew. *Vaxtskyddsnotiser* **64(3–4):** 37–43.

Parman, Price, T.V. and Lee, M. 1991. Studies on fungicidal control of lettuce anthracnose. *Aust. Pl. Pathol*. **20:** 103–107.

Parman and Price, T.V. 1991. Production of microsclerotia by *Microdochium panattonianum. Aust. Pl. Pathol.* **20:** 41–46.

Patterson, C.L. and Grogan, R.G. 1985a. Role of microsclerotia as primary inoculum of *Microdochium panattonianum*, incitant of lettuce anthracnose. *Plant Dis.* **75:** 134–138.

Patterson, C.L. and Grogan, R.G. 1985b. Differences in epidemiology and control of lettuce drop caused by *Sclerotinia minor* and *S. sclerotiorum. Plant Dis.* **69:** 766–770.

Patterson, C.L. and Grogan, R.G. 1991. Role of microsclerotia as primary inoculum of *Microdochium panattonianum* indicated of lettuce anthracnose. *Plant Dis.* **75:** 134–138.

Patterson, C.L., Grogan, R.G. and Campbell, R.N. 1986. Economically important diseases of lettuce. *Plant Dis*. **70:** 982–987.

Paul, Y.S. 1992. Studies on seed borne mycoflora of coriander with special reference to stem gall in Himachal Pradesh. *Pl. Dis. Res.* **7:** 19–23.

Pavgi, M.S. and Mukhopadhyay, A.N. 1969. Transmission of *Protomyces macrosporus* Unger, the incitant of stem gall of coriander. *Ann. Phytopathol. Soc. Jpn.* **35:** 265–270.

Pavgi, M.S. and Mukhopadhyay, A.N. 1972. Development of coriander fruit infected by *Protomyces macrosporus* Unger. *Cytologia* **37:** 619–627.

Prakasam, V., Vedamuthu, P.G.B., Khader, M.A. and Jeyarajan, R. 1987. Screening coriander lines for wilt resistance. *South Indian Hortic.* **35:** 258–259.

Prasad, B.K. 1982. Seedborne fungi of coriander, their pathogenicity and control. *Indian J. Mycol. Pl. Pathol.* **12:** 238–239.

Prasad, B.K. 1983. Influence of preharvest rain on the incidence of seed microflora of coriander. *Indian J. Mycol. Pl. Pathol.* **13:** 118.

Raabe, R.D. and Pound, G.S. 1952. Relation of certain environmental factors to initiation and development of the white rust disease of spinach. *Phytopathology* **42:** 448–452.

Rajeh, O.N. and Hamed, K. 2000. Soft rot diseases of vegetables in Jordan: host range, reaction of some potato cultivars to the infection and effect of planting date. *Dirasat Agric. Sci.* **27:** 149–157.

Richards, M.C. 1939. Downy mildew of spinach and its control. *Bull. Cornell Univ. Agric. Exp. Stn.* 718.

Sadravi, M., Etebarian, H.R. and Torabi, M. 2000. Downy mildew of spinach and possible mechanisms of resistance in some spinach cultivars. *Seed and Plant* **15:** 403–412.

Scherm, H. and van Bruggen, A.H.C. 1995. Comparative study of microclimate and downy mildew development in subsurface drip-and furrow-irrigated lettuce fields in California. *Plant Dis.* **79:** 620–625.

Schettini, T.M., Legg, E.J. and Michelmore, R.W. 1991. Insensitivity to metalaxyl in California populations of *Bremia lactucae* and resistance of California lettuce cultivars to downy mildew. *Phytopathology* **81:** 64–70.

Schnathorst, W.C., Grogan, R.G. and Bardin, R. 1958. Distribution, host range and origin of lettuce powdery mildew. *Phytopathology* **48:** 538.

Sehgal, S.P. and Mathur, R.L. 1959. Cercospora leaf spot of spinach. *Sci. and Cult.* **25:** 77.

Shimazaki, Y. and Uchiyama, F. 1985. Resistance of spinach cultivars to downy mildew. *Proc. Kanto Pl. Prot. Soc.* **32:** 110.

Shrestha A, Sultana R, Jong-Chan Chae, Kangmin Kim and Lee KJ. 2015. *Bacillus thuringiensis* C25 which is rich in cell wall degrading enzymes efficiently controls lettuce drop caused by *Sclerotinia minor*. *Euro. J. Plant Pathol.* **142:** 577–589.

Shukla, S., Saxena, R.P., Awasthi, L.P. and Singh, S. 2006. Comparative efficacy of some commercial fungicides and onion extract for the control of stem gall disease of coriander (*Coriandrum sativum* L.). Presented in 58th Annual meeting and National Symposium on "Emerging Plant Diseases their Diagnosis and Management" held at North Bengal University, Siliguri, w.e.f. Jan. 31–Feb. 2, 2006 (Abst.).

Singh, A.K. 2009. Integrated management of wilt, *Fusarium oxysporum* f.sp. *coriandrii* of coriander. *Indian J. Plant Prot.* **37:** 132–133.

Singh, B. and Gupta, V.C. 1951. Diseases of Spinach. *Curr. Sci.* **20:** 104–105.

Singh, H.B., Singh, A., Tripathi, A., Rai, S.K., Katiyar, R.S., Johri, J.K. and Singh, S.P. 2003.Evaluation of Indian coriander accessions for resistance against stem gall disease. *Genetic. Reso. Crop Evol.* **50:** 339–343.

Singh, J. and Kanwar, J.S. and Chadha, M.L. 1987. Infection rate of Cercospora leaf spot as affected by spacings and plant cuttings in spinach. *Punjab Veg. Grow.* **22:** 18–22.

Singh, S.P., Gupta, J.S. and Sharma, A.K. 1984. Disease appraisal and crop loss estimates in coriander attacked by *Protomyces macrosporus* Unger in Uttar Pradesh. *Geobios* **11:** 276–278.

Smith, A, Camilo, AB, Kusunoki, M, Alba, MC, Kondoo, T and Deguchi, M. 2013. Diversity of soil-dwelling *Trichoderma* in Colombia and their potential as biocontrol agents against the phytopathological fungus *Sclerotinia sclerotiorum* (Lib.) de Bary. *J. Gen. Plant Pathol.* **79(1):** 74–85.

Soylu, S., Soylu, E.M. and Kurt, S. 2003. First report of Cercospora leaf spot on Swiss chard caused by *Cercospora beticola* in Turkey. *Plant. Pathol.* **52:** 804.

Srivastava, U.S. 1971. Edaphic factors and wilt of coriander. *Indian Phytopath.* **24:**679–685.

Srivastava, U.S. 1972. Effect of interaction of factors on wilt of coriander caused by *Fusarium oxysporum* f.sp. *coriandrii* Kulkarni, Nikam & Joshi. *Indian J. Agric. Sci.* **42:** 618–621.

Srivastava, U.S., Rai, R.A. and Agrawat, J.M. 1971. Powdery mildew of coriander and its control. *Indian Phytopath.* **24:** 437–440.

Stevens, FL and Hall, JG. 1911. A serious lettuce disease (*Sclerotiniose*) and a method of control. North Carolina Agricultural Experiment Station of the College of Agriculture and Mechanic Arts.

Su, H., Bruggen, A.H.C.-Van, Subbarao, K.V. and Scherm, H. 2004. Sporulation of *Bremia lactucae* affected by temperature, relative humidity and wind in controlled conditions. *Phytopathology* **94:** 396–401.

Subbarao, K.V., Koike, S.T. and Hubbard, J.C. 1996. Effects of deep ploughing on the distribution and density of *Sclerotinia minor* sclerotia and lettuce drop incidence. *Plant Dis.* **80:** 28–33.

Sugiyama, J, Hosaka, K and Sung, OS. 2006. Early diverging Ascomycota: phylogenetic divergence and related evolutionary enigmas. *Mycologia* **98:**6, 996– 1005.

Sullivan, M.J., Damicone, J.P. and Payton, M.E. 2002. The effects of temperature and wetness period on the development of spinach white rust. *Plant Dis.* **86:** 753–758.

Sydow, H and Butler, EJ. 1911. Fungi Indiae orientalis. *Ann. Mycol.* **9:**372–421.

Thinggaard, K. 1985. Investigation of the Danish *Brèmia lectucae* (lettuce downymildew) population in the period 1979–1984. *Tidsskrift. Planteavl* **89(3):** 263–271.

Thomas, C.E. 1970. Epidemiology of spinach white rust in South Texas. *Phytopathology* **60:** 588 (Abstr.).

Tripahi, A.K., Pandya, R.K. and Tripathi, M.L. 2001. Effect of nitrogen, phosphorus and potassium on stem gall disease and yield of coriander. *Ann. Pl. Prot. Sci.* **9:** 337–339.

Tripathi, A.K. 2001. Evaluation of coriander cultivars for resistance against stem gall disease and yield potential. *Crop Res. Hisar* **22:** 485–488.

Tripathi, A.K. 2003. Effect of sowing dates and weather parameters on stem gall disease of coriander. *Indian Phytopath.* **56:** 191–193.

Tripathi, A.K. 2005. Efficiacy of fungicides andd plant production against stem gall disease of coriander. *J. Mycol. Pl. Pathol.* **35:** 388–389.

Tripathi, A.K. and Bartaria, A.M. 2002. Chemical control of stem gall disease (*Protomyces macrosporus*) of coriander and its economics. *Pl. Dis. Res.* **17:** 119–120.

Tripathi, A.K., Bartaria, A.M., Pandya, R.K. and Shashi Chauhan. 1998. Evaluation of coriander cultivars against stem gall disease. *Flora and Fauna, Jhansi* **4:** 98.

Tripathi, A.K., Bartaria, A.M., Pandya, R.K. and Tripathi, M.L. 2002c. Effect of cultural practices on stem gall disease of coriander. *Ann. Agric. Res.* **23:** 171–173.

Tripathi, A.K., Chauhan, R.K.S., Bartaria, A.M. and Chauhan, S. 2002b. Reaction of coriander cultivars to *Protomyces macrosporus. Ann. Pl. Prot. Sci.* **10:** 171–172.

Tripathi, A.K., Chauhan, R.K.S., Bartaria, A.M. and Chauhan, S. 2003. Quantitative and qualitative loss in coriander due to *Protomyces macrosporus. Indian Phytopath.* **56:** 451–452.

Tripathi, A.K., Pandya, R.K., Tiwari, R.J. and Tripathi, M.L. 2002a. Response of stem-gall disease (*Protomyces macrosporus*) and yield of coriander (*Coriandrum sativum*) to soil pH. *Indian J. Agric. Sci.* **72:** 128–129.

Unger, F. 1833. Die. Exanthame der Pflanzen: 344p.

Vanterpool, T.C. 1959. Oospore germination in *Albugo candida. Can J. Bot*. **37:** 169–173.

Verma, P.R. and Petrie, G.A. 1975. Germination of oospores of *Albugo candida. Can J. Bot.* **53:** 836–842.

Verma, P.R., Agnihotri, J.P. and Patel, P.N. 1975. Survival of *Cercospora cruenta* Sacc. and *Cercospora beticola* Sacc. through non-crop periods. *Phytopathol. Mediterr*. **14:** 23–24.

Verma, P.R., Dange, S.R.S. and Patel, P.N. 1968. Host range and epidemiology of Cercospora leaf spot disease of spinach-beet (*Beta vulgaris* L.). *Indian Phytopath*. **21:** 428–433.

Viswanath, S.M. and Nariani, T.K. 1981. Spinach leaf curl. *Curr. Sci*. **50:** 326–327.

Walkey, D.G.A. and Dance, M.C. 1979. High temperature inactivation of seed-borne lettuce mosaic virus. *Pl. Dis. Reptr*. **63:** 125–129.

Wiant, J.S. 1939. White rust of spinach. *Phytopathology* **29:** 616–623.

Wicks, T.G., Hall, B. and Pezzaniti, P. 1994b. Fungicidal control of metalaxyl – insensitive strains of *Bremia lactucae* on lettuce. *Crop Prot.* **13:** 617–623.

Wicks, T.J., Hall, B. and Pezzaniti, P. 1994a. Fungicidal control of anthracnose (*Microdochium panattonianum*) on lettuce. *Aust. J. Exp. Agric*. **34:** 277–283.

Wright, C.M. and Yerkes, W.D. 1950. Observations on the over-wintering of the pathogen causing downy mildew of spinach in the Walla-Walla area. *Pl. Dis. Reptr*. **34:** 28.

Wu, B.M., Subbarao, K.V. and Bruggan, A.H.C. Van. 2000. Factors affecting the survival of *Bremia lactucae* sporangia deposited on lettuce leaves. *Phytopathology* **90:** 827–833.

Yamaguchi, M. 1983. *World Vegetables: Principles, Production and Nutritive values.* AVI Publishing, Westport, CT.

■■■■

16

DISEASES OF GINGER

Ginger (*Zingiber officinale* Roscoe) is an important spice crop and India is a leading producer and exporter of ginger in the world. The production of ginger in India was 40,903 MT with an income of Rs. 44.04 crores in the year 2010–2011 (Dohroo *et al.*, 2012). In India, major ginger growing states are Kerala, Sikkim, Meghalaya, West Bengal, Orissa, Tamil Nadu, Karnataka, Andhra Pradesh, Maharashtra and Himachal Pradesh. During cultivation the crop is damaged by various diseases of fungal, bacterial and viral nature which reduce the potential yields drastically.

A. FUNGAL DISEASES

1. Rhizome Rot

This is a very serious disease of ginger crop throughout the world. In warm and humid conditions it may assume serious proportions and cause significant losses.

Geographical distribution: The disease was first recorded during the year 1907 from Surat in Gujarat, India (Butler, 1907). Thereafter, it was reported from Japan, China, Nigeria, Fiji, Taiwan, Australia, Hawaii, Sri Lanka and Korea. In India, the disease has been reported from almost all the states like Kerala, Rajasthan, Himachal Pradesh, Orissa, Maharashtra, Andhra Pradesh, etc. wherever ginger is grown (Dohroo, 1997).

Losses: Losses from this disease vary greatly in different localities and from year to year. The disease reduces potential yields to a greater extent in field, storage and market. Up to 40 per cent yield losses due to this disease has been reported (Dohroo *et al.*, 2012).

Symptoms: Ginger crop is susceptible to infection of this disease throughout the growing period. Almost all parts of the plant including sprouts, roots, developing rhizome and collar region of the pseudostem are vulenerable to infection. The disease manifests in field in the form of wet and dry rot which are commonly seen on the ginger rhizomes (Dohroo, 1979, Dohroo, 1982).

a. Wet rot

The general symptoms of the disease are first noticed as slightly pale leaves (Plate 1a). The yellowing starts from leaf tip and spread down the leaf blade, more along the margin than at the centre. The dead leaves droop and hang down along the pseudostem till the entire shoot becomes dry. The basal portion of the plant between the point of attachment with rhizome and collar region of pseudostem is also affected and exhibits a pale translucent colouration, later becoming water soaked and soft (Plate 1c). The whole shoot either topples down or can easily be pulled out from this point with a slight pressure. Soft rot extends from the collar region to the rhizomes enclosed by the tough rind of the rhizome (Plate 1d). The fibrovascular strands are not affected and remain isolated within the decaying mass (Dohroo, 1982).

b. Dry rot

In case of dry rot, though the leaves become pale but no soft rot of collar region is observed (Plate 1b). Such plants cannot be easily pulled out. Mycelial growth in the form of white, peach or buff coloured cushions can be seen on the surface of rhizomes (Dohroo, 1982).

In some cases, other fungi and bacteria follow and hasten the rotting process. Foul smell emits from the rotten rhizomes. The flies like *Mimegralla* and *Eumerus* lay eggs in the tissue and the maggots develop in the decomposed rhizomes (Iyer, 1988).

The Pathogen

The disease is caused by different species of *Pythium* and *Fusarium.* Several species of *Pythium* have been reported to cause the disease in different parts of the world. These are: *P. aphanidermatum* (Edson) Fitzp. (Mitra and Subramanian, 1928), *P. butleri* Subram. (Thomas, 1938), *P. complectens* Braun (Park, 1934), *P. deliense* Meurs (Haware and Joshi, 1974), *P. gracile* (de Bary) Schenk (Butler, 1907), *P. graminicolum* Subram. (Park, 1941; Bertus, 1942), *P. vexans* de Bary (Ramakrishnan, 1949), *P. pleroticum* T. Ito (Dohroo and Sharma, 1985), *P. zingiberum* (Ichitani and Shinsu, 1980) and *P. ultimum* Trow. (Dohroo 1987).

Different species of *Fusarium* such as *F. solani* (Mart.) Sacc. (Rath *et al.*, 1979; Dohroo, 1987; Chauhan and Patel, 1990), *F. equiseti* (Corda) Sacc. (Sharma and Dohroo, 1980; Bhardwaj *et al.*, 1988b; Dohroo *et al.*, 1987) and *F. oxysporum* Schlkechtend ex Fr. f. sp. *zingiberi* Trujillo (Teakle, 1965; Yang *et al.*, 1988) are associated with dry rot phase of this disease.

Combined infection of *Pythium* sp. and *Fusarium solani* in ginger rhizomes is common (Chauhan and Patel, 1990). In Korea, *F. oxysporum f. sp. zingiberi* and *P. zingiberum* caused rhizome rot (Yang *et al.*, 1988). In west coast of Madras, *Pythium* sp. in association with *Sclerotium rolfsii* Sacc. were considered as the cause of rhizome rot (Anonymous, 1953). In Himachal Pradesh, *P. pleroticum* was found more predominating and destructive than *F. equiseti* (Corda) Sacc. *Pythium pleroticum* causes wet rot, whereas *F. equiseti* is responsible for dry rot (Sharma and Dohroo, 1980). Bhardwaj *et al.* (1988b) reported five pathogens of rhizome rot of ginger in Himachal Pradesh. Out of these *P. aphanidermatum* causes wet rot and other fungi, such as *Fusarium equiseti, F. solani, Cladosporium cladosporioides* (Fresen.) De Vries and *Mucor hiemalis* Wehmer cause only dry rot in injured rhizomes. Besides, wilt and rhizome rot complex sysptoms, plants were also found infected with rhizome fly (*Mimegralla coeruleifrons*) or nematodes such as *Meloidogyne incognita* and *Pratylenchus coffeae.*

P. aphanidermatum, the mycelium is colourless, sometimes lustrous, occasionally slightly yellowish due to abundant oospores or hyphal swellings or grayish lilac

(Dohroo *et al.*, 2012). The main hyphae is up to 10 μm wide. Sporangia consisting of terminal complexes of swollen hyphal branches of varying length and up to 20 μm wide (Fig. 1). Oogonia terminal, globose, smooth, 20–25 μm in diameter. Anthredia mostly intercalary, sometimes, broadly sac shaped, 10–14 μm wide, oogonium, monoclinous or diclinous, oospores aplerotic (18 – 22 μm) in diameter, wall 1–2 μm thick. In *P. pleroticum* abundant sporangia are produced measuring 11.0 to 16.5 μm which liberate zoospores by the formation of evanescent vesicle (15.5 – 23.5 μm) through an emission tube. Zoospores are released either by a pore developed at the tip of sporangium or by bursting of vesicle. Sporangia are never detached from the sporangiophore. The oospores are smooth walled, plerotic (oospore wall fused with oogonial wall) and spherical in shape measuring 12.0–20.0 μm in diameter (Dohroo, 1982).

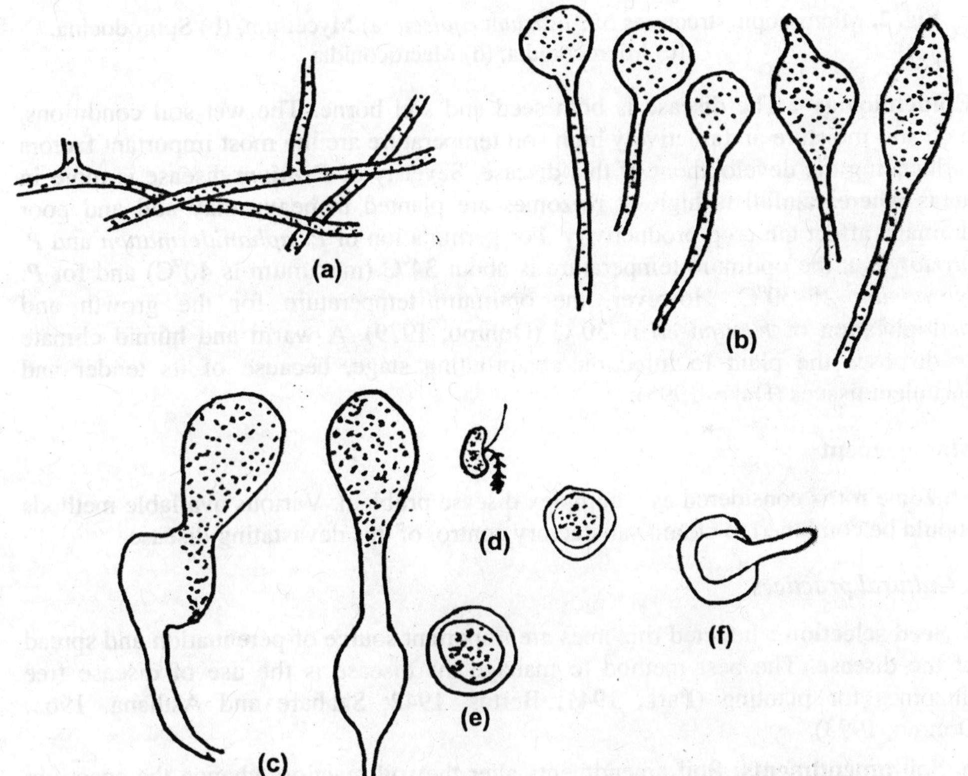

Fig. 1. Microscopic structures of *Pythium aphanidermatum* (a) Mycelium, (b) Sporangia, (c) Vesicle, (d) Zoospore, (e) Oospore, (f) Germination of oospore.

Fusarium equiseti (Forda) Sacc. produces microconidia and macroconidia. Microconidia are aseptate measuring 6.5 – 10.0 x 3.0 – 4.0 μm in size. Macroconidia are 1 to 3 septate, sickle to spindle shaped measuring 23.5–33.0 x 3.0–4.0 μm in size (Fig. 2). Sporodochial formation is rare (Dohroo, 1982).

Disease cycle: There are two ways in which the disease is carried over and perpetuated, firstly through diseased rhizomes as oospores in scales (Thomas, 1938) and secondly through oospores in soil. *Pythium* species are capable of saprophytic survival in plant debris. The infected plant debris remaining in the field forms an important source of primary infection. Such plant parts may contain large number of perennating oospores.

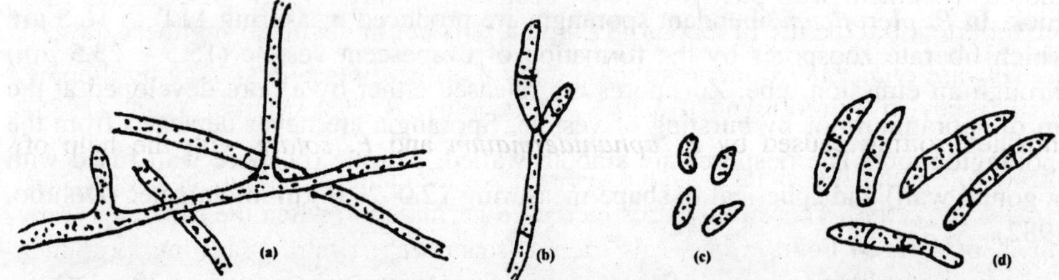

Fig. 2. Microscopic structures of *Fusarium equiseti* (a) Mycelium, (b) Sporodochia, (c) Microconidia, (d) Macroconidia

Epidemiology : The disease is both seed and soil borne. The wet soil conditions, high soil moisture and relatively high soil temperature are the most important factors influencing the development of this disease. Severity of *Pythium* disease is more in areas where rainfall is high or rhizomes are planted in heavy clay soil and poor drainage affect the crop productivity. For germination of *P. aphanidermatum* and *P. myriotylum,* the optimum temperature is about 34°C (maximum is 40°C) and for *P. pleroticum* 25–30°C. However, the optimum temperature for the growth and multiplication of *F. equiseti* is 30°C (Dohroo, 1979). A warm and humid climate predisposes the plant to infection at sprouting stage, because of its tender and succulent tissues (Dake, 1995).

Management

Rhizome rot is considered as a complex disease problem. Various available methods should be combined to obtain satisfactory control of this devastating disease.

i. Cultural practices

a. Seed selection : Infected rhizomes are important source of perennation and spread of the disease. The best method to manage the disease is the use of disease free rhizomes for planting (Park, 1941; Bertus, 1942; Shahare and Asthana, 1962; Dohroo, 1993).

b. Soil amendments: Soil amendments alter the soil reaction, change the spectrum of soil microflora and thus affect the pathogens existing in soil (Dohroo, 1993; Dohroo *et al.*, 1994; Dohroo and Pathania, 1997). Reduction in rhizome rot incidence after addition of soil amendments like oil seed cakes, neem cake and other organic matter from various plants in the field was recorded (Ghorpade and Ajri, 1982; Sadanandan and Iyer, 1986; Thakore *et al.*, 1987). Suppressive soils can also exploited for the control of rhizome rot (Lee *et al.*, 1990). Neem cake amendment resulted in the lowest disease incidence (22.92%) and highest yield (32.91 q/ha) followed by application of FYM.

ii. Use of chemicals

a. Seed treatment: The seed treatments have been tried prior to storing and planting. Treatment of rhizomes with 0.1 per cent mercuric chloride (0.1%) for 24 hours (Park, 1935) or 90 minutes was effective against soft rot caused by *Pythium* sp. Efficacy of rhizomes treatment with phenyl mercury acetate in controlling the disease has also been reported, but the use of mercurial fungicides is not in vogue during these days.

Haware *et al.* (1973) recorded good control of the disease in storage by dipping the rhizomes in Aureofungin (0.02%) and benomyl (0.2%). Mathur *et al.* (1985) controlled rotting caused by *P. aphanidermatum* and *F. solani* with the help of triadimefon. Other fungicides, such as Fenaminosulf, captafol, dodine and metalaxyl were also effective. These fungicides increased germination when the rhizomes were dipped for 4 to 24 hours. Chemicals such as mancozeb, ziram, guazatine, propineb and copper oxychloride effectively controlled rhizome rot when used as 30 minutes dip treatments for rhizomes (Dohroo and Sharma, 1986; Thakore *et al.*, 1988). Steeping of rhizomes in carbendazim (0.1%) for 60 minutes before storage also controlled storage rots and reduced the disease incidence from 71.4 to 18.2 per cent (Sharma and Dohroo, 1991).

b. Soil application: Different chemicals have also been tested by various workers as soil drench against rhizome rot of ginger. Rhizome rot was reduced by zineb, captafol, methyl bromide, mercuric chloride, thiram, phenyl mercury acetate, copper oxide and mancozeb (Shahare and Asthana, 1962; Kothari, 1966; Ichitani, 1980; Sharma *et al.*, 1980; Doshi and Mathur, 1987). Dohroo *et al.* (1984a) found metalaxyl application quite effective for the control of rhizome rot.

c. Seed and soil application: As the disease is both seed and soil borne, use of disinfected seed rhizomes and drenching of soil with chemical simultaneously have been proved more effective in comparison to either seed treatment or soil application.

Dipping the seed rhizomes in Bordeaux mixture (2:2:50) and applying this fungicide to soil eight days before sowing satisfactorily controlled the soft rot due to *P. myriotylum* (Bhagwat, 1960). In Philippines, sowing of seed rhizomes after disinfecting by immersion in 0.06 per cent mercury seed protectant for 90 minutes and addition of DDT or aldrin in soil against insects were recommended by Pordesimo and Raymundo (1963) for successful control of rhizome rot of ginger. Application of echlomezol as a drench around sources of primary infection prevented further spread of the disease (Ichitani, 1980). Blitane (zineb + copper oxychloride), zineb and captafol when used as drenches plus seed treatment also gave good results (Sharma *et al.*, 1980). Dipping of seed rhizomes one day before planting and soil drenching with a mixture of metalaxyl plus captafol three months after planting controlled rhizome rot of ginger (Rathaia, 1987). Fosetyl-Al, metalaxyl, oxadixyl, propamocarb and ethazole (epidiazole) were also evaluated against *P. aphanidermatum* (Ramachandran *et al.*, 1989). Of these fungicides, metalaxyl formulations (Ridomil 5G and Apron 35 WS) gave best control of the disease when used as soil and seed treatments. Rhizome rot caused by *F. oxysporum* was controlled by soil treatment with 4 per cent formaldehyde combined with treatment of rhizome planting material with thiophanate methyl. Rhizome treatment with 0.1 per cent carbendazim or 0.3 per cent mancozeb and soil treatment with formaldehyde were also found effective against the disease.

Srivastava (1994) managed soft rot (*P. aphanidermatum*) in Sikkim effectively by drenching the soil with zineb or mancozeb following rhizome treatment with carbendazim and incorporating Thiodan dust into the soil to control insect invasion. Nath (1993) suggested planting of ginger under shade after treating with 1 per cent formaldehyde. Treated rhizomes grown under shade had 19.4 per cent incidence of rhizome rot (*P. myriotylum*) as compared to 41.3 per cent incidence in treated rhizomes grown without shade.

iii. Biological control

Antagonistic mode of *Trichoderma* sp. against *P. aphanidermatum, F. equiseti, F. solani, Cladosporium cladosporioides* and *Mucor hiemalis* has been studied (Bhardwaj and Gupta, 1987) under *in vitro* conditions. *Trichoderma viride, T. harzianum* and *T. hamatum* inhibited the growth of all the pathogenic fungi tested. *Trichoderma harzianum* and *Gliocladium virens* are also known to inhibit the growth of *F. oxysporum* f.sp. *zingiberi*, causing rhizome rot in ginger (Sharma and Dohroo, 1991). Rathore *et al.* (1992) suggested that *T. viride* produced some volatile and non-volatile substances which inhibit growth of *P. myriotylum* and *F. solani*. Treatment of ginger rhizomes by steeping in spore suspension of *T. viride* or smearing with *T. hamatum* provided protection against dry and wet rot (Dohroo and Sharma, 1984; Bhardwaj *et al.,* 1988a).

iv. Resistant cultivars

Indrasenan and Paily (1974) reported Maran cultivar to be resistant against ginger rot caused by *Pythium aphanidermatum.* Setty *et al.* (1995 a) evaluated eighteen ginger cultivars against rhizome rot (*Pythium* sp.) under field conditions and none of which was resistant. However, the cultivar Supraba demonstrated less than 3 per cent disease incidence under Himachal Pradesh conditions.

v. Integrated disease management

Bandyopadhyay and Bhattacharya (2012) reported the management of this disease by integrating physical, chemical and biological methods like rhizome treatment with hot water at 51°C for 10 minutes followed by soil application of *Trichoderma harzianum* inoculum mixed with neem cake at the time of planting which resulted in lowering the disease incidence as compared to control. Similarly, Dohroo *et al.* (2015) advocated the treatment which included rhizome treatment in hot water at 47°C for 30 min and soil application of *T. harzianum* @ 2.5 kg/50 kg FYM/ha at the time of planting, followed by three drenching of mancozeb (0.25%) at 20 days interval was found highly effective in limiting the disease incidence besides having the significant effect in improving the growth and yield. Crop rotation with non-solanaceous crops, raised bed cultivation, proper drainage of the field, pre and post-harvest rhizome dip in combination of mancozeb (0.25%) and carbendazim (0.1%) for 1 h and drenching with copper oxychloride (0.3%) during monsoon season at an interval of 15 days provided effective control of this disease (Shukla and Gupta, 2015).

2. Yellows

Yellows disease of ginger was first described by Simmonds (1955) from Queensland and later from Hawaii (Trujillo, 1963) and India (Haware and Joshi, 1973). It causes significant losses in warm and humid climatic conditions.

Symptoms: Yellowing of the margins of the lower leaves is the initial symptom which gradually spreads, covering the entire leaves. Older leaves dry up first followed by the younger ones (Trujillo, 1963). Plants may thus show a premature drooping, wilting, yellowing and drying in patches or in whole bed. The basal portions of the affected plant become soft and watery, and the shoot can be easily pulled out from the mother rhizome. Plants may show stunting.

In rhizomes, a cream to brown discolouration accompanied by shriveling is commonly seen. Central rot is also prominent. Rotting of roots is common and the rhizome formation is affected. In final stages of decay, only the fibrous tissue remain within the rhizomes. A white cottony fungal growth may develop on the surface of stored rhizomes.

Causal organism: The disease is caused by *Fusarium oxysporum* Schlechtend ex Fr. *zingiberi* Truzillo. Other species of *Fusarium,* such as *F. solani* (Mart.) Sacc., *F. equiseti* (Corda) Sacc. and some unidentified *Fusarium* sp. were also reported to be associated with ginger rhizomes. Sharma and Dohroo (1990) reported *F. oxysporum* as the major cause of yellows from Himachal Pradesh. The second most frequently isolated species was *F. solani. Fusarium moniliforme* Sheld. (Tel.: *Gibbrella fujikuroi to in Ito & Kimura), F. graminearum* Achwabe (Tel.: *G. zeae* (Schwein) Petch) and *F. equiseti* were also found associated with the diseased plants. Isolates of *F. oxysporum* f. sp. *zingiberi* differed in their aggressiveness (Dohroo and Sharma, 1992b).

Epidemiology: For the development of yellows disease, a temperature range of 15 to 30°C is favourable (the optimum being 23–29°C) along with high humidity and continuous presence of free film of water (Sharma and Jain, 1978). Maximum disease incidence occurred when soil temperature ranged form 24 to 25°C and the soil moisture from 25 to 30 per cent (Sharma and Dohroo, 1989). Agrawal *et al.* (1974) found the medium amended with galled root (galls caused by *Meloidogyne incognita*) extract of ginger supported better growth of the pathogen than healthy root extract medium.

Disease cycle: The seasonal carryover of the fungus inoculum takes place through infected rhizomes and soil (Dohroo, 1989a). The fungus survives in soil as chlamydospores which may remain viable for many years in the field. Seed transmission of the fungus has also been reported by Dohroo (1989a). The secondary spread of the disease can also take place through irrigation water and by mechanical means. Sharma (1977) has also reported survival of *F. oxysporum* f.sp. *zingiberi* as a competitive saprophyte on agar plates.

Management

i. Seed selection: As the disease spreads through contaminated rhizomes, selection of healthy seed rhizomes has been found an effective control measure for the disease (Dohroo *et al.*, 1988; Rana, 1991; Dohroo, 1993).

ii. Cultural practices: In solarized plots, there was a considerable increase in seed germination, plant height and yield and greater reduction in disease incidence (88.18%) over unsolarized plots (Mohan *et al.*, 2004).

iii. Use of chemicals: Seed treatment with mercurial fungicides has been reported effective for the control of the disease. A dip treatment of seed rhizomes in ethoxyethyl $HgCl_2$ (6% Hg) at 1 lb/40 gallons for 30 minutes or 2 lbs/40 gallons for 10 minutes controlled the disease effectively (Teakle, 1965). In a field trial on the effect of different fungicides on *Fusarium* sp., Kothari (1966) observed better rhizome germination with 0.1 per cent $HgCl_2$ followed by thiram (0.5%), phenyl mercury acetate (0.5%) and zineb (0.2%) when the fungicides were applied as soil drenches or rhizome dips. In field conditions, application of Ridomil MZ resulted in the highest seed germination (96.50%) and yield (250.25 q/ha), and lowest disease incidence (5%) (Singh *et al.*, 2004).

Under storage conditions, postharvest dip treatment of aureofungin (0.02%) and benomyl (0.2%) provided better control of the disease (Haware *et al.*, 1973). Disease incidence under storage was reduced from 71.4 to 18.2 per cent by steeping the rhizomes in carbendazim (0.1%) for sixty minutes (Sharma and Dohroo, 1991).

iv. Biological control: Growth of *F. oxysporum* f.sp. *zingiberi* was inhibited by *Trichoderma harzianum* and *Gliocladium virens* (Sharma and Dohroo, 1991). Application of *Bacillus subtilis* reduced the yellows incidence (Sharma and Jain, 1979). Dohroo (1995) suggested an integrated approach to combat the yellows disease of ginger which includes treatment of seed rhizomes with mancozeb and carbendazim and use of biocontrol agents like *Trichoderma harzianum, T. hamatum and Gliocladium virens* as seed treatment and soil application.

v. Resistant cultivars: By using resistant or less susceptible cultivars of ginger the disease can be managed to a great extent. The incidence of yellows was slightly less in cultivars like SG 666 (Dohroo, 1989b) and Kerala local (Rana and Arya, 1991) in field trials under Himachal Pradesh conditions.

3. Phyllosticta Leaf Spot

Phyllosticta leaf spot disease is becoming increasingly important in many of the states due to severe leaf rot and blight it causes. The disease was first reported from Godavari district of Andhra Pradesh and Malabar area of Kerala, erstwhile Madras state (Ramakrishnan, 1942). Later it was reported from Himachal Pradesh (Sohi *et al.*, 1964), Madhya Pradesh (Shukla and Haware, 1972) and Maharashtra (Haware and Joshi, 1974).

Symptoms: Initial symptoms of the disease are small oval to elongated spots on the leaves, measuring 1-10 mm x 0.5–4 mm. later on, the spots show white papery centre and dark brown margins with a yellowish halo (Plate 2) surrounding it (Ramakrishnan, 1942). The spots when increase in size, coalesce to form larger lesions. The affected leaves become shredded and may suffer extensive desiccation. Symptoms appear first on younger leaves. As the plants put forth fresh leaves, these get infected subsequently (Shukla and Haware, 1972).

Causal organism: Phyllosticta leaf spot is caused by *Phyllosticta zingiberi* T.S. Ramakr. The fungus forms amphigynous, subglobose, dark brown ostiolate pycnidia

(Fig. 3a) on the host measuring 78 to 150 μm in diameter. On standard media, the fungus forms pycnidia having 100–270 μm diameter bearing hyaline, unicellular, oblong, biguttulate spores (Fig. 3a) measuring 3.7 to 7.4 × 1.2 to 2.5 μm (Ramakrishnan, 1942).

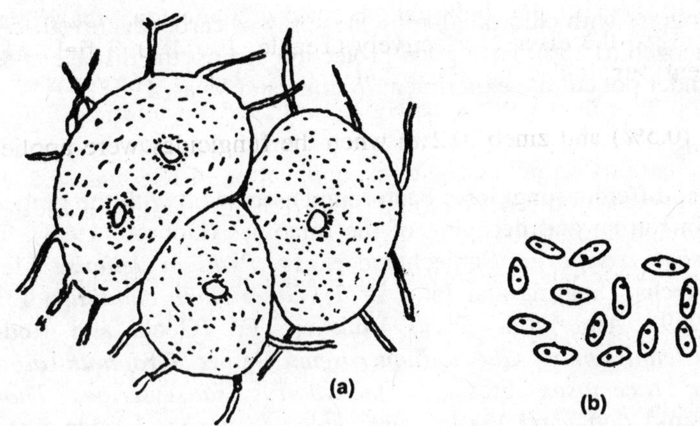

Fig. 3. Microscopic structures of *Phyllosticta zingiberi* (a) Pycnidia, (b) Conidia

Disease cycle and epidemiology: The infected debris or seed serve as primary inoculum for the disease. In leaf, pycnidiospors and mycelium remain viable for 14 months under laboratory conditions (Brahma and Nambiar, 1982). Pycnidia survive in the leaf debris throughout the summer having temperature range of 30 to 35°C. The pycnidiospores remain viable in soil even at 25 cm depth for 6 months. The optimum temperature range for mycelial growth of *Phyllosticta* was 25.0 to 27.5°C with maximum and minimum to be 32.5 and 10.0°C, respectively. At 5 and 35°C complete inhibition of mycelia growth was observed (Cerezine *et al.*, 1995).

The disease begins to appear towards the end of June. During this period, the temperature varies in between 23.4 to 29.6°C and relative humidity from 83.3 to 90.2 per cent. Later in July when the number of rainy days and total rainfall increase, the disease aggravates and spreads fast (Brahma and Nambiar, 1984). Ginger plants upto the age of 6 to 7 months are more susceptible to this disease.

The extent of dispersal of the causal fungus depends upon the intensity of precipitation. Higher intensity of rain accompanied by wind seems to exert greater impact on target leaf, so that spores are splashed to greater distances, resulting in liberation of greater amount of spores and increasing disease severity (Brahma and Nambiar, 1984).

Management

*i. **Resistant cultivars:*** None of the 18 cultivars tested in Karnataka was resistant to Phyllosticta leaf spot (Setty *et al.*, 1995b). However, the cultivars Narasapatom, Tura, Nadia, Tetraploid and Thingpani were classed as moderately resistant with a disease index less that 5 per cent. In Himachal Pradesh, none of the tested material of ginger was rated as resistant to *P. zingiberi,* however, 8 lines showed moderate resistance (Dohroo *et al.*, 1986b).

ii. Use of chemicals: Sprays of Bordeaux mixture, zineb and maneb have been reported effective in checking the disease (Ramakrishnan, 1942; Sohi *et al.*, 1973). Twelve sprays of mixture of benomyl (0.1%), mancozeb (0.2%) and soluble boron (0.1%) and iprodione (0.2%) alone were most effective in reducing the disease (Grech and Frean, 1988). In Brazil, Cerezine *et al.* (1995) found highest reduction in the disease progress with chlorothalonil. One spray of carbendazim (0.15%) and two sprays of mancozeb (0.25%) gave good protection against the disease and resulted in higher yield under pot culture experiment (Verma and Vyas, 1981).

4. Storage Rots

During storage, different fungi have been found associated with the ginger rhizomes, which result in rotting and decaying of the rhizomes (Dohroo, 1993). These fungi include *Fusarium oxysporum* Schlechtend ex Fr. *Pythium deliense* Meurs and *P. myriotylum* Drechs. (Sharma and Jain, 1977). *Geotrichum candidum* Link (Mishra and Rath, 1989), *Aspergillus flavus* Link ex. Fr. (Geeta and Reddy, 1990), *Cladosporium lennissimum, Gliocladium roseum* Bainer, *Graphium album* (Corda) Sacc., *Mucor racemosus* Fresen., *Stachybotrys sansevieriae, Thanatephorus cucumeris* (Frenk) donk and *Verticillium chlamydosporium* Goddard (Dohroo and Sharma, 1992a). *Geotrichum candidum* caused complete rotting of inoculated rhizomes in 15 days at 25°C and 100 per cent relative humidity (Mishra and Rath, 1989). The fungus *A. flavus* in association with ginger rhizomes was implicated in the production of carcinogenic aflatoxin (Geeta and Reddy, 1990). *Aspergillus niger* and *Rhizoctonia solani* which were associated with ginger rhizomes produced pectolytic and cellololytic enzymes during pathogenesis in inoculated rhizomes (Agrawal and Gupta, 1986).

Management

i. Use of chemicals: Okwuowulu and Nnodu (1988) suggested pre-storage chemical treatments of benomyl (750 ppm) and/or gibberellic acid (150 ppm) to reduce the incidence of storage rots. Dipping of rhizomes in imazalil or prochloraz at 0.8 g a.i. per litre and then storing at 10°C gave good protection against the infection with fungi, such as *Botryodiplodia, Aspergillus, Diplodia, Fusarium, Rhizoctonia* and *Pythium* in storage (Grech and Swarts, 1990). A combination of mancozeb and carbendazim treatment to rhizomes controlled storage rot of ginger (Dohroo *et al.*, 1986a; Dohroo and Malhotra, 1995). Under stored conditions, application of 0.3% Ridomil MZ resulted in the lowest incidence of the disease (Singh *et al.*, 2004).

Amongst various fungicides used for dipping the rhizomes before storage, mancozeb is known to persist longer than carbendazim (Sharma *et al.*, 1992). Mancozeb residues were observed even after 120 days of storage. However, the health risk in carbendazim treated rhizomes is low as compared to mancozeb if they are consumed after peeling.

ii. Modification of storage conditions: Rhizomes, which are harvested at least six months after planting are most suited to prolong storage in moist saw dust, which is effective in controlling dehydration (Okwuowulu and Nnodu, 1988). Storage of rhizomes under cooled conditions may prolong storability by reducing weight loss and sprouting, but may result in higher pathogen incidence than storage at room temperature. Packing of rhizomes in PVC film also reduces weight loss, but increases incidence of fungal infection (Lana *et al.*, 1993).

B. BACTERIAL DISEASES

1. Bacterial Wilt

Bacterial wilt is second most important disease of ginger after rhizome rot. Orian (1953) reported this disease for the first time from Mauritius. Later on the disease was reported from Hawaii (Ishii and Aragaki, 1963), Malaya and Australia (Hayward *et al.*, 1967). However, the disease occurred in India since middle of the century, but Mathew *et al.* (1979) reported it occurrence in 1979 from Kerala. Dohroo (1991) reported the occurrence of this disease from Himachal Pradesh.

Symptoms: On the collar region water soaked patches or linear streaks appear. These symptoms are followed by yellow to bronze colouration of margins of the lower-most leaves which gradually progresses upwards. At later stages, the leaves become flaccid with intense yellowish bronze colour and droop ultimately exhibiting typical wilt symptoms. In the infected plants, the ligules and leaf sheaths look yellowish to dull green. The leaves roll up and the whole plant dries up finally. The pseudostems can be easily separated with a gentle pull and can be broken off at the base. At advanced stage, the pseudostem appears slimy. The plants which are infested by the disease stand persistently and do not collapse. If the infected rhizomes are pressed, a milky bacterial exudates oozes out. When infected tissues are steeped in clear water for a while, the water turns cloudy and milky.

Causal organism: The disease is caused by a bacterium *Ralstonia solanacearum* (Smith) Yabuuchi *et al.* Three biotypes of this bacterium *R. solanacearum* have been described and out of these, biotype III causes the wilt in India (Dake *et al.*, 1988). Biotype III of the bacterium causes slow wilt, whereas biotype IV causes rapid wilting and death of infected plants (Pegg *et al.*, 1974). *R. solanacearum* biotype III has been found restricted only to ginger and its common weeds, whereas biotype IV has a wide host range including tomato, potato, *Zinnia elegans, Capsicum frutescens, Physalis peruviana* and eggplant. Biotype II of *R. solanacearum* was also isolated from potato plants (Hayward *et al.*, 1967).

In cross incoculation studies, Lum (1973) observed that a weakly virulent form of *R. solanacearum* isolated from ginger did not produce symptoms on tomato, tobacco and groundnut but was easily recovered from inoculation sites on these plants.

In Indonesia, the race 1 of biovar III of *R. solanacearum* is considered as the cause of ginger wilt (Mulya *et al.*, 1990). The disease incidence increases when abundant nematodes are there in the ginger soil (Samuel and Mathew, 1986).

Management

i. Cultural practices: Indrasenan *et al.* (1981) suggested selection of healthy seed rhizomes, eradication of weeds and adoption of an effective crop rotation as control measures for the disease. Almost all the cultivars are susceptible to wilt pathogen (Indrasenan *et al.*, 1982).

ii. Use of chemicals: Treatment of seed rhizomes with emisan plus plantomycin for 30 minutes followed by three sprayings, first at 30th day after planting and others at an interval of 15 days, also gives good control of the disease (Ojha *et al.*, 1986). Soil fumigation with methyl bromide is also effective in reducing the disease incidence (Ishii and Aragaki, 1963).

2. Bacterial Blight

The occurrence of bacterial blight was observed in the last week of October, 2012 in Solan area and up to 10 per cent yield reductions due to this disease have been recorded (Dohroo *et al.*, 2013).

Symptoms: The symptoms start appearing on the tip of the leaves which spread downwards giving them 'V' shaped appearance. In the favourable weather conditions, the disease further spread down to the tillers which later dry. Late in the season, the crop show blighted appearance.

Causal organism: The bacterium associated with this disease is *Pantoea ananatis* strain nvgl-01. The colonies of the test bacterium are yellow, raised and translucent with smooth margins. The bacterium is Gram negative, facultative anaerobes, rod shaped, motile and indole positive.

Management

No specific control measures have been suggested in literature, but can be kept in check by application of Streptocycline (100 ppm) sprays.

C. VIRAL DISEASES

1. Mosaic

The symptoms appear as yellowish and dark green mosaic patten on leaves. The affected plants show stunting.

The virus causing mosaic in ginger has spherical particles with diameter of 23 to 38 μm. It showed positive serological reaction with antiserum of *Cucumber Mosaic Virus* (CMV). The virus is known to be transmitted by sap to different plants known to be hosts of CMV (Su, 1980). Nambiar and Sarma (1974) reported that sap transmission from ginger to ginger, ginger to *Nicotiana tabacum* var. harrison special, *N. tabacum* var. *rustica, N. tabacum* var. *xanthii, N. glutinosa, Elettaria cardamomum, Curcuma longa* and *C. aromatica* gave negative results.

2. Chlorotic Fleck

The chlorotic fleck virus was characterized and described by Thomas (1986). He detected the ginger virus in ginger imported into Australia from a number of countries. The geographical distribution of the virus is uncertain, but is thought to include India, Malaysia and Mauritius.

The *Ginger Chlorotic Fleck Virus* (GCFV) has isometric particles about 30 nm diameter, with a sedimentation coefficient of 111s and readily purified from infected ginger leaf tissue. The purified preparations contained a major species of single stranded RNA MW 1.5 x 10^6 and major coat protein MW 29 x 10^3. At pH 7, the particles formed a single zone in both cesium chloride and cesium sulphate gradients, with buoyant densities of 1.355 g/cm^3 (fixed virus) and 1.297 g/cm^3 (unfixed virus), respectively. The virus particles migrated as two electrophoretic components and were liable when treated with 10mM EDTA, 1M NaCl, 10 mM tris pH 8.25 or when negatively stained with potassium phosphotungstate.

The virus is transmitted by *Myzus persicae, Pentalonia nigronervosa, Rhopalosiphum maidis* or *R. padi*. Possible affinities of GCFV with the subemovirus group has also been described by Thomas (1986).

3. Big Bud

The disease is caused by *Tomato Big Bud Virus* (Pegg *et al.*, 1974). The affected plants cease to grow and leaves become bunched at the top of the stem. As the disease advances, plants turn yellow and die. The pathogen has a wide host range and the disease is transmitted by leafhoppers.

REFERENCES

Agrawal, P.S., Joshi, L.K. and Haware, M.P. 1974. Effect of root knot extract of ginger on *Fusarium oxysporum* f.sp. *zingiberi* Trujillo causing yellows disease. *Curr. Sci.* **43**: 752.

Agrawal, R.D. and Gupta, M.N. 1986. Studeis on *in vitro* enzymes production by *Aspergillus niger* and *Rhizoctonia solani*. *Indian J. Mycol. Pl. Pathol.* **16**: 224–225.

Anonymous, 1953. Detailed report of the Subordinate Officers of Dep. Agric., Madras for 1943–44, 1944–45, 1948–49, 1949–50, 1950–51, 84pp.

Bandyopadhyay, S. and Bhattacharya, P.M. 2012. Management of rhizome rot of ginger using physical, Chemical and Biological Methods. *J. Mycol. Pl. Pathol.* **42(3)**: 314–316.

Bertus, L.S. 1942. Plant Pathology. *Adm. Rep. Dir. Agric. Ceylon,* 1941. p D5.

Bhagwat, V.Y. 1960. Control of soft rot of ginger. *Poona Agric. Coll. Mag.* **51**: 47–49.

Bhardwaj SS, Gupta, P.K., Dohroo, N.P. and Shyam, K.R. 1988b. An addition to fungi causing rhizome rot of ginger. *Pl. Dis. Res.* **3**: 66.

Bhardwaj, S.S. and Gupta, P.K. 1987. *In vitro* antagonism of *Trichoderma* species against fungal pathogens associated with the rhizome rot of ginger. *Indian J. Plant Pathol.* **5**: 41–42.

Bhardwaj, S.S., Gupta, P.K., Dohroo, N.P. and Shyam, K.R. 1988a. Biological control of rhizome rot of ginger in storage. *Indian J. Plant Pathol.* **6**: 56–58.

Brahma, R.N. and Nambiar, K.K.N. 1982. Survival of *Phyllosticta zingiberi* Ramakr., causal agent of leaf spot of ginger. Pp. 123–125. In: *Proceedings of National Seminar on Ginger and Turmeric.* (Eds. Nair, M.K., Prem Kumar, T., Ravindran, P.N. and Sarma, Y.R.) CPCRI, Kasargod.

Brahma, R.N. and Nambiar, K.K.N. 1984. Spore release and dispersal in ginger leaf spot pathogen *Phyllosticta zingiberi.* In: *Proceedings of National PLACROSYM-V 1982.* (Eds. Bavappa, K.V.A. *et al.*), Placrosym Standing Committee.

Butler, E.J. 1907. An account of genus *Pythium* and some chytrodiaceae. *Mem. Dep. Agrci. India (Bot. Ser.)* **1**: 70.

Cerezine, P.C., Olinisky, I.A., Bittencourt, M.V.L. and Valerio Folho, W.V. 1995. *Phyllosticta* leaf spot on ginger. Cultural characterization of the pathogen and effect of chemical treatment on disease control in Morrestes, Parana state, Brazil. *Pesqui. Agropecul. Bras.* **30**: 477–487.

Chauhan, H.L. and Patel, M.H. 1990. Etiology of complex rhizome rot of ginger (*Zingiber officinale*) in Gujarat and *in vitro* screening of fungicides against its causal agents. *Indain J. Agric. Sci.* **60**: 80–81.

Dake, J.N. 1995. Diseases of ginger (*Zingiber officinale* Rosc.) and their management. *J. Spices Arom. Crops* **4**: 40–48.

Dake, J.N., Ramachandran, N. and Sarma, Y.R. 1988. Strategies to control rhizome rot (*Pythium* sp.) and bacterial wilt (*Pseudomonas solanacearum*) of ginger. *J. Coffee Res.* **18**: 68–72.

Dohroo, N.P. 1979. Rhizome Rot of Ginger (*Zingiber officinale*). M.Sc. Thesis, HPKVV, Solan, HP.

Dohroo, N.P. 1982. Further Studies on Rhizome Rot of Ginger (*Zingiber officinale* Rosc.). Ph. D. Thesis, HPKVV, Solan, HP.

Dohroo, N.P. 1987. *Pythium ultimum* on *Zingiber officinale*. Indian Phytopath. **40**: 275.

Dohroo, N.P. 1989a. Seed transmission of pre-emergence rot and yellows of ginger. *Plant Dis. Res.* **4**: 73–74.

Dohroo, N.P. 1989b. Peroxidase and polyphenol oxidase activities in rhizome rot of ginger. *Indian Phytopath.* **42**: 167.

Dohroo, N.P. 1991. New record of bacterial wilt of ginger in Himachal Pradesh. 2nd Phytopatho. North Zone Meet, April 29–30, 1991. p. 16. (Abst.).

Dohroo, N.P. 1993. Final ICAR Report on Multilocational Project on Rhizome rot of ginger. VHF, Solan, 38 p.

Dohroo, N.P. 1995. Integrated management of yellows of ginger. *Indian Phytopath.* **48**: 90–92.

Dohroo, N.P. 1997. Global status on diseases of ginger (*Zingiber officinale* Roscoe). Directorate of Extension Education, UHF, Solan.

Dohroo, N.P. and Edison, S. 1989. Compendium of rhizome rot of ginger. *J. Hill. Res.* **2**: 4–13.

Dohroo, N.P. and Malhotra, R. 1995. Control of storage rot of ginger in Himachal Pradesh. Pp. 199–202. In: *Integrated Disease Management and Plant Health* (Eds. Gupta, V.K. and Sharma, R.C.) Scientific Publishers (India), Jodhpur.

Dohroo, N.P. and Pathania, N. 1997. Ginger diseases management. Pp. 247–250. In: *Neem in Sustainable Agriculture.* (Narwal, S.S., Tauro, P. and Bisla, S.S., Eds.). Scientific Publishers (India), Jodhpur.

Dohroo, N.P. and Sharma, M. 1992a. New host records of fungi from India. *Indian Phytopath.* **45**: 280.

Dohroo, N.P. and Sharma, S.K. 1992b. variability of *Fusarium oxysporum* f.sp. *zingiberi*, the incitant of ginger yellows. *Indian Pytopath.* **45**: 247–248.

Dohroo, N.P. and Sharma, S.L. 1984. Biological control of rhizome rot of ginger in storage with *Trichoderma viride. Indian J. Plant Pathol.* **2**: 185–186.

Dohroo, N.P. and Sharma, S.L. 1985. *Pythium pleroticum* on *Zingiber officinale. Indian Phytopath.* **38**: 391.

Dohroo, N.P. and Sharma, S.L. 1986. Evaluation of fungicides for the control of rhizome rot of ginger in storage. *Indian Phytopath.* **36**: 691–693.

Dohroo, N.P., Bhardwaj, S.S. and Shyam, K.R. 1986a. Management of rhizome rot of ginger in storage through fungitoxicants. *Pestology* **10**: 24–25.

Dohroo, N.P., Kansal, S. and Neha Ahluwalia 2015. Studies on eco-friendly practices for management of soft rot of ginger (*Zingiber officinale*). *Indian Phytopath.* **68:** 93–96.

Dohroo, N.P., Kansal, S., Ahluwalia, Neha and Shanmurugam, V. 2013. First report of *Pantoea ananatis* strain nogl-olon ginger. *Dis. Res.* **28(2):** 200–202.

Dohroo, N.P., Kansal, S., Mehta, P. and Ahluwalia, N. 2012. Evaluation of eco-friendly disease management practices against soft rot of ginger caused by *Phythium aphanidermatum. Pl. Dis. Res.* **27:** 1–5.

Dohroo, N.P., Korla, B.N. and Rattan, R.S. 1988. Effect of chemical seed treatments on pre-emergence rot of ginger. Pp. 153–155. In: *Proc. Nat. Sem. Chillies, Ginger and Turmeric,* Spices Board, and APAO, Hyderabad.

Dohroo, N.P., Sharma, Onkar, Sharma, Meera and Sarlach, R.S. 1994. Effect of organic amendments of soil on rhizome rot nematodes and rhizosphere mycoflora of ginger. *Ann. Biol.* **10:** 208–210.

Dohroo, N.P., Sharma, S.L. and Bhardwaj, S.S. 1984a. Efficacy of soil applied fungitoxicants against rhizome rot of ginger. *Indian J. Plant Prot.* **12:** 59–60.

Dohroo, N.P., Sharma, S.L. and Bhardwaj, S.S. 1984b. Comparative phenol and sugar contents in healthy and diseased ginger plants. *Biovigyanam* **10:** 85–86.

Dohroo, N.P., Shyam, K.R. and Bhardwaj, S.S. 1987. Distribution, diagnosis and incidence of rhizome rot complex of ginger in Himachal Pradesh. *Indian J. Plant Pathol.* **5**: 24–25.

Dohroo, N.P., Shyam, K.R., Bhardwaj, S.S. and Korla, B.N. 1986b. reaction of ginger germplasm to Phyllosticta leaf spot. *Indian Phytopath.* **39**: 650–606.

Doshi, Anila and Mathur, Sneh. 1987. Symptomatology, interaction and management of rhizome rot of ginger. *Xenobiotics* **26**: 261–265.

Geeta, G.S. and Reddy, T.K.R. 1990. *Aspergillus flavus* Link and its occurrence in relation to other mycoflora on stored spices. *J. Stored Prod. Res.* **26**: 211–213.

Ghorpade, S.A. and Ajri, D.S. 1982. Effectiveness of soil seed cakes in control of rhizome rot malady of ginger. *J. Maharashtra Agric. Univ.* **7**: 272–273.

Grech, N.M. and Frean, R.T. 1988. Ginger leaf spot: a cause for concern in the ginger industry. *Citrus Grow. Subtrop. Fruit J.* **644**: 14–15.

Grech, N.M. and Swarts, D.H. 1990. Postharvest application of fungicides for control of fungal decay of ginger rhizomes stored under stimulated low temperature shipping conditions. *Phytophylactica* **22**: 457–458.

Haware, M.P. and Joshi, L.K. 1974. Studies on soft rot of ginger from Madhya Pradesh. *Indian Phytopath.* **27:** 158–168.

Haware, M.P. and Joshi, L.K.. 1973. Basal rot of ginger (*Zingiber officinale*) caused by *Sclerotium rolfsii* from Madhya Pradesh. *Indian Phytopath.* **26**: 575–576.

Haware, M.P. and Joshi, L.K.. 1974. Studies on soft rot of ginger from Madhya Pradesh. *Indian Phytopath.* **27**: 158–168.

Haware, M.P., Joshi, L.K. and Sharma, N.D. 1973. Effect of postharvest treatment of aureofungin on rhizome rot and viability of ginger seed rhizomes. *Hind. Antibio. Bull.* **15**: 84–85.

Hayward, A.C., Moffett, M.L. and Pegg, K.G. 1967. Bacterial wilt of ginger in Queensland. *Queensl. J. Agric. Anim. Sci.* **24**: 1–5.

Ichitani, T. 1980. Control of rhizome rot of ginger cultivated successively and protectively for immature rhizome production in plastic house. *Proc. Kans. Plant Proc. Soc.* **22**: 7–11.

Ichitani, T. and Shinsu, T. 1980. *Pythium zingiberum* causing rhizome rot of ginger plant and its distribution. *Ann. Phytopahtol. Socl. Jpn.* **46**: 435–441.

Indrasenan, G. and Paily, P.V. 1974. Studies on the soft rot of ginger (*Zingiber officinale* Rosc.) caused by *Pythium aphanidermatum* (Edson) Fitz. *Agric. Res. J. Kerala* **11**: 53–56.

Indrasenan, G., Kumar, K.V., Mathew, J. and Mamen, M.K. 1981. The mode of survival of *Pseudomonas solanacearum* (Smith) Smith causing bacterial wilt of ginger (*Zingiber officinale* Rosc. *Agric. Res. J. Kerala* **19**: 93–95.

Indrasenan, G., Kumar, K.V., Mathew, J. and Mamen, M.K. 1982. Rection of different types of ginger to bacterial wilt caused by *Pseudomonas solanacearum* (Smith) Smith. *Agric. Res J. Kerala* **20**: 73–75.

Ishii, M. and Aragaki, M. 1963. Ginger wilt caused by *Pseudomonas solanacearum* E.F. Smith. *Plant Dis. Reptr.* **47**: 710–713.

Iyer, R. 1988. Diseases of ginger. pp. 251–258. In: *Review of Tropical Plant Pathology* Vol. **4**. (Eds. Raychaudhuri, S.P. and Verma, J.P.). Today and Tomorrow's Printers and Publishers, New Delhi.

Kothari, K.L. 1966. Effect of fungicides on germination and growth of ginger plants. *Sci. & Cult.* **32:** 48–49.

Lana, M.M., Casali, V.W.D., Finger, F.L. and Reis, F.P. 1993. Evaluation of postharvest storage of ginger rhizomes. *Hortic. Bras.* **11**: 139–141.

Lee, W.H, Cheong, S.S. and So, I.Y. 1990. Properties of suppressive and conducive soils of ginger rhizome rot *Korean J. Plant Pathol.* **6**: 338–342.

Lum, K.V. 1973. Cross inoculation studies of *Pseudomonas solanacearum* from ginger. *MARDI Res. Bull.* **1**: 15–21.

Mathew, J., Abraham, K., Indrasenan, G. and Samuel, M. 1979. A new record of bacterial wilt of ginger incited by *Pseudomonas solanacearum* E.F. Smith from India. *Curr. Sci.* **48**: 213–214.

Mathur, S., Thakore, B.B.L. and Singh, R.B. 1985. Effect of different fungicides on ginger rhizome rot pathogen and their effect on germination and rotting of rhizomes. *Indian J. Mycol, Pl. Pathol.* **14**: 155–157.

Mishra, B. and Rath, G.C. 1989. Geotrichum rot of stored ginger. *Indian J. Mycol. Pl. Pathol.* **18:** 213.

Mitra, M. and Subramanian, LS. 1928. Fruit rot disease of cultivated cucurbitaceae caused by *Pythium aphanidermatum* (Edson) Fitz. *Mem. Dep. Agric. India (Bot. Ser.)* **15**: 79–84.

Mulya, K., Shiomi, T. and Oniki, M. 1990. Bacterial wilt disease on industrial crops in Indonesia. *Ind. Crops Res. J.* **2**: 30–36.

Nambiar, K.K.N an Sharma YR. 1974. Mosaic diseases of ginger (*Zingiber officinale*). *Arecanut and Spices Bull.* **6**: 3–4.

Nath, P.D. 1993. Effect of shade and treatment for rhizome rot of ginger. *Ann. Agric. Res.* **14**: 327–328.

Ojha, K.L, Yadav, B.P. and Bhagat, A.P. 1986. Chemical control of bacterial wilt of ginger. *Indian Phytopath.* **39**: 600–601.

Okwouwulu, P.A. and Nnodu, E.C. 1988. Some effect of pre-storage chemical treatments and age at harvesting on the storability of fresh ginger rhizomes (*Zingiber officinale* Rosc.). *Trop. Sci.* **28**: 123–125.

Orian, G. 1953. Botanical Division. *Rep. Dep. Agric. Mauritius,* 1952, pp. 37–40.

Park, M. 1934. Report on the work of Mycological division Ceylon. *Admn. Rep., Rep. Dir. Agric.,* 1933, pp. D126–130.

Park, M. 1935. Report on the work of Mycological division Ceylon. *Admn. Rep., Rep. Dir. Agric.,* 1934, pp. D124–131.

Park, M. 1941. Report on the work of Mycological division Ceylon. *Admn. Rep., Rep. Dir. Agric.,* 1939, pp. D20–22.

Pegg, K.G., Moffett, M.L. and Colbran, R.C. 1974. Disease of ginger in Queensland. *Queensl. Agric. J.* **100**: 611–618.

Pordesimo, A.N. and Raymundo, S.A. 1963. Rhizome rot of ginger and its control. *Coffee Cacao J.* **5**: 240.

Ramachandran, N., Dake, G.N. and Sarma, Y.R. 1989. Effect of systemic fungicides on *in vitro* growth of *Pythium aphanidermatum,* the rhizome rot pathogen of ginger. *Indian Phytopath.***42**: 463–465.

Ramakraishanan, T.S. 1942. A leaf spot disease of *Zingiber officinale* caused by *Phyllosticta zingiberi* n.sp. *Proc. Inian Acad. Sci. Sect. B.* **20**: 167–171.

Ramakraishanan, T.S. 1949. Occurrence of *Pythium vexans* deBary in south India. *Indian Phytopath.* **2**: 27–30.

Rana, K.S. 1991. Effect of seed selection in the management of yellows disease of ginger. *Indian J. Mycol. Pl. Pathol.* **21**: 183–185.

Rana, K.S. and Arya, P.S. 1991. Rhizome rot and yellows disease of ginger in H.P. *Indian J. Mycol. Pl. Pathol.* **21**: 60–62.

Rath, G.S., Mishra, D. and Mishra, B. 1979. Fungi causing rhizome rot of ginger. *Indian Phytopath.* **31**: 387.

Rathaiah, Y. 1987. Control of soft rot of ginger with Ridomil. *Pesticides* **21**: 29–30.

Rathore, VRS, Kusum Mathur and Lodha, B.C. 1992. Activity of volatile and non-volatile substances produced by *Trichoderma viride* in ginger rhizome rot pathogen. *Indian Phytopath.* **45**: 253–254.

Sadanandan, A.K. and Iyer, R. 1986. Effect of organic amendments onrhizome rot of ginger. *Indain Cocoa Arecanut Spices J.* **9**: 94–95.

Samuel, M. and Mathew, J. 1986. Role and association of root-knot nematode *Meloidogyne incognita* in induction of bacterial wilt of ginger incited by *Pseudomonas solanacearum. Indian Phytopath.* **36:** 398–399.

Setty, T.A.S., Guruprasad T.R., Mohan, E. and Reddy, M.N.N. 1995a. Susceptibility of ginger cultivars to rhizome rot at West Coast conditions. *Environ. Ecol.* **13**: 242–244.

Setty, T.A.S., Guruprasad, T.R., Mohan, E. and Reddy, M.N.N. 1995b. Susceptibility of ginger cultivars to Phyllosticta leaf spot at West Coast conditions. *Environ. Ecol.* **13**: 443–444.

Shahare, K.C. and Asthana, R.P. 1962. Rhizome rot of ginger and its control. *Indian Phytopath.***15**: 77–78.

Sharma, I.D., Dohroo, N.P, Dubey JK an Korla BN. 1992. Monitoring mancozeb and carbendazim residues in ginger (*Gingiber officinale* Rosc.) following postharvest dip. *Pl. Dis. Res.* **7**: 13–15.

Sharma, M.P, Thakore, B.B.L. and Singh, R.D. 1980. Control of rhizome rot of ginger by seed and soil treatment.s *MACCO Agric. Dig. Res. Bull.* **5**: 7–9.

Sharma, N.D. 1977. Survival of *Fusarium oxysporum* f.sp. *zingiberi* as competitive saprophyte on agar plate. *Kavaka* **5**: 29–30.

Sharma, N.D. and Jain, A.C. 1977. A checklist and selected bibliography of ginger diseases of the world. *PANS* **23:** 474–481.

Sharma, N.D. and Jain, A.C. 1978. Studies in biocontrol of *Fusarium oxysporum* f.sp. *zingiberi,* the causal organism of yellows disease of ginger. *Indian Phytopath.* **31**: 260–261.

Sharma, N.D. and Jain, A.C. 1979. Studies on biological control of *Fusarium oxysporum* f.sp. *zingiberi*, the causal agent of yellows disease of ginger. *Indian Phytopath.* **31(2):** 260–261.

Sharma, S.K. and Dohroo, N.P. 1989. Effect of soil hydrothermal regimes on the development of ginger yellows. *Indian J. Plant Pathol.* **7**: 109–111.

Sharma, S.K. and Dohroo, N.P. 1990. Occurrence and distribution of fungi causing ginger yellows in H.P. *Pl. Dis. Res.* **5**: 200–202.

Sharma, S.K. and Dohroo, N.P. 1991. Postharvest management of rhizome rot (*Fusarium oxysporum* f.sp. *zingiberi* Trujillo) of ginger through chemical and antagonist. *Indian Cocoa, Arecanut Spices J.* **14**: 150–152.

Sharma, S.L. and Dohroo, N.P. 1980. Efficacy of chemicals in controlling rhizome rot of ginger and turmeric (ICAR) CPCRI, Calicut, 8–9 April, 1980, pp 65.

Sharma, S.L. and Dohroo, N.P. 1985. Studies on macerating enzymes produced by two rhizome rot causing fungi in ginger. *Indian J. Mycol. Pl. Pathol.* **15**: 87–89.

Shukla, Arti and Gupta, S.K. 2015. Management of rhizome rot of ginger under mid hill conditions of Himachal Pradesh. *J. Krishi Vigyan* **3(2)**: 102–104.

Shukla, B.N. and Haware, M.P. 1972. Phyllosticta leaf spot of ginger (*Zingiber officinale*) in Madhya Pradesh. *Indian J. Mycol. Pl. Pathol.* **2**: 93.

Simmonds, J.H. 1955. Science Branch, Plant Pathology Section. *Rep. Dep. Agric. Queensl.* 1953–54, pp 55–56.

Singh, A.K., Singh, S. and Edison, S. 2004. Effect of shading on the Phyllosticta leaf spot, sun burn of leaves and yield of ginger. *Indian Phytopath.* **57(2)**: 197–199.

Sohi, H.S., Jain, S.S., Sharma, S.L. and Verma, B.R. 1964. New records of plant diseases from Himachal Pradesh. *Indian Phytopath.* **17**: 35–41.

Sohi, H.S., Jain, S.S., Sharma, S.L. and Verma, B.R. 1973. Chemical control of Phyllosticta leaf spot of ginger (*Zingiber officinale*). *Pesticides* **7**: 21–22.

Srivastava, L.S. 1994. Management of soft rot of ginger in Sikkim. *Pl. Dis. Res.* **9**: 146–149.

Su, I.Y. 1980. Studies on ginger mosaic virus. *Korean J. Plant Prot.* **19**:67–72.

Teakle, D.S. 1965. Fusarium rhizome rot of ginger in Queensland. *Queensl. J. Agric. Sci.* **22**: 263–272.

Thakore, B.B.L., Sneh Mathur, Singh, R.B. and Chakravarti, B.P. 1987. Soil amendment with oil cakes in ginger field for rhizome rot control. *Korean J. Plant Prot.* **26**: 267–268.

Thakore, B.B.L., Sneh, Mathur and Singh, R.B. 1988. Effect of rhizome treatment with fungicides for economic control of rot. *J. Phytol. Res.* **1**: 83–84.

Thomas, J.E. 1986. Purification and properties of ginger chlorotic fleck virus. *Ann. App. Biol.* **108**: 43–50.

Thomas, K.M. 1938. Detailed Administration Report of the Government Mycologist, Madras, 1937–38, 21pp.

Trujillo, E.E. 1963. Fusarium yellows and rhizome rot of common ginger. *Phytopathology* **53**: 1370–1371.

Verma, R.K. and Vyas, S.C. 1981. Persistence and protective activity of some fungicides in relation to Phytllosticta leaf spot of ginger. *Indian J. Mycol. Pl. Pathol.* **11**: 14–16.

Yang, K.D., Kim, H.M., Lee, W.H. and So, I.N. 1988. Studies on rhizome rot of ginger caused by *Fusarium oxysporum* f.sp. *zingiberi* and *Pythium zingiberum. Korean J. Plant Pathol.* **4**: 271–277.

■■■■

17

POST-HARVEST DISEASES OF VEGETABLES

India is the second largest producer of vegetables after China but a considerable part of their production is spoiled by micro-organisms between harvest and consumption, particularly in tropical areas where high temperature and relative humidity favour rapid microbial growth. This is an important aspect because the value of the produce is determined by the appearance and quality of vegetables. Furthermore, ethylene produced by rotting vegetables can cause premature ripening of other produce in the same storage and transport environment, and it can contaminate healthy ones also. Apart from actual losses due to wastage, further economic loss occurs if the market requirements necessitate sorting and repacking of partially contaminated consignments.

Several microorganisms can cause post harvest spoilage of vegetables. However, it is well established that the major postharvest losses are caused by species of fungi like *Alternaria, Botrytis, Botryosphaeria, Colletotrichum, Diplodia, Monilinia, Penicillium, Phomopsis, Rhizopus* and *Sclerotinia* and of bacteria like *Erwinia* and *Pseudomonas*. Most of these pathogens are weak parasites and invade only damaged tissues of the produce. However, few pathogens like *Colletotrichum* spp. are able to penetrate the skin of healthy produce. Vegetables may also be subjected to mechanical damage (wounds, cuts and bruises) in which the air borne spores of the microbes are capable of causing infection (Sharma and Sumbli, 1992). Complete loss of the commodity occurs when one or a few pathogens invade and break down the tissues. The initial attack is followed by a broad spectrum of weak pathogens that magnify the damage caused by the primary pathogens. Sometimes, the appearance of vegetables may be marred by surface lesions caused by the pathogenic organisms, without the internal breakdown of the affected tissues.

Losses

Post harvest loss has been defined as "that weight of wholesome edible product that is normally consumed by humans and that has been separated from the medium and the sites of its immediate growth or production by deliberate human action with the intension of using it for human feeding but which for any reason fails to be consumed by humans". Not only quantity and quality but even appearance of vegetables may be affected and their market valueis reduced. Post harvest losses in vegetables may vary from season to season in different areas of production. Losses are much higher in less developed countries of the tropics, where transportation and refrigeration facilities are not adequate. Vegetables are tender and succulent with fair amount of water content and these are very much vulnerable to injury and spoilage by microorganisms. Moreover, as the vegetables are generally kept at high relative humidities to avoid shrinkage, this provides an excellent substrate for the attack of the microorganisms like fungi and bacteria. The loss due to post harvest pathogens is influenced by the growing method, harvesting stage and storage conditions of the vegetables (Dennis, 1985). It is very difficult to assess the exact extent of losses due to post harvest diseases owing to complicated chain of events in the handling of vegetables from the field to the consumer, because they vary from season to season and the area where they are produced. Losses are substantial in tropics and have been estimated to range between 5–50 per cent or more of the harvest (Salunkhe and Desai, 1984). Losses are still on higher side in the developed countries with more advanced technologies (Singer, 1980; Moline, 1984). In India, 20–30 per cent losses of fresh fruits and vegetables have been reported (FAO, 1977). Some of these pathogens, in addition to loss of nutritional elements, may produce mycotoxins and toxic stress metabolites in the colonized vegetables, which is emerging as a major dietary problem (Verma *et al.*, 1981).

Infection process

Post harvest pathogens cause rotting of the vegetables either by infecting the produce while still immature and attached to the plant or during the harvesting and subsequent handling and marketing operations. The infection process, particularly during post harvest stage is greatly influenced by mechanical injuries to the skin of the produce such as scratches and abrasions, rough handling, insect punctures and cut stems. It is always better to understand the pattern of infection process so that proper management strategies are adopted to check the menace of these pathogens.

(a) Pre-harvest infection

The infection of vegetables at pre-harvest stage may occur through several avenues, such as direct penetration of the skin, through natural openings and through mechanical injuries. Different post harvest pathogenic fungi are able to initiate an infection on the surface of floral parts and on developing fruit and such infections remain quiescent until the harvest, when the resistance of the host decreases and conditions become favourable for growth, such as when the fruit begins to ripen or the tissue senesces. For example, spores of *Colletotrichum* spp. germinate in free moisture on the surface of bell pepper fruit and form appressorium within several hours of germination that may or may not penetrate the skin before the infection is arrested. Weak parasitic fungi and bacteria also gain entry to immature vegetables through natural openings like stomata, lenticels and growth cracks. Again these infections may not develop until the host becomes less resistant to the invading organism or the fruits ripen.

(b) Post-harvest infection

Many postharvest parasitic fungi are unable to penetrate the intact skin of the harvested vegetables, but readily invade via any rupture in the skin. Though the damage is very little but it is sufficient for the pathogens present on the crop and in the packinghouse to gain entry into the produce. In addition, the cut stem is a frequent point of entry for microorganisms, and stem end rots are important forms of postharvest wastage of many vegetables like tomato. Postharvest infection can also occur through direct penetration of the skin by *Sclerotinia* and *Collettrichum* species.

Factors affecting development of infection

Surrounding environment is probably the most important factor affecting the development of postharvest damage of the produce. Generally high temperature and high humidity favour the development of postharvest decay and chilling injury generally predisposes the tropical and subtropical produce to postharvest decay. However, low temperature, low oxygen, and high carbon dioxide levels and the required humidity can restrict the rate of postharvest decay by either retarding the rate of ripening or senescence, depressing the growth of the pathogen or both.

Several other factors affect the rate of development of infection in vegetables. The pH of host tissue acts as a selective medium for the development of postharvest decay. Many vegetables have a pH above 4.5 and are largely attacked by bacterial pathogens.

A large number of postharvest diseases of vegetables have been reported from India (Sohi and Parkash, 1972; Prasad and Ambasta, 1988; Sharma and Sumbli, 1993). Some of the economically important diseases are described in detail.

1. ALTERNARIA ROT OF TOMATO

This is an important postharvest disease of tomato and about 9.9 per cent incidence of the disease has been reported from Himachal Pradesh (Sharma, 1994).

Symptoms

Lesions caused by *A. alternata* are found near the stem-scar or alternatively at the blossom end of the fruit (McColloch and Worthington, 1952). Decayed tissue is firm dry and brown to black. Internal cavities may form lines with dark grey mould.

The Pathogen

The disease is caused by *Alternaria alternata* (Fr.) Keissler. The description of the pathogen is given in the tomato chapter in the book. The pathogen is very common in the air and in soil, where it survives in plant debris. Infection of the fruits can occur via dying floral parts and also through insect injury, cracks at the stem end etc. (Halfon- Meiri and Rylski, 1983). The optimum temperature for the fungus growth is 28ºC but it can make significant progress at cold storage temperature. A toxin production by the fungus is also reported (Stinson *et al.*, 1981).

2. SOUR ROT OF TOMATO

This disease was first reported in 1922 (Jones *et al.*, 1991) and since then it is occurring in all tomato growing areas of the world. The disease causes losses in both fresh market and processing tomatoes. About 17.5 per cent incidence of this disease has been reported from different collection centers of Himachal Pradesh (Sharma, 1994).

Symptoms

On green fruits, sour rot lesions have a dull, greasy, water-soaked to bleached appearance. Although the decay may originate at cracks, cuts or skin punctures, it frequently starts at the edge of stem scar. From there, the decay may extend in a sector over the shoulder and down toward the blossom end of the tomato, or it may develop all around the stem scar and progress uniformly out over the shoulder area. The affected tissues remain fairly firm until the decay is quite advanced. Affected fruits have a pickled appearance and a definite sour odour. Such fruits either in transit or in the ripening rooms may show a dingy white, scum like growth of the causal fungus in the stem scar or at points where the skin is broken. In ripe fruits, the disease progresses fast, particularly under warm conditions and the epidermis covering the lesions usually cracks, allowing the watery contents to spill out. A white to dirty white, crusty, yeast like growth develops at the breaks in the fruit surface associated with the lesion.

The Pathogen

The fungus responsible for this disease is *Geotrichum candidum* Link. Its mycelium is hyaline and septate and branches dichotomously. Conidia are formed in chains by disarticulation at the septa of the lateral hyphae. The conidia are cylindrical, barrel shaped, ellipsoidal or sub-globose and measure 3–6 x 6–12 μm.

The fungus is a common inhabitant of soil and aquatic environments. Fruits may become contaminated with the initial inoculum through contact with fruit flies and other insects, machinery, splashing rainfall, decaying vegetation, harvesters and other sources. In processing plants and packing-houses, the fungus may grow on pulp residue remaining on equipment. Optimum temperature for the disease is 30ºC.

3. ANTHRACNOSE OF BELL PEPPER

The disease is present wherever pepper crop is grown throughout the world. Commonly bell pepper is marketed green and rarely shows anthracnose except at wounds or following other diseases like blossom end rot. In the collection centers and packinghouses of Himachal Pradesh, the incidence of this disease ranged from 13.2 to 19.4 per cent (Sharma, 1998).

Symptoms

Symptoms of the disease first appear as small water-soaked spots, which later turn dirty green to greenish brown and become slightly sunken. The margins of the spots are definite and circular in outline. The spots become covered with raised black pustules, which are sometimes arranged in concentric rings. Under moist conditions these pustules exude creamy to pink spore masses. The decay penetrates the fruit walls and discolour the seed cavity and seeds.

The Pathogen(s)

The fungi responsible for this disease are *Colletotrichum capsici* (Sydow) Butler and Bisby and *C. gloeosporioides* (Penz.) Penz. & Sacc. The description of both fungi is given in the chapter of chilli and bell pepper diseases in the book.

The fungi responsible for this disease are seed borne and also survive in the infected plant debris. These fungi can invade both green and ripe fruits without wounds. It is believed that early infections of fruit in the field, too small to be graded out at packing time are responsible for the disease spots found on the fruits in the market. Warm, humid or rainy weather favours the spread of the spores to the fruit in the field and the rapid development of decay.

4. PHYTOPHTHORA BLIGHT OF BEANS

This is an important disease of both French and lima beans and in favourable weather causes significant losses. After its first report in 1923, the disease has been reported from many countries of the world including India (Malik, 1992). The disease causes losses by rendering the green pods unfit for table purpose. The crop losses may range from 50 to 90 per cent under favourable conditions. A severe attack of *Phytophthora* was noticed on bush type bean cultivars in Himachal Pradesh, India and the incidence ranged from 40–74 per cent (Malik, 1992).

Symptoms

On lima beans, the fungus (*Phytophthora phaseoli*) occurs on the pods either in patches or covering them completely. The fungus grows through the pod wall into the grains resulting in drying and ultimate blackening of the pods.

In French beans, the rotting of the pods start from the distal end with conspicuous white cottony growth of the fungus on the affected portion. Usually one third to two third portion of the pods is rotten and the remaining portion remains healthy. The young green diseased pods show twisting and curling. In advanced stages of infection, the pods finally become dry and remained hanging to the plant.

The Pathogen

The disease is caused by several species of *Phytophthora* like *P. parasitica* Dast. and *P. nicotianae* var. *nicotianae* (Malik, 1992). Since *Phytophthora phaseoli* Thaxter is known to attack lima bean only (Ribeiro, 1978 and Erwin *et al*., 1983), here only *Phytophthora* blight caused by *P. nicotianae* var. *nicotianae* is described. The pathogen has been described in chilli and bell pepper diseases chapter in the book.

The pathogen being soil borne and seed borne in nature, persists in infected pods, seed or other plant debris. The pathogen overwinters in the form of oospores. Moderate temperatures and moist conditions favour disease development.

5. DRY ROT OF POTATO

This is an important disease of potato and causes extensive losses in storage and transit.

Symptoms

The disease causes dry rot of infected tubers, although a moist rot may occur if secondary infection with soft rot bacteria is involved. The surfaces of the infected tubers are sunken or wrinkled and rotten tissues are brown or grey to black. White or pink mold growth may be visible on tuber surfaces. Internal cavities often develop in rotten tissues and contain white, yellow or pink moulds that are visible when tubers are cut. In storage, blue black, purple, grey, white, yellow or pink spore masses may form in these cavities. At low storage temperatures, internal tissues often become firm and dry or even powdery.

The Pathogen

The disease is caused by species of *Fusarium*. The fungus is soil borne in nature and survives in the form of chlamydospores. Dry rot does not spread in the storage if the potatoes are properly healed. However, the disease is known to spread occasionally from diseased tubers to adjacent healthy tubers. Dry rot affected tubers are unfit for processing.

6. NECK ROT OF ONION

The disease is one of the most serious diseases affecting onions during transit, storage and marketing. Losses during storage are often extensive because the fungus is slowed down, but not inhibited at cold storage temperatures. White onions, however, are the most seriously affected. Coloured onions have a considerable measure of resistance because of their chemical composition. The disease can cause up to 90 per cent rotting of the stored onions under favourable conditions (Botter, 1973). The disease is caused by three different species of *Botrytis* each of which is responsible for this disease. However, they differ somewhat in symptoms. On the basis of symptoms produced by different species, the name was given as grey mould neck rot (*B. allii*), mycelial neck rot (*B. bysoidea*) and small sclerotial neck rot (*B. squamosa*). Among these grey mould neck rot is most common.

Symptoms

The first indication of the disease is water soaking and softening of affected scales. The water-soaked conditions give the affected scales a cooked appearance. The decayed tissues lose moisture and become sunken. Disease appears on the bulbs after harvest. Infection usually takes place through neck tissue and occasionally else where. The first sign is the softening of the affected tissue. A dense grayish mycelial mat later develops upon the decaying tissues of the scale bearing short conidiophores with conidia. Black sclerotial bodies are developed later on. When diseased bulbs are cut open, water soaked brown tissues are seen near the neck region.

The Pathogen

Three different species of *Botrytis* like *B. allii* Munn, *B. byssoidea* J.C. Walker and *B. squamosa* J.C. Walker cause this disease. Appearance of the conidiophore as grape bunch is the characteristic of *Botrytis*. The conidiophores are tall, erect, branching irregularly or dichotomously. They are dark and septate. The terminal cells swell to produce sporogenous ampullae. On each ampulla numerous conidia arise simultaneously on short denticles. The conidia are hyaline or tinted, aseptate and globose to ovoid. Colonies become grey in culture with the powdery mass of spores.

Although symptoms become apparent at or after harvest but the main source of infection is diseased seed (Trichelaar, 1970). *B. allii* is carried with in seed and remain viable for more than 3 years in seed stored at low temperature. The incidence of the disease is directly correlated to the percentage of the infected seed sown and to the weather conditions also during the growth period of the crop. The other sources of infection are diseased onion sets, soil borne inoculum like plant debris and sclerotia. The disease is prevalent when cool, moist weather prevails before and during harvest.

7. ONION SMUDGE

The disease is of common occurrence on white onions. The occurrence of the disease has been reported from Europe, north-eastern and mid western United States and India. Its importance consists chiefly in the reduction in market value due to blemishes on the outer scales. Under very moist conditions in transit and during storage scales are destroyed, bulbs shrink and often sets sprout pre-maturely. Smudge is confined almost entirely to the scales and lower portions of the unthickened leaves.

Symptoms

Characteristic symptoms of this disease appear as small, dark-green to black dots on the outer scales. These small dots may be grouped together to form concentric rings or blotches, giving a smudge appearance to the white onions. During moist conditions, acervuli of the fungus develop on the infected areas and spore masses and setae are easily recognized. The disease may spread to the underlying fleshy scales resulting in yellow depressions over bulb.

The Pathogen

The fungus responsible for this disease is *Colletotrichum circinans* (Berk.) Voglino. Mycelium is septate, hyaline, the walls becoming darker and thicker and cytoplasm becomes dense with age. In culture, the fungus produces dark brown colonies with vinaceous aerial mycelium through which abundant stroma protrude. Intercalary chlamydospores are also formed. The thick walled hyphae interwine and form dark green to black stromata. They sometimes coalesce. Acervuli are formed on stromata beneath the cuticle by formation of a palisade layer of short, hyaline conidiophores, which later rupture the cuticle. Scattered all over the acervulus are numerous thick-walled, dark coloured, zero to three-septate setae, which may be 80 to 315 μm long. The conidia are fusiform falcate, hyaline measuring 14–30 x 3–6 μm in diameter.

The fungus overwinters on onion scales in the soil or in bulbs and sets in storage. In field the disease appears shortly before harvest. Under favourable conditions the fungus attacks the outer scales, and spores that are borne on the infected areas, are carried by drops of water to other scales, where they germinate in few hours and cause infection. The disease develops and spread most rapidly in the field when the soil temperature is around 25°C alongwith abundance of rainfall.

8. BROWN ROT OF ONION

It is a bacterial disease which causes internal browning of onions and a very serious disease in storage particularly in subtropical countries including India (Gupta *et al.*, 1991; 1994).

Symptoms

Dark brown discolouration in bulb scales is the characteristic symptoms of the disease. In some bulbs, necrotic tissue may be completely surrounded by healthy scales. The necrotic scales are generally firm except in the neck region where a soft rot can be observed due to secondary infection. In advanced stages of the infection, the tissue collapse and disease spreads to the neighbouring bulbs.

The Pathogen

The disease is caused by bacterium *Pseudomonas aeroginosa* (Schroeter) Migula. Bacteria are motile with 1–4 polar flagella, pale dirty-white colony on nutrient agar, that turns the agar to reddish grey and the colonies become translucent and flat with irregular edges after 3 days. The bacterium is unable to utilize sucrose. An important characteristic of the bacterium is the ability to utilize erythritol and creatine (Cother *et al.*, 1976).

Bacterium remains viable in field soil and spreads through surface drainage water. The infection occurs through the wounds. The rot begins at the neck of the bulbs, which later gives foul smell through the neck when squeezed.

9. BLUE MOLD

It is a wide spread disease of bulbs throughout the world. The disease is caused by several species of *Penicillium* which are frequently isolated from diseased bulbs of local cultivars.

Symptoms

Early symptoms are visible as pale yellowish blemishes, watery soft spots or purplish red stains on the outer side of the bulb. Initial infections are visible at the base of the bulbs as masses of blue green spores. Later infection may be hidden within the tissue and therefore are not noticeable. Tissue decaying as a result of infection by *Penicillium* usually has a musty odour.

The Pathogen

The disease is caused by various *Pencillium* spp. Raju and Naik (2007) identified the causal organism as *P. digitatum* Sacc.

The pathogen survives and multiply on growing plants and senescing plant or within soil (Summer, 1995). *Penicillium* infects tissue through wounds but the scales that have healed over with callus tissue are not penetrated by the pathogen. Secondary infection takes place from the spores produced on infected bulbs which become air-borne. After infection, the mycelium grows throughout the fleshy tissues. The optimum temperature for conidia germination and growth is in between 21 to 25°C though the growth and germination can occur at temperature range of 15 to 32°C.

10. SOFT ROT

This is an important disease of vegetables and tuber crops and causes considerable losses during transit and storage. Lesions on the diseased fruits typically have an unpleasant aroma and texture.

Symptoms

Tomato

Symptoms of the disease first appear as small, sunken, light to dark coloured water soaked areas adjacent to or associated with injuries and natural openings. Bacterial ooze may flow from wounds, breaks in the epidermis over lesions, stem scars or open stylar pores. Stem scars and open stylar pores allow the development of lesions inside the fruit. Early external symptoms of internal lesions include softness on parts of the surface and dark, water soaked areas adjacent to the stem scar or stylar pores. Green fruits are more susceptible to decay from bacterial soft rot than ripe fruits.

Onion

The characteristic symptoms of the disease appear as water-soaked lesions on the outer scales which become soft in due course of time. Due to infection the inner scales of the bulb start emitting foul odour. Sometimes bulbs appear sound from outside but when cut open 1–2 scales are found infected.

The Pathogen

The disease is caused by bacterium *Erwinia carotovora* sub sp. *carotovora* (Jones) Bergey *et al*. The bacterium is described in other chapters in the book.

Soft rot bacteria are ubiquitous and survive apart from the host in a variety of ways. They often survive as epiphytes on various fleshy plant organs or as very low populations in soil or aquatic environments. When introduced into an infection court under favourable environmental conditions multiply rapidly to high numbers. Major outbreaks of bacterial soft rot in harvested fruit often accompany improper management of the water system in packinghouses. When fruits are dumped into or washed with recycled water, there is a accumulation of microbial population. If the water is not treated with suitable chemical than the microbes from water may cause infection of the fruits passing in it.

MANAGEMENT OF POST-HARVEST DISEASES

Since the post-harvest diseases cause significant losses, proper management practices should be followed in such a way that the practices may not aggravate the problems by causing bruises on the produce and leave any toxic residue until it is consumed. The post-harvest losses can be kept under check by using pre-harvest and post-harvest treatments including physical, chemical and biological methods.

Pre-harvest measures

Pre-harvest treatments are one of the best methods to combat the ravages of post – harvest diseases because several pathogens cause latent infections in the field, which may be difficult to control after harvest and affect the self-life of the produce. These incipient infections present before harvest may subsequently develop in to rots and spread to adjacent vegetables during transportation, storage and marketing. Therefore, prevention of pre-harvest infection appears to be more logical way of preventing post-harvest losses in fruits and vegetables (Pathak, 1995). Several fungal and bacterial pathogens attack vegetables during various handling operations required to move the harvested crop from the grower to the consumer. Vegetables may be subjected to mechanical damage in which the air-borne spores of microbes are capable of causing infection (Sharma and Sumbli, 1992). Wherever possible, sources of infection like seed borne inoculum, overwintering structures in plant debris etc. should be eliminated and sprays for the control of the causal agent be applied. Pre-harvest sprays of the fungicides should prevent decay in the field, reduce inoculum level and minimize quiescent infections that give rise to storage rots.

Although not much work has been done on the management of post-harvest vegetable rots through pre-harvest application of fungicides, but it would be much economical to check the invasion of several fungal pathogens before harvest so as to avoid tremendous losses during transit and storage. Seed treatment of onion seed with benomyl (0.1%) virtually eliminated the neck rot (*Botrytis allii*) in onion (Trichelaar, 1970) whereas considerable reduction of disease in stored bulbs was obtained by dusting sets with Rovral before planting. Potato seed treatment with benzimidazole fungicides reduced tuber diseases on the crop during storage (Cayley *et al.*, 1983). Pre-harvest sprays of metalaxyl formulations on potato crop contains sufficient amount of this fungicide to protect the harvested tubers from rotting due to *Phytophthora infestans* for 4 months of storage (Barak *et al.*, 1984). Similarly field applications of Ridomil MZ a combination product of metalaxyl with mancozeb, and other phenylamide fungicide mixtures have been observed to reduce tuber infection in field and their subsequent rotting after harvest (Thind *et al.*, 1989). Application of captafol, dimethyl sulphoxide, maneb and formaldehyde considerably reduced the population of viable *Alternaria* spores on the soil surface and prevented tuber infections (Lahman *et al.*, 1982). Pre-harvest sprays of benomyl, thiabendazole and dicarboximide have been reported to reduce *Botrytis* decay of cabbage during storage while benomyl and vinchlozoline reduced Sclerotinia rot of carrot (Tahvonen, 1985). Likewise, storage rots of brinjal caused by various fungi were completely controlled by pre-harvest sprays of carbendazim (Kaur and Bedi, 1989). Effect of pre- and post-harvest treatment on shelf life and quality of carrot was studied under storage conditions by Paul *et al.* (1991). Carrots (cv. Nantes) were sprayed with maleic hydrazide (MH) at 1000, 1500, 2000 and 2500 ppm at 25 days before harvesting. After harvest carrots were stored in a cool chamber. The shelf life of carrots was extended from 5 days at room temperature to 33 days in a cool chamber on application of bleaching powder as a pre-storage treatment with or without pre-harvest application of 1500 ppm of MH. The quality of the carrots after storage was highest with pre-harvest application of 1500 ppm of MH at storage in a cool chamber. Application of chlorothalonil to tomato crop 4 and 6 weeks prior to harvest in fields reduced the incidence of black mould to 50 per cent and increased marketable fruit yields and shelf life of the produce.

Careful handling during harvesting can minimize mechanical damage and reduce subsequent wastage due to microbe attack. Similarly, it is also unwise to harvest fruits of tomato and other vegetables immediately after rain or dew, as the peel is turgid and can be easily damaged.

Post-harvest measures

(i) Physical treatments: Several physical methods like low and high temperatures, modified atmosphere, correct humidity, ionizing radiations, good sanitation, development of wound barriers etc. may be effective in controlling postharvest wastage of fruits. Low temperature handling and storage is the most important physical method of post-harvest wastage control and the remaining methods can be considered as supplements to low temperature. The extent to which low temperature and other environmental modifications can be used to control wastage, depends on the tolerance of the tissue to that environment. For instance, most of the tropical and subtropical produce is susceptible to chilling injury and therefore cannot be subjected to very low temperature. Vegetables like brinjals, sweet peppers, tomatoes, squash, cucumbers and green beans should be stored above 7.5ºC as they are cold sensitive. The storage of tomatoes at 5ºC caused undue loss of fruit quality and should be stored between 10-15ºC because at this temperature the growth of *Alternaria* spp. was checked fairly well (Sood and Sharma, 2003). Manipulation of temperature to promote healing or curing of tissues is very effective in preventing losses in many vegetable crops. Curing yams at 25–40ºC and 80–90 per cent RH for 5-6 days protects them from subsequent post –harvest fungal infections (Been *et al.*, 1977). Similarly in carrots also, curing promoted healing of wounds (Garrod *et al.*, 1982). Heat treatments in the form of either moist hot air or hot water dips have some commercial applications for the control of post-harvest wastage of vegetables. The advantage of hot water dipping is that it can control surface infections as well as infections that have penetrated the skin, and it leaves no chemical residues on the produce. Exposing the mature tomato fruits to higher temperature of 38ºC for 24 or 48 h reduced the disease proportionately in the *Alternaria* spp. infected tomatoes while fruit dip at 52ºC for 5 min. appreciably checked the *A. alternata* (Sood and Sharma, 2003). They further reported that combined hot water-mancozeb (500 ppm) treatment at 52ºC for 5 minutes was most effective in eliminating *A. tomato* and significantly reduced *A. solani* and *A. alternata.*

Field heat of the vegetables like tomatoes or bell peppers should be immediately removed to prolong the life of the produce. If the produce is packed as such due to high temperature of the produce, respiration will be more and the life of the produce is reduced. Moreover, microorganisms, which are present at the time of harvesting, will also survive and hasten the decay. Field heat can be removed either by keeping the harvested produce under sheds overnight particularly in hilly areas where temperature is less during night or keep in cold storage at 5–10ºC temperature for some time. This treatment will reduce the load of microorganisms, which are carried on the fruit surface and reduce the transpiration, which will ultimately prolong the life of the produce.

In addition, moisture also affects development of post-harvest diseases, which are initiated before harvest. Geeson (1976) reported that in post-harvest rot of cabbage caused by *Phytophthora porri*, the fungus, which survives in the soil as oospores or chlamydospores, enters the cut stem of the cabbage during harvesting operations by soil splash. Correlation between prevalence of cabbage storage rot and wet field conditions at harvesting has also been confirmed by Derbyshire and Shipway (1978). Likewise, post-harvest drying of onions within 48 h of topping, reduces the incidence of neck rot caused by *Botrytis allii* (Maude *et al.*, 1984). In potatoes, the wound pathogen *Fusarium avenaceum* is favoured by the presence of moisture in the wound itself rather than RH alone, whereas, the other wound pathogen *F. coerulweum* is more tolerant to dry conditions and low RH (Logan, 1983). Removal of field heat is also found effective in the management of various fungal diseases of vegetables. The vegetables after harvest are kept in the field in open during nights and the incipient infections of the pathogens, which it is carrying at the time of harvesting, are killed due to low temperature during nights.

Controlled or modified atmospheres (CA) are useful in reducing post-harvest losses of especially those vegetables that undergo ripening after harvest e.g. tomato. Storage atmosphere containing 5–10 per cent carbon monoxide (CO) and 4 per cent oxygen (O_2) reduces *Botrytis* decay in tomatoes (Kader *et al.*, 1978). Carbon monoxide is highly fungistatic to fungal pathogens when the oxygen level of the atmosphere is less than 5 per cent (Sommer *et al.*, 1981). CA storage is beneficial in controlling fungal diseases of cauliflower, whereas, it did not affect the development of *Botrytis* on stored lettuce. Rangana *et al.* (1997) have reported that ultraviolet irradiation completely suppressed the development of both dry and soft rot in potatoes stored at 8ºC for 3 months. The amount of disease suppression increased as the dose of UV increased but decreased with an increase in the pre-treatment incubation period. Dry rot was completely suppressed at UV dose of 15.0 kJm^{-2} when tubers inoculated with *Fusarium solani* were incubated at 28ºC for one day but not when incubated for 2 days. The soft rot was suppressed at a UV dose of 15.0 kJm^{-2} when tubers inoculated with cells of *Erwinia carotovora* were incubated at 37ºC for 6 h but not for 12 h.

Packaging material also play important role in reducing the losses due to post-harvest decay. Bell pepper fruits packed in wooden boxes with iodine and diphenylamine impregnated liners reduced the soft rot (*Fusarium* spp.) infections by 80 and 70 per cent, respectively (Sharma and Shukla, 2003) as compared to unpacked without paperliners.

(ii) Chemical treatments: Chemical control of post-harvest wastage has become an integral part of the handling and successful marketing of vegetables in the last decade. The level of control of wastage depends on the marketing strategy for the commodity and the type of infection. There is no point in treating a short-life commodity with a fungicide that has a long residual activity. The success of a chemical treatment for decay control depends on several factors like the initial spore load, the depth of the infection within the host tissues, growth rate of the infection, prevalent temperature and humidity and the depth to which the chemical penetrate the host tissues. Moreover, the applied chemical must not be phytotoxic to the host tissues.

Various chemicals having antimicrobial properties have been used for the control of some post-harvest decay. The incidence of dry rot (*Fusarium coeruleum*) and soft rot (*Rhizopus arrhizus*) in potato were reduced by disinfecting the fresh tubers with formalin (Thakur and Chenulu, 1974). Effectiveness of sodium o-phenylphenate (SOPP) in preventing black rot (*Ceratostomella fimbriata*) and soft rot (*Rhizopus* spp.) of sweet potato has been reported. Post-harvest dip of SOPP proved beneficial in controlling storage rots of carrots also. However, Derbyshire and

Crisp (1978) observed that benomyl and thiabendazole sprays before storage gave better control of carrot storage rots than SOPP. Foam prepared from SOPP (0.5%) and sodium lauryl sulfate (0.3%) was highly effective in controlling wound infection of tomatoes caused by *Geotrichum* and *Rhizopus* (Krochta *et al.*, 1977).

A wax formulation containing dichloran protected squash against *Sclerotinia sclerotiorum* and tomatoes against Botrytis rot (Chastagner and Ogawa, 1979). Post-harvest treatment of mature green tomatoes with imazalil or etaconazole inhibited *Alternaria alternata* infection on tomatoes and bell peppers (Spalding, 1980; Spalding and King, 1980). Imazalil proved to be even more effective than dichloran against Botrytis rot of tomatoes (Manji and Ogawa, 1985). Benzimidazoles applied to the tubers soon after harvesting proved highly effective in controlling fungal diseases of potatoes during long-term storage (Hide and Cayley, 1983). Certain surfactants having fungistatic properties which make them useful in reducing post-harvest decay. Surfactants or surface active agents are those compounds that reduce the surface tension of the solutions. Santomerase F 85 has been reported to reduce post-harvest decays caused by *Erwinia carotovora, Aerobacter cloacae* and *Alternaria alternata* when added to chlorine for washing tomato fruit. Likewise, fungitoxicity of surfactant Nacconol to four common decay fungi (*Botrytis cinerea, Geotrichum candidum, Phytophthora parasitica* and *Rhizopus stolonifer*) of fresh market tomatoes has been reported (Hay and Ogawa, 1984).

Several antibiotics have also been used for the control of post-harvest decay of vegetables particularly for eradicating bacterial soft rot of leafy vegetables, where an extension of market life by few days would be of great economic significance. Post harvest dip treatment with Aureofungin reduced Alternaria rot of tomatoes (Dharam Vir *et al.*, 1967). However, continuous low level of antibiotics in human diet may result in possible harm and hence their use in post-harvest treatment of vegetables is not recommended. Chilli fruits dipped in indol-3 –butyric acid or naphthalic acid (200 mg/l) for 30 min. delayed the fruit rot development by six days (Datar, 1996).

Biological control

The chemical applied for the control of post-harvest decay of the vegetables are known to leave residues which are toxic to the consumers and have also been reported to be either mutagenic or carcinogenic (Norman, 1988). Some rot causing pathogens have also developed tolerance to post-harvest fungicides (Spotts and Cervantes, 1986). With the introduction of WTO regime, sale of fruits and vegetables with detectable residues of fungicides will be difficult. Keeping this in view, search for alternative means for controlling post-harvest diseases have been stimulated and biological control by microbial antagonists has emerged an alternative to the use of synthetic fungicides (Wilson *et al.*, 1991; Droby *et al.*, 1992). Pang *et al.* (2000) described the following ways in which biocontrol agents are acting:

(i) Effective antagonists may inhibit post-harvest pathogens by producing antibiotics,

(ii) By successfully competing with the pathogens for nutrients and space by interacting directly with the pathogen,

(iii) By producing exolytic enzymes or by inducing host resistance,

(iv) By enhancing the activity of the antagonists naturally existing on the surface of the vegetables and

(v) By introducing effective antagonists and applying the antibiotics produced by antagonists.

It is always better to use the resident microflora isolated from the surface of the produce and soil to treat the vegetables. Wisniewski and Wilson (1992) described that an ideal antagonist for post-harvest disease control should be genetically stable, effective at low concentrations in may host-pathogen combinations, survive in adverse environmental conditions, able to colonize wound surfaces, not fastidious in its nutrient requirements, inexpensive to produce, with a long shelf life, easy to dispense, resistant to pesticides, does not produce metabolites that are deleterious to human health and be non-pathogenic to host, other plants, animals and human beings. Application of biocontrol agents can be done either by drenching or by dipping fruits in a spore suspension of antagonists, by applying with waxes or pressure infiltration can be used to facilitate penetration of the antagonist in order to affect pathogens present under the cuticle or in lenticels. The frequency and rate of application of biocontrol agents is dependent on the target pathogen, growth and survivability of the control agent after application and also on the type of propagules used (Janisiewicz, 1991).

Bacterization of seed potatoes before harvest and of the tubers after harvest with *Pseudomonas putida* isolate antagonistic to *Erwinia* spp. resulted in 50 and 70 per cent reduction of rotting, respectively (Colyer and Mount, 1984). Elson *et al.* (1997) have reported effective control of silver scurf (*Helminthosporium solani*) of potato tuber through the use of bioagents. They screened microbiota of 47 agricultural sorts for biocontrol of silver scurf and selected 13 strains for further study on three different strains of *H. solani*, including one that was resistant to thiabendazole. Microbial strains that significantly inhibited *H. solani* were identified as *Pseudomonas putida, Nocardia globerula* and *Xanthomonas campestris*. Two antagonistic bacteria like *Pseudomonas aeruginosa* and *Acinetobacter genospecies* were isolated and both inhibited the growth of *Erwinia carotovora* sub sp. *carotovora*, the causal agent of soft rot of cabbage, both *in vitro* and *in vivo* conditions (Adeline *et al.*, 1999).

REFERENCES

Adeline, T.S.Y., Sijam, K., Hong, L.W., Sastroutomo, S.S., Caunter, I.G., Ali, J., Yang, L.K., Vijayasegaran, S. and Sen, Y.H. 1999. Biological control of bacterial soft rot of cabbage. *Proc. Symp. Biol. Cont. Tropics, held at MARDI Trg. Centre, Serdang, Malaysia* w.e.f. 18-19 March, 1999, pp. 133–134.

Barak, E., Edgington, L.V. and Ripley, B.D. 1984. Bioactivity of the fungicide metalaxyl in potato tubers against some species of *Phytophthora, Fusarium* and *Alternaria* related to polyphenol-oxidase activity. *Can. J. Pl. Pathol.* **6:** 304–308.

Been, B.O., Peskins, C. and Thompson, A.K. 1977. Yam curing for storage. *Acta Hortic.* **62:** 311–316.

Botter, H. 1973. On the occurrence of diseases during onion storage. *Naethrichtenbl. Pflanzenschutz.* **27:** 217–222.

Cayley, G.R., Hide, G.A., Read, P.J. and Dunne, Y. 1983. Treatment of potato seed and ware tubers with imazalil and thiabendazole for control of silver scurf and other storage diseases. *Potato Res*. **26:** 163–173.

Chastagner, G.A. and Ogawa, J.M. 1979. A fungicide-wax treatment to suppress *Botrytis cineria* and protect fresh market tomatoes. *Phytopathology* **69:** 59–63.

Cloyer, P.D. and Mount, M.S. 1984. Bacterization of potatoes with *Pseudomonas putida* and its influence on post-harvest soft rot diseases. *Plant Dis*. **68:** 703–706.

Cother, E.J., Darbyshire, B. and Brewer, J. 1976. *Pseudomonas aeruginosa*. Cause of internal brown rot of onion. *Phytopathology* **66:** 828–834.

DahramVir, Raychaudhuri, S.P. and Thirumalachar, M.J. 1967. Aureofungin as fruit dip and wrap treatment for the control of Diplodia rot of mango and Alternaria rot of tomato fruits during transit. *Indian Phytopath*. **20:** 301–303.

Datar, V.V. 1996. Efficacy of growth regulators and fungitoxicants on fruit rot of chilli. *J. Mycol Pl. Pathol*. **26:** 239–242.

Dennis, C. 1985. Fungi. In: *Post-harvest Physiology of Vegetables* (Ed. Weichmann, J.), New York, Marcel Dekker Inc.

Derbyshire, D.M. and Crisp, A.F. 1978. Studies on treatments to prolong the storage life of carrots. *Exp. Hortic*. **30:** 23–28.

Derbyshire, D.M. and Shipway, M.R. 1978. Control of post-harvest deterioration in vegetables in the UK. *Outlook on Agric* **9:** 246–252.

Droby, S., Chalutz, E., Wilson, C.L. and Wisniewski, M.E. 1992. Biological control of post-harvest diseases: a promising alternative to the use of synthetic fungicides. *Phyto-parasitica* **20(Suppl.):** 149–153.

Elson, M.K., Schisler, D.A. and Botrast, R.J. 1997. Selection of microorganisms for biological control of silver scurf (*Helminthosporium solani*) of potato tuber. *Plant Dis.* **81:** 647–652.

Erwin, D.C., Burtnicki-Garcia, S. and Tsao, R.H. 1983. *Phytophthora its Biology, Taxonomy, Ecology and Pathology*. APS Press, The American Phyto-pathological Society, St. Paul, Minnesota, 55121 USA, 391pp.

FAO, 1977. Analysis of an FAO survey of postharvest crop losses in developing countries. *AGPP Misc. 27, FAO*, Rome.

Garrod, B., Lewis, B.G., Brittain, M.J. and Davis, W.P. 1982. Studies on the contribution of lignin and suberin to the impedance of wounded carrot root tissue to fungal invasion. *New Phytologist* **90:** 99–108.

Geeson, J.D. 1976. Storage rot of white cabbage caused by *Phytophthoraporri. Plant Pathol.* **25:** 115–116.

Gupta, R.P., Srivastava, K.J. and Pandey, U.B. 1991. Management of onion diseases and insect pests in India. *Onion Newslett. Trop*. **3:** 15–17.

Gupta, R.P., Srivastava, K.J., Pandey, U.B. and Midmore, D.J. 1994. Diseases and insect pests of onion in India. International Symposium on *Alliums* for the Tropics. *Acta Hortic*. **358:** 265–269.

Halfon-Meiri, A. and Rylski, I. 1983. Internal mold caused in sweet pepper by *Alternaria alternata. Phytopathology* **73:** 63–70.

Hay, M.W. and Ogawa, J.M. 1984. Toxicity of the surfactant Nacconal to four decay causing fungi of fresh market tomatoes. *Plant Dis*. **68:** 699–702.

Hide, G.A. and Cayley, G.R. 1983. Effects of delaying fungicide treatment of wounded potatoes on the incidence of Fusarium dry rot in store. *Ann. App. Biol*. **102:** 107–115.

Janisiewicz, W.J. 1991. Control of post-harvest diseases with biocontrol agents. In: The Biological Control of Plant Diseases. *Proc. Int. Semin. Biol. Cont. Plant. Dis. Virus Vectors 2nd FFTC Book Ser.* 42pp.

Jones, J.B., Jones J.P., Stall, R.E. and Zitter, T.A. 1991. *Compedium of tomato diseases*. APS Press, St. Paul, Minnesota, U.S.A. 73 p.

Kader, A.A., Chastagner, G.A., Morris, L.L. and Ogawa, J.M. 1978. Effects of carbon monoxide on decay, physiological responses ripening and composition of tomato fruits. *J. Am. Soc. Hortic. Sci.* **103:** 665–670.

Kaur, S. and Bedi, P.S. 1989. Relative efficacy of different fungitoxicants in controlling fungal fruit rot of brinjal. *Indian Phytopath.* **42:** 326.

Krochta, J.M., Carlson, R.A., Ogawa, J.M. and Manji, B.T. 1977. Harvesting into foam reduces tomato losses. *Food Tech.* **31:** 42–46.

Lahman, L.K., Harrison, M.D. and Knutson, K.W. 1982. Pre-harvest chemical application for control of tuber infection by *Alternaria solani. Am. Potato J.* **59:** 523–532.

Logan, C. 1983. Potatoes. In: *Post-harvest Pathology of fruits and Vegetables* (Ed. Dennis, C.) pp. 179–217. Academic Press, London.

Malik, P.K. 1992. Studies on Phytophthorta blight of French bean (*Phaseolus vulgaris* L.). M.Sc. Thesis, Deptt. Pl. Pathol., HPKVV, Palampur. pp. 72.

Manji, B.T. and Ogawa, J.M. 1985. Suppression of *Botrytis cinerea* lesions in fresh market tomato fruits with fungicides. *Phytopathology* **75:** 1329 (Abstr.).

Maude, R.B., Bambridge, J.M. and Spencer, A. 1984. Storage rots of onions. In: *Rep. Int. Veg. Res. Stn., Wellesbourne*, pp.64–65.

McColloch, L.P. and Worthngton, J.T. 1952. Low temperature as a factor in the susceptibility of mature-green tomatoes to Alternaria rot. *Phytopathology* **42:** 425–427.

Moline, H.E. 1984. Post-harvest pathology of fruits and vegetables: Postharvest losses in perishable crops. *Univ. Calif. Agric. Exp. Stn. Bull.* 80pp.

Norman, C. 1988. EPA sets new policy on pesticides cancer risks. *Science* **242:** 366–367.

Pang, X.Q., Zhang, Z., Dong, C., Pang, X.Q., Zhang, Z.Q. and Dong, C. 2000. Biocontrol of post-harvest diseases of fruits and vegetables by antagonism. *Acta Hortic. Sinica* **27(Suppl.): 546–552.**

Pathak, V.N. 1995. Post-harvest disease management in developing countries. *Global Conf. Adv. Res. Pl. Dis. Mgt.* RCA, Udaipur, Feb. 12–17, 1995 (Abstr. 167).

Paul, R.K., Roy, K.S. and Wasker, D.P. 1991. Effect of pre and post-harvest treatments on shelf life and quality of carrot under different storage conditions. *Maharashtra J. Hortic.* **5:** 98–105.

Prasad, S.S. and Ambasta, K.K. 1988. Fruit rot of spongegourd. *Indian J. Mycol. Pl. Pathol.* **17:** 235.

Raju, K. and Naik, M.K. 2007. Survey and assessment for the post harvest diseases of onion in north eastern Karnataka. *Karnataka J. Agri. Sci.* **20:** 164–165.

Rangana, B., Kushalappa, A.C. and Raghvan, G.S.V. 1997. Ultraviolet irradiation to control dry rot and soft rot of potato in storage. *Can. J. Pl. Pathol.* **19:** 30–35.

Ribeiro, O.K. 1978. *A source book of the genus Phytophthora* pp. 14, Straws and Cramer Gmbh, D-6945 Hirschberg 11, Germany, 417pp.

Salunkhe, D.K. and Desai, B.B. 1984. *Post-harvest Biotechnology of fruits*. Vol. 1. Boca Raton, Fl. CRC. 168 pp.

Sharma, R.L. 1994. Prevalence of postharvest diseases of tomato in Himachal Pradesh. *Pl. Dis. Res.* **9:** 195–197.

Sharma, R.L. 1998. Prevalence of postharvest diseases of bell pepper in Himachal Pradesh. *Pl. Dis. Res.* **13:** 193–194.

Sharma, R.L. and Shukla, Arti, 2003. Effect of different packagings on soft rot (*Fusarium* spp.) of bell pepper in storage. *J. Mycol. Pl. Pathol.* **33:** 134–135.

Sharma, S.K. and Sumbli, G. 1992. Airborne mycoflora of vegetable shops and the occurrence of vegetable rots. *Geobios New Rep.* **11:** 59–62.

Sharma, S.K.and Sumbli, G. 1993. Post-harvest fungal decay of some vegetables from North India. *Indian J. Mycol. Pl. Pathol.* **23:** 241–246.

Singer, D.D. 1980. Post-harvest food losses-world overview.*Prog.Food Nutr. Sci.* **4:** 3–9.

Sohi, H.S. and Prakash, O. 1972. New records of fungal diseases from India. *J. Indian Instt. of Hortic. Res.* **2:** 139–142.

Sommer, N.F., Fortlage, R.J., Buchanan, J.R. and Kader, A.A. 1981. Effect of oxygen on carbon monoxide suppression of post-harvest pathogens of fruits. *Plant Dis.* **65:** 347–349.

Sood, Ruchi and Sharma, R.L. 2003. Efficacy of some physical methods in the management of fruit rot of tomato caused by *Alternaria* spp. *J. Mycol. Pl. Pathol.* **33:** 101–105.

Spalding, D.H. 1980. Control of Alternaria rot of tomatoes by post-harvest application of imazalil. *Plant Dis.* **64:** 169–171.

Spalding, D.H. and King, J.R. 1980. Inhibition of Alternaria rot of tomatoes and bell peppers by post-harvest treatment with CGA-64251 or imazalil. *Proc. Fla. State Hortic. Soc.* **93:** 307–308.

Spotts, R.A. and Cervantes, L.A. 1986. Populations, pathogenicity and benomyl resistance of *Botrytis* spp., *Penicillium* spp. and *Mucor piriformis* in packing houses. *Plant Dis.* **70:** 106–108.

Stinson, E.E., Osman, S.F., Heiser, E.G., Siciliano, J. and Bills, D.D. 1981. Mycotoxin production in whole tomatoes, apples, oranges and lemons. *J. Agric. Food Chem.* **29:** 790–792.

Summer, D.R. 1995. Penicillium decay of garlic and blue mold. In: "*Compendium of Onion and Garlic Diseases*, APS Press, Minnesota, pp. 27–28.

Tahvonen, R. 1985. The prevention of *Botrytis cinerea* and *Sclerotinia sclerotiorum* on carrots during storage by spraying the tops with fungicide before harvesting. *Ann. Agric. Fennial.* **24:** 89–95.

Thakur, D.P. and Chenulu, V.V. 1974. Chemical control of soft rot of potato tubers caused by *Rhizopus arrhizus. Indian Phytopath.* **27:** 375–378.

Thind, T.S., Mohan, C., Bedi, J.S., Grewal, R.K. and Sokhi, S.S. 1989. Role of application time of fungicides in the control of late blight of potato. *Pl. Dis. Res.* **4:** 113–117.

Trichelaar, G.M. 1970. Annual Report for 1969. *Inst. Phytopath. Res. Wageningen.* Pp. 30–31.

Verma, S.K., Verma, R.A.B. and Prasad, J.S. 1981. Possibilities of aflatoxicosis on the consumption of *Aspergillus flavus* infested fruits and vegetables in North Bihar. *J. Indian Bot. Sci.* **60:** 20–21.

Wilson, C.L., Wisniewski, M.E., Biles, C.L., McLaughlin, R., Chalutz, E. and Droby, S. 1991. Biological control of post-harvest diseases of fruits and vegetables: alternatives to synthetic fungicides. *Crop Prot.* **10:** 172–177.

Wisniewski, M.E. and Wilson, C.L. 1992. Biological control of post-harvest diseases of fruits and vegetables: Recent advances. *Hortic. Sci.* **27:** 94

■■■■

18

DISEASES OF VEGETABLES UNDER PROTECTED CULTIVATION CONDITIONS

Protected cultivation or greenhouse is the most contemporary approach to produce mainly high value vegetable crops qualitatively and quantitatively, and has spread extensively the world over in the last few decades (Sabir and Singh, 2013). This technology is suitable for those growers who have smaller land holdings as it is highly productive, conservative of water and also protective of the environment (Jensen, 2002). Vegetables like tomato, bell pepper and cucumber are raised in these structures at such a time when climate is not that much favourable for their production under open cultivation conditions as the environmental factors like temperature, humidity, light, soil, water, fertilizers etc. are manipulated in such a way to attain maximum produce as well as to maintaim regular supply of vegetables as an off-season produce. So the farmers can reap a better harvest from limited area and can obtain premium returns of the produce being an off-season.

Commercial vegetable production under protected condition is popular among 115 countries of the world. In India, the area under protected cultivation conditions is presently around 25,000 ha, while the greenhouse vegetable cultivation area is about 2000 ha. Since land holdings are shrinking due to rapid urbanization, declining crop production, declining biodiversity and ever increasing population, demand for food, especially vegetables has increased many folds and protected cultivation has offered a new dimension to produce more in a limited area.

Vegetables should be grown under protected conditions for following factors:

i. To protect the crops from biotic and abiotic stresses.

ii. Off-season cultivation of crops.

iii. For long duration cultivation of crops.

iv. Better quality of produce.

v. Very high productivity.

vi. Reduced use of pesticides.

vii. Virus free vegetable cultivation during rainy and post rainy season.

viii. High income generation from from small holdings.

Being an intensive activity, vegetable cultivation under protected conditions entails perfect planning and numerous phases of operations for its success. The design of protected structures varies depending on the location, whether in a desert, the tropics or in a temperate region (Jensen, 2002). The vegetables can be cultivated in the following types of protected structures:

i. Greenhouses:
   - Climate controlled
   - Semi-climate controlled
   - Naturally ventilated

ii. Net-houses
   - Insect-proof nets
   - Shade nets

iii. Walk-in-tunnels

iv. Plastic low tunnels

v. Floating mulch

vi. Plastic mulches

Earlier farmers used to cultivate vegetables under open field conditions, but due to stagnation in the income, the farmers ventured into new system of cultivating vegetables under above mentioned protected structures which enhanced the income of farmers in Himachal Pradesh and Punjab. Vegetables like tomato, capsicums (coloured like red, yellow, orange, purple) and cucumbers are commercially grown under protected structures. Indeterminate cultivars/hybrids of these vegetables are grown under protected conditions as they continue to grow and set fruits for longer periods till the conditions are favourable for their cultivation. During cultivation these crops are affected by various diseases of fungal, bacterial, viral and nematode natured organisms which not only reduce potential yields but also impair the quality of the produce. During summer, the temperature of polyhouses in mid hills of hilly areas reaches up to 30°C and above as about 20 per cent increase in temperature inside the polyhouse as compare to outside temperature (Spehia, 2011). Prevalence of high temperature and humidity leads to the development of various diseases in these greenhouses. Some diseases which were not that much important under open field conditions became important under protected structures. This new system of cultivation has also brought alongwith it a new pattern of diseases which are described in this chapter alongwith their management.

A. FUNGAL DISEASES

I. Damping –off

It is an important disease of tomato and capsicum nurseries which substantially impair the quality of the nursery. They are not only responsible for the poor seed germination and seedling stand in the nursery but also for carryover of the pathogens/ insect-pests to the protected structures where transplanting is done for raising of the crop.

Symptoms: The disease manifests in two phases *i.e.*, Pre-emergence and Post-emergence phase. In pre-emergence phase, there is failure of seedling emergence from the soil either due to seed rots or killing of young seedlings before their emergence from the soil, hence resulting in patchy appearance. In post-emergence phase, toppling over of infected seedlings at any time after their emergence from the soil results in mortality of the seedlings. In the nurseries, the disease usually radiates from initial infection points and thus gives the appearance of islands wherein almost all seedlings are killed.

The associated pathogen(s), disease cycle and predisposing factors are described in detail in nursery diseases chapter in the book.

Management: Disease can be kept in check by raising nursery in pro-trays or in beds under net house conditions (Anonymous, 2014). Treat the bed soil either with formalin (5%) or solarize it atleast 21 and 45 days, respectively before sowing treat the seed before sowing with captan (3g/kg seed). Apply Malathion dust (5–10g/m^2) around the bed for the control of ants and other insects. After seedling emergence from the soil, drench bed with the mixture of mancozeb (0.25%) and carbendazim (0.1%). For the control of insect-vectors, also apply 1–2 sprays of Metasystox or Confidor. Give light, but frequent irrigations.

II. Fusarium wilt

a. Tomato

Fusarium wilt is an important disease of tomato particularly in those parts of the world where soil temperature remains high during the period of plant growth under protected cultivation condition.

Geographical distribution: This disease was first reported by Massee (1895) from England and since then its occurrence has been reported from at least 32 countries and is particularly severe in countries with warm climate. Fusarium wilt spread throughout the U.S.A, the Middle East (Katan, 1971; Ismail and Abdulla, 1976), North Africa and Latin America (Gonzales and Galindo, 1974). In Europe, Fusarium wilt is widespread in areas of Southern France, Italy, Balkans and Bulgaria (Mitov and Nikolaeva, 1977). The disease is also reported from Taiwan in Kaohsiung in 1981 (Huang and Sun, 1982). Kapoor (1988) has observed widespread occurrence of the disease in various states in India. The disease was reported for the first time from Pusa (Bihar) by Butler (1918). Since then it has been reported from many states.

Losses: The disease inflicts heavy losses, particularly in susceptible cultivars especially when grown under favourable weather conditions. This disease nearly destroyed tomato production in parts of Florida and the southeastern states of United States (Strobel *et al.,* 1969, Dababat *et al.,* 2008). On an average, annual loss due to this disease is 10 to 20 per cent, which may go up to 80 per cent in rare cases. In India, losses ranging from 10 to 80 per cent have been reported (Kapoor, 1988). In Himachal Pradesh, 9.90 to 27.65 per cent incidence of this disease under protected cultivation conditions has been reported (Bharat and Sharma, 2014).

Symptoms: Two of the initial symptoms on young plants are clearing of veinlets and epinasty (drooping of the petioles). The disease first appears in the field as yellowing of the lower leaves. The affected leaves die prematurely and the symptoms continue to appear on successively younger leaves. One or more branches may be affected, while the others remain symptomless. On any given petiole, leaflets on one side may be affected showing yellowing. Browning of the walls of the vascular bundles follows rather promptly, and is best seen by cutting of stem or petiole and examining a cross section. Similar symptoms also appear in bell pepper.

The Pathogen(s)

The disease is caused by the fungus *Fusarium oxysporum* Schlechtend ex Fr. f.sp. *lycopersici* (Sacc.) Snyder and Hans. The old synonyms are *F. lycopersici*, *F. bulbigenum* and *F. bulbigenum* f.sp. *lycopersici*. The fungus is a facultative saprophyte, which grows readily on various solid or liquid media. The mycelium is hyaline at first becoming cream coloured with age and has branches and septations, and some isolates produce blue or red or pinkish pigments. The micro conidia are aseptate, hyaline, oval to round in shape and measured 2.2–4.8 × 1.4–2.4 μm in size. Macroconidia are 3–5 septate, often hooked at tapered tip and somewhat thicker at the upper third portion than in the central portion and measure 23.0–38.0 × 2.8–5.4 μm in size (Fig. 1). Three septate spores are more common. Chlamydospores, both rough and smooth walled, characterized by thick walls are terminal or intercalary serving as the chief means of long-term survival in soils in the absence of host (Singh, 1995). They are often solitary, but occasionally form in pairs or chains. Three physiological races *i.e.* race 1, 2 and 3 have been reported (Jones *et al.*, 1991). Race 1 is the most widely distributed and has been reported from most geographical areas. Although race 2 was first reported in Ohio in 1945 (Alexander and Tucker, 1945) and attacked all of the varieties of that time that has been bred for resistance to race 1. Race 2 did not become widespread or of economic concern until its discovery in Florida (Strobel *et al.,* 1969). In India, Pandey and Gupta (2013) collected 56 isolates from different states and out of these 31, 17 and 6 were grouped into race 2, race 1 and race 3, respectively. However, two isolates (Kr-F-3 and Mg-F-1) that infected only Assam Local Variety indicated the presence of another race in India and may be designated as race 4 or a variant till further confirmation.

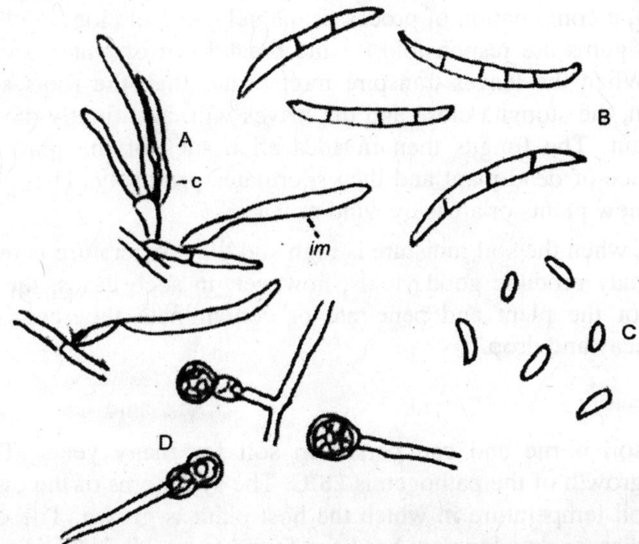

Fig. 1. Different spore stages of *Fusarium oxysporum* f.sp. *lycopersici*
(A) Sporodochia; (B) Macro-conidia; (C) Micro-conidia; (D) Chlamydospores

The fungus is regarded as an invader of the epidermis or hair of the fibrous root system of the host, finally reaching the vascular bundles where it is confined largely to xylem vessels. There is enough evidence that invasion is enhanced by wounding of the root system.

Host range: The fungus *Fusarium oxysporum* Schlechtend ex Fr. f.sp. *lycopersici* has been found to infect only species related to *Lycopersicon*, especially *L. esculentum* and *L. pimpinellifolium*, although other genera have also been shown to be latent hosts by harbouring the fungus as a facultative saprophyte in their fibrous root systems where chlamydospores are produced.

Disease cycle

The pathogen is soil borne in nature. Between crops, it survives in the infected plant debris and in the soil as mycelium and spore forms especially as chlamydospores. It spreads over small distances by means of water and contaminated farm equipments, and over long distances, primarily in infected transplants or in the soil carried with them. Usually, once an area becomes infested with *Fusarium*, it remains so indefinitely.

When healthy plants grow in contaminated soil, the germ-tube of spores or the mycelium penetrates root tips and the mycelium advances through the root cortex intercellularly reaching the xylem vessels through the pits. The mycelium then remains exclusively in the vessels and travels through, mostly upward in stem and crown of the plant. While in vessels, the mycelium branches produce microconidia, which are detached and carried upward in the sap stream. The microconidia germinate at the point where their upward movement is stopped. The mycelium penetrates the upper wall of the vessel and more microconidia are produced in the next vessel. The mycelium also advances laterally in to adjacent vessels, penetrating them through pits.

Presumably, a combination of processes, namely, vessel clogging by mycelium, spores, gels and gums are responsible for the breakdown of water economy of the infected plant. When the leaves transpire more water than the roots and stem can transport to them, the stomata close and the leaves wilt, and finally die followed by death of the plant. The fungus then invades all tissues of the plant extensively, reaches the surface of dead plant and then sporulates profusely. The spores may be disseminated to new plants or areas by wind or water.

Sometimes, when the soil moisture is high and the temperature is relatively low, infected plants may produce good yields, however, in such cases, the fungus may reach the fruit of the plant and penetrate or contaminate the seed. Usually, the infected fruits decay and drop.

Epidemiology

The fungus is soil borne and can persist in soil for many years. The optimum temperature for growth of the pathogen is 28°C. The symptoms of the disease vary in relation to the soil temperature in which the host plant is grown. The optimum soil temperature for disease development has been found to be 25–31°C, characterized by sudden wilting, usually without yellowing starting from the lower leaves. The rapid development of the disease occurred at 31°C. The temperature conditions of soil and air most favourable to the disease are 27 and 28°C, respectively. Hot dry weather favoured wild development. During summer, the temperature of polyhouses in mid hills of hilly areas reaches up to 30°C and above as about 20 per cent increase in temperature inside the polyhouse as compare to outside temperature (Spehia, 2011). Prevalence of higher temperature in polyhouses is favourable for the development and progress of this disease. The disease is more severe in acidic soils than in alkaline ones. Hydrogen ion concentration has been found to play a marked role in controlling the growth of causal organism. A pH of 4.0 to 4.5 is best for growth of the pathogen and the disease developed mostly in sandy soils and least in clay soils.

Management

i. Cultural practices: Fusarium wilt can be efficiently controlled by careful crop rotation and the use of healthy seed. Combined application of inorganic fertilizers and organic manures is considered effective in reducing disease incidence. Application of phosphate fertilizers like super phosphate helps to reduce wilt and increase yield. Nitrate fertilizers reduce disease severity and ammonium fertilizers increase disease development. Sarhan and Kiraly (1981) obtained complete control of *Fusarium oxysporum* f.sp. *lycopersici* in cultivar Primset of tomato by application of N as 420 ppm nitrate + raising the soil pH to 7.6 with $Ca(OH)_2$ + drenching the roots with benomyl @ 33 mg/l. Micronutrients like Cu, Zn, Fe, Mg, B and I have given encouraging results when sprayed at transplanting, flowering and fruit formation by increasing resistance in the plants to the disease (Ahmed *et al.*, 1987).

Nirwano *et al.* (1994) reported that application of leguminous foliage, such as groundnut and *Sesbania grandiflora* reduced the percentage of infected plants of tomato variety Money Maker to 22.17 and 11.68 per cent, respectively compared to control. Biofumigation of polyhouse soil with taramira crop residues for 30 days was also found effective in reducing the incidence of this disease (Sharma, 2013). Application of organic amendments (crab shell powder and chicken manure) greatly reduced the severity of *Fusarium oxysporum* f.sp. *lycopersici* through reduction in propagules and inhibition of their germination.

ii. Physical methods: Mulching of soil with transparent polyethylene sheet has proved effective against tomato wilt. A 70 per cent reduction in population of the pathogen at a depth of 5 cm, following 30 days of soil solarization in the mulched plot was recorded (Dwiwedi, 1991). The maximum temperature (55°C), recorded in the mulched plot was sufficient to kill the pathogen under Varanasi conditions. El Shami *et al.* (1990) reported that soil solarization by using single layer of transparent polyethylene was much more efficient in reducing the wilt than fumigation with methyl bromide. Soil solarization is thus recommended for management of tomato wilt under Egyptian conditions.

Hot water treatment of roots at 48–49°C for 30 seconds has proved beneficial in delaying the symptoms (Archisi *et al.*, 1985). Steam sterilization of soil is also quite effective under greenhouse conditions for the control of *Fusarium oxysporum* f.sp. *lycopersici*.

iii. Host resistance: Different cvs./lines like EC 26104, Gambdc, Globe, HS 101, HS 107, Improved Meerut, Marglobe and Pusa Early Dwarf (Chauhan, 1988) and wild sources like *Lycopersicon pimpinellifolium*: EC 65992, EC 65993, EC 86522 (USA);

EC 7353 (Australia) and EC 121453 (Netherlands), *L. hirsutum*: EC 486, EC 65999 and EC 127985 (USA), *L. peruvianum*: EC 2630, EC 65984, EC 86512 and EC 106293 (USA), *L. pimpinellifolium* x *L. hirsutum*: EC 66007, EC 86516, EC 106295, EC 116039 and EC 128013 (USA) (Bisht *et al.*, 1989) *L. peruvianum*: EC 148897 and EC 148808, *L. hirsutum* var. *glabratum*: EC 144975 and B 6013 (Banerjee *et al.* 1990), F_{1s} and backcrosses of Indian varieties (HS 101, HS 102, S 17 and Pusa Ruby) with *L. pimpinellifolium* cv. Pan American (Malhotra and Vashistha, 1993) and T 27 and BT 62 (Mishra and Mishra, 1993) were found resistant to highly resistant against this disease and can be used in breeding programme.

iv. Chemical control: Soil fumigation with chloropirin can provide an effective control of tomato wilt in highly susceptible varieties. Soil fumigation with methyl bromide (98%) + chloropicrin (2%), applied under polythene covers for 2 days proved very effective against *F. oxysporum* f.sp. *lycopersici* for two seasons.

In greenhouse tomatoes, Atkinson and Adamson (1977) observed consistent reduction in the number of wilted or dead plants by a single pre-plant drench of benomyl or thiophanate-methyl. Etebarian (1992) also reported good efficacy of benomyl, iprodione plus carbendazim and carbendazim alone against tomato wilt. Furadan (carbofuran) has been reported to reduce the incidence of Fusarium wilt when applied to soils at pH 6.5-7.5 @ 10 lb/acre (Overman and Jones, 1975). Since the presence of nematodes in wilt-affected field further aggravates the situation, it is better to apply some nematicide like Furadan in soil before transplanting. Chebil *et al.* (1980) found that soil disinfestation with Basamid (dazomet) ensured better plant growth in soils infested with *Fusarium oxysporum* f.sp. *lycopersici*. Seed treatment with thiram and ceresan @ 4 g/kg seed can also decrease infection by the wilt pathogen. Soil drenching with benomyl or thiophanate methyl was also found effective in the management of this disease under protected cultivation conditions.

v. Biological control: Several antagonistic microorganisms have been reported effective in suppressing *Fusarium* wilt of tomato. *Cephalosporium* species may result in the reduction of tomato wilt by reducing the severity of *F. oxysporum* f.sp. *lycopersici*. Severity of wilt symptoms was suppressed by non-pathogenic species of *Fusarium* like *F. oxysporum* f.sp. *melongenae*, *F. oxysporum* f.sp. *cucumerinum* and *F. oxysporum* f.sp. *batatas* as root dip treatment in spore suspension before transplanting (Amemiya *et al.*, 1986). Kroon *et al.* (1991) demonstrated that *F. oxysporum* f.sp. *dianthi* was able to induce resistance against the wilt disease. Additive effects were obtained in most cases when tomato plants were inoculated with non-pathogenic *F. oxysporum* str. F 047 at seeding stage (Fuchs and Defago, 1991).

Pseudomonas isolates from rhizosphere of different plants can reduce infection of *F. oxysporum* f.sp. *lycopersici* in tomato. Mutants from avirulent strain of *Pseudomonas solanacearum* are capable of detoxifying fusaric acid and thus possess the ability to protect tomato plants from wilt.

Coating of tomato seed with *Trichoderma viride*, *T. harzianum*, *Streptomyces griseus* or *Bacillus subtilis* prior to sowing in wilt affected soil significantly reduced damping-off, crown rot and increased seedling survival and yield (Yehia *et al.*, 1981, Sivan *et al.* 1987). Khakimov and Abdullaer (1992) also observed a good control of the pathogen through soil, seed or seedling treatment with trichodermin. Isolates of

T. viride, T. harzianum and *T. koningii* were inoculated into the growing medium at a concentration of 4.6 x 10^8 propagules/g of dry soil, and assessed for the control of FOL on tomato seedlings cv. Marmande in a greenhouse. All the isolates reduced disease incidence and increased crop yields significantly (Ercole and Nipoti, 1986). Under greenhouse, soil application with *T. harzianum* spores resulted in wilt disease reduction by 92 per cent (Ramezani, 2010). Isolates of non-pathogenic *F. oxysporum, T. hamatum, T. harzianum* T-22, *G. virens, P. fluorescens,* and *B. cepacia* applied alone or in combination significantly reduced Fusarium wilt of tomato in greenhouse tests (Larkin and Fravel, 1998, Yigit and Dikilitas, 2007). Reductions in disease incidence ranged from 35 to cent per cent compared to the pathogen-infested controls. Sivakumar *et al.* (2008) while studying bio-efficacy of antagonists against FOL applied *T. viride* at 10 g/kg of seed (ST)+5.0 kg/ha (SA); *Pseudomonas fluorescens* at 10 g/kg of seed (ST)+5.0 kg/ha (SA); *Paecilomyces lilacinus* at 10 g/kg of seed (ST)+5.0 kg/ha (SA); neem cake at 250 kg/ha. Combination of *T. viride, Pseudomonas fluorescens, Paecilomyces lilacinus* and neem cake as seed treatment plus soil application recorded the minimum disease incidence and significantly increased the shoot length, root length, biomass and fruit yield. Although all biocontrol agents applied individually reduced disease incidence, combination of *T. harzianum* T-22 + fluorescent *Pseudomonas* isolate gave the best control (70.2%). Application of coir pith compost and consortia application of *Pseudomonas fluorescens* and *Bacillus subtilis* as seed treatment, and soil drenching at 10th day after transplanting significantly reduced the wilt incidence and increased plant growth attributes like seed germination, shoot length, root length and ultimately fruit yield compared to controls besides inducing defense enzymes (Christopher *et al.*, 2013).

Arbuscular- mycorrhizal (AM) fungi

Frank (1885) found that roots of most of the plant are colonized by fungi and transformed in to fungus root organs, which he called "Mycorrhizal". Mycorrhizae results from a mutualistic symbiosis between plant roots and certain fungi. Tomato plants infected with *F. oxysporum* f.sp. *lycopersici* wilted less, if they had previously been inoculated with the endotrophic mycorrhizal fungus, *Glomus mosseae*. The fungus enhanced lignin deposition in the cell wall of the epidermis and stele, which is considered to increase resistance to the wilt pathogen. Other organisms like *Streptomyces* sp., *Penicillium* sp. and *Acremonium* sp. have also been tried in this direction. Sulfhydryl proteinases such as papain, bromelian, ficin and shoot extract of *Inula viscosa* have shown potential as biopesticides against this disease of tomato (Qusan *et al.*, 1995). Gardezi *et al.* (1999) found favourable effect of AMF on tomato plant growth, both in sterile soil and in pathogen-infected soil. It was concluded that *Glomus* resulted a decrease in disease severity and pathogen population as a result of the plants root colonization by mycorrhizae. Mwangi *et al.* (2011) reported significant reduction in disease severity and increase in plant growth parameters (height; shoot and root dry weight) compared to the control when tomato seedlings were applied with AM fungi culture under greenhouse condition.

Integrated disease management

Bio-fumigation of sick soil with taramira crop residues for 30 days, and application of formulation of *T. viride* along with FYM and inoculation of transplants with indigenous AM fungi resulted in to more than 80 per cent disease control under polyhouse conditions (Bharat and Sharma, 2015). This practice also increased the growth *viz.*, height, number of branches and plant dry weight as well as yield parameters like number of fruits per plant and fruit yield per plot.

b. Capsicum

Fusarium wilt is an important disease of capsicum both in open as well as in protected conditions. Of late, this disease is becoming the main disease of capsicum under protected cultivation conditions.

Geographical Distribution: The occurrence of this disease in capsicum was first reported from New Mexico in the year 1919 and named as rapid wilt. The disease is now occurring in almost all parts of the world wherever this crop is grown. The occurrence of this disease was first time recorded in Solan, Himachal Pradesh, during 2003 under naturally ventilated polyhouses (Gupta *et al.*, 2005). The pathogen attacks only chilli and capsicum.

Losses: The disease is becoming important and its incidence is on the rise in protected cultivation conditions being a soil borne disease. Gupta *et al.* (2005) reported its incidence upto 8.0 per cent in Solan area of Himachal Pradesh under naturally ventilated polyhouses.

Symptoms: The symptoms of the disease appear as slight yellowing of the lower leaves and drooping of the upper parts during hot of the day. The infected plants show predominant wilting within few days of infection. The vascular system of such plants becomes discoloured particularly in the lower stem and roots (Plate 1). The disease generally appears in patches in protected structures and can be recognized from a distance.

The Pathogen: The fungus responsible for this disease is *Fusarium oxysporum* Schlecht Fr.f.sp. *capsici.* Mycelium of the fungus is septate. Fungus produces microconidia, macroconidia and chlamydospores. The macroconidia are produced on branched conidiophores, which are multiseptate, whereas microconidia are oval to kidney shaped which are produced on short conidiophores. Chlamydospores are thick walled and are produced in adverse climatic conditions.

Disease cycle: The pathogen is both soil and seed borne (Vidyasekaran and Thiagarajan, 1981) in nature. It mostly survives in infected plant debris and in soil as mycelium or chlamydospores. Dispersal of the pathogen mostly takes by water, contaminated farm implements and through infected transplants or adhered soil to the roots. The infection takes place at the root tips. The mycelium advances through the root cortex intercellularly reaching the xylem vessels. The mycelium travels upwards in the vessels. Vessel clogging by mycelium, spores, gels and gums, is responsible for the breakdown of water movement in the infected plants and result in wilt symptoms.

Epidemiology: The pathogen is soil or seed borne in nature. In protected structures, mostly temperature and humidity remains high which favour its infection and spread. High temperature and wet soil conditions favour disease development. Disease is more severe in poorly drained soils. The fungus grows and sporulates best at 20–30°C and 5–8 pH.

Management

i. Cultural practices: Avoid cultivation of capsicum in the same structure year after year. Use healthy seed/ transplants and proper fertilization also reduce the incidence of this disease.

ii. Host resistance: Jones and Black (1992) identified some accessions of *Capsicum baccatum* which are resistant to this disease. The resistant accessions have smaller vessel cavities and thicker vessel walls.

iii. Biological Control: Increased population of antagonistic fungus *Trichoderma viride* was observed around the roots of the capsicum plants which were free from the disease (Naqvi and Chauhan, 1980). This suggests that the antagonist can be utilized in combatting the ravages of this disease.

iv. Chemical control: Seed treatment with fungicides like captan, carbendazim or carboxin has been reported to reduce the disease incidence and increase fruit yields. With the initiation of the disease drenching of plants with the mixture of carbendazim (0.1%) and mancozeb (0.25%) has also been found effective in reducing the disease incidence.

v. Integrated Disease Management: Use of healthy seed, seed treatment with fungicides like carbendazim or captan and soil application of *Trichoderma* formulation can keep this disease in check. Soil application of *Trichoderma harzianum* or *T. viride* multiplied on neem cake (500g/m^2) before transplanting was found effective and reduced the disease incidence (Gupta *et al.*, 2005).

III. Sclerotinia Rot of Capsicum

It is an important disease of Capsicum grown under protected structures in Punjab (Aujla *et al.,* 2010). The prevalence of cool and humid weather conditions in greenhouses is highly conducive for the development of this disease.

Geographical Distribution: The occurrence of this disease has also been reported from Ohio State of USA (Yanar *et al.,* 1996). Since then it has been reported from many parts of the world (Purdy, 1979; Boland and Hall, 1994). In India, the occurrence of this disease in protected structures have been reported from Punjab in 2010 (Aujla *et al.,* 2010).

Losses: About 30 to 40 per cent losses due to this disease has been reported from Ohio State of USA (Yanar *et al.,* 1996). The incidence of this disease varied from traces to as high as 87.4 per cent in different locations of Punjab (Kaur *et al.,* 2016). They further observed that the incidence of this disease was more in those greenhouses which are located in sub mountainous areas of Punjab due to prevailing cool and moist weather conditions.

Symptoms: The symptoms of this disease appear on the root and above ground parts of the plants. The first symptoms of this disease appear as dark brown lesions on the stem or just below the soil line. First foliar symptoms are progressive yellowing and wilting. Later the fungus producing whitish mycelium around the rotten stem and foliar parts of the plants including fruits under high moisture conditions (Jeon *et al.,* 2006). When the food is exhausted, the mycelium coagulate and start forming sclerotia which are initially white and then turned to black. On splitting open infected stems, a large number of sclerotia are seen in the embedded tissues.

The Pathogen: The fungus associated with this disease is *Sclerotinia sclerotiorum* (Lib.) de Bary. Mycelium in culture as well as on host tissue is hyaline, cottony, branched, septate hyphae. Microconidia are produced on short lateral branches of the mycelium in chains. The pathogen produces sclerotia when the food is exhausted. The sclerotia germinate both myceliogenically and carpogenically. In carpogenic germination, sclerotia germinate by producing stalked apothecia under suitable humidity and light conditions. Apothecia are brown in colour and are round or lobate type. Asci are cylindrical. Each ascus contains eight ascospores and released in clouds. Ascospores are elliptical in shape.

Disease Cycle: The pathogen is soil borne in the form of **sclerotia**. The fungus also has very wide host range affecting more than 400 plant species (Purdy, 1979, Boland and Hall, 1994). Sclerotia are capable of germination both carpogenically and myceliogenically and cause infection of the plants. After infection, the fungus produces oxalic acid which kills the cells in advance. A fluffy mycelium appears on the host surface during cool and moist weather conditions. During scarcity of food, the mycelium coagulates and starts forming sclerotia which are initially white in colour but later turns black.

Epidemiology

The fungus over winters in soil as sclerotia over a wide range of temperature i.e., 0 to 25°C. For infection optimum temperature is 15–21°C. High moisture and low temperature conditions are essential for infection, that's why the disease is more severe in sub mountainous regions of the Punjab State. In protected structures, more consistent environment; moderate temperatures, high humidity, less air movement, overcrowded conditions and rapid lush green growth highly favours the disease outbreaks (Aujla *et al.*, 2010). Soil borne diseases are more damaging when soil conditions are poor as a result of inadequate drainage, poor soil structure, low soil fertility and high soil compaction. The fungus can tolerate wide pH range, but is best adapted to acidic substrate.

Management

a. Cultural Practices: Collection and destruction of infected plant material, deep summer ploughing and paddy or maize rotation can reduce the sclerotial population in the infected soil of the protected structure.

b. Chemical Control: With the initiation of the disease, application of carbendazim or tiophenate methyl (0.1%) as spray drench proved most effective in checking the disease and in increasing the per cent marketable fruit yield (Kaur *et al.*, 2016).

IV. Sclerotium Stem rot of Capsicum

The disease is also known as southern blight or white rot. It is an important disease of the capsicum in protected cultivation conditions throughout the world and cause significant yield losses.

Geographical Distribution: The pathogen responsible for this disease has extremely wide host range and can infect more than 500 species of plants belonging to 115 families including vegetables like chilli, French bean, carrot, pea, cauliflower, cabbage, potato and arum (Punja, 1988). In Himachal Pradesh, this disease is widespread in the naturally ventilated polyhouses in Hamirpur and other parts of the state.

Losses: The disease is more severe in those structures which have sandy loam soils, particularly collected from riverbeds along with high soil moisture and temperature conditions. Plant mortality up to 50 per cent have been recorded under favourable disease conditions (Gupta *et al.*, 2014).

Symptoms: Initial symptom of this disease appears as foliar discolouration. Later on infected leaves may turn yellow. The cortical tissues at the base of the stem turn brown and decayed above and below the ground. Fan like white mycelial growth usually is visible at the base of the stem and on the soil around the base of the plant (Plate 2). Sclerotia of mustard seed size of tan to brown when mature are produced in the mycelial mat. The disease is generally characterized by sudden wilt of individual plants in patches scattered throughout the field. Branches or fruits may also become infected when they are in contact with soil.

The Pathogen: The fungus associated with this disease is *Sclerotium rolfsii* Sacc. In culture, initially the mycelium is silky but gradually looses its lustre and becomes somewhat dull in appearance. The hyphae are hyaline, thin walled, sparsely septate when young and measuring 60–350 × 2–8 μm in size. The broader hyphae show clamp connections, which is the characteristic feature of this fungus. Sclerotial initials are formed from hyphal strands that consist of 3–12 hyphae lying parallel. A spherical shape is soon assumed even if it is only a loose mass of hyphae. With further hardening, differentiation takes place and the sclerotium show an outerlayer of polyhedral cells, 3–4 μm in diameter, surrounding the compact hyphae. Mature sclerotia are dark brown, but variation from lighter brown to darker colour may be found. These are small resembeling to reddish seed in size, hard and usually round.The basidial stage of the fungus is *Athelia rolfsii* (Cruzi) C.C. Tu & Kimbr. The basidia are obvoid, 7–9 × 4–5 μm in size. Each basidium bears 2–4 parallel or divergent sterigma, which are 2.5–4 μm long. Basidiospores are unicellular, elliptical to obvate, sometimes rounded or pyriform, smooth walled hyaline and epiculate at the base measuring 6–7 × 3–5 μm in size. Genetic variation among isolates collected from different geographic regions also exists.

Disease Cycle

The pathogen is soil borne in the form of sclerotia and it can also survive as saprophyte on crop debris. Mostly sclerotia serve as the source of primary inoculum. These sclerotia immediately germinate at lower water potential and cause infection.

Epidemiology

The pathogen is soil borne in the form of sclerotia. High soil moisture and temperature favour disease development, although symptom expression is more during dry condition following a wet period. Local dissemination of the fungus is through irrigation water and intercultural operation tools.

Management

a. Cultural Practices: Collection and destruction of infected plant debris helps in reducing the initial inoculum of the pathogen. Deep summer ploughings of the protected structure soil not only burry sclerotia and debris deep in soil but also help to hasten in their decomposition. Solarization of the protected structure soils in raised beds for six weeks prior to transplanting has been found to reduce disease incidence.

b. Biological Control: Several fungi, bacteria and arbuscular mycorrhizal (AM) fungi are known to reduce the disease. Inoculation of plants with AM fungi like *Glomus fasciculatum* and *G. macrocarpum* caused disease reduction. Dual inoculation of AM fungi and antagonist *Trichoderma harzianum* was found more effective in reducing the disease and suppressing the pathogen (Sreenivasa, 1994). Soil application of *Trichoderma harzianum* and *T. viride* reduced disease incidence by 15.5 to 37.5 per cent and increased fruit yields by 14.8 to 30.3 per cent (Bhat *et al.*, 2015).

c. Chemical control: The growing medium of the protected structure should contain sufficient organic matter and must be either solarized at least for six weeks or sterilized with Formalin (5%) before filling in the beds. Drenching of the beds either with propineb (0.3%) or mancozeb (0.25%) or thiram (0.3%) can reduce infection of the plants.

V. Phyllosticta leaf Spot Tomato

Tomato is an important crop which is largely grown in different parts of Himachal Pradesh under low cost naturally ventilated greenhouses round the year. Due to the prevalence of high humidity and temperature under protected cultivation conditions, the crop is severely infected with various leaf spotting fungi of which Phyllosticta leaf spot is important which was noticed during August–September, 2009 on hybrid Naveen 2000. The severity of this disease was up to 40 per cent during 2009 crop season which reached up to 75 per cent in 2010 crop season in different polyhouses at Nauni and nearby polyhouses (Bharat and Gupta, 2013). The occurrence of this disease on tomato has also been reported from North American and African countries and in India from Maharashtra state (Rao, 1962).

Symptoms: The symptoms of the disease appear as small spots on the leaves which are amphigenous, conspicuous, circular to irregular, coalescing to form bigger spots (Plate 3). The centre of the spots becomes papery which ultimately fell off. When the intensity of the disease is very high, the symptoms of the disease also appear on fruits as light coloured spots with conspicuous dark coloured pycnidia at the point of their attachment with the plant. The spots on fruits turn dark in colour in later stages.

The Pathogen: The disease is caused by fungus *Phyllosticta lycopersici* Peck (Syn. *Phoma lycopersici* Cooke). The pycnidia of the fungus are dark coloured, amphigenous, scattered to gregarious, globose to subglobose, erumpent or submerged, ostiolated round, conspicuous and measured 90–240 μm in diameter. When ruptured these pycnidia release conidial mass in unique fashion forming a cirhus. Conidia are one celled, hyaline, oval to slightly elongated in shape, smooth and measured 3.0–6.0 x 2.0–3.0 μm in size.

Disease Cycle: The pathogen over winters in the infected plant debris as pycnidia. The pathogen requires moderate temperatures and high humidity which is prevalent in protected structures during August–September.

Management

The disease can be kept in check by collection and destruction of infected plant debris. With the initiation of the disease spray the crop with carbendazim (0.1%) or combination of carbendazim (0.05%) and mancozeb (0.25%) and repeat at 10–12 days interval.

VI. Late blight Tomato

Late blight is an important disease of tomato and potato throughout the world. Since the farmers are also taking cultivation of tomato and cherry tomato under protected conditions, this disease is also appearing in these crops due to dense planting, improper pruning and luxuriant plant growth leading to poor ventilation helps in providing favourable conditions for this disease in net houses and polyhouses (Aujla *et al.*, 2010). In Himachal Pradesh, Cherry tomato is also gaining importance and is grown during February to December in polyhouses. Though cherry tomato is less susceptible to different foliar diseases and it has been found resistant to buckeye rot under field conditions (Dodan *et al.*, 1995), however, it was found infected with late blight in October, 2010 in this crop grown under protected conditions at Kandaghat (Shukla *et al.*, 2013). Tomato crop grown in polyhouses around Shimla has also been infected with this disease.

Losses: The losses from this disease are immense. In Punjab, the disease started appearing in December and January and up to 84 per cent disease severity was recorded (Aujla *et al.*, 2010). Dense planting, improper pruning and luxuriant plant growth leading to poor ventilation helps in providing favourable conditions for the development of this disease. The severity of this disease ranged from 45 to 70 per cent in different localities of Solan district where cherry tomato was grown in polyhouses (Shukla *et al.*, 2013).

Symptoms: Symptoms of the disease appear on all above ground parts of the plant. On leaves, the symptoms appear as pale, green, irregular spots which in moist weather enlarge rapidly with central tissue turning necrotic and dark brown or black. On the lower surface of the leaves, a white downy growth of the fungus appears around the dead areas. In dry weather, the growth of the fungus is checked and the water-soaked areas dry up and turn brown. Brown streaks also develop on stems. Symptoms also appear on green fruits as olivaceous greasy spots which gradually cover the entire fruit surface.

The Pathogen: The fungus associated with this disease is *Phytophthora infestans* (Mont.) deBary. Sporangiophores are slender, hyaline, indeterminate, septate having side branches with swollen bases and measuring 114–140 x 4.0–5.4 μm in size. Sporangia are lemon shaped, papillate and hyaline measuring 28–50 x 20–27 μm in size (Fig. 2).

(The pathogen, disease cycle, favourable environmental conditions are described in detail in the chapter of potato diseases in this book).

Management

The disease can be kept in check by sprays of fungicides like Sectin 60 WG (0.12%), Ridomil Gold (o.25%), Curzate M-8 (0.25%) and Cabrio Top (0.2%) and repeat at 10 days interval (Shukla *et al.*, 2013). Do not repeat spray of same chemical to avoid development of resistance in the pathogen.

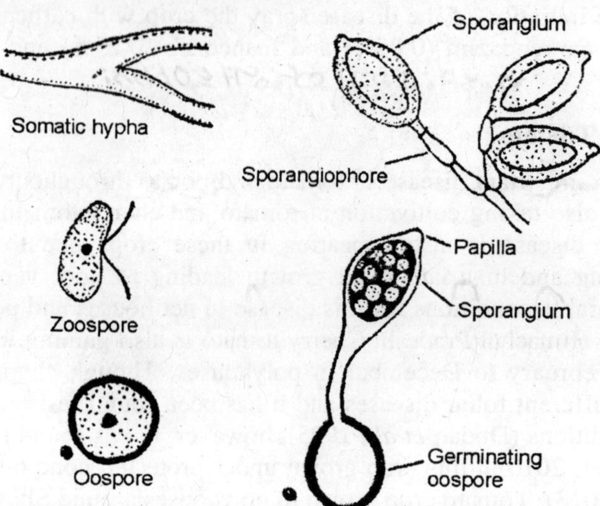

Fig. 2. Sporangiophore and sporangia of *P. infestan*

VII. Anthracnose and Ripe fruit Rot of Capsicum

It is an important disease of coloured capsicums in protected cultivation conditions because the climate inside the polyhouses especially in naturally ventilated polyhouses is generally warm and humid which is ideal for fruit rotting fungi which causes substantial yield reductions by impairing the quality of the fruits. Gupta and Gupta (2014) noticed the occurrence of anthracnose on coloured capsicum during 2012 and 2013 crop seasons in various polyhouses in the vicinity of Solan.

Symptoms: The characteristic symptoms of the disease appear as small circular or elongated water soaked spot which increase in long axis. Later these spots turn brown on the margin and centre become slightly sunken and bleached. These lesions bear light black, brown or orange coloured acervuli either in rings or in a scattered fashion (Plate 4).

The Pathogen: The disease is caused by fungus *Colletotrichum gloeosporioides* Penz. The pure cultures of the pathogen are raised. Colonies on PDA at 25±1°C are pinkish white to grey with reverse dark brown. The mycelium in culture forms brown to orange acervuli. Acervuli bearing setae are rarely observed in the culture, however on fruits the acervuli are subepidermal with light to dark brown setae disrupting outer epidermal cell walls of the host. The conidiophores are closely packed on the surface of the acervulus and were hyaline. The conidia are oblong to cylindrical with obtuse ends having one to three oil globules. Conidia measures 5.32–21.69 x 1.31–6.73 µm in size (Fig. 3).

Disease Cycle, favourable environmental conditions and management practices of this disease are same as described in detail in the chapter Diseases of bell pepper and chilli in this book.

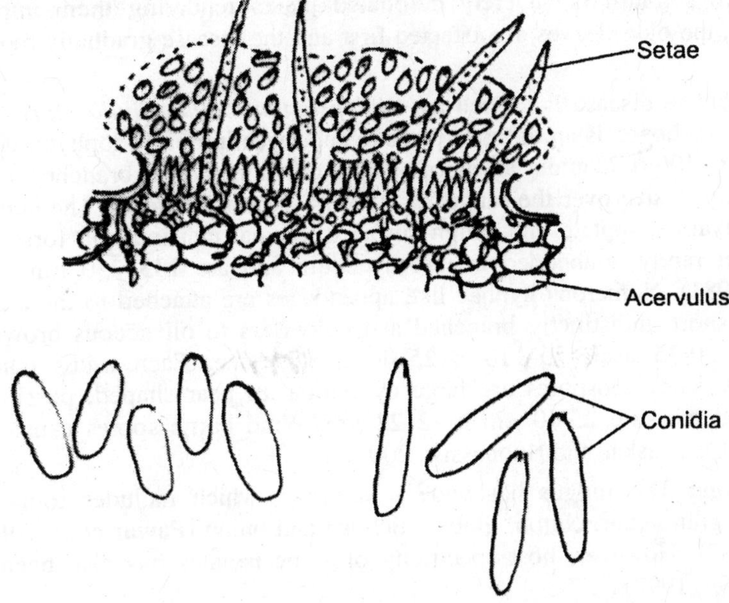

Fig. 3. Acervulus and conidia of *C. gloeosporoides*

VIII. Powdery Mildew of Capsicum

Under protected cultivation conditions, powdery mildew of Capsicum is an important disease which causes significant yield reductions by premature defoliation under warm and humid climatic conditions.

Geographical Distribution: This disease was first recorded in South West Florida in 1971 (Blazquez, 1976), since then it has been occurring in California on chillies and bell peppers (Smith *et al.*, 1999). By late 1990's it has spread to Arizona, New York, Oklahoma (Damicone, 1999), North Central Mexico (Velasquez, 1999), Canada (Cerkauskas and Brown, 1999). It was 1st detected in British Columbia in 2003 on greenhouse pepper crop and since then it has spread throughout the greenhouse pepper industry. In India, the occurrence of this disease was first recorded in Himachal Pradesh under naturally ventilated polyhouse during 2016 crop season (Gupta *et al.*, 2016).

Losses: Yield losses up to 10–15 per cent have been recorded (Cerkauskas and Brown, 1999). Premature defoliation is the most important phenomenon in this disease and about 50–75 per cent defoliation may occur (Damicone, 1999). Severity up to 65–80 per cent was recorded in different surveyed polyhouses (Gupta *et al.*, 2016). They further reported that the fruits were exposed to sunburn in severely defoliated plants which varied from 15–20 per cent rendering them unmarketable.

Symptoms: Symptoms of the disease first appear as light green to bright yellow, irregularly shaped blotches on upper leaf surfaces and the center of the lesion turn necrotic. The fungal growth is seen on the corresponding under surface of these blotches which sporulate profusely (Plate 5). Severely infected leaves turned yellow, rolled upward and dropped prematurely. No lesion appears on the stem. Fruits are exposed to sunburn in severely defoliated plants rendering them unmarketable. Generally, the older leaves are affected first and the disease gradually moved up the plant.

Pathogen: The disease is caused by *Oidiopsis taurica/Leveillula taurica.*(Smith *et al.*, 1999). Pathogen is an obligate parasite and mycelium is endophytic (Cerkauskas and Brown, 1999). Conidia are borne singly on short hyphal branches, hyaline and vary greatly in size over the range 25–95 x 14–20 µm (Plate 6). The conidiophores are also hyaline, septate and rough surfaced. Sexual cleistothecia form within the stroma but rarely, embedded in the mycelium and are 135–250 µm in diameter (Dixon, 1981). Numerous hyphae like appendages are attached to the cleistothecia, these are short, indistinctly branched and colourless to olivaceous brown. Usually there are 20-35 asci, 70–110 x 25–40 µm in size. Each ascus contains two ascospores. The ascospores are large cylindrical to pear shaped, possibly slightly curved and measure 25-40 µm x 12–22 µm. Wind borne spores cause secondary infection (Cerkauskas and Buonassist, 2003).

Host Range: The fungus has wide host range, which includes tomato, brinjal, coriander, gram, guar, cotton, globe artichoke and onion (Pawar *et al.*, 1985, Correl *et al.*, 1987). However, host specificity of some isolates has also been observed (Ayesu-offei, 1998).

Disease cycle: The fungus is an obligate parasite and perennates either through cleistothecia wherever they are produced or as mycelium and conidia on the cultivated or wild hosts. In areas where cleistothecia are not produced the fungus over winters on cultivated and wild hosts, which ensures the survival of the pathogen form one season to another. Conidia are wind borne or dispersed by splashing rain. The conidia on the host surface germinate and germ tubes enters the cuticle directly by perforating and also through stomata (Tafradjiiski *et al.*, 1975). The hyphae grow intercellularly and from haustoria in the spongy and palisade parenchyma. After appearance of symptoms, the conidia are produced which cause secondary infections.

Epidemiology: Powdery mildew has been more severe in shaded areas whether humid or dry. The disease is restricted to warm semiarid regions (Glawe, 2010). Temperature range for conidia germination is 8–35°C and 93–100 per cent RH (Reuveni and Rotem, 1973), but optimum temperature is 25–30°C. Close canopy, warmer, humid climate increase the disease incidences (Velasquez, 1999). Low relative humidity i.e.< 50 per cent favour disease spread (Diop Bruckler, 1989). Free moisture on leaves inhibits spore germination.

Management

i. Cultural control: Restrict visitors' access to the greenhouse and follow strict greenhouse hygiene throughout the growing season. Collect and destroy the infected plant parts. Practices like wider plant spacing, sprinkler irrigation, misting and air circulation are found effective in reducing the severity of the disease (Moens and Walvaert, 1981).

ii. Biological control: Mycoparasitic fungi *Ampelomyces* can be used as the natural antagonist against powdery mildew fungi (Kiss and Russell 2004).

iii. Host resistance: Few cultivars/lines like HV-12, S-32, Pondaicha, K. Red and Local (Pawar *et al.*, 1985a) have been reported as resistant to this disease while cultivars viz., B 15, R 7, Padasali, Tinwari, Vrias de California, Du Danube, Minis 27, Vidra 9 and Golden Superior (Singh *et al.*, 1985; Iosifescu, 1985) were moderately resistant to *L. taurica*. Tolerance in some cultivars has also been reported (Ullasa *et al.*, 1981).

Chemical control: Sulphur fungicide is most effective as preventive measure (Smith *et al.*, 1999) but not much recommended in poly-house conditions. Benomyl (0.1%) was outstanding fungicide when applied as spray either before or after disease appearance (Ondieki, 1973). Benomyl and Cercobin can also be used as soil drench before disease appearance. Foliar applications of Sodium or Potassium *Bi* Carbonate solutions have also been found effective in the management of this disease (Fallik *et al.*, 1997). Application of mono potassium phosphate (MKP) (KH_2PO_4 @ 1% (w/v) solution on the upper surface of lower leaves of greenhouse peppers induced local and systemic control of this disease (Reuveni and Dor, 1998). Various EBI fungicides like triadimefon and hexaconzole as foliar sprays have also been found effective against this disease (Negron *et al.,* 1991). Strobilurin fungicides have high efficacy against this disease but should not be applied more than once as foliar application due to resistance problem (Glawe *et al.*, 2010). Premature defoliation of leaves is an important phenomenon in this disease and can be markedly inhibited by application of IAA and IBA.

IX. Downy mildew of Cucumber

Cucumber is another crop which is popularly grown under protected structures to make the fruit available round the year and to substantiate the farmer's income. Cultivation of cucumber is being practiced using plastic (polythene) tunnel for getting early harvest in some districts of Punjab and in naturally ventilated polyhouses in hilly areas of the country. Plastic low tunnels provide the best way for off-season cultivation of cucumber during winter season by modifying the microclimate around the plants and thus protecting against cold during September to November in hills and December to February in plains.

Losses: Downy mildew leads to serious losses in cucumber production when the crop is grown under protected structures. These structures provide ideal temperature and promote high humidity for downy mildew development which has become the main limiting factor for cucumber production (Wehner and Shetty, 1997). The losses due to this disease are manifold as it decreased flower set and fruit development is hampered due to premature death of the foliage (Kristkova *et al.*, 2009). Up to 80 per cent disease severity was recorded in cucumber cultivation grown in low plastic tunnels in different districts of Punjab (Aujla *et al.*, 2010; Bhat *et al.,* 2013). Disease severity was maximum in Amritsar district while it was minimum in Ludhiana.

Symptoms: The initial symptoms of this disease appear on upper surface of the leaves as mosaic mottling comprising of light green, dark green and yellow patches which later turn yellow in colour (Plate 7). Corresponding under surface of these spots is covered with purplish downy growth of the pathogen particularly in the morning hours. The number and size of these spots increase and finally the leaves become chlorotic, brown and shrivelled. Leaves of 5–15 days are more susceptible. Severely infected vines wilt prematurely and die.

The Pathogen: The disease is caused by *Peronospora cubensis* (Berk & Curt.) Rostow.

Pathogen description, disease cycle and epidemiology is described in the Chapter "Diseases of Cucurbits" in this book.

Management

a. Cultural practices: Various cultural practices like removal and destruction of infected plant debris, cucurbitaceous weeds, separating mature and newly planted crops, avoiding dense crop density in protected structures for preventing the buildup of moist microclimates within the leaf canopy are important in keeping the disease under check. Early sown crops (15th December) recorded minimum disease severity on cv. Poinsett and Punjab Naveen compared to delayed sowing (30th January) where the disease severity increased manifold in these cultivars (Bhat *et al.*, 2013). In protected structures side ventilation plays an important role to make the microclimates favourable or unfavourable for disease development. Less disease severity was recorded when daily side ventilation was given for 4 h and it increased (54.49 and 75.88 %) considerably when no side ventilation was given in cvs. Poinsett and Punjab Naveen, respectively (Bhat *et al.*, 2013).

b. Chemical management: Bhat *et al.* (2013) evaluated 11 systemic and non-systemic fungicides against this disease in low plastic tunnels and found Ridomil MZ, Ridomil Gold (0.25%) and Amritsar 25SC (0.1%) as highly effective with minimum disease severity when repeated at 10 days interval followed by Curzate M-8 and Acrobat M2 (0.25%). A mixture composed of 5 parts of plant activator BABA (DL-3-aminobutyric acid) and 1 part of mancozeb was found highly effective in the management of downy mildew of cucumber in greenhouse conditions and also improved plant growth (Baider and Cohen, 2003).

B. BACTERIAL DISEASES

I. Bacterial wilt

Bacterial wilt is one of the most devastating bacterial diseases of solanaceous plants like tomato and capsicum particularly in many warm temperate, subtropical and tropical regions of the world under protected cultivation conditions.

Geographical Distribution: The occurrence of this disease was first reported in 1896 by Smith on potato, tomato and eggplant (Smith, 1896). Its origin and early dissemination remain undetermined, since the first reports were published towards the end of the 19th Century in Asia, South America, USA and Australia where it was already apparently well established. Then it disseminated to different susceptible hosts in tropical, subtropical and warm temperate zones of the world (Kelman, 1953). In India, the disease occurs in almost all the states wherever tomato and capsicum is cultivated (Khan *et al.*, 1980). It is well established in the mid hill sub-humid areas of Himachal Pradesh and appears endemically every year (Sood and Singh, 1993; Gupta *et al.*, 1998).

Losses: It is one of the most serious diseases, limiting the production of capsicum and tomato in protected cultivation conditions (Hayward, 2000, Ashlesha and Paul, 2012). The damage leads to large losses of yield and income as well as disease control is difficult (Hartman *et al.,* 1993; Doan and Nguyen, 2005).

Symptoms: Characteristic symptoms of bacterial wilt on most of the hosts are rapid and complete wilting of normal grown up plants. During wet and warm climatic conditions, the most conspicuous symptom seen on tomato plants is sudden drooping of leaves, without yellowing often accompanied by rotting of the stem. The roots appear healthy and often are well developed. Pathogen is mostly confined to vascular region. In advance cases, it may invade the cortex and pith, and cause yellow brown discolouration of tissues. Bacterial polysaccharides mechanically block the vascular system, which checks the translocation of water and minerals resulting in wilting of plants.

The Pathogen: The bacterium responsible for wilt is *Ralstonia solanacearum* (Yabuuchi *et al.*, 1995). The bacterium is gram negative, rod shaped, frequently occurs in pairs, motile with 1 to 4 polar flagella and measures 0.5–0.7 × 1.5–2.5 μm. Molecular based detection of *R. solanacearum* can be done by using specific primers Y2 and OLII amplified at 288bp. The primers were also able to distinguish the *R. solanacearum* from other saprophytic bacteria through colony PCR within 2 days (Singh *et al.*, 2010). Five races (race1, 2,3,4, 5) and four biovars (biovars I, II, III and IV) are known to occur in different parts of the world. In India race 1 and 3 and biovar II, III and IV are prevalent (Kalha and Sood, 1994).

Disease Cycle: The pathogen is soil borne and can survive at least for 2 years in the absence of any host in the diverse agro-climates of India (Shekhawat *et al.*, 1979). It survives in infected plant parts, crop debris, wild host plants and in weeds. The bacterium enters the plant through wounds or injuries made by transplanting, cultivation, insects, nematodes or natural wounds where secondary roots emerge (McCarter, 1991). Once the bacterium is inside the host, it has the affinity for the vascular system where it multiplies rapidly filling the xylem elements with bacterial cells and slime.

Epidemiology: Since the pathogen is soil borne, the decaying infected tissues release bacteria in the soil. It spreads through irrigation water, soil movement, implements and infected plant material. Root knot nematodes may pre-dispose tomato plants to bacterial wilt by interfering with nutrition of plants and root wounding to provide entry to the pathogen. Relatively high soil moisture and soil temperature favour the disease development. The disease mostly develops when mean temperature is >20ºC with more than 50 per cent soil moisture.

Management

The management of this disease in infested soil is very difficult, although resistant varieties are available, their resistance is often incomplete (Wydra and Semran, 2006). Cultural practices have been found to be ineffective due to high variability of the pathogen (Sequeira, 1993). Chemical control may not be suitable due to high cost and environmental risk associated with application. Keeping all this in view, the options available are described below:

a. Cultural Practices: The disease can be kept in check by adoption of suitable rotation in protected structure by avoiding capsicum after tomato or vice-versa or by planting non-solanaceous crops like maize or ragi (Sharma and Kumar, 2004). Although soil solarization has been exploited for the control of numerous soil borne pathogens, its role for the control of soil borne bacterial pathogens has not been much studied. However, bacterial wilt incidence was slightly reduced after 8 weeks of soil solarisation with 25 μm thick transparent polyethylene sheet (Sharma *et al.*, 2004). Solarization of growing media at least for eight weeks before filling in beds of protected structures can be of use in reducing the disease incidence. Movement of water from infected plants to healthy plants should be restricted. Changes in the planting dates should be so designed to avoid either periods of high temperature or heavy rainfall or both. Green manuring or biofumigation with *Brassica* sp. may reduce the bacterial inoculum in soil. Various soil amendments, such as saw dust, peat moss also decreased wilt incidence. Suppression of pathogen in amended soil was observed by stimulating the growth of saprophytic micro-organisms especially actinomycetes and *Bacillus* species or by the release of toxins during decomposition by soil microorganisms. When different nitrogen forms (NH_4Cl, $NaNO_3$ and $NaNO_2$) substituted urea in the components of S- H mixture, only the soil amended with $NaNO_2$ was found effective in suppressing the wilt pathogen.

b. Host Resistance: At least two types of resistance seem to exist in tomato; one is polygenic, originating from *Lycopersicon esculentum* var. *ceresiformae* and needs to be sustained by additional crop management measures. The second type is a dominant resistance present in cv. Hawaii, which is of simpler heredity, like the monogenic resistance described in Hawaii 7998. Most of the resistant sources against *R. solanacearum* are found in *L. pimpinellifolium* or ecotypes of *L. esculentum* var. *ceresiformae*, which are wild tomato germplasm with small fruits and resistant to elevated temperatures. The Asian Vegetable Research and Development Centre (AVRDC), Taiwan together with institutions in many countries including Australia, India, South Africa and U.S.A. have played a key role in the development of cvs. with stable resistance. Tomato lines/cultivars viz., EC 369060A (Swarna Naveen), BT-10, CKVT-17, Sikkam Local, BB-6, JC-8, Arka Keshav, Arka Nidhi, Singnath, BB-40, Green Round have been reported resistant to this disease (Singh *et al.*, 2014). Hybrid Sun 7711 is still performing good under open field conditions against this disease in Solan area of Himachal Pradesh, but it succumbed to the disease under protected structures due to root knot nematode attack. Some other hybrids like Naveen 2000^{+} and Manisha were tried and found less susceptible to root knot nematode attack and the disease.

c. Chemical Control: Chemical control is not effective in the management of bacterial wilt of tomato and capsicum, however, application of bleaching powder (15 kg/ ha) has also been found effective against this disease. Sprays of streptocycline sulphate (200 ppm) also provide good control of this disease (Mazumdar, 1998). The disease can also be controlled effectively if dazomet application is combined with soil solarization.

d. Biological Control: Application of biological control agents like *Bacillus* sp., *B. subtilis, Enterobacter* sp., *Pseudomonas fluorescens, P. aeruginosa*, AM fungi and *Trichoderma* sp. has been found effective in delaying the wilt development.

Arbuscular mycorrhizal fungi have also been observed to reduce the wilt incidence (Suresh and Rai, 1991).

e. Organic Inputs: Ashlesha and Paul (2012) evaluated seven organic inputs viz., biosol, matka khad, fermented butter milk, panchgavya, biodynamic compost tea, cow urine and vermiwash against *R. solanacearum* both *in vitro* and in polyhouse conditions. Combination of biosol, cow urine, fermented butter milk, biodynamic compost tea, vermiwash and homa ash enhanced the survivability of seedlings up to 13 days when tested at 1:10 ratio after 1 h dip duration and seedlings kept in Hoagland's solution. Drenching of mixture of biosol, fermented butter milk, cow urine, vermiwash and homa ash before and after transplanting along with capsicum seedling deep in biosol for 10 min before transplanting in polyhouse provided 86.64, 85.22 and 84.64 per cent disease control at Rajol, Palampur and Sungal areas, respectively.

f. Integrated disease management: Integration of soil solarization for 8 and 10 weeks during March to May along with incorporation of antagonistic rhizobacteria *i.e., Pseudomonas fluorescence* and *Bacillus cereus* in soil prior to solarization resulted in increased soil temperature at 5 cm depth in wilt sick plots by 5.2 to 6.2°C under Palampur (Himachal Pradesh) conditions and caused up to 73.4 per cent suppression of wilt incidence which was attributed to enhanced population of the antagonists and reduced *R. solanacearum* population (Kumar and Sood, 2005). No wilt incidence was observed when bacterized (*Pseudomonas fluorescens* and *Bacillus cereus*) tomato seedlings of moderately resistant line EC 191536 were planted in 10 weeks solarized soil.

II. Bacterial Canker of tomato

It is a serious disease of tomato both under open and protected conditions and cause substantial yield losses in favourable climatic conditions. Losses are associated with stunting, wilting, cankering, and sometimes death of systemically infected plants along with fruit spotting resulting from secondary infections.

Geographical Distribution: This bacterial disease was first reported by E. F. Smith in 1910 from Grand Rapids, Michigan (USA) as Grand Rapids disease, but later it was named as bacterial canker. Thereafter, McKeen (1973) and Chang *et al.* (1992a) reported bacterial canker in greenhouse and field grown tomatoes in other states of USA, like Ontario and Illinois. The disease was then reported from other countries like Italy (Scortichini, 1997), Cyprus (Ioannou *et al.*, 2000), Turkey (Basim *et al.*, 2004), Mexico (Holguin *et al.*, 2006), Iran (Nazari *et al.*, 2007), Korea (Myung *et al.*, 2008), Indonesia (Zainal *et al.*, 2008), Serbia (Milijasevic *et al.*, 2009), Syria (Ftayeh *et al.*, 2010) and Italy (Lamichhane *et al.*, 2011).

In India, the disease was first recorded during 1949 from Nagpur, Maharashtra state but the pathogen was not confirmed to be this bacterium (Jain, 1951). Later, confirmed reports of bacterial canker were received during surveys of tomato fields in the state of Karnataka, India with an incidence of 25–48 per cent by Sarala and Shetty (2005) and Umesha (2006).In Himachal Pradesh, the wide spread occurrence of the disease was recorded in 2013 (Singh *et al.*, 2015).

Losses: In literature, no report on exact losses due to this disease exist but the incidence ranging from 25–48 per cent have been reported from Karnataka state (Sarala and Shetty, 2005; Umesha, 2006). In Himachal Pradesh, India, incidence and severity of 30.10 and 23.68 per cent, respectively, have been reported (Singh *et al.*, 2015)

Symptoms: The disease is observed in midseason, especially when plants are in flowering and at maturity and ripe stage (Holguin *et al.,* 2006). The disease symptoms appear on almost all the above ground plant parts *viz.,* leaves, petioles, stem, branches and fruits. Young plants show poor growth and wilting of branches (Fig. 4). The lower leaves of affected plants become yellow, shrivelled with dark brown spots usually at the edges and surrounded by yellow halo. In some cases one half of the affected plant shows wilt symptoms while the other half appears healthy. The stem of the affected plant also develop scars or cavities at some locations. Sometimes splitting of the affected stem was also noticed with longitudinal cankers. The pith of affected stem appears reddish brown, especially at the nodes. The pith later turned somewhat mealy in appearance and hollow. Fruit symptoms appear as bird's eye lesions on severely affected plants. The lesions on fruits were yellow to brown, slightly raised, about 3 mm in diameter and surrounded by a persistent white halo.

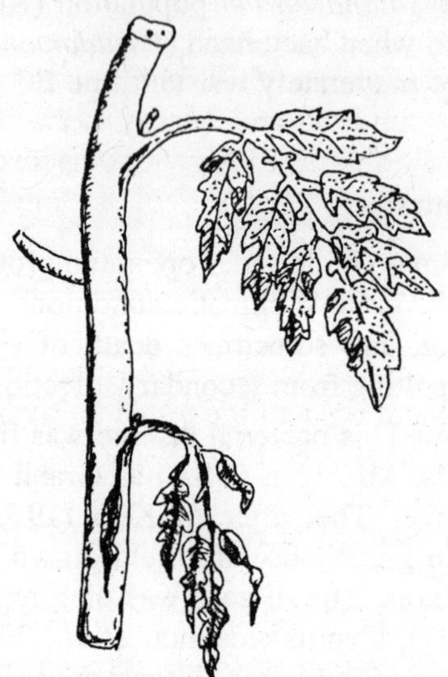

Fig. 4. Symptoms of bacterial canker caused by *C. michiganensis* pv. *michiganeusis*

The Pathogen: The disease is caused by bacterium *Clavibacter michiganensis* subsp. *michiganensis (Cmm).* The bacterium is aerobic non-sporulating, gram positive, rod shaped. The colonies of bacterium are mostly creamish to yellow in colour, roundish and semi fluidal. The colonies became deeper yellow and glistering with the period of incubation. All the bacterial isolates were found positive for Gram stain, esculin hydrolysis, gelatin liquefaction, catalase and 2,3,5-triphenyl tetrazolium chloride (TTC), and were negative for starch hydrolysis, oxidase, levan production and casein hydrolysis (Singh *et al.*, 2015).

Disease Cycle and Epidemiology: The bacterium is seed borne and known to spread through contaminated seed to new areas (Chang *et al.*, 1992b). Due to seed borne nature of the pathogen, disease spread to the areas where it was not reported earlier. The pathogen spread within plantings by splashing rain and human activities. Once the bacterium enters through leaf stomata and/ or small wounds, systemic infection occurs. Moisture condition of the plant and temperature are found to be important factors in the development of disease. During summers particularly in hilly areas, the congenial conditions prevail for the development of bacterial canker in the month of June when the temperature remains high and pre-monsoon showers provide sufficient humidity. Warm temperatures (23°C to 28°C) along with high moisture or relative humidity (>80%) favours disease development.

Management

Management of this disease is very difficult after the appearance of symptoms. Some practices are suggested to keep the ravages of this disease in check.

a. Cultural Practices: Once this disease appear in the field it is very difficult to keep in check. Field sanitation and preventive measures must be followed. Use only disease free seed collected from disease free fruits. At least three years crop rotation should be followed and remove symptomatic seedlings in the nursery or plantings as quickly as possible and destroy them. Stakes from infested fields should also be destroyed. If bacterial canker was a problem in the previous year, implements, tools, and cages should be sanitized prior to use. Collect and destroy the infected plants as soon as they are noticed in the field and destroy them. Deep summer ploughings to bury deep the infected plant debris. Eliminate solanaceous weeds like nightshades and horse nettles from in and around the field, as they are alternate hosts for this bacterium. Avoid working tomatoes when the foliage is wet. Irrigate protected structures and fields through drip irrigation to minimize the length of time so that foliage is not wet.

b. Host Resistance: Ulukus and Karaca (1985) studied the incidences of bacterial canker on tomato cultivars and none of the tested cultivars was found to be immune. Out of 24 tomato genotypes evaluated for resistance to systemic infection by *Cmm* in the glasshouse, for two consecutive crop seasons, lines CL1131-0-0-13-8 and MR8 were observed resistant in both the seasons (Quezado and Lopes, 1994). Meneisy *et al.* (2005) reported cvs. like Dora, Flora-Dade, KTM 141, Money Maker, Niagra, Super strain B hybrid and GS 12 as moderately resistant to this disease.

c. Biological Control: Seed treatment with antagonistic bacterium *Pseudomonas fluorescens* improved the seed quality under laboratory conditions and drastically reduced the disease incidence under field conditions (Grish and Umesha, 2005, Umesha, 2006). Seeds treated with PGPR viz., *Bacillus subtilis* GB03, *B. amyloliqenciens* IN937a and *Brevibacillus brevis* IPC11 enhanced seed quality parameters like seed germination and seedling vigour and maximize protection from the disease under greenhouse conditions (Umesha, 2006, Amkraz *et al.* 2010). They further reported that the rate of reduction in the disease incidence was directly proportional to the amount of increased level of PAL and total phenol content. Three *Streptomyces* sp. were evaluated for their microbial antagonism potential against bacterial canker (Abyad *et al.*, 1993; Zhang *et al.*, 2010). Seed-coating with antagonistic *Streptomyces* sp. significantly improved tomato growth and reduced the disease incidence.

d. Chemical Control: Applications of fixed copper plus either maneb or mancozeb and streptomycin may reduce epiphytic (superficial) populations of the bacterium before symptoms appearance, but they generally have little impact on disease control.

III. Bacterial spot of Bell pepper

This is a very important and common disease of Capsicum under protected cultivation conditions in those areas of the world where significant precipitation and high temperature occur during growth and fruiting stages and cause potential yield reductions due to premature defoliation. Same disease also occurs on tomato.

Losses: In polyhouses high temperature and relative humidity prevails which favours the development of this disease causing severe damage to the crop due to premature defoliation thus reducing potential fruit yields (Jindal and Malhotra, 1995, Sharma *et al.*, 2006).

Symptoms: The symptoms of the disease appear on all the above ground parts like leaves, stems and fruits of the plants. Initially, the lesions are water-soaked, circular or irregular that becomes necrotic with brown centres and thin chlorotic borders. Lesions may enlarge to 10 mm in diameter. These lesions are generally sunken on the top surface of the leaf and slightly raised on the bottom. In favourable environmental conditions these spots on leaves coalesce and have a blighted appearance. Such leaves turn yellow and fall down prematurely. Stem lesions are narrow, elongated and raised expanding up to 6 mm. The lesions become light brown and rough in appearance. Fruit lesions initially begin as green spots, which enlarge and become brown in colour. These spots are raised with a cracked, roughened wart like appearance and are up to 5 mm in diameter. During periods of high humidity, fruit around the lesions may start rotting. Similar spots may appear on floral parts, cotyledons and petioles.

The Pathogen: The bacterium responsible for this disease is *Xanthomonas vesicatoria* (ex Doidge) Vauterin *et al.* Association of *Xanthomonas euvesicatoria* with this disease has also been recorded (Ramyashreedevi *et al.*, 2017). The bacterium is motile, strictly aerobic, Gram-negative, rod shaped with round ends, non-spore forming and possess a single pollar flagellum. Each cell measures 1–1.8 x 0.5–0.7 μm. Bacteria occur singly, but occasionally in pairs and rarely in short chains. Its colonies on nutrient agar or on specific medium i.e., yeast dextrose agar (YDA) are lemon yellow, slightly convex, wet and shining. Eleven pepper races have been identified based on differential reactions to four resistance genes. Strains are pathogenic either on pepper only or tomato only or on both pepper and tomato.

Disease Cycle: The pathogen is both seed and soil borne. Dissemination of the bacterium in or on the seed and transplants is an important factor in disease cycle. The bacterium penetrates directly through the injuries created by wind-driven sand, insect punctures or mechanical means, hydathodes and stomata, and lesions develop more frequently on the wounded fruits than unwounded ones. After entry, the bacterium produces slimy masses of cells within the stomatal cavities and adjoining intercellular spaces. The adjoining cells swell and break the epidermis producing lesions. In infected fruits, the seed may be infected.

Epidemiology: The bacterium is both seed and soil borne. Moderate temperatures along with high precipitation and relative humidity favour disease development. The optimum temperature lies between 22 to 34ºC and maximum disease develops in between July and September. The bacterium is disseminated within a field by wind driven rain droplets, clipping of transplants and aerosols.

Management

a. Cultural Practices: Since the disease requires high precipitation and high relative humidity, protected structures should be well drained free of low-lying areas to minimize water-logged conditions. The field should be rotated with non-solanaceous crops and pepper should not be rotated with tomato. The structures should be kept free from weeds and other volunteer plants. Sprinkler or overhead irrigation of the polyhouses should be limited to keep the disease under check.

b. Biological Control: Prashad and Gupta (2012) reported the efficacy of *Pseudomonas fluorescens* and plant origin pesticides like extracts of *Allium sativum* and *Curcuma longa,* which can be used as an alternative to manage this disease particularly in organic production of capsicum.

c. Chemical Control: Use healthy seed and treat the seed with Streptocycline (100 ppm) for 30 minutes. Spray the crop with copper oxychloride (0.3%) or Bordeaux mixture (4:4:50) or combination of Streptocycline (00 ppm) and copper oxychloride (0.25%), or mancozeb (0.25%) and repeat at 10 days interval (Prashad *et al.,* 2011).

C. VIRAL DISEASES

During cultivation under protected structures, tomato, capsicum and cucumber are infected by many viral diseases. The important ones are described below:

i. Capsicum chlorosis

Capsicum chlorosis disease is emerging viral disease and is a serious threat to Capsicum cultivation both under protected conditions and open field conditions in Himachal Pradesh (Brakta *et al.,* 2014). The occurrence of this disease has already been reported from Bangalore, Haryana and Uttar Pradesh on chilli, capsicum and tomato (Kunkalikar *et al.*, 2010).

Losses: This is disease is emerging as a serious concern in commercial cultivation of bell pepper (Krishnareddy *et al.,* 2008, Kunkalikar *et al.*, 2010, Mandal *et al.,* 2012). The affected plants remain dwarf and the fruits produced on infected plants are mostly misshapen. In Himachal Pradesh, this disease is wide spread and reducing the marketable fruit yield drastically both under open and protected conditions (Brakta *et al.,* 2014).

Symptoms: Naturally infected plants remain dwarf, show stunted growth and exhibit ring spot symptoms on leaves which later turned necrotic and finally leading to chlorosis of the leaf. Fruits produced on infected plants are either misshapen with nail head symptoms or in case where fruits are healthy, the ripe fruits show pale red or yellow areas on the skin often appearing as ring spots of alternate colours. Similar symptoms also appear on tomato.

The Virus: The virus associated with this disease is *Capsicum chlorosis virus* (CaCV) type member of genus *Tospovirus* of family *Bunyaviridae,* belongs to the *Watermelon silver mottle virus* (WSMOV) serogroup (Yeh and Chang, 1995) and infects solanaceous and leguminous crops (Kunkalikar *et al.*, 2010, Mandal *et al.*, 2012). Virus represents a quasi-spherical enveloped particle of 80–110 nm in diameter constituted of a tripartite ssRNA genome which is segmented into large (L), medium (M) and small (S) RNA segments which are individually encapsidated. L-RNA consists of 8912 (nucleotides) nt, M-RNA consists of 4821 nt and S-RNA consists of 3105 nt. L-RNA encodes for RNA dependent RNA polymerase, hence plays role in virus replication. M and S-RNA encodes for non-structural proteins NSm and NSs responsible for cell to cell movement of virus and silencing of suppressor gene, respectively. S-RNA also codes for nucleocapsid (N) gene.

Transmission: The virus is typically transmitted by thrips vector mainly *Frankliniella* sp. and *Thrips palmi.*

Management

Collect and destroy the infected plant material regularly. Manage vector by application of suitable insecticide like Karate (0.05%) and repeat at 10 days interval.

ii. Tomato Spotted Wilt Virus

It is an important disease of tomato under protected cultivation conditions where it causes significant yield reductions.

Geographical distribution: *Tomato spotted wilt virus* was first reported in Georgia in 1986, but was not considered a problem until 1989. The disease infects many of the commercial crops in Georgia and also many of the weeds found in the Coastal Plain in Georgia. In India also, it is wide spread in different states of the country.

Losses: Losses from this disease are immense as it produce rings on the fruits rendering them unmarketable. In Georgia, the level of TSWV has been consistently high (>30%) and it has been common for entire fields to be destroyed by the virus.

Symptoms: Symptoms of this disease appear on affected plants as bronzing, curling, necrotic streaks and spots on the leaves. Dark brown streaks also appear on leaf petioles, stems and growing tips. The plants remain dwarf and exhibit stunted growth. The ripe fruits show paler red or yellow areas on the skin (Plate 8). Sometimes affected plants are killed by severe necrosis.

Virus: The virus responsible for the disease is *Tomato Spotted Wilt Virus* which belongs to genus *Tospovirus* and family *Bunyaviridae*.

Host range: This virus has one of the widest host range of any plant virus. Bell pepper, lettuce, pea, tobacco, potato, tomato and a large number of ornamental plant species are the main hosts of this virus.

Transmission: The virus is transmitted and spread in nature by thrips. The virus is acquired only by intensive feeding of the larvae and transmitted by adults. Once acquired by the larvae, the virus is passed transtadially, i.e., TSWV persists through insect molts from larval to adult stages. The virus replicates in thrips, and the thrips can transmit the virus within about 24 hours after acquisition as larva and during their entire adult life. There are at least 13 species of thrips that transmit *tospoviruses*. *Frankliniella occidentalis*, the western flower thrips (WFT), is considered to be the most important vector species because it is globally distributed and can transmit most *tospoviruses*.

Management

The presence of thrips in tomato fields can be monitored by using yellow sticky cards (Gupta and Handa, 2009). Consider planting of non-susceptible crops in the area if TSWV and thrips populations throughout the area are very high. Infected plants should be removed and destroyed immediately. Maintain seed beds away from cropped area as well as from other susceptible plants. Raise nursery in net house (40 mesh or higher netting) to exclude thrips. Maintain at least 10 m plant free border around the production fields to prevent the spread of *TSWV*. Remove crop debris, weeds and other sources of thrips at the end of each crop. Plough and keep the field fallow for 3–4 weeks to allow thrips to emerge and disperse.

REFERENCES

Abyad M S, Sayed M A, Shanshoury A R and Sabbagh S M. 1993. Towards the biological control of fungal and bacterial diseases of tomato using antagonistic *Streptomyces* sp. *Plant and Soil* **149(2):** 185–195.

Ahmed, B.Y., Sharif, F.M. amd Sarhan, A.R.T. 1987. Effect of certain micronutrients on Fusarium wilt of tomato. *I. J. Agric. Water Resour. Res. Pl. Food* **6:** 13–28.

Alexander L J and Tucker C M. 1945. Physiological specialization in the tomato wilt fungus *Fusarium oxysporum* f. sp. *lycopersici*. *J. Agric. Res.* **70**: 303–313.

Amemiya, Y., Yamaguchi, K., hirano, K. and Iida, W. 1986. Suppression of Fusarium wilt of tomato by use of crop protection. *Tech. Bull. Fac. Hortic., Chiba Univ.* **37:** 79–83.

Amkraz N, Boudyach E H, Boubaker H, Bouizgarne B and Aoumar A A B. 2010. Screening for *fluorescent pseudomonades*, isolated from the rhizosphere of tomato, for antagonistic activity toward *Clavibacter michiganensis* subsp. *michiganensis*. *World J. Microbiol. Biotechnol.* **26(6):** 1059–1065.

Anonymous, 2014. Package of practices of vegetable production, Directorate of Extension Education, YSP University of Horticulture and Forestry, Solan, 178 pp.

Archisi, M., Gennori, M., Matta, A. 1985. Retardations of *Fusarium* wilt symptoms in tomato by pre and post inoculum treatments of the roots and aerial part of host in hot water. *Physiol. Pl. Pathol.* **26(2):** 175–183.

Ashlesha and Paul, Y.S. 2012. Evaluation of organic inputs against *Ralstonia solanacearum*-a bacterial wilt pathogen of capsicum under protected cultivation. *Pl. Dis. Res.* **27(1):** 66–70.

Atkinson, R.G. and Adamson, R.M. 1977. Effects of fungicides on wilt and yield of green house tomatoes grown in *Fusarium* infected saw dust. *Can. J. Pl. Sci.* **57(3):** 675–680.

Aujla, I.S., Thind, T.S. and Kumar, Pardeep 2010. Disease scenario in vegetables under protected cultivation in Punjab. *Pl. Dis. Res.* **25(2):** 185–187.

Ayesu-offei, E.N. 1998. Formae speciales of *Leveillula taurica* (Lev.) Arn. infecting peppers and eggplants in Ghana. *Trop. Agric.* **66:** 355–360.

Baider, A. and Cohen, Y. 2003. Synergistic interaction between BABA and mancozeb in controlling *Phytophthora infestans* in potato and tomato and *Pseudoperonospora cubensis* in cucumber. *Phytoparasitica* **31:** 399–409.

Banerjee, M.K., Kalloo, G. and Jalali, I. 1990. Studies on Fusarium wilt resistance in *Lycopersicon* screening and method of assessment. *Veg. Sci.* **17(2):** 167–174.

Basim E, Basim H, Dickstein E R and Jones J B. 2004. Bacterial canker caused by *Clavibacter michiganensis* subsp. *michiganensis* on greenhouse-grown tomato in the Western Mediterranean Region of Turkey. *Plant Dis.* **88(9):** 1048.

Bharat, N. and Sharma, J. 2014. Occurrence of Fusarium wilt of tomato under protected conditions in Himachal Pradesh, India. *Int. J. Bio-resource Stress Mgt.* **5(2):** 285–287

Bharat, N. and Sharma, J. 2015. Management of Fusarium wilt of tomato under protected conditions using Brassica crop residues, AM fungi and biocontrol agents. *Indian Phytopath.* **68(3):** 287–292.

Bharat, N.K. and Gupta, S.K. 2013. Occurrence of Phyllosticta leaf spot of tomato under protected cultivation in Himachal Pradesh. *Pl. Dis. Res.* **28(2):** 203–204.

Bhat, J.H., Thind, T.S., Bala, Anju and Kumar, Pardeep. 2013. Factors affecting development of downy mildew of cucumber grown under plastic low tunnel and its management with fungicides. *Pl. Dis. Res.* **28(1):** 58–63.

Bhat, M.N., Sardana, H.R., Singh, D., Chitra Srivastava and Ahmand, M. 2015. Evaluation of chemicals and `bioagents against *Sclerotium rolfsii* causing Southern blight of bell pepper (*Capscium annuum*). *Indian Phytopath.* **68(1):** 97–100.

Bisht, I.S., Chandra, C., Thomas, T.A. and Chandra, I. 1989. Field reactions of wild tomato species to various diseases. *Veg. Sci.* **16:** 104–106.

Blazquez, C.H. 1976. A powdery mildew of chilli caused by *Oidiopsis* sp. *Phytopathology* **66(10):** 1155–1157

Boland, C.J. and Hall, R. 1994. Index of plant hosts of *Sclerotinia sclerotiorum*. *Can. J. Pl. Path.* **16:** 93–108.

Brakta, A., Bohra, Yogita, Handa, Anil and Thakur, P.D. 2014. First report of *Capsicum chlorosis virus* infecting capsicum in Himachal Pradesh. *Pl. Dis. Res.* **29(2):** 193–196.

Butler, E.J. 1918. *Fungi and Diseases in Plants.* Thacker Spink & Co. Calcutta, India. 300 pp.

Cerkauskas, R.F. and Brown, J. 1999. First report of powdery mildew of greenhouse pepper caused by *Leveillulla taurica* in Canada. *Plant Dis.* **83(8):** 781.

Cerkauskas, R.F. and Buonassist, A. 2003. Management of powdery mildew, *Leveillula taurica* in greenhouse peppers. *Plant Dis.* **87:** 1151.

Chang R J, Ries S M and Pataky J K. 1992a. Reductions in yield of processing tomatoes and incidence of bacterial canker. *Plant Dis.* **76(8):** 805–809.

Chang R J, Ries S M and Pataky J K. 1992b. Local sources of *Clavibacter michiganensis* sp. *michiganensis* in the development of bacterial canker on tomatoes. *Phytopathology* **82(5):** 553–560.

Chang R J, Ries S M and Pataky J K. 1992c. Effects of temperature, plant age, inoculum concentration, and cultivar on the incubation period and severity of bacterial canker of tomato. *Plant Dis.* **76(11):** 1150–155.

Chauhan, S.S. 1988. Reaction of varieties/lines of tomato to wilt. *Indian J. Mycol. Pl. Pathol.* **18:** 72–73.

Chebil, A., Margari, A. and Verlodt, H. 1980. Preliminary results with early tomato cultivation in plastic bags with a substrate of sea weed *Posscidonia oceanica*. *32nd Int. Symp. Phytopharm. Phytiatry Part II.* **45:** 775–781.

Christopher, D.J., Suthinraj, T. and Kumar, R.S. 2013. Role of defense enzymes induced by consortia of antagonists and decomposed coir pith in managing wilt of tomato caused by *Fusarium oxysporum* f.sp. *lycopersici*. *J. Mycol. Pl. Pathol.* **43(4):** 472–477.

Correl, J.C., Gordon, T.R. and Elliot, V.J. 1987. Host range, specificity and biometrical measurements of *Leveillula taurica* in California. *Plant Dis.* **71:** 248–251.

Dababat A A, Selim M E, Saleh A A and Sikora R A. 2008. Influence of Fusarium wilt resistant tomato cultivars on root colonization of the mutualistic endophyte *Fusarium oxysporum* strain 162 and its biological control efficacy toward the root-knot nematode *Meloidogyne incognita*. *J. Plant Dis. Protec.* **115 (6):** 273–278.

Damicone, J.P. 1999. First report of pepper powdery mildew caused by *Leveillula taurica* in Oklahoma. *Plant Dis.* **83(11):** 1072.

Diop Bruckler, M. 1989. Effect of climatic factors on development of *Leveillula taurica* and susceptibility of *Capsicum annuum* at different vegetative stages. *J. Phytopathol.* **126:** 104–114.

Dixon, G.R. 1981. *Vegetable Crop Diseases*. Avi Publishing Co. INC. Westpot, Connecticut. 404pp.

Doan, T.T. and Nguyen, T.H. 2005. Status of research on biological control of tomato and groundnut bacterial wilt in Vietnam. In: Wolfgang Zeller, Correlia Ulrich (eds.). 1st *Int. Symp. Biol. Cont. Bacterial Pl. Dis.*, Darmstadt, Germany 2005: 105–111.

Dodan, D.S., Shyam, K.R. and Bhardwaj, S.S. 1995. Relative response of tomato cultivars/ lines against buckeye rot under field conditions. *Pl. Dis. Res.* **10:** 135–136.

Dwivedi, S.K. 1991. Studies on population dynamics of *Fusarium oxysporum* f.sp. *lycopersici* in solar heated soil. *Natl. Acad. Sci. Newsl.* **14(6):** 235–237.

El Shami, L., Salem, D.E., Fadl, F.A., Ashour, W.E. and El Zayat, M.M. 1990. Soil solarization and plant disease management. II. Effect of soil solarization in comparison with soil fumigation on the management of Fusarium wilt of tomato. *Agric. Res. Rev.* **68(3):** 601–611.

Ercole D N and Nipoti P. 1986. Biological control of *Fusarium* and *Verticillium* infections in tomatoes under protected cultivation. *Colture Protette* **15 (3):** 55–59.

Etebarian H R. 1992. Studies on Fusarium wilt of tomato and its chemical control in Varamin area. *Iranian J. Agric. Sci.* **23**(1): 1–14.

Fallik, E., Ziv, O., Grinberg, S., Alkalai, S. and Klein, J.D. 1997. Bicarbonate solutions control powdery mildew (*Leveillula taurica*) on sweet pepper and reduce the development of postharvest fruit rotting. *Phytoparasitica* **25:** 41–43.

Frank A B. 1885. Veber nene Mykorriza-formen. *Berichte der Deutschen Botanischen Gesellschaft* **5**: 395.

Ftayeh R, Tiedemann A V, Koopmann B, Ghorrah M and Rudolph K. 2010. Occurrence of *Clavibacter michiganensis* subsp. *michiganensis*, the causal agent of bacterial canker of tomato, in Syria. *Phytopathol. Mediterr.* **49(2):** 172–178.

Fuchs, J. and Defago, G. 1991. Protection of tomatoes against *Fusarium oxysporum* f. sp. *lycopersici* by combining non-pathogenic Fusarium with different bacteria on untreated soil. *Bull. SROP* **14(8):** 51–56.

Gardezi A K, Espinoza G R, Cerrato F R and Saavedra L M. 1999. Effect of arbuscular mycorrhizae on tomato (*Lycopersicon esculentum* Mill) in naturally infested soil with *Fusarium oxysporum* f. sp. *radicis-lycopersici*. *Revista Mexicana de Fitopatologia* **17 (1):** 23–28.

Glawe, D.A, Barlow, T., Eggers, J.F. and Hamm, P.B. 2010. First report of powdery mildew caused by *Levillula taurica* of field grown sweet pepper in the Pacific Northwest. *Plant Health Prog.* **10:** 1094.

Gonzales G R and Galindo A J. 1974. Fusarium wilt of tomtato caused by *Fusarium oxysporum* f. sp. *lycopersici*, in the valley of Culiacan, Sinaloa. *Agrociencia* **18**: 97–104

Grish, N. and Umesha, S. 2005. Effect of plant growth promoting rhizobacteria on bacterial canker of tomato. *Arch. Phytopathol. Plant Protect.* **38**: 235–243.

Gupta, A.K., Chandel, Sunita and Bharat, N.K. 2014. Soil borne diseases of temperate and sub-temperate begetables under protected cultivation. In: *Diseases of Vegetable Crops: Diagnosis and Management* (Eds. D. Singh, P. Choudappa and Pratibha Sharma). Today and Tomorrow Printers and Publishers, New Delhi, pp. 709–723.

Gupta Meenu and Gupta, S.K. 2014. Anthracnose of coloured Capsicum under protected Cultivation conditions. *Indian Phytopath.* **67(3)**: 316–317.

Gupta, S.K., Dohroo, N.P. and Shyam, K.R. 1998. Occurrence of bacterial wilt of tomato in Himachal Pradesh. *Pl. Dis. Res.* **13:** 174.

Gupta, S.K., Bhardwaj, R., Sharma, M.K. and Gupta, Y.C. 2005. Diseases of bell pepper under protected cultivation conditions and their management. In: *Integrated Plant Disease Management* (Ed. R.C. Sharma and J.N. Sharma), Scientific Publishers (India), Jodhpur pp:183–186.

Gupta, S.K. and Handa, Anil. 2009. *Eco-friendly management of diseases of pea and other vegetables.* Department of Mycology and Plant Pathology, Dr. Y.S. Parmar University of Horticulture & Forestry, Nauni, Solan, 101 pp.

Gupta, S.K., Watpade, S., Kumar J., Sharma, Y.P. and Gupta, Meenu. 2016. Occurrence of capsicum powdery mildew under protected conditions in Himachal Pradesh. *Pl. Dis Res.* **31(2):** In Press

Hartman, G.L., Hong, W.F., Hanudin and Hayward, A.C. 1993. Potential of biological and chemical control of bacterial wilt. In Hartma, G.H. and Hayward, A.C. (eds.). Bacterial wilt, pp. 322-326. ACIAR, Canbera.

Hayward, A.C. 2000. *Ralstonia solanacearum. Encyclopedia of Microbiology*, Vol. 4, 2nd Ed. Pp. 32–42. Academic Press, London, UK.

Holguin P R J, Vazquez J R C and Rueda P E O. 2006. Bacterial canker caused by *Clavibacter michiganensis* subsp. *michiganensis* on tomato in the Baja California Peninsula of Mexico. *Plant Dis.* **90(12):** 1550.

Huang J W Sun S K. 1982. Tomato wilt, *Fusarium oxysporum* (Schl.) f.sp. *lycopersici* (Sacc.). *Plant Protec. Bull. Taiwan* **24 (4)**: 265–270.

Ioannou N, Psallidas P G and Glynos P. 2000. First record of bacterial canker (*Clavibacter michiganensis* subsp. *michiganensis*) on tomato in Cyprus. *J. Phytopathol.* **148(6):** 383–386.

Iosifescu, M. 1985. Aspects of the epidemiology of powdery mildew in sweet pepper, varietal response to it and its control. *Prod. Veg. Hortic.* **34:** 3–7.

Ismail A L S and Abdullah S K. 1976. Studies on Fusarium-wilt of tomatoes in Iraq. *Proc. Indian Nat. Sci. Acad.* **42**: 189–193.

Jain A C. 1951. A bacterial disease (canker) of tomatoes new to India. *Sci. Cult.* **17**: 46–47.

Jensen M H. 2002. Controlled environment agriculture in desert tropics and temperate regions- A world review. *Acta Hortic.* **578**: 19–25.

Jeon, Y.Y., Kwon, H.W., Nan, J.S. and Kim, S.H. 2006. Characterization of *Sclerotinia sclerotiorum* isolated from paprika. *Mycobiology* **34:** 154–157.

Jindal, K.K. and Mehrotra, R. 1995. *Pl. Dis. Res.* **10:** 149–152.

Jones, J.B., Jones J.P., Stall, R.E. and Zitter, T.A. 1991. *Compedium of tomato diseases*. APS Press, St. Paul, Minnesota, U.S.A. 73 p.

Jones, M.M. and Black, L.L. 1992. Sources of resistance among *Capsicum* sp. to Fusarium wilt of pepper. *Capsicum Newsl.* **11:** 33–34.

Kalha, C.S. and Sood, A.K. 1994. Occurrence of biovar III of *Pseudomonas solanacearum* on tomato in Himachal Pradesh. *ACIAR Bacterial Wilt Newsletter* **10:**10.

Kapoor I.J. 1988. Fungi involved in tomato wilt syndrome in Delhi, Maharashtra and Tamilnadu. *Indian Phytopath.* **41**: 208–213.

Katan J. 1971. Symptomless carriers of the tomato Fusarium-wilt pathogen. *Phytopathology* **61**: 1213–1217.

Kaur, Manpreet, Singh, A., Jain, S. and Mohan, C. 2016. Status of Sclerotinia rot of capsicum under protected conditions in Punjab and its management. *Pl. Dis. Res.* **31(1)**: 10–14.

Kelman, A. 1953. The bacterial wilt caused by *Pseudomonas solanacearum*. A literature review and bibliography. Releigh, N.C.: North Carolina State College, USA.

Khakimov, A.Kh. and Abdullaev, B.Ya. 1992. *Trichoderma* against fusariose of tomato. *Zaschita Rastenii Moskva* **8**: 25.

Khan, A.N.A.,Shetty, K.S. and Patil, R.B. 1980. Occurrence of bacterial wilt of chilli in Karnataka and its relationship to the wilt of other solanaceous vegetables. *Indian Phytopath*. **32**: 507–512.

Kiss, L. and Russell, J.C. 2004. Biology and biological potential of *Ampelomyces* mycoparasites, natural antagonist of powdery mildew fungi. *Biocont. Sci. Technol.* **14(7)**: 635–651.

Krishnareddy, M., Usha, R.R., Anil, K.K.S., Madhavi, R.K. and Pappu, H.R. 2008. *Capsicum chlorosis virus* (Genus *Tospovirus*) infecting chilli pepper (*Capsicum annuum*) in India. *Plant Dis*. **92**: 1469.

Kristkova, E., Lebeda, A. and Sedlakova, B. 2009. Species specra, distribution and host range of cucurbit powdery mildew in the Czech Republic, and in some other European and Middle Eastern countries. *Phytoparasitica* **37**: 337–350.

Kronn, B.A.M., Scheffer, R.J. and Elgersma, D.M. 1991. Induced resistance in tomato plants against Fusarium wilt induced by *Fusarium oxysporum* f. sp. *dianthi. Neth. J. Plant Path.* **97(6)**: 401–408.

Kumar, P. and Sood, A.K. 2005. An eco-friendly approach for the management of bacterial wilt of tomato. *Pl. Dis. Res*. **20**: 55–57.

Kunkalikar, S.R., Sudarsana, P., Rajagopalan, P. Zehr, U.B. and Ravi, K.S. 2010. Biological and Molecular characterization of *Capsicum chlorosis virus* infecting chilli and tomato in India. *Arch. Virol.* **155**: 1047–1057.

Lamichane, J.R., Balestra, G.M.a nd Varvaro, L. 2011. Severe out break of bacterial canker caused by *Clavibacter michigenesis* sub sp. *Michiganesis* on tomato in central Italy. *Plant Dis.* **95(2)**: 22–29.

Malhotra, S.K. and Vashistha, R.N. 1993. Genetics of resistance to *Fusarium* wilt Race 1 in Currant tomato (*Lycopersicon pimipinellifolium*). *Indian J. Agric. Sci.* **63**: 246–247.

Mandal, B., Jain, R.K., Krishnareddy, M., Kumar, K.K., Ravi, K.S. and Pappu, H.R. 2012. Emerging problems of tospoviruses (*Bunya viridae*) and their management in the Indian subcontinent. *Plant Dis.* **96**: 468–479.

Massee G E. 1895. *British fungus-flora*. (Vol. 4). 522 p.

Mazumdar, N. 1998. Managing *Ralstonia solanacearum* wilt of tomato. *J. Mycol. Pl. Pathol.* **28**: 189–192.

McCarter, S.M. 1991. Bacterial wilt. In: *Compendium of Tomato Diseases.* (Eds. Jones JB, Jones, J.P., Stall, R.E. and Zitter, T.A.) pp. 28–29. APS Press St. Paul, Minesota, USA.

McKean C D. 1973. Occurrence, epidemiology and control of bacterial canker of tomato in southwestern Ontario. *Plant Dis.* **53(3)**:127–130.

Meneisy Z A A, Ghafar N Y A, Wafaa M A E, Yazeed A A and Nagwa A M G. 2005. Susceptibility of some tomato cultivars to bacterial canker and spot diseases and the role of seeds in pathogen transmission. *Arab Univ. J. Agric. Sci.* **13(3)**: 917–926.

Milijasevic S, Todorovic B, Rekanovic E and Stepanovic M. 2009. Occurrence of bacterial canker of tomatoes in southern Serbia. *Acta Hortic.* **808**: 307–311.

Mishra, N.C. and Mishra, S.N. 1993. Perfomance of tomato varieties against wilt (*Fusarium oxysporum* f.sp. *lycopersici*) and fruit borer (*Helicoverpa armigera*) in the North Eastern Ghat Zone of Orissa. *Indian J. Pl. Prot.* **21(2):** 183–186.

Mitov N and Nikolaeva V. 1977. Physiologic races of the pathogen of Fusarium wilt of tomato in Bulgaria. *Gradinarska i Lozarska Nauka* **14 (1):** 110–114.

Moens, M. and Walvaert, W. 1981. Capsicum varietal susceptibility to *Leveillula taurica* and the influence of different irrigation methods. *Meded. Fac. Landbouwwet. Rijksuniv. Gent.* **46:** 805–811.

Mwangi M W, Monda E O, Okoth S A and Jefwa J M. 2011. Inoculation of tomato seedlings with *Trichoderma harzianum* and arbuscular mycorrhizal fungi and their effect on growth and control of wilt in tomato seedlings. *Brazilian J. Microbiol.* **42**: 508–513.

Myung I S, Kim D G, An S H, Lee Y K and Kim W G. 2008. First report of bacterial canker of tomato caused by *Clavibacter michiganensis* subsp. *michiganensis* in Korea. *Plant Dis.* **92(10):** 1472.

Naqvi, S.M.A. and Chauhan, S.K. 1980. Effect of root exudates on the spore germination of rhizosphere and rhizoplane mycoflora of chilli cultivars. *Plant Soil* **55:** 397–402.

Nazari F, Niknam G R, Ghasemi A, Taghavi S M, Momeni H. and Torabi S. 2007. An Investigation on Strains of *Clavibacter michiganensis* subsp. *michiganensis* in North and North West of Iran. *J. Phytopathol.* **155**: 563–569.

Negron, J.A, Rodrigues, R. and Aviles, L. 1991. Fungicide evaluation for control of powdery mildew, *Leveillula taurica* (Lev.) Arn. of peppers in Puerto Rico. *J. Agric. Univ. Puerto Rico* **75:** 119–123.

Nirwanto, H., Djajati Horjodi, S.S., Tjtrosomo, S., Horjadiw Widodo, W.D., Sudorsono. 1994. Study on the use of soil amendment with leguminous foliage and liming in effort to control *Fusarium* wilt disease of tomato. *Acta Hortic.* **369:** 155–162.

Ondieki, J.J. 1973. The control of powdery mildew of Capsicums with certain systemic and non-systemic fungicides. *Acta Hortic.* **33:** 137–142.

Overman, A.J. and Jones, J.P. 1975. Effect of Furadan on yield and expression of Fusarium wilt in tomato planted as bare root and containerised transplants. *Proc. Fl. State Hortic. Soc.* **88:** 214–217.

Ozdemir Z. 2005. First report of fruit infections of *Clavibacter michiganensis* subsp. *michiganensis*, on processing tomato in Turkey. *Plant Pathol. J.* **4(2):** 143–145.

Pandey, K.K and Gupta, R.C. 2013. Virulence analysis of *Fusarium oxysporum* f.sp. *lycopersici* causing tomato wilt in India. *J. Mycol. Pl. Pathol.* **43:** 409–413.

Pawar, R.N., Utikar, P.G. and Shinde, P.A. 1985a. A survey of powdery mildew of chilli. *J. Maharashtra Agric. Univ.* **10:** 218.

Pawar, R.N., Utikar, P.G. and Shinde, P.A. 1985b. Field evaluation and economics of fungicidal control of powdery mildew of chilli. *Pesticides* **19:** 69–70.

Prashad, D. and Gupta, S.K. 2012. *In vitro* efficacy of biocontrol agents and plant origin pesticides against *Xanthomonas campestris* pv. *vesicatoria*, the incitant of bacterial spot of bell pepper. *Pl. Dis. Res.* **27**(2): 234–236.

Prashad, D., Gupta, S.K. and Shukla, Arti. 2011. Management of bacterial spot of bell pepper (*Capsicum annuum*) through chemicals and antibiotic. *Indian Phytopath*. **64:** 204–205.

Punja, Z.K. 1988. *Sclerotium* (*Athelia*) *rolfsii*, a pathogen of many plant species. *Adva, Plant Pathol.* **6:** 523–534.

Purdy, L.H. 1979. *Sclerotinia sclerotiorum*: History, diseases and symptomatology, host range, geographic distribution and impact. *Phytopathology* **69:** 875–880.

Quezado A M and Lopes C A. 1994. Reaction of tomato genotypes to bacterial canker. *Hortic. Brasil.* **12(2):** 154–160.

Qusan, J.R., Al Abed, A.S. and Abu Blan, H.A. 1995. Antifungal activity of clanny mula (*Inula viscose*) on *Helminthosporium sativum* and *Fusarium oxysporum* f. sp. *lycopersici*. *Phytopathol. Meditr.* **34(1):** 7–14.

Ramezani H. 2010. Antagonistic effects of *Trichoderma* sp. against *Fusarium oxysporum* f.sp. *lycopersici* causal agent of tomato wilt. *Plant Protec. Journal* **2** (1): 167–173.

Ramayashreedevi, G.S., Singh, D., Srivastava, Arpita, Biswas, K.K. and Gupta, A.K. 2017. Characterization of *Xantomonas* species causing bacterial leaf spot disease of pepper (*Capscium annuum*) in India. *Indian J. Agric. Sci.* **87(12):** 1679–86.

Rao, V.G. 1962. Sydowia **16:** 275–283.

Reuveni, R. and Dor, G. 1998. Local and systemic control of powdery mildew (*Leveillula taurica*) on pepper plants by foliar spray of mono-potassium phosphate. *Crop Prot.* **17(9):** 703–709.

Reuveni, R. and Rotem, J. 1973. Epidemics of *Leveillula taurica* on tomatoes and peppers as affected by conditions of humidity. *Phytopathologische-Zeitschrift* **76(2):** 153–157.

Sabir N and Singh B. 2013. Protected cultivation of vegetables in global arena: a review. *Indian J. Agric. Sci.* **83** **(2):** 123–135.

Sarala I S and Shetty S. 2005. *Clavibacter michiganensis* subsp. *michiganensis* in tomato fields of Karnataka, India. *Advances in Plant Sci.* **18(2):** 581–589.

Sarhan, A.R.T. and Kiraly, Z. 1981. Control of Fusarium wilt of tomato with an integrated nitrate lime fungicide regime. *Acta Phytopathol. Acad. Sci. Hung.* **16(1-2):** 9–14.

Scortichini M. 1997. Early infections of *Clavibacter michiganensis* subsp. *michiganensis* on tomato cultivated in unheated greenhouse in central Italy. *J. Pl. Pathol.* **79(3):** 223.

Sequeira, L. (1993). Bacterial wilt: past, present and future. In: Hartman, G.L. and Hayward, A.C. (eds.) Proc. Int. Conf. Bacterial wilt Kaohsiung, Taiwan, 28-31, Oct., 1992:12-21.

Sharma, H.R., Gupta, S.K., Thakur, A.K. and Jarial, Kumud. 2006. Evaluation of bell pepper germplasm for resistance to bacterial spot leaf blight and fruit rot diseases. *J. Mycol. Pl. Pathol.* **36:** 219–223.

Sharma, J. 2013. Biointensive management of fusarium wilt of tomato under protected cultivation. M.Sc. thesis, submitted to Dr. Y.S. Parmar University of Horticulture & Forestry, Nauni, Solan, 58 pp.

Sharma, S.K., Kumar, R., Dohroo, N.P. and Kaushal R. 2004. Effect of soil solarization on *Ralstonia solanearum* population and bacterial wilt of tomato in Himachal Pradesh. *Indian Phytopath.* **57(2):** 200–204.

Sharma, J.P. and Kumar, S. 2004. Effect of crop rotation on population dyanamics of *Ralstonia solanacearum* in tomato wilt sick soil. *Indian Phytopath.* **57:** 80–81.

Shekhawat, G.S., Kishore, V., Singh, D.S., Khanna, R.N., Singh, R. and Bahal, V. K. 1979. Survival of *Pseudomonas solanacearum* under diverse agroclimates in India. *Indian J. Agric. Sci.* **49:** 736–738.

Shukla Arti, Gupta, S.K. and Sharma, H.R. 2013. Occurrence of late blight in cherry tomato grown under protected conditions in Himachal Pradesh. *Pl. Dis. Res.* **28(2):** 205–206.

Singh, D., Chowdappa, P. and Sharma, Pratibha (2014). *Diseases of Vegetable Crops: Diagnosis and Management*. Indian Phytopathological Society, IARI, New Delhi and Today and Tomorrow's Printers and Publishers, New Delhi, 734pp.

Singh, D., Sinha, Shweta, Yadav, D.K., Sharma, J.P., Srivastava, D.K., Lal, H.C., Mondal, K.K. and Jaiswal, R.K. 2010. Charaterization of biovar/races of *Ralstonia solanacearum*, the incitant of bacterial wilt in solanaceous crops. *Indian Phytopath.* **63(3):** 261–265.

Singh, G., Bharat, N.K. and Sharma, Monica 2015. Occurrence of bacterial canker of tomato in Himachal Pradesh, India: Identification and Molecular characterization of the pathogen. *The Bioscan* **10(4):** 1753–1757.

Singh, R.S. 1995. *Diseases of Vegetable Crops.* 3rd ed. New Delhi: Oxford and IBH Publ. Pvt. Ltd. pp. 107–109.

Singh, S., Lodha, S. and Lodha, S. 1985. Varietal reaction and evaluation of fungicides *in vitro* and *in vivo* against powdery mildew of chilli. *Indian J. Agric. Sci.* **55:** 85–87.

Sivakumar T, Eswaran A and Balabaskar P. 2008. Bioefficacy of antagonists against for the management of *Fusarium oxysporum* f. sp. *lycopersici* and *Meloidogyne incognita* disease complex of tomato under field condition. *Plant Arch.* **8 (1):** 373–377.

Sivan A, Ucko O and Chet I. 1987. Biological control of Fusarium crown rot of tomato by *Trichoderma harzianum* under field condition. *Plant Dis.* **71**: 587–592.

Smith, E.F. 1896. A bacterial disease of tomato, eggplant and Irish potato (*Bacillus solanacearum* nov.sp.). *Div. Veg. Phys. and Path. Bul.* 12: USDA.

Smith, R.F, Koike, S.T., Davis, M., Subbarao, K. and Laemmlen, F. 1999. Several fungicides control powdery mildew in peppers. *California Agric.* **53(6):** 40–43.

Sood, A.K. and Singh, B.M. 1993. Prevalence of bacterial wilt of solanaceous vegetables in mid hill sub humid zone of Himachal Pradesh (India). In: Hartman, G.L. and Hayward, A.C. (eds.). *Bacterial wilt: Proc. Int. Conf. held at Kaohsing, Taiwan* 28-31 Oct., 1992. *ACIAR Proc. No.* **45:** 358–361.

Spehia R S. 2011. Polyhouse introduction, history and important aspects. *In: protected cultivation for sustainable horticulture*, (eds. Spehia R S and Sharma I P). Dehradun, UK, pp. 3–26.

Sreenivasa, M.N. 1994. Biological deterrent activities of VA mycorrhiza and *Trichoderma harzianum* on *Sclerotium rolfsii* in Chilli. *Environ. Ecol.* **12:** 319–321.

Strobel J W, Hayslip N C, Burgis D S and Everett P H. 1969. Walter, a determinate tomato resistant to race 1 and 2 of the Fusarium wilt pathogen. *Florida Agric. Exp. Sta. Cir.* University of Florida, Gainesville, 202p.

Suresh, C.K. and Rai, P.V. 1991. Interaction of *Pseudomonas solanacearum* with antagonistic bacteria and VA-mycorrhiza. *Curr. Sci.* **20:** 26–37.

Tafradjiiski, I, Elenkov, E. and Neshev, G. 1975. On the pathogenesis of *Leveillula taurica. Arn. C.R. Acad. Agric. Georgi Dimitrov.* **8:** 65–67.

Ullasa, B.A., Rawal, R.D., Sohi, H.S., Singh, D.P. and Joshi, M.C. 1981. Reaction of sweet pepper genotypes to anthracnose, Cercospora leaf spot and powdery mildew. *Plant Dis.* **65(7):** 600–601.

Ulukus I and Karaca I. 1985. Investigations on disease rates, causal agents and symptoms of bacterial diseases on tomatoes and capsicums in Elazg, Diyarbakr and Mardin provinces, and control measures against the most important disease. *Ege Universitesi Ziraat Fakultesi Dergisi* **22(1):** 61–88.

Umesha S. 2006. Occurrence of bacterial canker in tomato fields of Karnataka and effect of biological seed treatment on disease incidence. *Crop Protect.* **25**: 375–381.

Velasquez-Valle, R. 1999. First report of powdery mildew of pepper in North Central Mexico. *Plant Dis.* **83(3):** 302.

Vidyasekaran, P. and Thiagarajan, C.P. 1981. Seed borne transmission of *Fusarium* in chilli. *Indian Phytopath.* **34:** 211–213.

Whener, T.C. and Shetty, N.V. 1997. Downy mildew resistance of the cucumber germplasm collection in North Carolina field tests. *Crop Sci.* **37:** 1331–1340.

Wydra, K. and Semran, J. 2006. Phenotypic and molecular characterization of the interaction of antagonistic bacteria with *Ralstonia solanacearum* causing tomato bacterial wilt. In: Wolfgang Zeller, Cornella Ulrich (ed.). *1st Int. Symp. Biol. Cont. of Bacterial Pl. Dis.*, Darmstadt, Germany 2005: 112–118.

Yabuuchi, E., Kosako, Y., Yano, I., Hota, H. and Nishiuchi, Y. 1995. Transfer of two *Burkholderia* and an *Alcaligenes* species to *Ralstonia* gen. nov. Proposal of *Ralstonia picketi* (Ralston, Palleroni and Doudoroff, 1973). Comb. Nov. and *Ralstonia eutropha* (Davis, 1969) Comb. *Nov. Microbiol. Immunol.* **39:** 897–904.

Yanar, Y., Sahin, F. and Miller, S.A. 1996. First report of stem and fruit rot of pepper caused by *Sclerotinia sclerotiorum* in Ohio. *Plant Dis.* **80:** 342.

Yeh, S.D. and Chang, T.P. 1995. Nucleotide sequence of the N gene of *Watermelon silver mottle virus*, a proposed new member of the genus *tospovirus*. *Phytopathology* **85:** 58–64.

Yehia, A.H., Fl Hassan, S.A., Ismail, F.K. 1981. Studies on damping-off disease of tomato seedlings and its biological control. *Meso. J. Agric.* **16(2):** 115–124.

Yigit F and Dikilitas M. 2007. Control of Fusarium wilt of tomato by combination of fluorescent *Pseudomonas*, non-pathogen *Fusarium* and *Trichoderma harzianum* T-22 in greenhouse condition. *Plant Pathol.* **6**: 159–163.

Zainal A, Anwar A, Khairul U and Sudarsono. 2008. Distribution of *Clavibacter michiganensis* subsp. *michiganensis* in various tomato production centres in Sumatra and Java. *Microbiol. Indonesia* **2(2):** 63–68.

Zhang W H, Yang W X, Meng Q F, Li, Y N and Liu D Q. 2010. Screening and identification of antagonistic *Streptomyces* spp. against *Clavibacter michiganensis* subsp. *michiganensis* from tomato rhizosphere. *Front. Agric. China* **4(2):** 159–164.

■■■■

19

NEMATODE PROBLEMS IN VEGETABLE PRODUCTION

Vegetables are attacked by different plant parasitic nematodes both under field and in protected (glasshouse and polyhouse) conditions and cause significant losses. There are numerous nematode species associated with vegetable crops (Goodey *et al.*, 1965; Potter and Olthol, 1993). A total of 176 nematodes belonging to 72 genera have been reported to be associated with vegetable crops in India (Mahajan *et al.*, 1986). Chandel *et al.* (2010) carried qualitative and quantitative analysis of major plant parasitic nematodes in 214 greenhouses associated with main crops (Sweet pepper, cucurbits, tomato and cauliflower) and 19 nurseries (various vegetables) in Himachal Pradesh during 2006-09 crop seasons and reported the presence of three plant parasitic nematodes including *Meloidogyne incognita*, *Helicotylenchus dihystera* and *Pratylenchus* spp. Besides above three nematodes, one more additional nematode i.e. *Rotylenchulus reniformis* was also encountered in nurseries, however, *M. incognita* was observed to be the most alarming. Nematode problems are now receiving the attention of both scientists and farmers in view of the direct and indirect losses caused by them. The most important nematode species causing economically important diseases in vegetables are described in this chapter.

1. ROOT KNOT NEMATODE

Root knot nematodes (*Meloidogyne* spp.) rank first among the nematodes that cause damage to vegetables. It is a serious problem both under protected and open cultivation conditions.

Geographical distribution

Root knot disease of vegetables is a common disease of many vegetables and has global distribution. Berkeley first recognized the root knot disease on glasshouse cucumbers in England (Berkeley, 1855). Since then it has been reported from almost all the countries of the world.

Losses

The yield losses due to root-knot nematodes have been found to be 28.0 – 47.5, 26.2–50.0, 19.7–33.0, 6.0–90.9, 38.0–47.2 and 18.0–33.0 per cent in tomato, brinjal, chillies, okra, bittergourd and melons, respectively (Sakhuja and Jain, 2001). Since the nematodes are soil borne, these sometimes provide avenues for the entry of other soil borne fungal (*Fusarium* spp., *Rhizoctonia solani, R. bataticola* etc.), bacterial and viral pathogens in the root system, thus causing indirect losses by helping in disease causation (Sakhuja, 1984). Growth parameters like height, number of fruits, fruit length, yield, root length and fresh and dry weight of roots were reduced more in okra when *M. incognita* and yellow vein mosaic virus were present together compared with the plants infected by either pest (Neog *et al.*, 2000). Avoidable losses due to this nematode in greenhouse tomato were estimated to be 11.31 per cent during 2008–09 crop season (Chandel *et al.*, 2010).

Symptoms

The symptoms of the root knot nematode appear in the form of galls/ knots on the roots (Fig. 1). Root-knot affected plants look pale, undernourished and water deficient (Plate 1). The affected plants bear undersized and lesser number of fruits and sometimes mature plants show wilting in the day. The size and number of galls depend upon inoculum load, crop and nematode species. Large sized galls are often formed in crops like cucurbits, brinjal, okra and tomato (Plate 2) while small to medium galling occurs in chillies and potato. Severely infested nursery plants sometimes fail to establish in the field after transplanting.

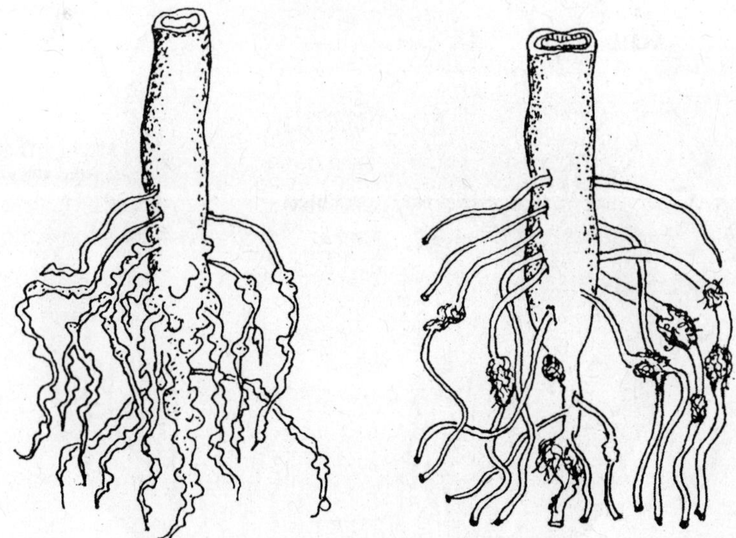

Fig. 1. Symptoms of root knot of tomato (*Meloidogyne incognita*)

The Pathogen

More than 80 species of the genus *Meloidogyne* have been reported to be associated with this disease, the world over but in India species like *M. incognita, M. javanica, M. arenaria, M. hapla, M. africana, M. brevicauda, M. thamesi, M. graminicola, M. indica, M. exigua, M. lucknowica, M. graminis* and *M. triticoryzae* have been reported. Among these, *M. incognita, M. javanica* and *M. arenaria* are the most predominant. Root knot nematodes are sedentary, endoparasitic and gall producing nematodes. The infective stage is the second stage juveniles, which have lightly sclerotized cephalic framework. The stylet in juveniles and adult female is weak. Third and fourth stage juveniles lack stylet and the body is saccate. Adult females are swollen, pyriform, saccate with short neck, vulva is terminal, perinealcuticular pattern of striae is present. Tail is absent. Didelphic, prodelphic ovaries are coiled. Excretory pore is present near stylet base. Large rectal glands produce gelatinous matrix expelled through the anus. Males are vermiform with single or paired teste. Mail tail end is twisted and rounded. Bursa is absent. Reproduction is parthogenic.

Host range

The occurrence and prevalence of root-knot nematodes are more in tropical and subtropical climatic conditions, where summers are longer than the winter. Being polyphagus in nature, this nematode parasitizes most of the crops grown all over the world. However, in India, more than 232 and 144 genera of plants are infected by *M. incognita,*and *M. javanica,* respectively (Krishnappa, 1985).

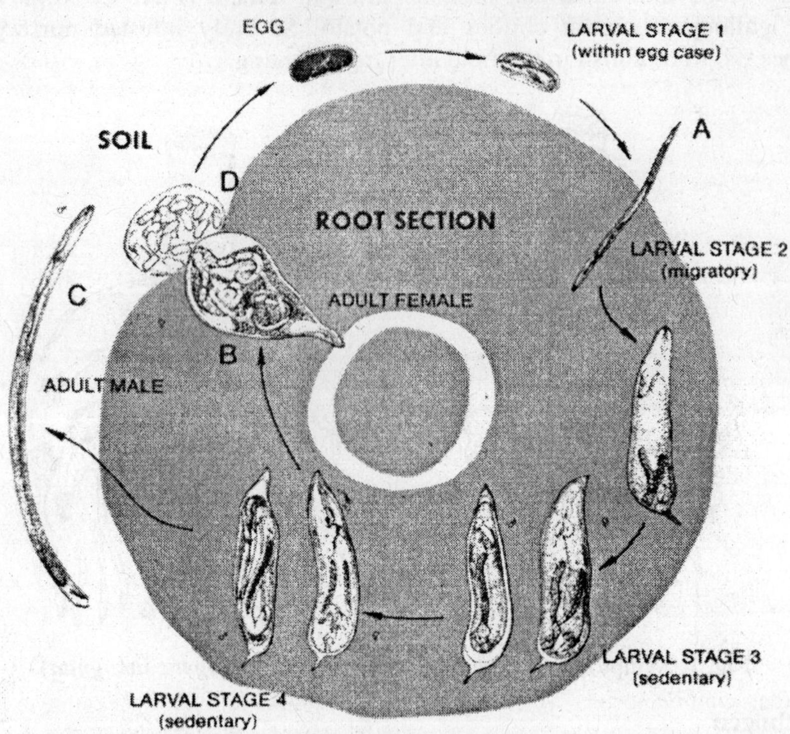

Fig. 2. Life cycle of a root-knot nematode, *Meloidogyne* sp.

Disease cycle

The second stage juveniles in the soil after searching a suitable site, normally behind the root cap, start feeding on epidermal cells and penetrate in cortical layers of root. They reach up to stellar region, where they form giant cells for feeding. Second stage juveniles initiate gall formation and feed for 7–15 days, undergo three moults to become adults (Fig. 2). The total time taken for completing one life cycle under optimum conditions (temp. 27–30°C) is 3-4 weeks in most species depending upon the host and other weather conditions thus leading to completion of several generations in a year. Generally, inoculum level varying from 0.5 to 1 J^2/cc soil in different vegetable crops *viz.*, tomato, brinjal, okra, cucurbits, radish, turnip, pea etc. has been found to be damaging threshold level (Mangat and Bhatti, 1987). Four races of *M. incognita* have been reported and all the four races occur in India (Bajaj *et al.*, 1986; Sakhuja and Singh, 1993). Patel *et al.* (1991) have identified two races in *M. javanica* on the basis of groundnut cultivar differentials. All the races of *M. incognita* and *M. javanica* attack vegetable crops. Moderate temperature (29–30°C) and higher relative humidity (79–80%) favoured *M. incognita* reproduction and spread (Sharma and Bhatia, 2002).

Management

Use of several management practices like cultural methods, host-plant resistance, biocontrol, chemicals and integrated approaches can contain the root knot disease.

(i) Cultural practices: The major source of spread of this nematode into the non-infested fields is through the use of infected transplants. Therefore, the nursery beds should be prepared on such sites, which do not have nematode infestation. Before sowing the soil of the bed should be either solarized for at least 35–40 days or dry paddy husk or saw dust etc. may be burnt at the site of nursery raising. Rotation of the fields with certain non-hosts, graminaceous poor hosts and a few antagonistic crops for one or two years has been reported to be effective in reducing the population of root-knot nematode (Jain and Bhatti, 1985). Groundnut-mustard rotation was found most effective in reducing the population of *M. incognita* (Sharma *et al.*, 1980). Sharma (2007) evaluated different crop sequences and found pea, wheat and maize beneficial against root knot nematode which is a very serious problem in *kharif* vegetables.

The incorporation of organic material and their decomposition products are known for helping in reducing the incidence of root-knot nematodes, either through promotion of natural enemies or release of certain toxic compounds during the course of decomposition (Bhattacharya and Goswami, 1987). These soil amendments (dry or green crop residues, oil cakes, meals, sawdust, FYM etc.) are allowed to decompose in the nematode infested field, which in turn helps in alteration of physical, chemical and biotic conditions of the soil. Amending the potting mixture with composted dry cork reduced the root galling and final populations of *M. incognita* race 1 and *M. javanica* in tomato (Nico *et al.*, 2004). The suppression of nematode (*M. javanica*) development, reproduction, and number of root knot index in tomato plants was observed when foam sulfur and chicken faeces were applied (Ami and Al-Sabie, 2004). Neem formulations like Neemax and Nemadol have been found to inhibit hatching of *M. incognita* and thereby check the multiplication of population (Randhawa *et al.*, 1998). Addition of neem (*Azadirachta indica*) cake @ 40-80 q/ha, mahua, castor and groundnut cake @ 110 q/ha have been found to control root-knot nematodes in a few vegetable crops (Alam, 1991; Agbakli *et al.*, 1992). Incorporation of organic amendments also leads to the production of a variety of phenolics, which may in turn cause toxicity to the nematodes. Among the different phenolics tested as root dip treatment, hydroquinone was most effective against root-knot nematode (Mehmood and Siddiqui, 1993).

Soil amendment with the tissues of non-traditional legumes like *Canavaliaensi formis, Crotalaria retusa, Indigofera hirsute, I. nummulariifolia, I. spicata, I. suffruticosa, I. tinctoria* and *Tephrosia adunca* reduced the gall numbers when applied in *M. incognita* infested soil (Morris and Walker, 2002). These non-traditional legumes, known to contain bioactive phytochemicals, may offer considerable promise as soil amendments for control of plant parasitic nematodes. Besides nematode control they also enhance plant height and dry weight of the plants. Reduction in population dynamics of *M. incognita* was observed by application of mustard cake (5 t/ha) and marigold leaves (10 t/ ha) besides improving the root and shoot length of brinjal (Stuti and Dhillon, 2016).

Altering the date of sowing/planting of vegetable crops can also reduce damage due to root-knot nematode. Ahuja and Singh (1990) reported that mid and late varieties of cauliflower escaped infestation of *M. incognita* while early-planted varieties suffered much damage. Flooding of root-knot infested fields for 4–5 months also helps in reducing the population of nematodes in the field. Deep ploughing of infested fields during summer not only leads to disturbance and instability in the nematode community but also causes mortality by exposing the nematodes to solar heat and desiccation (Raghavan, 1964; Vadhera *et al.*, 2001). Soil tarping with transparent polyethylene sheet during summer months also results in simultaneous control of weeds, fungi, bacteria and nematodes (Sakhuja *et al.*, 1997).

Planting of short-cycle trap crops like lettuce, radish, Chinese cabbage (*Brassica rapa* subsp. *perkinensis*) or Chinese Salt-word (*Brassica rapa* subsp. *chinensis*) is recommended to remove nematodes prior to planting a susceptible crops such as tomato, cucumber or cowpea (Cuadra *et al.*, 2000). The nematode numbers in the soil decreased by 50 per cent and crop yields increased by 10–20 per cent within 3–5 months. This type of treatment is effective only when the roots of these trap crops are completely removed from the field prior to transplanting/sowing of susceptible crops.

Green manuring is one practice which not only adds nutrients to soil but roots of certain plants when incorporated into soil and their decomposition products help in reduction of root knot nematode. Dhillon and Sharma (2009) evaluated guar and sorghum varieties as green manuring crops against root knot nematode of brinjal and tomato. These crops were incorporated into the soil 50 days after their sowing and left as such for 15 days. After 15 days, seedlings of brinjal and tomato (susceptible to root knot) were transplanted and a reduction in the incidence of root knot was observed.

(ii) Host resistance: Krishnappa (1985) has listed a number of varieties in tomato, potato, brinjal, cucurbits, sugarbeet, chillies, cowpea etc. reported to be resistant to *Meloidogyne* spp. In Haryana also, a number of varieties of different vegetables have been screened (Jain and Bhatti, 1984; 1986). The success, however, has been achieved in tomato only. Cultivation of *M. javanica* resistant tomato cultivars in rotation with the susceptible variety has led to reduction in *M. javanica* population and increase in tomato yield (Jain and Bhatti, 1985). Resistance breeding in some other vegetable crops like brinjal, okra, cucurbits etc. has not met with success. However, several wild species in these crops have been found to possess resistance. But the genetic barriers make it difficult to introduce resistance of the wild species in the cultivated ones. Modern techniques like protoplast culture and somatic hybridization and other biotechnological techniques may help to create viable hybrids and other genotypes resistant to root-knot nematode.

Kaur *et al.* (2013) evaluated forty tomato genotypes against root knot nematode (*M. incognita*) and found four lines viz., Hisar Lal, Line 82125, Line 1-6-1-4 and EC 119197 as resistant while EC521056 and EC 521079 were moderately resistant. Biochemical analysis further revealed high total phenols, orthodihydroxy phenols, flavonols, phenyl ammonia lyase (PAL) and peroxidise (PO) activity in leaves and roots of resistant genotypes than the susceptible ones.

(iii) Chemical methods: The chemical nematicides used for controlling root-knot nematodes are either soil fumigants or non-fumigant granular or liquid compounds, which are soluble in water. Fumigants like Basamid have been found to provide excellent control of root-knot nematode and increased the yield by 265 per cent over control when applied in nursery beds @ 40 g/m^2 20 days before sowing (Sakhuja and Singh, 1994). Soil fumigants are not recommended during these days because of their high cost, hazardous nature and toxicity to non-target organisms. Chemicals belonging to organophosphate and carbamate groups have been evaluated by various workers and are being preferred over fumigants because of low doses, easy application and systemic action. Carbofuran @ 0.2–03 g a.i./ m^2 has been recommended for control of this disease for adoption by the farmers.

Bare root dip treatment with dimethoate @ 500 ppm for 6 h and a number of other chemicals have been found to be effective in suppressing root-knot nematode populations (Tiagi *et al.*, 1986). Bare root dip treatment of tomato in 0.03 or 0.06 per cent asataf or monocrotophos for 4 h gave significant control of *M. incognita* and increased the yield (Sakhuja and Singh, 1992). Soaking of okra seeds in trizophos 1000 ppm for 12h also provided protection against *M. javanica* (Jain and Gupta, 1986).

(iv) Biological control: Nematodes in soil are subject to interactions with numerous types of fungi and bacteria. This creates the possibility of using soil microorganisms to control plant parasitic nematodes. Since the first report of parasitism of females of *Heterodera schachtii* by *Tarichum auxilare*, a large number of organisms, such as fungi, bacteria, viruses, predatory nematodes, insects, mites and some invertebrates have been tested and found to parasitize nematodes.

Paecilomyces lilacinus was found to control *M. javanica* on okra when applied either before planting or as seed treatment (Walia *et al.*, 1991). Further studies revealed that wood charcoal powder could serve as a good carrier of this bioagent for storage and field application (Bansal *et al.*, 1992). Beta-exotoxin produced by *Bacillus thuringiensis* killed 95 per cent of larvae of *Meloidogyne* spp. (Rai and Rana, 1979). *Pasteuria penetrans*, an important bacterial strain, is also being investigated throughout the world for its nematophagous capabilities (Walia and Dalal, 1994). This is an obligate parasite, which does not produce any toxin but kills the nematode by mechanical action. Efficacy of *Pseudomonas fluorescens* in reducing the soil population of *M. incognita* was reported when treated plants were planted in infested soil (Jothi *et al.*, 2003).

Arthrobotrys oligospora,' Dactylaria brochopaga, Aspergillus terrus and *Paecillomyces lilacinus* reduced the number of knots of tomato when applied in sterilized and unsterilized soils (Bandyopadhyay *et al.*, 2001; Goswami and Sharma, 2001). Nematophagous fungus *Pochonia chlamydosporia* var. *catenulata* when applied in root knot infested soils reduced nematode infestations significantly in the following tomato crop (Atkins *et al.*, 2003). They further studied the survival of the fungus in soil in controlled conditions in which it remained in soil in densities significantly greater than its original application rate for at least 5 months and this effect can be utilized in regulation of root knot nematode populations in the infested fields.

Tomato seedlings from nursery treated with VAM fungi like *Glomus mossae* (2.5 kg/M^2) when planted in nematode infested soil gave maximum fruit yield and was more than carbofuran (1 kg/ha) treatment (Babu and Suguna, 2000).

Sharon *et al.* (2010) evaluated two *Trichodema* isolates viz. *T. hamatum*-382 and *T. asperellum* in *M. incognita* and *M. javanica* inoculated sandy soil and potting-mixes and reported that *Trichoderma* treatments reduced root galling and nematode reproduction besides improving plant growth parameters of seedlings. El-Hahad *et al.* (2010) isolated 165 bacterial cultures and evaluated under *in vitro* as biocontrol agents against the second stage juvenile (J2s) of the *M. incognita* andthe highest mortality percentages were recorded by cultures of *Paenibacillus polymyxa*, *Bacillus megaterium* and *B.circulans* (100% at dilution 1/10), while it was 99.3, 99 and 97.8% at dilution 1/100, respectively. *P. polymyxa* exhibited a high nitrogen fixation rate (4.2 micro mole N2/mL/h), while *B. megaterium* and *B. circulans* effectively solubilized phosphate and potassium comparing with the control treatments (1.94 fold of available phosphate and 2.0 fold of available potassium, respectively). All bacterial isolates also showed the highest protease, gelatinase and chitinase activities which were thought to be responsible for their nematicidal effect.

Singh *et al.* (2016) evaluated nematicidal efficacy of a *Streptomyces* based bionematicide Actinovate AG against *M. incognita* in tomato and found Actinovate (840g/ ha) application as effective in reducing the nematode population in soil (>42%) besides improving plant growth parameters like shoot and root length and weight.

Plant extracts

Plant parts, their products and extracts have also been reported to possess nematicidal properties. Leaf extracts of *Allium sativum, Ricinus communis, Xanthium strumarium, Azadirachta indica, Eucalyptus* sp., *Tagetes* sp., *Mentha viridis, Casia fistula, Cordiamyza, Dalbergia sisoo* etc. were found to have nematicidal properties (Sakhuja and Jain, 2001). Water extracts of *Calotropis gigantea* significantly reduced the *M. incognita* population both at 45 days after transplanting (87.30%) and at harvest (89.96%) over the control and increased fruit yield in tomato (Saravanapriya *et al.*, 2003). The population of root knot nematode was significantly reduced in soil when treated with neem oil (Shah *et al.*, 2003).

Integrated control

Marigold planting combined with carbofuran application @ 1.0 kg a.i./ha was most effective in controlling *M. javanica* infecting brinjal (Singh, 1991). Root-knot nematode infested plots treated with karanj cake, neem cake, groundnut cake and carbofuran gave significantly higher okra yield and reduced root knot index. Sivakumar *et al.* (1993) have found that the combined application of *P. lilacinus* and carbofuran controlled root-knot nematode and produced much healthier seedlings of brinjal, than with either alone. Soil solarization during summer months and planting of resistant cv NR-7 provided maximum yield of tomato followed by carbofuran treatment (Vadhera *et al.*, 2001). *Verticillium chlamydosporium* plus cabofuran application to the root knot infested field resulted in superior plant growth of tomato and lowest number of galls and egg masses (Gopinatha *et al.*, 2002). Greater parasitization by *V. chlamydosporium* was recorded when carbofuran was integrated with the fungus than when the fungus was applied alone or in combination with neem cake. Application of soil amendment (mustard oil cake, poultry refuse) in combination with biological agents like *Trichoderma* spp. drastically reduced the severity of root knot disease and increased plant growth and fruit yield of tomato (Faruk *et al.*, 2002). Soil solarization and amendment with chicken litter significantly reduced the root knot disease in tomato and stimulated a shift in the soil microbial population dynamics (Stevens *et al.*, 2003). Rhizobacteria viz., *Bacillus* spp. and fluorescent pseudomonads increased significantly in the rhizosphere, rhizoplane and interior root tissues of tomatoes grown in solarized soil compared to nonsolarized soil. These microorganisms may have contributed to the increased growth response of vegetables and some were probably suppressive to soil borne diseases.

Riga (2011) evaluated the effects of *Brassica* green manures of their own and in combination with reduced rates of synthetic nematicides on plant-parasitic nematodes and free living nematodes both under green house and field conditions and observed that *Brassica* green manures in combination with half the recommended rate of 1,3-dichloropropene (1,3-D, Telone) reduced root knot nematode (*Meloidogyne chitwoodi*) to below detection levels, and reduced lesion nematodes (*Pratylenchus penetrans*) and stubby root nematodes (*Paratrichodorus allius*) to below economic thresholds. The combination treatments did not affect the beneficial free-living nematode populations and the non-pathogenic *Pseudomonas*. The total cost of growing and soil-incorporating of *Brassica* crops as green manures in combination with reduced rates of 1,3-D was approximately 35 per cent lower than the present commercial costs for application of the full rate of this fumigant.

2. RENIFORM NEMATODE

It is the second most important nematode of vegetable crops. The occurrence of this nematode on cowpea was first time reported from Hawaii (Linford and Oliviera, 1940). It is a limiting factor in vegetable production but is often neglected or remained unnoticed due to the absence of clear-cut symptoms on the infected plants. Its occurrence has been reported from at least 36 countries of the world. It is also known to be prevalent in almost all states of India.

Symptoms

The affected plants show unthrifty plant growth, yellowing and stunting. It mostly feeds on the secondary or tertiary roots and produces necrotic lesions in the root area around its head. Severe necrosis, discoloration and destruction of feeder roots also occur. The presence of partly protruding reniform females covered with a gelatinous matrix and adhering soil particles on the poorly developed root system, specially secondary and tertiary roots, is unmistakable evidence of the attack of reniform nematode.

The Pathogen

The nematode responsible for this disease is *Rotylenchus* spp. Ten species of *Rotylenchus* are known the world over (Gaur and Perry, 1991) but *R. reniformis* is the most prevalent species. It is a semi-endoparasitic nematode and has well developed cephalic framework. Juveniles, males and young females are vermiform. Mature female is kidney-shaped, with an irregular, less swollen neck, paired ovary, a post median vulva and a short pointed tail. Orfice of dorsal oesophageal gland is usually about one stylet length, behind stylet base. Stylet and oesophagus are regressed in males. Eggs are laid in gelatinous matrix that covers the entire protruding portion of the body.

Host range

Rotylenchus reniformis attacks over 200 plant species including many vegetable crops. This nematode is an important pest of tomato, brinjal, okra, cowpea, cucumber, pea, lettuce, radish, turnip, bean, carrot and sugarbeet. Roy *et al.* (2007) surveyed different leguminous crops in 7 districts of West Bengal and found reniform nematode infecting pea and cowpea only.

Disease cycle

The immature females are infective and penetrate the roots both close to root tip and in the entire root system. The female traverses its head through the cortex and rests in an epidermal cell, which becomes hypertrophied and assumes dome-shaped appearance with adjacent cells remaining normal. The pericycle cells adjacent to endodermal cells undergo a similar change. They become hypertrophied, with thickened cell walls, enlarged nucleus and densely granulated cytoplasm. These hypertrophied cells form a semilunar sheet surrounding vascular region on both sides of nematode head. Males do not penetrate or feed. The cylindrical shape of female nematodes changes to kidney shape. Life cycle of *R. reniformis* is completed in 24–29 days (Sivakumar and Seshadri, 1971). Development of *R. reniformis* is temperature dependent. At 21.5°C, life cycle on tomato is completed in 25–26 days while at 27.6°C it takes only 14 days. Reproduction is mostly through amphimixis and rarely through parthenogenesis. High population in rhizosphere occurs in June-October. The survival of the nematode in the absence of host was observed even after seven months in moist soil and after six months in dry soil (Sivakumar and Seshadri, 1976).

Vats and Dalal (1998) reported one larva per CC soil as the damaging threshold level of *R. reniformis* in pea. Existence of two distinct races of *R. reniformis*, i.e. race 'A' multiplying on cowpea, castor and cotton and race 'B' on cowpea has been reported (Dasgupta and Seshadri, 1971). Prasada Rao and Ganguly (1996) reported the occurrence of 4 physiological races based on differential reproduction capabilities on cotton, pearl millet, cowpea, castor and mustard.

R. reniformis has been found to interact with many other micro-organisms, like *Rhizoctonia solani, Macrophomina phaseoli, Fusarium oxysporum* f.sp. *vasinfectum* and species of *Verticillium* on several vegetable crops that often aggravated the disease.

Management

(i) Cultural practices: Deep ploughing, fallowing and soil solarization have been found to reduce the nematode population density below threshold levels at places having high temperature i.e. >35°C during summers. Crop rotation is also helpful in reducing the population of this nematode if the crops like marigold, pangola grass, peanut, chillies, sugarcane, maize, sorghum etc. are included in rotation (Bridge, 1983). Certain organic amendments in the form of green manures, as oil cakes or as leaf extracts of certain plants are also effective against *R. reniformis* (Alam *et al.*, 1977).

(ii) Host resistance: Patel *et al.* (1986) tested 60 lines of cowpea and found 11 to be moderately resistant to *R. reniformis*. High degree of resistance in Kalyanpur selection and Kalyanpur III of tomato and CA 121 of *Lycopersicon pimpinellifolium* against this nematode has been observed (Balasubramanian and Ramakrishnan, 1983). Cowpea varieties V-16, Phalguni, RC-48 also possess resistance to this nematode (Thakar and Patel, 1984; Gaur, 1986).

(iii) Biological control : Limited work has been done in this direction. *Paecilomyces lilacinus* has been found effective against *R. reniformis* in tomato and brinjal (Parvatha Reddy and Khan, 1989). More work needs to be done on this aspect of plant disease management.

(iv) Chemical control: Application of DBCP @ 50 l/ha, metham sodium @ 250 l/ha, copper oxychloride @ 50 kg/ha, aldicarb/thionazin/monocrotphos/dimethoate @ 4-10 kg/ha were found effective against *R. reniformis* infecting tomato (Sivakumar *et al.*, 1976; 1977). Seed treatment of okra with carbofuran (3 and 6 % a.i. w/w) is also effective in reducing the nematode attack (Sivakumar *et al.*, 1973).

(v) Integrated management: Sivakumar *et al.* (1979) obtained 89 per cent reduction in the population of *R. reniformis* by growing *Capsicum annum* and applying carbofuran/aldicarb/phorate/disulfotan @ 1 kg a.i./ha 15 days after transplanting. Deep ploughing for one month integrated with seed treatment of okra with carbofuran or soil application of aldicarb @ 0.8 kg a.i./ha also proved effective against *R. reniformis* (Reddy, 1983).

3. CYST NEMATODES

Cyst nematodes like *Globodera pallida* and *G. rostochiensis* are the most important nematodes in potato and are causing huge losses to the growers. These nematodes are endoparasites. The remarkable ability of these nematodes to survive as cysts under unfavourable conditions, remaining dormant in the absence of host for years and occurrence of biotypes pose serious challenge to their control strategies.

Globodera spp.

Golden cyst nematode, *G. rostochiensis* has been reported from 60 countries of the world while yellow cyst nematode, *G. pallida* is present in 27 countries (Krishna Prasad and Singh, 1998). In India, both these species are found in Tamil Nadu state where they are considered serious pests of potato. These nematodes have also been encountered in Kerala and Karnataka in recent years.

In all 5 races of *G. rostochiensis* and 3 of *G. pallida* are generally recognized the world over. In India, race Ro1 of *G. rostochiensis* and Pa 2 of *G. pallida* are most prevalent. Other prevalent races are Ro 2, Ro 5 of the former and Pa 1 and Pa 3 of the later species.

Management

This nematode can be managed by using cultural practices and host resistance.

(i) Cultural practices: Crop rotation with non-solanaceous crops like beet, cabbage, carrot, cauliflower, French bean, garlic, radish and turnip during autumn reduces the cyst population density. Presently, potato-cabbage-carrot rotation is being followed extensively in Nilgiri Hills of Tamil Nadu for the management of cyst nematode.

(ii) Host resistance: Very high degree of resistance in *Solanum ajanhurri, S. bulbocastanum, S. gandarlassi, S. tuberosum* sub sp. *andigena, S. vernei* and *S. spegazzinii* has been recorded against these nematodes. Different races of *Globodera* spp. prevalent in India can be best managed with resistance from *S. vernei* as it combines well to all these populations (Krishna Prasad and Singh, 1998).

***Heterodera* spp.**

In temperate regions, *Heterodera schachtii* and *H. cruciferae* are known to affect vegetable crops. *H. schachtii* causes significant losses in cruciferous vegetable crops in Mexico, Canada, U.S.A., Libya, Senegal and Zambia. The nematode can be managed by long rotations or with nematicides (Netscher and Sikora, 1990).

The cabage cyst nematode, *H. cruciferae* causes damage to cruciferous vegetables in California, Libya etc. (Edongali and Dabaj, 1982). General symptoms include off-colour foliage, stunting of plants and sometimes even wilting. Root development may also be affected with reduction in size and secondary root formation.

In India, cowpea, guar and *Dolichos lablab* are affected by pigeonpea cyst nematode, *H. cajani*. However, not much work has been done on its prevalence on these crops and its management, though lot of information has been generated on pulse crops affected by it. High degree of control of *H. cajani* by the combined application of *Pasteuria penetrans* and carbofuran in cowpea has been reported (Singh and Dhawan, 1998).

4. LESION NEMATODE

The nematodes of this group cause severe necrotic lesions on the roots. In *Pratylenchus* genus, there are more than 68 species and these are migratory endoparasites. All juvenile stages and adults invade the roots and sometimes other underground parts also. They feed and reproduce on the root cortex and cause necrosis of the infected and surrounding cells. Irregular, brown to black longitudinal lesions can be seen on the infected roots. They have a wide host range. *P. penetrans, P. minor, P. brachyurus, P. coffeae, P. vulnus, P. pratensis, P. scribneri, P. thornei* and *P. capitatus* are found on vegetable crops like tomato, cauliflower, cabbage, okra, potato, radish, pea, beans, pepper, onion and sugarbeet (Khan and Khan, 1993).

The yield of potatoes up to 50 per cent was reduced by this nematode in the affected fields of Italy (Holgado *et al.*, 2009). Transect-sampling showed plant growth to be negatively correlated with densities of *P. penetrans* and suggested a damage threshold of potato to the nematode of 100 specimens per 250 g of soil. *P. penetrans* was present in roots, underground stems, stolons and tubers. In tubers, nematodes were detected inside cross-lesions typical of common scab, and occurred also in the outermost 0.5 mm of tissue associated with such lesions. On potato cv. Saturna grown in a greenhouse, *P. penetrans* alone induced tuber lesions similar to those of common scab. *P. penetrans* survives storage of potatoes, and new infections may develop from infected tubers used as seed. Hence, potato tubers do appear to be an important means for the spread of *P. penetrans* to new areas.

Life cycle is completed normally within 30–90 days depending upon the temperature, host and nematode species.

The affected plants show a gradual decline in vigour and become stunted. They develop chlorosis and often wilt during daytime. The lesions on the roots provide favourable entry sites to other disease causing microorganisms. The interaction of *Verticillium dahliae* and *P. penetrans* is well known on crops like brinjal, tomato, pepper and tomato (Khan and Khan, 1993).

Management

Soil amendments with carbonaceous and nitrogenous material have been found to reduce the population of *Pratylenchus* spp. (Gupta, 1990). Intercropping of bell pepper with ginger reduced the population of *P. penetrans* and increased the yield of ginger (Sharma and Bajaj, 1998). Inclusion of crops like tomato, onion and pea in rotation were found to be of high significance in the management of this nematode (Sharma, 2007). Use of resistant varieties (Armstrong and Jensen, 1978) and nematicides has also shown potential against lesion nematode.

REFERENCES

Agbakli, R. Maraite, H. and Colin, J.E. 1992. Use of neem extract (*Azadirachta indica*) to control root-knot nematode (*Meloidogyne* spp.) in leafy vegetable crops in South Benin. *Mededelingen-Van-de-Faculteit-Landobouwweteis, Chappen Rijksunivers: teit-Genet.* **57(3A):** 913–918.

Ahuja, S. and Singh, J. 1990.Variations in susceptibility of some cole crops to root knot nematode, *Meloidogyne incognita* at different soil temperatures. *Indian J. Ent.* **52:** 702–704.

Alam, M.M. 1991. Control of plant parasitic nematodes with oil seed cake on some vegetables in field. *Pakistan J. Nematol.* **9(1):** 21–30.

Alam, M.M., Saxena, S.K. and Khan, A.M. 1977. Influence of interculture of Merigold and Margosa with some vegetable crops on plant growth and nematode population. *Acta Botanica Indica.* **5:** 33–39.

Ami, S.N. and Al-Sabie, R.F. 2004. Application of foam sulfur and chicken faeces for controlling root-knot nematode *Meloidogyne javanica* on tomato plant. *Dirasat Agric. Sci.* **31:** 115–129.

Armstrong, J.M. and Jensen, H.J. 1978. Indexed Bibliography of nematode resistance in plants. *Sta. Bull. Agri. Expe. St., Oregon State Uni., Corvallis* No. **639:** 246 pp.

Atkins, S.D., Hidalgo-Diaz, L., Kalisz, H., Mauchline, T.H., Hirsch, P.R. and Kerry, B.R. 2003. Development of a new strategy for the control of root-knot nematodes (*Meloidogyne* spp.) in organic vegetable production. *Pest Mgt. Sci.* **59:** 183–189.

Babu, R.S. and Suguna, N. 2000. Management of root-knot nematode (*Meloidogyne incognita*) on tomato with Vesicular Arbuscular Mycorrhiza. *Int. J. Trop. Pl. Dis.* **18:** 47–50.

Bajaj, H.K., Jain, R.K. and Gupta, D.C. 1986. Races of root-knot nematode *Meloidogyne incognita* prevalent in Haryana. *Haryana Agric. Univ. J. Res.* **16:** 399–400.

Balasubramanian, P. and Ramakrishnan, C. 1983. Resistance to the reniform nematode, *Rotylenchus reniformis* in tomato. *Nematol. Medirt.* **1:** 203–204.

Bandyopadhyay, P. and Kumar, D., Singh, V.K. and Singh, P.K. 2001. Eco-friendly management of root-knot nematode of tomato by *Arthrobotrys oligospora* and *Dactylaria brochopaga. Indian J. Nematol.* **31:** 153–156.

Bansal, R.K., Walia, R.K. and Bhatti, D.S. 1992. Wood charcoal powder, a carrier of *Paecilomyces lilacinus* spores. *Nematol. Meditr*. **20:** 5–7.

Barber, C.A. 1901. Teaeelworm diseases in South India. *Bull. Govt. Madras, Dep. Land Rec. Agric.* **45:** 227–234.

Berkeley, M.J. 1855. Vibrio forming cysts on the root of cucumbers. *dnrs' Chron.* 220.

Bhattacharya, D. and Goswami, B.K. 1987. A study on the comparative efficacy of neem and groundnut oil cakes against root-knot nematode *Meloidogyne incognita* as influenced by microorganisms on sterilized and unsterilized soil. *Indian J. Nematol.* **17:** 81–83.

Bridge, J. 1983. Nematodes In: *Pest control in tropical tomatoes*. Centre for overseas Pest Research, Overseas Development Administration, London, pp. 9–84.

Chandel, Y. S., Kumar, Sunil, Jain, R. K. and Vashisth, S. 2010. An analysis of nematode problems in green house cultivation in Himachal Pradesh and avoidable losses due to *Meloidogyne incognita* in tomato. *Indian J. Nematol.* **40(2):** 198–203.

Cuadra, R., Cruz, X. and Fajardo, J.L. 2000. The use of short cycle crops as trap crops for the control of root-knot nematodes. *Nematropica* **30:** 241–246.

Dasgupta, D.R. and Seshadri, A.R. 1971. Races of the reniform nematode, *Rotylenchusreniformis*, Linford & Oliviera, 1940. *Indian J. Nematol.* **1:** 21–24.

Dhillon, Napinderjeet Kaur and Sharma, S.K. 2009. Effect of green manuring with non-conventional crops on root knot nematode in brinjal and tomato. *Pl. Dis. Res.* **24(1):** 78 (Abstr.).

Edongali, E.A. and Dabaj, M. 1982. Cauliflower (*Brassica oleraceacv. capitata*) cyst nematode in Libya. *Libya J. Agric*. **11:** 205.

El-Hadad, M.E., Mustafa, M.I., Selim, S.M., Mahgoob, A.E.A., El-Tayeb, T.S. and Aziz, N.H.A. 2010. *In vitro* evaluation of some bacterial isolates as biofertilizers and biocontrol agents against the second stage juveniles of *Meloidogyne incognita. World J. Microbiol. Biotechnol.* **26(12):** 2249–2256

Faruk, M.I, Rahman, M.L. and Bari, M.A. 2002. Management of root-knot nematode of tomato using *Trichoderma harzianum* and organic soil amendment. *Bangladesh J. Pl. Pathol.* **18:** 33–37.

Holgado, R., Skau, K.A.O. and Magnusson, C. 2009. Field damage in potato by lesion nematode *Pratylenchus penetrans*, its association with tuber symptoms and its survival in storage. *Nematologia-Mediterr*. **37(1):** 25–29

Gaur, H.S. 1986. Invasion and reproduction of the reniform nematode on some cultivars of cowpea and green gram. *Int. Nematol. Netw. Newsl.* **3:** 8–10.

Gaur, H.S. and Perry, R.N. 1991. The biology and control of plant parasitic nematode, *Rotylenchus reniformis. Agril. Zoology Rev.* **4:** 177–211.

Goodey, J.B., Franklin, M.T. and Hooper, D.J. 1965. *The nematode parasites of plants catalouged under their hosts*. Thirsedn. Common Wealth Agricultural Bureaux, Farnham, Royal.

Gopinatha, K.V., Gowda, D.N. and Nagesh, M. 2002. Management of root knot nematode, Meloidogyne incognita, on tomato using the biocontrol agent *Verticillium chlamydosporium*, neem cake, marigold and carbofuran. *Indian J. Nematol.* **32 :** 179–181.

Goswami, B.K. and Sharma, S.B. 2001. Application of *Aspergillus terreus* and *Paecilomyceslila cinus* for the management of *Meloidogyne incognita* on tomato. *Int. J. Nematol.* **11:** 270–273.

Gupta, M.C. 1990. Response of *Meloidogyne* spp. and *Pratylenchus* spp. in soil amended with carbonaceous and nitrogenous materials. *Indian J. Mycol. Pl. Pathol.* **20:** 38–40.

Jain, R.K. and Bhatti, D.S. 1984. Screening of germplasm of some vegetable crops for resistance to root-knot nematode, *Meloidogyne javanica. Haryana J. Res.* **15:** 261–262.

Jain, R.K. and Bhatti, D.S. 1985. Evaluation of resistant v/s susceptible tomato varieties in reduction of root-knot nematode (*Meloidogyne javanica*) population. *Int. Nematol. Netw. Newsl.* **2:** 17-18.

Jain, R.K. and Bhatti, D.S. 1986. Screening of cowpea germplasm against root-knot nematode *Meloidogyne javanica. Indian J. Nematol.* **16:** 113–114.

Jain, R.K. and Gupta, D.C. 1986. A note on the screening of some chemicals as seed treatment for control of root-knot nematode (*Meloidogyne javanica*) on okra. *Haryana J. Hortic. Sci.* **15(3-4):** 288–290.

Jothi, G., Sivakumar, M. and Rajendran, G. 2003. Management of root-knot nematode by *Pseudomonas fluorescens* in tomato. *Indian J. Nematol.* **33:** 87–88.

Kaur, Maninder, Sharma, S. and Kaur, Sukhjeet. 2013. Screening of tomato germplasm against root knot nematode (*Meloidogyne incognita*) and biochemical characteristics of selected tomato genotypes. *Pl. Dis. Res.* **28(2):** 191–196.

Khan, M.W. and Khan, A.M. 1993. Nematode problems of vegetables. In "*Handbook of Economic Nematology*" (Eds. R.S. Singh and K. Sitaramaiah), Cosmo Pub. New Delhi.

Krishna Prasad, K.S. and Singh, D.B. 1998. Management of potato nematode parasites. *Group meet of Nematologists of Hortic. Crops. C.P.C.R.I. Reg. St. Kayangulam, Kerala*, Jan. 16–18, 1998.

Krishnappa, K. 1985. Studies on races of root-knot nematode *Meloidogyne incognita* in India with particular reference to Karnataka. Ph.D. thesis, Univ. Agril. Sci. Bangalore, India, pp. 197.

Linford, M.B. and Oliveira 1940. *Rotylenchus reniformis*, nov.gen. N. sp., a nematode parasite of roots. *Proc. Helminth. Soc. Wash.* **7:** 35–42.

Mahajan, R., Mangat, B.P.S. and Nandpuri, K.S. 1986. *Plant parasitic and soil nematodes associated with vegetable crops in India - A checklist*, Department of Vegetable Crops, PAU, Ludhiana pp. 1–48.

Mangat, B.P.S. and Bhatti, D.S. 1987. Pathogenicity of *Meloidogyne javanica* on *Brassica compestris* ssp. *penkiensis. Int. Nematol. Netw. Newsl.* **4:** 14–15.

Mehmood, I. and Siddiqui, Z.A. 1993. Effects of phenolics on the growth of tomato and reproduction of *Rotylenchus reniformis. Nemato. Meditr.* **21(1):** 97–98.

Morris, J.B. and Walker, J.T. 2002. Non-traditional legumes as potential soil amendments for nematode control. *J. Nematol.* **34:** 358–361.

Neog, D.J., Pathak, J.J., Bhagbati, K.N. and Sinha, A.K. 2000. Interaction between yellow vein mosaic virus and *Meloidogyne incognita* on growth of okra plants. *J. Interacademicia* **4:** 376–381.

Netscher, C. and Sikora, R.A. 1990. Nematode parasites of vegetables. In: *Plant Parasitic Nematodes in Sub-tropical and Tropical Agriculture*. (Eds. Luc, M., Sikora, R.A. and Bridge, J.). CAB International, U.K. pp. 237–283.

Nico, A.I., Jimenez-Diaz, R.M. and Castillo, P. 2004. Control of root-knot nematodes by composted agro-industrial wastes in potting mixtures. *Crop Prot*. **23:** 581–587.

Prasada Rao, G.M.V. and Ganguly, S. 1996. Host preference of six geographical isolates of reniform nematode, *Rotylenchulus reniformis. Indian J. Nematol.* **26(1):** 19–22.

Parvatha Reddy and Khan, R.M. 1989. Evaluation of *Paecilomyceslila cinus* for the control of *Rotylenchulus reniformis* infecting tomato, compared with carbofuran. *Nematol. Meditr.* **16:** 113–115.

Patel, B.A., Chawda, J.C., Patel, S.T. and Patel, D.J. 1986. Relative susceptibility of certain cowpea lines to *Rotylenchus reniformis. Int. Nematol. Netw. Newsl.* **3:** 16–18.

Patel, D.J., Patel, H.V. and Patel, S.K. 1991. Comparative efficacy of the LDPE film covers for soil solarization in control of root-knot nematode tomato nurseries. *Plasticulture* **91:** 33–36.

Potter, J.W. and Olthol, T.H.A. 1993. Nematode pests of vegetable crops. In: *Plant Parasitic Nematodes in Temperate Agriculture* (Eds. Evans, K., Trudgill, D.L. and Webster, J.M.). CAB International, Wallingford.

Raghwan, D. 1964. *Agriculture in ancient India*. ICAR Publ. New Delhi.

Rai, G.P. and Rana, R.S. 1979. Effectiveness of beta-exotoxin of *Bacillus thuringiensis* var. *thuringiensis* on the ability of *Meloidogyne* sp. from brinjal (*Solanum melongena*) to survive. *J. Biosci.* **1:** 271–278.

Randhawa, N., Sakhuja, P.K. and Singh, I. 1998. Effect of different neem extracts on the hatching of root knot nematode *Meloidogyne incognita. Nat. Symp. Mgt. seed soil Borne Pl. Pathogens, MPKV, Rahuri*, Feb. 9–10, 1998: 82.

Reddy, P.P. 1983. *Plant Nematology*. Agricole Publishing Academy, New Delhi, India. 287p.

Riga, F. 2011. The effects of Brassica green manures on plant parasitic and free living nematodes used in combination with reduced rates of synthetic nematicides. *J. Nematol.* **43(2):** 119–121.

Roy, K., Mukhopadhyay, A.K. and Paramanik, A. 2007. Occurrence, distribution and community analysis of plant parasitic nematodes associated with leguminous vegetable crops in West Bengal. *Indian J. Nematol.* **37(1):** 58–62.

Sakhuja, P.K. 1984. Studies on the inter-relationship between root-knot nematode, *Meloidogyne javanica* and root rot fungi on groundnut. Ph.D. thesis, IARI, New Delhi.

Sakhuja, P.K. and Jain, R.K. 2001. Nematode diseases of Vegetable crops and their management. In: *Diseases of fruits and vegetables and their management* (Ed. Thind, T.S.), Kalyani Publishers, Ludhiana, pp. 439–459.

Sakhuja, P.K. and Singh, I. 1992. Chemotherapy evaluation of some pesticides against *Meloidogyne incognita* in tomato. *Afro-Asian Nem. Netw.* **1(2):** 67.

Sakhuja, P.K. and Singh, I. 1993. Occurrence of races of *Meloidogyne incognita* in Punjab.*Pl. Dis. Res.* **8(2):** 208.

Sakhuja, P.K. and Singh, I. 1994. Evaluation of basamid granules for the control of root-knot nematode *Meloidogyne incognita* in tomato. *Pl. Dis. Res.* **9(2):** 241–242.

Sakhuja, P.K., Singh, I. and Chhabra, H.K. 1997. *Two decades of Nematology in Punjab*, Department of Plant Pathology, PAU, Ludhiana. p. 58.

Saravanapriya, B., Sivakumar, M., Singh, R.V., Pankaj, Dhawan, S.C. and Gaur, H.S. 2003. Effect of plant extracts on tomato against *Meloidogyne incognita. Proc. Natl. Symp. Biodiv. Mgt. Nematodes Cropping Sys. Sust. Agric., Jaipur, India*, Nov. 11–12, 2002. pp. 125–128.

Shah, G.S., Pathan, M.A., Lodhi, A.M. and Rajput, M.A. 2003. Effect of different management practices on *Meloidogyne incognita*. *Pakistan J. Nematol*. **21:** 25–30.

Sharma, G.C. and Bajaj, B.K. 1998. Effect of inter-cropping bell pepper with ginger on plant parasitic nematode populations and crop yields. *Ann. App. Biol.* **133:** 199–205.

Sharma, G.C. and Bhatia, M. 2002. Effect of cultural practices on nematode populations and yield of French beans. *Nematol. Meditr*. **30:** 23–25.

Sharma, G.G. 2007. Impact of different cropping sequences on the population density of phytoparasitic nematodes in the farmer's field. *Indian Phytopath*. **60(1):** 95–98.

Sharma, S.K., Singh, I. and Sakhuja, P.K. 1980. Influence of different cropping sequences on the population of root-knot nematode *Meloidogyne incognita* and the performance of subsequent mungbean crop. *Indian J. Nematol*. **10(1):** 53–58.

Sharon, E., Mor, M., Bar-Eyal, M., Maghodia, A.B., Spiegel, Y., Oka, Y. Kerckhove, S.van, Vanachter, A. and Chet, I. 2010. Tricho Nema: a *Trichoderma*-based project to develop a commercial bio-agent product against phyto-nematodes. *Acta Hortic*. **833:** 223–227.

Singh, B. and Dhawan, S.C. 1998. Compatibility of *Pasteuria penetrans* with carbofuran in controlling *Heterodera cajani* on cowpea. *Indian J. Nematol*. **28(2):** 110–113.

Singh, D. 1991. Studies on effect of *Tagetis* sp. on *Meloidogyne javanica* infecting brinjal (*Solanum melongena*) Ph.D. thesis, H.A.U. Hisar, Haryana, India p.83.

Singh, J., Dhillon, N.K. and Singh, H. 2016. Nematicidal efficacy of bionematicide Actionovate against *Meloidogyne incognita* in tomato. *Pl. Dis. Res.* **31(2):** 215 (Abstr.)

Sivakumar, C.V. and Seshadri, A.R. 1971. Pathogenicity of reniform nematode, *Rotylenchulus reniformis* to castor. *Indian J. Nematol*. **1:** 227–236.

Sivakumar, C.V. and Seshadri, A.R. 1976. Longevity of the reniform nematode, *Rotylenchulus reniformis* in host-free soil. *Indian J. Nematol*. **6:** 138–144.

Sivakumar, C.V., Bhaskaran, M.D. and Balasasaswathy, S. 1993. Control of root-knot nematode, *Meloidogyne arenaria* in brinjal nursery with the *fungus Paecilomyces lilacinus* and the systemic nematicide carbofuran. *J. Biol. Control* **7(1):** 49–50.

Sivakumar, C.V., Kuppuswamy, S. and Meerzainuddin, M. 1973. Evaluation of carbofuran seed treatment for the control of root-knot of okra. *Indian J. Nematol*. **3:** 71–72.

Sivakumar, C.V., Palanisamy, S. and Balasubramanian, M. 1977. Control of *Rotylenchulus reniformis* in tomato nursery. *Indian J. Nematol*. **7:** 78–80.

Sivakumar, C.V., Lakshmanan, P.L. and Balasubramanian, M. 1979. Interaction of a non-host plant (*Capsicum annum* L.) and certain systemic nematicides in population decline of *Rotylenchulus reniformis. Revue. Nematol*. **2:** 165–168.

Sivakumar, C.V., Lakshmanan, P.L. and Palanisamy, S. 1976. Control of root-knot and reniform nematodes with low doses of granular systemic nematicides. *Indian J. Nematol*. **6:** 78–80.

Stevens, C., Khan, V.A., Rodriguez, K.R., Ploper, L.D., Backman, P.A., Collins, D.J., Brown, J.E., Wilson, M.A. and Igwegbe, E.C.K. 2003. Integration of soil solarization with chemical, biological and cultural control for the management of soilborne diseases of vegetables. *Plant and Soil* **253:** 493–506.

Stuti and Dhillon, N.K. 2016. Management of root knot nematode (*Meloidogyne incognita*) in brinjal (*Solanum melongena*) with non-chemical methods. *Pl. Dis. Res.* **32(2):** 213 (Abstr.).

Thakur, N.A. and Patel, C.C. 1984. Screening of cowpea varieties for resistance against root-knot nematodes Meloidogyne spp. In *Nat. Seminar on Breeding Crop Plants for Resistance to Pests and Diseases*. May 25–27, 1983.

Tiagi, S.A., Siddiqqui, M.A. and Alam, M.M. 1986. Control of *Meloidogyne incognita* on tomato by bare root dip treatment in dimethoate. *Int. Nematol. Netw.Newsl*. **3:** 4–5.

Vadhera, I, Tiwari, S.P. and Shukla, B.N. 2001. Comparative efficacy of soil solarization, summer ploughing, resistant variety and nematicide treatment for the management of root knot nematode infecting tomato. *JNKVV Res. J.* **35:** 79–81.

Vats, R. and Dalal, M.R. 1998. Pathogenicity of the reniform nematode, *Rotylenchulus reniformis* on pea (*Pisum sativum* L.). *Indian J. Nematol.* **23:** 22–23.

Walia, R.K. and Dalal, M.R. 1994. Efficacy of bacterial parasite *Pasteuria penetrans* applications as nursery soil treatment in controlling root-knot nematode, *Meloïdogyne javanica* on tomato. *Pest Mgt. Econ. Zoo.* **2**(1): 19–21.

Walia, R.K., Bansal, R.K. and Bhatti, D.S. 1991. Effect of *Paecilomyces lilacinus* application time and method in controlling *Meloidogyne javanica* on okra. *Nematol. Meditr.* **19:** 247–249.

■■■■